Springer Natural Hazards

The Springer Natural Hazards series seeks to publish a broad portfolio of scientific books, aiming at researchers, students, and everyone interested in Natural Hazard research. The series includes peer-reviewed monographs, edited volumes, textbooks, and conference proceedings. It covers all categories of hazards such as atmospheric/climatological/oceanographic hazards, storms, tsunamis, floods, avalanches, landslides, erosion, earthquakes, volcanoes, and welcomes book proposals on topics like risk assessment, risk management, and mitigation of hazards, and related subjects.

Ikuo Towhata · Gonghui Wang · Qiang Xu ·
Chris Massey
Editors

Coseismic Landslides

Phenomena, Long-Term Effects and Mitigation

Springer

Editors
Ikuo Towhata
Department of Civil Engineering
Kanto Gakuin University
Yokohama, Kanagawa, Japan

Gonghui Wang
Disaster Prevention Research Institute
Kyoto University
Uji, Kyoto, Japan

Qiang Xu
The State Key Laboratory of Geohazard
Prevention and Geoenvironment Protection
Chengdu University of Technology
Chengdu, Sichuan, China

Chris Massey
GNS Science
Lower Hutt, Wellington, New Zealand

ISSN 2365-0656 ISSN 2365-0664 (electronic)
Springer Natural Hazards
ISBN 978-981-19-6596-8 ISBN 978-981-19-6597-5 (eBook)
https://doi.org/10.1007/978-981-19-6597-5

This Springer imprint is published by the registered company Springer Nature Singapore Pte Ltd.
The registered company address is: 152 Beach Road, #21-01/04 Gateway East, Singapore 189721, Singapore

Preface

Fifty years ago, one of the great dreams of human beings was that the development of science and technology would solve calamities so that the twenty-first century might be full of glory. In spite of the development of science and technology as expected, however, the human still suffers from many types of calamity today. To say nothing of pandemics, natural disasters still occur frequently worldwide every year. Unfortunately, it appears that the power of natural disasters in the new century is more serious than in the late twentieth century as exemplified by the 2004 tsunami disaster in the Indian Ocean and the 2008 Wenchuan earthquake in China.

Many landslides devastated communities during the 2005 earthquake in Northern Pakistan and the 2008 Wenchuan earthquake. Landslide disasters are triggered by strong earthquakes and claim a substantial number of lives. We have to agree, therefore, that the human dream of safe glorious century has failed.

There are numerous reasons for the failure of dream. In the second half of the twentieth century, the human activity became very powerful and expanded into regions where landslide risk was too high. At the same time, the global climate change increased the risk of heavy rain and consequent disasters as well. Noteworthy is that a greater number of extremely strong earthquakes (magnitude > 8) have started to rise in the twenty-first century from the low level in the preceding period (1980–2000); see Fig. 11.244 in my article in this book. These negative situations work together and increase the risk of earthquake-induced landslide, which is the main theme of this book. New problems have emerged recently. Landslide dams, long runout of landslide mass and compound effects of different disaster mechanisms, such as rainfall + earthquake, are taken more seriously than in the previous century.

To cope with this adversity, all that we can do is to understand the nature of many aspects of landslide disasters in more details and propose reasonable mitigation measures. With this viewpoint, the authors of this book made substantial efforts to collect and present information from real disasters.

The idea of this book on earthquake-induced landslide was originally suggested to me several years ago by Prof. Luciano Picarelli who was the chair of the Joint Technical Committee One (JTC1) on landslide issues under the auspices of Federation of International Geo-engineering Societies (FeDIGS) that is an international umbrella organization of four geo-engineering societies. Because the range of topics to be studied was so vast, writing articles took a long time. It is my greatest pleasure today that this book is published internationally now as a product of JTC1.

I should stress that the problem of earthquake-induced landslide has not been solved. Ambitious people are invited to take part in more studies. To facilitate further development of the study, I would point out that there are two information lacks. The one is that seismology cannot predict earthquakes. Therefore, we cannot predict earthquake-induced landslides. Hence, evacuation principle does not work in contrast to mitigation of conventional gravity-induced landslides. The other lack of information is that we have to assess the likelihood of earthquake-induced landslides in slopes without knowing the subsurface conditions. We do not know mechanical properties of heterogeneous mountain body. We do not know geohydrology under the slope surface. Obviously, funding is too little to run subsurface investigation and provide local communities with life safety. It is my hope, therefore, that this book with ample information will become the first great step towards the development of safety against earthquake-induced disasters, including landslides.

Yokohama, Japan Ikuo Towhata

Contents

Report on Landslides Triggered by the 2008 Wenchuan Earthquake

Qiang Xu, Weile Li, Chuan Tang, Xuanmei Fan, Yonghong Luo, Jing Zhu, Chong Xu, Aiguo Xing, Yu Zhuang, and Ikuo Towhata

Abstract The devastating 2008 Wenchuan earthquake (Mw 7.9) was the largest seismic event in China in the past more than 50 years and is characterized by extremely high level of induced acceleration. For example, the strong ground motion recorded at the Wolong seismic station registered the EW, NS, and UD peak accelerations being 957.7, 652.9, and 948.1 Gal, respectively. Accordingly, this earthquake triggered nearly 200,000 coseismic landslides over an area of 110,000 km^2, out of which more than 800 dammed rivers, posing severe threats to downstream settlements. The total area of the coseismic landslides was about 1160 km^2 and the

Q. Xu (✉) · W. Li · C. Tang · X. Fan · Y. Luo · J. Zhu
State Key Laboratory of Geohazard Prevention and Geoenvironment Protection, Chengdu University of Technology, Chengdu, China
e-mail: xq@cdut.edu.cn

W. Li
e-mail: liweile08@mail.cdut.edu.cn

C. Tang
e-mail: Tangc@cdut.edu.cn

X. Fan
e-mail: fanxuanmei2014@cdut.edu.cn

Y. Luo
e-mail: luoyonghong2012@cdut.edu.cn

J. Zhu
e-mail: zhujing@cdut.edu.cn

C. Xu
National Institute of Natural Hazards, Ministry of Emergency Management, Beijing, China
e-mail: xuchong@ies.ac.cn

A. Xing · Y. Zhuang
State Key Laboratory of Ocean Engineering, Shanghai Jiao Tong University, Shanghai, China
e-mail: xingaiguo@sjtu.edu.cn

Y. Zhuang
e-mail: zyzhuangyu@sjtu.edu.cn

I. Towhata
Kanto Gakuin University, Yokohama, Japan

I. Towhata et al. (eds.), *Coseismic Landslides*, Springer Natural Hazards,
https://doi.org/10.1007/978-981-19-6597-5_1

total volume was about 10 km^3. These landslides are characterized by their great number, high density, large scale, widespread distribution, and clustering of occurrence. Analysis shows that they are positively correlated with the slope angle, the seismogenic fault distance and seismic intensity, and have dominant orientations, i.e., east, southeast, and south. Meanwhile, most landslides took place in the areas with sandstone, siltstone, and granitic rocks. Large-scale landslides (with an area of more than 50,000 m^2) were concentrated in the areas, where fault slip rates were highest near the intersections and junctures of individual segments of the faults. The head scarps of the large coseismic landsides are serrated, rough, and steep due to the tensile stress, and differ from the smooth arc-shaped scarps of gravitational landslides caused by shear stress. Aftershock monitoring has been carried out to better understand the slope response to seismic shaking. After the earthquake, the regions of strong shaking experienced several heavy rainstorms that induced many catastrophic landslides and debris flows. In this paper, the distribution and evolution of landslides in the post-Wenchuan earthquake area are discussed. Case studies of representative post-seismic landslide areas are presented as well. Predicting the post-earthquake landslides and assessing their long-term effect and risk are still of great concern and remain a challenge, due in part to lack of empirical data with sufficient temporal resolution.

1 Introduction

On May 12, 2008, a powerful earthquake (Mw 7.9) struck Sichuan Province in China. It resulted in 69,255 deaths, with 17,939 people missing, and 374,640 injured and was the country's deadliest earthquake in the past more than 50 years. The earthquake occurred in the NE-trending Longmenshan Thrust Fault zone, exhibiting the focal depth of circa 14–19 km. This zone separates the steep and strongly dissected eastern margin of the Tibetan Plateau from the Sichuan Basin (Xu et al. 2009a).

In addition to causing devastation through earthquake shaking, the earthquake triggered numerous landslides over a broad area. It was these landslides that caused about one-third of the total number of fatalities (Yin et al. 2009). Researchers have prepared a range of landslide inventory maps for the Wenchuan earthquake (Huang and Li 2009; Yin et al. 2009; Qi et al. 2010; Dai et al. 2011; Gorum et al. 2011; Parker et al. 2011; Xu 2012; Xu et al. 2014). Xu et al. (2014) compared four different maps and reported major differences in the number of landslides mapped. Xu et al. (2014), Dai et al. (2011), Gorum et al. (2011), and Huang and Li (2009) mapped about 196,000, 56,000, 60,000, and 11,300 landslides. The inventory of Xu et al. (2014) is largely an updated version of Dai et al. (2011)'s inventory. The occurrence of landslides during the earthquake was probably controlled by the strong and long-lasting ground shaking, steep, rugged topography, and the fragile and densely jointed bedrock (Huang and Fan 2013).

The frequency and magnitude of post-earthquake debris flows induced by subsequent rainfalls (i.e., Tang et al. 2009, 2012; Xu et al. 2012) were increased by the huge deposits from these coseismic landslides (estimated at 5–15 km^3 by Parker et al. 2011). Some of these landslides and debris flows blocked rivers, creating landslide dams and barrier lakes that threatened people and property by inundating areas upstream, and flooding areas downstream if the dam breached (Cui et al. 2009; Fan et al. 2012a). The coseismic landslides produced a tremendous amount of loose material on the hillslopes, promoting debris flows. Based on records in Beichuan (Tang et al. 2011a), the threshold of hourly rainfall for triggering debris flow was around 40% lower after the earthquake than before.

The post-earthquake landslides may last for many years; for example, at least 10 years for the 1999 Chi-Chi earthquake in Taiwan (Lin et al. 2006). Huang and Li (2014) predicted that the Wenchuan earthquake landslides will last 20–25 years. After the Wenchuan earthquake, post-seismic landslides have posed a danger to local people, and a lot of effort has been put into understanding the characteristics and mechanisms of their debris flows. Tang et al. (2011a) studied the landslides and debris flows that were induced by the first heavy rainfall in Beichuan region four months after the earthquake. Xu et al. (2012) also investigated the debris flows triggered by the heavy rain on 12–14 August 2010 in the Wenjia Gully and Zoumaling Gully in Qingping catchment, those in the Hongchun Gully near Yingxiu, and the Longchi debris flow near Dujiangyan City.

The sediment disasters induced by the Wenchuan earthquake have received much attention from geoscience communities, such as engineering geology, structural geology, earthquake geology, and hazard geology. The published reports address seismic response of slopes during aftershocks, coseismic landslides, subsequent debris flows induced by rainstorms, and the evolution of landslides in the area affected by the Wenchuan earthquake.

2 Geological and Tectonic Environments

The Longmenshan Mountain is a steep range located between the eastern Tibetan Plateau and the Sichuan Basin. The Indian Plate continues to underthrust beneath the Eurasian plate at a speed of 40 mm/yr, resulting in the formation of the Roof of the World, the Tibetan Plateau (Fig. 1a). The eastward and northward diffusion of the internal materials of the plateau lead to the growth of crust in two directions (Molnar et al. 1993; Molnar 2005), while the crustal materials migrate eastward along several large strike-slip fault systems (Molnar and Tapponnier 1975; Tapponnier and Molnar 1977; Tapponnier et al. 1982). In the eastern margin of the plateau, the eastward extrusion of the plateau materials is blocked by the rigid Sichuan basin, creating local nappe thrust tectonic belts and associated foreland basins (Tapponnier and Molnar 1977; Deng et al. 1994; Xu and Deng 1996), typically the Longmen thrust fault, which is the boundary between the Bayankala block and the Yangtze block. The regional lithology of the Longmen Mountain area consists of mixed Proterozoic basement rocks, a sequence of Paleozoic passive margin rocks, Mesozoic foreland basin rocks, and Cenozoic sediments (Burchfiel et al. 1995). The bedrock of the

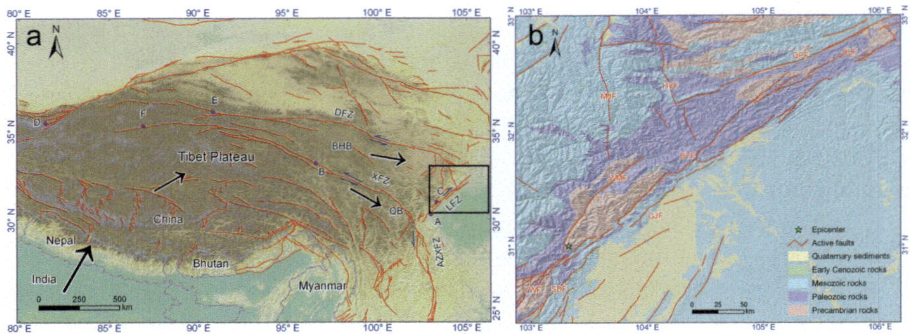

Fig. 1 Tectonics in Longmenshan and adjacent regions. **a** Tectonic setting of the Tibetan plateau. LFZ Longmenshan fault zone, XFZ Xianshuihe fault zone, AZXFZ Anninghe-Zemuhe-Xiaojiang fault zone, DFZ Dongkunlun fault zone, BHB Bayan Har block, and QB Qiangtang block. Earthquake epicenters: A Lushan event; B Yushu event; C Wenchuan event; D Yutian event, E Kunlun event; F Manyi event. **b** Tectonic setting of Longmenshan area. BYF Beichuan-Yingxiu Fault; GJF Guanxian-Jiangyou Fault; QPF Qingchuan-Pingwu Fault; CLF Chaba-Linansi Fault; HYF Huya Fault; MJF Minjiang Fault; WMF Wenchuan-Maowen Fault; YWF Yanjing-Wulong Fault; SDF Shuangshi-Dachuan Fault. The box in Fig. 1a shows the area shown in Fig. 1b. Active faults are drawn after Xu et al. (2016a, b)

Longmenshan Mountains ranges from Precambrian to Cretaceous in age, consisting of basalt, granite, phyllite, dolomite, limestone, sandstone, shale, and other types of rocks (Chigira et al. 2010).

The Longmenshan fault zone is about 500-km long and 40–50-km wide (Fig. 1b). This fault zone is extremely complex, being composed of three major reverse imbricate faults trending N45° E and dipping northwest at 50°–75°, i.e., the Back Range Fault (Wenchuan-Maoxian Fault), the Central Range Fault (Yingxiu-Beichuan Fault), and the Front Range Fault (Guanxian-Jiangyou Fault) (Deng et al. 1994; Wang and Meng 2009) (Fig. 2). The three faults converge into a single shear belt about 20-km below the surface, which is the main control structure of the Tibetan Plateau pushing over the Sichuan Basin (Burchfiel et al. 1995, 2008; Densmore et al. 2007). The present tectonic deformation rate is less than 2 mm/yr as shown by GPS observations across the Longmenshan fault zone on a decadal timescale (Zhang et al. 2008). Historical documents and modern instruments recorded only three earthquakes with magnitude greater than 6 in the central and southern segments of the Longmenshan Fault before the 2008 Wenchuan earthquake. They were the M6½ event on April 21, 1657, the M6¼ event on February 8, 1958, and the one with M6.3 on February 24, 1970.

The Longmenshan mountain area has a steep topographic escarpment and ranges in elevation from 500 to 5000 m over its 50-km width. Its ridges and valleys generally trend NE and run parallel to the geologic structures in the area. Crossing these trends, the Minjiang, Fujiang, Jinjiang, and Bailong Rivers flow from the north or northwest to the south or southeast. Within the deeply dissected topography, hillslope gradients are steep, particularly in the Longmenshan Mountains, where slopes are commonly steeper than 30° (Chigira et al. 2010; Ouimet 2010).

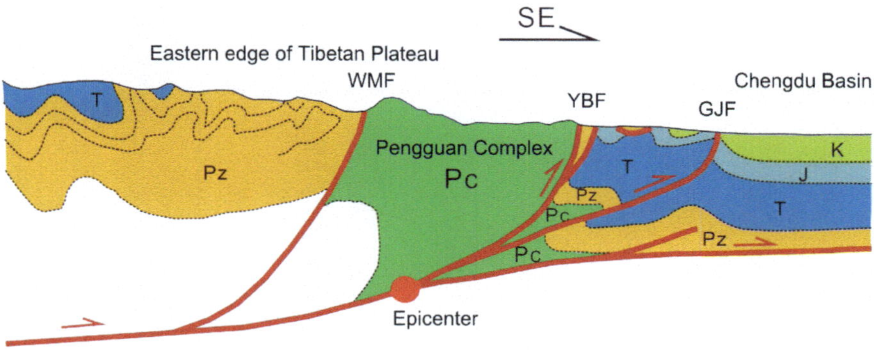

Fig. 2 Cross section of the Longmenshan and the adjacent areas. Pc: Precambrian; Pz: Paleozoic; T: Triassic; J: Jurassic; K: Cretaceous; WMF: Wenchuan-Maoxian fault; YBF: Yingxiu-Beichuan fault; GJF: Guanxian-Jiangyou fault. Modified from Zhang et al. (2008) and Burchfiel et al. (2008)

3 Earthquake Characteristics

The 2008 Wenchuan earthquake had a focal depth of 14–19 km, started near the base of the Yingxiu-Beichuan Fault, and propagated upwards. In more details, the seismological data suggests that the first rupture occurred in the southern Longmenshan Mountains and propagated toward the northeast over the distance of about 320 km, along the northwest dipping fault (Xu et al. 2009a). Its shaking was felt as far as in Beijing (1500-km distance) and Shanghai (1700 km), where high-rise office buildings were shaken. The induced tremor was also felt in nearby countries. Both of the 240-km long Yingxiu-Beichuan and 72-km long Jiangyou-Guanxian Faults underwent rupture. These faults are linked by a short northwest-striking rupture zone at the southern end through a lateral ramp (Fig. 3). The rupture process lasted for nearly 120 s, whereas most of the rupture energy was released during the first 80 s. Starting from Yingxiu, the rupture propagated at an average velocity of 3.1 km/s toward northeast. The seismic intensity of up to IX was registered on Modified Mercalli scale. Xu et al. (2009c) obtained the maximum vertical and horizontal displacement of 6.5 m and 4.9 m, respectively (Fig. 3). The largest surface slip, which was 11.0 m in the vertical direction and 4.5 m in the right-lateral direction, occurred near Beichuan Town (Liu-Zeng et al. 2009). Until 11:00 (local time), March 15, 2010, 72,046 aftershocks had been recorded. During the main shock, the maximum peak ground acceleration (PGA) was recorded at the Wolong seismic station (31.035° N, 103.182° E) with the EW, NS, and UD PGA components being 957.7 Gal, 652.9 Gal, and 948.1 Gal, respectively.

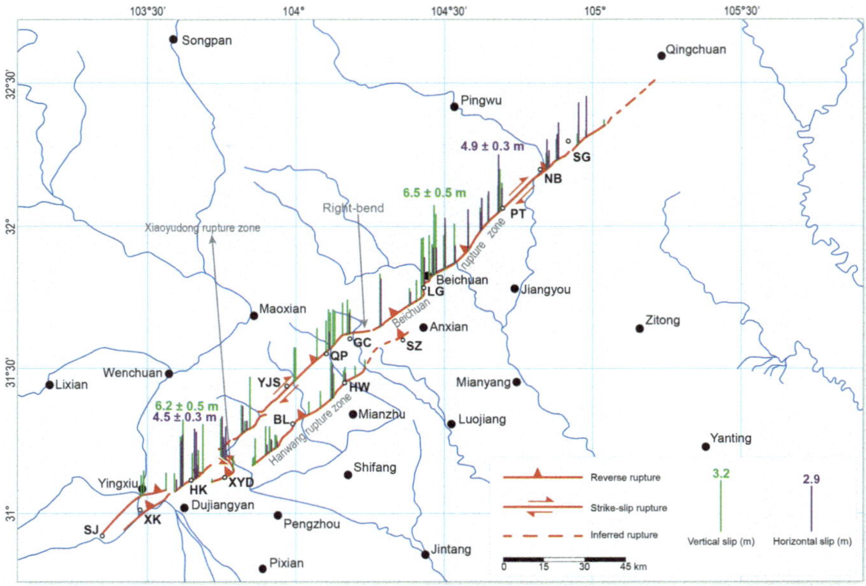

Fig. 3 Vertical and horizontal slip (*m*) during the Wenchuan earthquake (Revised after Xu et al. 2009b)

4 Slope Seismic Response Characteristics of the Aftershocks

The seismic response of a hill slope is a complex phenomenon, and the dynamic characteristics of seismic wave propagation affect the slope motion. The topography, geological structure, lithology, elevation, and many other factors influence the dynamic response of a slope. In this regard, the present section studies the recorded seismic motion upon different field profiles to demonstrate the characteristics of local seismic responses to a certain extent.

In order to understand the amplification effect in the seismic response of a slope response, the State Key Laboratory of Geohazard Prevention and Geoenvironment Protection (SKLGP) selected in early 2009 the Mt. Dong-Mt. Shizi area near Qingchuan County in the northern section of the Longmenshan fracture zone, and started a long-term observation of the aftershocks of the Wenchuan earthquake. Following the Wenchuan main shock, many aftershocks occurred along the Qingchuan-Pingwu Fault, which is the north section of the Longmenshan fracture. Three branch faults of the Qingchuan-Pingwu fracture cross the Qingchuan County (Fig. 4), Mt. Dong, and Mt. Shizi are located between the central and south branches of this fracture. The exposed strata therein are mainly Sinian and Silurian bedrocks with a lithology of dolomite and phyllite. During the Wenchuan earthquake, those mountains underwent a very strong dynamic response, and many seismic cracks were found in Weigan hill, Mt. Shizi, and Mt. Dong. In order to determine the seismic response characteristics of mountain slopes on the basis of the aftershocks, 12 seismic monitoring stations were placed in Weigan hill, Mt. Dong, and Mt. Shizi (Fig. 4). The basic information at each monitoring station is given in Table 1.

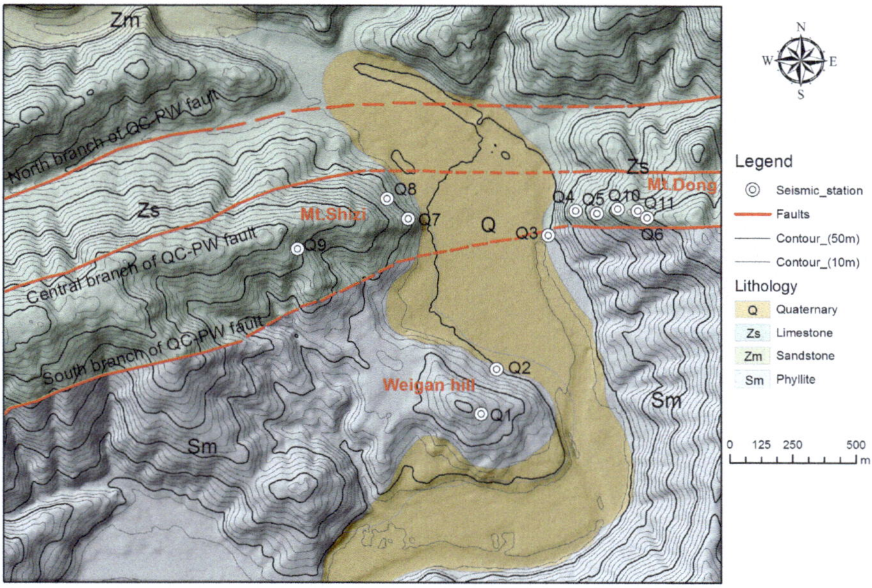

Fig. 4 Geological sketch map of Qingchuan county (Reproduced from Luo et al. 2020 with permission of Elsevier)

Table 1 Information on seismic monitoring stations

ID	Elevation (*m*)	Geological condition	Location
Q1	875	Bedrock (Phyllite)	Weigan hill
Q2	805	Bedrock (Phyllite)	Weigan hill
Q3	788	Bedrock (Phyllite)	Mt. Dong
Q4	871	Bedrock (Phyllite)	Mt. Dong
Q5	921	Bedrock (Dolomite)	Mt. Dong
Q6	962	Bedrock (Dolomite)	Mt. Dong
Q7	902	Bedrock (Dolomite)	Mt. Shizi
Q8	893	Bedrock (Dolomite)	Mt. Shizi
Q9	962	Bedrock (Dolomite)	Mt. Shizi
Q10	960	Bedrock (Dolomite)	Mt. Dong
Q11	981	Bedrock (Dolomite)	Mt. Dong

More than 200 aftershocks were recorded from 2009 to 2016, with magnitudes ranging between Ms = 1.3 and 5.2. Their focal depth was less than 30 km, and the epicentral distance from the monitoring stations was between 2.0 and 46.0 km.

In Mt. Dong, six stations (Fig. 4) were distributed from the slope foot to the mountain top. Among them, three stations (Q3, Q4, Q6) were put in three short tunnels (<20 m), respectively, while other stations (Q5, Q10, Q11) were installed on the slope surface. Because all these stations were set at different times, only four typical aftershocks were recorded by five stations at the same time. Figure 5 presents the amplification characteristics of PGA at Q4, Q5, Q10, and Q11, using Q3 as the reference. It can be observed that both horizontal and vertical PGAs are amplified between 1.08 and 4.64, and Q5 shows the strongest amplification. And the horizontal PGA of north–south component shows bigger amplification than the east–west and vertical components. It implies that the east–west elongation of Mt. Dong ridge (Fig. 4) makes for transverse direction of the ridge producing stronger amplification effects.

On the other hand, between 2009 and 2010, the Q3, Q4, Q6, Q8, and Q9 stations monitored dozens of earthquakes. Figure 6 presents the PGA amplification features of typical recordings in Mt. Dong and Mt. Shizi. In comparison with Q3, Q4 shows

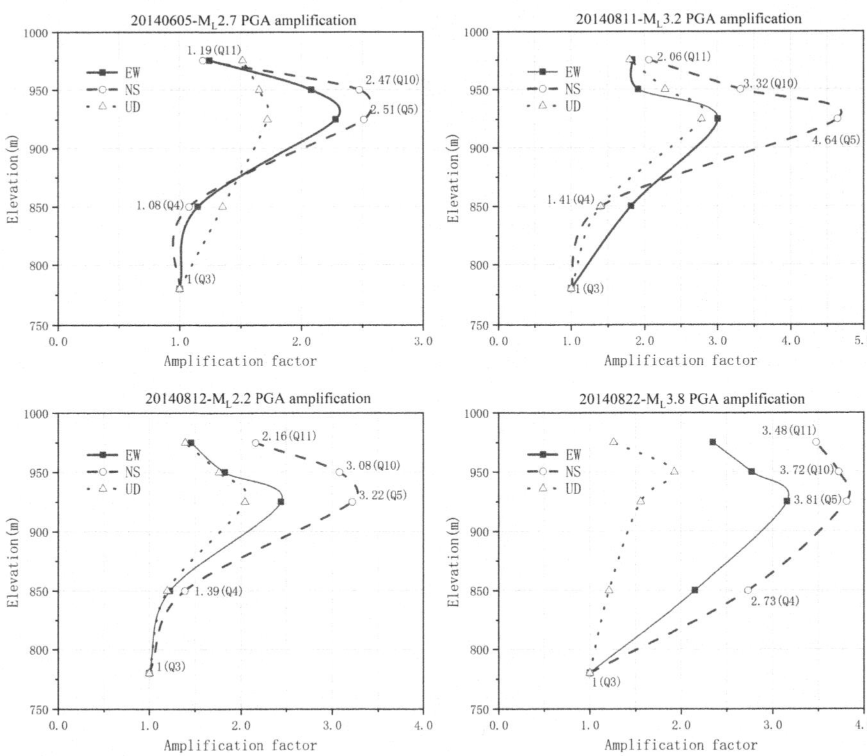

Fig. 5 PGA amplification factor of Q4, Q5, Q10, and Q11

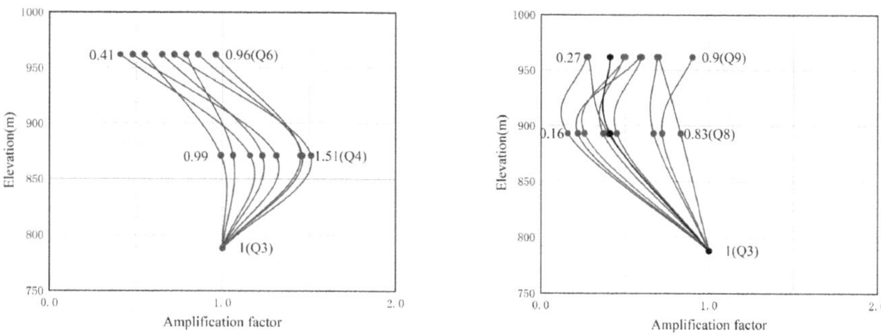

(a) Combined PGA amplification of horizontal and vertical components in Mt. Dong

(b) Combined PGA amplification of horizontal and vertical components in Mt. Shizi

Fig. 6 PGA amplification factor of Q4, Q6, Q8, and Q9

the amplification to be between 0.99 and 1.51, but Q6 shows attenuation of dynamic responses (the amplification factor less than 1.0) in Mt. Dong. In Mt. Shizi, Q8 shows the amplification factor less than 0.83, and the Q9 shows a little bigger amplification factor than Q8. The PGA response characteristics shows an opposite amplification trend at the similar elevation of Q4 and Q8. Although the topography features (such as the steep slope, big curvature and high elevation) alone can promote the amplification effect, and the combined effects of topography and layer should enhance the extent of amplification (Luo et al. 2020), it seems that the abundant groundwater at Q8 site may reduce the strong dynamic response at Mt. Shizi slope.

5 Characteristics and Spatial Pattern of the Coseismic Landslides

5.1 Coseismic Landslide Inventory

After the Wenchuan earthquake, many researchers produced inventories of landslides triggered by this earthquake. Table 2 lists ten typical inventories out of them. These inventories can be used to analyze information on the spatial patterns, and the hazards and mechanisms of the earthquake-triggered landslides. However, most of these inventories have shortcomings such as being incomplete or being subject to errors related to automatic extraction. Moreover, some inventories indicate landslide boundaries by points in place of polygons, or group many individual landslides into one landslide. Consequently, significant differences occur among the inventories. It seems likely that the differences arose from the research aims, seriousness, and professional knowledge of the researchers. In this section, the author presents his own inventory (Xu et al. 2014). It strictly follows the criteria for landslide inventories (Xu 2015) and is the most detailed and complete inventory of landslides triggered by

Table 2 Inventories of landslides triggered by the Wenchuan earthquake

Method	Inventory type	Landslide number	Landslide area	Integrity of coverage	Source
Visual interpretation	Point	11,308		Near complete	Huang and Li (2009)
Visual interpretation	Polygon	43,842	632	Complete	Li et al. (2013)
Visual interpretation	Point, Polygon	13,085	418.85	Near complete	Qi et al. (2010)
Visual interpretation	Point	60,104		Near complete	Gorum et al. (2011)
Automatic extraction	Polygon	73,367	565.8	Near complete	Parker et al. (2011)
Visual interpretation	Polygon	48,007	712	Near complete	Xu et al. (2009c)
Visual interpretation	Polygon	56,000	811	Near complete	Dai et al. (2011)
Visual interpretation	Polygon	5154	1139	Complete	Guo et al. (2010)
Automatic extraction	Polygon	57,150	396	Complete	Li et al. (2014)
Visual interpretation	Polygon	197,481	1160	Complete	Xu et al. (2014)

the quake, including those landslides of small scales or in areas far from the human settlement.

Figure 7 shows the author's inventory map of landslides triggered by the 2008 Wenchuan earthquake (Xu et al. 2014). Herein, 197,481 landslides were delineated throughout the greater elliptical area of about 110,000 km^2 as shown in the map. The total area occupied by these landslides is about 1160 km^2. This elliptical boundary is located far away from the Yingxiu-Beichuan Fault and coseismic landslides are quite rare, whereas the employed satellite images are of low resolutions. Accordingly, most of the landslides were identified in the smaller area of about 44,000 km^2. In this area, 196,007 landslides were mapped and occupy the total area of about 1150.6 km^2. The landslide area percentage (LAP) in this landslide-prone area is 1150.622 km^2/44,031.130 km^2 $\times$ 100% = 2.613%, and the landslide number density (LND) is 196,007 sites/44,031.130 km^2 = 4.452 km^{-2}. The total volume of the landslides is estimated to be about 10 km^3 (Xu et al. 2016a), based on an empirical correlation between the individual landslide volume (V) and the area (A), $V = 1.3147 \times A^{1.2085}$. This landslide inventory is the largest, detailed, and most complete inventory related to an individual earthquake ever recorded in the world.

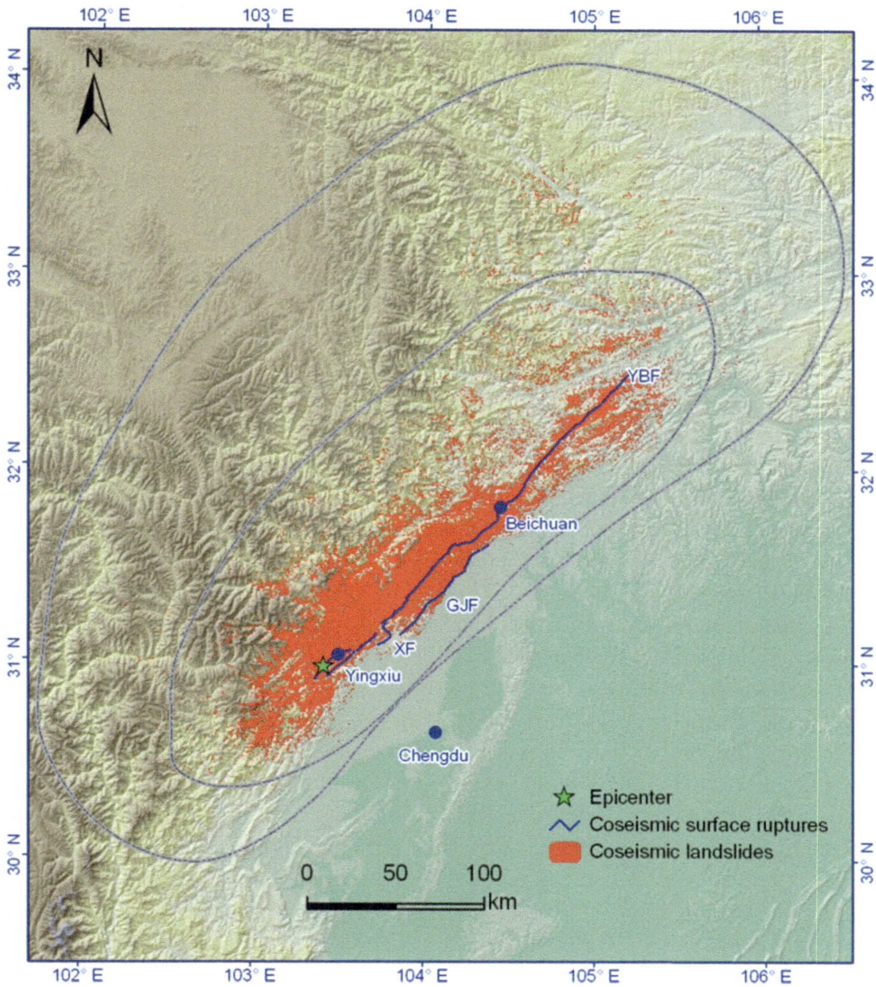

Fig. 7 Coseismic landslides induced by the 2008 Wenchuan earthquake (Reproduced from Xu et al. 2014). The two elliptical areas represent the approximate limit (larger) and the main distribution area (smaller) of the landslides, respectively. YBF, Yingxiu-Beichuan fault; GJF, Guanxian-Jiangyou fault; XF, Xiaoyudong fault

5.2 General Characteristics and Distribution Pattern of Coseismic Landslides

The Wenchuan earthquake-triggered landslides that are characterized by great number, high density, large scale, wide distribution, and clustering of occurrence. These slope failures are mainly shallow disrupted landslides, but also include other types such as rock falls, landslides, rock avalanches, and soil avalanches. Most of them have long runout distances, had high speeds, short occurrence times, large energy, broken landslide bodies, complex mechanisms, and remarkable jumping phenomenon. An air cushion effect was recognized during the occurrence of the large landslides, resulting in high speed and long runout distance, which caused serious disasters (Xu et al. 2014).

The spatial distribution of the landslides shows an ellipse, of which the long axis is aligned with the strike of the seismogenic faults (Yingxiu-Beichuan and Guanxian-Jiangyou Faults), with obvious asymmetry to the seismogenic faults. Most of the coseismic landslides concentrated along the faults, indicating a strong control by the causative faults. Most of them occurred to the northeast of the epicenter and extended as long as 300 km. The elongation of spatial distribution of the landslides is consistent with the spread and propagation of the seismic waves. The landslides also are concentrated on the hanging wall of the Yingxiu-Beichuan surface rupture, especially the section between Yingxiu and Beichuan, dominated by reverse faulting with a dextral strike-slipping component. In spite of the concentration of coseismic landslides along the Yingxiu-Beichuan fault, the difference in landslide abundance along the fault is quite significant. The area controlled by reverse faulting with a dextral strike-slip component between Beichuan and Yingxiu registered many more coseismic landslides than the area northeast to Beichuan County controlled by a dominant dextral strike-slipping fault. Overall, the places with the highest abundance of landslides are mainly located on the hanging-wall side of the fault segment between Yingxiu and Beichuan of the Yingxiu-Beichuan Fault, the area between the Yingxiu-Beichuan Fault and the Guanxian-Jiangyou Fault, the stripe area extending northeast to Beichuan along the Yingxiu-Beichuan Fault, and the distant area along rivers on the hanging wall of the Yingxiu-Beichuan Fault. Although the area between the Yingxiu-Beichuan and Guanxian-Jiangyou surface ruptures registered a large number of coseismic landslides, the landslide abundance of this area is less than in the hanging wall area of the segment of the Yingxiu-Beichuan Fault.

The local topography also controls the location of coseismic landslides, and the topographic amplification of the seismic wave triggered many landslides on the middle and upper slopes. The coseismic landslides also tend to occur along rivers in some areas. The slopes facing southeast; i.e., the movement direction of the hanging wall during the quake has a higher landslide abundance than the slopes with other slope aspects. In general, the landslides have positive correlations with the slope angle, distance from the seismogenic faults, and seismic intensity. The slopes with sandstone, siltstone, and granitic rocks bear more landslides. The earthquake not only triggered massive landslides, but also deteriorated the mechanical strength of rocks in hillslopes in the affected area. Such affected slopes are more likely to fail during strong rainfall or aftershocks in the future (Xu et al. 2014).

6 Large-Scale Landslides Induced by the Wenchuan Earthquake

6.1 Characteristics and Distribution Pattern of Large-Scale Landslides

112 large-scale landslides with areas greater than 50,000 m^2 were identified by interpretation of remote sensing imagery and field investigations (Fig. 8, Xu et al. 2011). Their deformation and failure are distinct from those of general landslides that were induced by gravity (Xu et al. 2016b). Under the influence of gravity, sliding on a rock slope is often associated with a gentle tension crack at some depth on their rear side (Fig. 9a). This crack is connected to a shear plane in the lower part of the slope to form a failure surface. In contrast, seismic shaking tends to produce deeper and steeper tension cracks on the rear side, promoting shear deformation in the lower part of the slope (Fig. 9b).

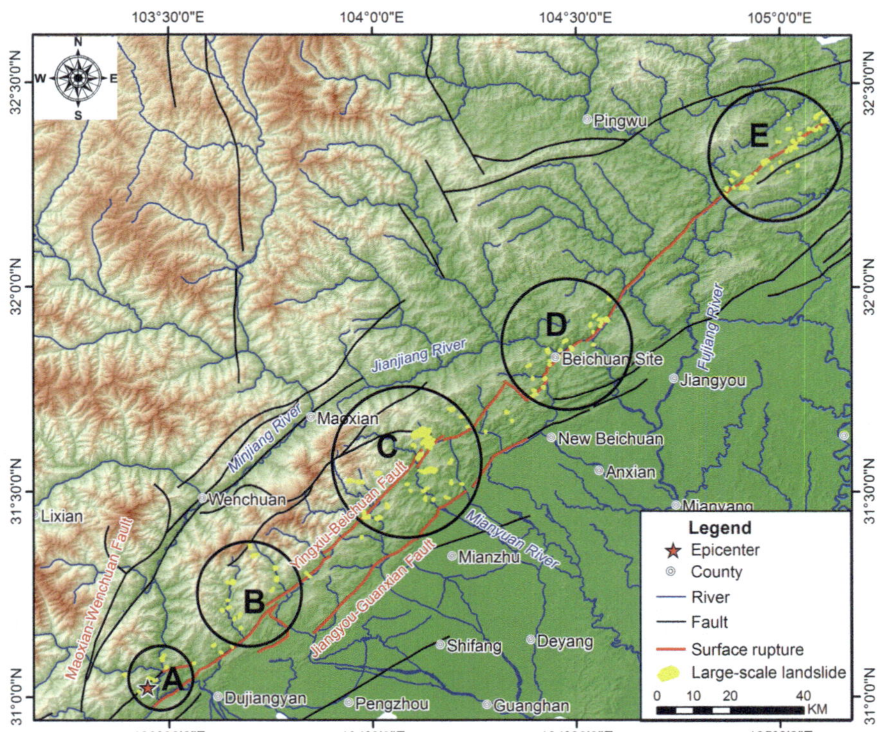

Fig. 8 Distribution map of large landslides triggered by the 2008 Wenchuan earthquake (Huang and Li 2014)

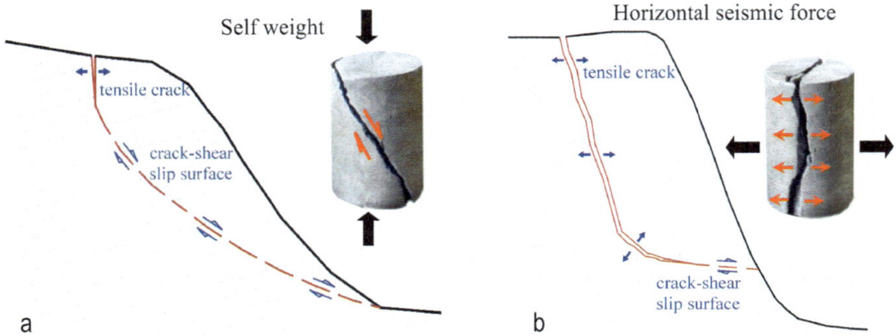

Fig. 9 Development of a sliding surface in a rock slope **a** induced by gravity and **b** due to the effect of earthquake shaking (Reproduced from Xu et al. 2016b)

The present study analyzed the distribution of large-scale landslides along the Yingxiu-Beichuan Fault, which extends over a rupture length of 240 km (Xu et al. 2009a; Liu-Zeng et al. 2009; Gorum, 2013). In this area, two distinct types of fault occurred along the entire length of the fault near Beichuan City. Herein, the dominant type of fault changes from reverse to strike-slip types. The analysis suggested that the distribution of large landslides was affected by the following four major factors.

(1) Distance from the fault: Large landslides were situated within a limited distance from the surface manifestation of the seismic fault; about 40% of them occurred within 1 km, 70% within 3 km, and more than 80% within 5 km of the surface manifestation of the fault. No large landslides occurred beyond 13 km.
(2) The hanging wall: Among the large landslides, more than 70% occurred on the hanging-wall side of the fault presumably because the peak acceleration was higher there.
(3) Locked segments of the fault: Large landslides were concentrated in the areas A, B, C, D, and E (see Fig. 8). The C area was associated with the highest concentration, as shown by the two largest landslides at Daguangbao and Wenjiagou. The zones with high landslide concentrations occurred along the locked segments of the Yingxiu-Beichuan Fault, which showed major dislocations and rupture phenomena. Due to the rupture and displacement of the seismic fault, a large amount of seismic energy was released in these areas.
(4) Direction of slopes: The face slopes (slopes dipping toward the earthquake wave source) had fewer large landslides than the slopes that dipped in the same direction as the seismic wave traveled (back slope). In each section, the large landslides slid in the direction of rupture propagation of the seismic fault. In the Qingchuan area (area E), where the main fault activity was strike-slip movement, many landslides appear to have slid and moved in the NE direction, which is the direction of rupture propagation.

6.2 Daguangbao Landslide, the Largest Event Induced by the Wenchuan Earthquake

The Daguangbao landslide, which was the largest landslide induced by the Wenchuan earthquake, is located in Anxian County, Sichuan Province. It covered an area of 7.8 km^2, had the maximum width of 2.2 km and the debris volume estimated to be 1200 million m^3 (Fig. 10). Daguangbao landslide is on the hanging-wall side of the Yingxiu–Beichuan Fault.

The debris of this landslide traveled about 4.5 km and blocked the Huangdongzi Valley, where it formed a landslide dam which was nearly 600-m high (Huang et al. 2012). The source area was about 2.4-km long and 1.2-km wide, and the deposit is approximately 3.0-km long and 2.2-km wide. The head scarp of the Daguangbao landslide is serrated and nearly vertical (70°–90°), with the maximum height of 700 m. This landslide is located at an elevation between 2200 and 2900-m asl (Fig. 11a). Many landslides triggered by the Wenchuan earthquake have very steep and high head scarps, with tensile features generated by the long-lasting and strong shaking. The southern boundary of the Daguangbao landslide also has a steep and high cliff similar to the head scarp, with an average slope angle of 55°–70°. The cliff strikes to the NE–SW direction, and is controlled by two joint sets. The evidence of shattering and cracking is still visible at the northern boundary (Fig. 11b) and the rear side of the landslide (Fig. 11c).

Fig. 10 Spot-5 satellite image of Daguangbao landslide

Fig. 11 Steep head scarp (**a**) the tension cracks and fractured rock mass remaining at the northern boundary and (**b**) and the rear side of the Daguangbao landslide (**c**)

6.3 The Second Largest Landslide Induced by the Wenchuan Earthquake—Wenjiagou Landslide

More than 100 rock avalanches with volumes greater than 10 million m^3 were triggered by the 2008 Wenchuan earthquake. Among these massive landslides, the rock avalanche with the longest runout occurred at Wenjiagou, Mianzhu, Sichuan (Fig. 12). Approximately, 27.5 million m^3 of sandstone, coming from the source area, was included in this landslide. This material traveled about 4170 m, and its descending was about 1360 m in elevation. This movement is equivalent to a fahrböschung angle (Heim 1932; Hsü 1975) of 16.9°. The debris covered an area of 1.5 million m^2 and had a volume of approximately 49 million m^3. This event was the second largest landslide induced by the Wenchuan earthquake, and destroyed the village of Yanjing, killing 48 people and burying houses at the entrance of the Wenjiagou Valley.

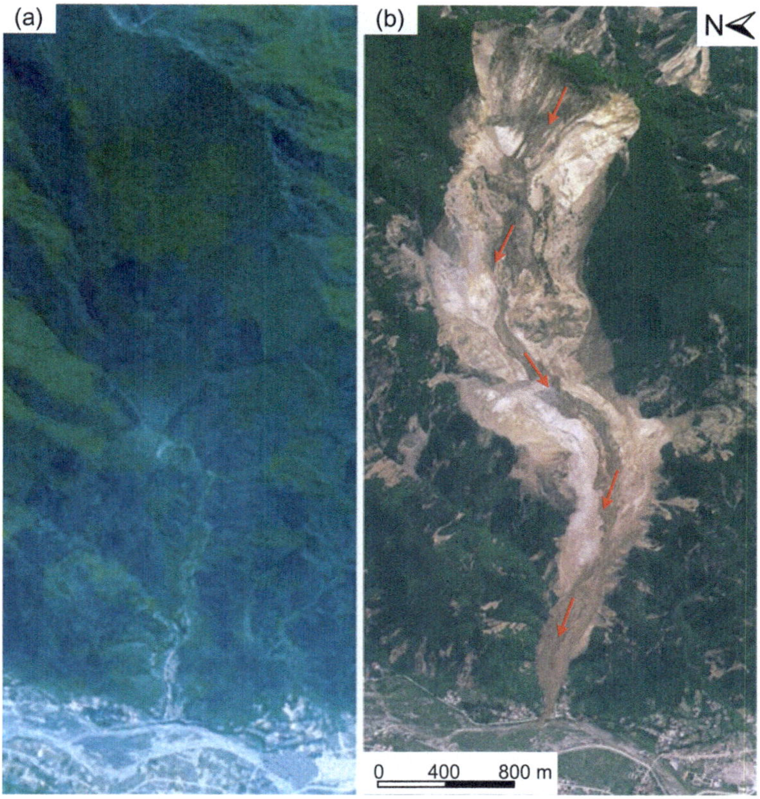

Fig. 12 Aerial images of the Wenjiagou rock avalanche prior to (**a**) and after the landslide (**b**) (Reproduced from Xing et al. 2017 with permission from Springer)

The Wenjiagou landslide occurred on the left bank of the Mianyuan River near Qingping in Mianzhu County, Sichuan, at about 80-km northeast of the epicenter of the Wenchuan earthquake. This landslide occurred in a region of low to medium relief within a moderately incised valley. The elevation of this landslide varied from 2402 m at the summit of Mount Jiudingshan to 883 m at the entrance of the Wenjia Valley. The valley had been incised to the depth of 30–50 m. The valley bottom has a gradient from 150 to 180‰, and the gradient of the side slopes lies between 35° and 55°.

The landslide movement resumed as a rock slide in a mass of limestone. Undergoing strong earthquake motion, this rock mass detached from its source, and on the left side, ran down rapidly in the direction of 283°. This landslide mass struck the Hanjia Platform and was then transformed into a debris flow. Crossing a 280-m high scarp at the elevation of 1599 m, the debris ran up on the opposite (right bank) slope of the valley at location A (see Fig. 13). The debris then collided with the right bank and changed direction to cross the platform at an elevation of 1300 m. At the valley

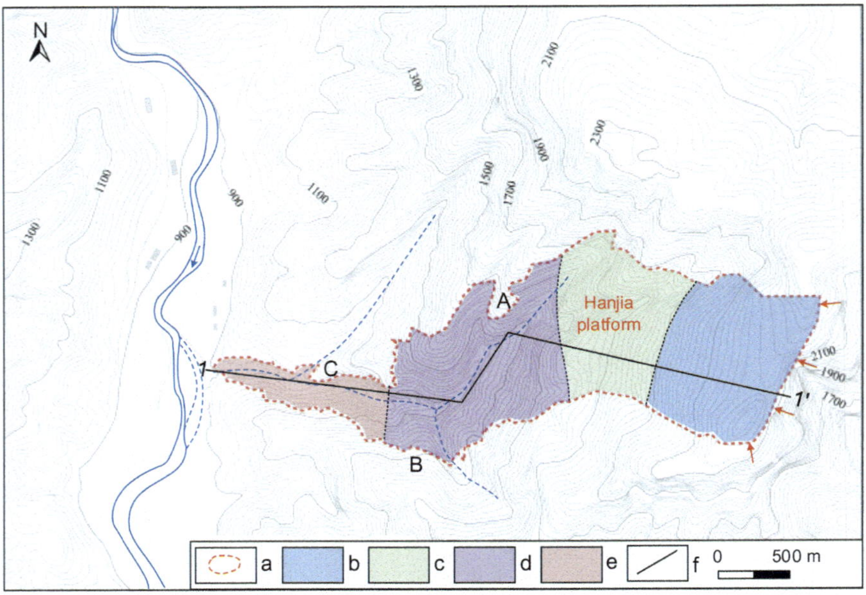

Fig. 13 Topography of the Wenjia valley. **a** Landslide boundary. **b** Source area. **c** Hanjia platform (Zone I). **d** Zone II. **e** Zone III. **f** Cross section line. Red arrows indicate the direction of the movement of the sliding mass (Reproduced from Xing et al. 2017 with permission from Springer)

bend at location B in Fig. 13, the debris flow became superelevated and changed direction again, traveling along the valley in the direction of 278°. It then ran up a ridge with a trend of 070° (location C). After colliding with the right bank, the debris continued to move along the valley and came to rest at the entrance of the valley. The landslide created an air blast that was powerful enough to knock down trees along its sides and front margin.

The flow of the rock avalanche after onset of the landslide was simulated using the numerical model (DAN3D) (McDougall and Hungr 2004, 2005). This numerical tool suggested the duration of the movement to be 240 s as well as the average flow velocity to be 17.4 m/s (Fig. 14). In the three major impact zones, the simulated run-ups and the superelevations were compared against the measured trimline. It was then indicated that the flow velocities are reasonably reproduced by the numerical tool. Figure 15 shows the maximum simulated flow velocities along the runout path. The greatest value of the velocity was worked out near the toe of the rupture surface, which was 56 m/s. The calculated velocities were also examined against the run-ups and super elevations. After passing the cliff at an elevation of 1599 m, the debris ran up the opposite slope on the right bank of the valley at location A. Then, the debris flows superelevated at the valley bend at location B and ran up the ridge at location C.

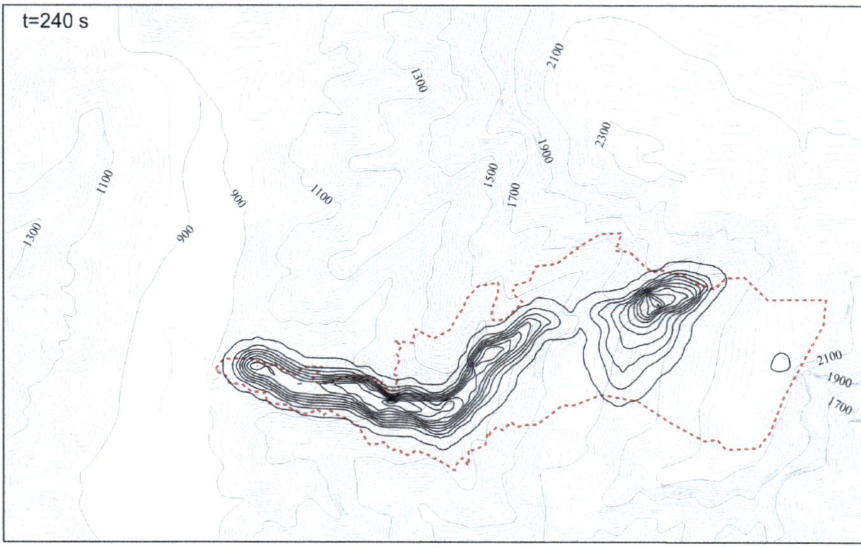

Fig. 14 Time-lapse image of the DAN3D simulation at 240 s. The contours of the deposit depth are drawn with 10-m intervals and the elevation contours are with 20-m intervals (Reproduced from Xing et al. 2017 with permission from Springer)

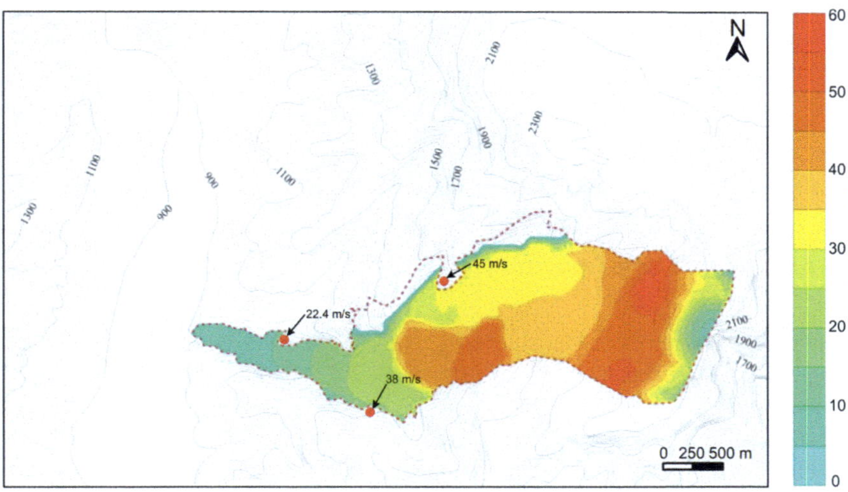

Fig. 15 Distribution of the maximum simulated flow velocities along the runout path (DAN3D program). The maximum velocity contours are at 5-m/s intervals and the elevation contours with 20-m intervals (Reproduced from Xing et al. 2017 with permission from Springer)

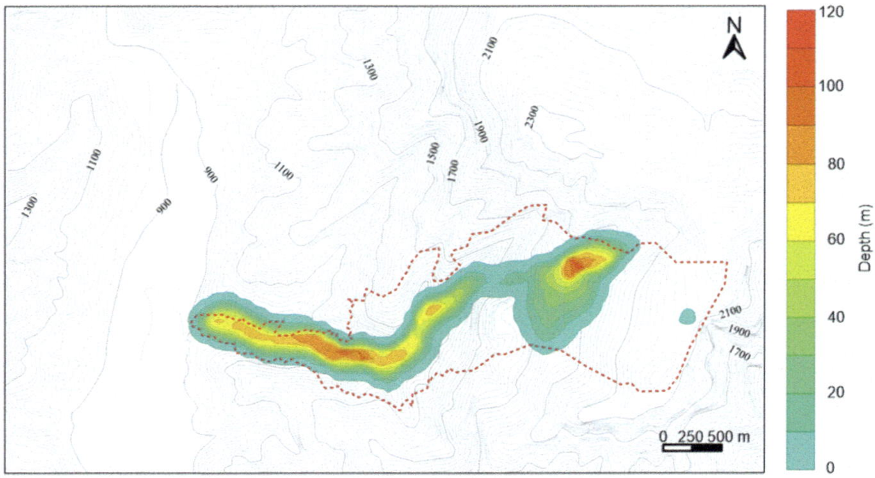

Fig. 16 Final distribution of the debris at 10-m intervals (obtained by DEM) (Reproduced from Xing et al. 2017 with permission from Springer)

The distribution of the debris obtained by the DAN3D simulation is reasonably consistent with what was obtained by DEMs constructed from 1:50,000-scale topographic maps (Fig. 16). The final volume of the deposit was estimated to be 49.0 million m^3, which is different from the aforementioned volume because materials were entrained along the runout path.

7 Landslide Dams Induced by the Wenchuan Earthquake

7.1 Landslide Dam Inventory

The unprecedent number of landslides triggered by the 2008 Wenchuan earthquake (Huang and Li 2009; Dai et al. 2011; Gorum et al. 2011) blocked rivers and produced landslide dams (Cui et al. 2009; Fan et al. 2012a, b). Among those dams, 32 were breached intentionally to drain water and to reduce the risk of catastrophic floods (Xu et al. 2009b). This event promoted the studies on coseismic landslide dams, in which their causal factors, spatial distribution, dynamic decay, and impacts were focused on.

The detailed interpretation of remote sensing imageries produced a database including 828 landslide dams as a consequence of the Wenchuan earthquake. This number comprises approximately 1.4% out of more than 60,000 coseismic landslides mapped and attributed to during this earthquake. This database also uses data in common with Gorum et al. (2011) who mapped landslides in the same affected region, using 52 satellite imageries before and after the earthquake. Furthermore, the present data made uses of the EQ-1 (10-m spatial resolution) in July 2008 and ASTER imageries (15-m resolution) in July 2010 for mapping landslide dams and reservoirs. Then, the identified dams were examined virtually by comparing the pre- and post-earthquake false color composites or panchromatic imageries supplemented by field reconnaissance within the accessible areas. This field check took care of 60% of the dams. Note that the limitation of the resolution of imageries did not allow to capture the landslides and reservoirs smaller than 200 m^2. Consequently, the landslide dams were classified into two groups, which are (1) full blockage of rivers by landslide debris and (2) partial blockage that could not produce water impoundment. Among the 828 river blockings contained in the author's inventory, 501 of them (61%) completely dammed rivers, whereas 327 (39%) were partial damming of river flow. The complete damming landslides had the area of 1249 m^2 to 7.1×10^6 m^2, while the partial damming events were slightly smaller with the area of 768 m^2 to 1.3×10^6 m^2. The total area of the landslide dams was approximately 54 km^2, and the produced lake area varied from 217 m^2 to 6.5×10^6 m^2.

Figure 17 shows the locations of landslide dams in the area of high landslide distribution density suggested by Gorum et al. (2011). The classification of the river-blocking dams in this figure follows the terminology by Cruden and Varnes (1996). Rock slides are the majority among them because the fractured rocks slopes in this area are highly prone to seismic shaking (Chigira et al. 2010). Thus, it occupies 35% of the entire river blockage. First, the aforementioned Daguangbao landslide was the largest coseismic landslide, and its area as well as the estimated volume were 7.8 km^2 and 1.17×10^9 m^3, respectively (Huang and Fan 2013). Second, the dam produced by the Tangjiashan landslide impounded, at maximum, 3×10^8 m^3 of water. The problem herein was that the second largest city of Sichuan Province, the Mianyang City having the population of about 1.2 million, was located at the distance of approximately 85 km in the downstream direction. In fear of its possible devastating breaching, the Tangjiashan natural dam was artificially breached after evacuating the vulnerable people in the city (Liu et al. 2009) and the lake water was drained.

7.2 Spatial Distribution of Landslide Dams

A moving circular window of 1 km^2 in area was employed to assess the spatial density of the coseismic landslides (Gorum et al. 2011) and landslide dams. As shown in Fig. 18, their maximum densities were 54.1 landslides/km^2 and 4.5 landslide dams/km^2, respectively. The profiles of the landslide and dam densities along the fault were illustrated by picking up the events within a rectangular swath which was 240-km long and 25-km wide (Fig. 18c) and calculating the mean normalized point densities. Further, four smaller and perpendicular swarths of 25 km by 20 km were employed for details (Fig. 18c 1–4). The swath width of 25 km was thought relevant because it is equivalent with the surface projection of the hanging wall of the Yingxiu-Beichuan Fault. Both profiles of the landslides and dams in Fig. 19c follow the same

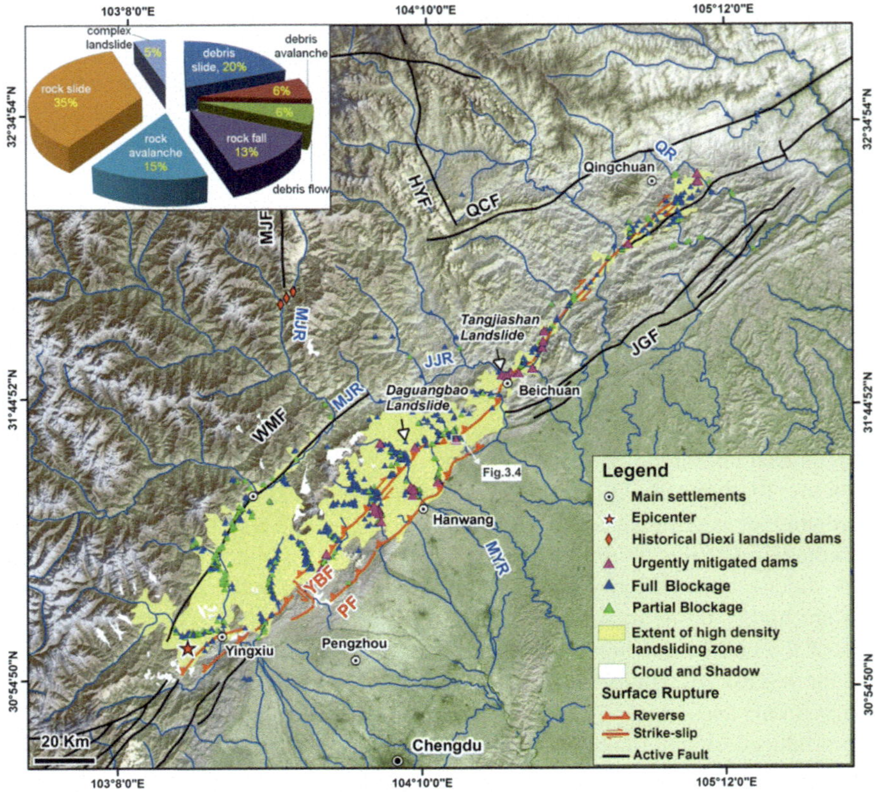

Fig. 17 Distribution of landslide dams triggered by the Wenchuan earthquake (drawn after Fan et al. 2012a). Yellow color shows the high landslide area density exceeding 0.1 km^{-2}. The epicenters of historical earthquakes and historical Diexi landslide dams are illustrated. White polygons show unmapped areas due to clouds and shadows in the post-earthquake imagery. The Pengguan Massif bounded by the white dashed curve has catchments such as the Min river, Jian river, Fu river, Qing river, and 7 smaller rivers (P1–P7) in. WMF: Wenchuan-Maowen Fault; YBF: Yingxiu-Beichuan Fault; PF: Pengguan Fault; JGF: Jiangyou-Guanxian Fault; QCF: Qingchuan Fault; HYF: Huya Fault; MJF: Minjiang Fault (after Xu et al. 2009a). The source of this figure is Fan et al. (2012a) with permission from Elsevier

trend, exhibiting the maximal in the SW section between Yingxiu and Beichuan as well as near the fault tip near Qingchuan. The maximum density of the landslide dams occurred at several river junctions because of the abundance of coseismic landslides there. The landslide dams are most abundant in steep watersheds (P1–P7) of the Pengguan Massif and along the thrust segment of the Yingxiu-Beichuan Fault (YBF in Fig. 18b). Thereafter, the density decreases to the NE direction, extending from Beichuan Town to the tip of the fault.

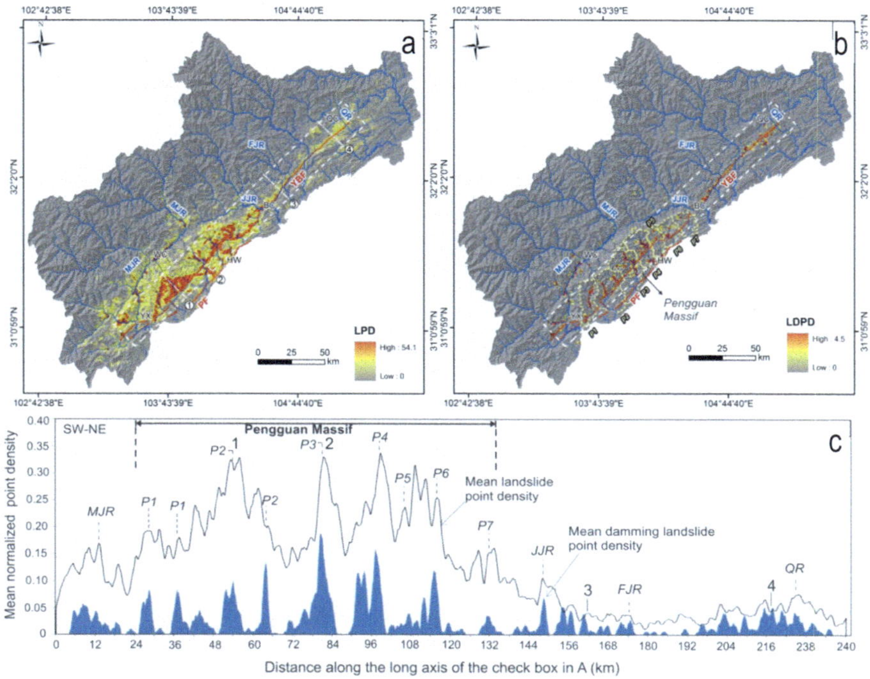

Fig. 18 Densities of river blocking and non-blocking landslides (modification of Fan et al. 2012a with permission from Elsevier). **a** and **b** show the landslide and landslide dam point density. White dashed lines illustrate 240 km by 25-km swath profiles and cross Sects. (1–4). LPD and LDPD stand for the landslide and landslide dam point density. **c** Mean normalized landslide and landslide dam densities along the SW-NE profile. Red lines are Yingxiu-Beichuan Fault (YBF) and Pengguan Fault (PF). Yellow dash lines are the boundary of the P1–P7 watersheds in the Pengguan Massif. YX, WC, BC, and QC are the cities of Yingxiu, Wenchuan, Beichuan, and Qingchuan. MJR, JJR, FJR, and QR represent the Min, Jian, Fu, and Qing rivers

7.3 Simulation of Dam Breaching and Flood, and Prediction on the Tangjiashan Landslide Dam

Floods caused by breaching of landslide dams are a profound threat to people and property in the downstream region. Therefore, it is essentially important to urgently assess the risk of dam breaching and to reduce the potential risk, if there is any. The city of Beichuan is situated at about 5-km downstream of the Tangjiashan landslide dam that is located at 31.84° N, 104.43° E (Fig. 19). This city was completely devastated by the Wenchuan earthquake followed by the large coseismic landslides. It was later decided that this city would be preserved in memory of the disaster without reconstruction. The landslide blocked the Tongkou River (otherwise called Jianjiang), which has an upstream catchment area of 3550 km^2.

Chinese authorities evacuated partially Beichuan until the Tangjiashan landslide dam was artificially breached and the lake was drained through a spillway (Liu

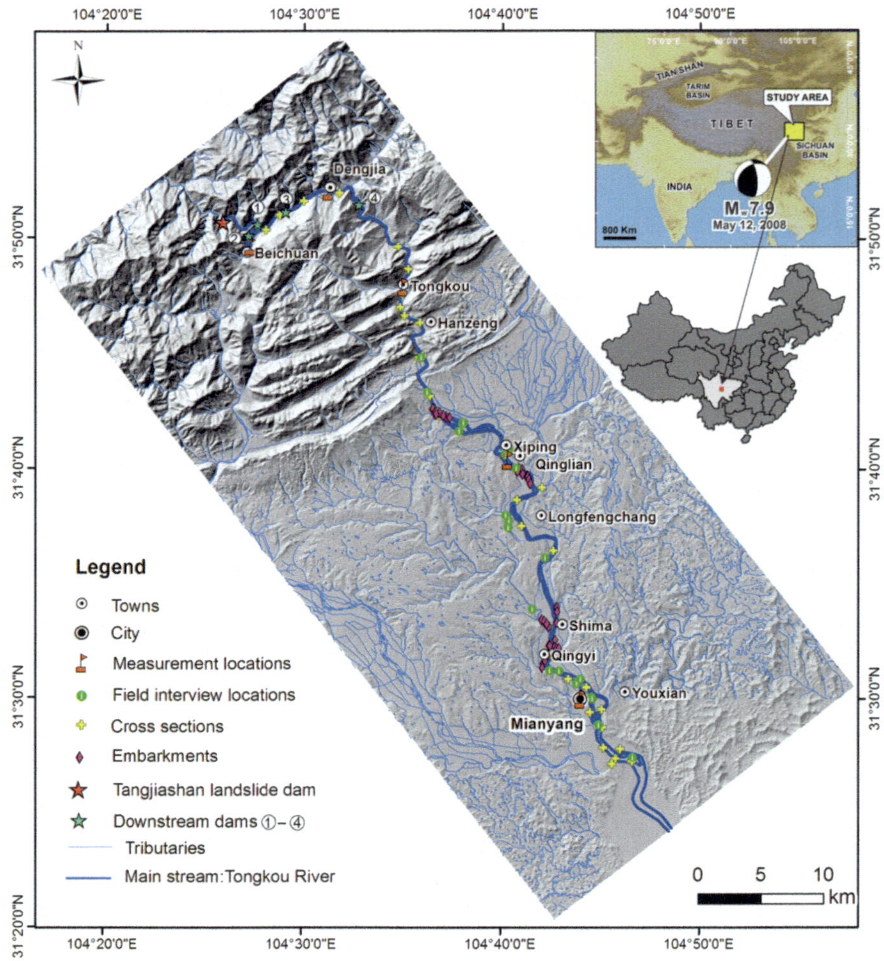

Fig. 19 Locations of the Tangjiashan landslide dam and the field investigation in the Beichuan hydraulic station (after Fan et al. 2012c)

et al. 2009). To simulate the breaching-induced flood, the BREACH model and the SOBEK 1D-2D model developed by Deltares were integrated. The BREACH model can simulate the outflow hydrograph from a dam and quantify the dam breach. The base scenario for calibration of the model was based on the 2008 event together with the spillway. The outputs of the BREACH model under different scenarios were used in the SOBEK model.

There were four different scenarios; refer to Fan et al. (2012c) for the modeling parameters and settings for different scenarios).

Scenario 1: assuming that the dam would breach naturally without the spillway under the same hydrological condition as in the base (calibrated) scenario; aiming to evaluate the effects of the spillway.

Scenario 2: being similar to Scenario 1, but considering that the Tangjiashan dam breach initiates cascading breaches of the four smaller downstream dams.

Scenario 3: assuming that the dam would breach naturally during a heavy rainstorm whose flood discharge of 5-year return period is 2190 m^3/s as inflow of the lake. The consequence of the 5-year return period is substantially different from those of other scenarios.

Scenario 4: assuming the unfavorable geotechnical parameters of dam material that lead to the total breaching of the dam, and complete releasing of the water in the barrier lake. This is the most catastrophic scenario.

According to the BREACH analyses, the most catastrophic scenario (Scenario 4) gives the highest flood peak discharge of about 75,500 m^3/s However, the peak duration time was shortest, in which the discharge exceeded 1000 m^3/s (Fig. 20a). On the other hand, the peak water flow arrived within the shortest time for Scenario 3 which had higher inflow rate and impounded the lake faster, thereby accelerating the process toward the dam breach. Compared to Scenario 1, the base calibrated scenario reduced the peak discharge significantly and delayed the arrival time of the peak flow by nearly 32 h, proving that the spillway is crucial in mitigating the hazard of dam-break outburst and flood. The difference between Scenarios 1 and 2 stems from the cascading breach of the downstream dams, which can be analyzed only by the SOBEK model.

Figure 20b illustrates the assessment of the flood area, which is mainly constrained in the floodplain surrounded by higher terrains. The flood area obtained by the Scenarios 1 and 2 is virtually identical (about 62 km^2) due to the limited effect

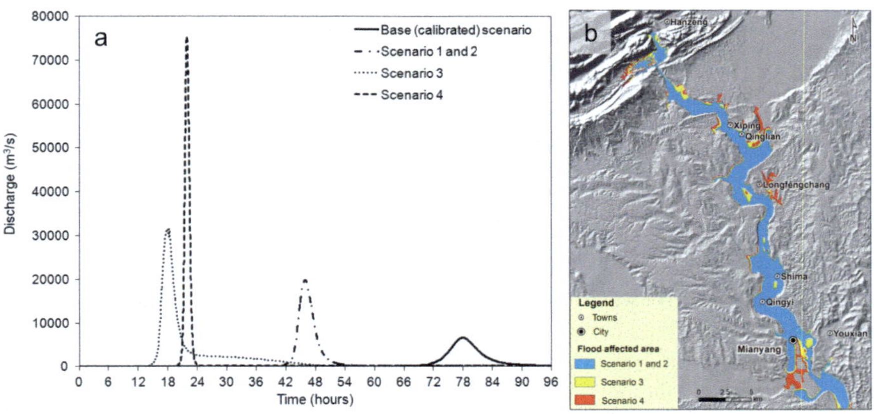

Fig. 20 **a** Output hydrographs of the BREACH model, and **b** modeled flooded area under different scenarios. The yellow and red areas show the increase in flooded area obtained from Scenarios 3 and 4 and are modified based on Fan et al. (2012c)

of the cascading breaches of the downstream dams. Scenarios 3 and 4 infer flooded areas to be 70.5 and 78.7 km^2, respectively, including most parts of Mianyang City.

8 Slope Instability After the Wenchuan Earthquake

After the earthquake, the regions of strong shaking had several heavy rainstorms that induced a number of catastrophic landslides and debris flows. Lacking empirical and temporal data on post-earthquake landslide activity makes it difficult to plan reconstruction, and more knowledge is needed on how geo-environmental systems destabilized by large earthquakes return to the original state of equilibrium.

8.1 Development of Landslides After the Wenchuan Earthquake

To understand the changes in landslide activity after the Wenchuan earthquake, Tang et al. (2016) generated five landslide inventories for different years for the same area. Their landslide interpretation area, Yingxiu and Longchi, covers approximately 179 km^2. The evolution of post-earthquake landslides was investigated using remote sensing imageries in different years since 2009. The landslide inventory that was made shortly after the earthquake is the basis of the inventory of the post-earthquake landslides in the following years. In each year, the landslide activities were classified into the following levels;

- level 0: there is no landslide activity and the landslide is dormant;
- level 1: less than 1/3 of the area of a landslide is active;
- level 2: about 1/3 to 2/3 of the area is active;
- level 3: more than 2/3 of the area is active or new landslide is formed.

These landslide activity levels were determined referring to the changes in diagnostic features between remote sensing images taken in different periods. The inventory in the previous year was the basis to build the next-year version. In more details, the landslide polygons were classified for each year with their aforementioned activity level. Any enlargement of existing landslides and newly triggered landslides was put in the digital inventories, and the changes in the landslide classifications were recorded in the attribute tables. In case that the landslide activity is confined within the same polygon as in the previous year's landslide, no new digitization of image was performed and the polygon of the previous year's inventory was assigned with the landslide activity level that was relevant for the new year.

A summary of the statistics is shown in Fig. 21. After the earthquake, from 2008 until April 2015, the number of landslides (including dormant landslides) in the study area increased by 660, to 7387. The total landslide area increased from 54.6 to 58.3

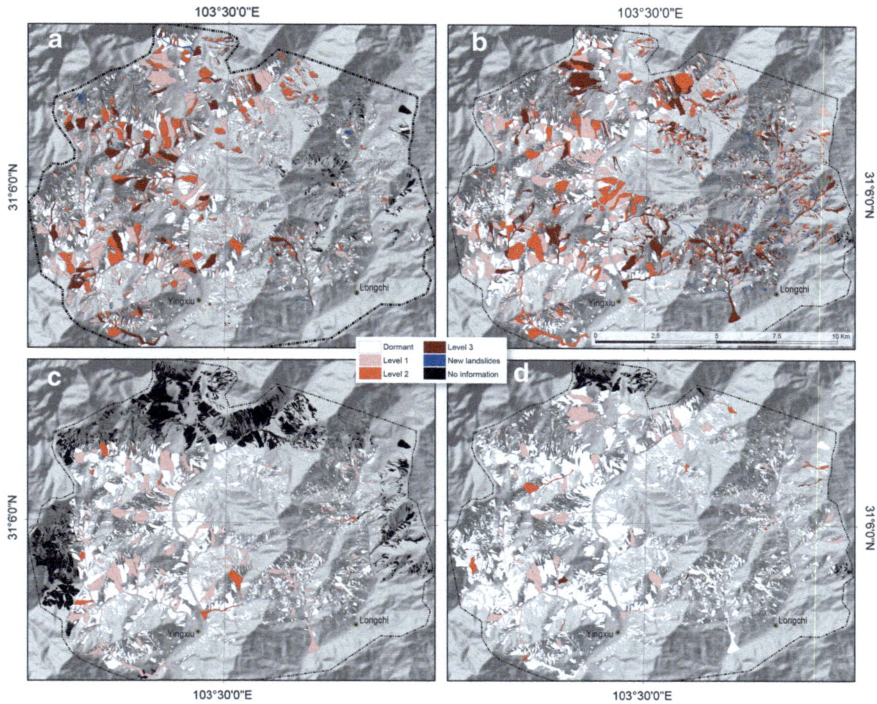

Fig. 21 Landslide activity maps for the areas between Yingxiu and Longchi for different years: **a** 2009; **b** 2011; **c** 2013; **d** 2015. Over 5 view statistics of the number of landslides within the various activity classes (Revised after Tang et al. 2016)

km^2. Most of the new landslides occurred during 2009 to 2011. The most frequent landslide type for the new landslides was slide (60.4%), followed by flow (36%) and fall. The average area of the new landslides is 1339 m^2, and ranged from 34 to 31,356 m^2. The large new landslides are mainly debris flows with long run-outs. Overall, the new landslides are mainly small and medium-sized: 90% of the new landslides are less than 3023 m^2.

8.2 Major Debris Flow Events After the Wenchuan Earthquake

The coseismic landslides produced huge amounts of loose material on the hillslopes, promoting the occurrence of debris flows during the aftermath. Accordingly, the rainfall threshold in terms of hourly precipitation for triggering debris flows decreased to around 40% lower than the original level according to the records in Beichuan (Tang et al. 2011a). The Land and Resources Department of Sichuan Province recorded

2333 debris flows between the 2008 earthquake and 2012, excluding small ones in remote regions. This number is significantly greater than 758 debris flows in a comparable time period between 2003 and 2007 prior to the earthquake (Huang and Fan 2013).

One of the important lessons from the Wenchuan earthquake was that the post-earthquake geohazards can last for many years. In line with this, Lin et al. (2006) mentioned the continuation of geohazards for at least 10 years after the 1999 Chi-Chi earthquake in Taiwan. As for the Wenchuan earthquake, 20–25 years have been predicted by Huang and Li (2014). Thus, the Wenchuan earthquake posed serious threats of geohazards (landslides, debris flows, and sediment increase) to local people and authorities. To shed further light to this issue, Tang et al. (2011a) studied the landslides and debris flows induced by the first heavy rainfall four months after the earthquake in the Beichuan region. Similarly, Xu et al. (2012) studied debris flows triggered by a heavy rainstorm during August 12–14, 2010, including the debris flows in the Wenjia Gully and Zoumaling Gully in the Qingping catchment, the debris flows in the Hongchun Gully near Yingxiu and the Longchi debris flows near Dujiangyan City.

Heavy rainfalls induced massive debris flows many times after the Wenchuan earthquake (Fig. 22, Table 3). For example, a severe rainfall on September 24, 2008, in Beichuan County caused 72 gully debris flows and claimed 42 deaths, inducing serious damage of roads and areas where people had been relocated after the Wenchuan earthquake (Tang et al. 2009). The old town of Beichuan County was completely buried by 6 to 10-m thick debris. Another rainfall on August 13 to 14, 2010, caused 80 debris flows in Qingping, Yingxiu and Longchi towns that had been reconstructed after the main earthquake. It was unfortunate that those communities were partially or totally destroyed by the debris flows (Tang et al. 2011b; Xu et al. 2012). In addition, heavy rainfall on July 10, 2013, triggered more than 200 debris flows along the Yingxiu-Wenchuan section of the Minjiang River. This disaster claimed 29 lives and caused serious damage to the reconstructed highway between Dujiangyan and Wenchuan.

8.3 Catastrophic Debris Flows Triggered by the September 24, 2008 Rainfall in the Beichuan Area

On September 24, 2008, the Beichuan County had numerous debris flows as a consequence of rainstorm after the substantial damage induced by the earthquake in May of the same year. The death toll due to this rain was 42 in Beichuan County and many roads were affected. A debris flow, consisting of soil, debris and boulders, came from the Xishanpo Gully, and most part of the city was buried underneath (Fig. 23). It was fortunate, however, that this debris flow did not claim life in the city center because all the residents had already been evacuated in safer places after the Wenchuan Earthquake. On 24th and 25th of September 2008, geohazard scientists

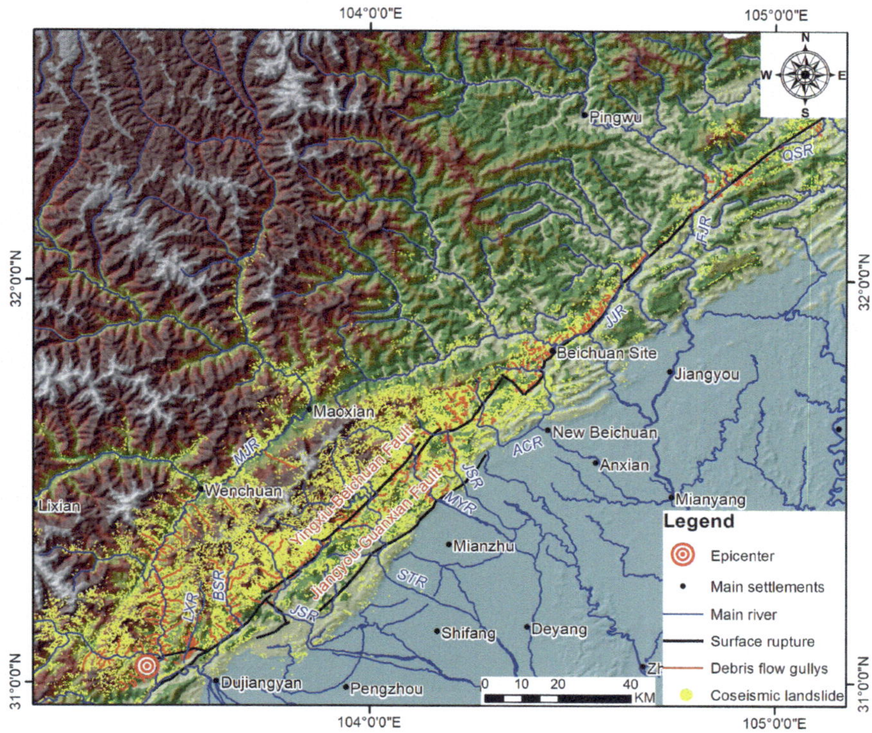

Fig. 22 Distribution of debris flows after the Wenchuan earthquake. MJR: Minjiang River; LXR: Longxi River; BSR: Baisha River; JSR: Jinsha River; STR: Shiting River; MYR: Mianyuan River; JSR: Jushui River; ACR: Anchang River; JJR: Jiangjiang River; FJR: Fujiang River; QSR: Qingshui River

of SKLGP investigated the source area of the debris flow, the transport paths and the deposits in Beichuan, aiming to capturing the rainfall threshold that triggered the debris flow and the processes leading to the earthquake-related debris flows. The short interval of four months between the earthquake and the debris flow made it possible to examine the role played by the Wenchuan earthquake in the onset of debris flows and their triggering mechanisms.

The geology in this area comprises the outcrops of Lower Paleozoic materials accompanied by Cambrian sandstones and argillaceous limestones, Silurian slates and phyllites, and limestones of Devonian and Carboniferous age (Fig. 24). The terraces and alluvial fans are covered by the quaternary unconsolidated deposits. The orientation of the geological structure is NE–SW here. The causative Yingxiu-Beichuan Fault is situated to the southeast of the study area and is forming a NW-dipping thrust fault with an angle of 60°–70°. The hanging wall is made of the Cambrian sandstones, while the footwall consists of Silurian and Devonian limestones. Beichuan County, which is basically a mountainous region, lies in the transition between the Sichuan Basin and the Western Sichuan Plateau. Its northwestern

Table 3 Catastrophic debris flow events after the 2008 Wenchuan earthquake on May 12

Location	Date	Type of disaster	Damage
Beichuan county	September 24, 2008	Debris flows occurred in 72 gullies simultaneously. The largest one was Weijia gully debris flow with its volume of more than one million m^3	42 people were killed. The old town of Beichuan county was almost buried by 6 to 10-m thick debris
Gaochuan, Anxian county	July 18, 2009	Debris flows occurred in ten gullies. The largest one had a volume of more than 3×10^5 m^3	Houses and roads were damaged
Qingping, Mianzhu county	August 13, 2010	Debris flows occurred in 20 gullies, with the total volume of 10×10^6 m^3. The volume of the largest Wenjia gully debris flow was 4.5×10^6 m^3	14 died, 33 injured, and 379 houses damaged. Direct economic loss = 600 million RMB
Yingxiu, Wenchuan county	August 14, 2010	Debris flows occurred in 21 gullies, with the total volume of 2×10^6 m^3. The largest one was Hongchun gully debris flow, with the volume of 7.5×10^5 m^3	32 deaths. Villages were ruined, and barrier lakes were formed, leading to secondary flood hazards
Longchi, Dujiangyan county	August 13, 2010	Debris flows occurred in 44 gullies, with the total volume of 3×10^6 m^3	161 houses damaged. Direct economic loss = 400 million RMB
Yinchang gully, Pengzhou county	August 18, 2012	Debris flows occurred in 12 gullies, with a total volume of 6×10^5 m^3	2 deaths. Direct economic loss = 500 million RMB
Yingxiu-Wenchuan county	July 10, 2013	Debris flows occurred in more than 200 gullies. The largest one was the Qipan Gully debris flow, with a volume of 1.0×10^6 m^3	29 deaths, 80 km of highway damaged. Direct economic loss = 4.6 billion RMB

part has high mountains (generally 1200–3100 m-asl), the central part has medium elevations consisting of rugged mountains and deeply incised valleys, and the south-eastern part has a lower relief (700–1400-m asl). The highest place in this area is Mt. Chaqi (4769-m asl) and the lowest elevation is only 540-m asl in the Jian River Valley.

Fig. 23 Photographs showing a debris flow channel deeply incised in an old alluvial fan (**a**) and debris flows inundating old Beichuan city (**b**) (Reproduced from Tang et al. (2009) with permission from Springer)

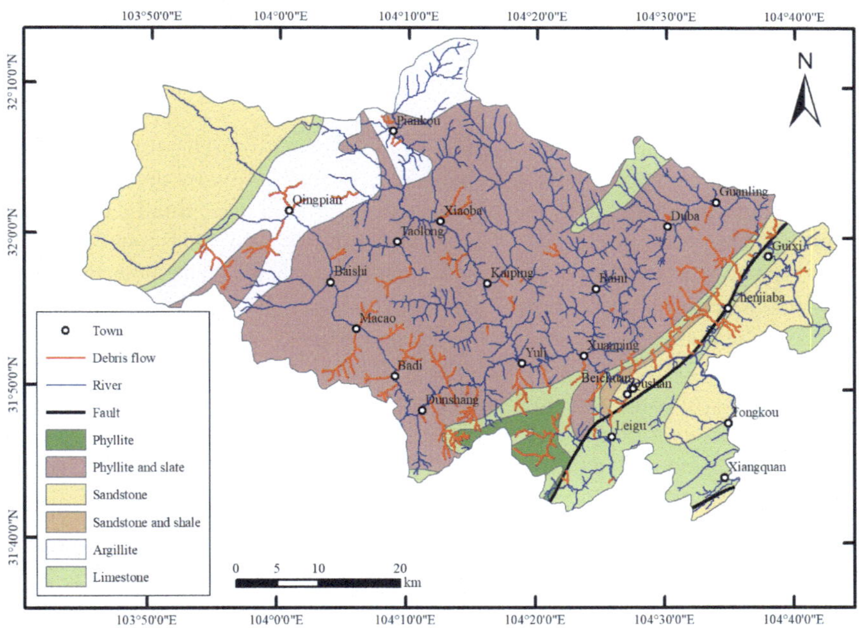

Fig. 24 Simplified geological map and debris flow inventory of Beichuan county, Sichuan province (Revised after Tang et al. 2009)

The studies on the debris flow events on September 24, 2008 revealed the following points:

(1) The heavy rainfall on September 24, 2008 not only triggered 72 debris flows in Beichuan County, resulting in 42 deaths and serious damage to roads, but

also threatened the areas designated for relocation of the earthquake-stricken people.

(2) The abundant rock falls and landslides caused by the Wenchuan earthquake provided a source of loose material that can be washed away as debris flows during subsequent rainfalls.

(3) The combination of the abundant antecedent rainfall and the heavy precipitation in September contributed to the onset of the post-earthquake debris flows. The post-earthquake threshold of precipitation and rainfall intensity that trigger debris flows were different from the pre-earthquake values. The threshold accumulated rainfall was reduced by 15–22%, while that of the hourly rainfall intensity was reduced by 25–32%, compared with the data in 2003.

(4) Lithology substantially influences the risk of debris flows because slopes made of the Silurian slates and phyllites are particularly prone to landsliding. These rocks have developed interbed shears and other fractures during tectonic folding and faulting, and this geological setting is in favor of the progress of weathering. Because clay minerals are soured and, hence, bonding is lost during weathering, the debris flow and the transport of large rock blocks become more likely to occur. The proximity between active faults and major rivers influences the spatial distribution of debris flows.

(5) The debris flows are initiated as a shallow movement toward the drainage channels, and then the loose debris in the channel is flushed out by the flow, which is sometimes referred to as the "fire hose effect." Moreover, small dams in the gully made of loose materials forming dams in the gully, that may block the mass movement for a short time, are eventually breached, and the flow discharge is increased instantaneously. The abundance of the loose material in the source area of the debris flow increases the discharge speed, volume, and distance of flow unexpectedly in spite of the relatively small drainage area. Accordingly, there is a significant need to consider the additional risk of debris flows after big earthquakes for the safety of infrastructure, residential areas, and roads.

(6) The rainfall-induced debris flows on 24th of September indicated that the Wenchuan Earthquake areas are particularly susceptible to frequent debris flows during the next 5–10 years. To cope with the debris flow risk, the geohazard management of cities and towns in the heavily affected areas has to be reinforced. The installation of slope monitoring and early warning systems in potentially high-risk areas is needed together with the construction of engineered safety structures in order to reduce the occurrence of and damage caused by the debris flows.

Fig. 25 Aerial photograph of the debris flow in August 2010 between Yingxiu and Laohuzui in the Minjiang river watersheds

8.4 Catastrophic Debris Flows Triggered by the 14 August 2010 Rainfall in the Yingxiu Area

At 3:00 a.m. on August 14, 2010, the area around Yingxiu Town was affected by many debris flows caused by intense rainfall. The G213 national road and the Dujiangyan-Wenchuan highway (under construction) were covered by deposits of mud, debris, and rocks. This debris flow finally moved down into the Mingjiang River. At the confluence of main and tributary channels, debris dams were formed, and entirely or partially, blocked the Mingjiang River. This resulted in severe flooding in Yingxiu Town. After the event, the inspection of aerial photographs identified 21 gullies along the Mingjiang River, wherein debris flows had occurred during this intense rainfall. Figure 25 addresses the damage extent in roads, bridges, and buildings in Yingxiu.

The debris flow coming from the Hongchun Gully was extremely catastrophic, traveling across the Mingjiang River and creating a debris dam that was 10-m high, 100-m long across the river, and 150-m long along the river (Fig. 26). Because of this dam, the river changed its course and flooded the newly reconstructed Yingxiu. The maximum discharge of the river through the town was estimated to be 570 m^3/s shortly after the onset of the debris flow on August 14th and increased to 1400 m^3/s on August 15th. The flood water was 2.5–3.5-m deep and the flood lasted for 7 days, until the debris dam was finally removed on August 20th. 62 people were killed by this disaster and more than 8000 residents were forced to evacuate.

8.5 Catastrophic Debris Flows Triggered by Rainfall on 10 July 2013 in the Wenchuan Area

On July 10, 2013, a heavy rainstorm around the Wenchuan town induced massive, devastating debris flows. Consequently, 29 people died and more than 8000 people were forced to evacuate. Aerial photographs identified 105 gullies along the

Fig. 26 Depositional fan of the Hongchun Gully before and after the debris flow in August 2010; **a** debris dam after the debris flow that blocked the Mingjiang river's course and buried the road, **b** old alluvial fan before the debris flow occurred

Mingjiang River (Figs. 27 and 28). 28 debris dams completely or partly blocked the Minjiang River. Moreover, the debris flow from Qipan catchment killed 15 people and more than 900 houses were destroyed on the alluvial fan at the exit of the catchment (Fig. 29).

The Qipan Gully is 15.1-km long and its catchment area is 54.2 km^2 in size. It is the first tributary of the Minjiang River and is located at about 5-km southwest of Wenchuan City. The geology in this catchment is chiefly underlain by granitic and carbonate rocks. The volume of debris from the landslides and sediments in the channels was estimated to be 22 million m^3. After the disaster, the field reconnaissance detected at least three failed landslide dams along the passage of the torrent. These dams were 20–30-m high and formed barrier lakes. The breaching of these landslide dams aggravated the effects of the debris discharge, and the transporting capacity increased drastically such as a domino effect. This accounts for the transport of huge quantities of sediments and blockage of the Minjiang River. The field investigation found that the debris came mainly from the coseismic landslide deposits. The breaching of the landslide dams instantaneously increased the flow discharge and the channel erosion was intensified. The run-off of water in the catchment areas mobilized a large amount of loose debris from the widespread coseismic landslides, and those debris was transformed into huge debris flows. In this manner, the likelihood of the debris flow disaster has been very high for many years in this earthquake-hit area. It is thus necessary to improve the debris flow risk management of the villages and towns in this area. Note that the possibility of debris flows had been recognized in the region prior to the Wenchuan earthquake, but their potential for widespread and devastating impacts had not been fully understood. The assessment of the risk of this type in future is required together with the estimation of the volume of the debris flow in order to quantitatively evaluate the hazard. The risk of debris flow deserves serious attention because many pre-existing fans are being used or considered as settlement areas in the region.

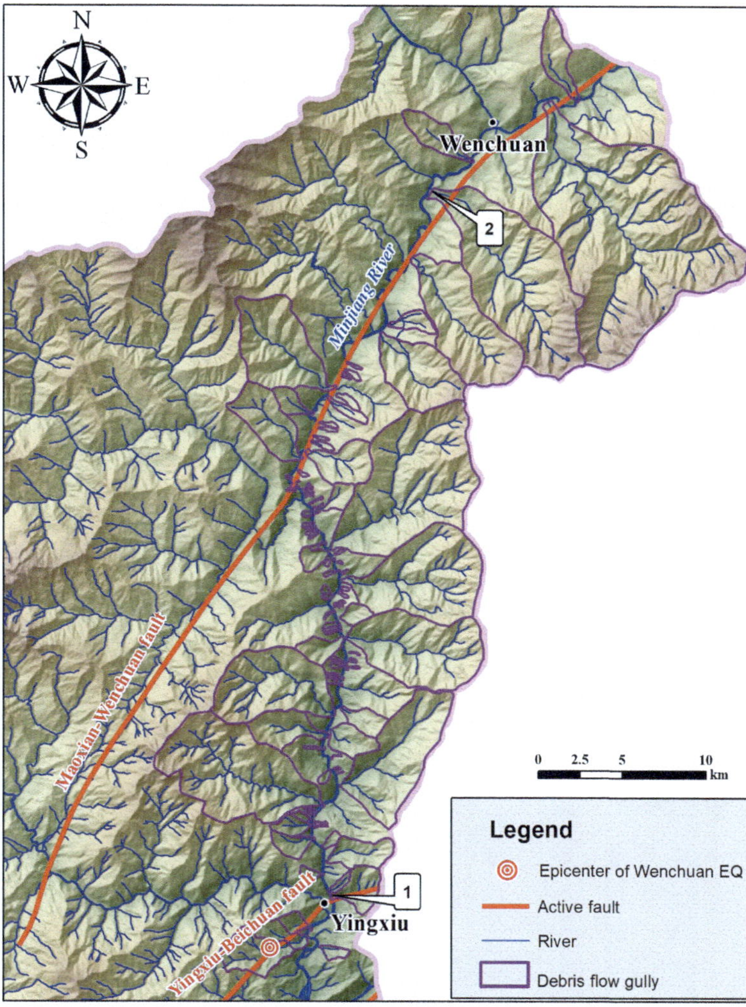

Fig. 27 Distribution of debris flows following the Wenchuan earthquake in the Yingxiu-Wenchuan section of Minjiang river. (1) Hongchun Gully, (2) Qipan Gully

9 Conclusion

This report summarizes the regional geology and tectonic setting, earthquake characteristics, slope seismic response of the aftershocks, and characteristics and spatial patterns of the coseismic and post-seismic landslides related with the 2008 Wenchuan earthquake of Mw = 7.9. The main shock generated two surface ruptures along existing faults and induced very high peak accelerations. The major conclusions drawn from this study are summarized in what follows.

Fig. 28 Aerial photograph of debris flows along the Minjiang river taken on July 20, 2013

Fig. 29 Aerial photograph of the Qipan Gully debris flow taken on July 20, 2013

(1) The earthquake triggered at least 197,481 landslides in an elliptical area that measures about 110,000 km^2 in size. Those landslides occupied the total area of about 1160 km^2, and the total volume of the landslides was about 10 km^3.
(2) The extent of this landslide disaster was probably affected by the strong and long-lasting ground shaking, steep and rugged topography of the mountains, and the fragile and densely jointed lithology.
(3) The fault type and the slip rate during the rupture also affected the distribution of the coseismic landslides.

(4) The large landslides were concentrated in the areas, where the fault slip rates were highest, near the intersections and junctions of the segments of the fault.
(5) The head scarps of the large-scale landsides are serrated, rough, and steep due to the tensile stress in situ.
(6) The earthquake left a tremendous amount of loose material on the hillslopes, and this material is prone to erosion during heavy rain and transportation toward the downstream areas.
(7) The threshold hourly rainfall for triggering debris flows was reduced by around 40% by the earthquake as compared with the pre-earthquake level, according to the records in Beichuan.
(8) During the decade after the Wenchuan earthquake, heavy rains produced many eroded gullies in the earthquake-hit area, resulting in many casualties and property losses.
(9) The distribution of post-quake debris flows is closely related with the seismic faults. Although much work has been carried out, the risk assessment of post-seismic geologic hazards is insufficient partially because of the lack of data on post-seismic evolution of landslides and debris flows with sufficient temporal resolution.

Acknowledgements The work was financially supported by the Science Fund for Creative Research Groups of the National Natural Science Foundation of China (Grant No. 41521002) and the National Key Research and Development Program of China (Grant No. 2021YFC3000401).

References

Burchfiel BC, Chen Z, Liu Y, Royden LH (1995) Tectonics of the Longmenshan and adjacent regions. Int Geol Rev 37:661–735

Burchfiel BC, Royden LH, van der Hilst RD, Eric K (2008) A geological and geophysical context for the Wenchuan earthquake of 12 May 2008, Sichuan, People's Republic of China. GSA Today 18(18):4–11

Chigira M, Wu X, Inokuchi T, Wang G (2010) Landslides induced by the 2008 Wenchuan earthquake, Sichuan, China. Geomorphology 118(3–4):225–238

Cruden DM, Varnes DJ (1996) Landslide types and processes. In: Turner AK, Schuster RL (eds) Landslides, investigation and mitigation: special report, Transportation Research Board, National Research Council, vol 247, pp 36–75

Cui P, Zhu YY, Han YS, Chen XQ, Zhuang JQ (2009) The 12 May Wenchuan earthquake-induced landslide lakes: distribution and preliminary risk evaluation. Landslides 6(3):209–223

Dai FC, Xu C, Yao X, Xu L, Tu XB, Gong QM (2011) Spatial distribution of landslides triggered by the 2008 Ms 8.0 Wenchuan earthquake, China. J Asian Earth Sci 40(4):883–895

Deng Q, Chen S, Zhao X (1994) Tectonics, seismicity and dynamics of Longmenshan mountains and its adjacent regions. Seismol Geol 16(4):389–403

Densmore AL, Ellis MA, Li Y, Nick JR (2007) Active tectonics of the Beichuan and Pengguan faults at the eastern margin of the Tibetan plateau. Tectonics 26(4):171–178

Fan X, van Westen CJ, Korup O, Görüm T, Xu Q, Dai F, Huang R, Wang G (2012a) Transient water and sediment storage of the decaying landslide dams induced by the 2008 Wenchuan earthquake, China. Geomorphology 171–172:58–68

Fan X, van Westen CJ, Xu Q, Görüm T, Dai F (2012b) Analysis of landslide dams induced by the 2008 Wenchuan earthquake. J Asian Earth Sci 57:25–37
Fan X, Tang CX, van Westen CJ, Alkema D (2012c) Simulating dam-breach scenarios of the Tangjiashan landslide dam induced by the Wenchuan earthquake. Nat Hazards Earth Syst Sci 12:3031–3044
Gorum T (2013). Toward a better understanding of earthquake triggered landslides: an analysis of the size, distribution pattern and characteristics of co-seismic landslides in different tectonic and geomorphic environment. PhD Thesis, Twente University, Enschede, The Netherlands, 154 p
Gorum T, Fan XM, van Westen CJ, Huang RQ, Xu Q, Tang C, Wang GH (2011) Distribution pattern of earthquake-induced landslides triggered by the 12 May 2008 Wenchuan earthquake. Geomorphology 133(3–4):152–167
Guo Z, Zhou C, Sun X, Zhang J (2010) The distribution of landslide triggered by Wenchuan earthquake and its causative factors. Earth Sci Front 17(5):234–242
Heim A (1932) Bergstürz und Menschenleben. Fretz and Wamuth Verlag Zürich, pp 114–119 (in German)
Hsü KJ (1975) Catastrophic debris streams (Sturzstroms) generated by rockfalls. Geol Soc Am Bull 86:129–140
Huang RQ, Li WL (2009) Analysis of the geo-hazards triggered by the 12 May 2008 Wenchuan earthquake, China. Bull Eng Geol Env 68:363–371
Huang RQ, Fan XM (2013) The landslide story. Nat Geosci 5:325–326
Huang RQ, Li WL (2014) Post-earthquake landsliding and long-term impacts in the Wenchuan earthquake area, China. Eng Geol SI:111–120
Huang RQ, Pei XJ, Fan XM, Zhang WF, Li SG, Li BL (2012) The characteristics and failure mechanism of the largest landslide triggered by the Wenchuan earthquake, May 12, 2008, China. Landslides 1:131–142
Li WL, Huang RQ, Tang C, Xu Q, van Westen CJ (2013) Co-seismic landslide inventory and susceptibility mapping in the 2008 Wenchuan earthquake disaster area, China. J Mountain Sci 10(3):339–354
Li G, West AJ, Densmore AL, Jin ZD, Parker RN, Hilton RG (2014) Seismic mountain building: landslides associated with the 2008 Wenchuan earthquake in the context of a generalized model for earthquake volume balance. Geochem Geophys Geosyst 15(4):833–844
Lin CW, Liu SH, Lee SY, Liu CC (2006) Impacts on the Chi-Chi earthquake on subsequent rainfall-induced landslides in central Taiwan. Eng Geol 86:87–101
Liu N, Zhang J, Lin W, Cheng W, Chen Z (2009) Draining Tangjiashan Barrier lake after Wenchuan earthquake and the flood propagation after the dam break. Sci China Ser e: Technol Sci 5(4):801–809
Liu-Zeng J, Zhang Z, Wen L, Tapponnier P, Sun J, Xing X, Hu G, Xu Q, Zeng L, Ding L, Ji C, Hudnut KW, Woerd JVD (2009) Co-seismic ruptures of the 12 May 2008, Ms 8.0 Wenchuan earthquake, Sichuan: East-west crustal shortening on oblique, parallel thrusts along the eastern edge of Tibet. Earth Planetary Sci Lett 286(3–4):355–370
Luo YH, Fan XM, Huang RQ, Wang YS, Yunus AP, Havenith HB (2020) Topographic and near-surface stratigraphic amplification of the seismic response of a mountain slope revealed by field monitoring and numerical simulations. Eng Geol 271:105607
McDougall S, Hungr O (2004) A model for the analysis of rapid landslide motion across three-dimensional terrain. Can Geotech J 41(6):1084–1097
McDougall S, Hungr O (2005) Dynamic modelling of entrainment in rapid landslides. Can Geotech J 42(5):1437–1448
Molnar P (2005) Mio-Pliocene growth of the Tibetan plateau and evolution of East Asian climate. Palaeontol Electron 8(1):1–23
Molnar P, Tapponnier P (1975) Cenozoic tectonics of Asia: effects of a continental collision. Science 189(4201):419–426
Molnar P, England P, Martinod J (1993) Mantle dynamics, uplift of the Tibetan plateau, and the Indian monsoon. Rev Geophys 31(4):357–396

Ouimet WB (2010) Landslides associated with the May 12, 2008 Wenchuan earthquake: implications for the erosion and tectonic evolution of the Longmen Shan. Tectonophysics 491(1–4):244–252

Parker RN, Densmore AL, Rosser NJ, De Michele M, Li Y, Huang RQ, Whadcoat S, Petley DN (2011) Mass wasting triggered by the 2008 Wenchuan earthquake is greater than orogenic growth. Nat Geosci 4(7):449–452

Qi SW, Xu Q, Lan HX, Zhang B, Liu JY (2010) Spatial distribution analysis of landslides triggered by 2008.5.12 Wenchuan earthquake, China. Eng Geol 1–2:95–108

Tang C, Zhu J, Li WL, Liang JT (2009) Rainfall-triggered debris flows following the Wenchuan earthquake. Bull Eng Geol Env 2:187–194

Tang C, Zhu J, Qi X, Ding J (2011a) Landslides induced by the Wenchuan earthquake and the subsequent strong rainfall event: a case study in the Beichuan area of China. Eng Geol 1–2:22–33

Tang C, Zhu J, Ding J, Cui XF, Chen L, Zhang JS (2011b) Catastrophic debris flows triggered by a 14 August 2010 rainfall at the epicenter of the Wenchuan earthquake. Landslides 4:485–497

Tang C, van Asch TWJ, Chang M, Chen GQ, Zhao XH, Huang XC (2012) Catastrophic debris flows on 13 August 2010 in the Qingping area, southwestern China: the combined effects of a strong earthquake and subsequent rainstorms. Geomorphology 139–140(2):559–576

Tang CX, van Westen CJ, Tanyaş H, Jetten VG (2016) Analysing post-earthquake landslide activity using multi-temporal landslide inventories near the epicentral area of the 2008 Wenchuan earthquake. Nat Hazard 16(12):1–26

Tapponnier P, Molnar P (1977) Active faulting and tectonics in China. J Geophys Res 82(20):2905–2930

Tapponnier P, Peltzer G, Le Dain AY, Armijo R, Cobbold P (1982) Propagating extrusion tectonics in Asia: new insights from simple experiments with plasticine. Geology 10(12):611–616

Wang EC, Meng QR (2009) Mesozoic and Cenozoic tectonic evolution of the Longmenshan fault belt. Sci China, Ser D Earth Sci 52(5):579–592

Xing AG, Yuan XY, Xu Q, Zhao QH, Huang HQ, Cheng QG (2017) Characteristics and numerical runout modelling of a catastrophic rock avalanche triggered by the Wenchuan earthquake in the Wenjia valley, Mianzhu, Sichuan, China. Landslides 14:83–98

Xu C (2012) Detailed inventory of landslides triggered by the 2008 Wenchuan earthquake and its comparison with other earthquake events in the world. Sci Technol Rev 30(25):18–26

Xu C (2015) Preparation of earthquake-triggered landslide inventory maps using remote sensing and GIS technologies: principles and case studies. Geosci Front 6(6):825–836

Xu X, Deng Q (1996) Nonlinear characteristics of paleoseismicity in China. J Geophys Res: Solid Earth 101(B3):6209–6231

Xu C, Dai FC, Chen J, Tu XB, Xu L, Li WC, Tian W, Cao YB, Yao X (2009a) Identification and analysis of secondary geological hazards triggered by a magnitude 8.0 Wenchuan earthquake. J Remote Sens 13(4):745–762 (in Chinese)

Xu Q, Fan XM, Huang RQ, van Westen CV (2009b) Landslide dams triggered by the Wenchuan earthquake, Sichuan Province, south west China. Bull Eng Geol Env 68:373–386

Xu X, Wen X, Yu G, Chen G, Klinger Y, Hubbard J, Shaw J (2009c). Coseismic reverse- and oblique-slip surface faulting generated by the 2008 Mw 7.9 Wenchuan earthquake, China. Geology 37(6):515–518

Xu Q, Zhang S, Li WL (2011) Spatial distribution of large-scale landslides induced by the 5.12 Wenchuan earthquake. J Mt Sci 2:246–260

Xu Q, Zhang S, Li WL, van Asch TWJ (2012) The 13 August 2010 catastrophic debris flows after the 2008 Wenchuan earthquake, China. Nat Hazard 1:201–216

Xu C, Xu XW, Yao X, Dai F (2014) Three (nearly) complete inventories of landslides triggered by the May 12, 2008 Wenchuan Mw 7.9 earthquake of China and their spatial distribution statistical analysis. Landslides 11(3):441–461

Xu C, Xu XW, Shen L, Yao Q, Tan X, Kang W, Ma S, Wu X, Cai J, Gao M, Li K (2016a) Optimized volume models of earthquake-triggered landslides. Sci Rep 6:29797

Xu Q, Li YR, Zhang S, Dong XJ (2016b) Classification of large-scale landslides induced by the 2008 Wenchuan earthquake, China. Environ Earth Sci 75(1):1–12

Yin Y, Wang F, Sun P (2009) Landslide hazards triggered by the 2008 Wenchuan earthquake, Sichuan, China. Landslides 6(2):139–152

Zhang PZ, Xu XW, Wen XZ, Ran YK (2008) Slip rates and recurrence intervals of the Longmen Shan active fault zone, and tectonic implications for the mechanism of the May 12 Wenchuan earthquake, 2008, Sichuan, China. Chinese J Geophys 51(4):1066–1073

Earthquake-Induced Landslides and Ground Failure in Chile: The Aysen 2007 and Maule 2010 Earthquakes

Sergio A. Sepúlveda, Felipe Ochoa-Cornejo, and Alejandra Serey

Abstract The 2007 Aysen earthquake, of magnitude Mw 6.2, was a shallow crustal earthquake in the large, regional Liquiñe-Ofqui Fault Zone. It triggered a total of 541 landslides (reported) classified as rock slides and avalanches, rock falls, shallow soil and soil-rock slides, and debris flows. The largest landslides induced a local tsunami in the Aysen fjord, causing severe damage to aquaculture facilities and ten fatalities along the fjord coast. The 2010 Maule earthquake, of magnitude Mw 8.8, was the second largest subduction earthquake to have occurred in Chile, after the 1960 Valdivia earthquake (Mw 9.5). The Maule earthquake was followed by a destructive tsunami that affected an extensive area, causing significant damage. This earthquake and tsunami caused more than 500 deaths, and an economic loss close to 20% of the GDP of Chile. The damage was partly attributed to ground failure. Among the most significant effects associated with ground failure were the liquefaction of soils and landslides. The triggering of liquefaction caused the failure of urban, civil, and industrial infrastructures such as buildings, bridges, ports, and mining facilities. Liquefaction cases were reported even in locations far away from the rupture area. Likewise, 1160 landslides were induced, including rock falls, rock, and soil slides. Over 800 landslides were observed in areas with lower shaking intensities, occurring in clusters. This paper presents the earthquake-induced damage related to the liquefaction and landslides cases reported after these seismic events. In the case of the coseismic landslides, a significant difference is found in the number of landslides

S. A. Sepúlveda (✉) · A. Serey
Departamento de Geología, Universidad de Chile, Santiago, Chile
e-mail: sergio_sepulveda@sfu.ca

A. Serey
e-mail: alejandra.serey@uoh.cl

S. A. Sepúlveda
Department of Earth Sciences, Simon Fraser University, Burnaby, B.C., Canada

F. Ochoa-Cornejo
Departamento de Ingeniería Civil, Universidad de Chile, Santiago, Chile
e-mail: fochoac@uchile.cl

S. A. Sepúlveda · A. Serey
Instituto de Ciencias de la Ingeniería, Universidad de O'Higgins, Rancagua, Chile

I. Towhata et al. (eds.), *Coseismic Landslides*, Springer Natural Hazards,
https://doi.org/10.1007/978-981-19-6597-5_2

generated by subduction earthquakes and shallow crustal earthquakes, a critical point to consider in future seismic landslide hazard analyses.

1 Introduction

The 2007 Aysen earthquake, of magnitude Mw 6.2, in the Chilean Patagonia fjordland, was the first documented large-impact shallow crustal earthquake in Chilean history. In turn, the Mw 8.8 Maule 2010 earthquake, in the south-central area of Chile, was the second largest earthquake ever recorded in the country, and the fifth most massive earthquake in recorded history to date. Tsunami waves followed both earthquakes; in the Aysen earthquake, a tsunami was triggered by large rock slope failures in a fjord, while the 2010 Maule tsunami in the Pacific Ocean destroyed part of some coastal cities, causing many casualties. In both cases, hundreds of landslides were induced by the earthquakes, with a very distinct size and geographic distribution closely related to the earthquake type.

Landslides are common natural phenomena in Chile caused by different triggers such as rainfall, human intervention, volcanic eruptions, and earthquakes. They can cause significant economic losses and fatalities. Despite their social and economic impact, earthquake-induced landslides in the country are still poorly understood. There is no full understanding regarding how earthquake magnitude, intensity, and the number of landslides correlate, and more efforts are required to determine if the various types of landslides correlate with different types of earthquakes. An understanding of how geology, geotechnics, and seismicity interact to define the occurrence of landslides is required, as well as how the occurrence of one landslide could influence the occurrence of another. In this context, comprehensive inventories of landslides triggered by earthquakes are an essential part of seismic landslide hazard analyses.

Before the 2007 Aysen earthquake, there were no comprehensive inventories of landslides induced by earthquakes in Chile. In this report, a review of the first coseismic landslide inventories is presented. These inventories of the 2007 and 2010 earthquakes represent exceptional cases to study the occurrence of landslides generated by earthquakes originating in subduction tectonic environments.

The infrastructure of Chile had, in general, performed well in the latest large earthquakes, despite the damage observed. In 2010, the cases where moderate-to-high levels of damage were reported resulted from poor construction practices related to either construction requirements, code misinterpretation, or the occurrence of massive ground displacement due to liquefaction or landslides (Bray and Frost 2010).

Liquefaction is a large-strain phenomenon that occurs in saturated loose sand deposits during earthquakes (Ochoa-Cornejo et al. 2016). The rapid ground shaking caused by the earthquake induces undrained cyclic shearing that generates excess pore pressures due to the inability of the water to flow out of the grain pores rapidly. As a consequence, the excess pore pressure increases, decreasing the effective stresses, eventually to zero, triggering liquefaction, inducing large ground deformations and damage to urban and industrial facilities (Ishihara 1996).

Soil liquefaction was widely observed along the rupture length of the 2010 Maule earthquake, and caused significant tilting and settlement of various types of infrastructure. Effects such as lateral spreading in sloping grounds and ground failure, including large displacement due to the liquefied soil beneath the foundations of buildings, as well as coastal structures like piers and ports, were observed too. Infrastructure such as bridges, abutments, roads, and other transportation infrastructure also had severe liquefaction damage, disrupting traffic, and transportation (Bray and Frost 2010).

This document provides information on ground failure observed during the 2007 and 2010 earthquakes in Chile, and mainly details the occurrence of landslides and liquefaction.

2 Landslides Induced by the 2007 Aysen Earthquake, Patagonia

2.1 The April 21, 2007 Mw 6.2 Aysen Shallow Crustal Earthquake

After an unusual seismic activity over four months in the Aysen fjord area in the Chilean Patagonia (45.3° S, 73.0° W), a major Mw 6.2 strike-slip earthquake struck the region on April 21, 2007. The earthquake source was the Liquiñe-Ofqui Fault Zone, a major active transpressional, NNE trending intra-arc fault system that accommodates the parallel component of the Nazca plate oblique subduction beneath the South American plate (Cembrano et al. 1996, 2007; Legrand et al. 2011; Vargas et al. 2013). The earthquake occurred in the Punta Cola fault, a secondary branch of the Liquiñe-Ofqui Fault Zone, in a transference zone between several N–S trending regional branches of the fault zone (Vargas et al. 2013).

The main shock (Mw 6.2, Ms 6.3; Global CMT Catalog, 2008; NEIC, 2008) was recorded at 13:50 local time. The reported intensity was VII in the nearby towns of Puerto Aysén and Puerto Chacabuco (over 20 km epicentral distance), causing moderate structural damage. Unfortunately, no local instrumental records of the mainshock and aftershocks were retrieved from a purpose-installed local network, as most of the instruments were either damaged or destroyed by a local landslide-induced tsunami. Thus, precise localization of the main shock is unclear, and there are no local acceleration records for the earthquake (Sepúlveda et al. 2010).

2.2 Landslide Inventory and Distribution

Sepúlveda et al. (2010) published an inventory of landslides induced by the 2007 Aysen fjord earthquake. In this inventory, a total of 541 landslides were mapped on the basis of field observations and remote sensing analysis. Therein, those landslides were classified into rock slides and avalanches, rock falls, shallow soil and soil-rock slides, and debris flows (Figs. 1 and 2). The majority of landslides were triggered on the steep slopes of the fjord and adjacent glacial valleys. The geology of the slopes is dominantly composed of strong intrusive rocks of the North Patagonian Batholith, which are covered by thin pyroclastic soils. Some slides also occurred in the slopes of small monogenetic volcanoes that are that located along the fault branches of the Liquiñe-Ofqui Fault Zone (Sepúlveda et al. 2010).

The largest landslide was a rock avalanche in the fjord lateral creek of Punta Cola, which is located very close to the epicenter. Redfield et al. (2011) and Oppikofer et al. (2012) studied this landslide in detail and described the structure of the rock slide area and the geomorphology of the rock avalanche, respectively. Oppikofer et al. (2012) assessed the landslide volume to be 22 mm^3, of which around 55–65% reached the fjord located at about 1 km away from the source (Oppikofer et al. 2012; Vollmer 2017). Numerical modeling of the rock avalanche suggests velocities to be up to 20–30 m/s (Vollmer 2017). Topographic amplification effects played a major role in the triggering of the largest rock slope failures (Sepúlveda et al. 2010). This hypothesis was examined by preliminary analyses on slope stability (Serey 2011) together with numerical modeling for the Punta Cola landslide by

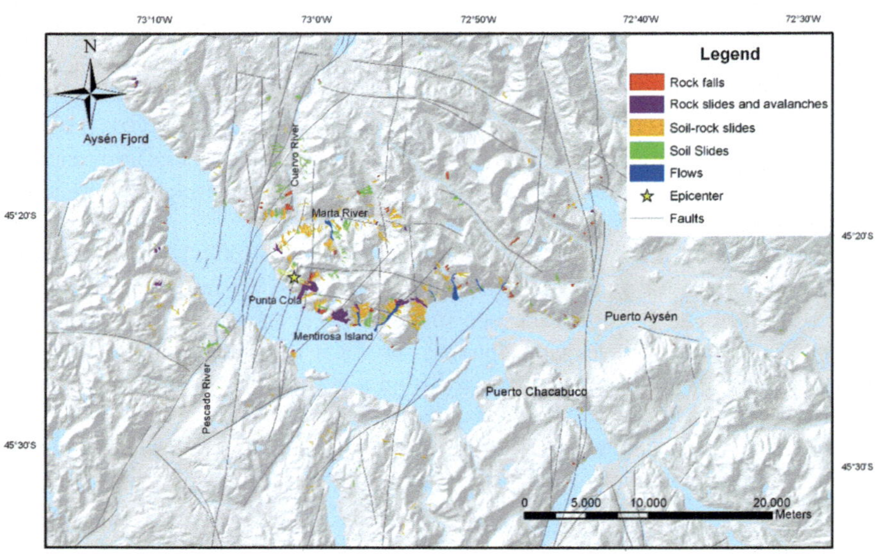

Fig. 1 Landslide inventory of the 2007 Aysen earthquake in southern Chile (Reproduced from Sepúlveda et al. 2010)

Fig. 2 Earthquake-induced landslides that triggered a local tsunami in the Aysen fjord, April 2007

Escudero (2017) and Aravena (2017). These studies confirmed the chief roles played by the topographic site effects as well the importance of local faults in the slope destabilization. The largest landslides, including the Punta Cola rock avalanche and a rock slide complex in front of Mentirosa Island, along with several smaller coastal rock slope failures and some debris flows, triggered displacement waves in the fjord. As a result, a local tsunami was produced and killed ten people, causing major damage in salmon farm infrastructure along the fjord shores (Naranjo et al. 2009; Sepúlveda and Serey 2009). Furthermore, oceanographic campaigns in the fjord after the earthquake (Van Daele et al. 2013; Hermanns et al. 2014; Lastras et al. 2016) showed how the largest landslide caused significant geomorphological changes in the fjord sea bottom; with the formation of spectacular ring-shaped structures.

3 Landslides Induced by the 2010 Maule Earthquake, Central-Southern Chile

3.1 *The February 27, 2010 Mw 8.8 Maule Megathrust Earthquake*

The Maule region, in the south of Chile, was severely hit by a strong earthquake of moment magnitude Mw 8.8 on February 27, 2010. It is the second largest earthquake in the documented history of Chile, and the largest seismic event ever instrumentally recorded in the country. After the main earthquake shock, a destructive tsunami was triggered, with waves exceeding 5 m in height, covering an extensive area of land, and causing significant damage. The 2010 Maule earthquake caused nearly 600 deaths. About one-third of the deaths were due to the tsunami. The economic loss was close to USD 30 Billion, close to 20% of the GDP of Chile (Verdugo and González 2015).

This earthquake was of a subductive nature, and due to the convergence of the Nazca Plate beneath the South American Plate, which occurs at velocities ranging between 6 and 8 cm/yr. The source of this earthquake was located in Cobquecura, 60 km to the north of Concepción, at a depth of 37 km, presenting a fault displacement of close to 3 m (Ruiz et al. 2012; Fig. 3). The rupture length was larger than 500 km, and the width approximately 160 km, developing from south to north, connecting the rupture areas of the earthquakes of Valdivia (1960) and Valparaíso (1985) (Campos et al. 2002). According to the acceleration records, the duration of the ground shaking was approximately two minutes, and it was felt within 1000 km of the epicenter. Field observations suggested that tectonic displacement of the hanging wall produced both uplifts of over 2 m and subsidence of up to 1 m in coastal regions (Bray and Frost 2010).

The aftershocks of the Maule earthquake were of magnitudes close to 7.0, and most of them were shallow, at a depth of approximately 10 km. The most important was on March 11, 2010, of magnitude 6.9, in Pichilemu, 450 km north of the epicenter. The largest one, an interplate earthquake, was on March 25, 2012, of magnitude 7.1, in Constitución, 120 km to the north of Cobquecura.

The main Maule shock was recorded by approximately 40 seismic stations installed on both rock sites and soil deposits. The highest horizontal peak ground acceleration registered on the rock was 0.32 g in the center of Chile, 500 km to the north of the epicenter. The highest horizontal peak ground acceleration registered on soil was 0.94 g and 250 km to the south of the epicenter (Verdugo and González 2015; Verdugo et al. 2019).

Among the most significant effects associated with ground failure was the liquefaction of soils and landslides. Severe liquefaction was reported during the 2010 Maule Earthquake. The large displacements induced by the triggering of liquefaction caused the failure of urban, civil, and industrial facilities. The earthquake-induced liquefaction affected residential buildings, bridges connecting important cities, ports of great economic importance, and mining facilities such as tailings dams, vital for

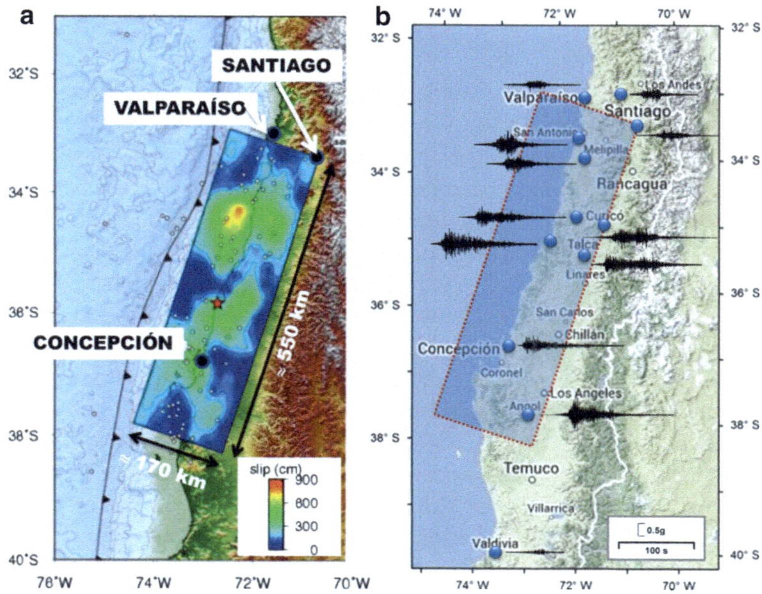

Fig. 3 **a** Rupture area associated with the 2010 Maule earthquake (Reproduced from Verdugo et al. 2019 with permission from Elsevier). **b** Time histories of some of the seismic stations that recorded the earthquake (Reproduced from Verdugo and González 2015 with permission from Elsevier)

ore production. The 2010 Mw 8.8 Maule earthquake also triggered over a thousand slope failures, including rock falls, rock and soil slides, flows and lateral spreads, among others (Fig. 4). Most of the landslides (over 800) are located in Principal Cordillera, a mountain area with strong relief but where lower intensities were registered due to a larger distance from the epicenter. The landslides tended to occur in clusters.

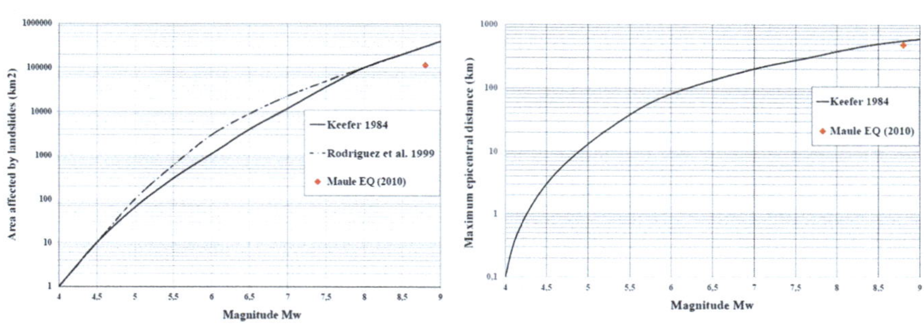

Fig. 4 Comparison with curves of Keefer (1984) and Rodriguez et al. (1999) of maximum landslide area and epicentral distance for worldwide earthquake-triggered landslides and the 2010 Maule earthquake. Left. Maximum landslide area. Right. Maximum epicentral distance for disrupted landslides (After Serey et al. 2017)

3.2 Landslide Inventory and Distribution

A landslide inventory of the 2010 Maule earthquake was started by Escobar (2013) in the coastal range and completed by Serey et al. (2017, 2019), who presented the full inventory between 32.5° and 38.5° S in Chilean territory, although a few landslides (mainly rock fall activity) have been reported in the Argentinean part of the Andes Main Range (Wick et al. 2010). The inventory is mainly based on satellite imagery revision, plus local reports from the National Geological and Mining Survey of Chile (SERNAGEOMIN), the Ministry of Public Works, academic works (Sepúlveda et al. 2012; Mardones and Rojas 2012; Moya 2016), and an inventory of lateral spreads provided by Verdugo et al. (2012).

Serey et al. (2017, 2019) produced landslide maps by interpreting Landsat satellite images before and after the earthquake using Google Earth, employing the mapping scales between the 1:2000 and 1:10,000. The location of any landslide that was identified in the post-seismic image was examined by referring the latest pre-seismic image without cloud or snow cover. Moreover, field validation was performed mainly in the coastal regions where the density of landslides was high to facilitate identification, classification, and description of the landslides previously mapped from the viewpoint of their failure modes. The minimum size of the landslides in the mapping was 30 m^2, whereas field inspections showed an indefinite number of small mass movements that were not recognized in the satellite images. Thus, the inventory is not complete for small-size landslides.

The 2010 landslide inventory mapped by Serey et al. (2017) and analyzed in detail by Serey et al. (2019) counted 1226 landslides in an area of about 120,500 km^2; they include 1059 disrupted slides, 110 flows, 49 lateral spreads, and eight coherent slides (Fig. 5). The maximum distance to the epicenter is 487 km. Despite the lower earthquake intensities, most of the landslides (over 800, mainly disrupted shallow slides and falls) are located in the farther Andes Main Range that has greater relief and steeper slopes than the coastal range. A larger number of landslides (387) address those events having the volume range of 1000–5000 m^2, while just a few (29) are more than 50,000 m^2 (Serey et al. 2017). Furthermore, many landslides in the central depression are lateral spreads caused by soil liquefaction according to the bibliographic reviews by Verdugo et al. (2012) and Serey et al. (2017).

The compiled dataset was compared with the curves from the main literature regarding the maximum landslide area and the epicentral distance (Fig. 4). The geographical distribution of landslides is in agreement with the predictions defined for an earthquake of magnitude 8.8 (Serey et al. 2017). However, the events are not evenly distributed in the study area. Serey et al. (2017, 2019) highlighted the presence of landslide clusters. The most important landslides (127 failures) are located in the Arauco Peninsula, Biobio Region (Figs. 5 and 6), and they were mainly situated in the low-strength Neogene marine sedimentary rocks, suggesting an important lithologic control as a major factor in the generation of landslides (Moya et al. 2015; Moya 2016). These rocks have been tested by Moya (2016) and Brain et al. (2021) using shear devices equipped with back-pressure channels. Those rocks exhibited different

Fig. 5 Photographs of earthquake-induced landslides triggered by the 2010 Maule earthquake. Above, landslides in the Arauco peninsula, southern Chile (Moya 2016). Below, landslides in the coastal area of central Chile (Reproduced from Sepúlveda et al. 2012 with permission from Springer)

stress–strain behaviors, depending on the testing conditions. The increase in the shear strength under cyclic testing deserves attention.

Serey et al. (2019) present a comprehensive inventory of landslides induced by the 2010 Chile megathrust earthquake and their correlations with geological (slope, lithology) and seismic factors (rupture distance, PGA, PGV). Their comparisons found no statistically significant correlation between landslide intensity and ground shaking (PGA and PGV) across the study area. For the Maule megathrust earthquake, this suggests that geology was a controlling factor more relevant to coseismic landsliding than ground motion.

4 Liquefaction During the 2010 Maule Earthquake

The southern part of Chile, particularly to the south of the epicentral area, has a vast number of lakes, rivers, and beaches. Therefore, a significant number of sites are underlain by saturated sandy soil deposits, with varying percentages of fines, where liquefaction phenomena are likely to occur during a massive earthquake.

After the 2010 Maule earthquake, Verdugo and González (2015) carried out a detailed field reconnaissance focused on liquefaction phenomena triggered by the

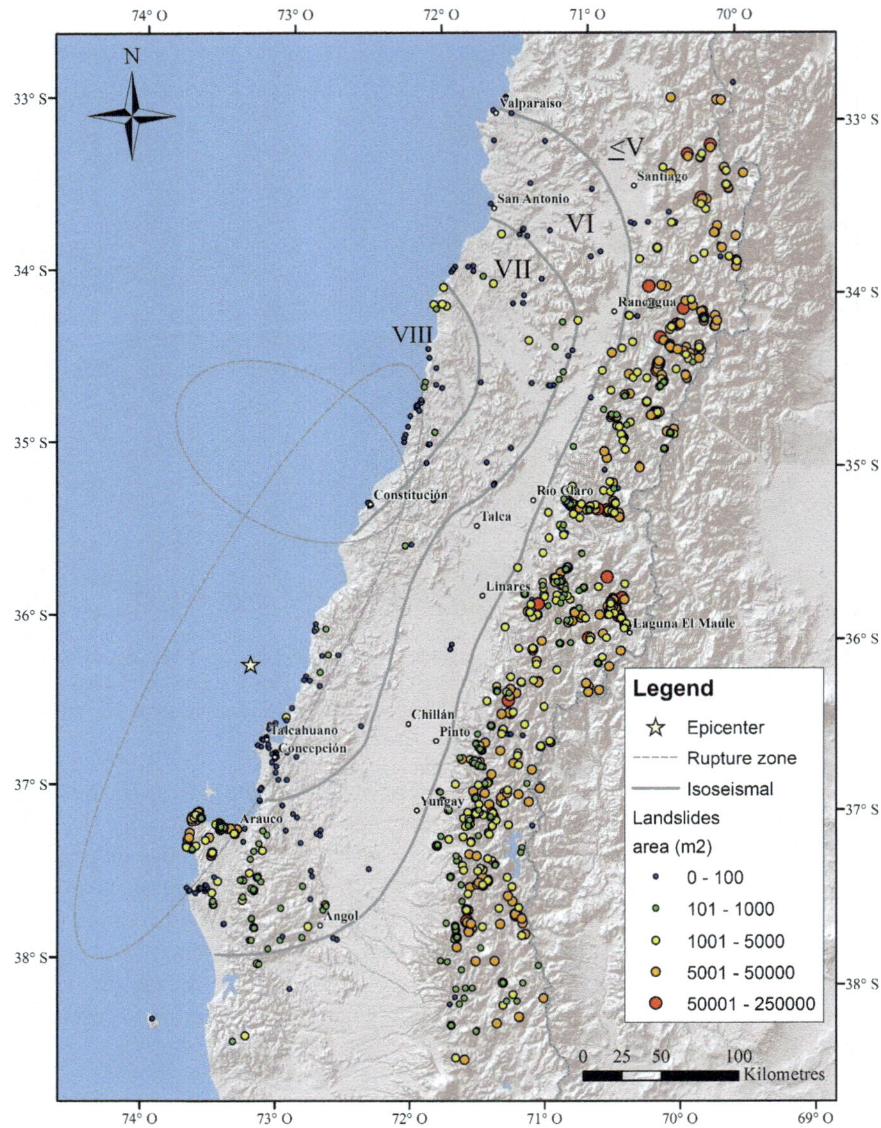

Fig. 6 Landslide inventory for the 2010 Maule earthquake (After Serey et al. 2017). Intensity data from Astroza et al. (2012) and rupture zone from Ruiz et al. (2012)

seismic event. Liquefaction was widely observed in an area of 1000 km length from north to south, in more than 150 spots. The reported cases were observed in both horizontal and gently sloping grounds. The farthest site presenting liquefaction was the Llanquihue Lake (41° S), more than 500 km south of the epicenter. The liquefaction observed in this area is difficult to explain, given the considerable distance

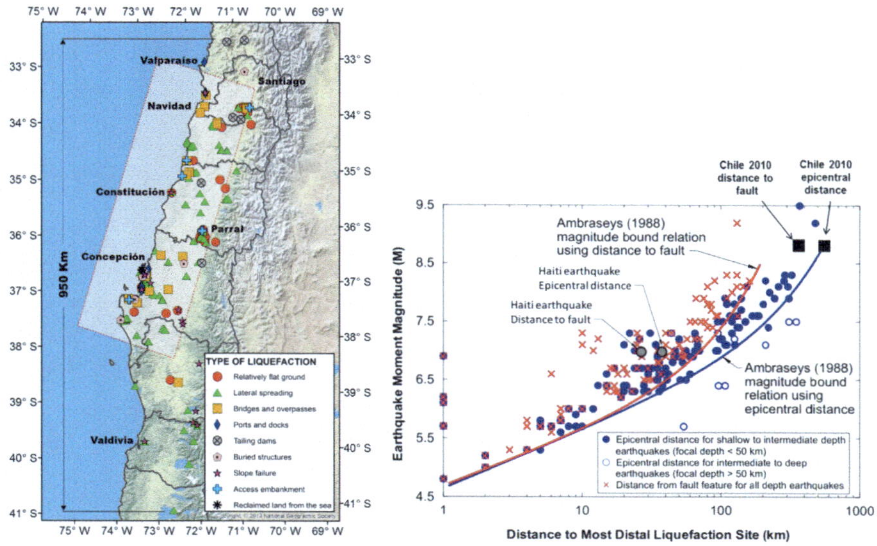

Fig. 7 Left: Spatial distribution of liquefaction cases observed in central Chile, between Valparaiso region and Llanquihue lake (southern edge of the map) (González 2015); Right: Farthest distance of liquefaction observed versus magnitude (Reproduced from Verdugo and González 2015 with permission from Elsevier)

between this area and the epicenter and also the 0.14 g of acceleration recorded, which is somewhat moderate. This area was severely affected by the Mw 9.5 1960 Valdivia earthquake, inducing liquefaction at several sites. The spatial distribution of liquefaction cases observed, as well as the plot of magnitude versus the farthest distance of liquefaction observed, are presented in Fig. 7.

4.1 Description of Reported Liquefaction Cases

Several liquefaction cases were reported, among which the most important were: (1) flat areas in where significant settlements were induced, (2) tailing dams, (3) lateral spreading, and (4) failure of pile foundations due to lateral spreading. Examples of liquefaction phenomena are presented in Fig. 8, and are described in the paragraphs below.

Liquefaction in flat grounds induced significant settlements, such as the one depicted in Fig. 8a, once the excess pore pressures from ground shaking were entirely dissipated. Likewise, buried geostructures were observed to rise to the ground surface.

Close to the epicentral area, there were several tailing dams of medium to large size. Only four of them showed limited damage. One of them, Las Palmas, experienced earthquake-induced flow failure, with the material flowing for several meters, causing four casualties and severe contamination. Figure 8b shows the position of

Fig. 8 Liquefaction cases reported: **a** flat areas with significant settlements, **b** before and after Las Palmas dam breach, **c** lateral spreading, **d** failure of pile foundations due to lateral spreading (a from González 2015; b and c modified from Bray and Frost 2010, d from Verdugo and González 2015)

the tailings material before and after the occurrence of the flow of the material. Las Palmas is an old geostructure that was constructed using the upstream construction method, the most common construction technique in the past, which is currently forbidden in seismic countries due to the several catastrophic flow failures that occurred in the last decades. Nowadays, old abandoned upstream dams imply a significant risk to the local population due to the possibility of a large earthquake inducing flow failure.

Evidence of lateral spreading was evident through the development of cracks in the ground along shorelines. Factors such as the size, the orientation of the cracks, sloping angle, and extension have a strong influence in the geometry of the liquefied area affected by lateral spreading. Figure 8c shows the lateral spreading observed in a gently sloping ground, in which the shallow crust was displaced over the liquefied ground, experiencing significant large deformations.

Figure 8d shows the effect of lateral spreading on pile foundations of port facilities, which has been widely observed in other seismic events, resulting in significant operation and maintenance costs due to the shutdown and reduced capability of operations.

4.2 *Liquefaction Effect on Bridges*

The 2010 Maule Earthquake proved that the Chilean bridge engineering is safe and robust. In general, the performance of bridges during this seismic event was quite satisfactory. Some bridges had damage due to lateral spreading or excessive settlement, particularly those crossing rivers where several bridge spans became unseated. Strong shaking also induced significant damage not only along shorelines of the river but also in the inland area of Concepcion, and even in Santiago. Effects like those observed in Santiago suggest that local soil conditions and topography set the conditions for more concentrated damage in areas with no other damaged structures (e.g., the northern area of Santiago). Although few bridges collapsed in the interior areas of Chile, those that did were attributed to poor detailing and construction. Figure 9 shows some of the effects of liquefaction on bridges.

5 Discussion and Concluding Remarks

The subduction 2010 Maule earthquake, which occurred on February 27, 2010, was caused by a rupture of more than 500 km in length, at a depth of approximately 40 km. The rupture area connected the seismic gap produced by the earthquakes of 1960 in Valdivia and 1985 in Valparaiso.

Evidence of liquefaction was observed throughout the area affected by the 2010 Maule earthquake in saturated granular materials, and sediments in rivers and shorelines. Liquefaction occurrence was reported over an extent of over 1000 km, with some occurring more than 300 km from the epicenter. The most affected sites were flat fields, bridge piles and abutments, port facilities, and buried structures, with lateral spreading affecting piles and the failure of a tailing dam, resulting in significant repair costs. Although no significant concentrated damage was reported, it was widely distributed.

The Maule earthquake triggered 1160 landslides, mainly reported as rock falls, rock and soil slides, flows and lateral spreads. More than 800 landslides were located in Principal Cordillera, an area with stronger relief but with lower recorded intensities. In general, the landslides occurred in clusters. The 2007 Mw 6.2 Aysen earthquake, which affected a much smaller area, of about 30 km around the epicenter, triggered

Fig. 9 Juan Pablo II bridge. Top left: Evidence of bridge deck settlement, Top right: Lateral spreading on a bridge abutment, Bottom left and right: a column and pier affected by liquefaction (Bray and Frost 2010)

close to half of the number of landslides induced by the much larger 2010 subduction earthquake, including large volume (in the order of 10^6 to 10^7 m^3) rockslides and rock avalanches not seen in the 2010 Maule event.

A comparison of the number of landslides during the 2007 Aysen and the 2010 Maule earthquakes with the number of landslides from other recent earthquakes worldwide (Tohoku, Chi-chi, Wenchuan, Kashmir) suggests that the number of landslides generated by subduction earthquakes is at least one order of magnitude lower than events of similar magnitude triggered by shallow crustal earthquakes. This is critical in future seismic landslide hazard analysis (Serey et al. 2019). Larger focal distance and attenuation of high frequency waves may explain at least in part this large difference. Figure 10 shows the frequency distribution curves of landslides (see Malamud et al. 2004 for details) in the Aysen and Maule earthquakes, compared with other important coseismic landslides inventories, such as the 1994 Mw = 6.7 Northridge and the 2011 Mw = 9.0 Tohoku earthquakes.

As observed in Fig. 10, while the Chilean earthquake inventories curves are relatively similar to the general laws proposed by Malamud et al. (2004), and they differ in the part for smaller areas before the inflection point. However, they show similar decay for larger landslides and are above the curve of the Tohoku earthquake. This may suggest a similar frequency distribution curve for large landslides in both Chilean earthquakes, independent of their earthquake source type, although the shallow crustal Aysen earthquake reaches larger areas.

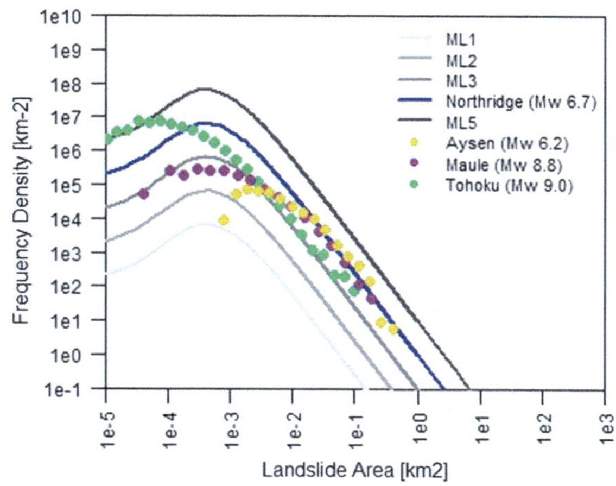

Fig. 10 Frequency density plot of Aysen and Maule earthquakes in Chile, compared with Northridge, Tohoku (Wartman et al. 2013) and Malamud et al. (2004). ML refers to landslide magnitudes by Malamud et al. (2004) (after Serey 2020)

Acknowledgements This work is a literature review constructed upon the text and images of the references indicated in this document, and information obtained by the authors in projects Fondecyt 11070107, Fondecyt 1140317, Fondecyt 11181252, NERC-Newton fund grant NE/N000315/1, UInicia 24/2018 and startup funds from the Research Dean and the School of Civil Engineering of Universidad de Chile.

References

Aravena N (2017) Análisis 3D del deslizamiento de roca de Punta Cola generado por el terremoto de Aysén del año 2007. Undergraduate dissertation, Dept of Geology, University of Chile (in Spanish)

Astroza M, Ruiz S, Astroza R (2012) Damage assessment and seismic intensity analysis of the 2010 (Mw 8.8) Maule earthquake. Earthq Spectra 28(S1):S145–S164

Brain MJ, Moya S, Kincey ME, Tunstall N, Petley DN, Sepúlveda SA (2021) Controls on post-seismic landslide behavior in Brittle rocks. J Geophys Res-Earth Surf 126:e2021JF006242

Bray J, Frost D (2010) Geo-engineering reconnaissance of the 2010 Maule, Chile earthquake. Report No. GEER-022 of the NSF Sponsored GEER Association Team, 1

Campos J, Hatzfeld D, Madariaga R, Lopez G, Kausel E, Zollo A, Lyon-Caen H (2002) A seismological study of the 1835 seismic gap in south-central Chile. Phys Earth Planet Inter 132(1-3):177–195

Cembrano J, Hervé F, Lavenu A (1996) The Liquiñe Ofqui fault zone: a long-lived intra-arc fault system in southern Chile. Tectonophysics 259:55–66

Cembrano J, Lara L, Lavenu A, Hervé F (2007) Long-term and short-term kinematic history of the Liquiñe-Ofqui fault zone: a review and implications for geologic hazard assessment. In: Proceedings Geosur 2007, Santiago de Chile, pp 30

Escobar P (2013) Inventario de remociones en masa desencadenadas por el sismo del 27 de febrero de 2010 en Chile central. Memoria de título, Universidad de Chile, Departamento de Geología: 109 p (in Spanish)

Escudero I (2017) Modelamiento dinámico mediante elementos discretos del deslizamiento de roca de Punta Cola generado por el terremoto del fiordo Aysén del 2007. Undergraduate dissertation, Dept of Geology, University of Chile (in Spanish)

González J (2015) Estudio del fenómeno de licuefacción en Chile para el terremoto del Maule, 2010. MSc. thesis, Dept of Civil Engineering, University of Chile (in Spanish)

Hermanns RL, Sepúlveda SA, Lastras G, Amblas D, Canals M et al (2014) Earthquake-triggered subaerial landslides that caused large scale fjord sediment deformation: combined subaerial and submarine studies of the 2007 Aysén Fjord event, Chile. In: Lollino G, Manconi A, Locat J, Huang Y, Canals M (eds) Engineering geology for society and territory—Volume 4 marine and coastal processes, Chap. 14. Springer, pp 67–70

Ishihara K (1996) Soil behaviour in earthquake geotechnics, 1st edn. Clarendon Press, Oxford, UK

Lastras G, Amblas D, Canals M, DETSUFA Shipboard Party (2016) Fjord flank collapse and associated deformation in Aysén fjord, Chile. In: Dowdeswell JA, Canals M, Jakobsson M, Todd BJ, Dowdeswell EK, Hogan KA (eds) Atlas of submarine glacial landforms: modern, quaternary and ancient. Geological Society, London, Memoirs, p 46

Legrand D, Barrientos S, Bataille K, Cembrano J, Pavez A (2011) The fluid-driven tectonic swarm of Aysen Fjord, Chile (2007) associated with two earthquakes (Mw = 6.1 and Mw = 6.2) within the Liquiñe-Ofqui fault zone. Cont Shelf Res 31:154–161

Malamud BD, Turcotte DL, Guzzetti F, Reichenbach P (2004) Landslides, earthquakes, and erosion. Earth Planet Sci Lett 229:45–59

Mardones M, Rojas J (2012) Procesos de remoción en masa inducidos por el terremoto del 27F de 2010 en la franja costera de la Región del Biobío, Chile. Revista de Geografía Norte Grande 53:57–74

Moya S (2016) Comportamiento monotónico y cíclico de suelos y rocas blandas afectados por remociones en masa cosísmicas. MSc Thesis, Dept of Geology, Universidad de Chile (in Spanish)

Moya S, Sepúlveda SA, Serey A, García M (2015) Remociones en masa generadas por el terremoto del Maule del 2010 en la Península de Arauco. In: Proceedings, XIV Congreso Geológico de Chile, La Serena

Naranjo J, Arenas M, Clavero J, Muñoz O (2009) Mass movement-induced tsunamis: main effects during the Patagonian Fjordland seismic in Aysén (45° 25′ S), Chile. Andean Geol 36:137–145

Ochoa-Cornejo F, Bobet A, Johnston CT, Santagata M, Sinfield JV (2016) Dynamic properties and small strain behavior of sands containing super-plastic fines. Soil Dyn Earthq Eng 88(9):265–279

Oppikofer T, Hermanns R, Redfield T, Sepúlveda S, Duhart P, Bascuñan I (2012) Morphologic description of the Punta Cola rock avalanche and associated minor rockslides caused by the 21 April 2007 Aysén earthquake (Patagonia, Southern Chile). Revista de la Asociación Geológica Argentina 69(3):339–353

Redfield TF, Hermanns RL, Oppikofer T, Duhart P, Mella M, Derch P, Bascuñan I, Fernández J, Arenas M, Sepúlveda S, Rebolledo S, Loew S, Yugsi Molina F, Abächerli A, Henderson I, Jaboyedoff M, Kveldsvik V (2011) Analysis of the 2007 earthquake induced Punta Cola rockslide and tsunami, Aysén Fjord, Patagonia, Chile (45.3 S, 73.0 W). In: Proceedings 5th international conference on earthquake geotechnical engineering, Santiago de Chile, January 2011

Ruiz S, Madariaga R, Astroza M, Saragoni R, Lancieri M, Vigny C, Campose J (2012) Short-period rupture process of the 2010 Mw 8.8 Maule earthquake in Chile. Earthquake Spectra 28(S1):S1–S18

Sepúlveda SA, Serey A (2009) Tsunamigenic, earthquake-triggered rock slope failures during the 21st of April 2007 Aisén earthquake, Southern Chile (45.5° S). Andean Geol 36(1):131–136

Sepúlveda SA, Serey A, Lara M, Pavez A, Rebolledo S (2010) Landslides induced by the 2007 Aysen Fjord earthquake, Chilean Patagonia. Landslides 7(4):483–492

Sepúlveda SA, Rebolledo S, Farías M, Vargas G, Arriagada C (2012) Efectos geológicos. In: Mw = 8.8: Terremoto en Chile, 27 de febrero 2010. Departamento Ingeniería Civil, FCFM, Universidad de Chile

Serey A (2020) Coseismic landslides during strong shallow crustal and megathrust earthquakes: controlling factors and conceptual hazard geomodels. PhD. Thesis, Dept of Geology, Universidad de Chile (in Spanish)

Serey A (2011) Análisis de las remociones en masa generadas por el terremoto del fiordo Aysén de 2007 (45.5° S). Tesis.Uchile.Cl. Universidad de Chile. Retrieved from http://www.tesis.uchile.cl/tesis/uchile/2007/celis_c/sources/celis_c.pdf

Serey A, Escobar P, Moya S, Sepúlveda SA, Petley DN (2017) Landslide inventory of the 2010 Mw 8.8 Maule earthquake, Central Chile. In: 16th World conference on earthquake engineering, Santiago, Chile, paper 1873

Serey A, Piñero-Feliciangeli L, Sepúlveda SA, Poblete F, Petley DN, Murphy W (2019) Landslides induced by the 2010 Chile megathrust earthquake: a comprehensive inventory and correlations with geological and seismic factors. Landslides 16:1153. https://doi.org/10.1007/s10346-019-01150-6

Van Daele M, Versteeg W, Pino M, Urrutia R, De Batist M (2013) Widespread deformation of basin-plain sediments in Aysén fjord (Chile) due to impact by earthquake-triggered, onshore-generated mass movements. Mar Geol 337:67–79

Vargas G, Rebolledo S, Sepúlveda SA, Lahsen A, Thiele R, Townley B, Padilla C, Rauld R, Herrera MJ, Lara M (2013) Submarine earthquake rupture, active faulting and volcanism along the major Liquiñe-Ofqui fault zone and implications for seismic hazard assessment in the Patagonian Andes. Andean Geology 40(1):141–171

Verdugo R, González J (2015) Liquefaction-induced ground damages during the 2010 Chile earthquake. Soil Dyn Earthq Eng 79:280–295

Verdugo R, González J, González V, Torres A (2012) Características y efectos del fenómeno de licuefacción. In Mw = 8.8: Terremoto en Chile, 27 de febrero 2010. Departamento Ingeniería Civil, FCFM, Universidad de Chile, pp 63–106

Verdugo R, Ochoa-Cornejo F, Gonzalez J, Valladares G (2019) Site effect and site classification in areas with large earthquakes. Soil Dyn Earthq Eng 126:105071

Vollmer M (2017) Simulación numérica de la avalancha de roca del 21 de abril de 2007 en Punta Cola, Región de Aysén, Chile. Undergraduate dissertation, Dept of Geology, University of Chile (in Spanish)

Wartman J, Dunham L, Tiwari B, Pradel D (2013) Landslides in Eastern Honshu induced by the 2011 off the Pacific coast of Tohoku earthquake. Bull Seismol Soc Am 103(2B):1503–1521

Wick E, Baumann V, Jaboyedoff M (2010) Report on the impact of the 27 February 2010 earthquake (Chile, Mw 8.8) on rockfalls in the Las Cuevas valley, Argentina. Nat Hazard 10:1989–1993

Earthquake-Induced Landslides in the Nepal Himalaya

Ranjan Kumar Dahal

Abstract Himalayan belt is well known for seismic risks and Himalayan populations are in higher risk of coseismic landslides. The earthquake-induced landslides are possessing severe damages to settlements, civil structures, resource management system, and tourism industries in the Nepal Himalaya. Landslides associated with the past earthquakes in the Himalaya are not well documented, but few authors have described some cases of landslides in past earthquakes. In this report, six major earthquakes and related earthquake-induced landslides are documented. Out of six major earthquakes, two earthquakes (2011 Sikkim-Nepal Boarder Earthquake and 2015 Gorkha Earthquake) were investigated in more detail. To assess the slope damages after these earthquakes, many field visits were conducted and landslide inventory maps were prepared field and remote sensing data. The landslides were evaluated in terms of characteristics and mode of occurrences along with hazard assessment. In the case of the Himalayan coseismic landslides, a significant discriminant equation is also developed to evaluate future seismic landslide hazard in Nepal.

1 Introduction

The Himalayan belt is associated with active subduction of Indian plate beneath the Eurasian plate and has been well recognized as an area of high seismic hazard (Molnar and Deng 1984; Bilham 2004; Bollinger et al. 2014). The subduction rate is believed to be about 45 mm/yr (Sella et al. 2002) which produced many thrust and fault systems in the Himalaya. As a result, the rocks in the Himalaya are uplifting in the vertical direction as well as in the horizontal direction towards south along the major thrusts (Meigs et al. 1995), which are, namely, main central thrust (MCT), main boundary thrust (MBT), and main frontal thrust (MFT). Among these thrusts, MFT is believed to be highly seismogenic because it accommodates most of the accumulated strain induced by subduction of the Indian plate (Lavé and Avouac 2000; Bilham et al. 2001; Bilham 2004). Figure 1 illustrates the active faults and

R. K. Dahal (✉)
Central Department of Geology, Tribhuvan University, Kirtipur, Kathmandu, Nepal 44618
e-mail: rkdahal@gmail.com; ranjan.dahal@cdgl.tu.edu.np

I. Towhata et al. (eds.), *Coseismic Landslides*, Springer Natural Hazards,
https://doi.org/10.1007/978-981-19-6597-5_3

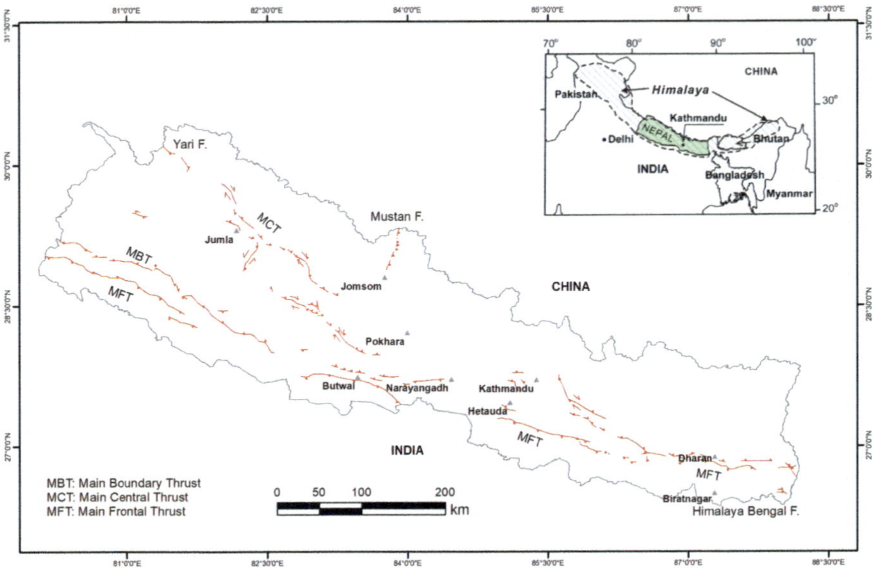

Fig. 1 Active faults in Nepal, HFF (Himalayan frontal fault) and MFT are synonym (Modified after Nakata 1982 and Nakata et al. 1984)

thrust systems in Nepal, which might have affected either directly or indirectly the past earthquakes and will do so in future.

The Himalayan belt has extreme threat from both rainfall-induced and earthquake-induced landslides (Dahal and Hasegawa 2008; Korup et al. 2010 and Dahal et al. 2013, Dahal 2015a). These landslides cause severe damages to settlements, civil structures, resource management system, and tourism industries. Nevertheless, the landslides associated with the past earthquakes in the Himalaya are not well documented, but a few authors who recorded some cases of landslides during some past earthquakes such as the 1934 Mw = 8 Bihar-Nepal earthquake (Pandey and Molnar 1988), the 1950 Mw = 8.5 Assam earthquake (Mathur 1953), and the 2005 Mw = 7.6 Kashmir earthquake (Owen et al. 2008). Such gigantic seismic events in Pleistocene and Holocene epochs triggered a few large-scale landslides as well as sediment yields from landslide site that blocked river channels in Central Nepal (Hasegawa et al. 2009; Schwanghart et al. 2016). Seismicity in the Himalayan region has been well studied from with regard to the relation with known geological faults and tectonic activities, but study on earthquake-induced landslides is missed.

During the past 100 years, three great earthquakes (Table 1) occurred along the Himalayan belt. After the 1934 Bihar-Nepal Earthquake in Nepal, eight major earthquakes hit Nepal, and the latest earthquake was the 2015 Gorkha Earthquake. Therein, the earthquake-induced landslides have been one of the most damaging natural disasters and sometime the damage was bigger than the damage related to ground shaking and surface rupture of the earthquake itself. There were many earthquake-induced landslides in Nepal, but they are not well documented. In this regard, this report was prepared to give concise information about the earthquake-induced landslides in Nepal. For this purpose, earthquakes of last 100 years were considered and coseismic landslides related to various major and great earthquakes are discussed. Most of the information presented in this report for the Nepal Himalaya has been cited from Roback et al. (2018), Dahal et al. (2013), Dahal (2015a, b, c), Dahal (2016), Shrestha et al. (2016), and Tsou et al. (2018).

2 The 1934 Bihar-Nepal Earthquake (Mw = 8.1)

Among the great earthquakes that struck the Himalayan belt in the last 100 years, the 1934 Bihar-Nepal Earthquake was one of the most pernicious earthquakes in the belt. This earthquake occurred on 15 January 1934 and it is still remained in ambiguity in terms of its location, nature, and intensity (Bhattarai et al. 2018). According to Rana (1935) and Dunn et al. (1939), the damage of this earthquake was very uneven and much of the destruction was due to slumping, fissuring, and tilting of the ground and mainly extended mainly in Midlands, Siwaliks and Gangatic Plain zones (Fig. 2). These damage descriptions clearly suggest that the earthquake triggered many landslides in mountainous area of central and eastern Nepal. But, there are very few documentations regarding the landslide's location and proximity to the epicenter. According to the report, this earthquake destroyed about 19% and damaged about 38% of the buildings in the valley (Pandey and Molnar 1988; Rana 1935). Rana (1935) clearly mentioned that there were many landslides in Pokhara area. One of such incidents has been figured out. Villagers of Jogithok of Syagja district confirmed that the debris flow deposit in their village was a result of the 1934 Bihar-Nepal Earthquake (Fig. 3). The landslide location and recent condition of the terrain are shown in Fig. 3. The landslide blocked the creek for seven days and a consecutive rainfall of 3 days one week after the earthquake-induced a huge debris flow. The explanations made by a senior citizen in the village about the debris flow clearly suggested that the flow was a result of a Landslide Dam Outburst Flood (LDOF).

Although the available damage reports on the 1934 Bihar-Nepal Earthquake unanimously address house damages and land subsidence in India and Nepal, the extensive damage on slopes in the mountain area in such strong shake cannot be eliminated from further study.

3 The 1980 Bajhang Earthquake—(ml = 6.5)

On July 29, 1980, western Nepal was hit by an earthquake named 1980 Bajhang Earthquake. In total, 178 people were killed by this earthquake and 40,000 houses were destroyed. It affected nearly 40,000 km^2 area of western Nepal (Fig. 4). A countless number of landslides and ground subsidences were also observed in this area. The earthquake intensity shows that most of the severe damages were confined

Table 1 Mega and major earthquakes in Nepal in last 800 years (modified after Chaulagain et al. 2018)

Year	Epicenter	Magnitude	Casualties	Structural and geological damages
1255	–	Intensity MMIBX	One third of the Kathmandu valley population along with the then King Abhay Malla	Severe damage in residential buildings, monumental and architectural heritage
1260	–	–	Many people killed by the earthquake as well as the famine followed by earthquake	Damages in residential buildings, monumental and architectural heritage
1408	–	–	No records available	Severe damage in residential buildings, temples Lateral spreading/soil liquefaction
1681	–	–	No records available	Damage on residential buildings
1810	–	–	Some casualties in Bhaktapur	Significant damage in residential buildings and monumental constructions
1823	–	–	No	Some residential buildings damaged
1833	–	ML7.7	414 deaths in and around Kathmandu valley	In total 18,000 buildings damaged Around 4000 in Kathmandu valley and Banepa
1834	–	–	No records available	Many residential as well as monumental constructions were damaged
1837	–	–	No records available	Damage only in Indian state of Bihar
1869	–	–	No records available	No records available
1897	–	–	No records available	No records available
1917	–	–	No records available	No records available

(continued)

Table 1 (continued)

Year	Epicenter	Magnitude	Casualties	Structural and geological damages
1934	Eastern Nepal	MW 8.1	8519 total deaths in Nepal 4296 within Kathmandu valley	More than 200,000 residential buildings, monuments and historical constructions damaged About 81,000 buildings collapsed Almost 55,000 buildings damaged in Kathmandu valley 12,397 buildings collapsed in Kathmandu valley Soil liquefaction/lateral spreading observed in central part of Kathmandu valley and many other locations of central and eastern plains of Nepal cascading effects like landslides, floods due to blockade in river course aggravated the damage
1936	Annapurna	ML 7.0	No records available	No records available
1954	Kaski	ML 6.4	No records available	No records available
1965	Taplejung	ML 6.1	No records available	No records available
1966	Bajhang	ML 6.0	24	6544 buildings damaged 1300 buildings collapsed
1980	Chainpur	ML 6.5	103	25,086 buildings damaged 12,817 collapsed
1988	Udaypur	MW 6.5	721	66,382 buildings damaged Several cases of liquefaction reported in eastern Nepal
2011	Sikkim-Nepal border	MW 6.9	Six deaths and 30 injuries in Nepal side (damage was intense in Indian side) 2 casualties in Kathmandu valley	14,554 buildings damaged 6435 buildings collapsed

(continued)

Table 1 (continued)

Year	Epicenter	Magnitude	Casualties	Structural and geological damages
2015	Barpak, Gorkha	MW 7.8	8790 deaths and 22,300 injuries 8 million people displaced	498,852 buildings collapsed 256,697 buildings partly damaged As many as 6000 landslides (polygons) and avalanches, Roback et al. (2018) have identified 24,915 landslides with a total landslide area of 87 km^2. Many cases of liquefaction and lateral spreading in Kathmandu valley Severe damage in infrastructures and lifelines

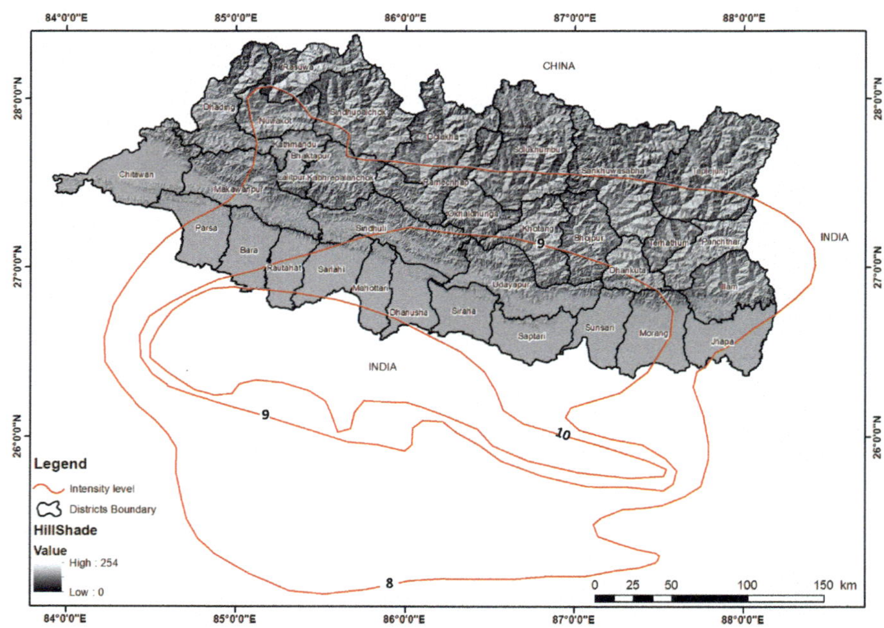

Fig. 2 Observed Rossi-Forel seismic intensity scale during the 1934 Bihar-Nepal earthquake (after Pandey and Molnar 1988)

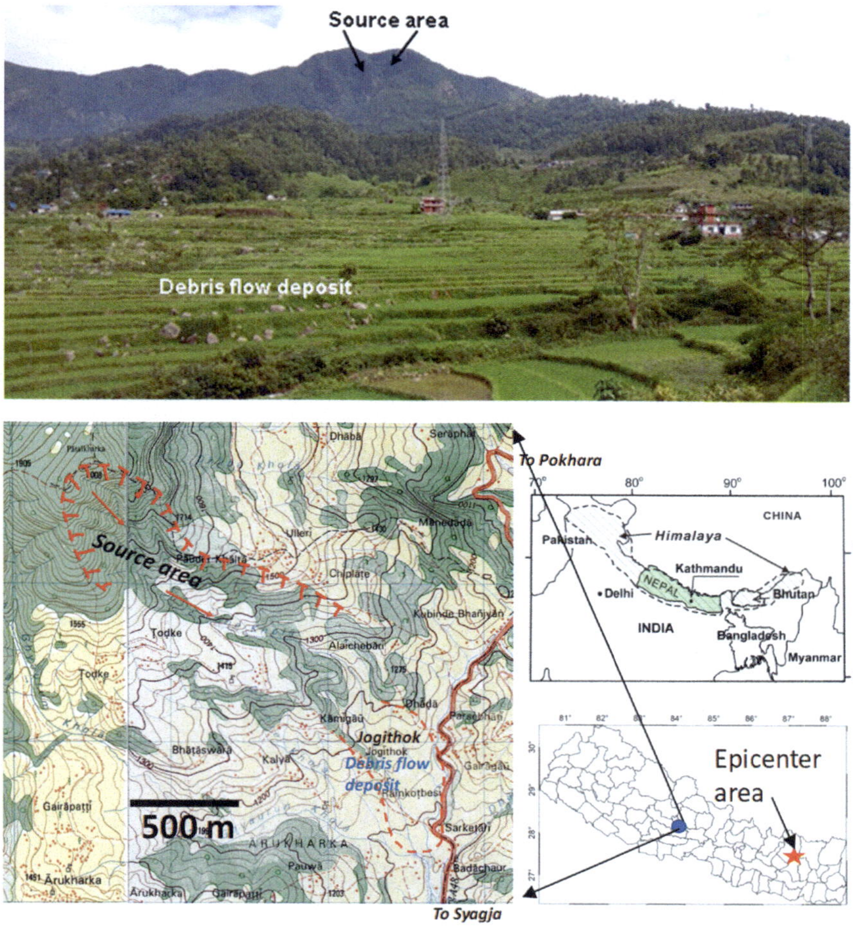

Fig. 3 Earthquake-induced landslide in Pokhara area of Nepal after the 1934 Bihar-Nepal earthquake, inset shows the location of site in the map of Nepal along with tentative location of the earthquake epicenter

to the Darchula and Bajhang Districts where the modified Mercalli Intensity Scale of VIII was registered (Singh 1985).

4 The 1988 Udayapur Earthquake (Mw 6.6)

A 54 years after the 1934 Bihar-Nepal earthquake, Nepal was hit again by another seismic event of medium size (Mw = 6.6) at 04:55 local time on August 20, 1988. Its epicenter was located at Udaypur, about 160 km southeast of Kathmandu. The shaking lasted for about 40 s, affecting 22 districts in central and eastern Nepal. It

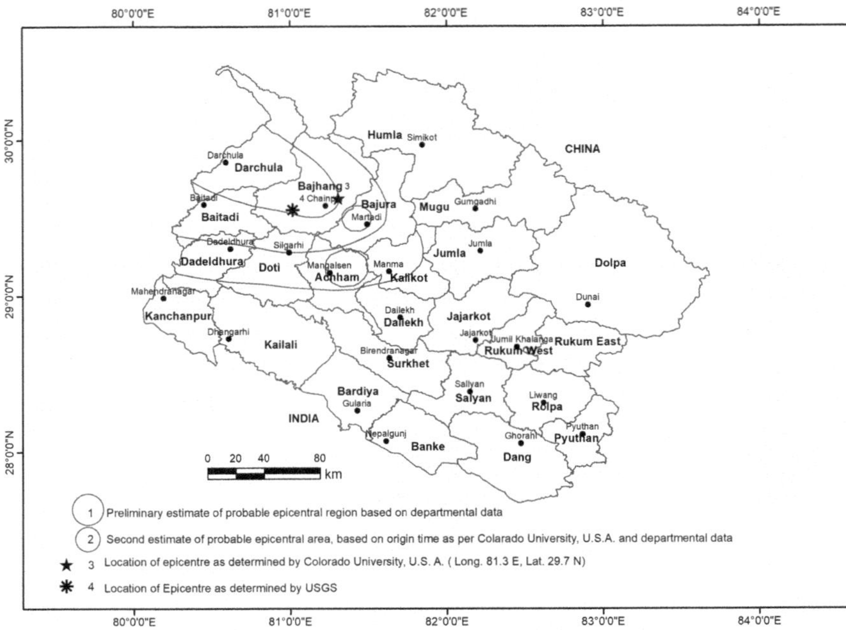

Fig. 4 Area affected by the 1980 Bajhang earthquake (After Singh 1985)

killed 717 people and more than 23,000 people lost their homes completely. Likewise, 65,145 homes severely damaged during this earthquake. Several landslides were triggered in the mountainous region. The epicentral intensity (Fig. 5) was found to be VIII in MMI scale in Udaypur and its surrounding area which suffered a higher order of damages (Dixit 1991). Structural damage and many other effects occurred in the zone of intensity = VII and VI. Six major aftershocks occurred with Richter scale greater than (Chaulagain et al. 2018). The common ground effects that were observed are surface collapsing and ground fissuring as well as shallow landslides, liquefaction, lateral spread, and sand boils. The roads in eastern Nepal were also severely damaged by earthquake-induced landslides.

5 The 2011 Sikkim/Nepal Border Earthquake (Mw = 6.2)

The 2011 Sikkim/Nepal Border Earthquake (Mw 6.2) occurred on September 18, 2011, in eastern Nepal. Although this earthquake was named Sikkim/Nepal Border Earthquake by USGS, its epicenter was actually located in the Nepalese territory; in the northeast part of Taplejung District near the Kanchenjunga Base Camp. This earthquake affected 14,544 houses (6435 completely destroyed), claimed six casualties, and injured 30 people in Nepal (Dahal et al. 2013). The shake map that was produced by USGS immediately after the earthquake (Fig. 6) indicates that the

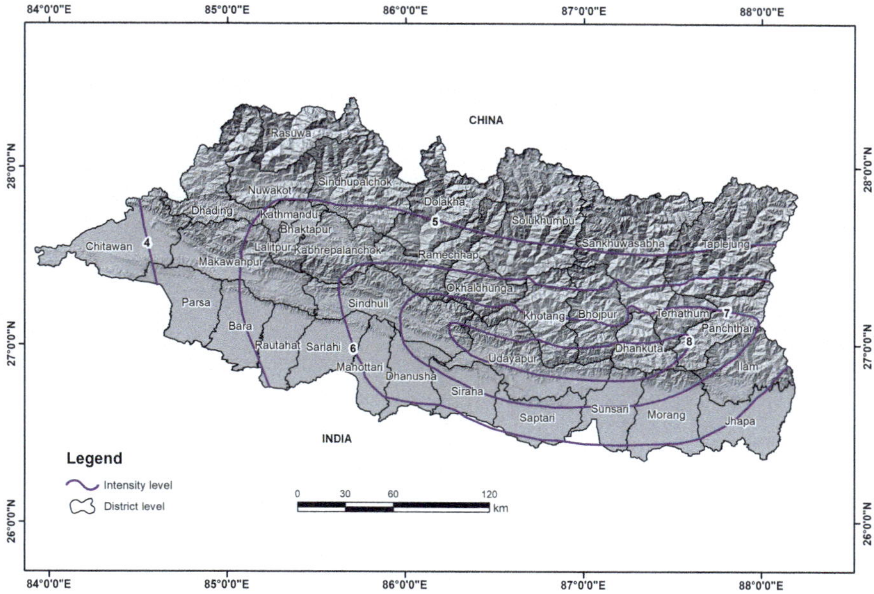

Fig. 5 Observed intensity for the 1988 Udayapur earthquake (After Dixit 1991)

seismic ground shaking was moderate to strong in east Nepal, Sikkim (India), and a few parts in south Tibet (China). During this earthquake, the peak ground accelerations (PGA) measured jointly by USGS and Department of Mines and Geology of Nepal ranged between 38 and 90 Gal in the Koshi Highway region, while being 90–177 Gal in the Mechi Highway region. This strong shaking triggered more slope failures along the Mechi Highway as compared with the Koshi Highway (Fig. 7, Plate 1 and Fig. 8). The damage reconnaissance found many longitudinal as well as transverse cracks in the ridge part of the roadside slopes. Moreover, a number of earthquake-induced landslides were found in the epicentral region. Noteworthy was that the epicenter was located in a glaciated Himalayan region at high altitude, and, therefore, the landslides in the epicentral area did not cause any socio-economic loss. In contrast, the smaller slope failures along the major highways, which occurred at the epicentral distance ranging from 70 to 100 km drew a commanding attention of the Nepalese planners about the consequences of earthquake-induced landslides. Rockfalls damaged the Koshi Highway and huge boulders deposited on the roadside slope. The geological map (modified after DMG (2011)) suggested that the area consists of Higher Himalayan rocks, namely gneiss, quartzite, and biotite schist in the landslide affected area.

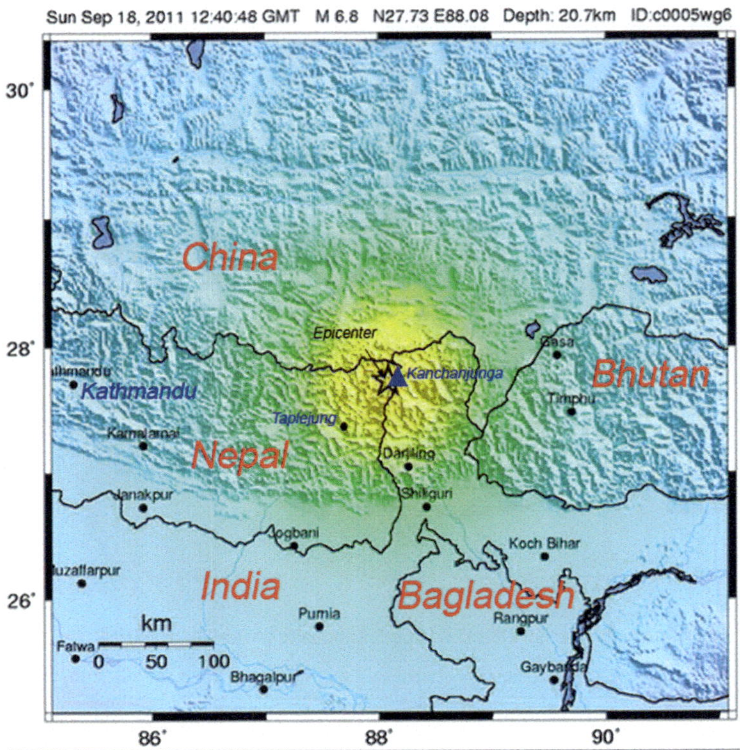

PERCEIVED SHAKING	Not Felt	Weak	Light	Moderate	Strong	Very strong	Severe	Violent	Extreme
POTENTIAL DAMAGE	None	None	None	Very light	Light	Moderate	Moderate /Heavy	Heavy	Very heavy
PEAK ACC (%g)	<0.17	0.17-1.4	1.4-3.9	3.9-9.2	9.2-18	18-34	34-65	65-124	>124
PEAK VEL (cm/s)	<0.1	0.1-1.1	1.1-3.4	3.4-8.1	8.1-16	16.31	31-60	60-116	>116
INSTRUMENTAL INTENSITY	I	II-III	IV	V	VI	VII	VIII	IX	X+

Fig. 6 Shake map of the 2011 Sikkim-Nepal Border Earthquake of September 18, 2011 (Modified after USGS 2011) (From Dahal 2015a with permission from Journal of Nepal Geological Society)

5.1 Hazard Zoning After the 2011 Sikkim/Nepal Border Earthquake

A GIS-based susceptibility analysis was performed on earthquake-induced slope failure by following the method proposed by Uchida et al. (2004) (Dahal 2015a). After extensive studies on the landslide damages in granitic Rokko Mountains induced by the 1995 Kobe earthquake, Uchida and his team (Uchida et al. 2004) derived a landslide probability function based on discriminant analysis using slope, average curvature, and maximum ground acceleration. This method is characterized by not using geological and other intrinsic factors. In their method, the average curvature is defined by the mean of the maximum and the minimum curvatures of the geodesic data on curved slopes. In raster GIS, this average curvature, ε, can be derived from

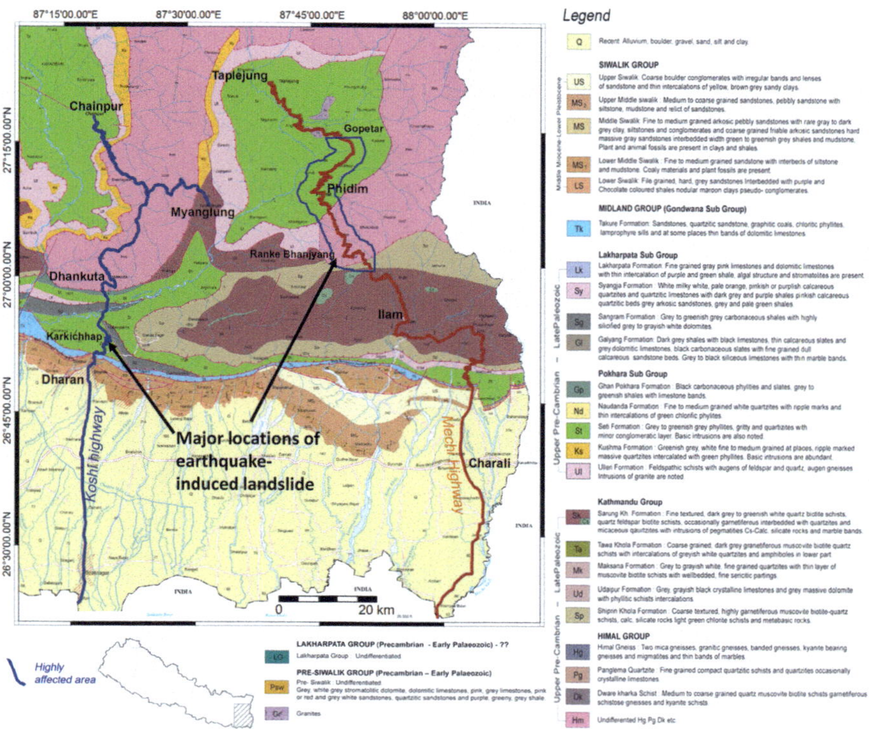

Fig. 7 Rock falls area after 2011 Sikkim/Nepal Border Earthquake at the Koshi highway (Revised after Dahal 2015a with permission from Journal of Nepal Geological Society)

the following relationship:

$$\varepsilon = \frac{\frac{\partial^2 y}{\partial x^2}\left\{1+\left(\frac{\partial f}{\partial y}\right)^2\right\}+\frac{\partial^2 f}{\partial y^2}\left\{1+\left(\frac{\partial f}{\partial x}\right)^2\right\}-2\frac{\partial f}{\partial x}\frac{\partial f}{\partial y}\frac{\partial^2 f}{\partial x \partial y}}{\sqrt{2\left(1+\left(\frac{\partial f}{\partial x}\right)^2+\left(\frac{\partial f}{\partial x}\right)^2\right)^3}} \tag{1}$$

in which f designates the pixel value of a DEM generated by the contour map, while x and y stand for the local coordinates. A concave slope has a positive curvature value, whereas a convex slope is assigned with a negative curvature. By using the mean, ε, of the curvatures defined by Eq. (1), Uchida et al. (2004) obtained the landslide probability function on the basis of a discriminant analysis as followings:

$$F = 0.075[\theta] - 8.9[\varepsilon] + 0.0056[a_{\max}] - 3.2 \tag{2}$$

where F is the landslide probability function or the discriminant score, θ is the slope angle in degrees, ε is the mean slope curvature, and $a_{\max}$ is the maximum ground

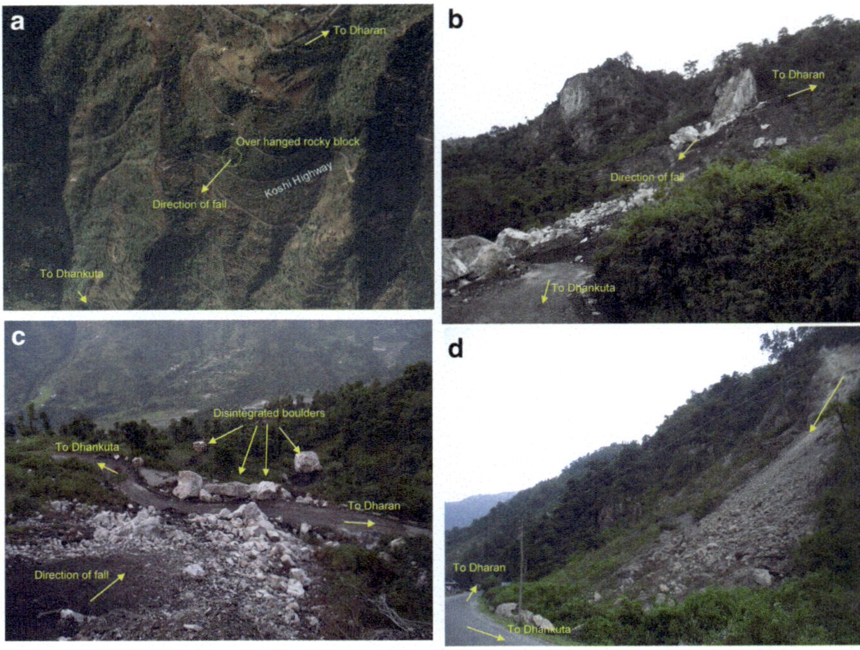

Plate 1 Rock fall caused by the 2011 Sikkim/Nepal border earthquake at the Koshi highway

acceleration in Gal. If F is positive, the corresponding pixels have a potential to fail during an earthquake. In contrast, a negative F value implies that the slope is unlikely to fail during an earthquake. Equations (1) and (2) are easily applicable to DEM in GIS platform.

The main limitation of this landslide risk assessment is that the effect of lithology, other topographic factors, and the amplification effect of earthquake motion cannot be taken into account of this simplified method. Nevertheless, once this equation is validated from one seismic event, it can be applied to other earthquake-prone regions.

It is important that the calculation of F value (Eq. 2) in any selected area does not use any data on the earthquake-induced slope failures. In this regard and, also, to validate the predictive ability of F for likelihood of slope failure, the F-values were qualitatively examined with the help of success rate curves. According to a statistical analysis on landslide susceptibility, the success rate is a measure of good fitting and, for the present research, the success rate is a measure of the predictive power of the landslide susceptibility values, i.e., the value of F. This is because the calculation of susceptibility values employs only three parameters on topography and shaking intensity (see Eq. 2) and do not use any data obtained from reported earthquake-induced slope failures. The calculated index values in all the pixels in the map were then sorted in a descending order in order to obtain a success rate curve in terms of F. The ordered pixel values were categorized into one hundred classes with 1% cumulative intervals and an F-value map was drawn. This F-value map was

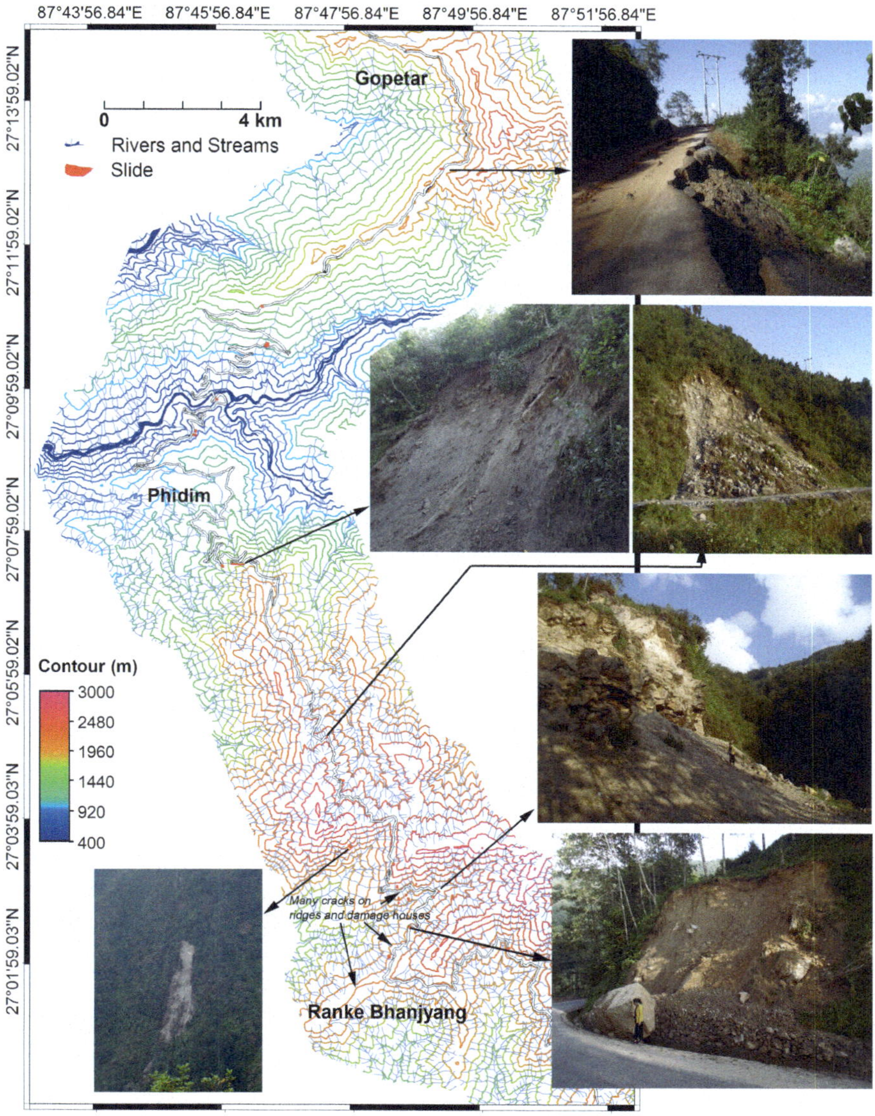

Fig. 8 Earthquake-induced landslide inventory (After Dahal 2015a) in a stretch of the Mechi highway after the 2011 Sikkim-Nepal border earthquake (With permission from Journal of Nepal Geological Society)

cross-examined against a slope failure inventory map and the success rate curve was prepared from cross table values.

An example case history was taken from the earthquake-induced slope failures that were induced by the September 18, 2011, Sikkim/Nepal border earthquake. The success rate reveals that F-values were high in 20% of the studied area, and was able

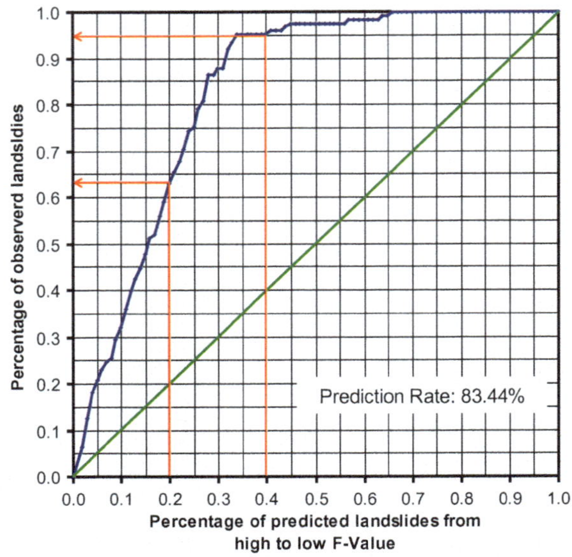

Fig. 9 Success rate curve and prediction rate of susceptibility map for the selected small area

to account for 64% of total landslides (Dahal 2015a). Similarly, 40% of the higher *F*-values corresponded to 95% out of all the existing landslides. Figure 9 illustrates the percent coverage of slope failures as a function of *F*. The predictive capability of *F*-value was further examined by measuring the area under the curve in the success rate diagram (Fig. 9). The area under the curve designates the success rate of the prediction by means of *F*. If the total area is equal to unity, the prediction "landslide is likely if *F* is positive" has a perfect accuracy. On the contrary, if the area under the curve is smaller than 0.5000, the analysis is judged to be invalid. In the present study, the area under the curve in Fig. 9 is 0.8344, showing 83.44% success of the prediction method. Thus, the examined method of prediction is considered valid.

To construct a classified susceptibility map, the reference success rates were defined in terms of *F*-values of 40, 60, 80, and 90% ranging from low to high, are calculated. Accordingly, four slope susceptibility classes were established as stable zone (less than 40% of *F*-value), Moderately stable zone (40–60% of *F*-value), quasi stable zone (60–80% of *F*-value), and unstable one (80–100% of *F*-value). The susceptibility map of the selected area prepared based on these intervals is shown in Fig. 10.

Finally, the zonation map was cross-examined against a slope failure inventory map of the same area. Accordingly, it was found that 96% of earthquake-induced slope failures induced by the Sikkim-Nepal Border Earthquake of September 18, 2011, are located in unstable and quasi stable zones (Fig. 11) (Dahal 2015a). Thus, the methodology proposed by Uchida et al. (2004) in Japan was judged applicable to susceptibility assessment of the earthquake-induced landslides in the mountainous terrain of eastern Nepal. It appears possible that the hazard of earthquake-induced landslides will be evaluated in the Nepal Himalaya where the risk of coseismic

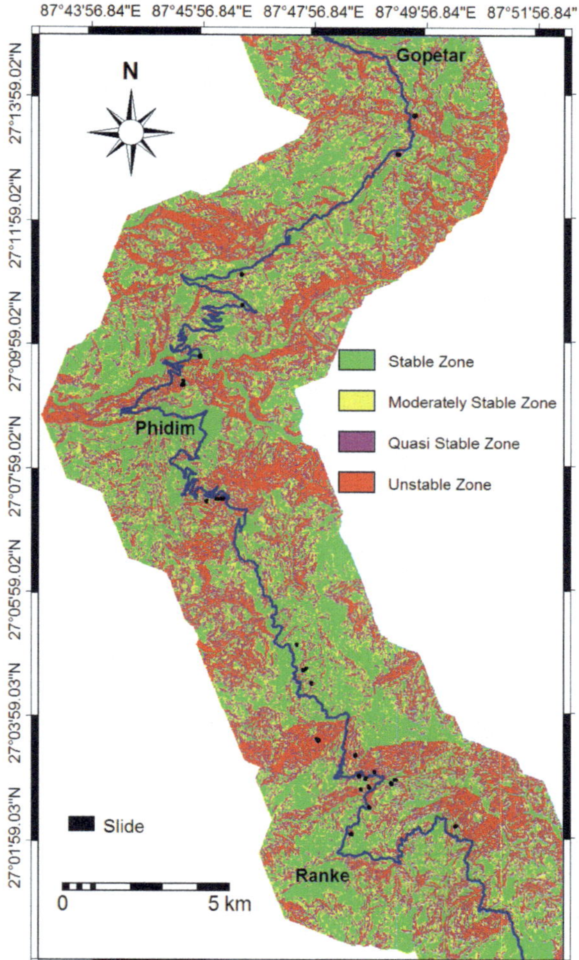

Fig. 10 Susceptibility map of a small selected area in Ranke-Phidim area. Slope failures after the September 18 2011 Sikkim/Nepal Border Earthquake are also presented in this map

slope failures is high and the mountainous roads passing through the highly elevated Himalayan ridges are always vulnerable to slope disasters during the events of great earthquakes.

6 The 2015 Gorkha Earthquake

The 2015 Gorkha Earthquake of Mw = 7.8 occurred at 11:56 AM Nepal Standard Time on April 25, 2015, with its epicenter located in Barpak village at 77 km northwest of Kathmandu. This earthquake was one of the most devastating events that had hit Nepal since the 1934 Nepal-Bihar earthquake. The main shock of this earthquake

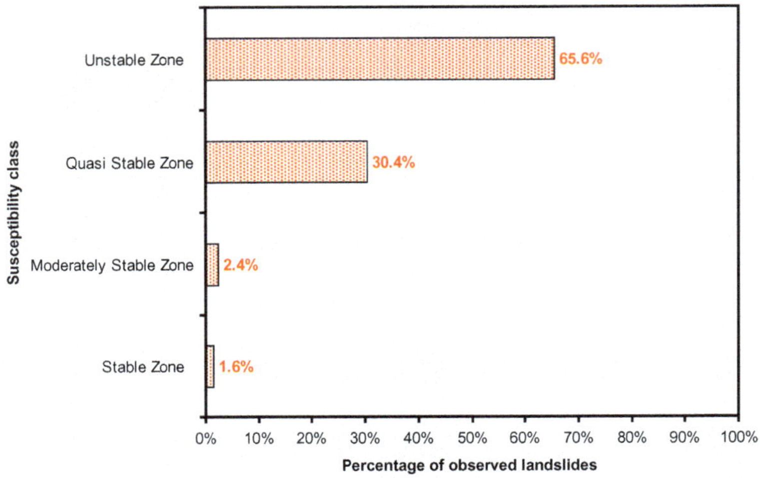

Fig. 11 Distribution of existing slope failures in a small selected area. In total, 96% of earthquake-induced slope failures after Sikkim-Nepal border earthquake of September 18, 2011 are found in unstable and quasi stable zones

was followed by hundreds of aftershocks, including an event of Mw = 7.3 on May 12, 2015 (USGS 2015). More than eighty-five million people have been affected by the Gorkha Earthquake in Nepal. This number corresponds to one quarter of the total population of the country (Dahal 2015b).

The Gorkha Earthquake triggered many landslides in the major hit areas and they concentrated, in particular, in the region from Barpak to Dolakha, which are mainly in the northern part of the Kathmandu Valley. After this disaster, several studies have been made of the distribution of the landslides using field reconnaissance data and satellite images, and analyzed their controlling factors (Kargel et al. 2015; Avouac et al. 2015; Dahal 2015b, 2016; Martha et al. 2017; Roback et al. 2018; Tiwari et al. 2017). It was found that most of the landslides were shallow and disrupted landslides and are distributed across the Fore Himalaya and the Higher Himalaya (Fig. 12). A total of 13,097 new landslides and 750 enlarged landslides were identified (Tsou et al. 2018). Roback et al. (2018) identified 24,915 landslides with a total landslide area of 87 km^2. Whatever the case during the 2015 Gorkha Earthquake, many coseismic landslides occurred in areas of forest and shrubland originating from the upper slopes (near ridges) and initiated by rock falls (Geli et al. 1988; Shrestha et al. 2016); Wang et al. (2016); Roback et al. (2018) have pointed out the seismic energy emitted from the deepest part of the rupture patch played a chief role in landslide occurrence along the Fore Himalayan part along the direction of the rupture propagation.

The author was also busy to collect data on earthquake-induced landslides after the 2015 Gorkha earthquake. More than 6000 landslides were recognized as fresh failures after the earthquake. A discriminant analysis has been done to explore the relationship of earthquake-induced landslides with ground acceleration, slope and

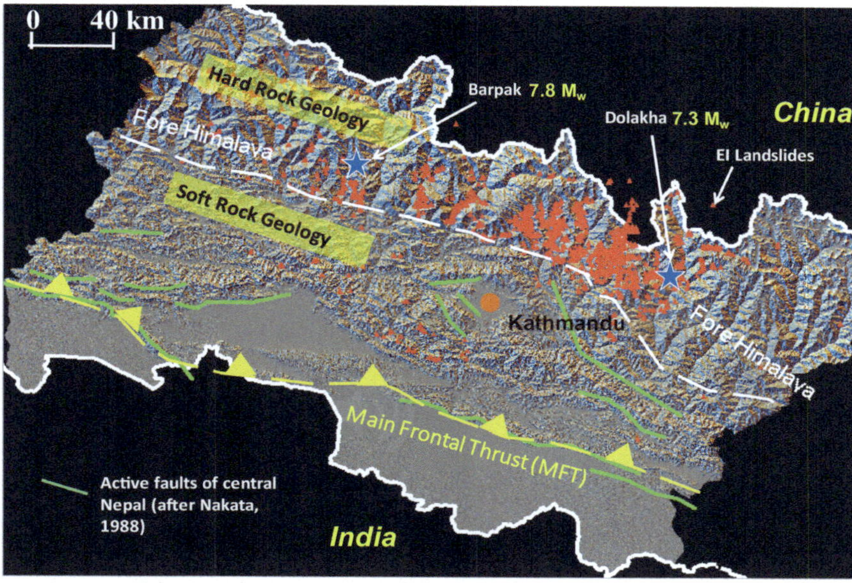

Fig. 12 Landslide distribution in central Nepal

slope curvature of the affected area. The resulted discriminant function is given by Eq. 3;

$$F = 4.536 \times (\text{Curvature}) + 0.004 \times (\text{Slope}) + 0.005 \times (\text{PGA}) - 2.766 \quad (3)$$

Equation 3 can be used to investigate earthquake induced landslide hazards in the given area of the Nepal Himalaya, especially in Fore Himalayan and Higher Himalayan region.

7 Classification of the Earthquake-Induced Landslides for the Nepal Himalaya

In line with the landslides in other parts of the world, the field observations and measurements are able to classify the earthquake-induced landslides after the 2015 Gorkha Earthquake into six categories such as rock fall, shallow landslides and dry debris fall, deep seated landslides, debris flow and mud flow, collapse of fill in valleys, and cut-and-fill failure. These classifications are more or is similar to Category II and Category I slides proposed by Keefer (2000).

1. **Rock fall**
   Similar to the definition by Varnes (1984), rock fall in this article comprises a detachment of rock materials from a steep slope or cliffs followed by more or less free and extremely rapid descent of the material. The falling rock travels through the air, occasionally bouncing or rolling on the slope. In general, rocky slopes of 40°–90° are subject to coseismic rock fall. During major earthquakes, rock fall is a very common and significant problem along transportation routes in rocky terrain (Fig. 13).
2. **Shallow landslides and dry debris fall**
   Shallow landslides is defined as failures of slope in which shear failure occurs within the depth of 2 m or less. Although depending to a certain extent on soil thickness on slopes, most of the earthquake-induced shallow landslides in Nepal are of soil thickness less than 2 m. The shallow landslides comprise a detachment of slope materials from a gentle or steep slope (Fig. 14). Generally, the sense of movement of slope materials is translational or semi-translational along a slightly curved failure surface.

Fig. 13 Schematic illustration of rock fall and rock fall at Mailung Bazar, Nepal after the 2015 Gorkha earthquake

Fig. 14 Illustration of shallow-seated debris slide in profile and shallow landslides on thin soil cover on bedrock after the 2015 Gorkha earthquake

3. **Deep-seated landslides**
   Deep-seated landslides are associated with the failure plane at depth of more than 2 m. According to Chigira and Yagi (2006), deep-seated landslides are coherent landslides. They stated that most of the deep-seated landslides triggered by the 2004 Niigata-Chuetsu earthquake, Japan, had the failure at depth exceeding 5 m. Deep-seated landslides also include large topples and translational slides as well (Varnes 1984). Usually, both rock and soil are involved in deep-seated landslide and it has considerable areal extension. If large in scale, deep-seated landslides often block river and create lakes. Breaching or collapse of landslide dams lead to devastating flooding in the downstream area. At the time of the Gorkha Earthquake, most of the deep-seated landslides occurred in pre-existing large-scale landslide topography. It was interesting that deep valley was collapsing in association with wide or open cracks (Fig. 15).
4. **Debris flow and avalanche**
   Debris flow is a rapid movement of loose soils, rock fragments, and organic materials combined with entrapped air and water. Thus, slurry is formed and is able to flow downslope rapidly. Generally, earthquake-induced landslides are not debris flow at the moment when sliding starts. However, landslides, whether shallow or deep-seated, landslides may block river flows partly or totally and, eventually, the landslide dam is overtopped and eroded, leading to a disastrous debris flow in the downstream of the river. In the case of the 2015 Gorkha Earthquake, such debris flow problems were not noticed probably because it happened in relatively dry season of April. On the other hand, the rock fall coming down from the snow-capped peaks of Langtang Valley created a huge snow avalanche and debris flow, and buried the village of Langtang (Fig. 16). This event claimed more than 220 lives including many tourists from different countries.
5. **Failure of fill and embankment in valleys**
   Construction of road and building in mountainous terrains sometimes require earth filling over small creeks or across valleys. Earth filling is the common practice for road to cross a zero-order valley. Therein, the filling materials are

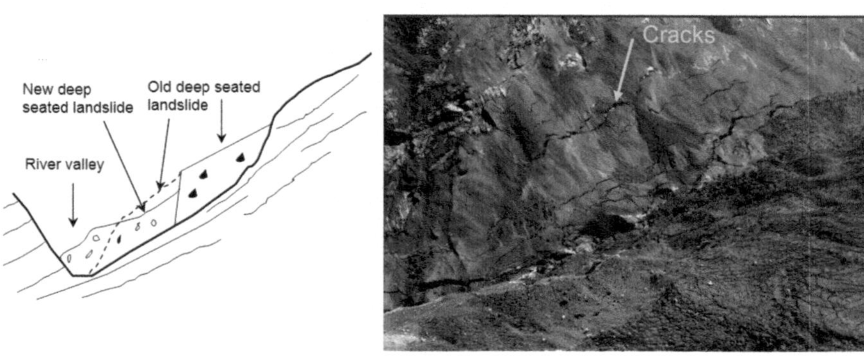

Fig. 15 Typical deep-seated translational slides profile of a landslide in Langtang valley that occurred after the 2015 Gorkha earthquake. The cracks were more than 30 m wide in many cases

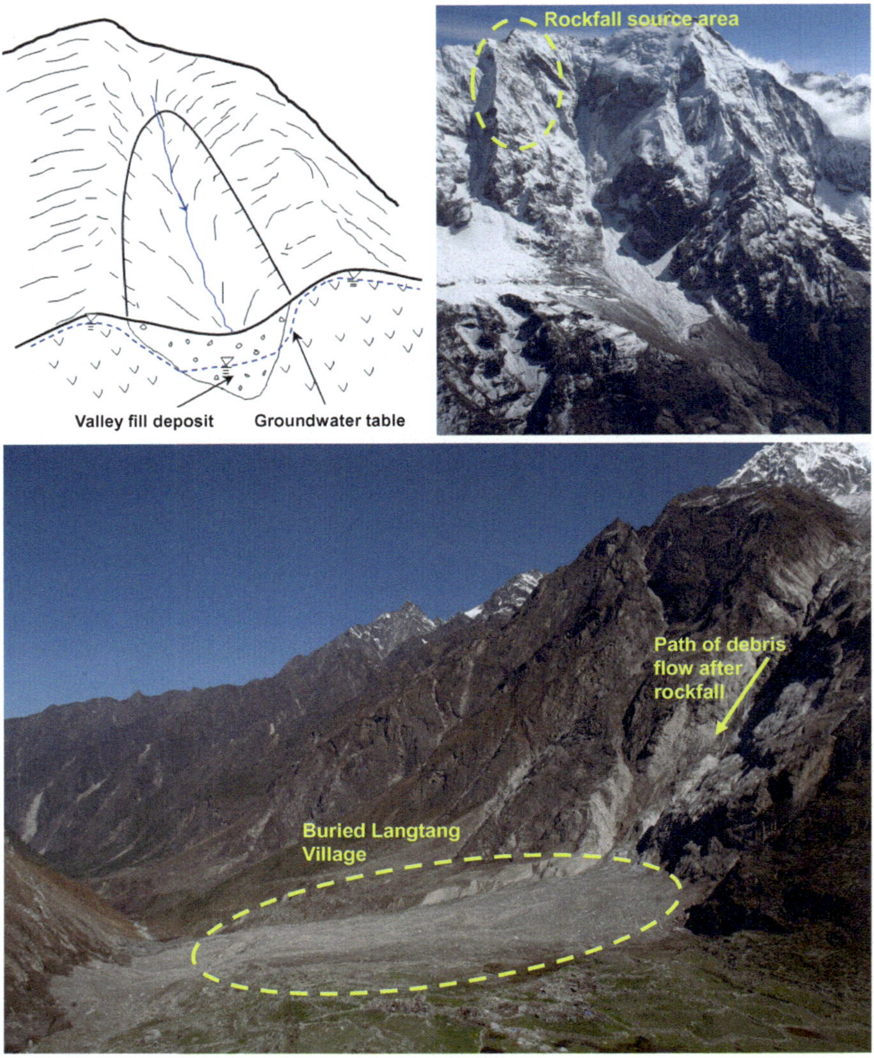

Fig. 16 Formation of debris flow and avalanche on Himalayan glacial valley during the 2015 Gorkha Earthquake and buried Langtang Village

mostly debris collected from river channels or weathered rock and soil obtained in the vicinity. Undergoing earthquake shaking, such materials are extremely prone to failure. Such a failure of fill in valley was found during the 2015 Gorkha Earthquake. About 100 m stretch of Kathmandu Airport Road collapsed during the earthquake. Accordingly, Department of Roads, Nepal is still facing serious problem from this landslide. This road collapse was of a lateral-spread type of failure and continued flowing.

6. **Cut-and-fill failure**

Cut and fill is very common in mountain road construction, preparation of terrace for farming, and construction of houses on mountain slopes. Unfortunately, the fill part is always more vulnerable to seismic effects than the cut part. In line with this, road embankments and river levees are also prone to failure under strong ground motion. In Nepal, most of the roads in agricultural area as well as more important highways are constructed by cut-and-fill methods. So, those roads are substantially subject to cut-and-fill failures during strong earthquake. Such problems were found at many places during the 2015 Gorkha earthquake (Fig. 17). The houses constructed on base of hills around the Kathmandu Valley faced serious damages associated with the cut-and-fill problems. Many rural roads in Central Nepal faced such failure problems as well (Fig. 18).

Fig. 17 Valley filling for road construction and collapsed Kathmandu airport road during the 2015 Gorkha earthquake

Fig. 18 **a** Cut-and-fill method used for road construction or housing and **b** Cut-and-fill failure at earthen rural road during the 2015 Gorkha earthquake in central Nepal

8 Conclusion

This report summarizes the earthquake-induced landslides in the Nepal Himalaya. Six major earthquakes of last 100 years including the recent 2015 Gorkha Earthquake are taken into consideration for this report. Although earthquake-induced landslides were not well documented during previous earthquakes, but for this report, author has attempted to explore few earthquake-induced landslides resulted from previous earthquakes and described. Earthquake-induced landslides of two earthquakes (the 2011 Sikkim-Nepal Boarder Earthquake and 2015 Gorkha Earthquake) were investigated by author and described in this report. Although a lot of work has been carried out by various scientists after the 2015 Gorkha Earthquake, the risk assessment of post-seismic landslide hazard remains not well understood due to a lack of field data. This report also describes an approach of earthquake-induced landslide hazard assessment in the Nepal Himalaya along with earthquake-induced landslide classification scheme to understand the earthquake-induced landslide process in the Nepal Himalaya.

Acknowledgements The author wishes to thank many scientists who have published journal papers and technical reports regarding earthquake-induced landslides after the 2015 Gorkha Earthquake. This report would not be possible without citing their works. Dr. Manita Timilsina and Ms. Chandani Bhandari of Geotech Solutions International (GSI) are sincerely acknowledged for their supports in field works and preparation of this report.

References

Avouac JP, Meng L, Wei S, Wang T, Ampuero JP (2015) Lower edge of main Himalayan thrust unzipped by the 2015 Gorkha earthquake. Nat Geosci 8:708–711

Bhattarai G, Chamlagain D, Rajaure S (2018) Seismic hazard assessment for eastern Nepal using 1934 and 1988 earthquakes. J Nepal Geol Soc 42:85–93

Bilham R (2004) Earthquakes in India and the Himalaya: tectonics, geodesy, and history. Ann Geophys 47:839–858

Bilham R, Gaur VK, Molnar P (2001) Himalayan seismic hazard. Science (Washington) 293(5534):1442–1444

Bollinger L, Sapkota SN, Tapponier P, Klinger Y, Rizza M, Van der Woerd J, Tiwari DR, Pandey R, Bitri A, Bes de Berc S (2014) Estimating the return times of great Himalayan earthquakes in eastern Nepal: evidence from the Patu and Bardibas strands of the Main Frontal Thrust. J Geophys Res Solid Earth

Chaulagain H, Gautam D, Rodrigues H (2018) Revisiting major historical earthquakes in Nepal: overview of 1833, 1934, 1980, 1988, 2011, and 2015 seismic events. In: Impacts and insights of the Gorkha Earthquake. Elsevier, pp 1–17

Chigira M, Yagi H (2006) Geological and geomorphological characteristics of landslides triggered by the 2004 Mid Niigata prefecture earthquake in Japan. Eng Geol 82:202–221

Dahal RK (2015a) Earthquake-induced slope failure susceptibility in eastern Nepal. J Nepal Geol Soc 49:49–56

Dahal RK (2015b) Engineering geological issues after Gorkha earthquake 2015b in Nepal—a preliminary understanding. In: Proceedings of 10th Asian regional conference of IAEG, Kyoto, Japan, September 26–27, 2015, 7p

Dahal RK (2015c) Consequences of large-scale landslide dams in the Nepal Himalaya. In: Proceeding of EGNM 2015, Indian Society of Engineering Geologists, October 2015c, 10p

Dahal RK (2016) Lessons learnt from the 2015 Gorkha earthquake in Nepal on the engineering geological perspectives. In: Singh SP, Khanal SC, Joshi M (eds) Lessons from Nepal's earthquake for the Indian Himalayas and the Gangetic plains. Central Himalayan Environment Association (CHEA), India, pp 15–29

Dahal RK, Hasegawa S (2008) Representative rainfall thresholds for landslides in the Nepal Himalaya. Geomorphology 100(3–4):429–443

Dahal RK, Bhandary NP, Timilsina M, Yatabe R, Hasegawa S (2013) Earthquake induced landslides in the roadside slopes of East Nepal after recent September 18, 2011. In: Ugai K, Yagi H, Wakai A (eds) Earthquake induced landslides. Springer, Berlin, Heidelberg, pp 149–157

Deuja BB, Dhital MR, Manandhar KC, Wagner A (1991) Risk engineering in the Himalaya. International Centre for Integrated Mountain Development (ICIMOD), p 117

Dixit AM (1991) Geological effects and intensity distribution of the Udayapur (Nepal) earthquake of August 20, 1988. J Nepal Geol Soc 7:1–17

DMG (2011) Geological map of Eastern Nepal, Department of Mines and Geology, Lainchaur, Kathmandu, Nepal

Dunn JA, Auden JB, Ghosh AMN, Wadia DN (1939) The Bihar-Nepal earthquake of 1934. Geol Surv India, Mem 73

Geli L, Bard PY, Jullien B (1988) The effect of topography on earthquake ground motion: a review and new results. Bull Seismol Soc Am 78(1):42–63

Hasegawa S, Dahal RK, Yamanaka M, Bhandary NP, Yatabe R, Inagaki H (2009) Causes of large-scale landslides in the Lesser Himalaya of central Nepal. Environ Geol 57(6):1423–1434

Kargel JS, Leonard GJ, Shugar DH, Haritashya UK, Bevington A, Fielding EJ, Fujita K, Geertsema M, Miles ES (2015) Geomorphic and geologic controls of geohazards induced by Nepal's 2015 Gorkha earthquake. SIAS Faculty Publications. 342. https://digitalcommons.tacoma.uw.edu/ias_pub/342

Keefer DK (2000) Statistical analysis of an earthquake-induced landslide distribution—the 1989 Loma Prieta, California event. Eng Geol 58(3–4):231–249

Korup O, Montgomery DR, Hewitt K (2010) Glacier and landslide feedbacks to topographic relief in the Himalayan syntaxes. Proc Natl Acad Sci USA 107:5317–5322

Lavé J, Avouac JP (2000) Active folding of fluvial terraces across the Siwaliks Hills, Himalayas of central Nepal. J Geophys Res Solid Earth 105:5735–5770

Martha TR, Roy P, Mazumdar R, Govindharaj KB, Kumar KV (2017) Spatial characteristics of landslides triggered by the 2015 Mw 7.8 (Gorkha) and Mw 7.3 (Dolakha) earthquakes in Nepal. Landslides 14:697–704

Mathur LP (1953) The Assam earthquake of 15th august, 1950: a short note on factual observations. In: Rao MBR (ed) A compilation of papers on the Assam earthquake of August, 15, 1950, pp 56–60

Meigs AJ, Burbank DW, Beck RA (1995) Middle-Late Miocene (>10 Ma) formation of the main boundary thrust in the western Himalaya. Geology 23(5):423–426

Molnar P, Deng Q (1984) Faulting associated with large earthquakes and the average rate of deformation in central and eastern Asia. J Geophys Res Solid Earth 89(B7):6203–6227

Nakata T (1982) A photogrammetric study on active faults in the Nepal Himalaya. J Nepal Geol Soc 2:67–80

Nakata T, Iwata S, Yamanaka H, Yagi H, Maemoku H (1984) Tectonic landforms of several active faults in the western Nepal Himalaya. J Nepal Geol Soc 4:177–200

Owen LA, Kamp U, Khattak GA, Harp EL, Keefer DK, Bauer MA (2008) Landslides triggered by the 8 October 2005 Kashmir earthquake. Geomorphology 94:1–9

Pandey MR, Molnar P (1988) The distribution of intensity of the Bihar-Nepal earthquake of 15 January 1934 and bounds on the extent of the rupture zone. J Nepal Geol Soc 5(1):22–44
Rana BSJB (1935) The great earthquake of Nepal. Jorganesh Press, Kathmandu, Nepal
Roback K, Clark MK, West AJ, Zekkos D, Li G, Gallen SF, Chamlagain D, Godt JW (2018) The size, distribution, and mobility of landslides caused by the 2015 Mw 7.8 Gorkha earthquake, Nepal. Geomorphology 301:121–138
Schwanghart W, Bernhardt A, Stolle A, Hoelzmann P, Adhikari BR, Andermann C, Tofelde S, Merchel S, Rugel G, Fort M, Korup O (2016) Repeated catastrophic valley infill following medieval earthquakes in the Nepal Himalaya. Science 351
Sella GF, Dixon TH, Mao A (2002) REVEL: a model for recent plate velocities from space geodesy. J Geophys Res Solid Earth 107
Shrestha AB, Bajracharya SR, Kargel JS, Khanal NR (2016) The impact of Nepal's 2015 Gorkha earthquake-induced geohazards. ICIMOD, Kathmandu, p 48
Singh V (1985) Earthquake of July 1980 in far Western Nepal. J Nepal Geol Soc 2(2):1–11
Tiwari B, Ajmera B, Dhital S (2017) Characteristics of moderate- to large-scale landslides triggered by the Mw 7.8 2015 Gorkha earthquake and its aftershocks. Landslides 14(4):1297–1318
Tsou C-Y, Higaki D, Chigira M, Yagi H, Dangol V, Amatya S, Hayashi K, Hiroki K (2018) Topographic characteristics of landslides induced by the 2015 Gorkha earthquake. Nepal J Nepal Geol Soc 55(Sp. Issue):69–75
Uchida T, Kataoka S, Iwao T, Matsuo O, Terada H, Nakano Y, Sugiura N, Osanai N (2004) A study on methodology for assessing the potential of slope failures during earthquakes. Technical note of national institute for land and infrastructure management, p 91 (in Japanese with English summary)
USGS (2011) Summary of Sikkim/Nepal boarder earthquake, 2011, September 18 available in http://earthquake.usgs.gov/earthquakes/pager/events/us/c0005wg6/index.html#summary. Last accessed: 27 June 2018
USGS (2015) The April-May 2015 Nepal earthquake sequence. available in http://earthquake.usgs.gov/learn/topics/Nepal-slides.pdf Last accessed: 15 July 2015
Varnes DJ (1984) Landslide hazard zonation: a review of principles and practice. Natural Hazards. UNESCO, Paris
Wang F, Miyajima M, Dahal R et al (2016) Effects of topographic and geological features on building damage caused by 2015.4.25 Mw7.8 Gorkha earthquake in Nepal: a preliminary investigation report. Geoenviron Disasters 3:7. https://doi.org/10.1186/s40677-016-0040-2

Landslides Triggered by the 2002 M 7.9 Denali Fault Earthquake, Alaska, USA

William Schulz

Abstract The 2002 M 7.9 Denali earthquake in Alaska, USA, was the largest inland earthquake in North America in nearly 150 years. The earthquake involved oblique thrusting but mostly strike-slip motion, and faults ruptured the ground surface over 330 km. Fault rupture occurred in a rugged, mountainous, subarctic environment with extensive permafrost and variable glaciation, geology, and groundwater presence, and many triggered landslides mobilized into avalanches that traversed varied physiographic settings, some moving farther than 11 km. However, only several thousand landslides were triggered, and these occurred in a narrow zone along the fault rupture. These characteristics of the event provide ample opportunities to improve understanding of controls on coseismic landsliding and avalanching mechanisms. The paucity and limited extent of landslides likely resulted from high directivity of seismic energy and relatively low levels of high-amplitude, high-frequency ground motion; glacial damping of seismic energy was likely not a factor. Landslides preferentially occurred on hillslopes steeper and higher than average. Meteorological conditions were near historical averages at the time of the event, although the region has experienced gradual warming historically that appears to have resulted in increased landslide occurrence in other parts of Alaska during recent years. Inherent susceptibility of geological formations to landsliding was not apparent with the available data, although discontinuities created dip-slope conditions for the five largest slides. Historical thinning of glaciers and consequent slope debuttressing may have been a factor in aiding occurrence of some of the earthquake-induced landslides, particularly some of the largest. Mobility of the largest avalanches was above average compared to global data, and mobility of all sizes of avalanches was apparently aided by movement over glacial ice. Avalanche deposits displayed characteristics indicative of turbulent flow on steeper slopes and laminar plug flow in flatter areas, where sliding also likely occurred.

W. Schulz (✉)
U.S. Geological Survey, Denver, CO, USA
e-mail: wschulz@usgs.gov

I. Towhata et al. (eds.), *Coseismic Landslides*, Springer Natural Hazards,
https://doi.org/10.1007/978-981-19-6597-5_4

1 Introduction

At 22:12:41 UTC on November 3, 2002, the largest inland earthquake to occur in North America in nearly 150 years shook southeastern Alaska, USA, with moment magnitude **M** 7.9 (Alaska Earthquake Information Center 2017). The earthquake ruptured 330 km of the Denali, Totschunda, and Susitna Glacier Faults during a ~100 s period (Fig. 1) and triggered thousands of landslides in the Alaska Range (e.g., Eberhart-Phillips et al. 2003; Harp et al. 2003). The earthquake also caused seiches as far away as Louisiana (~5000 km to the southeast) and triggered seismicity in Washington, California, and Utah (Alaska Earthquake Information Center 2017). The event primarily involved strike-slip displacement, making it one of the largest strike-slip earthquakes in history (e.g., Haeussler et al. 2004). Fault rupture occurred in a largely glaciated, high mountain setting, with much of the rupture occurring beneath and through glacier ice. Impacts of the earthquake on humans were relatively minor due to the sparse population of the region; nearly all landslides occurred in glaciated or formerly glaciated terrain nearly entirely absent of development. The variety of terrain, geology, climate, and glacier cover over which strong shaking occurred makes the event useful for evaluating conditions that influenced landslide occurrence and character. Different from many landslide-triggering earthquakes, the Denali Fault earthquake also permits evaluation of potential effects of glacier buffering on ground motions and related landsliding (McColl et al. 2012), as well as potential effects of the general lack of rainfall and snowmelt (e.g., Gorum et al. 2014) and widespread permafrost on coseismic landsliding. The paucity of near-field seismic ground motion records for the event renders the triggered landslides important for estimating coseismic ground motions (e.g., Jibson et al. 2004). The range of mobility and variety displayed by triggered landslides provide opportunities to study mechanisms controlling their movement, particularly those of rock avalanches. Finally, the event also is useful for studies of the effects of landslides on glacier movement and ablation, as many landslides were deposited on glaciers. Herein, I summarize previous studies of the event and provide data and simple analyses related to landslide causation, all of which are potentially useful for future, more detailed studies.

2 Setting

The earthquake occurred along the ~2000-km-long Denali strike-slip fault system, which is one of the longest in the world (Harp et al. 2003). This fault system accommodates collision of the Yakutat terrain into the North American Plate (e.g., Matmon et al. 2006), which is occurring at ~47 mm/yr (Fletcher and Freymueller 2003). Stout and Chase (1980) concluded that more than 90 km of slip occurred along the fault during the past 38 Myr, while its tertiary displacement was approximately 300 km. Offsets of Quaternary features suggest a slip rate of 8–13 mm/yr (Plafker et al.

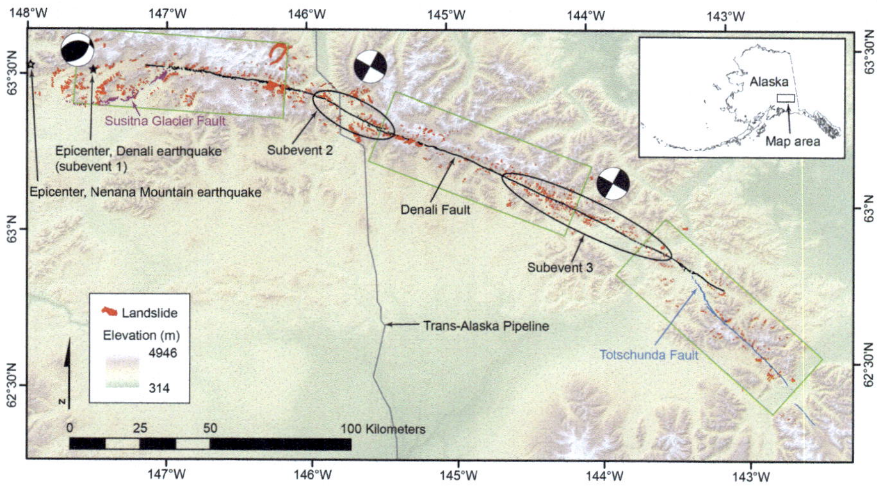

Fig. 1 Shaded relief map showing landslides (Gorum et al. 2014; Schmitt et al. 2017a, b), fault rupture (Haeussler 2009), earthquake epicenters, subevent locations, and lower-hemisphere projections of focal mechanisms (Frankel 2004). Green rectangles delineate areas quantitatively compared herein. Elevation data from U.S. Geological Survey 3D Elevation Program (3DEP), 1 arc second

1977), whereas exposure age dating and measurements of offset features suggest a late Pleistocene-Holocene slip rate of 12.1 mm/yr along the segment of the fault that ruptured during 2002 and 6 mm/yr along the Totschunda Fault (Matmon et al. 2006). No significant historical earthquakes occurred along the Denali Fault prior to 2002 (EERI 2003).

As a result of the great right-lateral displacement that has occurred along the Denali Fault, geological formations vary across it. On its north side in the vicinity of the 2002 rupture, bedrock primarily consists of deformed and metamorphosed sedimentary, volcanic, and plutonic rocks of middle Paleozoic age, whereas on its south side, bedrock is generally Paleozoic and Mesozoic sedimentary rock (Nokleberg et al. 1992 and references therein). On both sides of the fault, generally Mesozoic granitic to gabbroic plutons intrude older formations (Nokleberg et al. 1992).

Displacement along the Denali Fault system resulted in formation of the Alaska Range (e.g., Csejtey 1976; Ridgway et al. 2002; Matmon et al. 2006). These mountains extend over ~900 km from the US-Canadian border on the east to the Alaskan Peninsula on the west. The range's subarctic location (~62°–63° N) and elevation reaching 6190 m make it one of the coldest mountain ranges on Earth, resulting in widespread permafrost, precipitation that generally falls as snow in the higher elevations, and formation of glaciers. The earthquake occurred in the eastern part of the Alaska Range, which is relatively narrow, measured normal to the fault, and boasts valley glaciers as long as 65 km (Figs. 1 and 2).

The area in which fault rupture and nearly all landsliding occurred can be readily subdivided into three physiographic regions based on the character of the mountains, glaciation, and climate (Figs. 1, 2, 3, 4 and 5). The characteristics of these regions

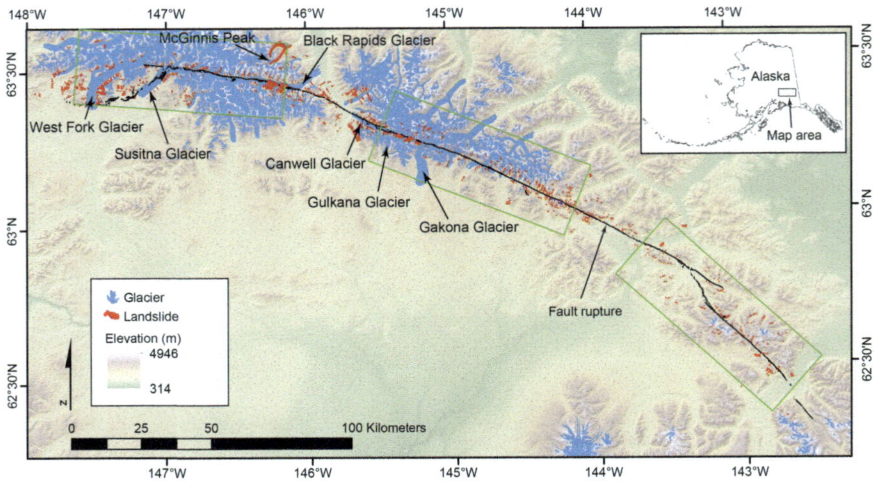

Fig. 2 Shaded relief map showing landslides (Gorum et al. 2014; Schmitt et al. 2017a, b), fault rupture (Haeussler 2009), and glaciers (Wilson et al. 2015; 1:500,000 data). Green rectangles delineate areas quantitatively compared herein. Elevation data from U.S. Geological Survey 3DEP, 1 arc second

likely partly control the distribution and nature of the landslides triggered by the earthquake. For quantitative comparison purposes, rectangles with areas of 2000 km^2 each were drawn approximately centered on the fault rupture in these western, central, and eastern regions (green rectangles in Figs. 1, 2, 3, 4 and 5) and statistics were determined within each.

The western region is the most rugged; it has the highest overall and mean elevations, is steepest, and displays greatest local relief (Table 1). Largely due to its higher elevation, this region also is coldest and wettest (Figs. 3 and 4, Table 2), although nearly all precipitation falls as snow that rarely melts. The estimated annual temperature during 1971–2000 averaged −6.5 °C (Gibson 2009a; Fig. 3), and permafrost is nearly continuous (Fig. 5), as indicated by the region's average permafrost zonation index (PZI; Gruber 2012a, b) of 0.92 (Table 2). A PZI of 0.9–1.0 indicates continuous permafrost; 0.5–0.9 indicates extensive, discontinuous permafrost; 0.1–0.5 indicates sporadic, discontinuous permafrost; and less than 0.1 indicates isolated patches of permafrost. In sum, these conditions result in the western region having the greatest glacier coverage (Figs. 2 and 6, Table 2), with glaciers covering 1197 km^2 within the 2000 km^2 rectangle. Most glaciers here comprise a vast ice field, although isolated valley glaciers also occur (Molnia 2008). The majority of the exposed geologic material consists of Tertiary calcareous sedimentary rock, and, to lesser extent, granitic plutons and unconsolidated Quaternary sediment (Figs. 6 and 7). Metamorphic rocks, mainly phyllite, underlie the majority of the remaining area.

The central region is generally lower in elevation, less steep, and has less local relief than the western region (Fig. 1, Table 1). Correspondingly, the central region is also warmer and drier than the western region (Figs. 3 and 4, Table 2), although

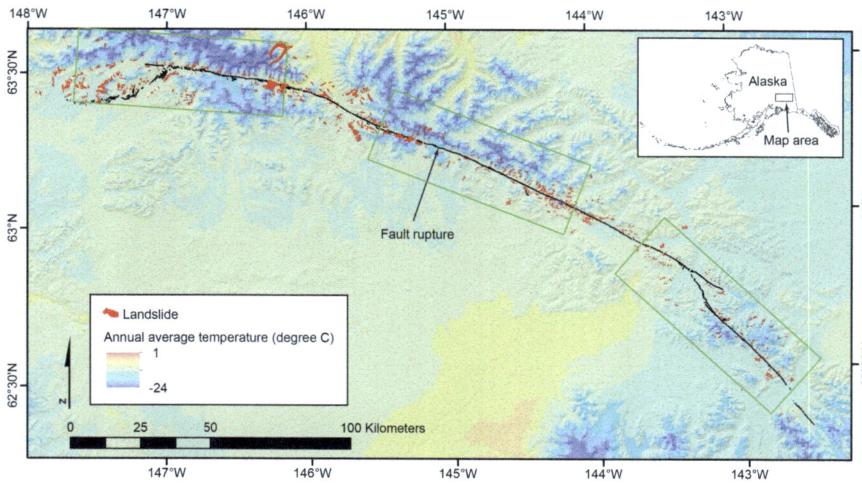

Fig. 3 Shaded relief map showing landslides (Gorum et al. 2014; Schmitt et al. 2017a, b), fault rupture (Haeussler 2009), and annual average air temperature estimated for 1971–2000 from parameter-elevation regressions on independent slopes model (PRISM) modeling results (Gibson 2009a). Green rectangles delineate areas quantitatively compared herein. Elevation data from U.S. Geological Survey 3DEP, 1 arc second

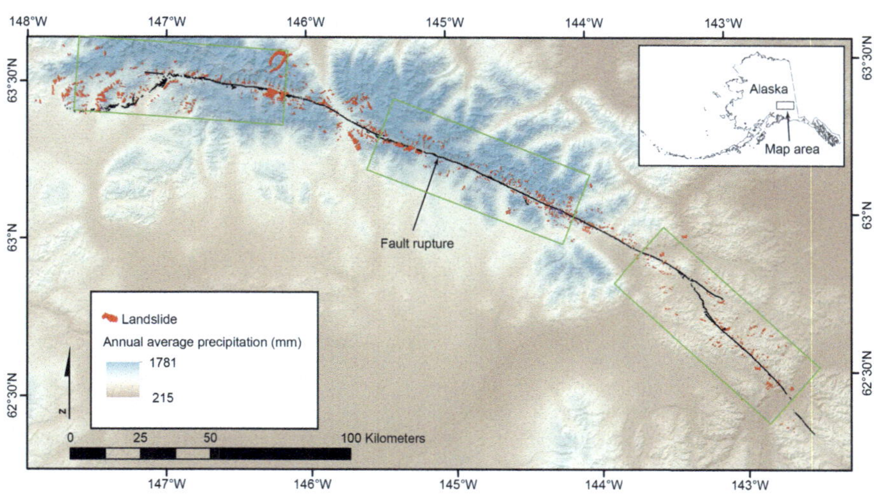

Fig. 4 Shaded relief map showing landslides (Gorum et al. 2014; Schmitt et al. 2017a, b), fault rupture (Haeussler 2009), and annual average total precipitation estimated for 1971–2000 from PRISM modeling results (Gibson 2009b). Green rectangles delineate areas quantitatively compared herein. Elevation data from U.S. Geological Survey 3DEP, 1 arc second

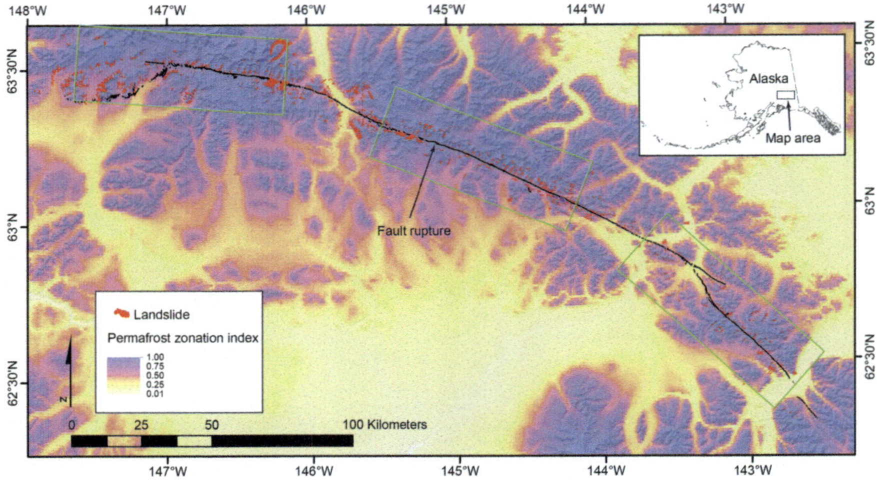

Fig. 5 Shaded relief map showing landslides (Gorum et al. 2014; Schmitt et al. 2017a, b), fault rupture (Haeussler 2009), and permafrost zonation index (PZI) modeling results (Gruber 2012a, b). A PZI of 0.9–1.0 indicates continuous permafrost, 0.5–0.9 indicates extensive, discontinuous permafrost, 0.1–0.5 indicates sporadic, discontinuous permafrost, and less than 0.1 indicates isolated patches of permafrost. Green rectangles delineate areas quantitatively compared herein. Elevation data from U.S. Geological Survey 3DEP, 1 arc second

Table 1 Topographic metrics and landslide statistics within west, central, and east regions

Region	Elevation[a] (m)			Slope[a] (degrees)		Relief[b] (m)		Landslides[c]		
	Min	Max	Mean	Max	Mean	Max	Mean	Total area (km^2)	Number	Average area (km^2)
West	794	4075	1802	76	14.4	1056	110	54.433	412	0.132
Central	867	3037	1612	73	14.3	590	104	22.876	430	0.0532
East	632	2493	1250	70	13.8	446	96	8.405	247	0.034

[a]Topographic metrics calculated from U.S. Geological Survey 3DEP 1 arc second elevation data
[b]Relief calculated within 300-m diameter circles
[c]Landslide area and number calculated from Schmitt et al. (2017a, b) using data from Gorum et al. (2014)

the estimated average annual temperature during 1971–2000 (Gibson 2009a; Fig. 2) remained well below freezing (−5.4 °C; Table 2). Permafrost is generally extensive but discontinuous in the region, as indicated by its average PZI of 0.86 (Table 2, Fig. 5). Glaciers still cover a large part of the region (752 km^2), although less than in the western region (Figs. 2 and 6, Table 2), and the primary ice field is smaller, with isolated valley glaciers becoming more abundant than in the western region. Unconsolidated Quaternary sediments compose most of the geology exposed in the region (Figs. 6 and 7), with metamorphic rocks, mainly phyllite, also being abundant. Basalt,

Table 2 Climate, glacier, and landslide statistics within west, central, and east regions

Region	Annual average temperature[a] (°C)			Annual average precipitation[a] (mm)			Permafrost zonation index[b]			Glacier area[c] (km^2)	Landslides[d]		
	Min	Max	Mean	Min	Max	Mean	Min	Max	Mean		Total area (km^2)	Number	Average area (km^2)
West	−19	−3	−6.48	540	1573	1116	0.40	1	0.92	1197	54.433	412	0.132
Central	−12	−4	−5.37	638	1525	1098	0.38	1	0.86	752	22.876	430	0.053
East	−9	−3	−4.94	271	883	471	0.18	1	0.65	8	8.405	247	0.034

[a]Temperature and precipitation metrics calculated from 1971 to 2000 PRISM models (Gibson 2009a, b)
[b]Permafrost zonation index metrics calculated from Gruber (2012a, b)
[c]Glacier area calculated from 1:500,000 version of Wilson et al. (2015)
[d]Landslide area and number calculated from Schmitt et al. (2017a, b) using data from Gorum et al. (2014)

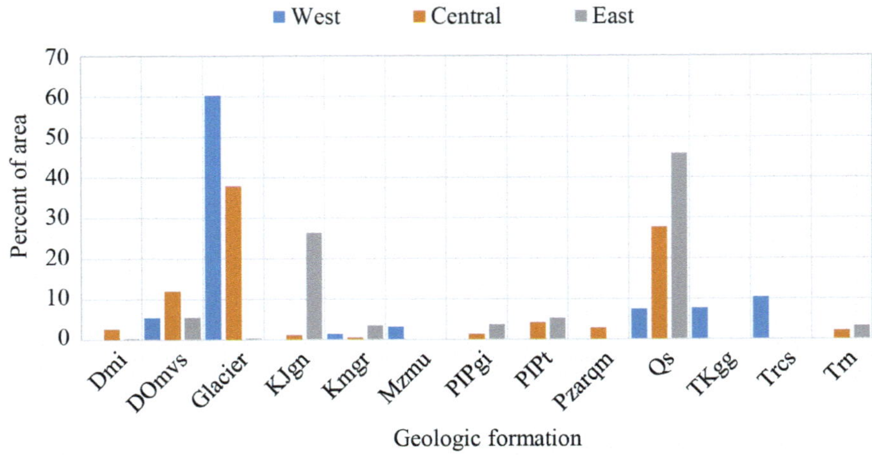

Dmi - Metamorphosed mafic igneous rocks (MDmg)
DOmvs - Metamorphic rocks, mainly phyllite (DCsp)
KJgn - Marine graywacke and mudstone
Kmgr - Granodiorite and lesser quartz diorite, granite, quartz monzonite
Mzmu - Metamorphic rocks (TMzmb)
PIPgi - Granodiorite, syenite, other granitic rocks
PIPt - Submarine andesite and basalt flows, volcanic graywacke (TrIPms)
Pzarqm - Schist and phyllite (DCsp)
Qs - Unconsolidated surficial deposits; alluvium, colluvium, glacial (QTs)
Tkgg - Gneissose granitic rocks (TMzmb)
Trcs - Calcareous sedimentary rocks
Trn - Massive basalt and greenstone (Trmb)

Fig. 6 Chart showing the percent of area each geologic formation covers in each of the three regions (see orange boxes on Fig. 7). Geology is from 1:500,000 scale version of Wilson et al. (2015); formation abbreviations that differ from those on the 1:1,584,000 scale version are provided in parentheses and these correlate with Fig. 7. Geologic formation percentages sum to 93% or more for each region; only the most widespread formations are included in the chart

greenstone, andesite, granitic rocks, and a small amount of marine sedimentary rocks are exposed in the remaining area.

The eastern region is generally lowest in elevation, least steep, and has the lowest local relief of the three regions (Fig. 1, Table 1). It also is warmest and driest (Figs. 3 and 4, Table 2), and is nearly free of glaciers, with only 8 km^2 of the 2000 km^2 rectangle covered by glaciers (Table 2, Figs. 2 and 6). Permafrost is also much more limited in the region, with the average PZI indicating extensive but discontinuous permafrost, in general, but with some of the region experiencing only sporadic permafrost (Table 2, Fig. 5). Corresponding with the present-day limited glacier cover, unconsolidated Quaternary sediments are exposed in nearly half of the region and many of these are glacial deposits (Figs. 6 and 7). In contrast to the western and central regions, marine greywacke and mudstone are very abundant here, followed

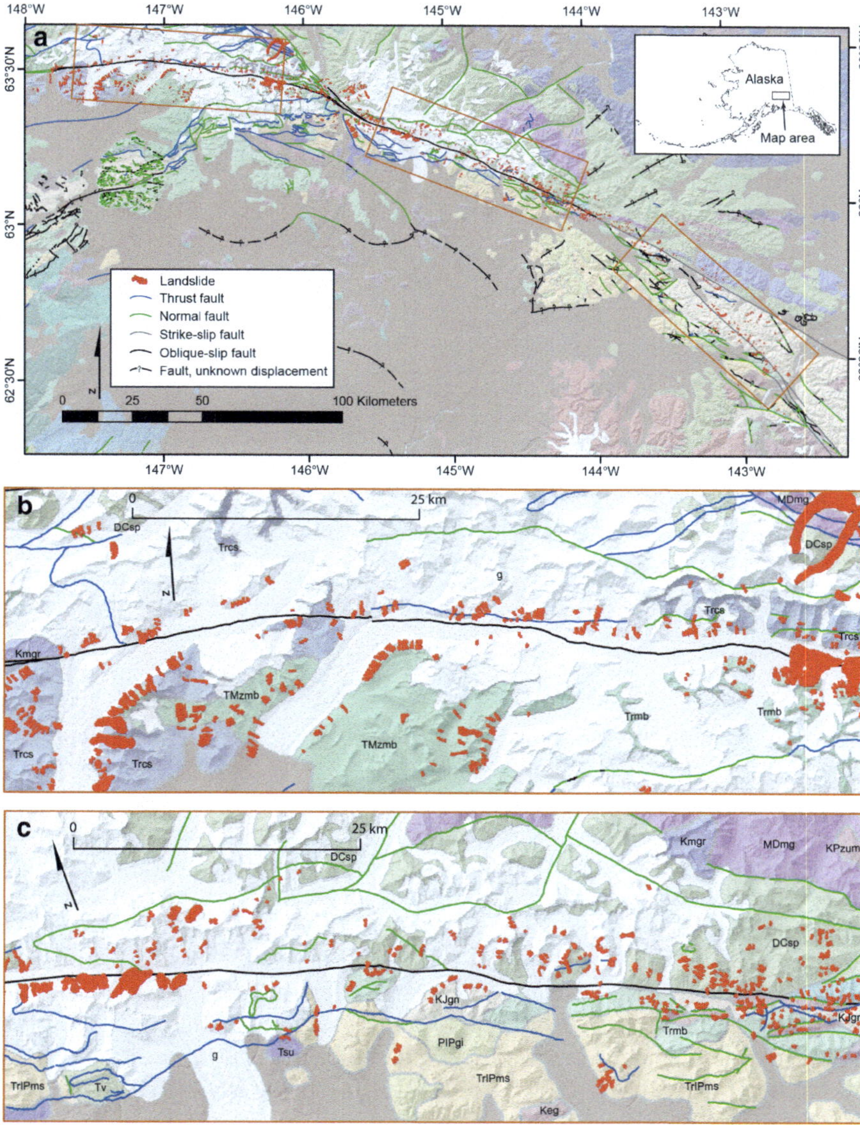

Fig. 7 Shaded relief map showing landslides (Gorum et al. 2014; Schmitt et al. 2017a, b), fault rupture (Haeussler 2009), and geology (1:1,584,000 scale version of Wilson et al. 2015). Elevation data from U.S. Geological Survey 3DEP, 1 arc second. Panels **b**, **c**, and **d** show enlarged views of the western, central, and eastern regions, respectively, which are outlined by the orange rectangles in **a**; abbreviations refer to geologic formation

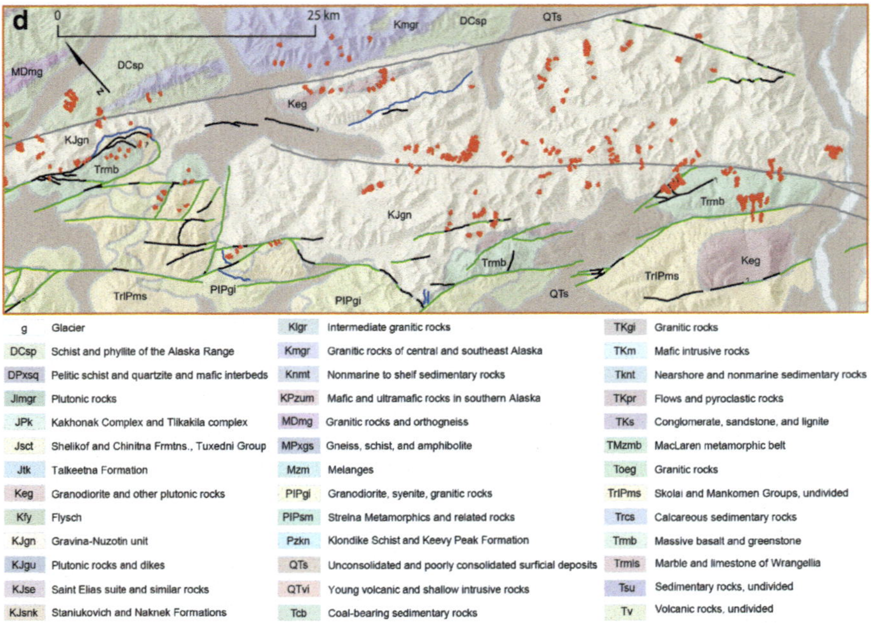

Fig. 7 (continued)

by phyllite, granodiorite, and other granitic rocks, and some andesite, basalt, and greenstone.

The variable glaciation in the area of coseismic landsliding during 2002 potentially affected the distribution and character of triggered landslides, considering the variability of glacial incision, deposition, and slope buttressing, and also potential effects of glaciers on coseismic ground accelerations. Many of the area's glaciers have been observed to significantly change historically. Most glaciers are stagnant, thinning, and retreating in the Alaska Range (Molnia 2008). In Denali National Park, ~100 km west of the earthquake-affected area, elevations measured along nine glaciers using lidar between 1994 and 2010 generally revealed up to ~1 m water equivalent/yr of thinning, whereas the combined area covered by glaciers declined by 8% between ~1953 and ~2009 (Loso et al. 2014). In the area of coseismic landsliding, the Gulkana Glacier (Fig. 2), one of the largest of the glaciers in the range, retreated more than 4 km between 1910 and 1952 (Molnia 2008), while it thinned by an average of ~11 m between 1957 and 1993 (Echelmeyer et al. 1996). Similarly, the nearby West Gulkana Glacier was observed to thin by an average of 24 m between 1957 and 1993 (Molnia 2008). In contrast, numerous surging glaciers (i.e., glaciers that experience periods of relatively rapid, long-distance motion separated by periods of slow movement) are present in the central part of the range, which scour and buttress slopes as they advance, but they also generally thin during surges. One of these surging glaciers is the Black Rapids Glacier, onto which three of the largest landslides triggered by the earthquake were emplaced (Figs. 2 and 8). This glacier retreated ~5 km between 1912 and 1936 (Molnia 2008), only to advance ~5 km during the 1936–1937 winter (Post 1960). Since then, the glacier has been stagnant and thinned while several of its tributary glaciers have retreated significantly; Black Rapids thinned on average by ~0.7 m/yr between the mid-1990s and 1999 (Molnia 2008). Susitna Glacier (Fig. 2), which shares a head with the Black Rapids glacier (the two move in diverging directions), surged ~4 km during 1952 or 1953 (Post 1960). Surging of both glaciers resulted in pronounced (but unmeasured) lowering of their upper parts; measured lowering during the 1956–1957 surging of the nearby Muldrow Glacier was ~170 m (Post 1960). Although reduced precipitation can cause glacier thinning, observed historical thinning probably did not result from reduced precipitation, as ice cores from Mt. Hunter in the Alaska Range suggest doubling of annual snowfall since ~1840 (Winski et al. 2017).

Although rainfall and melting of snow and ice are rare in much of the high mountains of the Alaska Range that were affected by coseismic landsliding during 2002, long-term climatic changes can strongly affect glaciation and permafrost, both of which can affect slope stability. Coe et al. (2018) showed that rock avalanches in other glaciated mountains of Alaska have recently become more frequent, larger, and more mobile, possibly due to permafrost degradation and slope debuttressing as glaciers thin. They found that avalanches were especially frequent during prolonged, abnormally warm periods and when the annual mean air temperature trended to above freezing. The continuously operating meteorological station nearest to the

Fig. 8 Oblique aerial photograph of the largest avalanche deposits on Black Rapids glacier. Photograph taken by Ed Harp (U.S. Geological Survey), November 18, 2002. Glacier on which deposits were emplaced is ~2 km wide

2002 epicenter is the McKinley Park station (USC00505778), which is ~80 km WNW of the epicenter and at an elevation of 631 m (NOAA National Climate Data Center, https://www.ncdc.noaa.gov/cdo-web/), or 618–1170 m lower than the mean elevations of the three regions affected by coseismic landsliding (Table 1). Records for 1923–2002 (Fig. 9) indicate that temperature has remained nearly steady at McKinley Park and, on average, below freezing (−2.3 °C average), although the average for water year 2000 (1 October 2000 to 30 September 2001) was 0.6 °C. The average was again below freezing (−1.8 °C) for water year 2001, which ended about one month prior to the earthquake. Considering also the elevation difference, it is therefore likely that average temperatures have remained below freezing in the area of coseismic landsliding during historic times. Precipitation increased by ~1.4 mm/yr at the McKinley Park station since 1923 and water year 2001 was especially wet, with 141% of the 1923–2002 average falling. Water year 1999 was similarly wet, although water year 2000 experienced 79% of the long-term average precipitation. How much precipitation falls as rain or snow that melts within the area of coseismic landsliding remains unclear. From May through September 2002, corresponding with above-freezing temperatures, precipitation at the McKinley Park station fell mostly as rain (Fig. 10). However, this station is 923 m lower than the average elevation of the three landslide-affected regions. Within those regions, modeled temperatures between 1971 and 2000 (Figs. 3 and 11; Gibson 2009a) suggest that average temperatures are above freezing during May through August. Hence, although temperatures in some areas were likely below freezing prior to the earthquake, precipitation at least in lower elevations probably fell as rainfall that infiltrated the ground for several months prior to the earthquake, similar to other years.

3 The 2002 Denali Earthquake

The moment magnitude **M** 7.9, November 3, 2002, Denali earthquake was preceded by the **M** 6.7 Nenana Mountain earthquake on October 23, 2002, (Kikuchi and Yamanaka 2002), located ~20 km west of the Denali quake epicenter (Fig. 1). Aerial evaluation immediately after the Nenana Mountain earthquake revealed that it triggered abundant small rockfalls and snow avalanches in the vicinity of the epicenter (Eberhart-Phillips et al. 2003; Haeussler et al. 2004). The Denali earthquake involved three subevents that propagated eastward (Kikuchi and Yamanaka 2002; Eberhart-Phillips et al. 2003; Choy and Boatwright 2004) at rupture speeds averaging above 2.8 km/s (Ellsworth et al. 2004); modeling suggests a rupture speed of ~3.5 km/s in the vicinities of subevents 1 and 2 (Fig. 1) and ~5 km/s near subevent 3 (Frankel 2004). The earthquake began with an oblique thrust event (**M** 7.3; Choy and Boatwright 2004) on the previously unknown Susitna Glacier Fault (Fig. 1), with failure occurring on a 48°, northward-dipping plane and rupturing 40–50 km of the ground surface with average displacement of 3.3–4 m (Eberhart-Phillips et al. 2003; Crone et al. 2004). Dextral strike-slip displacement then occurred along ~220 km of the Denali Fault (Fig. 1), with an average offset of ~5 m and a maximum of 8.8 m near the

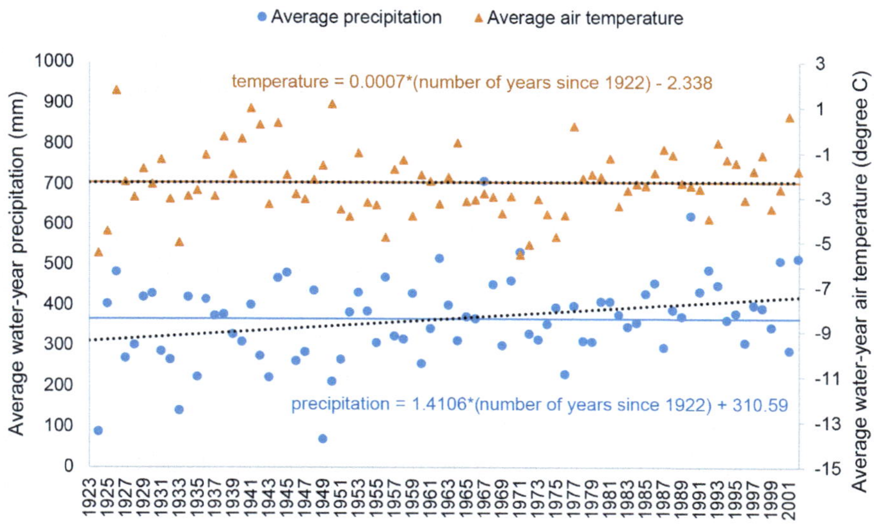

Fig. 9 Average annual precipitation and air temperature at the McKinley park weather station (USC00505778), located ~80 km northwest from the Denali earthquake epicenter and at an elevation of 631 m. Long-term averages are given by the colored lines, whereas best linear fits to data are given by the dashed black lines

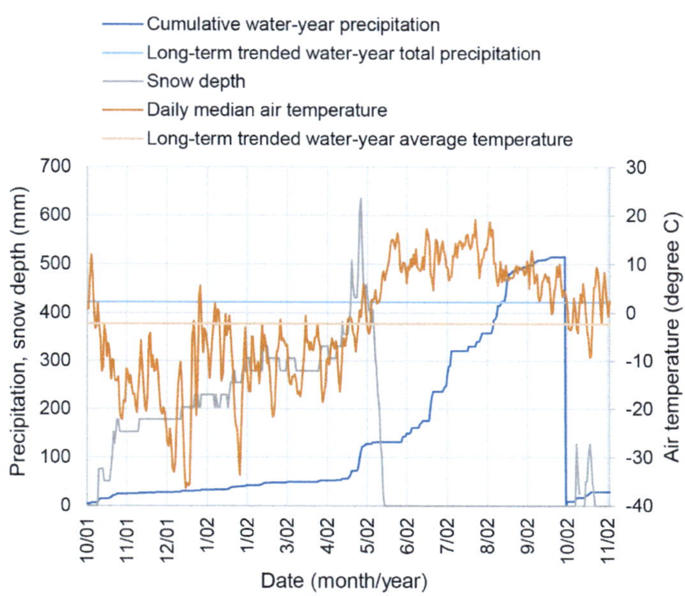

Fig. 10 Meteorological observations from the 13 months prior to the Denali earthquake at the McKinley park weather station (USC00505778), located ~80 km northwest from the earthquake epicenter and at an elevation of 631 m. Long-term trended annual average precipitation and temperature are from records from 1923–2002 (Fig. 9)

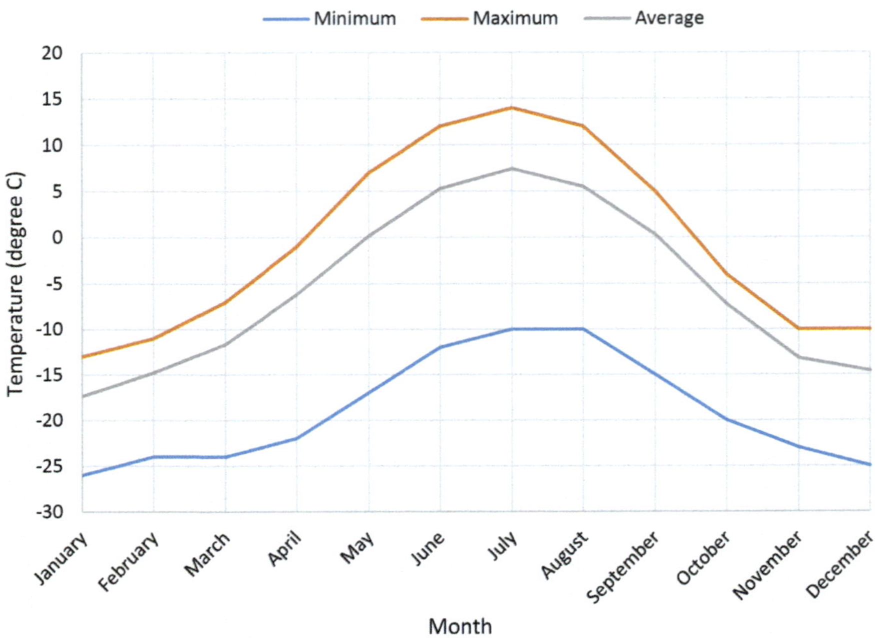

Fig. 11 Minimum, maximum, and average temperature within the three boxes (Fig. 3) estimated for 1971–2000 from PRISM modeling results (Gibson 2009a)

eastern limit, ~190 km east from the epicenter (Eberhart-Phillips et al. 2003; Haeussler et al. 2004). Subevent 2 (**M** 7.3; Choy and Boatwright 2004) occurred near where The Trans-Alaska Pipeline, one of the world's largest petroleum pipelines, crosses the Denali Fault (Fig. 1) and where average surface offsets increased to ~6 m (Eberhart-Phillips et al. 2003). Subevent 3 (**M** 7.6; Choy and Boatwright 2004) had the highest moment magnitude and was distributed along the eastern part of the Denali Fault rupture (Frankel 2004), generally correlating with the region of maximum surface offset and propagation speed (~5 km/s), which was supershear (Frankel 2004). The latter part of the earthquake involved dextral strike-slip rupture of the Totschunda Fault, which subparallels the Denali Fault (Fig. 1) and which ruptured the ground surface over ~70 km (Eberhart-Phillips et al. 2003; Haeussler et al. 2004) with an average offset of 1.7 m (Haeussler et al. 2004). Aftershocks were concentrated along the trace of the surface rupture and were generally located above 10 km depth; few were large (**M** 5.8 was the largest) and the sequence decayed quickly (Eberhart-Phillips et al. 2003).

Ground motions associated with the earthquake are poorly constrained, given that only one near-field ground-motion record exists. This record (Fig. 12) is from The Trans-Alaska Pipeline, Alyeska Pump Station 10, located ~3 km from the Denali Fault and 85 km east of the epicenter, and showed a peak horizontal ground acceleration of 341–360 cm/s^2 (Ellsworth et al. 2004; Martirosyan et al. 2004) and velocity of 180 cm/s (Ellsworth et al. 2004). Peak horizontal acceleration exceeded 0.1 g for only 10 s, probably due to strong directivity of the earthquake (Ellsworth et al. 2004). The station record shows only subevent 2 clearly, likely also due to directivity effects (Frankel 2004). Temporary seismometer deployments were made west of the Denali earthquake epicenter after the Nenana Mountain earthquake and recorded maximum horizontal accelerations of 107 cm/s^2 and 98 cm/s^2 49 km and 59 km from the epicenter, respectively (Martirosyan et al. 2004). Peak horizontal accelerations at the next closest permanent installations were 47 cm/s^2 and 8–22 cm/s^2 150 km and 270 km away from the epicenter, respectively (Martirosyan et al. 2004). Modeling by Frankel (2004) suggests that high-frequency energy release was greatest for the thrust subevent 1, whereas subevent 2 was very localized spatiotemporally. Subevent 3 produced the highest moment magnitude but lowest frequency energy of the subevents, and little high-frequency energy was generated during rupture of the Totschunda Fault (Frankel 2004). However, in contrast, from broadband teleseismic studies, Choy and Boatwright (2004) concluded that subevents 2 and 3 produced greater high-frequency energy release than did subevent 1. They also concluded that much longer period ground motion accompanied subevents 2 and 3 compared to subevent 1.

Ground motion records and modeling suggest that seismic energy release was highly directive. For example, ground motion stations at azimuths more than 90° from the fault rupture propagation direction did not record the third subevent (e.g., Eberhart-Phillips et al. 2003) and forward modeling indicated that ground motions should have been relatively large near the rupture and to the east in the direction of rupture propagation (Asano et al. 2005). Velasco et al. (2004) concluded that strong directivity was responsible for enhancing surface wave amplitudes by 2 orders of

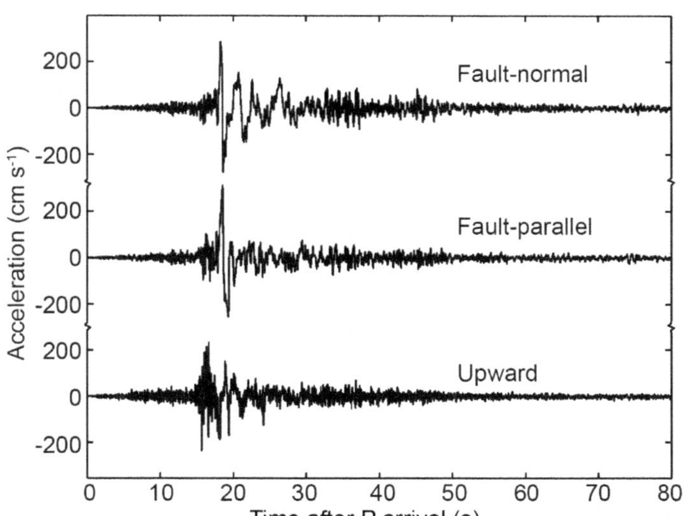

Fig. 12 Ground accelerations recorded at Pump Station 10 and corrected and projected by Ellsworth et al. (2004). Modified from Ellsworth et al. (2004)

magnitude. Field evidence also suggests that ground motions were highly directive, with lake ice cracking only within 40 km west of the fault rupture but 100 km away to the east (Harp et al. 2003), and increased occurrence of liquefaction to the east (Eberhart-Phillips et al. 2003; EERI 2003; Harp et al. 2003; Kayen et al. 2004).

3.1 Earthquake Effects

Because the earthquake occurred in a region with little development, effects on humans were relatively limited with no fatalities occurring, one injury, and about $40 million in damage (Kayen et al. 2004) and $90 million in total direct losses (EERI 2003). Additionally, most buildings in the affected area were ductile wood and log structures that accommodated ground motions relatively well (EERI 2003). Fault rupture propagated beneath The Trans-Alaska Pipeline with 5.8 m of right-lateral slip occurring, which damaged some pipeline support structures within a few kilometers of the rupture (Kayen et al. 2004). However, the pipeline was designed to accommodate 6.1 m of horizontal and 1.5 m of vertical offset in that location (EERI 2003), which rendered the damage to within acceptable limits. Some liquefaction was observed near the pipeline but caused no deformation, and no landslides impacted the pipeline or related infrastructure (EERI 2003). Fault offset destroyed ~670 m of the Richardson Highway and the Tok Cutoff where it intersects the fault, whereas many bridges were damaged and several required replacement, but all remained operational after the earthquake (EERI 2003). The airport runway at Northway located ~80 km from the Totschunda Fault rupture was rendered unusable by lateral spreading (Kayen et al. 2004).

The earthquake's effects on the natural environment were far reaching, in some cases. The earthquake triggered small earthquakes (Velasco et al. 2004) and seiches (EERI 2003) more than 3000 km away. Groundwater level and turbidity changed as much as ~5000 km away (EERI 2003). Nearer to the fault rupture, liquefaction was widespread within ~180 km of the location of maximum fault displacement (Harp et al. 2003) but was concentrated to the east as described previously. Although earthquake-induced landsliding and liquefaction typically are collocated (Keefer 1984), following the Denali earthquake, liquefaction was observed ~ 100 km beyond the limits of landsliding (Eberhart-Phillips et al. 2003; Harp et al. 2003), possibly due to a general absence of high-frequency energy thought responsible for most landslide triggering in rock and due to the sustained shaking that aids liquefaction (Harp et al. 2003).

4 Landslides Triggered by the Denali Earthquake

The earthquake triggered thousands of landslides ranging in size from a few cubic meters to several million cubic meters (Harp et al. 2003; Jibson et al. 2004, 2006;

Gorum et al. 2014), most of which were shallow rockfalls and slides (Jibson et al. 2006). Most hillslopes in the affected area comprise bare rock or very shallow colluvium decimeters to a few meters thick (Jibson et al. 2004, 2006); hence, slides primarily occurred in rock.

4.1 Spatial Distribution

Primarily from 30-m resolution satellite imagery, 1580 landslides were identified within an area of 7150 km^2 (Gorum et al. 2014; Schmitt et al. 2017a, b) comprising a band ~30 km wide centered along the Denali and Totschunda Fault ruptures, and on the hanging wall of the Susitna Glacier Fault (Eberhart-Phillips et al. 2003; Harp et al. 2003). The number of landslides mapped (Gorum et al. 2014; Schmitt et al. 2017a, b), and probably their distribution, are underestimates of those actually triggered, given the resolution of the imagery and that some of the mapped landslides are amalgamations of more than one slide (Tanyaş et al. 2017). Eighty percent of the mapped landslides occurred within 7 km from the Denali and Totschunda Fault ruptures, whereas the region of triggered slides near the Susitna Glacier Fault was wider and asymmetric and nearly all occurred on the fault hanging wall (Gorum et al. 2014). Landslides in potentially sensitive clay were observed ~150 km from the earthquake epicenter and a rockfall was observed ~250 km away (Harp et al. 2003; Jibson et al. 2004), but causation of these landslides cannot be definitively ascribed to the earthquake. Imagery available from Google Earth for years prior to and following the earthquake show repeated landslides in the potentially sensitive clay.

Landslides triggered by the earthquake were fewer in number and more limited in spatial extent than generally would be expected for an earthquake of this size (Eberhart-Phillips et al. 2003; Harp et al. 2003; Jibson et al. 2004, 2006). Global earthquake magnitude-landslide distribution relations developed by Keefer (1984, 2002) suggest that as many as 80,000 landslides should have been observed in a 300–500 km wide band with a mean total area of 28,000 km^2 (Harp et al. 2003; Jibson et al. 2004; Gorum et al. 2014), all of which exceed observations by an order of magnitude. Marc et al. (2017) provide an approach for estimating the area affected by coseismic landsliding considering seismic moment, source depth, focal mechanism, ground shaking, and fault rupture length. Their approach yields an affected area prediction for the Denali earthquake that is more than double the observed area.

The relatively small number of landslides triggered by the earthquake has been ascribed to a general absence of high-frequency, high-amplitude ground accelerations generally believed to be responsible for causing slope failures in rock (Eberhart-Phillips et al. 2003; Harp et al. 2003; Jibson et al. 2004, 2006). Strong directivity and associated narrow focusing of seismic energy along the direction of fault rupture (Eberhart-Phillips et al. 2003; EERI 2003; Harp et al. 2003; Choy and Boatwright 2004; Ellsworth et al. 2004; Kayen et al. 2004) may also have been responsible for the limited distribution and number of landslides triggered by the earthquake

(Harp et al. 2003; Jibson et al. 2004; Gorum et al. 2014), although Marc et al. (2017) concluded that directivity effects were small based on the symmetrical nature of landslide distributions beyond the western and eastern limits of fault rupture. Directivity effects on landslide causation should be greatest in the rupture direction (east-southeast), and Gorum et al. (2014) observed that more than 65% of triggered landslides occurred on east-southeast-facing hillslopes where landslide concentration was greatest near subevent 3 (Fig. 1), supporting the conclusion that directivity effects were important. Potential directional components to ground motions also were suggested by five of the largest rockslides having occurred from north-northeast-facing hillslopes (slides onto Black Rapids Glacier and from McGinnis Peak, Figs. 2, 8, and 13). These slides occurred between subevents 1 and 2, suggesting a north-northeast component to either or both of these events (Jibson et al. 2004, 2006).

Landslide concentration varied across the affected area and the nature of coseismic ground motions certainly contributed to this variability. Many landslides occurred in the vicinity of the earthquake epicenter near the Susitna Glacier Fault (32% of all triggered landslides) with 94% of these being on the hanging wall (Gorum et al. 2014). Landslide concentration is commonly observed to be greatest on fault hanging walls (e.g., Kieffer et al. 2006; Sato et al. 2007; Yagi et al. 2009; Chigira et al. 2010), possibly due to increased high-amplitude, high-frequency ground motion there (Harp et al. 2003). However, in terms of number of landslides per fault rupture length, the densest concentration of landslides occurred near subevent 3 (Fig. 1), ~30–80 km west of the Denali-Totschunda Fault junction. Subevent 3 had the highest moment magnitude, and modeling suggests (Frankel 2004) that it had the highest rupture

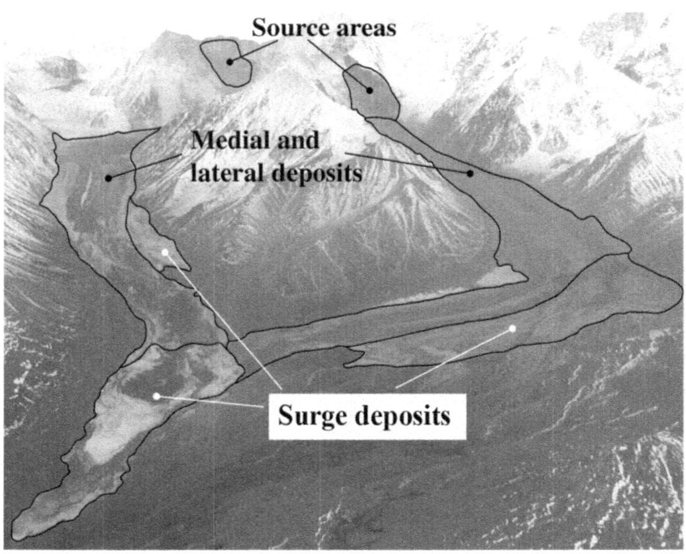

Fig. 13 Oblique aerial photograph of avalanches at McGinnis peak. Overall deposits are 11–12 km long. Modified from Schulz et al. (2008)

speed, which may have been partly responsible for this concentration. Sackungen (i.e., landforms including ridgetop grabens and uphill-facing escarpments suggestive of deep-seated slope failure) also were common near subevent 3 (Jibson et al. 2004). Coseismic ground motions may be damped by the presence of glaciers, especially at ridge crests between glaciers (McColl et al. 2012). However, Gorum et al. (2014) compared ground motions and landslide characteristics from Denali to those from other recent strike-slip earthquakes in periglacial and formerly glaciated terrain and found no conclusive evidence that landslide-inducing ground motions were affected by glacier cover.

Properties of geologic materials may control inherent susceptibility to landsliding in some instances. To evaluate potential relations between the Denali landslides and geologic material, I used the polygons delineating landslide-affected areas (including source, transport, and deposit areas; Gorum et al. 2014; Schmitt et al. 2017a, b) to identify points approximating each landslide source. This was done by using tools available in Esri ArcGIS (v. 10.4) to create a point at the highest elevation of each polygon, which generally resulted in point creation near the center of each landslide's headscarp. However, some individual polygons encompass amalgamations of multiple landslides (Tanyaş et al. 2017), so the resulting landslide points underrepresent the actual number of landslides. Table 3 displays the number and density of landslides within the most prevalent geologic formations (Figs. 6 and 7) mapped at 1:500,000. On average, landslide density varied within these units from as low as 0.05 landslide/km^2 to as high as 0.93 landslide/km^2. Nearly, 1 landslide/km^2 occurred in formations KJgn and TKgg, which comprise marine sedimentary rock and granitic rock, respectively. Formations Trcs (calcareous sedimentary rocks), Trn (basalt and greenstone), and Kmgr (granodiorite) displayed intermediate landslide density (0.66, 0.59, and 0.45 landslide/km^2, respectively). Lowest landslide density (0.05–0.07 landslide/km^2) was observed where rock was covered by glaciers, and in formations Dmi (metamorphosed igneous rocks) and PIPgi (submarine volcanic rocks). Given the lack of clear relations between material and landslide density (for example, intrusive igneous rocks composed of granite and granodiorite had highest and relatively low landslide density, respectively) ascribing variable landslide occurrence to inherent material variability is difficult for this event.

Variability in slope distributions and seismic energy are at least partly responsible for the inability to identify clear relations between geological materials and propensity for landsliding. However, inherent susceptibility to landsliding by material may be revealed by considering the spatial variability of landsliding within given geologic formations. For example, landslide density was generally 2 to 5 times higher in the central area compared to the western and eastern areas in all formations except for Qs (unconsolidated surficial deposits), which had highest landslide density in the west. Qs is more widespread toward the east and occurs mostly as valley-fill glacial and fluvial deposits, whereas in the western area, it occurs as steep lateral moraines very close to the Denali Fault rupture and also as flatter deposits along the Susitna Glacier Fault and its hanging wall. These topographic and seismic settings almost certainly contributed to a higher landslide density in Qs in the west. As another example, KJgn (marine graywacke and mudstone) had the highest landslide density

Table 3 Landslide distribution by geologic formation

Geologic formation[a]	Number of landslides[b]			Landslide density[b,c] (number of landslides/km^2)			
	West	Central	East	West	Central	East	Average
Dmi	1	0	0	0.07	NA	NA	0.07
DOmvs	20	144	11	0.18	0.60	0.10	0.30
Glacier	28	57	0	0.02	0.08	NA	0.05
KJgn	0	39	154	NA	1.57	0.29	0.93
Kmgr	16	8	9	0.55	0.68	0.13	0.45
Mzmu	8	0	0	0.13	NA	NA	0.13
PIPgi	0	0	5	NA	NA	0.07	0.07
PIPt	0	11	7	NA	0.13	0.07	0.10
Pzarqm	0	6	0	NA	0.10	NA	0.10
Qs	32	29	21	0.21	0.05	0.02	0.10
TKgg	127	0	0	0.82	NA	NA	0.82
Trcs	137	0	0	0.66	NA	NA	0.66
Trn	0	37	21	NA	0.85	0.32	0.59

[a]Geologic formations from 1:500,000 scale version of Wilson et al. (2015)
[b]Landslides from Schmitt et al. (2017a, b) using data from Gorum et al. (2014)
[c]Landslides modified from affected-area polygons to points located near headscarp centers

of all formations and displayed 5 times greater density in the central area compared to the eastern area (it doesn't occur in the western area). This probably results from it being limited in spatial extent to within 4 km of the fault rupture in the central area and extending farther than 15 km away in the eastern area; as noted previously, landsliding was densest nearest the fault rupture and decayed rapidly with distance normal to the fault. Similarly, Trn (massive basalt and greenstone) had higher density in the central area compared to the east and is located as much as 5 km and 10 km from the fault rupture in the two areas, respectively; hence, its nearer proximity to the fault in the central area likely contributed to its higher density there. Additionally, subevent 3 occurred at the eastern end of the central area, and landslide density was particularly high near this subevent (Fig. 1). As a final comparison, Kmgr (primarily granodiorite) displayed similar landslide density in the western and central areas but lower density by 4–5 times in the eastern area. The formation mostly occurs on the hanging wall of the Susitna Glacier Fault in the western area, very near the Denali Fault rupture in the central area and occupies a large part of the eastern area 1–6 km from the Denali rupture near and beyond its eastern end. Hence, it appears that variable coseismic ground motions associated with characteristics of the rupture at least partly caused the disparity in landslide density for this and the other formations, which along with topographic variability, makes evaluation of inherent material susceptibility to landsliding difficult with the data available.

Other factors, in addition to material properties and ground shaking characteristics, also were important for determining the distribution of landslides, but identifying the magnitude of their contributions is not readily tractable. Gorum et al. (2014) observed that landsliding predominantly occurred from slopes higher and steeper than the modes and means of overall relief and slope inclination near the fault, regardless of proximity to rupture ends or to the epicenter. The western region of landsliding is highest in elevation and displays steepest slopes and greatest relief, followed by the central and eastern regions, implying decreasing landslide susceptibility from west to east, but landslide numbers decreased from the central region to the western and eastern regions (Figs. 1, 6, and 7; Tables 1 and 2). The lack of positive correlation between topographic metrics and landslide numbers may partly result from the presence of groundwater in susceptible slopes, which is more likely in the eastern part of the landslide-affected area because temperatures are higher and permafrost less extensive. Hence, the potential for coseismic pore-pressure elevation and resulting landsliding should have increased toward the east. Finally, debuttressing of hillslopes by deglaciation and glacier thinning may increase landslide susceptibility. Considering that the degree of glaciation decreases from west to east, such debuttressing should be more pronounced toward the east. However, as noted previously, many of the existing glaciers in the central and western regions have thinned historically, especially those that have experienced surges, so slope debuttressing by deglaciation is more recent in those regions, although having occurred to a lesser extent than in the east. Hence, it is likely that these and other factors variably affected landslide susceptibility during the Denali earthquake, but disentangling their contributions requires more detailed study.

4.2 Landslide Size

Landslides ranged in size from small rockfalls (Harp et al. 2003) with volumes of a few cubic meters to slides mostly in rock that transitioned to avalanches as large as 20.4×10^6 m^3 (Jibson et al. 2006). Tables 1 and 2 show that, although the number of landslides was nearly equal in the western and central areas, the total area covered by landslides in the western area was more than double that in the central area. Hence, the largest landslides occurred in the west, and average surface areas were an order of magnitude smaller in the central and eastern areas. Large landslides may have occurred preferentially in the western part of the affected area due to subevent 1 being a thrust event, presumably having higher amplitude high-frequency ground motion than the strike-slip events (Harp et al. 2003; Jibson et al. 2006). The five largest landslides triggered by the earthquake occurred in the eastern part of the western area (Figs. 1, 2, 8, and 13), ~25–50 km west of subevent 2. These slides occurred from slopes above the Black Rapids Glacier and from McGinnis Peak, all from north-northeast-facing hillslopes, suggesting a north-northeast component to subevents 1 and/or 2 (Jibson et al. 2004, 2006). Worth noting is that laterally extensive discontinuities that paralleled source hillslopes created dipslope conditions for the bases of these rockslides (Jibson et al. 2006; Schulz et al. 2008).

Landslide areas mapped by Gorum et al. (2014; Schmitt et al. 2017a, b) include sources, transport zones, and deposits; and the western area has the greatest glacier coverage, with many of the landslides having moved onto glaciers. This apparently aided runout and spreading, thereby increasing landslide-affected area (Table 1, 2, 4, and 5). Gorum et al. (2014) observed that average landslide runout on glaciers was 127% of that in ice-free areas, and Jibson et al. (2006) found that the seven largest avalanches had runout longer than average when compared to global landslide runout data (Fig. 14). These seven landslides moved mostly over glaciers. This includes the two largest landslides, which descended from McGinnis Peak (Figs. 2 and 13) and which skew results for landslide size toward the largest having occurred in the western region. Discounting landslides from areas mapped as glaciers, the western area still had an average landslide size that was 2–3 times the average sizes for the central and eastern areas. This largely results from landslides in metamorphosed mafic igneous rocks (Dmi) being an order of magnitude larger on average than landslides from nearly all other geologic formations (Table 4). Dmi is of limited extent throughout the earthquake-affected area (Figs. 6 and 7), and landslides occurred in this formation only in the western area, also near McGinnis Peak. An additional cluster of particularly large landslides occurred near subevent 2 (Fig. 1), and two very large sackungen also occurred in this area, whereas another sackungen occurred near subevent 3 (Jibson et al. 2004). The three large landslides that occurred from slopes above Black Rapids Glacier (Figs. 2 and 8) also created sackungen on the mountainsides opposite their source areas and left several million cubic meters of landslide debris perched high on the mountains (Jibson et al. 2004).

4.3 Large Rock Avalanches

The seven largest earthquake-triggered landslides transformed into avalanches with long runouts (Table 5, Fig. 14). The settings in which these avalanches occurred varied, which permits comparison of deposit features potentially indicative of controls on avalanche movement. Three of the avalanches primarily traversed open slopes and the approximately flat surface of Black Rapids Glacier (Figs. 2 and 8; Table 5), two were confined to glacial valleys on the flanks of McGinnis Peak (Figs. 2 and 13, Table 5), and two were initially confined to glacial valleys before moving across short open slopes and then onto the relatively flat surface of the West Fork Glacier (Figs. 2 and 15, Table 5). The mobility of each of the avalanches was greater than average for rock avalanches worldwide (Fig. 14).

At Black Rapids Glacier, three large rockslides (and many smaller slides) occurred on north-facing hillslopes above the south side of the glacier (Figs. 2 and 8). These rockslides involved Tertiary-Late Cretaceous gneissose granitic rock (TKgg, Fig. 6; TMzmb, Fig. 7), were 30–50 m thick (Jibson et al. 2004, 2006), and failed along a series of joint surfaces that formed dip-slope conditions with source hillslopes. Source

Table 4 Average landslide size by geologic formation

Geologic formation[a]	Average landslide size[b] (km^2)			
	West	Central	East	Overall average
Dmi	0.34	NA	NA	0.34
DOmvs	0.06	0.02	0.04	0.04
Glacier	1.05	0.24	NA	0.64
KJgn	NA	0.02	0.03	0.02
Kmgr	0.02	0.02	0.03	0.02
Mzmu	0.02	NA	NA	0.02
PIPgi	NA	NA	0.02	0.02
PIPt	NA	0.03	0.02	0.03
Pzarqm	NA	0.04	NA	0.04
Qs	0.12	0.06	0.06	0.08
TKgg	0.07	NA	NA	0.07
Trcs	0.05	NA	NA	0.05
Trn	NA	0.01	0.06	0.04

[a]Geologic formations from 1:500,000 version of Wilson et al. (2015)
[b]Landslides from Schmitt et al. (2017a, b) using data from Gorum et al. (2014)

Table 5 Average geometric characteristics of large rock avalanches triggered by the earthquake[a]

Landslide	Vertical drop	Length	Width	Length per	Source area	Deposit area	Depth	Volume
	(km)	(km)	(km)	vertical drop	(km^2)	(km^2)	(m)	(10^6 m^3)
McGinnis peak								
North	1.65	11.0	1.5	6.67	0.40	10.21	2	20.4
South	1.80	11.5	0.7	6.39	0.48	5.71	2	11.4
Black Rapids glacier								
East	0.98	4.6	1.6	4.69	0.66	4.64	3	13.9
Middle	0.80	4.5	1.2	5.63	0.79	4.55	3	13.6
West	0.73	3.2	1.0	4.38	0.52	3.24	3	9.7
West Fork glacier								
North	0.76	3.3	0.7	4.34	0.42	1.37	3	4.1
South	0.90	4.1	0.8	4.56	0.55	1.47	3	4.4

[a]Modified from Jibson et al. (2006)

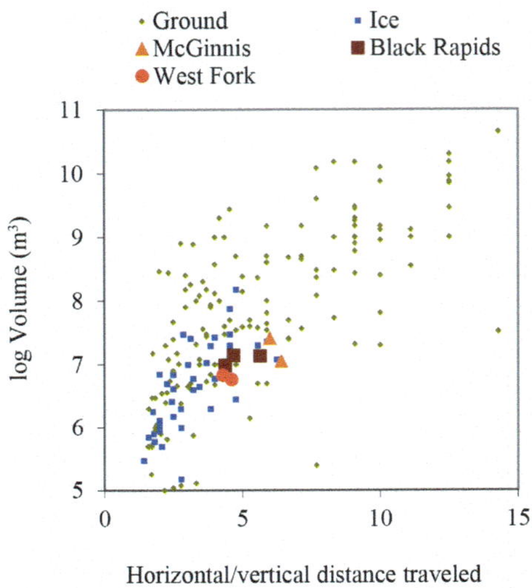

Fig. 14 Plot comparing volumes of rock avalanches with the ratio of their horizontal to vertical travel distance. Rock avalanches traveling on ice (glacial surfaces) and ground generally cluster together. Rock avalanches triggered by the Denali earthquake plot within, but on the long-runout side of, the worldwide data. Modified from Jibson et al. (2006), which was modified from Evans and Clague (1988) and Shaller (1991)

Fig. 15 Oblique aerial photograph of the northern West Fork glacier avalanche deposit. Photograph taken by Randy Jibson (U.S. Geological Survey). Peak at right rises ~1100 m above the glacier in the foreground

slopes were inclined 28°–35° and extended from the mountaintops to the valley bottom adjacent to the Black Rapids Glacier (Schulz et al. 2008). The landslides spawned rock avalanches with volumes of 9.7×10^6 m^3 to 13.9×10^6 m^3 (Table 5) that crossed the relatively flat glacier surface below (Figs. 2 and 8), whereas larger volumes of material slid but remained perched high on source mountainsides (Harp et al. 2003; Jibson et al. 2004, 2006). The avalanches left deposits generally 2–3 m thick, 3.2–4.6 km long, and 1.0–1.6 km wide on the glacier (Schulz et al. 2008). The deposits were ~1 m thicker on the proximal side and ~1 m thinner on the distal side (Schulz et al. 2008) of a 50-m-high medial moraine located near the center of the glacier, and the avalanches were moving faster than 36 m/s to cross this moraine (Jibson et al. 2004).

At West Fork Glacier (Figs. 2 and 15), two rockslides with volumes of 4.1–4.4 $\times$ 10^6 m^3 (Table 5) occurred at the heads of adjacent hanging glacial valleys within Triassic calcareous sedimentary rock (Trcs, Figs. 6 and 7). The slides transitioned to avalanches that traversed the ~1.5–2 km long, 10–12° inclined surfaces of the hanging valleys before moving down the 26° inclined main wall of the West Fork Glacier valley and then onto the relatively flat glacier surface. Runup on walls of the hanging valleys suggested speeds in excess of 31 m/s (Jibson et al. 2004). Avalanche deposits were 3.3–4.1 km long and 0.7–0.8 km wide, and generally 2–4 m thick, excluding blocks as large as 20 m across that were sporadically distributed within the deposits (Fig. 16). A few large blocks were observed that ran out farther than the main deposit and gouged the glacier surface as they moved (Jibson et al. 2004). Both avalanches left lateral deposits 2–6 m thick along parts of their paths above the West Fork Glacier, particularly on the main valley wall (Fig. 15).

The two avalanches with the longest runout occurred at the heads of ~5° inclined glacial valleys on McGinnis Peak (Figs. 2 and 13). The northern of these was the largest triggered by the earthquake with a volume of 20.4 $\times$ 10^6 m^3, about 10% of which was ice from a hanging glacier at the valley head (Harp et al. 2003), and it occurred on a source slope inclined 47°. The southern rockslide occurred on a source slope inclined 62° and it had a volume of 11.4 $\times$ 10^6 m^3, about 30% of which was ice also from a hanging glacier (Harp et al. 2003; Jibson et al. 2006; Table 5). Devonian to Ordovician metamorphic rock, primarily phyllite (DOmvs, Fig. 6; DCsp, Fig. 7) was involved in the slides, which failed along several series of slope-parallel

Fig. 16 Photograph showing large boulders near the distal end of the deposit shown in Fig. 15, some of which are coated by finer material. Note author (circled) for scale. Photograph taken by Ed Harp (U.S. Geological Survey)

discontinuities. The larger, northern avalanche had a deposit 11.0 km long, while the smaller, southern avalanche had an 11.5-km long deposit (Jibson et al. 2006); the mobility of the larger slide may have been aided by its higher ice content and steeper source area. Both avalanches produced overriding dust clouds with deposits reaching 50–70 m above the valley bottoms (Harp et al. 2003). Superelevation and ridge overtopping of the northern avalanche suggest speeds greater than 40 m/s and 54 m/s, respectively, 4–5 km down valley from its source (Jibson et al. 2004). Deposit characteristics suggest that both avalanches developed multiple surges as they flowed, and these surges produced medial and lateral deposits along their paths until ceasing movement and forming surge deposits (Schulz et al. 2008; Fig. 13). Individual medial, lateral, and surge deposits generally were 0.5–2 m, 1–3 m, and 2–5 m thick, respectively, with overlapping of multiple deposits being common and resulting in thicker, overall deposits (Schulz et al. 2008).

Deposits of each of the avalanches comprised subangular clay to boulder-sized rock fragments displaying no evidence of size gradation with distance from source slopes (Schulz et al. 2008). Anomalously large rock blocks were randomly distributed within the deposits as well. The lower ~two-thirds of the deposits displayed relatively uniform size distribution, whereas the upper third consisted of ~80% cobbles through boulders and ~20% pebbles through clay. Particle-size distributions were very consistent with distance from source areas, suggesting that particle comminution and mixing mostly occurred during initial failure and movement along steeper surfaces where deposition was limited. The avalanches flowed as indicated by deposit geometries bearing no resemblance to source geometries, and by parallel banding and zonation of different rock types in the deposits (Schulz et al. 2008; Figs. 8, 17, and 18). Although flow occurred, many of the large rocks on the deposit surfaces were coated with smaller rock fragments, suggesting little turbulence during at least the latter stages of flow; some of these coated large rocks reached diameters of ~20 m (Fig. 16). Sliding and/or laminar and plug flow also occurred, as suggested by likely compressional features (folds) resembling wrinkles on deposit surfaces, particularly near the proximal sides of obstacles (Schulz et al. 2008; Fig. 17). Sliding was suggested during the latter stages of movement for the avalanches on the Black Rapids Glacier surface by strike-slip faults within the deposits that offset compositional bands and zones (Schulz et al. 2008; Fig. 18). Ice from glacier surfaces and soil and vegetation in unglaciated areas were locally plowed by the leading edge of each of the large avalanches. Such plowing was apparent along ~20% of deposit margins (Schulz et al. 2008). Avalanche movement also left some overrun substrate undisturbed, including where the northern McGinnis Peak avalanche overran a 150-m-high ridge at speeds estimated to be greater than 54 m/s (Jibson et al. 2004, 2006; Schulz et al. 2008). Altogether, substrate and deposit characteristics suggest turbulent flow down source slopes involving disaggregation and local mixing of avalanche material, followed by laminar flow with overriding plugs in deposit areas, and, finally, localized sliding as avalanches halted, forming faults within deposits and gouging and plowing of substrate.

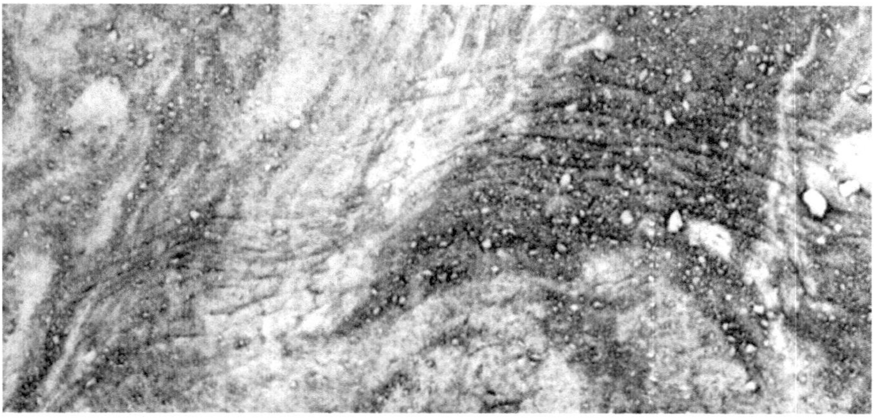

Fig. 17 Vertical aerial photograph of apparent folds in an open-slope avalanche deposit. View is 225 m wide. Avalanche movement was from the lower part of the view to the upper part of the view. Modified from Schulz et al. (2008)

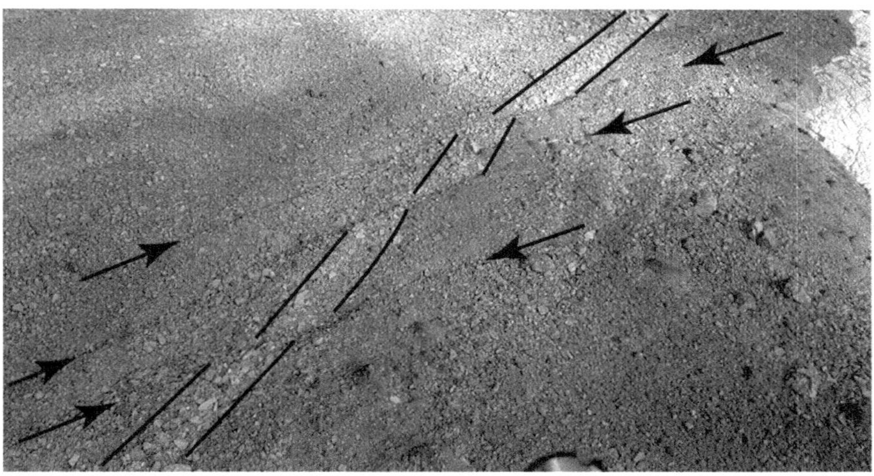

Fig. 18 Oblique aerial photograph showing faults (between arrows) offsetting rock type zones near the distal end (right side) of an open-slope avalanche deposit. One zone is outlined to illustrate the offset. Avalanche movement was from lower left to upper right and the central part of the deposit is toward the upper left. Center of view is about 350 m wide. Modified from Schulz et al. (2008)

5 Conclusion

On November 3, 2002, the **M** 7.9 Denali earthquake occurred in southeastern Alaska, USA, which was the largest inland earthquake in North America in nearly 150 years (e.g., Alaska earthquake Information Center 2017). The event started with thrusting displacement but primarily involved strike-slip displacement (e.g., Choy and Boatwright 2004), making it one of the largest strike-slip earthquakes in history. The earthquake triggered thousands of landslides in the rugged, glaciated Alaska range (e.g., Harp et al. 2003; Jibson et al. 2004). The variety of terrain, geology, climate, and glacier cover over which strong shaking occurred make the event useful for evaluating conditions resulting in landslide occurrence and character. Presented herein is a summary of previous work and simple analyses potentially useful for more detailed studies.

Landslides occurred mostly in rock and ranged from a few cubic meters to 20.4 $\times 10^6$ m^3 (e.g., Harp et al. 2003). The greatest concentration of landslides and sackungen occurred near subevent 3 (the most eastern of the three), possibly because this subevent had the highest moment magnitude and rupture speed (Frankel 2004). Landslides also were densely concentrated near the earthquake epicenter (subevent 1), with nearly one-third of all landslides produced by the earthquake having occurred on the hanging wall associated with this thrust event. However, the overall number and spatial distribution of landslides was an order of magnitude lower than previous studies (Keefer 1984, 2002) would suggest (e.g., Eberhart-Phillips et al. 2003). The limited number and distribution may have resulted from a general lack of high-amplitude, high-frequency ground motion, which is supported by the high concentration of landslides on the hanging wall of subevent 1, where such ground motions would have been most likely (Harp et al. 2003). Strong directivity of coseismic ground motion may have contributed to landslides being concentrated very near the fault and is supported by liquefaction occurring farthest from the epicenter in the rupture propagation direction (e.g., Harp et al. 2003). Directivity also was indicated by the absence of subevent 2 from seismic records for stations at azimuths greater than 90° from the rupture propagation direction (e.g., Eberhart-Phillips et al. 2003). The abundance of slides from slopes facing the propagation direction also supports directivity control (Gorum et al. 2014). The five largest landslides triggered by the earthquake occurred from north-northeast-facing hillslopes ~25–50 km west of subevent 2, suggesting a north-northeast component to subevents 1 and/or 2 (Jibson et al. 2004, 2006). Although a study suggests that damping of coseismic ground motion by glacier loads may reduce the potential for coseismic landsliding (McColl et al. 2012), another study found no evidence that this contributed to the limited number and distribution of the Denali landslides (Gorum et al. 2014).

The area affected by coseismic landsliding was subdivided into three physiographic regions that are progressively less rugged, less glaciated, and warmer from west to east. Landslides preferentially occurred on slopes steeper and higher than average and consistently did so as mean and mode slopes varied along the length of the affected area (Gorum et al. 2014). Landslides were most abundant in the central region and especially so if the relative overabundance of landslides on the hanging wall of the epicentral event are discounted because of the more likely landslide-producing energy from that thrust event (western region), followed by the western and then the eastern regions. The reduced permafrost and higher likelihood of groundwater in susceptible hillslopes in the central and eastern regions may have importantly contributed to landsliding there. Climatic conditions prior to the earthquake were about average for the historical record, and, although historically observed rising temperatures may be increasing landslide susceptibility as observed elsewhere in Alaska (Coe et al. 2018), average annual temperatures in the affected area remained below freezing prior to the earthquake.

Glaciers in the area have been observed historically to thin by tens of meters to more than 100 m from melting, ablation, and surging (e.g., Molnia 2008). Such thinning should debuttress potentially susceptible hillslopes, which may have affected landslide occurrence. Three of the five largest triggered landslides occurred adjacent to the Black Rapids Glacier, and many other large slides occurred adjacent to the Gulkana and Gakona Glaciers, all of which were observed to thin historically.

The landslide-affected area displays geologic variability, with bedrock units including intrusive and extrusive volcanic rocks, marine and terrestrial sedimentary units, and a range of metamorphic rock types. No clear patterns emerge with respect to inherent susceptibility to landsliding by geological formation, partly because coseismic ground motions are poorly constrained. Spatially extensive discontinuities (foliation and jointing) that created dip-slope conditions appeared to predispose some slopes to releasing the largest rockslides of the event.

Many landslides transformed into rock avalanches. Avalanche runout was apparently aided by movement over glacial ice and, for the largest avalanches, was slightly greater than average when compared to global data from both glaciated and non-glaciated terrain (Jibson et al. 2006). The seven largest avalanches occurred in variable topographic settings, making their evaluation potentially useful for gaining insight into rock avalanching mechanisms. Where data permitted quantitative estimation, avalanches locally moved at speeds exceeding 50 m/s and ran out as much as 11.5 km over gently inclined ($\leq 5°$) glacier surfaces (Jibson et al. 2006). Comminution of source rock appeared to occur during turbulent flow as avalanches proceeded down source slopes before significant deposits formed on flatter surfaces, as indicated by spatially consistent particle-size distributions. Avalanche material flowed during the majority of movement, and flow appeared to be laminar and plug in depositional areas, as suggested by folding of deposits on proximal sides of obstacles. Sliding also occurred, as was revealed by faulting of deposits and localized plowing of substrate (Schulz et al. 2008).

Acknowledgements Randy Jibson and Ed Harp (U.S. Geological Survey) are thanked for our collaborative studies of landslides triggered by the Denali Fault earthquake. Yao Jiang, Erin Bessette-Kirton, Jeffrey Coe, Jill McCarthy, and Ikuo Towhata are thanked for their constructive reviews of this paper. Any use of trade, firm, or product names is for descriptive purposes only and does not imply endorsement by the U.S. Government.

References

Alaska Earthquake Information Center (2017) M 7.9 Denali fault earthquake of November 3, 2002. http://www.aeic.alaska.edu/Denali_Fault_2002/. Last accessed 15 Dec 2017

Asano K, Iwata T, Irikura K (2005) Estimation of source rupture process and strong ground motion simulation of the 2002 Denali, Alaska, earthquake. Bull Seismol Soc Am 95(5):1701–1715

Chigira M, Wu X, Inokuchi T, Wang G (2010) Landslides induced by the 2008 Wenchuan earthquake, Sichuan, China. Geomorphology 118(3–4):225–238
Choy GL, Boatwright J (2004) Radiated energy and the rupture process of the Denali fault earthquake sequence of 2002 from broadband teleseismic body waves. Bull Seismol Soc Am 94(6B):S269–S277
Coe JA, Bessette-Kirton EK, Geertsema M (2018) Increasing rock-avalanche size and mobility in Glacier Bay national park and preserve, Alaska detected from 1984 to 2016 Landsat imagery. Landslides 15(3):393–407
Crone AJ, Personious SF, Craw PA, Haeussler PJ, Staft LA (2004) The Susitna Glacier thrust fault–characteristics of surface ruptures on the fault that initiated the 2002 Denali fault earthquake. Bull Seismol Soc Am 94(6B):S5–S22
Csejtey B Jr (1976) Tectonic implications of a late Paleozoic volcanic arc in the Talkeetna Mountains, south-central Alaska. Geology 4(1):49–52
Eberhart-Phillips D, Haeussler PJ, Freymueller JT, Frankel AD, Rubin CM, Craw P, Ratchkovski NA, Anderson G, Carver GA, Crone AJ, Dawson TE (2003) The 2002 Denali fault earthquake, Alaska: a large magnitude, slip-partitioned event. Science 300(5622):1113–1118
Echelmeyer KA, Harrison WD, Larsen CF, Sapiano J, Mitchell JE, De Mallie J, Rabus B, Adalgeirsdottir G, Sombardier L (1996) Airborne surface profiling of glaciers: a case-study in Alaska. J Glaciol 42(142):538–547
EERI (Earthquake Engineering Research Institute) (2003) Preliminary observations on the November 3, 2002 Denali fault, Alaska, earthquake. EERI special earthquake report, February 2003. Available at https://www.eeri.org/2002/11/denali-alaska/. Last accessed 15 Dec 2017
Ellsworth WL, Celebi M, Evans JR, Jensen EG, Kayen R, Metz MC, Nyman DJ, Roddick JW, Spudich P, Stephens CD (2004) Near-field ground motion of the 2002 Denali fault, Alaska, earthquake recorded at pump station 10. Earthq Spectra 20(3):597–615
Evans SG, Clague JJ (1988) Catastrophic rock avalanches in glacial environments. In: Bonnard C (ed) Proceedings, 5th international symposium on landslides, Lausanne, Switzerland, vol 2, pp 1153–1159
Fletcher HJ, Freymueller JT (2003) New constraints on the motion of the Fairweather fault, Alaska, from GPS observations. Geophys Res Lett 30(3):1139
Frankel A (2004) Rupture process of the M 7.9 Denali fault, Alaska, earthquake: subevents, directivity, and scaling of high-frequency ground motions. Bull Seismol Soc Am 94(6B):S234–S255
Gibson W (2009a) Mean average temperature for Alaska 1971–2000. National Park Service, Alaska Regional Office GIS Team. Available from https://irma.nps.gov/DataStore/Reference/Profile/2170516. Last accessed 7 July 2018
Gibson W (2009b) Mean precipitation for Alaska 1971–2000. National Park Service, Alaska Regional Office GIS Team. Available from https://irma.nps.gov/DataStore/Reference/Profile/2170508. Last accessed 7 July 2018
Gorum T, Korup O, van Westen CJ, van der Meijde M, Xu C, van der Meer FD (2014) Why so few? Landslides triggered by the 2002 Denali earthquake, Alaska. Quaternary Sci Rev 95:80–94
Gruber S (2012a) Derivation and analysis of a high-resolution estimate of global permafrost zonation. Cryosphere 6:221–233
Gruber S (2012b) Global permafrost zonation index map. http://www.geo.uzh.ch/microsite/cryodata/pf_global/. Last accessed 7 July 2018
Haeussler PJ (2009) Surface rupture map of the 2002 M7.9 Denali fault earthquake, Alaska; digital data. U.S. Geological Survey Data Series 422
Haeussler PJ, Schwartz DP, Dawson TE, Stenner HD, Lienkaemper JJ, Sherrod B, Cinti FR, Montone P, Craw PA, Crone AJ, Personius SF (2004) Surface rupture and slip distribution of the Denali and Totschunda faults in the 3 November 2002 M 7.9 earthquake, Alaska. Bull Seismol Soc Am 94(6B):S23–S52
Harp EL, Jibson RW, Kayen RE, Keefer DK, Sherrod BL, Carver GA, Collins BD, Moss RES, Sitar N (2003) Landslides and liquefaction triggered by the M 7.9 Denali fault earthquake of 3 November 2002. GSA Today, vol August 2003, pp 4–10

Jibson RW, Harp EL, Schulz WH, Keefer DK (2004) Landslides triggered by the 2002 Denali fault, Alaska, earthquake and the inferred nature of the strong shaking. Earthq Spectra 20(3):669–691

Jibson RW, Harp EL, Schulz WH, Keefer DK (2006) Large rock avalanches triggered by the M 7.9 Denali fault, Alaska, earthquake of 3 November 2002. Eng Geol 83:144–160

Kayen R, Thompson E, Minasian D, Moss RE, Collins BD, Sitar N, Dreger D, Carver G (2004) Geotechnical reconnaissance of the 2002 Denali fault, Alaska, earthquake. Earthq Spectra 20(3):639–667

Keefer DK (1984) Landslides caused by earthquakes. Geol Soc Am Bull 95(4):406–421

Keefer DK (2002) Investigating landslides caused by earthquakes—a historical review. Surv Geophys 23(6):473–510

Kieffer DS, Jibson R, Rathje EM, Kelson K (2006) Landslides triggered by the 2004 Niigata Ken Chuetsu, Japan, earthquake. Earthq Spectra 22(S1):47–73

Kikuchi M, Yamanaka Y (2002) Source rupture processes of the central Alaska earthquake of Nov. 3, 2002, inferred from teleseismic body waves (+ the 10/23 M6.7 event). EIC Seismological Note, No 129

Loso M, Arendt A, Larsen C, Rich J, Murphy N (2014) Alaskan national park glaciers—status and trends. Natural resource technical report NPS/AKRO/NRTR—2014/922. National Park Service, Fort Collins, Colorado

Marc O, Meunier P, Hovius N (2017) Prediction of the area affected by earthquake-induced landsliding based on seismological parameters. Nat Hazard 17:1159–1175

Martirosyan A, Hansen R, Ratchkovski N (2004) Strong-motion records of the 2002 Denali Fault, Alaska, earthquake. Earthq Spectra 20(3):579–596

Matmon A, Schwartz DP, Haeussler PJ, Finkel R, Lienkaemper JJ, Stenner HD, Dawson TE (2006) Denali Fault slip rates and Holocene–late Pleistocene kinematics of central Alaska. Geology 34(8):645–648

McColl ST, Davies TRH, McSaveney MJ (2012) The effect of glaciation on the intensity of seismic ground motion. Earth Surf Proc Land 37:1290–1301

Molnia BF (2008) Satellite image atlas of glaciers of the world. In: Williams RS, Ferrigno JG (eds) U.S. geological survey professional paper 1386-K, 525p

Nokleberg WJ, Aleinikoff JN, Lande IM, Silva SR, Miyaoka RT, Schwab CE, Zehner RE (1992) Preliminary geologic map of the Mount Hayes Quadrangle, Eastern Alaska Range, Alaska. U.S. Geological Survey Open-File Report 92-594, 39p

Plafker G, Hudson T, Richter DH (1977) The United States geological survey in Alaska: accomplishments during 1976. In: Blean KM (ed) USGS circular 751-B, pp B67–B69

Post AS (1960) The exceptional advances of the Muldrow, Black Rapids, and Susitna Glaciers. J Geophys Res 65(11):3703–3712

Ridgway KD, Trop JM, Nokleberg WJ, Davidson CM, Eastham KR (2002) Mesozoic and Cenozoic tectonics of the eastern and central Alaska range: progressive basin development and deformation in a suture zone. Geol Soc Am Bull 114(12):1480–1504

Sato HP, Hasegawa H, Fujiwara S, Tobita M, Koarai M, Une H, Iwahashi J (2007) Interpretation of landslide distribution triggered by the 2005 Northern Pakistan earthquake using SPOT 5 imagery. Landslides 4(2):113–122

Schmitt RG, Tanyas H, Nowicki J, Zhu J, Biegel KM, Allstadt KE, Jibson RW, Thompson EM, van Westen CJ, Sato HP, Wald DJ, Godt JW, Gorum T, Xu C, Rathje EM, Knudsen KL (2017a) An open repository of earthquake-triggered ground-failure inventories. U.S. Geological Survey Data Series 1064, 17p

Schmitt RG, Tanyas H, Nowicki J, Zhu J, Biegel KM, Allstadt KE, Jibson RW, Thompson EM, van Westen CJ, Sato HP, Wald DJ, Godt JW, Gorum T, Xu C, Rathje EM, Knudsen KL (2017b) An open repository of earthquake-triggered ground-failure inventories. U.S. Geological Survey data release collection. https://doi.org/10.5066/F7H70DB4. Last accessed 29 Dec 2017b@@

Schulz WH, Harp EL, Jibson RW (2008) In: Chen Z, Zhang J, Li Z, Wu F, Ho K (eds) Landslides and engineered slopes, from the past to the future, vol 2. Taylor & Francis Group, London, pp 1447–1453

Shaller PJ (1991) Analysis and implications of large Martian and terrestrial landslides. Ph.D. Dissertation, California Institute of Technology, 586 p

Stout JH, Chase CG (1980) Plate kinematics of the Denali fault system. Can J Earth Sci 17:1527–1537

Tanyaş H, Westen CJ, Allstadt KE, Anna Nowicki Jessee M, Görüm T, Jibson RW, Godt JW, Sato HP, Schmitt RG, Marc O, Hovius N (2017) Presentation and analysis of a worldwide database of earthquake-induced landslide inventories. J Geophys Res: Earth Surf 122(10):1991–2015

Velasco AA, Ammon CJ, Farrell J, Pankow K (2004) Rupture directivity of the 3 November 2002 Denali fault earthquake determined from surface waves. Bull Seismol Soc Am 94(6B):S293–S299

Wilson FH, Hults CP, Mull CG, Karl SM (2015) Geologic map of Alaska. U.S. Geological Survey Scientific Investigations Map 3340, 197p

Winski D, Osterberg E, Ferris D, Kreutz K, Wake C, Campbell S, Hawley R, Roy S, Birkel S, Introne D, Handley M (2017) Industrial-age doubling of snow accumulation in the Alaska range linked to tropical ocean warming. Sci Rep 7:17869

Yagi H, Sato G, Higaki D, Yamamoto M, Yamasaki T (2009) Distribution and characteristics of landslides induced by the Iwate-Miyagi Nairiku earthquake in 2008 in Tohoku district, Northeast Japan. Landslides 6(4):335–344

Recent Earthquake-Triggered Landslide Events in Central Asia, Evidence of Seismic Landslides in the Lesser Caucasus and the Carpathians

Hans-Balder Havenith

Abstract This chapter presents an overview of earthquake-triggered landslide hazards in Central Asia, with a focus on the Tien Shan and Pamir Mountain Ranges. It essentially compiles information presented by the author in his previous publications as well as other published results. This overview will show that in Central Asian mountain areas, most giant mass movements are likely to have a seismic origin—though it cannot be proved for most of them (see companion report by A. Strom). Here, we will briefly introduce those paleo-cases to establish a link with the Carpathian Mountains and the Lesser Caucasus, where old massive landslides could have a seismic origin as well and thus could provide information on ancient high-magnitude earthquakes for which no written information is available. In recent history, large earthquakes triggered only minor slope failures in the Carpathians and the Lesser Caucasus, while almost all $M > 7$ events that occurred last century in Central Asian mountain ranges triggered numerous landslides, including at least one two-billion m^3 mass movement. However, none of these events can be compared with the 1999 Chi-Chi, the 2005 Kashmir or the 2008 Wenchuan earthquakes, which had triggered many thousands of landslides. The question is then—are such events impossible in Central Asia (maybe due to the much dryer climate than in the regions affected by the cited earthquakes) even though high-magnitude earthquakes are relatively common in the Tien Shan and Pamir? One possible response will be provided by a detailed outline of the largest earthquake disaster that had affected the Central Asian mountain regions in historic times: the Khait event in 1949. Some additional notes will outline the potential for future massive landslide activation during earthquakes in Armenia and Romania.

1 Introduction

Many disastrous earthquakes occurred in the past twenty years in the mountain regions of Taiwan (1999), El Salvador (2001), Pakistan (2005), China (2008), and

H.-B. Havenith (✉)
Liege University, Liège, Belgium
e-mail: hb.havenith@uliege.be

I. Towhata et al. (eds.), *Coseismic Landslides*, Springer Natural Hazards,
https://doi.org/10.1007/978-981-19-6597-5_5

Nepal (2015). As a consequence, these events increased the general attention to earthquake-induced landslides.

In the previous times, landslides triggered by earthquakes did not attract much attention, being regarded as a less important issue as compared with the impact of the ground shaking itself. According to Schuster and Highland (2001), this perception came from the relatively small impact of earthquake-triggered mass movements as compared with the substantial damages induced directly by earthquake shaking. The new idea that earthquake-induced landslides and other ground failures are globally of the 'secondary' importance was established by a study undertaken by Bird and Bommer (2004) who also admitted that consideration on seismic ground failures is essential for some earthquake risk scenarios for correct estimation of total losses. Explicitly, the two El Salvador earthquakes in 2001 ($M_w = 7.7$ in January and 6.6 in February) promoted the idea that 'the most devastating impact … has been the triggering of hundreds of landslides in volcanic soils, …' (Bommer et al. 2002).

Bird and Bommer (2004) completed their study before the $M = 7.6$ earthquake in the Kashmir Mountains on October 8, 2005. For this event, 26,500 people, which is about 30% of the total number of victims (officially 87,350 victims) were killed by coseismic landslides (Petley et al. 2006). Within less than three years after this disaster, the Wenchuan earthquake on May 12, 2008 hit the Sichuan and neighbouring provinces of China and claimed 'more than 15,000 geohazards in the form of landslides, rockfalls, and debris flows, which resulted in about 20,000 deaths' (Yin et al. 2009). This number of landslide victims corresponds again to almost 30% of the official number of total fatalities (69,197).

With these in mind, the present focuses on the mountainous regions in Central Asia, which are the Tien Shan and Pamir where records of case histories are available related to the landslides during the following earthquakes (see locations in Fig. 1): $M = 6.9$ Belovodsk, 1885, $M \sim 7.3$ Verniy, $M = 8.2$ Kemin, 1911, $M = 7.6$ Sarez, 1911, $M = 7.4$ Khait, 1949, $M = 5.5$ Gissar, 1989, and $M = 7.4$ Suusamyr, 1992 (Mushketov 1890; Leonov 1960; Havenith and Bourdeau 2010). Furthermore, the mid-term effect on slope stability and related hazards exerted by small and medium-sized earthquakes is discussed by adding one more landslide case history in the Tien Shan Mountains. Although three more earthquakes with $M > 7$ occurred in the Tien Shan Mountains, they are excluded from the discussion in the paper because very little is known about their impacts. Those events are the 1889 $M > 8.0$ Chilik earthquake north-east from Issyk-Kul Lake, the 1946 $M7.5 \pm 0.2$ Chatkal earthquake in the western Tien Shan, for which Leonov (1965) provided some data about triggered mass movements and landslide dams, and the $M = 7.3$ Markansu earthquake that affected the remote areas of the southern Tien Shan, for which Nikonov et al. (1983) indicated some data about induced ground failures.

The long-term effects of earthquakes in mountain regions deserve special attention. Examples will be shown of a substantial delay of slope failures after earthquakes as well as the post-seismic increase in landslide activity, as was reported after the 2008 Wenchuan earthquake in Sichuan, China. In this perspective, analyses will be made of what were called 'secondary or tertiary effects' of earthquakes in the previous times, such as natural dams and related flooding hazards. In the Tien Shan,

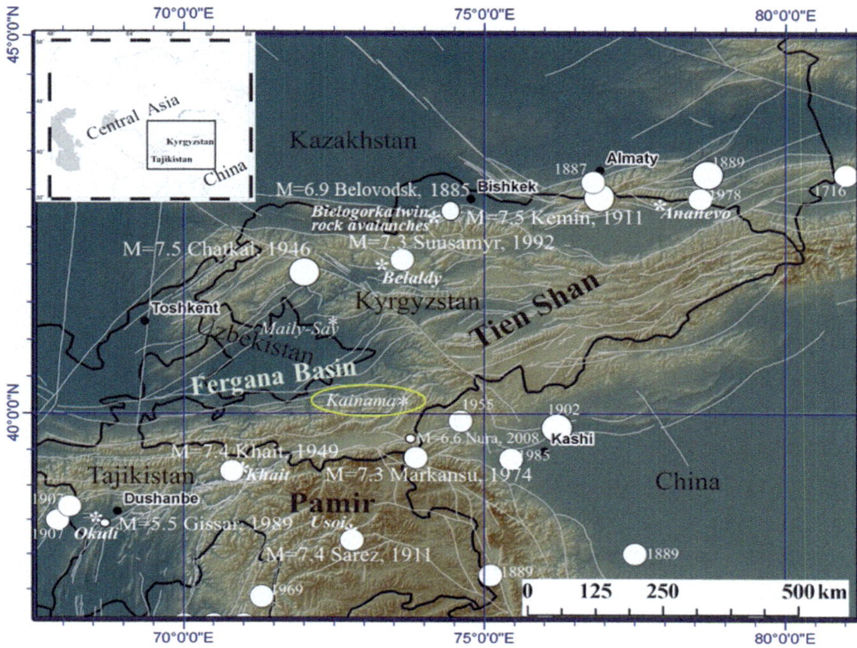

Fig. 1 Map of the Tien Shan and Pamir mountains in Central Asia, with locations of major faults and earthquakes (circles show all $M \geq 6.9$ earthquakes recorded prior to 2010 with the year of occurrence; the magnitude is indicated for analysed events) and related major mass movements (stars) (modified from Havenith and Bourdeau 2010)

furthermore, the most recent massive natural dam was the one formed after the 1992 Suusamyr earthquake. This dam partly failed in 1993 and triggered a long-runout debris flow in the downstream areas (Havenith and Bourdeau 2010).

Massive rockslides that have characteristics of seismic slope failures (initiation high on slopes, very deep scarps, anti-dip failure) can also be found in the Carpathians and the Lesser Caucasus—where they are far less studied (e.g., in comparison with the Greater Caucasus and the Tien Shan–Pamir). Especially in the Carpathians, such ancient massive rockslides could play a major role in seismic hazard assessment, as little historical information is available for the time before the largest event that is known for the Vrancea seismic region (SE Carpathians, Romania), the $M7.8$–7.9 'Great Earthquake' in 1802 (Havenith et al. 2016). Since then, at least 4 other $M > 7$ earthquakes occurred in the same area—but none had triggered landslides that are comparable to the large old deep-seated mass movements that our team is now studying in the SE Carpathians—they could be markers of ancient (>300 years) $M \geq 8$ earthquakes. Below, we will provide some data on two of those massive landslides.

Surprisingly, also relatively little historical information is available for landslides triggered by earthquakes in the Lesser Caucasus—even though Armenia and the neighbouring regions have a rich and long well-described earthquake history. At least, two massive slope failures are more-or-less well documented, one triggered by an earthquake in 1840 from Mount Ararat (formerly located in Armenia, now in Turkey), as well as the rock avalanche of Kamaz Mountain, triggered in 1139 by the Ganja earthquake, damming several lakes, including the well-known Goy-Gol Lake, in Azerbaijan (both case histories are briefly reviewed below). Other well-known giant mass movements are generally much older—but are also worth study (e.g. in Garni), for the same reasons as those presented for Romania: they can help assess seismic hazards, especially for long return periods, notably by setting the maximum possible magnitude.

Most of the information presented here for Central Asia has been extracted from Havenith et al. (2015) and Havenith and Bourdeau (2010), for Armenia from Matossian (2017), and for Romania from Mreyen et al. (2017).

2 Case Histories from Central Asia

The worldwide assessment of the probability of landslides and avalanches by Nadim et al. (2006) is based on the morphological, geological, meteorological, and seismological data. It was evident in their study that all landslide hotspots are located in seismically active mountain ranges. In Central Asia (Fig. 2), the estimated landslide hazard is rated as medium to very high. Further notice was made of some areas in Tajikistan that are prone to the highest mortality risk due to landslides.

In what follows, we will present a short overview of earthquake events that have (or are believed to have) triggered landslides in Central Asia, focusing on effects in the Kyrgyz Tien Shan and two major events in Tajikistan, in 1911 and 1949. However, it should be noted that the 1887 Verniy and 1889 Chilik events also caused widespread and large rock slope failures, mostly in the northeastern Kazakh part of the Tien Shan (some exceeding several tens of millions of cubic metres in volume, according to Mushketov 1890 and 1891); we will not review those in detail as we did not directly study their effects.

2.1 1885 Belovodsk Earthquake

The $M = 6.9$ Belovodsk earthquake in Kyrgyzstan that ruptured the Issyk-Ata thrust fault provided the first documented observation of seismic effects on surface (see Fig. 1). This earthquake-triggered instabilities in the northern slopes of the Kyrgyz Range (Ignatiev 1886; Mushketov 1890) together with the region around the capital Bishkek (formerly called Pishpek). Both contemporary reports and eyewitness testimonies stated that this earthquake caused many slope failures. Two neighbouring rock avalanches occurred near the Bielogorka Village at the epicentral distance of less than 15 km and at about 25 km south of the activated Issyk-Ata fault. Besides the geographic relationship, no direct (temporal) links exist, however, between the

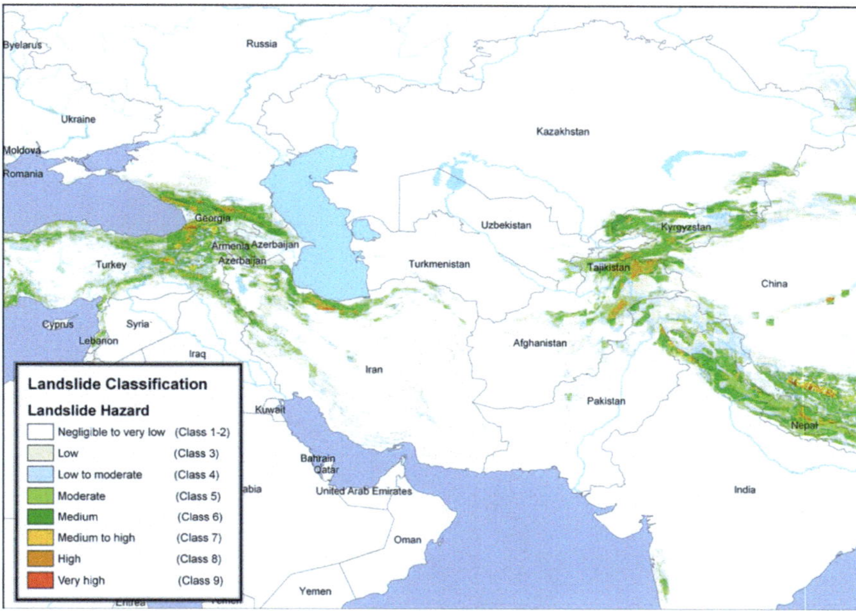

Fig. 2 Landslide hazard map of Central Asia and Middle East (from Nadim et al. 2006)

earthquake and the rockslides. Anecdotal evidence is only provided by an old man who claims that his ancestors were killed by the larger rock avalanche in 1885 (K. Abdrakhmatov, personal communication). Further discussion on the seismic origin of the rock avalanches is presented in the next section.

In August 2017, we had completed a new survey with a drone of the Bielogorka site and a 3D model has been constructed (Fig. 3). In 1998, we had tried to provide a minimum date to those rock avalanches through dendrochronological analyses applied to Archa trees growing on the deposits of the rock avalanches. We had hoped that this minimum age would be close to 110 years, thus providing the missing temporal link between the 1885 earthquake and the occurrence of those massive failures. However, those analyses provided maximum ages of only 50 to 60 years for the Archa trees, which is much more recent than the seismic event. One explanation for this difference is that those trees did not start growing on the deposits (which are purely made of bare rocks) right after failure, but that first some local soils had to develop before trees could grow on them. Another reason advanced by Strom and Abdrakhmatov (2018) is that all trees in that region had been cut by local people to get fuel during the difficult times of the Second World War (as all trees, also outside the rockslide zone, were found to be not older than 55 years in 1998). Anyway, our data showed that the rock avalanches are not 'ancient' and must be younger than a few hundreds of years. We, therefore, concluded that the only event that could have triggered those 'twin rock avalanches' must have been the 1885 event.

Fig. 3 Bielogorka twin rock avalanches (coordinates: N42° 38′ 26″, E74° 16′ 54″). Top left: Google Earth view (2017 image, see also scale and view orientation); other views of 3D drone image model (completed in 2017)

2.2 *1911 Kemin Earthquake*

The Kemin earthquake of Ms = 8.2 in 1911 (January 3) was one of the largest events ever recorded in the Tien Shan, and Bogdanovich et al. (1914) were the first that analysed it. Extensive landsliding is known to have occurred along the activated fault segments over a length of 200 km. Among many, the largest mass movements were three rockslides, first, in the Kemin Valley, second, to the north of Lake Issyk-Kul, and, third, in the upper reaches of the Chilik River. The first rock avalanche (about 15×10^6 m^3) composed of limestone material occurred along the activated Chon Kemin Fault at about 60 km west of the epicentre. This event is known to have buried a village of yourts (traditional tent houses) with 38 inhabitants. The second 'Ananevo' rockslide occurred to the north of Lake Issyk-Kul (at approximately 80 km east of the presumed epicentre) is one of the most prominent consequences of the Kemin earthquake (Fig. 4). This slope failure was triggered at the southern end of a mountain ridge, just above the discontinuous Chon Aksu Fault that was also activated by the Kemin earthquake (see also a lake produced by reactivated fault scarp at about 30 km to the west of Ananevo rockslide in Fig. 5). Delvaux et al. (2001) stated that this section of the Chon Aksu fault is a thrust gently dipping towards the northeast into the collapsed slope. The presence of the fault is verified by the related scarp with a height of 1 m at 3 km WNW of the site; this height increased up to nearly

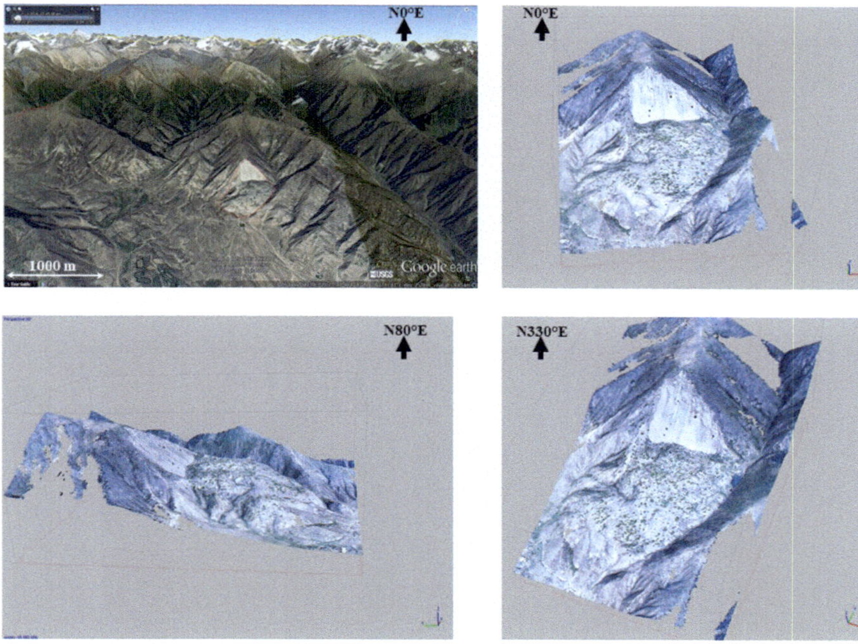

Fig. 4 Ananevo rockslide (coordinates: N42° 48′ 15″, E77° 37′ 40″). Top left: Google Earth view (2017 image, see also scale and view orientation); other views of 3D drone image model (completed in 2017)

10 m at 12 km to the WNW. On the site itself, disintegrated and weathered granitic rocks was found as outcrops at the foot of the southwest-oriented slope within a 100-to-200-m-wide fault zone.

2.3 1911 Sarez Earthquake

The 1911 Sarez earthquake with Ms = 7.6 struck the central Pamir Mountains, Tajikistan, on February 18, 1911.Although earthquakes of this magnitude likely triggered hundreds or thousands of mass movements, only one is well documented. It is the giant Usoi rockslide (Fig. 6). This rockslide fell down from a 4500-m-high mountain down to an elevation of 2700 m in the valley (Schuster and Alford 2004) and formed in the Murgab River a natural dam whose volume was about 2×10^9 m^3. Schuster and Alford (2004) reported that the location of the slide is related to 'a high degree of rock fracturing from previous tectonic activity…a major thrust fault with an unfavourable orientation …and… a series of intensively sheared zones forming geometric setting for a typical wedge failure.' In addition to this, one more slope failure, although being

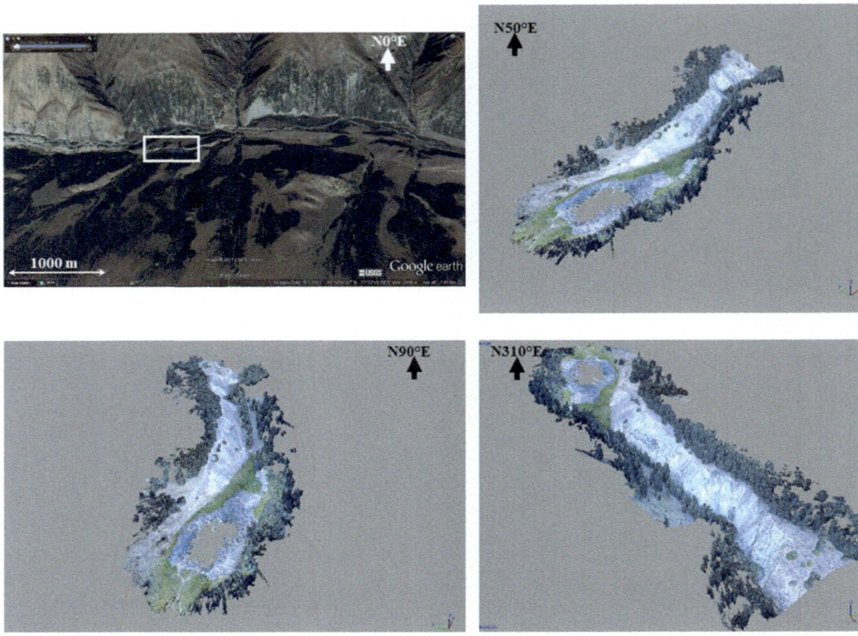

Fig. 5 Chon Aksu fault scarp dammed lake (coordinates: N42° 50′ 15″, E77° 21′ 54″). Top left: Google Earth view (2017 image, see also scale and view orientation); other views (of rectangle shown in Google Earth image) of 3D drone image model (completed in 2017)

several orders of magnitude smaller, occurred on the opposite slope of the ridge in the Kudara River valley.

Lake Sarez and the smaller Lake Shadau had been impounded behind this 600-m-high natural dam, which is the highest dam in the world (see the lower part of Fig. 6).

Various scenarios have been considered of the risk of possible breaching of the rockslide dam due to internal erosion or overtopping as well as to tsunami wave induced by the fall of mass movement into the lake. By investigating those scenarios, Schuster and Alford (2004) stated that the worst-case scenario is the flood wave that could affect more than 5 million people living in the Amu Darya River Basin.

2.4 1949 Khait Earthquake

The 1949 Khait earthquake of $M = 7.4$ hit northern Tajikistan on July 10 and triggered one of the most destructive coseismic landslides in the human history (Leonov 1960). Little, if any, has been known about this disaster out of the former Soviet Union because most information has been published in Russian. It is therefore worthwhile for this article to describe this event.

Fig. 6 Above: Usoi rockslide (coordinates: N38° 17′ 10″, E72° 36′ 20″) outlined in red with dammed Sarez (right) and Shadau (left) lakes (compiled Google earth imagery). Below: field view of dam and source area (behind)—(Photo from: http://wikimapia.org/1739699/Sarez-Lake#/photo/584782)

First, a massive rock avalanche embedded Khait and Kusurak Villages where up to one thousand inhabitants were living (Fig. 7). The exact number of fatalities will be unknown forever because 'during the formidable rule of Joseph Stalin, information about accidents and natural catastrophes was suppressed unless special permission was granted' (Yablokov 2001). This rock avalanche started from the Borgulchak Mountain at an elevation of about 2950 m and moved over more than 6 km until arriving at the inhabited valley at an altitude of 1550 m. The volume of this avalanche was initially estimated to be more than 200×10^6 m^3 (Leonov 1960). Nevertheless, Evans et al. (2009) and Strom and Abdrakhmatov (2018) re-examined this event to state that the total volume was much lower to be about 75×10^6 or 100×10^6 m^3, respectively. Moreover, Evans et al. (2009) observed that a noticeable part of this mass movement consisted of loess, which promoted the mobility and the long runout of this rockslide.

Second, Evans et al. (2009) investigated the situation in the Yasman Valley that is located on the opposite side of the Khait rock avalanche where hundreds of loess earth slides coalesced to form one massive earth flow. It is believed that this loess slide buried about 20 villages. In total, the Khait rock avalanche together with the Yasman loess flows presumably claimed more than 5000 lives during the 1949 disaster.

In addition to the details of catastrophes, Russian geologists described the general conditions of the earthquake-triggered slope failures. For example, Leonov (1960) stated '...involved are also amplification effects that can explain landsliding far from the epicentre...' (translation from the Russian original). We have visited the Khait rock avalanche and Yasman earth flow sites in summer 2010. We also climbed to the

Fig. 7 Khait rock avalanche (coordinates: N39° 11′ 13″, E70° 53′ 05″): viewing to the east from Yasman valley (a coloured version of this photograph in 2005 provided by A. Ischuk, and published in Evans et al. 2009)

scarp area of the rock avalanche that initiated at an altitude of almost 3000 m. There, we found a characteristic feature of massive rock slope failures triggered by seismic shaking: the scarp formed behind the crest (almost 100 m behind the original crest line), thus increasing the total volume of failed material. Such a feature confirms the severe shaking that was responsible for 'behind-crest' ruptures, which was likely due to topographic amplification effects.

2.5 1989 Gissar Earthquake

This earthquake occurred on January 23, 1989, to the south of Dushanbe in Gissar, Tajikistan, and registered Ms = 5.5. It triggered a series of earth flows in loess terrain. The number of victims was at least 250, whereas tens of houses were buried under soil; mostly under Sharora landslide. According to Ishihara et al. (1990), Zerkal (1996) and Voznesensky and Zerkal (1997), those slides were all related to extensive liquefaction, under the effects of the horizontal acceleration being about 0.15 G. All these literatures unanimously attribute this liquefaction to the 'collapsible nature' of the highly porous loess material (a silt-sized deposit with an average content of clay of 15% and a low plasticity).

Among three landslides triggered by this earthquake, the largest landslide was called 'Okuli-bolo' (Fig. 8) and had an estimated volume of 20×10^6 m^3. Ishihara (2002) indicated that 'at least two slides seem to have been triggered independently from the hillsides on the north, which then merged into the main stream of the mudflow.' The sliding plane was located at about 15 m below the ground surface, which was within the saturated part of the 30-m-thick loess stratum. It was also noted by Ishihara et al. (1990) that the scarps of the landslides were closely associated with an irrigation channel situated on the shoulder of the hills. It was discussed, accordingly, that 'water in the channel had been infiltrating in the loess over years, leading to final failure during earthquake due to liquefaction of water-bearing loess layer'. This point is supported by their observation of muddy water oozing from the earth flow.

2.6 1992 Suusamyr Earthquake

More recently, the Ms = 7.3 Suusamyr earthquake on August 19, 1992, triggered a variety of ground failures in the northern-central Tien Shan (Bogachkin et al. 1997). Most of the 50 victims occurred in the remote areas as a consequence of mass movements. Korjenkov et al. (2004) described the types of slope disasters as sagging of mountain slopes, rockfalls, landslides, soil avalanches and flows and mud/debris flows as well as a great variety of gravitational cracks. Along the crest and southern slopes of the Chet-Korumdy Ridge, substantial ground failures occurred.

Fig. 8 Okuli-bolo loess earth flow (coordinates: N38° 29′ 01″, E68° 38′ 12″) (photographs by Ishihara 2002)

Therein, most landslides developed from existing ground instabilities (see views of Chet-Korumdy landslides in Fig. 10).

The induced ground and slope failures were distributed widely over an area of 4000 km^2 around the Suusamyr Basin, whereas the failures were relatively small in the neighbouring mountain ranges; their volume being substantially less than 10^6 m^3. Figure 9 presents a map of all the captured landslides in the Suusamyr region in association with ground failures induced by the Suusamyr earthquake (modified from Korjenkov et al. 2004).

The Belaldy rock avalanche was the only large rock slope failure that occurred on the southern slope of the Suusamyr Range (Bogachkin et al. 1997 and Korjenkov et al. 2004). This avalanche covered a family of a shepherd and a flock of sheep. The pre- and post-earthquake situations at this rockslide sites are illustrated in Fig. 11. This mass movement blocked the Jalpaksu River to form a natural dam whose thickness was about 80 m associated with the width of 500 m and the volume of about 30–35 $\times$ 10^6 m^3. Two more small lakes were impounded behind this dam (area = 200–300 m^2 as of September 1992). Within one year after the earthquake, the water level in the lakes had risen enough to partially breach the dam (Korjenkov et al. 2004). This dam breaching resulted in mud and debris flows over a distance of 20 km, causing a profound extent of damage to infrastructures in the Toktogul Region. A satellite image in 1993 of the Belaldy site (Fig. 9) indicates the upper part of the debris flow just below the dam. Currently, the dam still has a volume of about 25–30 $\times$ 10^6 m^3, as also estimated by Strom and Abdrakhmatov (2018).

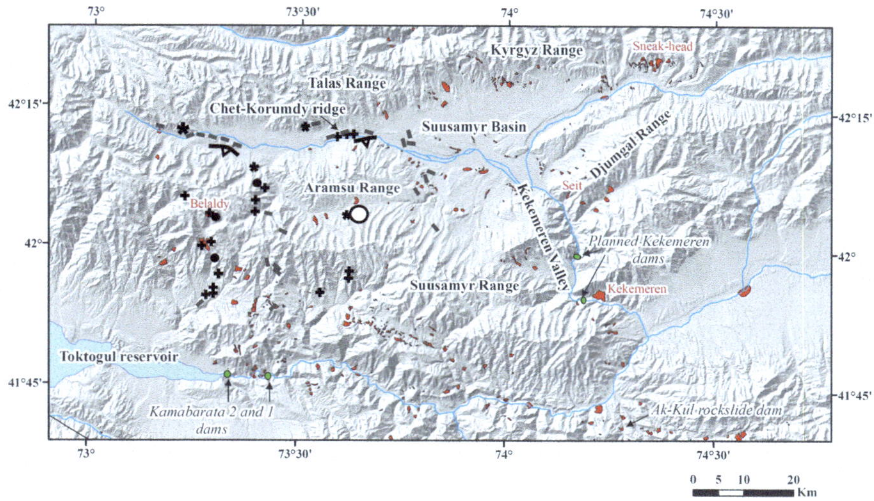

Fig. 9 Location of all detected landslides in the Suusamyr region (red), including several ground failures (partly hidden by the black symbols that specifically identify 1992 ground deformations) as a consequence of the Suusamyr earthquake (modified from Korjenkov et al. 2004—circle: $M = 7.3$ epicentre; black lines with triangle: 1992 fault scarps; crosses: rockslides and landslides; stars: mud eruptions; grey bars: ground and slope fractures, black dots: rock falls)

2.7 2004 Kainama Landslide—Delayed Effect of Local $M = 4.5$ Earthquakes?

The Kainama earth flow of 220,000 m^3 in volume occurred in the southern part of the Fergana Basin (location highlighted in Fig. 1) on April 26, 2004, and destroyed 12 houses, while claiming 33 lives. In total, 64 families lost their homes due to entirely or partially destroyed houses. This landslide was generated within a loess layer and developed into a very rapid flow with a long runout. It seems important that, in 2001, there was no clear sign of instability as illustrated in Fig. 12. In spring of 2003, several cracks appeared in the upper part of the slope.

The main landslide in April 2004 was preceded by an evolution of seismic activity. For example, on 2004 March 26, an earthquake of Ms = 4.7 occurred at 40 km SW of the Kainama site, associated with an estimated PGA at the site being 0.05 G). Two more smaller earthquakes (Ms = 4.2) were registered on April 8th and 9th at about 50 km NW of Kainama. After these earthquakes, new slope fractures were found. Thus, the time lapse between seismic events and the eventual slope failure was more than two weeks.

The delay of the massive slope failure after the earthquakes is presumably related with the changes in the local environments. At the end of March 2004, the mountains were still covered by snow and the groundwater level was low. In April, the melting snow and spring rains initiated a rapid rising of groundwater level. Note that the water infiltration into slopes had been probably accelerated through the new subsoil

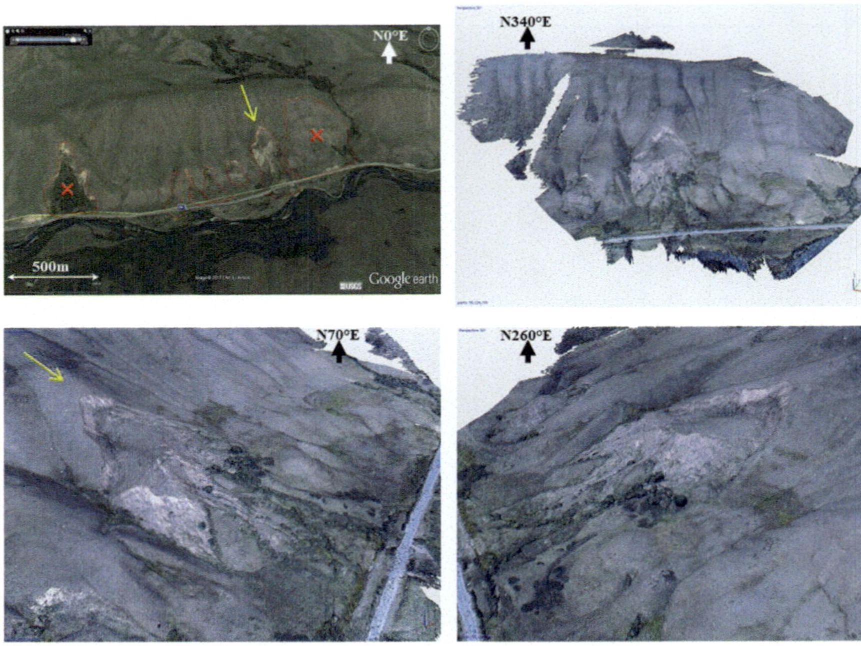

Fig. 10 Suusamyr landslide (coordinates: N42° 12′ 25″, E73° 36′ 33″) on Chet-Korumdy ridge (below yellow arrow). Top left: Google Earth view (2013 image, see also scale and view orientation); others views of 3D drone image model (completed in 2017). Note, the left and right landslides (marked by a red cross) in the top left view were not triggered by the 1992 earthquake; they existed before and have not been reactivated

fractures created by the earthquakes. To confirm the important role of this sequential process that promoted many landslides in this seismic area, numerical studies have been undertaken. Related results were published in Danneels et al. (2008).

2.8 The Most Recent Large Earthquakes in the Tien Shan and Pamirs

For more recent large earthquakes, we have almost no information on triggered landslides. For the 2008 Nura earthquake ($M = 6.6$) in south Kyrgyzstan, we just know that there had been no failures on major highways reaching the area—as we had overviewed the region using satellite imagery taken just after the event—to indicate to authorities if emergency help could be brought by trucks to a village where 75 people had been killed (and 150 were injured) or if helicopters were needed (notably due to obstruction of main roads by rockfalls). We had confirmed that no major rockfalls had occurred along the main roads reaching the remote large Alay Mountain Valley hit by the earthquake.

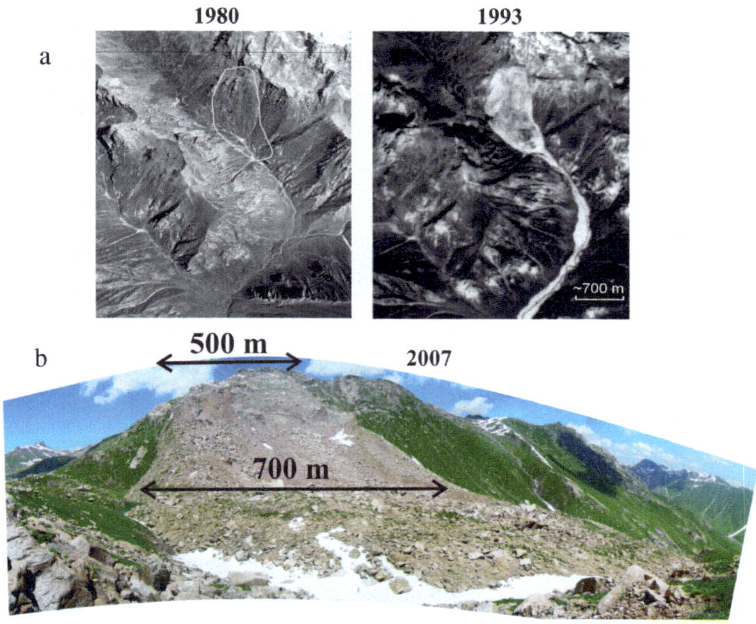

Fig. 11 Belaldy rock avalanche (coordinates: N42° 03′ 39″, E73° 16′ 44″). **a** Aerial photograph of the site acquired in 1980 (left), before the 1992 Suusamyr earthquake and satellite image of 1993 (right), after the earthquake. **b** Photograph of the rock avalanche made in 2007 (by the author)

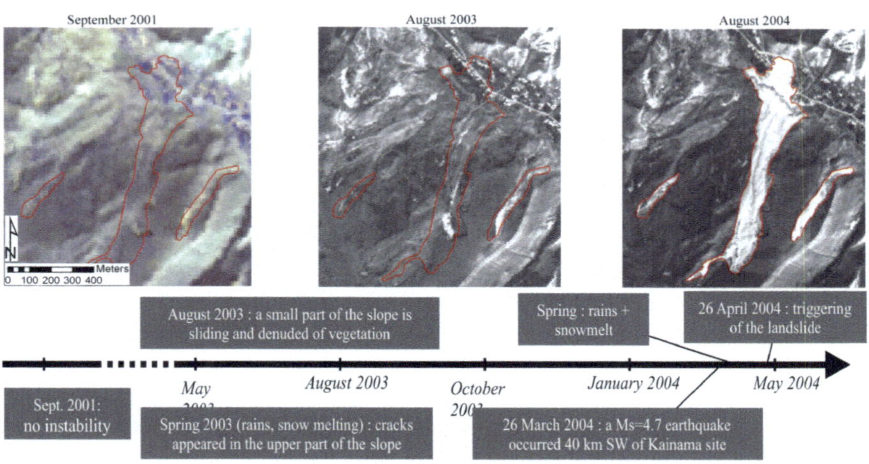

Fig. 12 Evolution of the Kainama landslide (coordinates: N40° 16′ 22″, E73° 33′ 51″) from 2001 (ASTER satellite image, left), to 2003 (SPOT image, middle) and 2004 (SPOT image, right)

Furthermore, some information may be added on the last $M = 7.2$ earthquake that occurred in the Pamirs in December 2015. This event triggered a few minor or partial slope failures around Lake Sarez, but massive rockslides or any serious effect on Usoy dam (besides a general lake level rise of about 20 cm) could not be identified by Ischuk and Strom (2016).

3 Case Histories of Possible Earthquake-Triggered Landslides in the Lesser Caucasus (Armenia and Neighbouring Regions)

3.1 1679 Garni Earthquake

The 1679 Garni earthquake was one of the most destructive earthquakes to have occurred in Armenia. According to historical sources, the intensity was between VIII and X and its magnitude was estimated, accordingly, between 5.5 and 7 (Guidoboni et al. 2003). It is likely that this earthquake reactivated a series of large landslides, including the giant ancient Garni landslide. The scarp is composed of breccia with some fractured large pebbles (Fig. 13, left). Those neatly fractured pebbles within the breccia inside the scarp hint at a rapid massive triggering of slope failure, as can be expected during seismic shaking. According to the observations in the field, our primary interpretation and explanations by Dr. Arkady Karakhanian in September, 2016, the landslide evolution is likely to be related to local/regional extensional tectonic deformation. The (possible seismic) triggering of the main part of the landslide, with development of rotational movements and the formation of gravitational grabens, might have occurred 2000 to 3000 years ago. The age of the landslide is quite difficult to determine due to the arid climate, which preserves the landslide morphology (the age could be easily underestimated). However, horizontally lying foundations of an ancient farm of Roman times (Lat: 40.1513—Long: 44.6860), dated near 0 AD (that apparently did not move) are still visible on the landslide body (A. Karakhanian, personal communication). They indicate that the main mass movement most likely predates the Roman era. The presence of a small basin between the gravitational graben in the breccia and the lower main part of the old landslide might be the consequence of local extensional tectonic movements that also favoured slope instability. The flat surface of this basin might be due to the filling of the basin over the last 2000 years (or more). This could explain the structure of the Garni landslide presenting a marked relief in the upper part and a smooth topography in the lower part.

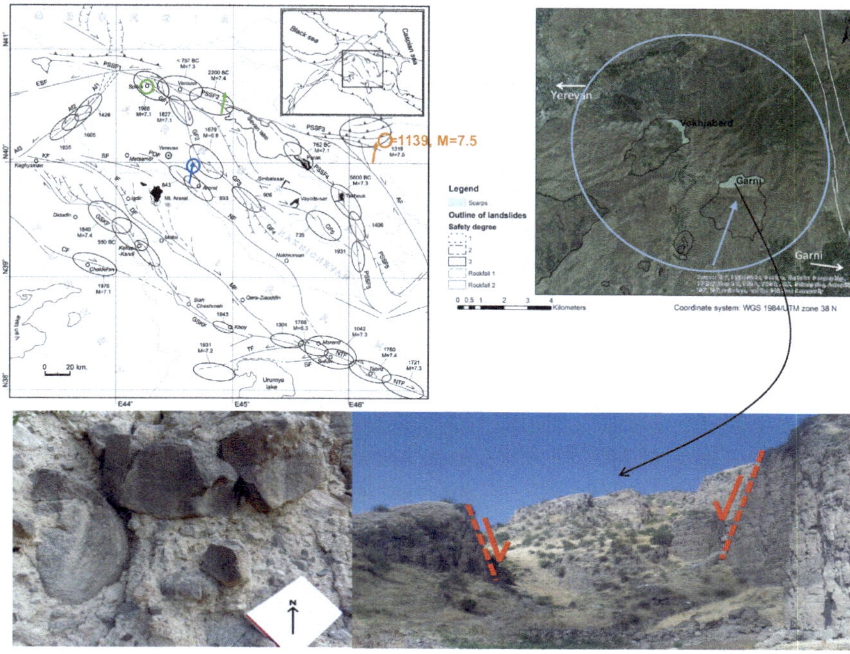

Fig. 13 Top left: Garni earthquake and landslide location (see blue circle and arrow); see also approximate location of the 1139 Ganja earthquake epicentral area (orange circle) and of the related seismically triggered Kapaz mountain rock avalanche (orange arrow) that dammed several lakes (map by Karakhanian et al. 2004. Note that the map probably has the wrong year for the $M = 7.5$ event in the eastern Lesser Caucasus, marked as 1319, which should be 1139 for the Ganja earthquake in that location). The green circle marks the epicentral area of the 1988 Spitak earthquake and the green arrow the approximate location of the Dilijan landslides. Top right: Voghjaberg and Garni landslides, likely initially triggered by earthquakes; the Garni landslide (coordinates: N40° 09′ 03″, E44° 41′ 09″) was apparently reactivated during the 1679 event. Below: volcanic breccia (left) and graben in the scarp area

3.2 *1840 Ararat Earthquake*

The 1840 Ararat earthquake, with an estimated magnitude of 7.4, occurred in the Ararat depression (Ambraseys and Melville 1982) and created a 72-km-long surface rupture. This earthquake killed 10,000 people. A landslide was triggered from the summit of Ararat, mainly destabilized by the earthquake and, possibly, by an eruption (Stepanian 1964; Karakhanian et al. 2002). According to historical sources and Karakhanian et al. (2002), an explosive Bandai-type phreatic eruption occurred near the Ararat summit. A cloud and pyroclastic flows would have been induced by that eruption. Some secondary volcanic effects were observed: an eruptive rain which left a blue liquid, with a harsh smell, and lahars which were triggered from the summit and travelled along the Arkory Canyon with a speed of 175 m/s (see Fig. 14). Added to this, the landslide destroyed the monastery of St. James (St. Jakob) and the Arkory village. The movement stopped at an elevation of 900 m and formed a natural dam. The dam broke a few days later and the resulting flood destroyed other surrounding villages. Other landslides dammed the Arax and also created floods. Liquefaction and subsidence of the ground were also generated in the Ararat depression, along the Arax, destroying some villages (Karakhanian et al. 2002).

The most damage was caused by secondary effects: the volcanic eruption triggering lahars, the landslide dams followed by floods, and earthquake-induced effects such as ground subsidence and liquefaction. The secondary effects killed 4000 people (Karakhanian et al. 2004). This earthquake destroyed many villages near the Ararat. Other cities and villages were slightly damaged, as Etchmiadzin (Armenia), Yerevan and Garni and the cities of Van (eastern Turkey), Tbilisi (capital of Georgia), Tabriz (northern Iran), and Gyumri (northern Armenia) felt the event (Ambraseys and Melville, 1982).

A very impressive historical event—but for which, surprisingly, little information is available in the literature—is the Kapaz Mountain rock avalanche (actually, two avalanches) triggered by the 1139 Ganja earthquake. Its location is shown by an

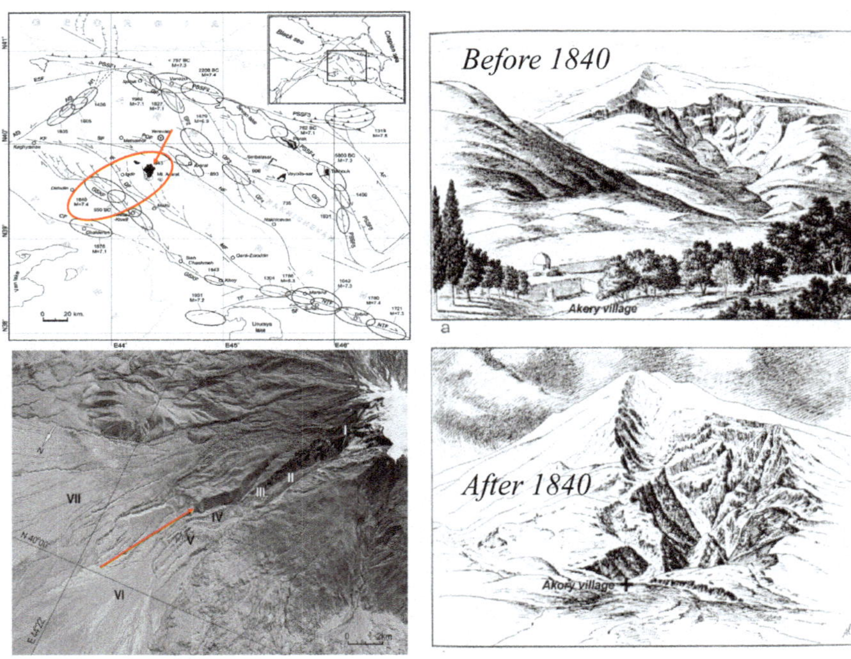

Fig. 14 Top left: 1840 earthquake and Arkory canyon failure on Ararat (see red arrow). Right: Drawings of Arkory canyon before and after the 1840 event. Lower left: Corona image of NE part of the Greater Ararat with Arkory canyon (coordinates: N39° 44′ 20″, E44° 19′ 53″). See: II—2, deposits of the first flow-landslide of July 2, 1840; III—St. Jakob's Monastery; IV: old village of Akory; V: today's village; VI—deposits of the July 6, 1840, flow-landslide; VII—deposits of the earlier flow-landslides (from Karakhanian et al. 2002)

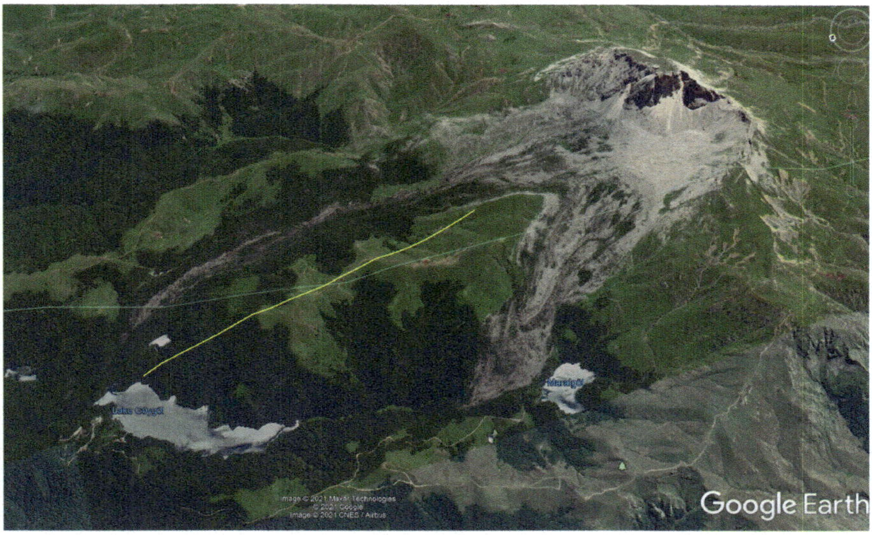

Fig. 15 Google Earth view from northwest of the Kapaz mountain rock avalanche(s) (coordinates: N40° 23′ 30″; E46° 19′ 40″) triggered by the 1139 Ganja earthquake in the southeastern Lesser Caucasus, which dammed several lakes, including the two larger ones of Goy-Gol (lower left corner) and the Maral-Gol (middle lower right). See also yellow line of 4 km length (for scale)

orange arrow and circle, respectively, in Fig. 13); it is considered the most powerful catastrophic event in the Lesser Caucasus (Nikonov and Nikonova 1986). The two rock avalanches dammed in total eight still existing lakes, including the two larger ones of Goy-Gol and Maral-Gol. The Google Earth view in Fig. 15 shows that the eastern rock avalanche part (damming Goy-Gol) is about 7 km long, while the western part (damming Maral-Gol) is almost 4 km long. The total volume of both parts would be about 500×10^6 m^3 (at least).

The most recent disastrous earthquake in the Lesser Caucasus, the 1988 Spitak earthquake with a magnitude of 6.8 had also (re)activated larger landslides. According to Boynagryan (2009), several landslides were (re)activated in the Dilijan area (see small green arrow in upper left part of Fig. 13, at a distance of about 50 km from the epicentre); he also indicates that the large Voghjaberg landslide (volume of more than 100×10^6 m^3) shown in the Google Earth view in the right upper part of Fig. 13 was reactivated during the 1988 event. Boynagryan (2009) also considered the activation of the Nubarashen landslide (in the south of the capital Yerevan, directly downstream from Garni landslide) soon after the 1988 earthquake, a delayed consequence of the seismic event (see destroyed graves on the landslides in Fig. 16).

Fig. 16 Destroyed graves on the Nubarashen landslide, activated soon after the 1988 Spitak earthquake (from Boynagryan 2009)

4 Case Histories of Possible Earthquake-Triggered Landslides in the Carpathians (Romania)

4.1 Recent M > 7 Earthquakes

The Vrancea seismic region was investigated intensely by Havenith et al. (2016). The southeastern Carpathian Mountains is the area where the largest deep-focus earthquakes occur in Europe. In the last two centuries, four earthquakes with $M_w >$ 7.4 occurred in this region. Georgescu (2003) stated that the 1802 earthquake (‘*the Great Quake*’) was presumably the most severe event ever recorded in this area in terms of magnitude ($M_w = 7.8$–7.9 at 150-km depth). This event was also one of the most important one in terms of MSK intensity scale ($I_{0I} = \mathrm{IX} - \mathrm{X}$). Moreover, the historical earthquakes in Vrancea triggered landslides at very large distances (250–300 km). For example, the November 1940 earthquake ($M_w = 7.7$ at 150-km depth) induced numerous geohazards such as ground fracturing and at least 40 shallow and medium-seated landslides accompanied by groundwater level disturbances (Radu and Spânoche 1977). This disaster proneness was further aggravated by the wet subsoils as a consequence of the overwhelmingly humid climate in 1937–1940 when annual mean precipitation was exceeded by 13–38%. Most of the above-mentioned disasters were triggered within an elliptical zone (300 by 200 km) which had its longer axis oriented in the NE-SW direction and was marked by the 8 degrees MSK intensity isoline. According to Angelova (2003), all the Vrancea earthquakes with $M > 5.5$ affected the ground and produced damage throughout Bulgaria. In particular, the 1977 event ($M_w = 7.4$, 94 km depth) is known by a number of coseismic phenomena such as cracks and fissures at epicentral distances of 160–290 km, falls from coastal cliffs, caves, steep river banks, limestone ridges at epicentral distances of up to 270–380 km and slides at epicentral distances of 210–260 km. Twelve coseismic slope failures were detected showing their surface area of 1–10 ha, and most of them occurred during the earthquake or immediately after it, while only a few (namely 4) were post-seismic failures 24 h to 5 weeks after the earthquake. Thus, the 1977 earthquake triggered a relatively small number of landslides in spite of its great magnitude (7.7). This was attributed to the relatively dry climatic conditions during that period, marked by a pluviometric record (global rainfall characteristics) that was 40% lower than the monthly average of 30–90 mm.

4.2 Ancient Possible M8+ events

As indicated above, recent earthquakes in Vrancea triggered landslides at very large epicentral distances (250–300 km), e.g., the events in 1940 and 1977, but none of the landslides exceeded 1 million m^3 in volume. The Balta and Eagle’s Lake landslides (both with a volume of about ~5–10 million m^3) that we have been investigating for several years represent deep-seated mass movements with the most pronounced

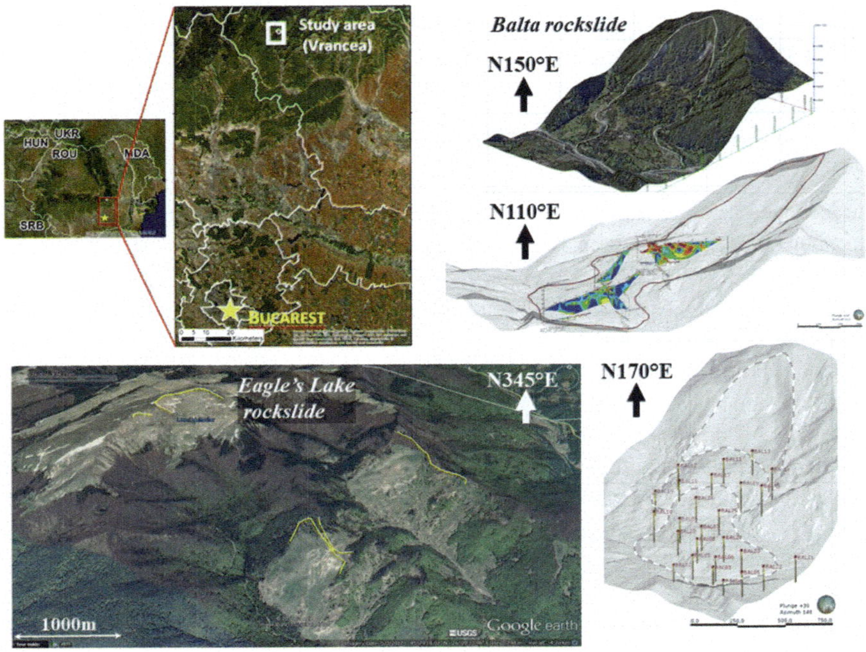

Fig. 17 Top left: Vrancea-Buzau seismic region location. Right part: Balta rockslide views and geophysical models (from Mreyen et al. 2017). Lower left: Google Earth view of Eagle's Lake rockslide and other ancient, likely earthquake-triggered slope failures in the Siriu Mountains (same area)

concave and steep scarps in this area. Furthermore, they are marked by a massive body, rock structures (rock layering) favouring static stability, a scarp initiation from the mountain top and with no river at the base of the initial slope that could have eroded the mountain toe before massive failure. Figure 17, we present some views of those two ancient rock slope failures which have most likely been triggered by very large, possibly $M \geq 8$, earthquakes in the SE Carpathian Mountains (see e.g. Mreyen et al. 2021). For these landslides, we have also collected drone images, and 3D models are now being processed.

5 Discussion and Conclusion

Extensive studies of earthquake-triggered landslides have been carried out in the Central Asian mountain regions. Those show that related events occur on a regular basis—with a return period that is close to the one of $M > 7$ earthquakes (as very large landslides in these areas are typically triggered by those earthquakes). If we combine the Tien Shan and Pamir Mountains, $M > 7$ events potentially triggering landslides occur once every 15 years. Over the past 130 years of well-known seismic hazard history in Central Asia (in practice starting with the 1885 Belovodsk event), just several large earthquakes (in 1887, in 1911, in 1949) caused massive activation of landslides. Among them, the $M = 7.4$ Khait event in 1949 was the most disastrous one, as it had triggered numerous large slope failures that killed several thousands of people. All other cases are marked by only a few larger (> 10^7 m^3) seismically triggered landslides. The main reason for this is that the areas that are most prone to landslide activation, i.e., the areas covered by thick loess deposits are generally not marked by the highest seismic hazard. However, there are a few exceptions: the seismically very active southern border of the Tien Shan where thick loess deposits also can be found (and where the Khait earthquake occurred) and the northern and eastern border of the Fergana Basin. Around the Fergana Basin, no large earthquakes occurred in recent historic times (since 1885); the largest known events had magnitudes smaller than 6.5—though our former seismic hazard estimates (Abdrakhmatov et al. 2003) clearly showed that NE Fergana is marked by a high seismic hazard. The probably most severe event would be an $M > 7$ earthquake along the central segment of the Talas-Fergana Fault, in the east of Babash-Ata Mountain (to the SE of the site where a prehistorical earthquake had, most likely triggered the Karasu rock avalanche directly within the fault zone). This densely populated eastern rim of Fergana Valley is covered by thick loess deposits and extremely prone to landslides. It is also relatively close to the Tala-Fergana Fault (~40 km) and thus could be hit by a massive activation of landslides in loess deposits that may have extreme impacts (similar to those of the 1949 Khait earthquake).

Currently, loess landslides are mainly triggered by climatic influences around the Fergana Basin and in the Tajik Depression south from the Gissar Range. We also know of earthquake effects on slope stability that prepare a later massive failure—see the Kainama case. There are also proven and monitored seismic activations of slope fractures that many years later were at the origin of large landslides. Earthquakes along the Talas-Fergana Fault are extremely rare (return period of >1000 years for $M \gg 7$ earthquakes). More frequent activations of loess landslides (combined with other soft materials) by earthquakes can also be expected from local $M \sim 6$ events. Even though less massive and extended, the Gissar case history showed that $M < 6$ earthquakes also can trigger widespread liquefaction and slope failure in loess deposits. Furthermore, Niyazov and Nurtaev (2009) showed that loess landslide activation along the western border of the Tien Shan could be triggered by distant deep foci earthquakes (in the Hindukush or Pamir). In 1997, there is also a proved distant activation of a landslide in the Kyrgyz part of the Fergana Basin rim. Therefore, we believe that most final climatically triggered slope failures in loess deposits are just the result of accumulated strain caused by previous seismic shaking (combined with long-term climatic influences).

Finally, it is just a question of time that a Khait-type event will also occur in southern Kyrgyzstan, possibly affecting areas in Uzbekistan as well. Along the eastern Fergana Basin rim, we have identified at least 3 slopes that could fail massively during an $M > 7$ event earthquake (with volumes exceeding 100 million m^3) which could occur along the central Talas-Fergana Fault segment. A massive rock slope failure could be triggered on Babash-Ata Mountain where a prehistoric > 1 km^3 failure has already occurred (coordinates: N41° 23′ 41″ E72° 56′ 12″), and one > 200 million m^3 soft sediment landslide (coordinates: N41° 14′ 41″ E72° 03′ 12″) could be triggered from an unstable slope in front of the village of Oogan-Talaa, and one from a landslide-prone slope in the east of Taran-Bazar (approximate coordinates: N41° 06′ E73° 25′). An example of giant mass movements likely to have been triggered by a Talas-Fergana Fault earthquake is the prehistoric Sary-Chelek rockslide with a volume of >5 km^3 damming Sary-Chelek Lake.

In addition to the possibly large impacts of a $M \geq 7$ earthquake in eastern Fergana within the densely populated loess-covered area, what would be the consequences of an earthquake and related triggered extremely large rock slope failures in the higher mountains (e.g., Babash-Ata in the Tien Shan or a new 'Usoy-type' rockslide in the Pamirs)? The direct consequences would probably be more limited as fewer people live in the higher mountains. But nobody could predict the later consequences if a large dam forms on a river, just as now for Usoy.

For the Tien Shan and the Pamirs, future research should focus on the potential of massive loess landslide activations by earthquakes near the northern and eastern Fergana Basin rim and along the southern border of the Tien Shan. In the higher mountain areas (with minor loess cover), it cannot be easily predicted where the next Usoy-type rockslide failure would occur (in the Tien Shan or Pamirs). Therefore, research should first focus on the total hazard assessment for the already existing Usoy-type rockslide—thus, on the landslide dam stability! At the same time, other large rockslide dams should be studied in these regions, such as Sary-Chelek, Kara-Kul, and Kutman-Kul (located at the head of Mailuu-Suu Valley). Serious hazard studies (including extensive dating of ancient events and identification of zones prone to large future massive failures) would certainly require investment of more than 10 million USD. The two other regions presented here, the Lesser Caucasus and the SE Carpathians are clearly less affected by earthquake-triggered landslides than the Tien Shan and Pamirs. However, such phenomena also exist there and should not be neglected. The particularity of the Lesser Caucasus is the presence of ancient volcanoes. At present, there is no sign that those structures increase regional landslide susceptibility—but, the 1840 Ararat earthquake showed that a massive slope failure can occur at least on that volcano (the highest mountain in that region that presents also the steepest slopes). For instance, the Armenian part of the Lesser Caucasus (most studied by our team) is covered by hundreds of large rockslides, but none occurred in recent history as a consequence of an earthquake. The Garni and Voghjaberg rockslides, which have most likely been triggered by an earthquake more than 1000 years ago (together or by different events?), and certainly also the Kapaz Mountain collapse in 1139 (in the Azerbaijani part of the Lesser Caucasus), indicate that case histories similar to those known from Central Asian mountain regions could also occur in the Lesser Caucasus—though less frequently. Here, more focus should also be on landslides that could be triggered along the Pambak-Sevan-Syunik Fault bordering Sevan Lake in eastern Armenia. In Kyrgyzstan, there is practically no hazard related to subaquatic landslides in Issyk-Kul Lake or to subaerial landslides impacting the lake (due to the gentle slopes bordering the lake). However, there is clearly a much higher hazard related to such phenomena in the eastern part of Sevan Lake, where it is bordered by the Pambak-Sevan-Syunik Fault that creates very steep subaerial and subaquatic slopes. Here, full cascading hazard scenarios should be studied, including the triggering of tsunami waves by subaerial or subaquatic mass movements.

An interesting aspect of studying possibly earthquake-triggered landslides in the SE Carpathian Mountains is that, compared to other regions in Europe affected by similar high seismic hazards, very little is known about the impacts of very large earthquakes in this region. The region is known to produce the largest intracontinental earthquakes in Europe, but their effects are mainly studied for the city of Bucharest, not for the mountainous areas which are actually relatively densely populated. We have mainly highlighted the importance of ancient large rockslides for estimating the maximum possible earthquake magnitude for the SE Carpathians, as the studied rockslides are much larger than the landslides known to have been triggered by the last century's M7.4–7.7 events. High uncertainties exist around the maximum possible earthquake magnitude, due to the fact that classical paleoseismology cannot be applied in the SE Carpathians, as the activated faults are too deep and do not create surface ruptures. Therefore, an adapted paleoseismological approach is needed—working with giant rock slope failures—if it can be proven that they really had been triggered by seismic shaking: this is what we are doing now. And, once this work is advancing well, the effects of $M \sim 8$ earthquakes should also be estimated for the mountain regions (in addition to the areas bordering the mountains, and to Bucharest). Inside the area marked by the highest seismic hazard is the city of Nehoiu, as well as the Siriu dam directly upstream. In the past, a large rockslide has already occurred within the dammed lake area. An $M \sim 8$ earthquake could cause problems to both the dam and to slope stability in the lake area, both representing an extreme risk source for the city of Nehoiu and the neighbouring relatively densely populated downstream areas.

References

Abdrakhmatov K, Havenith H-B, Delvaux D, Jongmans D, Trefois P (2003) Probabilistic PGA and arias intensity maps of Kyrgyzstan (Central Asia). J Seismolog 7:203–220

Ambraseys NN, Melville CP (1982) A history of Persian earthquakes. Cambridge University Press, New York, p 219

Angelova D (2003) Influence of the Vrancea earthquake from 04 03 1997 on terrains with established paleoseismic events in Bulgaria (preliminary data). In: Lungu D et al (eds) Proceedings of the international conference on earthquake loss estimation and risk reduction, Oct 24–26 2002, Bucharest, Romania, pp 303–309

Bird JF, Bommer JJ (2004) Earthquake losses due to ground failure. Eng Geol 75(2):147–179

Bogachkin BM, Korzhenkov AM, Mamyrov E, Nechaev YuV, Omuraliev M, Petrosyan AE, Pletnyov KG, Rogozhin EA, Charimov TA (1997) The structure of 1992 Susamyr earthquake source based on its geological and seismological manifestations. Izvestiya, Phys the Solid Earth 33(11):867–882

Bogdanovich MMC, Kark J, Korolkov B, Muchketov D (1914) Earthquake of the 4 January 1911 in the northern districts of the Tien Shan. Tr Geol Com Ser 89:1–270

Bommer JJ, Benito MB, Ciudad-Real M, Lemoine A, Lopez-Menjivar MA, Madariaga R, Mankelow J, Mendes de Hasbun P, Murphy W, Nieto-Lovo M, Rordiguez-Pineda CE, Rosa H (2002) The El Salvador earthquakes of January and February 2001: context, characteristics and implications for seismic risk. Soil Dyn Earthq Eng 22:389–418

Boynagryan V (2009) Landslides in Armenia. Rom J Geogr 53:197–208

Danneels G, Bourdeau C, Torgoev I, Havenith H-B (2008) Geophysical investigation and numerical modelling of unstable slopes: case-study of Kainama (Kyrgyzstan). Geophys J Int 175:17–34

Delvaux D, Abdrakhmatov KE, Lemzin IN, Strom AL (2001) Landslide and surface breaks of the 1911 M 8.2 Kemin earthquake. Landslides 42(10):1583–1592

Evans SG, Roberts NJ, Ischuk A, Delaney KB, Morozova GS, Tutubalina O (2009) Landslides triggered by the 1949 Khait earthquake, Tajikistan, and associated loss of life. Eng Geol 109(3):195–212

Georgescu ES (2003) The partial collapse of the Coltzea Tower during the Vrancea earthquake of 14–26 October 1802: the historical warning of long-period ground motions site effects in Bucharest. In: Lungu D, Wenzel F, Mouroux P, Tojo I (eds) Proceedings of the international conference on earthquake loss estimation and risk reduction, Oct 24–26 2002, Bucharest, Romania, pp 303–309

Guidoboni E, Haroutiunian R, Karakhanian A (2003) The Garni (Armenia) large earthquake on 14 June 1679: a new analysis. J Seismolog 7:301–328

Havenith H-B, Bourdeau C (2010) Earthquake-induced hazards in mountain regions: a review of case histories from Central Asia. Geol Belg 13:135–150

Havenith H-B, Strom A, Torgoev I, Torgoev A, Lamair L, Ischuk A, Abdrakhmatov K (2015) Tien Shan geohazards database: earthquakes and landslides. Geomorphology 249:16–31

Havenith H-B, Torgoev A, Braun A, Schlögel R, Micu M (2016) A new classification of earthquake-induced landslide event sizes based on seismotectonic, topographic, climatic and geologic factors. Geoenviron Disast 3(6):2–24

Ignatiev IV (1886) Earthquake in the Tokmak district in 1885. Proc Imp Russ Geogr Soc 22(2):150–164 (in Russian)

Ischuk N, Strom A (2016) Geological effects of the earthquake at the Sarez Lake on the 7th of December 2015. Georisk, 47–56 (in Russian)

Ishihara K (2002) Failure of slopes and soil property characterisation. Department of Civil Engineering, HKU and the Geotechnical Division, HKIE: The Second Lumb Lecture October 24, 2002

Ishihara K, Okusa S, Oyagi N, Ischuk A (1990) Liquefaction-induced flow slide in the collapsible deposit in the Soviet Tajik. Soils Found 30:73–89

Karakhanian A, Djrbashian R, Trifonov V, Philip H, Arakelian S, Avagian A (2002) Holocene-historical volcanism and active faults as natural risk factors for Armenia and adjacent countries. J Volcanol Geoth Res 113(1):319–344

Karakhanian AS, Trifonov VG, Philip H, Avagyan A, Hessami K, Jamali F, Bayraktutan S, Bagdassarian H, Arakelian S, Davtian V, Adilkhanyan A (2004) Active faulting and natural hazards in Armenia, eastern Turkey and northwestern Iran. Tectonophysics 380(3–4):189–219

Korjenkov AM, Mamyrov E, Omuraliev M, Kovalenko VA, Usmanov SF (2004) Rock avalanches and landslides formed in result of strong Suusamyr (1992, M = 7,4) earthquake in the Northern Tien Shan. Test structures for mapping of paleoseismic deformations by satellite images. In: Buchroithner MF (ed) High mountain remote sensing cartography VII (HMRSC VII), Institute for Cartography of the Dresden University of Technology. Kartographische Bausteine, Band 23, Dresden, 19 p

Leonov NN (1960) The Khait, 1949 earthquake and geological conditions of its origin. In: Proceedings of academy of sciences of the USSR, Geophysics, vol 3, pp 409–424 (in Russian)

Leonov NN (1965) Role of geological structure in the formation of residual deformations of soils during earthquakes. Seismic Microzoning, Trans Inst Phys Earth RAS 36(203):132–136 (in Russian)

Matossian A (2017) Identification of giant mass movements in the Lesser Caucasus and assessment of their spatial relationship with major fault zones and volcanoes. MSc thesis in Geology, University of Liege, Belgium, 108p

Mreyen AS, Micu M, Onaca A, Cerfontaine P, Havenith H-B (2017) Integrated geological-geophysical models of unstable slopes in seismic areas. In: Mikos et al (ed) Proceedings advancing culture of living with landslides. Springer, pp 269–279

Mreyen A-S, Cauchie L, Micu M, Onaca A, Havenith H-B (2021) Multiple geophysical investigations to characterize massive slope failure deposits: application to the Balta rockslide, Carpathians. Geophys J Int 225(2):1032–1047. https://doi.org/10.1093/gji/ggab028

Mushketov IV (1890) The Verniy earthquake of 28/05 (9/06) 1887. In: Proceedings of geological committee, St. Petersburg, 10 (in Russian)

Mushketov IV (1891) The Chilik earthquake. In: The materials for study of earthquakes in Russia, St. Petersburg, 1, pp 11–42 (in Russian)

Nadim F, Kjekstad O, Peduzzi P, Herold C, Jaedicke C (2006) Global landslide and avalanche hotspots. Landslides 3:159–173

Nikonov AA, Nikonova KI (1986) The strongest earthquake in Transcaucasia on September 30, 1139. In: Reconstruction by historical, historical-architectural and archaeological materials—Topics of engineering seismology, 27, pp 152–183 (in Russian)

Nikonov AA, Vakov AV, Veselov IA (1983) Seismotectonics and earthquakes in the Pamir—Tien Shan convergence zone. Nauka, Moscow (in Russian)

Niyazov RA, Nurtaev BS (2009) Landslides in Uzbekistan caused by earthquakes in Pamir-Hindukush zone. In: Proceedings of international conference "mitigation of natural hazards in mountainous areas", 15–18 September 2009, Bishkek, Kyrgyzstan, pp 136–139

Petley D, Dunning S, Rosser N, Kausar AB (2006) Incipient landslides in the Jhelum valley, Pakistan following the 8th October 2005 earthquake. In: Proceedings disaster mitigation of debris flows, slope failures and landslides. Universal Academy Press Inc., Tokyo, Japan, pp 47–55

Radu C, Spânoche E (1977) On geological Phenomena associated with the 10 November 1940 earthquake. Rev Roum Geol Geophys et Geogr—Geophysique, vol 21, pp 159–165

Schuster RL, Alford D (2004) Usoi landslide dam and lake Sarez, Pamir mountains, Tajikistan. Environ Eng Geosci 2:151–168

Schuster RL, Highland LM (2001) Socioeconomic and environmental impacts of landslides in the western hemisphere. U.S. Geological Survey, Open-File Report 01-0276

Stepanian V (1964) Earthquakes in the Armenian upland and adjacent areas. Publishing House Hayastan, Yerevan (Armenia), p 247p

Strom A, Abdrakhmatov K (2018) Rockslides and rock avalanches of Central Asia: distribution, morphology, and internal structure. Elsevier; Candice Janco, Amsterdam, The Netherlands. ISBN 978-0-12-803204-6

Voznesensky EA, Zerkal OV (1997) Dynamic instability of saturated loesses. In: Proceedings of international symposium on engineering geology and the environment. A.A. Balkema, pp1125–1130

Yablokov A (2001) The tragedy of Khait: a natural disaster in Tajikistan. Mt Res Dev 21(1):91–93

Yin Y, Wang F, Sun P (2009) Landslide hazards triggered by the 2008 Wenchuan earthquake, Sichuan, China. Landslides. https://doi.org/10.1007/s10346-009-0148-5

Zerkal OV (1996) Mechanism of formation and development of deep seismogenous landslides due to sudden liquefaction of loessal soils. In: Proceedings of the seventh international symposium on landslides, Trondheim, A.A. Ballkema, pp 1055–1059

Some Earthquake-Induced Rockslides in the Central Asia Region

Alexander Strom and Gonghui Wang

Abstract This chapter presents a brief description of rockslides triggered by the historical earthquakes in the Central Asia region. Based on field survey and interpretation of space images, the location, scale, geological and topographic features of these rockslides, and some characteristics of related earthquakes are introduced. Several rockslides, whose triggering earthquakes had not been clarified, are also presented as controversial case studies. Further, a study conducted on the eastern part of the Tien Shan region was introduced to show that strong earthquakes can occur in the same area repeatedly triggering landslides. Identifying the features and locations of these coseismic landslides can help distinguish the return period of strong earthquakes along the same faults and can provide information for estimating the location and timing of these earthquakes.

1 Introduction

Earthquakes are the obvious and one of the most important triggers of landslides in the Central Asian region. Besides slope failures associated with local crustal earthquakes with medium to large magnitudes, landslides in this part of Asia with a predominantly arid climate often occur due to the combined effect of the mantle earthquakes from the Hindukush focal zone (Fig. 1) and precipitation. Such earthquakes, despite the rather low intensity, can trigger landslides more than 500 km far from the epicenter, due to long duration and low-frequency shaking (Niyazov 2015).

Most of the earthquake triggered landslides occur in Quaternary sediments, especially in loess and loess-like loam, weak poorly lithified Mesozoic, and Cainozoic deposits that are widespread in the Fegrana Valley rim in Uzbekistan, Kyrgyzstan,

A. Strom (✉)
Geodynamics Research Center LLC, Moscow, Russia
e-mail: strom.alexabdr@yandex.ru

G. Wang
DPRI, Kyoto University, Kyoto, Japan
e-mail: wang.gonghui.3r@kyoto-u.ac.jp; gonghui.wang@gmail.com

I. Towhata et al. (eds.), *Coseismic Landslides*, Springer Natural Hazards,
https://doi.org/10.1007/978-981-19-6597-5_6

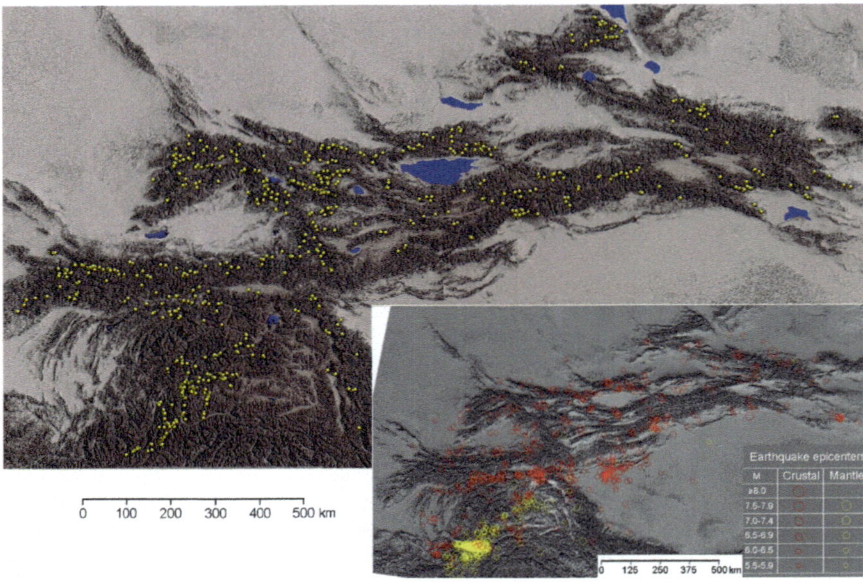

Fig. 1 Spatial distribution of rockslides in the Central Asian region with volumes exceeding 1 million cubic meters (yellow dots) and of the crustal and mantle earthquake epicenters (in the inset)

and Tajikistan and at other foothill areas. Some historical earthquakes, however, triggered large-scale rockslides that pose a special threat to population and infrastructure due to their extreme size and high mobility as well as the ability to form large natural dams. Analysis of such features that could remain in the landscape for Centuries and Millennia (see Fig. 1) helps the selection of the criteria supporting well-grounded identification of past (prehistoric) seismically induced landslides.

Most of the known large-scale historical rockslides in the Central Asia region including mountain systems of Pamir, Tien Shan, and Djungaria have been triggered by large ($M \geq 7.0$) crustal earthquakes that repeatedly affect this part of the continent (Table 1). Among them are the Akjar rockslide triggered by the 1887 Vernyi earthquake in Kazakhstan (Mushketov 1890); the Ananievo, the Chilik, and the Kaindy rockslides in Kazakhstan and Kyrgyzstan induced by the 1911 Kemin earthquake (i.e., Bogdanovich et al. 1914; Delvaux et al. 2001; Havenith et al. 2003, among others); the famous Usoi rockslide in Tajikistan, which was triggered by the 1911 Sarez earthquake with formation of the deepest dammed lake (Sarez lake) in the world, and had been widely reported (i.e., Bukinich 1913; Shpilko 1914; Preobrajensky 1920; Lange 1926, 1929; Selivanov and Andreev 1959; Sheko 1968; Sheko and Lekhatinov 1970; Fedorenko et al. 1981; Gaziev 1984; Fedorenko 1988; Agakhanjanz 1989; Alford and Schuster 2000; Ischuk 2011; Ambraseys and Bilham 2012; Kulikova et al. 2016, among others), the Sarykamish landslide, which was triggered by the 1946 Chatkal earthquake and had blocked the Naryn River valley

(Matveev 1968), the catastrophic rock avalanche in Tajikistan, which was induced by the 1949 Khait earthquake (Panasenko 1952; Semionov and Semionova 1958; Gubin 1960; Leonov 1960; Evans et al. 2009), the Belaldy rockslide, which was triggered by the 1992 Suusamyr earthquake in Kyrgyzstan (Bogachkin et al. 1997; Ghose et al. 1997; Korjenkov et al. 2004).

Data on other rockslides that, likely, had been triggered by the same or some other earthquakes, such as the 1812 Nilki, the 1885 Belovodsk, the 1889 Chilik, the 1902 Kashgar, the 1906 Manas, the 1907 Karatag, the 1946 Chatkal earthquakes, are either scarce or controversial. Detailed discussion on some contradictory information concerning the direct trigger and time for the Bielogorka, the Kaindy, and the Lower Kiol-Sai rockslides (all of them are in Northern Tien Shan), and for the Yashinkul rockslide in Southern Tien Shan can be found in (Strom and Abdrakhmatov 2018) and will be summarized in brief hereafter.

Most of the historical case studies described or discussed hereafter are from the western (former Soviet) part of the study region. A brief introduction on some landslides triggered by earthquakes in its eastern (Chinese) part is also presented to show the frequent occurrence of big earthquakes and resultant landslides in this area, although no evidence shows whether these landslides are rockslides or not.

It is also noted that not all known large earthquakes had triggered large-scale rock slope failures. For example, the M7.2 2015s Sarez earthquake did not produce such features (Ischuk and Strom 2016). On the other hand, some large Central Asian rockslides occurred without seismic triggering. This could be exemplified by the 1964 Aini rockslide in Tajikistan (Fedorenko 1988; Neshikhovskiy 1988; Strom and Abdrakhmatov 2018) and the 2008(?) Djamansu rockslide in Kyrgyzstan (Strom and Abdrakhmatov 2018). This highlights the complexity and ambiguousness of identification of the prehistoric seismically induced landslides and the necessity of further study of this controversial problem.

2 Some Landslides Triggered by the Historical Earthquakes

2.1 The 1887 Vernyi Earthquake

This earthquake with estimated $M_s \approx 7.3$, was the first one who's geological and geomorphic effects were studied by professional team led by Prof. I. Mushketov. Results of their reconnaissance were presented in a detailed report published by Russian Geological Survey (Mushketov 1890). The earthquake triggered numerous slope failures that affected slopes within a belt that stretched for about 35 km along the foothills of the Transili Range. Quite widespread were large earthflows up to millions cubic meters in volume that moved rather slowly along galleys within the epicentral zone.

Table 1 Largest rockslides in Central Asia triggered by historical earthquakes

Year, earthquake name, and magnitude	Rockslide name	Rock type	Approximate volume (Mm^3)	Comments
1887, Vernyi, 7.3	Akjar	Granite, schist	7–10	Very irregular and complex headscarp
1911, Sarez, 7.7	Usoi	Terrigenous, carbonate, gypsum, anhydrite	2200	Very deep wedge failure whose crown is located at a prominent topographic bench level. The upper part of the rockslide body is featured by a block sized ~1 × 1 km; rockslide buried Usoi village with all inhabitants except 3 men who left for Sarez village. It dammed the Murgab River valley and formed the 500 m deep Sarez Lake
1911, Kemin, 8.2	Ananievo	Granite	30	Failure destroyed rock spur with its length and width being several hundred meters just at the hanging wall of the surface rupture. The headscarp has a nearly planar back wall
	Kaindy	Metasediments	5	The failure occurred just at the hanging wall of the surface rupture. Rockslide killed several families in yurts

(continued)

Table 1 (continued)

Year, earthquake name, and magnitude	Rockslide name	Rock type	Approximate volume (Mm^3)	Comments
	Chilik	Metasediments and granites	8	Rockslide blocked the river, forming a natural dam, which was breached later. Some displaced materials are still left on the slope
1946, Chatkal, 7.5	Saryka-mysh	Carbonate	2 (4–5)	Large entrainment of the material occurred. a dam was temporarily formed. Nowadays the deposits are submerged by the Toktogul reservoir
1949, Khait, 7.4	Khait	Gneiss, schist, granites	100	The rockslide left a very deep irregular headscarp; the stability of the entire slope was affected by earthquake. Rockslide converted into long runout rock avalanche that buried the Khait town and killed several thousand people
1992, Suusamyr, 7.3	Belaldy	Granite	20	Rockslide dammed the valley and killed shepherds' family. About one year later the devastating debris flow occurred

One of the largest slope failures triggered by this earthquake was the Akjar rockslide in the Aksai River basin at 43.103° N, 76.802° E, about 10 km southwest of the present-day Almaty City (Mushketov 1890; Bogdanovich et al. 1914). The slide occurred on the 150–400 m high slope of 150–400 m at the contact zone among granite, diorite and schist. Rockslide had a very irregular and complex headscarp (Fig. 2). Though Khromovskikh (1984) estimated the volume of the collapsed rock as 40 Mm^3, Strom and Abdrakhmatov (2018) estimated it as much less—7 to 10 Mm^3 only.

This rockslide, however, was not the first one in the study catchment. The previous rock avalanche, whose age and exact source zone are still unknown, had formed a lake that was silted before its breach (Fig. 3). It allows assumption on the recurrent large earthquakes that had affected this area.

Fig. 2 3D Google Earth view of the 1887 Akjar rockslide induced by the M ~ 7.3 Vernyi earthquake. The yellow arrow indicates the position of the outcrop shown in Fig. 3

Fig. 3 Fine laminated lacustrine deposits in the Aksai River valley that had accumulated in the lake dammed by the prehistoric rock avalanche remnant (RA)

2.2 *The 1911 Kemim Earthquake*

One of the largest and, up to now, the best studied seismic catastrophe in the entire Central Asia region was the 1911 Kemin earthquake with estimated magnitude 8.2 (Kondorskaya and Shebalin 1982). More recent studies (Kulikova and Krüger 2015) gave lower, but still very large value (M = 8.0). This earthquake was studied by group of experts led by Prof. Bogdanovich (Bogdanovich et al. 1914). Earthquake was accompanied by a series of surface ruptures and triggered numerous slope failures including at least 4 large-scale rockslides. Despite extremely detailed (for that period) description, some severely affected areas were not visited by the reconnaissance team and, therefore, some earthquake triggered features could be missed.

The most prominent and impressive seismically induced rock avalanche in the eastern part of the epicentral zone of the Kemin earthquake, in the Chilik River valley, collapsed from the ~800 m high slope composed of weathered Paleozoic metasediments and granites at 43.02° N, 78.161° E. Its volume was about 8 Mm^3 and it had blocked the river. The valley here is rather wide and rock avalanche debris had spread over its bottom forming low dam that had been overtopped shortly after the event, so that earthquake reconnaissance team photographed the stream flowing over it and eroding it gradually (Fig. 4). Due to the low gradient of the overflow, there was no resultant catastrophic breach and outburst flood. At present, the Chilik River has almost returned to its former profile, though both the headscarp and the body of the rock avalanche are still visible.

Thorough check of space images revealed several rock slope failures not far from the 1911 Chilik rockslide that look quite "fresh". Being located in the hardly attainable tributary valleys of the Chilik River, they were not mapped and described by the team led by Prof. Bogdanovich. Nevertheless, they could be triggered by the same earthquake. Rather large rock avalanche, about 7 Mm^3 in volume, accompanied by much smaller rock slope failures is located at 42.913° N, 78.158° E, in the Taldy River valley, 12 km south of the Chilik rockslide (Fig. 5). Its body rests at the valley bottom below the distinct headscarp. Microrelief of these deposits differs significantly from that of the moraines and rock glaciers visible upstream, and this body has not been

Fig. 4 Photo of the Chilik rockdslide and temporarily lake. After Bogdanovich et al. (1914)

affected by any processes related to glaciation. Judging from its freshness it can be assumed that rock avalanche age should not be older than a century approximately. Therefore this slope failure could be triggered either by the 1911 Kemin or by the preceding $M \geq 8.0$ 1889 Chilik earthquake which epicentral zone partially coincides with that of the Kemin earthquake.

About 20 km west-north-west from this site another rock avalanche about 2 Mm^3 in volume, with a significant runup, also looking quite "fresh" blocked the Bolshoi (Large) Oriuktu River valley at 42.988° N, 77.935° E and formed a dammed lake that still exists (Fig. 6). Besides this rock slope failure that was released completely, there are evidence of destabilization of about 50 Mm^3 of the Ordovician terrigenous rocks that subsided and formed the sackungen-like landforms. Such combination of the evidence of long-term slope deformations and of the catastrophic rock slope failures was found at some other parts of the study region.

Probably the most impressive rock slope failure triggered by the 1911 Kemin earthquake is the Ananievo rockslide (Fig. 7). It is located at the southern foot of the Kungey-Alatau Range at 42.803° N, 77.629° E. The rocky spur up to 450 m high composed of granite and raising just above the surface rupture of the 1911 Kemin earthquake—the thrust fault dipping under the ridge (Delvaux et al. 2001), collapsed completely. The failure of the granite with a volume of 30 Mm^3 affected the most prominent part of the spur and left the headscarp with a nearly planar backwall about 200 m high and more than 600 m wide.

Fig. 5 Recent rock avalanche in the Taldy River valley. H–S: headscarp; R–A: rock avalanche deposits that differ significantly from moraine deposits and rock glaciers at the valley bottom upstream. 3D Google Earth view produced from the November 2014 space image

Fig. 6 Rock avalanche with ca. 70 m high runup that collapsed from almost 1 km high slope and blocked the Bolshoi Oriuktu River valley. Besides, distinct fractures can be seen on top of the ridge above and close to rock avalanche headscarp. The dam has been partially eroded. 3D Google Earth view

Fig. 7 Front view of the 1911 Ananievo Rockslide

Though the M8.2 earthquake profuced great effects over wide area, rockslides occurred just at this particular site probably due to a favorable combination of several factors: (1) topographic amplification of coseismic response (Havenith 2002; Havenith et al. 2003), (2) closeness to the seismic surface rupture with thrust kinematics that could "knock this spur up", additionally to intensive shaking produced by seismic waves, and (3) structure of rock massif—the latter had weak fault zone at the base of the spur overlain by less fractured and less deformable granite that outcrops at the ridge crest. These weak zone stretches in such way that westward from this spur it rise and cross the ridge more than 100 m above its feet, while east of it the weaker zone is hidden below the daylight surface.

Another large rockslide triggered by the 1911 Kemin earthquake and described by the reconnaissance team (Bogdanovich et al. 1914), is located at 42.838° N, 77.568° E, about 6 km north-west of the Ananievo rockslide in the Sutty-Bulak River valley within the same granite massif. It blocked the small valley with a dam that spread for 750 m along the valley.

One more recent rock avalanche identified in the Kemin Earthquake epicentral zone that likely was triggered by this event is located in the Chon-Uriukty River basin at 42.813° N, 77.8286° E, 16 km east of the Ananievo rockslide. It was about 3 Mm^3 in volume and occurred on the 700 m high slope composed of weathered granite. The headscarp is at the upper part of the slope (Fig. 8). Dry rockslide debris flowed down along rather steep gulley, passed 2.12 km, and terminated at 2200 m a.s.l. Freshness of the associated landforms and location within the area severely affected by the 1911 Kemin earthquake are the main reasons in favor of the assumption that this rock avalanche could be triggered by the 1911 earthquake.

In the western part of the 1911 Kemin earthquake epicentral zone, Kaindy rockslide was the largest coseismic one, which originated in the south-eastern corner of the Chon-Kemin intermontane depression (42.718° N, 76.205° E). It was composed of ca. 5 Mm^3 of debris of the Proterozoic metasediments, that collapsed from the 350 m high slope, also just above the surface rupture of this earthquake, similarly to the Ananievo rockslide. However, here surface rupture had a predominantly left-lateral style of motion (Delvaux et al. 2001). Landslide with a long runout of about 1.14 km entrained a large amount of loose material from the slope foot (Bogdanovich et al. 1914) (Fig. 9).

2.3 *The 1911 Sarez Earthquake*

Usoi rockslide, the most well-known seismically induced landslide in the Central Asian region, was triggered by the 1911 M7.3–7.7 Sarez earthquake (Golitzin 1915; Ambraseys and Bilham 2012; Kulikova et al. 2016). It blocks the Murgab River valley at 38.27° N, 72.6° E and forms the presently world-highest natural dam up to 670 m high. The Sarez Lake formed behind this dam is up to 500 m deep, 55.8 km long, and stores about 16–17 km^3 of water (Ischuk 2011). Many aspects of its structure, composition, and further evolution remain unknown due to the enormous size of the dam and the remoteness of the site, which makes its study quite challenging.

As a wedge failure, the slide occurred on the ca. 2.5 km high right-bank slope of the Murgab River valley. About 2/3 of its entire height below the prominent bench in the relief collapsed leaving the impressive headscarp (Fig. 10). Glacial valleys, whose frontal parts were truncated by the headscarp backwall, dissect the upper part of the southern slope of the Muzkol Range with its crest at an elevation of up to

Fig. 8 Recent rock avalanche in the Chon-Uriukty River basin. Google Earth image

Fig. 9 The 1911 Kaindy rockslide soon after its formation. After Bogdanovich et al. (1914)

5500 m a.s.l. Moraines from these glacial valleys provide material for debris flows fed by water from melting ice and snow and their deposits accumulated in the deep proximal lowering of this outstanding blockage. The sandstone, schist, and quartzite of the Carboniferous Sarez Formation, overlain by the Permo-Triassic dolomite and limestone with gypsum and anhydrite in its upper part, outcrop in the headscarp walls. The Carboniferous and Permo-Triassic units are divided by the Usoi thrust.

Preobrajensky (1920) presented one of the most detailed and accurate descriptions of the Usoi blockage, and estimated the volumes of the displaced rock mass and the rockslide body as 2.2 km^3 and ca. 2.7 km^3, respectively. The increase in the volume might result from the rock mass decompaction.

The carapace facies resting on top of the Usoi rockslide body is represented by a giant block of the terrigenous rocks of the Sarez series about 1.0 km^2 in size with well-preserved N–W striking bedding and intensive transverse (N–E) fracturing that is most likely originated during its motion. The bouldery alluvium that forms the narrow isthmus between the Sarez and the Shadau Lakes in the frontal part of the landslide body was scraped by rapidly moving rockslide from the valley bottom and bulldozed at its present position. The lower part of the natural dam up to ca. 300 m thick is almost impermeable. It was found that the filtration takes place through the upper part of the dam only. At present, the water level of the lake fluctuates at 3260 m a.s.l. ± 6 m approximately with the measured outflow being 47.1 m^3/s in maximum. The natural dam has a minimum freeboard of 38 m above the lake. The long-term observations indicate that the mean lake level increases gradually (Ischuk 2011). It should be pointed out that judgments on the long-term stability of the Usoi blockage are quite uncertain since neither structure, nor grain-size composition of the Usoi blockage interiors are known with confidence (Strom 2014).

Fig. 10 Wedge-like headscarp of the 1911 Usoi rockslide

Despite large magnitude of the 1911 Sarez earthquake and extremely rugged topography of the area, practically nothing is known about other slope failures that could be triggered by this earthquake. Most likely that on the opposite (northern) slope of the Muzkol Range some failure occurred whose traces can be seen near the body of the prehistoric Kudara-Pasor rock avalanche (Fig. 11) but its volume was incomparable with that of the Usoi rockslide.

Recent slope deformations at 38.248° N and 72.66° E (about 5 km upstream off the Usoi blockage) were evidenced. There, the lower part of the right-bank slope of the valley up to 340–400 m above the mean lake level is affected by the blockslide about 50 Mm^3 in volume (including its submerged part). The affected slope is 1 km long and the eastern half of this rockslide had collapsed already (Fig. 12). No data on the timing of this failure is available in the literature. Nevertheless, the possibility of occurring simultaneously with the Usoi rockslide cannot be excluded.

On December 7, 2015, the M7.2 earthquake occurred near the eastern part of the Sarez Lake (Kulikova et al. 2016). Its source was associated with submeridional fault zone that crosses the Murgab River valley. This earthquake, however, did not produce any voluminous slope failures, except rather limited subsidence on the left bank of the lake at the edge of the so-called "Mardjanai Plateau" (Ischuk and Strom 2016).

Fig. 11 Dark-gray fan in the Kudara River valley—trace of slope failure likely of the same age as the 1911 Usoi rockslide, which headscarp is behind the Muzkol range at the background. Just right of it is the body of the prehistoric Kudara-Pasor rock avalanche. 3D Google Earth view

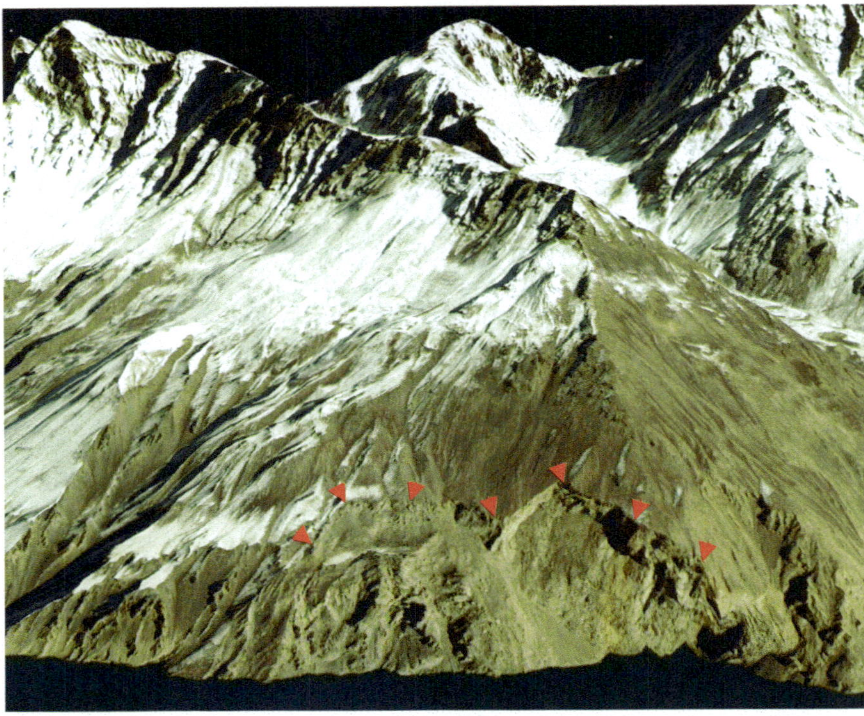

Fig. 12 3D Google Earth view of the slide on the base of the ca. 1300 m high right-bank slope of the Sarez Lake that is located about 5 km east of the Usoi blockage. Red triangles mark the well-defined headscarp

2.4 The 1946 Chatkal Earthquake

This very large (M7.5) earthquake was rather poorly investigated. Most of its environmental effects were studied decades later and published in early 70th (Leonov 1970; Kuchai 1971). The most studied feature was the rockslide that blocked the Naryn River at 41.745° N, 72.701° E (the so-called Sarykamysh rockslide) (Matveev 1968). According to Matveev (1968), about 1.2–1.8 $\times$ 10^6 m^3 of Carboniferous limestone caved from the right slope (more than 1 km high) of the Naryn River valley, fell on the talus accumulations at the central part of the slope, and finally rushed downslope with entrainment of these loose materials that were probably saturated by the autumn rains. Part of the displaced materials jumped from the rocky bench at the central part of the slope, while some other parts overlaid on the talus accumulations at the foot of this slope. Finally, a natural dam about 50–60 m high was formed with a lake volume of about 4–5 Mm^3 and resulted in valley inundation. To prevent possible further disaster, the dam was blasted and, later, naturally lowered by river erosion. It is noted that at present, the remnants of this rockslide body are submerged by the Toktogul Reservoir.

2.5 The 1949 Khait Earthquake

This M7.4 ± 0.1 earthquake that occurred on July 10, 1949 was one of the most disastrous natural phenomena in the entire Central Asia. Besides the devastating rock avalanche that buried small Khait town and a village and killed several thousand people (Evans et al. 2009), it triggered at least one more rock avalanche at the Turpi Range (Gubin 1960; Leonov 1960) and also numerous flow-like landslides in the loess-like loam in the Yasman River valley that reached the valley bottom, merged and formed the earthflow that buried several kishlaks (local villages) with their inhabitants (Gubin 1960).

The most spectacular was the 7.4 km long Khait rock avalanche ca. 100 Mm^3 in volume (Fig. 13). The accounts of eyewitnesses of this disastrous were summarized in Semionov and Semionova (1958) and are presented in English in Strom and Abdrakhmatov (2018). The rock mass of giant volume was dispatched from the north-western slope (more than 1 km high) of the Chokrak mountain and resulted in an impressive irregular headscarp (about ~300 m deep) at 39.188° N, 70.929° E. According to Leonov (1960) failure occurred along the old tectonic contact between gneiss, schist, and granites. Rockslide was triggered during the mainshock of this earthquake and transformed into an extremely mobile rock avalanche. The mainshock followed a series of foreshocks occurring on July 8, 1949, among them two foreshocks had magnitudes 5.1 ± 0.7 and 5.6 ± 0.3, respectively. Their intensity near the Khait town was estimated to be V–VI points of the MM scale, and the most affected areas (intensity: ~VIII points) were located about 15–20 km west of Khait in the Yasman River valley. It is noted that the 1949 earthquake was preceded by the 1941 Garm earthquake with M6.4 ± 0.2 7, which intensity at Khait town was about VII–VIII points a (Gubin 1943; Kondorskaya and Shebalin 1982). All these preceded earthquakes may in some extent have destabilized the slope that collapsed on July 10, 1949, although it is difficult to find direct evidence to make a conclusive remark on this inference.

Some effects of this earthquake were described in (Semionov and Semionova 1958) (some comments to their description are in italics). They wrote: "on July 10, 1949, in the morning, the Khait town and the surrounding area were affected by

Fig. 13 Vies of the 1949 Khait rock avalanche 70 years after the disaster

severe vertical shock accompanied by noisy boom resembling the roar of guns and followed by strong horizontal shocks. People could not remain standing and some of them were even thrown up (*indicating acceleration exceeding 1 g*). All buildings in the town collapsed. Along with the continuous shaking and endless boom some additional noise resembling, as one of the eyewitnesses said "sound of the stone mill", started far away and increased rapidly. Suddenly dark "wave" of debris up to 100–150 m high appeared from the Obi-Khauz-Dara River valley (*such height was indicated according to the emotional description of eyewitnesses, and, judging from the traces of rock avalanche in the upper narrow part of the valley could be less in fact. Probably its height could be enlarged visually by the dust storm, accompanying rapidly moving rock avalanche*). It moved as fast as an express train and buried the town. This surge was accompanied by very strong wind (*air blast*) that rooted trees out and threw them at a large distance. The same was observed even in the gardens that have not been reached by rock avalanche directly". It was also reported that: "while frontal parts of this avalanche stopped, the following surges of debris continued moving above and their deposits overlaid each other".

The displaced debris crossed the Yarhych River, blocked it with its toe part being spread for about 1.5 km over the opposite bank. The debris burring the Khait town was muddy and well saturated such that it was impossible to walk on it soon after the event (Semionov and Semionova 1958). It could be caused by the entrainment of rather a large amount of loess-like loam that had been saturated by the preceding rains (N. Leonov, personal communication). Besides, the rock avalanche splashed out several small ponds, mixing the water with debris.

The surface of the deposits is covered by numerous low conical hills (molards?). The Khait rock avalanche had a runout distance of about 7.41 km and covered an area of about 5.045 km^2. The volume of this rock avalanche estimated by earlier and later studies presented great differences. For example, following the first description of Panasenko (1952), Leonov (1960) and, later, Nikonov and Shebalina (1986) estimated a total volume of 400–500 Mm^3 approximately. However, this estimate seems to be significantly exaggerated. Evans et al. (2009) provided a much more reliable estimate of ca. 75 Mm^3. According to Strom and Abdrakhmatov (2018), the volume of rock avalanche deposits could be of the order of 100 Mm^3. Such an estimate is based on the area of the deposits (~5 km^2) and their mean thickness (about 20 m).

It is interesting that traces of intensive shaking visible on the Obi-Khauz-Dara River valley slopes along which the rock avalanche had passed are distributed quite asymmetrically. Most of them are on the left bank slope that was affected, aside from the main deep headscarp, by numerous fractures that create prerequisites for future slope failures. In contrast, the right-bank slope is almost intact. It indicates that seismic waves propagated from the south.

2.6 The 1992 Suusamyr Earthquake

Several landslides were triggered by the M7.3 Suusamyr earthquake on August 19, 1992. The only large rock slope failure associated with this earthquake (Fig. 14) was originated in the upper reaches of the Belaldy River valley at 42.06° N, 73.279° E, which is approximately 35 km far from the epicenter and 13 km far from the surface rupture of this earthquake (Bogachkin et al. 1997; Ghose et al. 1997). The rockslide with a volume of ca. 20 Mm^3 buried the shepherd's family who tended a herd, and spread over the bottom of a rather steep but wide glacial valley, and formed two 200–300 m high natural dams. In June 1993, ten months later after the formation of the dam, the dams were breached probably due to the intensive snowmelt at elevations exceeding 3000 m a.s.l. Powerful debris flow resulting from the dam breach partially devastated two villages (Belaldy and Torkent) that were located 17 km and 30 km downstream of the damming site (Korjenkov et al. 2004). It was assumed (Strom and Abdrakhmatov 2018) that debris flow was caused by the outburst of these lakes. However, no trace of such an outburst is visible and, most likely, it was triggered by intensive seepage through the dam and remobilization of the saturated rock avalanche debris (see Fig. 14, right part).

It is noted that for this area, aerial photographs of 1961 and the KFA-1000 space images taken in 20th Century 70 s–80 s, are available. However, analysis of these photos did not present any abnormal evidence of instability that would allow identifying this site as being a high potential of catastrophic failure (Strom and Abdrakhmatov 2018).

Besides this single large-scale rock slope failure, the Suusamyr earthquake triggered several landslides in loose Quaternary and poorly lithified Neogene sediments

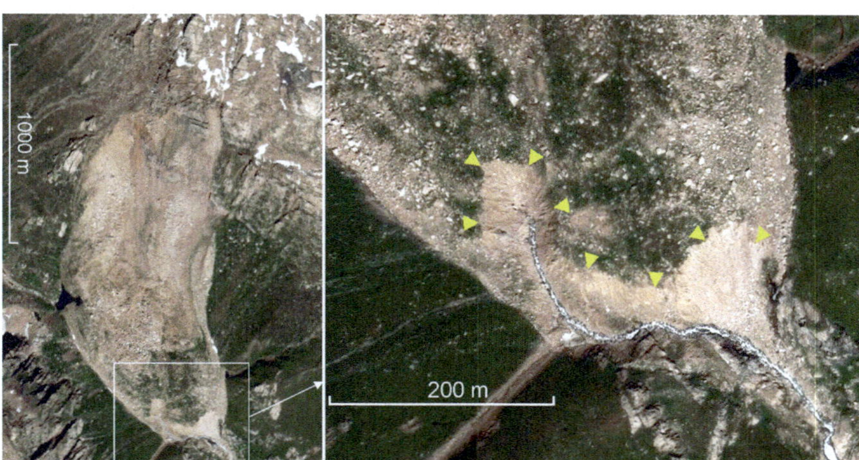

Fig. 14 Rockslide (rock avalanche) at the upper reaches of the Belaldy River triggered by the 1992 Suusamyr earthquake. Left—overview; right—frontal part of rock avalanche body with evidence of erosion (yellow triangles) that could be the source of the 1993 catastrophic debris flow

near the surface rupture of this earthquake at the steep southern slope of the small Chet-Korumdy ridge (Havenith et al. 2003). It is necessary to point out that all these slope failures inherited older landslides, and, at the same time that some parts of the slope with distinct scarps on top of the ridge were not activated by this earthquake.

One more area strongly affected by landslides during this earthquake was rather far, about 14–20 km from the epicenter and from the surface rupture—at the southern slopes of the Suusamyr Range. Most of them occurred on east-facing slopes. Some of these landslides killed people that did the haymaking. There were also unchecked reports that earthquake caused short-term blocking of the Naryn River, more than 40 km south from its surface rupture. No exact locality of this blockage is known, however.

3 Some Controversial Case Studies

Controversies in identifying earthquake-induced landslides exist both for slope failures that had occurred before the end of the XIXth Century when more or less regular reporting of natural disasters in the study region had started, and even for some rock slope failures of the early stage of the historical period.

An example could be raised by the Yashinkul rockslide dam in Southern Kyrgyzstan (Krukovskiy 1969; Rezvoi and Rezvoi 1969; Rezvoi et al. 1971; Reizvikh et al. 1971; Pushkarenko and Nikitin 1984; Strom 2010, 2013; Strom and Abdrakhmatov 2018). This dam was breached catastrophically in 1966. Data on the age of this dam formation and on its association with particular earthquakes are controversial. Some researchers inferred that the rockslide forming the dam was triggered by a strong earthquake. However, Pushkarenko and Nikitin (1984) mentioned that the dam had been formed in 1835, while Rezvoi and Rezvoi (1969), referring to local people's memorials, ascribed the rockslide to an earthquake that occurred about 300 years before the breach of the dam—in the middle of the 17th Century. Moreover, no data about strong earthquakes in this region in both time intervals have been presented in earthquake catalogs available (Mushketov and Orlov 1893; Kondorskaya and Shebalin 1982). On the other hand, Semionov (1958) and Djanuzakov et al. (2003) reported some historical data on large earthquakes in southern part of the Ferghana valley in 1822 or 1823. However, in Djanuzakov et al. (2003), the reported effects were localized in the area (40.3° N, 71.5° E) about 80 m north-west from the Yashinkul rockslide site. It is difficult to believe that such a large rockslide resulted from such a distant shock.

Quite controversial data are available also for much more recent rock slope failures such as well-known simultaneous Bielogorka rock avalanches not far from the Bishkek City (Fig. 15). Traditionally these very impressive features have been linked with the 1885 Belovodsk earthquake. However, in the description of this seismic catastrophe made by Ignatiev (1886) who inspected settlements affected by this earthquake, these features were not mentioned, despite they are clearly visible from the populated planes at the foot of the Kyrgyz Range visited by him.

On the other hand, Mushketov (1890, pp. 112–115) did not provide any indication of these rock avalanches in the Annex to the report on the 1887 Verniy earthquake concerning the 1885 event. However, at the same time, several other much smaller rock falls, earth flows, and debris avalanches were described in both publications. In the reconnaissance report made by Ignatiev (1886), who paid attention to the structural damage mainly, it was noticed that just after the earthquake water flow had stopped temporarily in the adjacent Aksu River. Most likely it was blocked by a landslide. It is hypothesized that such blockage might occur at 42.542° N, 74.005° E where the erosional canyon incised in the body of the gigantic Late Pleistocene Aksu rockslide was blocked temporarily by a landslide from the slope composed of rockslide debris (Fig. 16).

Fig. 15 Overview of the two Bielogorka rockslides

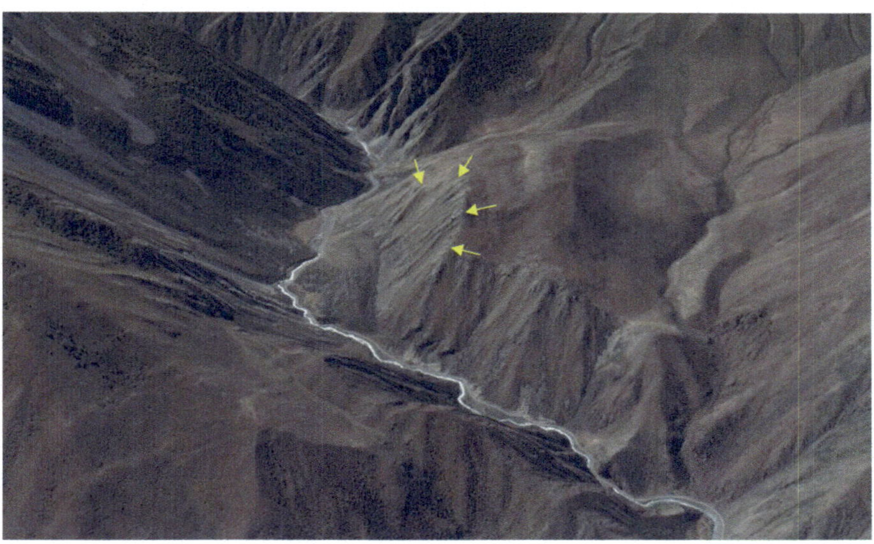

Fig. 16 Assumed most recent landslide in the Aksu River valley. Headscarp is marked by yellow arrows. 3D Google Earth image

Fig. 17 Submerged trees in the Kaindy Lake

It was hypothesized that the Bielogorka rockslides might result from some older earthquake (Strom and Abdrakhmatov 2018), though we could not provide strong evidence to make conclusive remarks on this assumption at present.

A similar controversy is the formation of several large rockslide dams at the north-eastern part of the Kungey-Alatau Range. In most popular resources like Wikipedia, these dams are considered as being triggered either by the 1889 Chilik or by the 1911 Kemin earthquakes, both of which had magnitudes more than 8.0 and severely affected this region. However, dating on the samples taken from the submerged trees collected by V.A. Gapich in the Kaindy Lake at 42.99° N, 78.469° E (Fig. 17) and in the Kiol-Sai Lower Lake at 42.995° N, 78.325° E (both lakes have been dammed by rockslides) shows that the ages are 430 ± 60 BP (1406–1635 AD) and 400 ± 30 BP (1436–1625 AD), respectively. Thus, these slope failures could be much older than the Chilik and the Kemin earthquakes (Strom and Abdrakhmatov 2018).

It should be pointed out that the most recent slope failure resulting in the formation of the Kaindy Lake had occurred within the headscarp of a much older and much larger rockslide (Fig. 18). The submerged trees had been growing on the flat valley bottom that can be derived from almost the same height of the dead trees throughout the entire lake (see Fig. 17). It means that the valley had been dammed here twice—the first lake had been silted completely forming a flat valley bottom and the forest that had been growing there was submerged after the more recent river-damming event. Such recurrent slope failures allow assessment of their recurrence interval and, thus, of the recurrence interval of the triggering phenomena—most likely of large earthquakes.

Fig. 18 Position of the headscarp of the most recent rockslide that has been damming the Kaindy Lake within the muck larger circus-like depression—the headscarp of the much larger older rockslide. The tourist camp is located on its deposits

4 Landslides and Paleo-Earthquakes in Xinjiang, China

It is well-known that the eastern (Chinese) part of Tien Shan region also suffers from earthquake disaster frequently. According to Yang (1988), since 1716, there were more than 19 great earthquakes with $M \geq 7.0$ in this region, and among them at least 11 earthquakes had been identified to have trigged landslides.

By now, studies on these landslides triggered by the earthquakes in the eastern Tien Shan had been rarely reported, so that the information on the locations, geological and topographic features of these coseismic landslides is very limited. Nevertheless, it is deemed that earthquake can occur along the same faults repeatedly, resulting in large number of landslides every time. As an example, in the following, we introduce a study conducted by Yang (1988).

On March 8, 1812, a strong earthquake ($M \sim 8$) occurred in Nilka (in the eastern part of the Northern Tianshan), Xinjiang, China. This earthquake had also been reported as Nilk earthquake (Yang et al. 1985; Brown et al. 2008), and Nilek earthquake (Yang 1988). Because many earthquakes had been reported in the same region, hereafter we call the earthquake event as 1812 Nilka earthquake.

The 1812 Nilka earthquake resulted from the rupture associated with a fault passing along the Kashi River (shown in Fig. 19) with its epicenter being located at 43.78° N and 83.47° E and focal depth of 25 km (Yang et al. 1985). Although the earthquake epicenter area was sparsely populated, this earthquake caused 58 fatalities. It was also reported that most of the fatalities were herders living in Yurt and loggers living in wooden houses.

On the survey to the 1812 Nilka earthquake area, Yang (1988) found a large number of landslides that were distributed intensively (Fig. 19), and inferred that these landslides might have been triggered by different paleo-earthquakes. Based

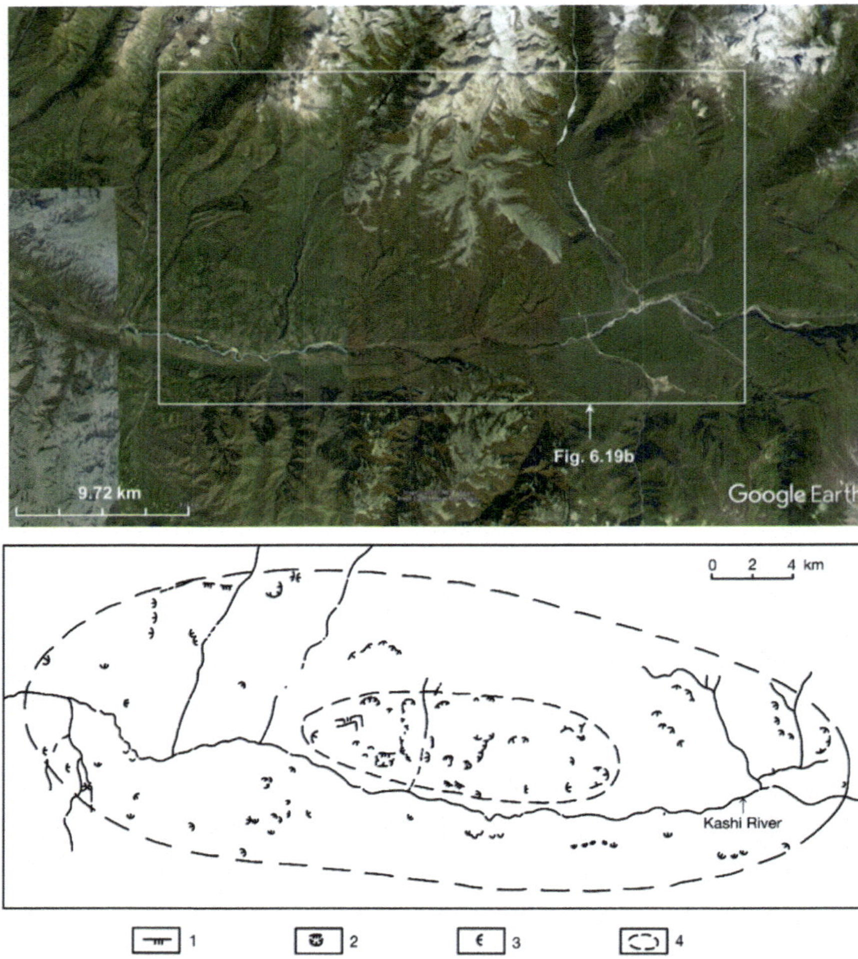

Fig. 19 The distribution of landslides triggered by paleo-earthquake Event A (after Yang 1988). Legend: (1) old coseismic landslide; (2) fault of paleo-earthquake; (3) ground settlement resulting from earthquake; (4) different areas of densely populated landslides resulting from different paleo-earthquakes

on the features of these landslides, such as the weathering state of the outcropped bedrocks, vegetation on the landslide area, and slope angles, these landslides were further divided into two zones (Fig. 19), which were inferred to be the results of two paleo-earthquake events (hereinafter called Event A and Event B), respectively.

The relics (as shown in Fig. 19) of Event A include a lot of old landslides, ground settlements, and faults of paleo-earthquake that had been identified by the excavated trenches. The landslide distributed area is approximately the same as the epicenter area of the 1812 Nilka earthquake. According to the width of the back scarp of the landslide cluster (about 4 km wide), it is estimated that Event A might be as strong

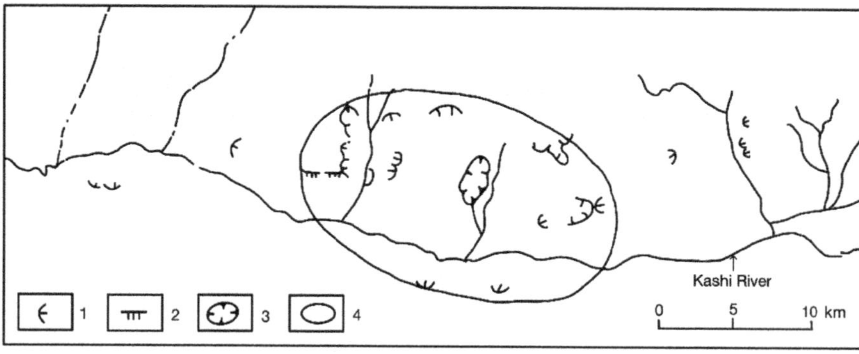

Fig. 20 The distribution of landslides triggered by paleo-earthquake Event B (after Yang 1988). Legend: (1) old coseismic landslide; (2) outcropped fault of paleo-earthquake; (3) ground settlement resulting from earthquake; (4) area with densely distributed landslides

as the 1812 Nilka earthquake. Based on the geometric center of the locations of these coseismic landslides, the macro-epicenter of Event A is estimated to be at 43.8° N, and 83.6° E.

The number of landslides triggered by Event B (Fig. 20) is less than that by Event A, probably because Event B was more distant, and it becomes difficult to identify them through their topographic features. The landslides are concentrated within an area about 20 km long. According to the back scarp of the cluster of landslides (about 4 km wide) and the area of ground settlement (about 4 km^2), the magnitude of Event B is also estimated to be the same as that of the 1812 Nilka earthquake. From the geometric center of the locations of these coseismic landslides, the macro-epicenter of Event B is estimated to be at 43.7° N, and 83.7° E.

Subsurface exploration on the toe part and source area of these landslides was conducted and carbonized wood fragments were obtained below the slide masses. Radiocarbon dating (C^{14}) indicates that the interval between the 1812 Nilka earthquake and the earthquake Event A is about 2713 years, and that between Event A and Event B is about 2448 years. Therefore, the return period of strong earthquake in this region is estimated to be approximately 2600 years.

5 Summary

Based on the field survey and examination of space images, the basic information on some rockslides triggered by the historical earthquakes in the Central Asia region had been presented. Some rockslides, whose triggers had not been clearly identified, had also been introduced as examples of controversial case studies. Our results show that earthquakes are the main, but not the only trigger of large landslides in Central Asia. Detailed complex study of such features, both newly formed and past, will provide information useful both for landslide and for seismic hazard assessment of the study region.

Studies on landslides occurring in the same areas as the epicentral zones of the 1812 Nilka, and the 1887 Vernyi earthquakes revealed that large earthquakes can occur in the same area repeatedly and result in densely populated landslides every time. Detailed examination of the features and locations of these coseismic landslides together with radiocarbon dating can help identify the macro-epicenters of paleo-earthquakes and distinguish the return period of strong earthquakes resulting from the activities of the same fault.

Acknowledgements We thank Jiangsu Provincial Seismological Bureau, China, for allowing us to introduce the research results and use of figures in the paper by Yang (1988) published on Journal of Seismology (now: Journal of Disaster Prevention and Mitigation Engineering).

References

Agakhanjanz OE (1989) Sarez. Leningrad Press, Leningrad, p 112 (in Russian)
Alford D, Schuster R (eds) (2000) Usoi Landslide Dam and Lake Sarez. An assessment of Hazard and Risk in the Pamir Mountains, Tajikistan. UN Publication. Sales No. E.00.III.M.1
Ambraseys N, Bilham R (2012) The Sarez-Pamir earthquake and landslide of 18 February 1911. Seismol Res Lett 83:294–314
Bogachkin BM, Korjenkov AM, Mamyrov E, Nechaev YV, Omuraliev M, Petrosian AE, Pletniov KG, Rogozhin EA, Charimov TA (1997) Structure of the 1992 Suusamyr earthquake source on the basis of its geological and seismological effects analysis. Phys Earth 11:3–18 (in Russian, English translation)
Bogdanovich MC, Kark J, Korolkov B, Muchketov D (1914) Earthquake of the 4th January 1911 in the northern districts of the Tien Shan. Trans Geol Com New Series 89:270 (in Russian with extended French abstract)
Brown ET, Bourles DL, Burchfiel BC, Deng QD, Li J, Raisbeck GM, Yiou F (2008) Estimation of slip rates in the southern Tien Shan using cosmic ray exposure dates of abandoned alluvial fans. GSA Bull 110(3):377–386
Bukinich DD (1913) Usoi earthquake and its consequences. Russian Records 187, Wednesday, August 14, 1913 (In Russian)
Delvaux D, Abdrakhmatov KE, Lemzin IN, Strom AL (2001) Landslides and surface breaks of the 1911, Ms 8.2 Kemin earthquake, Kyrgyzstan. Russ Geol Geophys 42:1167–1177
Djanuzakov KD, Omuraliev M, Omuralieva A, Iliasov BI, Grebennikova VV (2003) Strong Earthquakes in Tien Shan within the territory of Kyrgyzstan and the adjacent regions of Central Asia. Bishkek, Ilim, p 201 (in Russian)
Evans SG, Roberts NJ, Ischuck A, Delaney KB, Morozova GS, Tutubalina O (2009) Landslides triggered by the 1949 Khait earthquake, Tajikistan, and associated loss of life. Eng Geol 109(3–4):195–212
Fedorenko VS (1988) Rockslides and rock falls and their prediction. Moscow State University Publishing House, Moscow, p 214 (in Russian)
Fedorenko VS, Denikaev ShSh, Lim VV (1981) Main engineering-geological aspects of the Sarez Lake problem. Eng Geol 3:70–80 (in Russian)
Gaziev E (1984) Study of the Usoi Landslide in Pamir. In: Proceedings of 4th international symposium on landslides, Toronto, vol 1, pp 511–514

Ghose S, Mellors RJ, Korjenkov AM, Hamburger MW, Pavlis TL, Pavlis GL, Mamyrov E, Muraliev AR (1997) The Ms = 7.3 1992 Suusamyr, Kyrgyzstan earthquake: 2. Aftershock focal mechanisms and surface deformation. Bull Seismol Soc Am 87:23–38

Golitzin PB (1915) Sur le tremblement de terre du 18 de février 1911. Comptes Rendus De L'academie Des Sciences De Paris 160:810–813

Gubin IE (1943) The Garm Earthquake in 1941. Published by the Tajik branch of the Academy of Sciences of the USSR, p 132 (in Russian)

Gubin IE (1960) Patterns of seismic manifestations of the Tajikistan territory (geology and seismicity). Academy of Sciences of USSR Publishing House (in Russian), Moscow

Havenith H-B (2002) Landslides triggered by earthquakes. Experimental studies in the Tien Shan mountains (Central Asia) and dynamic modelling. Ph. D. Thesis. Université de Liège, Faculté des Sciences Appliquées, p 239

Havenith H-B, Strom A, Jongmans D, Abdrakhmatov K, Delvaux D, Trefois P (2003) Seismic triggering of landslides, Part A: field evidence from the Northern Tien Shan. Nat Hazard 3:135–149

Ignatiev IV (1886) The earthquake in the Tokmak district in the 1885. Proc Russian Imperial Geograph Soc 22 (2) (in Russian)

Ischuk AR (2011) Usoi Rockslide Dam and Lake Sarez, Pamir Mountains, Tajikistan. In: Evans SG, Hermanns R, Strom AL, Scarascia-Mugnozza G (eds) Natural and artificial rockslide dams. Lecture Notes in Earth Sciences, vol 133. Springer, Heidelberg, pp 423–440

Ischuk NR, Strom AL (2016) Geological effects of the earthquake at the Sarez Lake on the 7th of December, 2015. GeoRisk 3:56–74 (in Russian)

Khromovskikh VS (1984) Rocky dragon. Mysl Publishers, Moscow (in Russian)

Kondorskaya NV, Shebalin NV (1982) New catalog of strong earthquakes in the U.S.S.R. from ancient times through 1977. NOAA, Boulder

Korjenkov AM, Mamyrov E, Omuraliev M, Kovalenko VA, Usmanov SF (2004) Rock avalanches and landslides formed in result of strong Suusamyr (1992, M = 7,4) earthquake in the Northern Tien Shan—test structures for mapping of paleoseismic deformations by satellite images. In: Buchroithner MF (ed) High mountain remote sensing cartography VII (HMRSC VII). Kartographische Bausteine, Band 23, Dresden, pp 117–136

Krukovskiy GL (1969) On the role of slope failure phenomena in formation of high mountain Yashinkul and Kokkul lakes located in the southern part of Ferghana valley in the Tegermach River Basin. Hydrogeology and engineering geology of the arid zones of the USSR. Tashkent 9(II):25–31 (in Russian)

Kuchai VK (1971) Use of paleoseismic dislocations for studying seismic regime (exemplified by the pleistoseismal zone of the 1946 Chatkal earthquake). Geol Geophys 4:124–129 (in Russian)

Kulikova G, Krüger F (2015) Source process of the 1911M8.0 Chon-Kemin earthquake: investigation results by analogue seismic records. Geophys J Int 201:1891–1911. https://doi.org/10.1093/gji/ggv091

Kulikova G, Schurr B, Krüger F, Brzoska E, Heimann S (2016) Source parameters of the Sarez-Pamir earthquake of 1911 February 18. Geophys J Int 205:1086–1098

Lange OK (1926) Expedition on the survey of the Usoi blockage and the Sarez Lake. Nat Econ Central Asia 10:76–79 (in Russian)

Lange OK (1929) Modern state of the Usoi blockage. Proc Central Asian Geograph Soc 19 Tashkent (in Russian)

Leonov NN (1960) The Khait 1949 earthquake and geological conditions of its origin. Proc Acad Sci USSR Geophys Series 3:409–424 (in Russian)

Leonov NN (1970) The 1946 Chatkal earthquake. In: Problems of the engineering seismology, Issue 13, Moscow, Nauka, pp 64–77 (in Russian)

Matveev YD (1968) Regulationg of the colossal rock falls and landslides in the Toktogul dam construction region. In: Problems of engineering geology and soil science. Issue 2. Moscow, pp 245–258 (in Russian)

Mushketov IV (1890) The Vernyi earthquake of 28/05 (9/06) 1887. In: Proceedings of geological committee, vol 10, issue 1, St. Petersburg, p 154 (in Russian)
Mushketov IV, Orlov AP (1893) Catalogue of Earthquakes of Russian Empire. In: Proceedings of Russian geographic society, p 26 (in Russian)
Neshikhovskiy RA (1988) Floods on rivers and lakes. Hydrometeoizdat, Leningrad, p 184 (in Russian)
Nikonov AA, Shebalina TY (1986) Lichenometric method of seismic dislocations dating. Institute of Physics of Earth, Moscow, p 185 (in Russian)
Niyazov RA (2015) Landslides triggered by the Pamir-Hindukush earthquakes. Tashkent, ISBN 978-9943-4519-2-6, p 223 (in Russian)
Panasenko GD (1952) Collapse of the Chokurak Mountain. Transactions of the Academy of Sciences of the USSR, 35, No 3 (in Russian)
Preobrajensky IA (1920) The Usoi Landslide. Geol Comm Papers Appl Geol 14:21 (in Russian)
Pushkarenko VP, Nikitin AM (1984) Experience in the regional investigation of the state of mountain lake dams in Central Asia and the character of breach mudflow formation. Landslides and Mudflows (in two volumes). UNESCO/UNEP. Moscow (in Russian)
Rezvoi DP, Rezvoi PD (1969) Lake that vanished .... Priroda (Nature) 7:81–83 (in Russian)
Reizvikh VN, Krukovskiy GL, Nikitin AM (1971) Causes and conditions of the Yashinkul Lake outburst. Collect Papers Tashkent Hydro-Meteorol Observatory 4:49–95 (in Russian)
Rezvoi DP, Alexeenko AV, Bolgar BD, Ermolenko YA, Marushkin IA, Rezvoi PD, Soloshenko II (1971) Rockfall-dammed lakes of the Alai Range—important factor of the high mountain relief formation. Reports and News of the L'vov branch of the Ukraine Geographical Society (1968–1969). Lvov University Publishing House, pp 71–73 (in Russian)
Selivanov PI, Andreev VI (1959) About water level behavior and possible evolution of the Sarez Lake. Meteorol Hydrol 3 (in Russian)
Semionov AA (1958) List of earthquakes in Central Asia and adjacent countries from ancient times till 1830. Proc Inst Seismol, Acad Scie Tajik SSR 94:37–52 (in Russian)
Semionov PG, Semionova VA (1958) Catalogue of earthquakes felt on the territory of Tajikistan during the periods (of time) 1865–1940 and 1941–1952. Proc Inst Seismol Acad Sci Tajik SSR 86(3):143 (in Russian)
Sheko AI (1968) Assessment of the Usoi blockage stability and possibility of the Sarez Lake breach. Bull Moscow Soc Nat Scient. Sect Geol 43(4):151–159 (in Russian)
Sheko AI, Lekhatinov AM (1970) Present date state of the Usoy blockadge and tasks of future investigations. In: Materials of scientific-technical meeting on the methodical problems of mud flows, rockfalls and landslides investigation and forecast, Dushanbe, Donish, pp 219–223 (in Russian)
Shpilko GA (1914) The 1911 earthquake in Pamirs and its effects. Proc Russian Geograph Soc 3:69–94 (in Russian)
Strom AL (2010) Landslide dams in Central Asia region. Landslides J Japan Landslide Soc 47(6):309–324
Strom A (2013) Use of indirect evidence for the prehistoric earthquake-induced landslide identification. In: Ugai K, Yagi H, Wakai A (eds) Earthquake-induced landslides. Springer, Heidelberg, New York, Dordrecht, London (ISBN 978-3-642-32237-2), pp 21–30
Strom A (2014) Sarez Lake problem: ensuring long-term safety. In: Sassa K, Canuti P, Yin Y (eds) Proceedings of world landslide forum 3, 2–6 June 2014, Beijing, vol 3, pp 633–639
Strom A, Abdrakhmatov K (2018) Rockslides and rock avalanches of Central Asia: distribution, morphology, and internal structure. Elsevier, p 450
Yang Z (1988) Preliminary study landslip of paleoearthquake in Xinjiang. J Seismol 3(27):1–6
Yang Z, Zhang Y, Li J, Yi GH (1985) A preliminary study on the original tectonics of the Nilk Quake (M = 8) in Xinjiang Occurred on Mar. 8, 1812. Acta Seismologica Sinica 7(4):434–444

Some Recent Coseismic Landslides in Japan

Gonghui Wang

Abstract Earthquakes occur frequently in Japan, resulting in many casualties and great damage to properties. Landslides triggered by the earthquakes are a great threat to local properties and the lives of residents. In many cases, most of the casualties during the earthquake resulted from the coseismic landslides. In this report, we introduced those landslides triggered by some recent earthquakes that have occurred in Japan. A study of the 2004 Mid-Niigata Prefecture earthquake revealed that a huge number of landslides were triggered and most of the large-scale landslides are reactivated ones, but the displaced materials suffered long runouts and rapid movement. Rainfall preceding the earthquake may have favored the occurrence of landslides in large numbers and rapid movement of the displaced landslide materials. Some catastrophic landslides on tephra slopes triggered by the 2011 Tohoku Earthquake suggest that slopes that are undercut at the toe in part have lower slope stability, and that paleosols sandwiched on the tephra slopes favor the instability of the slope during earthquakes. Field monitoring of the ground motion indicates that the landslide area may have suffered very strong motion, although the seismic station nearby recorded motion weaker than other stations. Further experimental examination using the estimated ground motion during the main shock in the landslide area suggests that the ground motion of tephra slopes with different water moisture could be very complicated and further study on estimating the coseismic response of tephra slopes with high precision will be needed. The landslides triggered by the 2016 Kumamoto earthquake give another example showing the diversity of coseismic landslides; their initiation and movement are greatly dependent on the geological, geomorphological and also hydrologic features of the slopes Landslides in tephra slopes are probably not very large in size, but are normally catastrophic, because the displaced materials can be fluidized and undergo long runout and rapid movement, resulting great damage to local residential areas. This kind of disaster resulting from the landslides occurring on tephra slopes occurred again during the 2018 Hokkaido Eastern Iburi earthquake.

G. Wang (✉)
Disaster Prevention Research Institute, Kyoto University, Kyoto, Japan
e-mail: wang.gonghui.3r@kyoto-u.ac.jp

I. Towhata et al. (eds.), *Coseismic Landslides*, Springer Natural Hazards,
https://doi.org/10.1007/978-981-19-6597-5_7

1 Introduction

Japan is an island country located along the Pacific Ring of Fire, which is known as the most active earthquake belt in the world, mainly due to the subduction of the Philippine Sea Plate beneath the continental Amurian Plate and Okinawa Plate to the south, and that of the Pacific Plate under the Okhotsk Plate to the north. Due to this special geological feature, earthquakes strike Japan very frequently. Because Japan is also one of the most densely populated countries in the world, these earthquakes in Japan are also normally very devastating, resulting in great damage to the local properties and great loss of lives.

During an earthquake, the collapse of buildings or other types of construction is a great threat to human life, but in most cases, the resultant landslides played a key role in the loss of lives. This could be exemplified by some landslides that were triggered by the 2004 Mid-Niigata Prefecture earthquake, the 2011 Tohoku earthquake, and the 2016 Kumamoto earthquake. This report outlines the basic pictures of these landslides and also introduces some typical landslides in each earthquake event.

2 Landslides Triggered by the Mid-Niigata Prefecture Earthquake in 2004

On October 23, 2004, a strong earthquake (M6.8) hit a wide area in Niigata Prefecture, Japan, which was named the 2004 Mid-Niigata Prefecture Earthquake (Japan Meteorological Agency; JMA) (Fig. 1). This earthquake resulted in severe damage and isolating villages in the mountainous area near the epicentral region. During the earthquake, more than 1600 landslides were triggered, among them 362 landslides that are wider than 50 m and 12 landslides are $>1 \times 10^6$ m^3 in volume (Ministry of Land, Infrastructure and Transport 2005; Sassa et al. 2007). The distribution of these landslides is shown in Fig. 2. In many locations, the displaced landslide materials blocked the rivers and formed natural dams. Totally, forty-five natural dams had been formed during this earthquake, and five of them in Higashi Takezawa, Terano, Dainichisan, Haguro, and Myoken areas (Fig. 1) were the most catastrophic. For a detailed introduction to these landslides refer to Sassa (2005), Sassa et al. (2006), and Chigira and Yagi (2006), among others.

Chigira and Yagi (2006) detailed the geological setting of the epicentral area (as shown in Fig. 3). As reported, this area is within the Higashiyama Hills with an NNE–SSW trend, which is subparallel to the axes of several anticlines and synclines that control the regional geologic structure (Fig. 3). Underlaying the area are Pliocene and Pleistocene formations, which consist mainly of volcanic rocks, mudstones, sandstone, and alternating beds of sandstone, siltstone, and mudstone. Several active faults and folds exist in the Higashiyama Hills and nearby areas, among them the Yukyuzan fault is very close to the 2004 epicenter (Research Group for Active Faults of Japan 1991). Landslides were densely triggered in the Imo River basin along the

Fig. 1 Location of the study area for the 2004 Mid-Niigata Prefecture Earthquake (M6.8)

Kajigane Syncline, with the sliding direction being along the dip direction of the geologic formations.

As an example of the most catastrophic one occurring in this earthquake, the landslide in the Higashi-Takezawa area (hereinafter termed the Higashi-Takezawa landslide) had been surveyed and investigated in detail, and the results will be presented here.

The Higashi-Takezawa landslide (Fig. 4) occurred on the left bank slope of the Imo River. From Fig. 4, the scarp of an old landslide can be identified. The plan of the landslide area is depicted in Fig. 5, while a cross-section along line I–I' shown in Fig. 5 is presented in Fig. 6. The landslide originated at an elevation of about 200 m, and had a length of about 365 m, width of about 250 m, and a maximum thickness of about 30 m. The displaced landslide materials with an estimated volume of 1.3×10^6 m^3 traveled a horizontal distance of 100 m after a descent in elevation of about 60 m.

The Higashi-Takezawa landslide was a dip-slope one. The main scarp is approximately 50 m high and sloped 30–35 degrees. Outcropped on the main scarp is siltstone with a blue-gray color (Fig. 7). The sliding surface on the source area presents a strike of N5° W–N10° E with a dip angle of 18–22° W. The displaced landslide materials crossed the river and crashed on the opposite band, resulting in the formation of an uplifted hill 20 m high.

Immediately after the earthquake, many borings were drilled on the landslide body for the construction of counter measurements to stabilize the natural dam. These borings showed that the landslide body mainly consists of sandy silt-silty sand, silty sandstone, and siltstone/sandy mudstone (Yamabe et al. 2010). The sliding surface

Fig. 2 Distribution of landslides triggered by the 2004 Mid-Niigata Prefecture Earthquake (M6.8) (after Geospatial Information Authority of Japan)

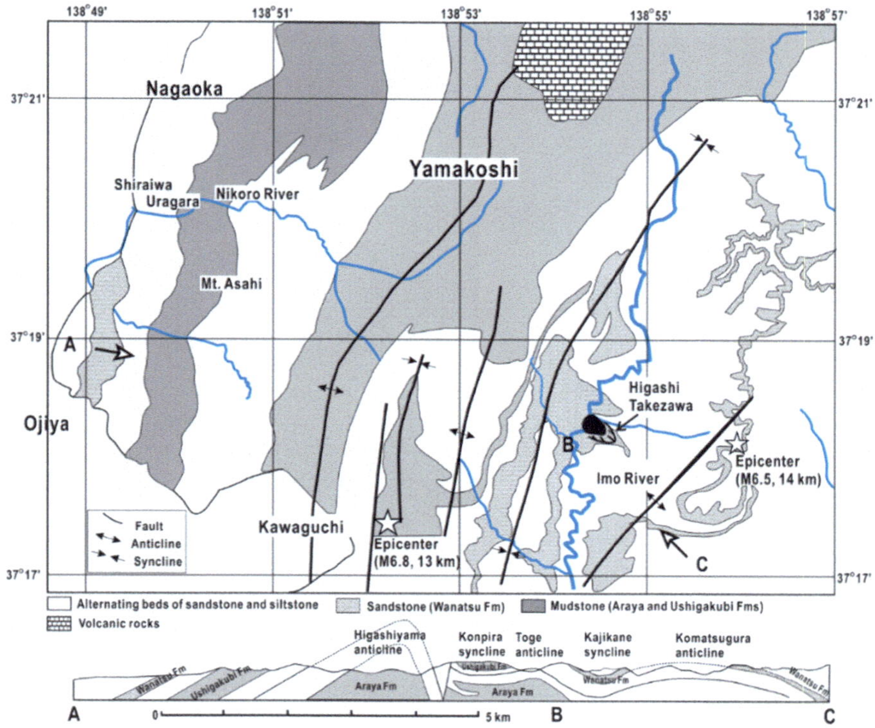

Fig. 3 Geologic outline (after Chigira and Yagi 2005) and location of the Higashi-Takezawa landslide (from Wang et al. 2016 with permission from Elsevier)

underlain the landslide body on the upper part was about 8.5 m deep. Above this sliding surface, the soil layer consists of mainly soft/extremely loose sandy silt and silty sand, while below which is siltstone or sandy mudstone. The sliding surface in the middle part of the landslide body appeared at the depth of about 28.0 m, and was formed within a loose, water-laden sandy layer at the base of the silty sandstone. The sliding surface in the toe part of the landslide body reached a depth of 31.6 m, and is located at the base of a 25-m thick rock stratum of weak siltstone/sandy mudstone. Therefore, the average and maximum thicknesses of the displaced landslide materials forming the natural dam were estimated to be about 26 m, and 32 m, respectively. The length of the landslide dam (along the river direction) was approximately 350 m, and its width was about 300 m. The barrier lake formed by this landslide dam is estimated to have a capacity of about 3.3×10^6 m^3 (Yamabe et al. 2010).

To prevent a potential catastrophic breaching, emergency countermeasures were conducted to some large-scale landslide dams through preventing possible overtopping and/or slope failure within the dams. For the Higashi-Takezawa landslide dam, immediately after the earthquake, sandbags transported by helicopters were piled up on the lower part of the dam crest, while water drainage was conducted at the same time by means of pumping the water out from the barrier lake and constructing a

Fig. 4 Oblique view of the Higashi-Takezawa landslide (photos taken by K. Sassa on 6 November 2004)

sluiceway. Finally, as the permanent countermeasures, a concrete sluiceway on the former right bank of the Imo River and two check dams downstream of the landslide dam were constructed. The dam crest had also been flattened. Further countermeasure works were conducted on the landslide source area to stabilize the slope. The plan of the landslide dam after the complement of all these countermeasures is shown in Fig. 8. Further information on the Higashi-Takezawa landslide dam can be obtained in Wang et al. (2016).

To examine the initiation and movement mechanisms of this landslide, we took samples from the sliding surface (sandy material) as shown in Fig. 7, and performed seismic ring shear tests. The seismic loadings (normal stress and shear stress) were inferred by using the seismic waves that were recorded during the Mid-Niigata earthquake in an earthquake observation station (K-NET) close to the landslide area. It is noted that K-NET is the strong earthquake motion monitoring network operating by the National Research Institute for Earth Science and Disaster Resilience (NIED). It is noted that K-NET is a system sending strong-motion data. There are about 1000 observatories (25 km mesh) located all over Japan, and all the data are available immediately through the internet once the earthquake has occurred. The nearest monitoring station to the Higashi Takezawa landslide area is NIG019 at Ojiya, which is located 10 km west of the Higashi Takezawa landslide, and 7 km WNW of the earthquake epicenter. It is noted that the location of the epicenter is between NIG019 and the landslides, but closer to the landslide. Using the record, we estimated the possible dynamic normal stress and shear stress loading (shown in Fig. 9) on the

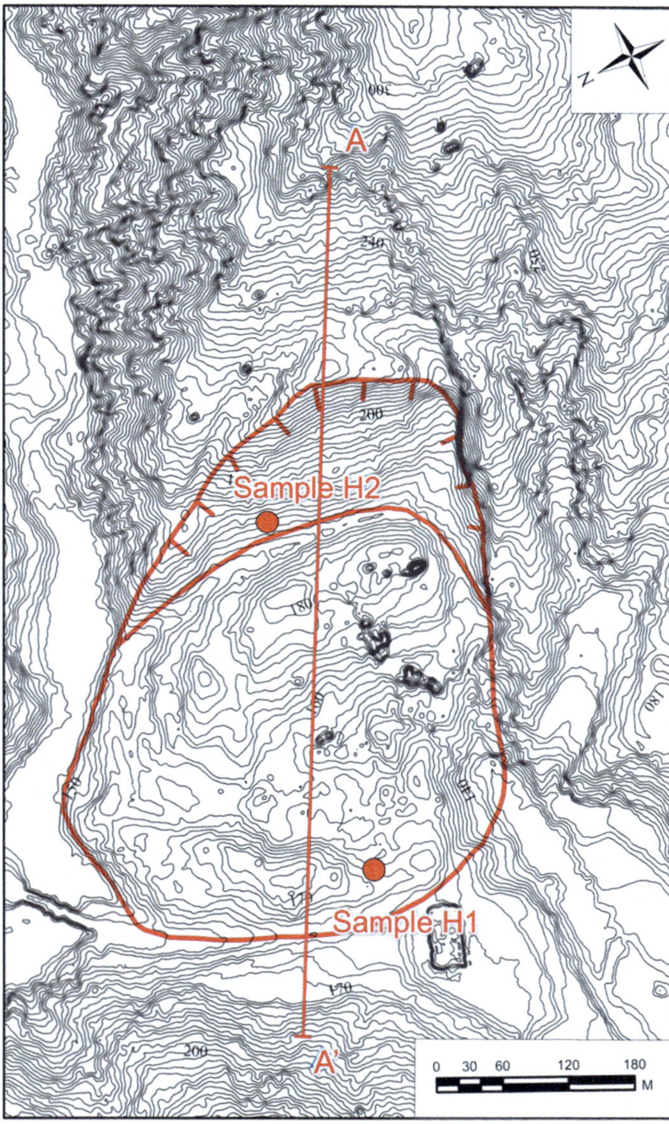

Fig. 5 Plan of Higashi-Takezawa landslide area (the map was made by the airborne laser scanning in the courtesy of Aero Asahi Corporation. Contour lines are 2 m spacing)

sliding surface due to the introduction of the earthquake. After the sample was fully saturated and normally consolidated, the estimated dynamic stresses were applied simultaneously to the sample under undrained conditions. An example of the results is shown in Fig. 10, in which it is seen that shear failure is triggered with the progress of seismic loadings. High pore-water pressure is generated, and shear strength lowers

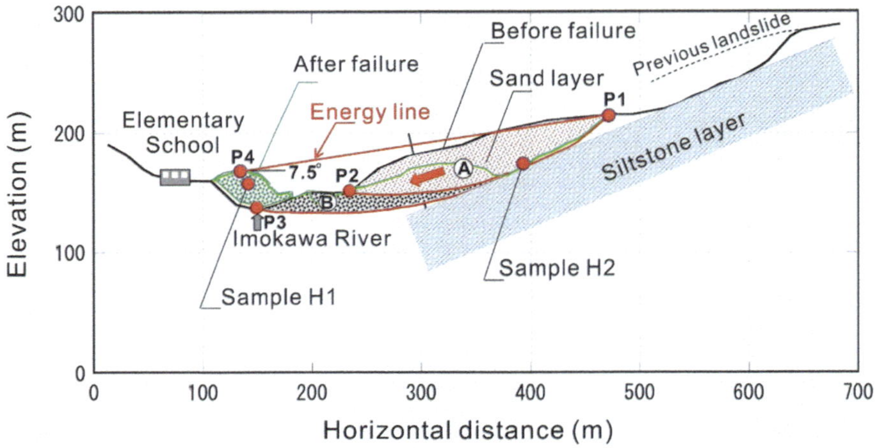

Fig. 6 Section of the landslide along the A–A' line in Fig. 5

Fig. 7 Outcrop of siltstone on the main scarp (taken on November 15, 2004)

Fig. 8 Higashi-Takezawa landslide dam after permanent countermeasures had been performed

to a very small value. These data provide evidence for the rapid movement featured in this landslide.

From Fig. 2, it is notable that a huge number of landslides were triggered by the earthquake within a relatively small area. Compared to other strong earthquakes occurred in Japan, the number triggered by this earthquake is obviously too big. The differences in the geological and geomorphological features could be the reasons, but an important reason for the occurrence of these landslides might be the rainfall before the earthquake. Three days before this earthquake, this area suffered from heavy rainfall accompanying the typhoon Tokage (also called Typhoon No.23 in Japan). Figure 11 plots the rainfall data that were obtained by Japan Meteorological Agency at four observation stations in Nagaoka, Tochio, Sumon, and Oide, which are located approximately 18 km north-west, 21 km north-east, 13 km north-east, and 10 km south-east of the Higashi-Takezawa landslide, respectively. All these stations recorded a daily precipitation >90 mm on October 20. The heavy rainfall before the earthquake may have elevated the ground water table or saturated the sliding surface, such that the shear resistance could be lowered after the occurrence of slope instability (as revealed by Fig. 10). It is noted that in July, heavy rain fell on the wide area of the Hokuriku region, resulting in devastating flooding and landslides. The daily precipitation reached 421 mm at Tochio station (Fig. 11b). Further studies will be needed to unravel the influence of the antecedent precipitation on the occurrence of coseismic landslides.

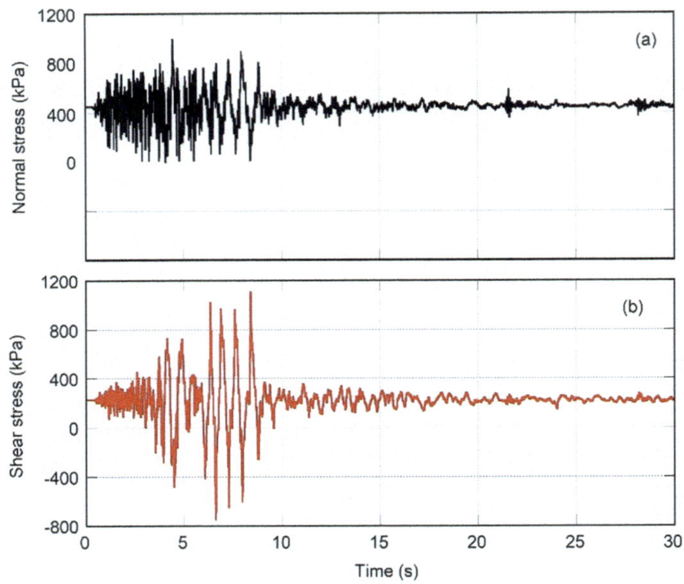

Fig. 9 Estimated dynamic normal and shear stresses due to the introduction of an earthquake for the Higashi-Takezawa landslide

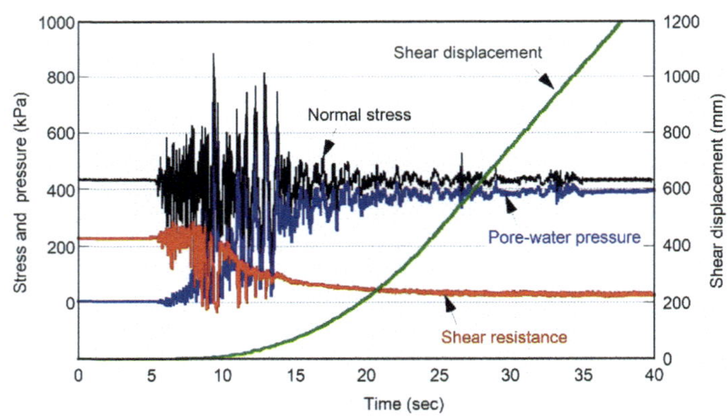

Fig. 10 Coseismic simulating test on a sample taken from the soil layer near the sliding surface of the Higashi-Takezawa landslide

3 Landslides Triggered by the 2011 Tohoku Earthquake

On March 11, 2011, a mega-earthquake (M_w9.0) occurred with the epicenter being offshore of the Pacific coast of the Tohoku Region, Japan (Fig. 12). This earthquake named the "2011 Tohoku Earthquake" is the strongest one known to have hit Japan. The earthquake was followed by an extremely powerful tsunami, which resulted in

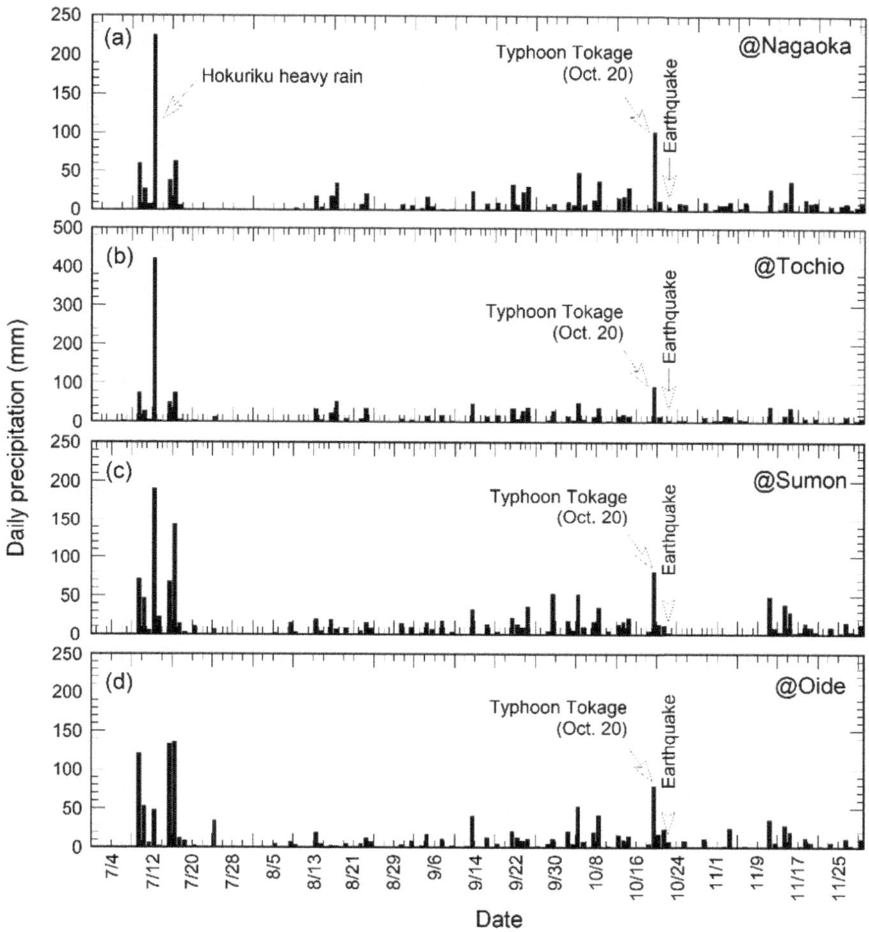

Fig. 11 Rainfall data for the period of July to November of 2004 for the Higashi Takezawa study area

a huge number of casualties and severe destruction to the infrastructure in Tohoku Region.

Besides the huge tsunami, this earthquake also triggered many landslides in the wider Tohoku Region. Both the number and size of these landslides are small compared to the scale of this earthquake, and also compared to the landslides induced by those $M_j = 7$–8 earthquakes that occurred before in the same area (Doshida and Uchiyama 2012). However, many landslides triggered by this earthquake presented high mobility with features of fluid-like motion of granular material. This kind of landslide is normally termed flowslide, according to Bishop (1973). Because these flowslides were scattered in different regions while these regions had a small amount of preceding precipitation, and also because almost all of these landslides occurred

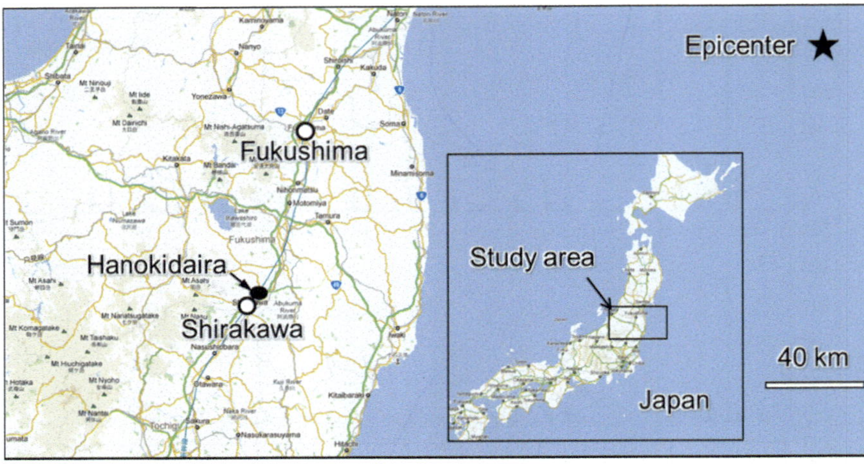

Fig. 12 Locations of the epicenter of the 2011 Tohoku earthquake and Hanokidaira landslide (after Google Map). ○: Location of Shirakawa city; ●: Location of Hanokidaira landslide

with their sliding surface located above the groundwater table, both geologists and geotechnical engineers had paid great concern to them. To understand the initiation and movement mechanisms of these landslides, we conducted a reconnaissance field examination of many of these flowslides and studied seven of them in detail. Here one landslide that occurred in the Hanokidaira area (hereinafter called Hanokidaira landslide) in the south part of Fukushima Prefecture will be described. The possible generation and movement mechanisms are discussed. Through monitoring aftershocks in the landslide source area, the seismic response of the landslide area during the mainshock was estimated. The shear behaviors of the landslide materials with different moisture contents were examined and then the post-failure movement of the landslide was discussed. It is noted that details on some other landslides occurring on tephra slopes triggered by the 2011 Tohoku earthquake are presented in Wang et al. (2014).

3.1 Hanokidaira Landslide

Most of the flowslides triggered by the earthquake are distributed in northern Tochigi Prefecture and southern Fukushima Prefecture. As samples of these flowslides, the landslide occurring in the Hanokidaira area, Shirakawa City, Fukushima Prefecture (Fig. 12) will be detailed. A geologic map of the landslide area is shown in Fig. 13. This map was based on the 1:200,000 Seamless Digital Geological Map of Japan (Geological Survey of Japan 2003). The study area mainly consists of Quaternary fluvial deposits, Neogene volcanic rocks and sedimentary rocks, Paleozoic and Mesozoic sedimentary/metamorphic rocks, and Mesozoic granite. The landslide was triggered in the soil layer of Pleistocene tephra and highly weathered welded tuff. In the strata of the landslide areas, there are 12 layers of tephra that are deposited by the eruptions of Nasu Volcano at different times (Suzuki 1992).

Figure 14 shows the topography of the areas with the Hanokidaira, which was based on the urban plan map (1/2500, with a contour interval of 2 m) of Shirakawa City, Fukushima Prefecture. The after-event topography was measured by using a laser rangefinder. The topographies of the landslide area before and after the earthquake were examined by using Google Earth images.

The Hanokidaira landslide (Fig. 15) was triggered on a southeast-facing slope of a hill, which is located approximately 1 km north of the JR Shirakawa Station in Shirakawa City. The displaced landslide materials killed thirteen people and destroyed 10 houses. Some of the dispatched material run into the golf driving range and formed a thin deposit.

Stratigraphically from top to bottom, the slope consists of forest soil, weathered pumice and scoria partly with a palaeosol, a palaeosol, and a basal deeply weathered welded tuff. The bedded texture is approximately parallel to the slope surface. There are visible striations on the surface of outcropped primary palaeosol that overlies the

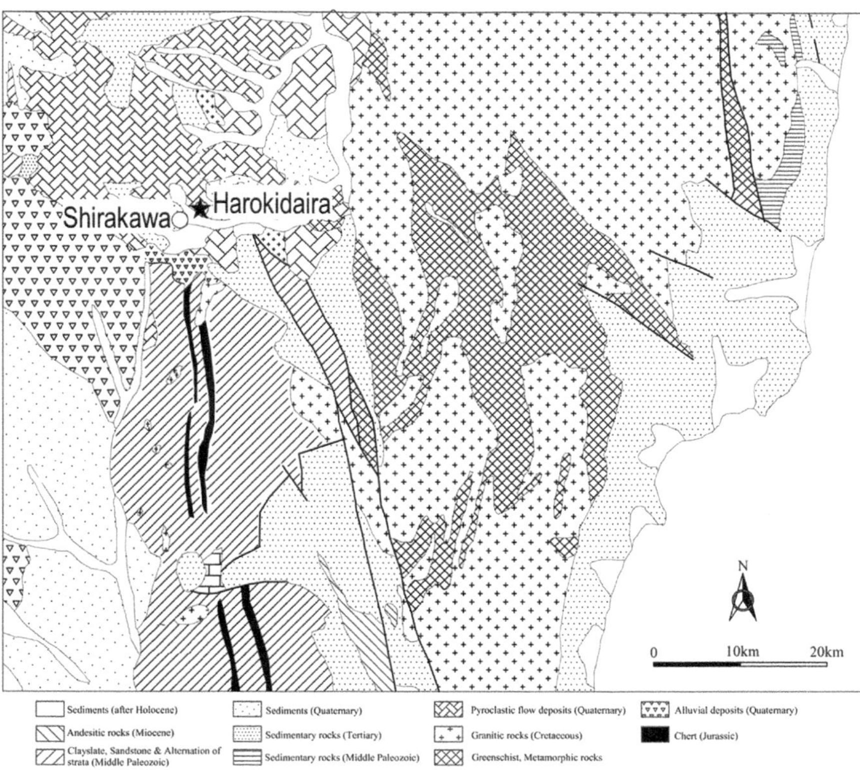

Fig. 13 Geological map of the study area (after geological survey of Japan 2003)

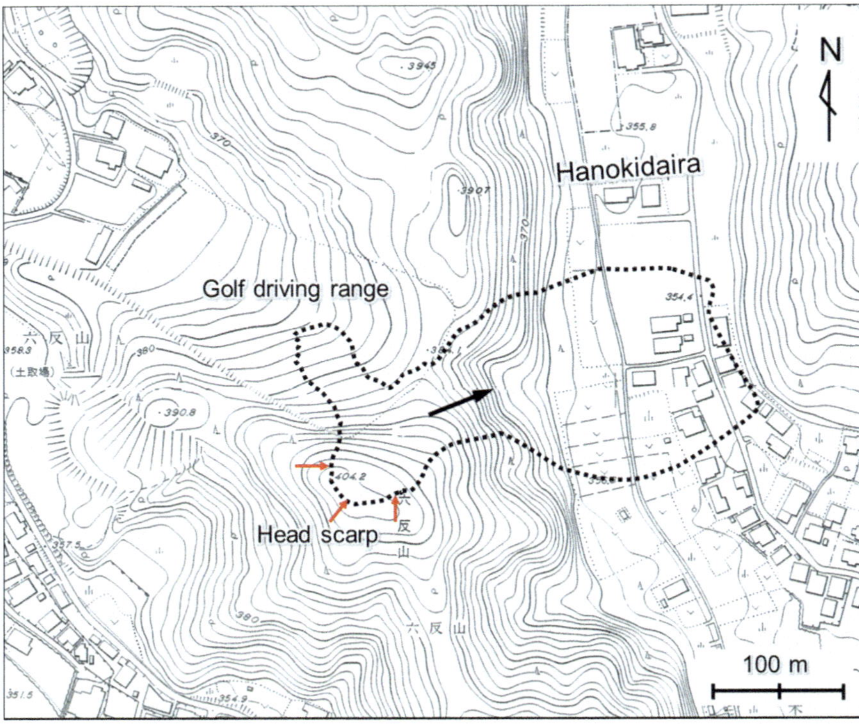

Fig. 14 Topographic map of Hanokidaira landslide area

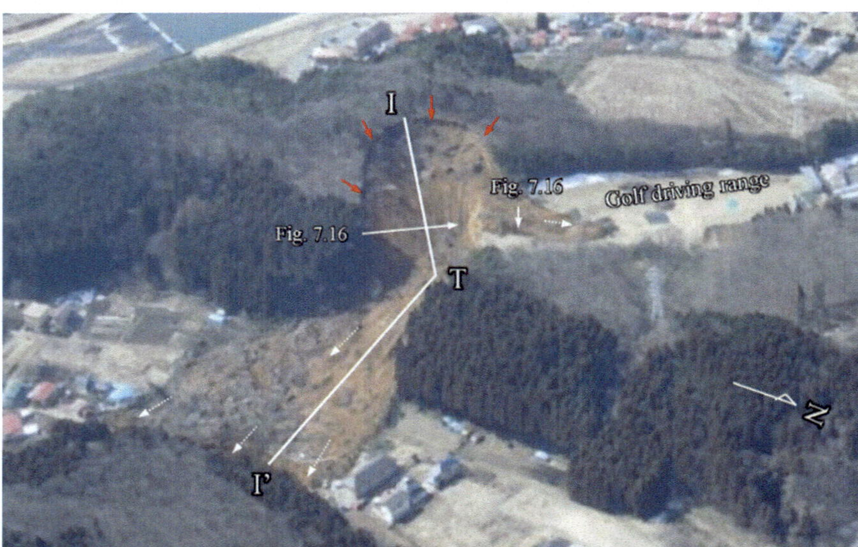

Fig. 15 Oblique view of the landslide on the Hanokidaira area (after Takeshi 2011)

welded tuff (Fig. 16a). An observation pit was dug into the deposits on the golf driving range, and it was found that grass was knocked down and buried by the displaced debris, but no grass had been cut or displaced from its original place (Fig. 16b). Above the grass was the palaeosol layer of ~10 cm thick, which was overlain by loose pumice. It is interesting to note that many trees stood almost vertically even after a long distance of travel with the displaced landslide maters. On the left flank of the landslide, muddy striae outcropped (Fig. 16c). On the high position above the sliding surface, splashes of mud were also visible (Fig. 16d). Therefore, it is inferred that the sliding movement was at first initiated and the liquefaction phenomenon occurred within the displaced landslide materials with the progress of downslope sliding, and the liquefied materials (mud) were splashed during the movement. Therefore, according to Cruden and Varnes (1996), the movement of this landslide can be categorized as earth *slide-flow*.

Based on the topographic measurements by means of a laser rangefinder, the longitudinal section of the landslide was obtained (Fig. 17). The main part of the displaced materials smashed into the residential area at the foot of the hill and presented typical features of fluid-like flow with spreading. Measuring from the point I to I' in Fig. 15, the displaced landslide material had a horizontal travel distance of about 290 m and a total relief of about 50 m, which gave an apparent mobilized friction angle of about

Fig. 16 **a** View of sliding surface and the left side margin; **b** grass buried on the golf driving range and the palaeosol layer above the grass; **c** muddy striation on the left-side margin; **d** view of the splashed mud on the left-side margin

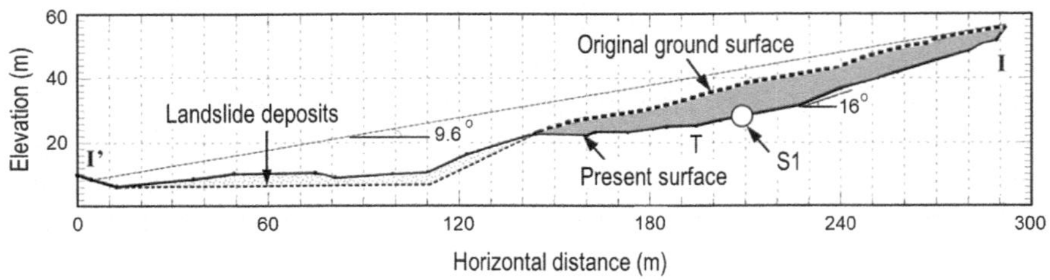

Fig. 17 The longitudinal section along the line I–I' shown in Fig. 15

9.6°, Based on the measured topography, the volume of the displaced material was estimated to be approximately 100,000 m^3.

At the first time of our field survey, we did not find standing groundwater or groundwater seepage on the landslide source, however, we were told that there was a spring at the toe part of the slope, which had continuous water flow before the earthquake. Late, with the progress of stabilization countermeasures on the landslide area, the debris on the lower part of the slope (near point T in Fig. 15) has been removed and some small springs with continuously flowing water outcropped on the slope (observation in July 2012). Therefore, it is reasonable to infer that there was standing groundwater within the soil layers of the slope before the earthquake.

With the progress of stabilization countermeasures, tephra layers on the source area were exposed, and a paleo-valley had also been identified (as shown in Fig. 18). From the upper surface of the welded tuff, it is inferred that the hilly area was substantially dissected with a paleo-valley, which was aligned approximately east–west and sloping to the east (PV–M in Fig. 18). A tributary of the paleo-valley is also noticed, which slopes up towards point PV-T as shown in Fig. 18. The palaeosol was mainly developed on this hilly paleo-surface and subsequently mantled by thick layers of air-fall volcanic ash. Through long-term erosion and deposition of tephra newly coming from interval eruption of volcanos, the thickness of deposited tephra on the flanks of the paleo-valley has become shallow than that along the axis of the paleo-valley. All these tephra layers above the palaeosol developed on the weathered welded tuff had slid out of the source area during the earthquake.

3.2 *Seismic Motion*

As well known, Japan has a very dense earthquake observation network operated by JMA (the Japan Meteorology Agency), NIED, Japan Agency for Marine-Earth Science and Technology (JAMSTEC), Port and Airport Research Institute, universities, and local government, and other communities. The observation system is under working continuously for 24 h per day, and all the observed data will be available immediately after each earthquake event. Around the Hanokidaira landslide area, there are two permanent strong-motion stations, one is the K-NET Shirakawa station

Fig. 18 View of paleo-valley on the Hanokidaira landslide source area exposed by the construction of slope stability measures. S1 and S1': Locations for disturbed and intact samples of palaeosol, respectively; PV–M and PV–T: the main paleo-valley and a tributary of the main paleo-valley, respectively

of NIED and another one is the JMA Shirakawa-Shi Kakunai station. The recorded seismic waves at these two stations during the mainshock are shown in Figs. 19a, b, respectively. At both stations, the N–S component of motion presented the greatest values, while the vertical component was the smallest. However, it is notable that the peak values in both stations showed a remarkable difference. The JMA station had a peak acceleration of 371.5 gals, while the K-NET station recorded a maximum of 1295.2 gals, about three times the former one. Considering that the JMA station is very close to the landslide area (about 800 m south), and the K-NET station is about 2.8 km far, it is reasonable to infer that during the mainshock the landslide area may have had a seismic response similar to those observed at the JMA station. However, as well known, the local characteristics of soil/rock layers and topography of a site may greatly affect the seismic responses. Therefore, it is necessary to further understand the local response of the landslide area for a better understanding of the seismic motion of the landslide area during the mainshock. To achieve this, we installed a seismometer on the upper part of the landslide area on April 11, 2011, and succeeded in recording aftershocks. Using these newly obtained seismic recordings and those records obtained by the JMA Shirakawa-Shi Kakunai station, the seismic response of the landslide area during the mainshock is estimated through a site-effect substitution method. Figure 20 shows the estimated seismic wave, which presents a peak acceleration of 778 gals in the E–W direction, and about 907 gals in the N–S direction. These values are approximately about 2.3 and 2.4 times greater than those records at the

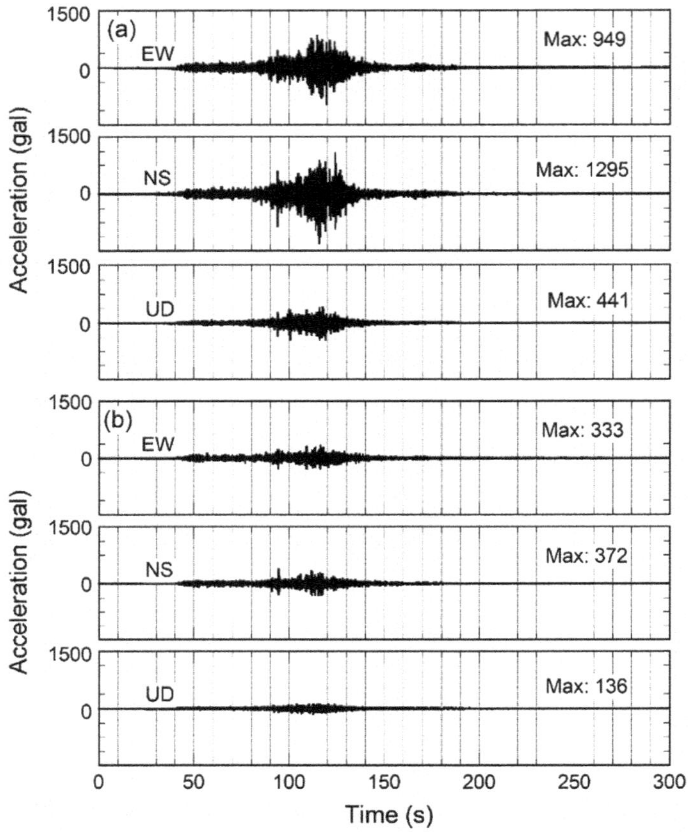

Fig. 19 **a**, **b** Seismic records and maximal values in each component that were observed at the K-NET Shirakawa station and the JMA Shirakawa-shi Kakunai station, respectively. EW, NS, UD: seismic motion in the direction of east–west, north–south, and up-down, respectively

JMA station, respectively. It is worth noting that the estimated vertical acceleration showed a peak value of 463 gals, which is approximately 3.4 times greater than that recorded at the JMA station, indicating that the landslide area may have a stronger site effect on the vertical motion. From Fig. 20, we infer that the peak seismic motion on the slope at Hanokidaira might be at least greater than 907 gals (NS component).

3.3 Geotechnical Aspect

Field evidence suggests that the boundary between the overlying pumice layer and the palaeosol layer on the slope might have played a key role in the initiation and movement of the landslide. Therefore, understanding the geotechnical properties of those materials distributed along the sliding surface will be of great importance. To

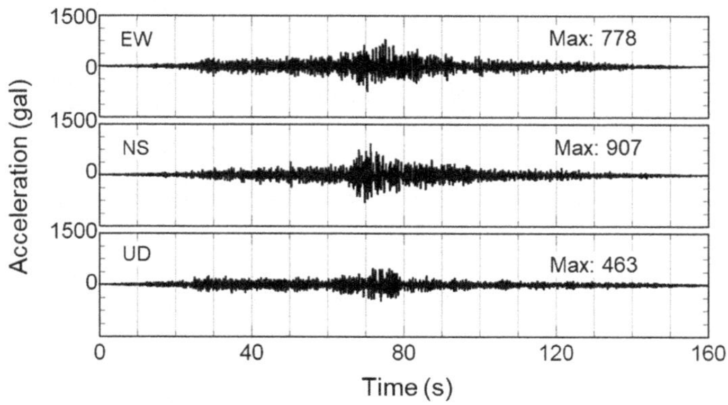

Fig. 20 The estimated seismic waves on the Hanokidaira area

achieve this, in 2011, disturbed samples were taken from the palaeosol (S1) and pumice with scoria (S2) on the source area (about 40 kg for each), respectively. In July 2012, intact samples (two blocks sized 40 × 40 × 30 cm approximately for each) were also taken from the palaeosol layer exposed during the construction of countermeasures (S1' in Fig. 18).

Palaeosol S1 had a specific gravity of approximately 2.70 g cm^{-3}, natural moisture content of 94–160%, dry density of 0.48–0.66 g cm^{-3}, and void ratio of 3.10–4.18, the pumice and scoria sample S2 had a specific gravity of 2.68 g cm^{-3}, natural moisture content of 94–151%, dry density of 0.51–0.64 g cm^{-3}, and void ratio of 3.25–4.25, while the intact blocks of S1' had a natural moisture content of 114% with a dry density of 0.636 g/cm^3.

Direct shear box tests were at first conducted on palaeosol sample S1 under the normal stresses of 84, 64, and 41 kPa, respectively. The results are presented in Fig. 21. In all the tests, the samples were at first sheared under their natural moisture contents. Some tests were also conducted on fully saturated intact samples. From Fig. 21, it is seen that the intact samples sheared under their natural moisture content have a friction angle (φ') of 27° and an apparent cohesion (c) of 22 kPa. For the remolded samples, the friction angle showed a lower value of 14°, while the cohesion was 26 kPa, presenting no significant change. The friction angle and cohesion for the intact sample under fully saturated conditions are 26° and 19 kPa, respectively. This may result from the fact that the intact sample originally had a very high natural moisture content, which gave a saturation degree was 94.3%.

Undrained cyclic shear behaviors of both samples under natural moisture contents were examined by using a ring shear apparatus. The moisture contents are 114% and 140% for palaeosol S1 and pumice with scoria S2, respectively. During the test, the samples were at first placed into the shear box and then were normally consolidated under a given normal stress. After the consolidation, cyclic shear stress was applied to the samples under undrained conditions. The applied cyclic shear stress has an amplitude of 24 kPa. A low frequency of 0.5 Hz was selected for better monitoring response of pore-water pressure, which could be built up within the shear zone with

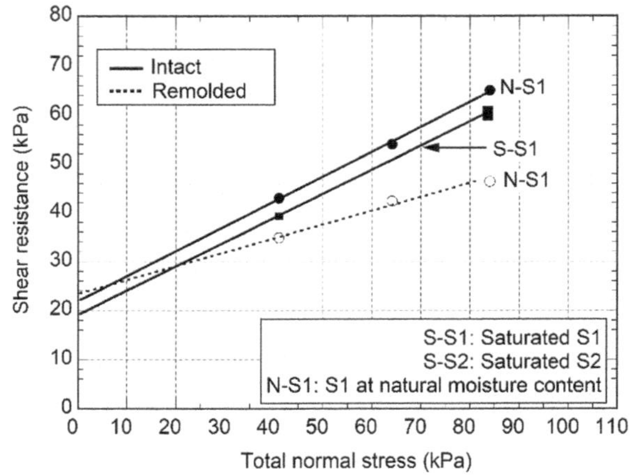

Fig. 21 Shear strength for intact and remolded sample S1 under different water contents measured by direct shear box tests

the progress of shearing. Figures 22 and 23 present the results for both samples, in the forms of time series data and variation of shear strength and monitored pore-water pressure with the progress of shear displacement, respectively. As shown in the figure, with the application of the cyclic shear loading, shear failure was triggered in both tests, shear displacement increased continuously even after the cyclic shear loading was stopped, and the shear resistance lowered further to approximately zero after the initiation of shear failure. Although the monitored pore water pressure is very small and may not represent that within the shear zone, probably due to the inclusion of pore air or problem in the PWP monitoring system (Wang et al. 2016), it is reasonable to infer that liquefaction phenomenon had been triggered within the soil layers along the sliding surface, because the landslide suffered from strong-motion with long duration during the mainshock of the earthquake, and the water content of the soil layers along the sliding surface was also very high (almost close to fully saturation degree).

Our field survey revealed that the sidewall of the landslide mass would have suffered from static shearing resulting from the movement of displaced landslide material. To understand the static shear behavior, static shearing tests were conducted on sample S2 with a natural moisture content of 140%. One of the static shear tests is exampled in Fig. 24. After the sample was placed into the shear box, it was normally consolidated under normal stress of 84 kPa, and then loaded by an initial shear stress of 24 kPa (corresponding to the gravitational stress of a soil layer on a slope of 16 degrees). After the consolidation, the shear box was switched into undrained condition, and then the sample was sheared by continuously increasing the shear stress. As shown in Fig. 24, excess pore water pressure was built up gradually with the increase of shear stress at the initial shear stage. However, after the shear failure was initiated, shear resistance became small gradually and finally lowered to zero approximately. The vertical displacement showed a continuous increase, indicating that the sample suffered consolidation with the progress of shearing. It is worth noting that when the shear box was switched into drainage condition after the undrained

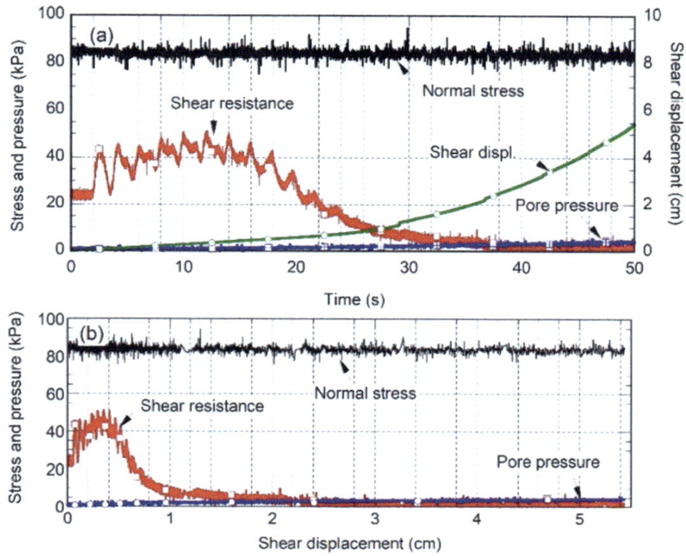

Fig. 22 Undrained cyclic shear tests on palaeosol sample S1 at natural water content (114%). **a** Time series data; **b** variation of stresses and pore water pressure against shear displacement. The dry density (after consolidation) was 0.70 g/cm^3

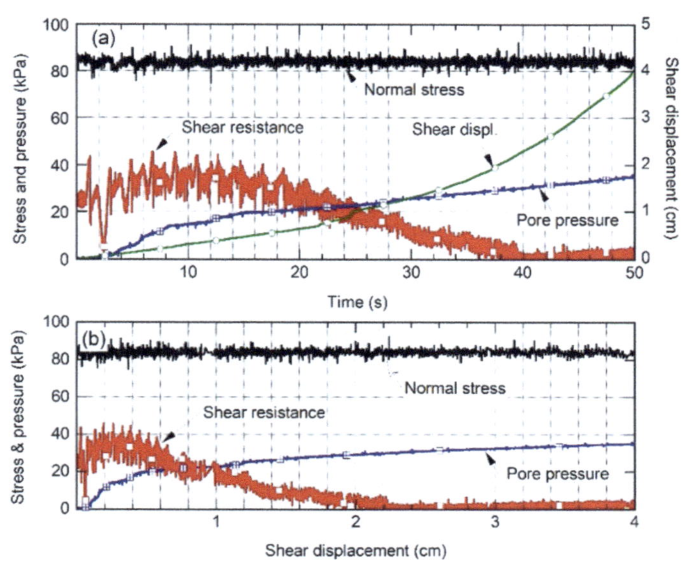

Fig. 23 Undrained cyclic shear tests on pumice with scoria samples S2 at a natural water content of 140%. **a** Time series data; **b** variation of stresses and pore water pressure against shear displacement. The dry density (after consolidation) was 0.51 g/cm^3

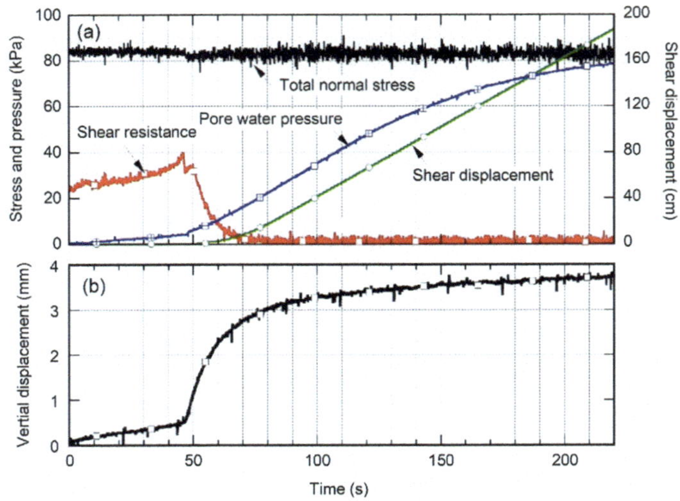

Fig. 24 Time series data for the undrained shear test on sample S2 (pumice with scoria, with a moisture content of 140%). **a** For normal stress, shear stress, shear displacement, and buildup of pore-water pressure against time; **b** for vertical displacement (positive indicates sample consolidation). The dry density of the sample after consolidation: 0.55 g/cm^3

shear test, water flew out from the drainage tubes. Therefore, it is inferred that even though the sample was not fully saturated, the high water content of the sample enabled the buildup of high pore-water pressure within the sample when shear failure was triggered.

4 Landslides Triggered by the 2016 Kumamoto Earthquake

The 2016 Kumamoto earthquakes started on 14 April 2016 with a Mj6.5 foreshock (epicenter: 32° 44.5'N, 130° 48.5'E, depth: 11 km) at the local time of 21:26, and then occurred on 16 April 2016 with a Mj7.3 mainshock (epicenter: 32° 45.2'N, 130° 45.7'E, depth: 12 km). Figure 25 presents the epicenter, distribution of the source faults of the earthquake and the location of study area. These earthquakes caused widespread damage throughout a wide area in Kumamoto and neighboring prefectures, resulting in the collapse of numerous buildings, and 50 deaths (Doi et al. 2019).

A huge number of landslides were triggered by the earthquake in and around Minamiaso Village located in the western part of Aso caldera, Kumamoto Prefecture. Figure 26 presents the distribution map of the landslides triggered by the earthquake, which was made by NIED (2016).

There are mainly two different types of landslides, based on their locations. One type occurred on the inner wall of the caldera, and the other type occurred on the

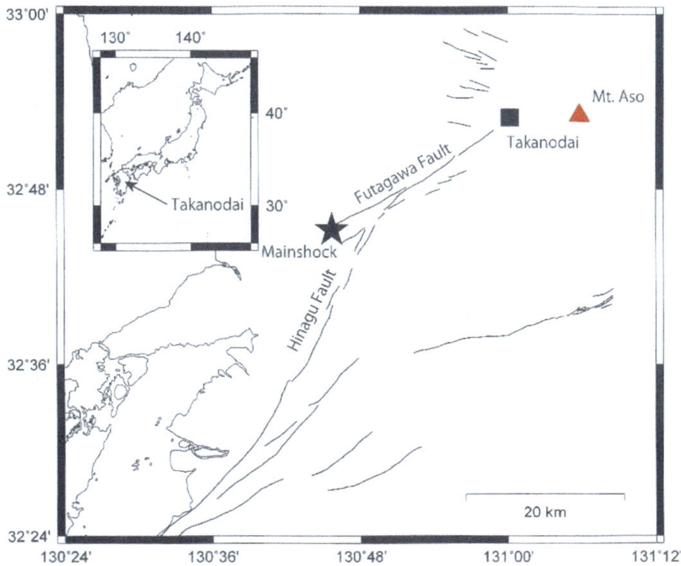

Fig. 25 Epicenter and the distribution of the source faults of the 2016 Kumamoto earthquake as well as the location of the Takanodai landslide area. The fault traces indicated by black lines are from the Research Group for active faults of Japan (1991)

hillslopes inside the caldera. Among all these landslides, the biggest one occurred on the inner wall of the caldera in front of the Aso Bridge (Fig. 27). This landslide is about 2.5 million m^3 in volume. The displaced landslide materials (mainly originating from volcanic rocks and pyroclastic flow deposits) destroyed the highway on the toe part of the slope, and further caused the collapse of the Aso Bridge across the river, directly resulting in the death of one person who was driving the car crossing the bridge.

During the earthquake, one of the devastating landslides occurred in the Takanodai area (Figs. 28 and 29). From the scarp of the landslides, it is identified that the soil layers above the sliding surface (about 8 m thick) are tephra deposits with several layers of andosol between them. The landslide materials were dispatched into three main different parts (termed R, M, and L parts, respectively), and each of them showed a different sliding direction (Fig. 28). The displaced landslide materials in Part R were fluidized with long and rapid movement, resulting in 5 fatalities and severe damage to the houses downstream. The debris in Part M also fluidized and finally rested on a gentle slope after traveling a long distance. However, Part L showed less mobility. The displaced materials in Part L were not completely scattered, and there are a lot of huge blocks (~5 m^2) near the source area with the original lawn on the surfaces of these blocks. These features indicate that Part L can be categorized as a spread flow, following the clarification given by Cruden and Varnes (1996). As shown in Fig. 30a, the sliding surface (exposed on the middle part of the source area) has a whitish soil layer on the surface, below which is volcanic soil with dark

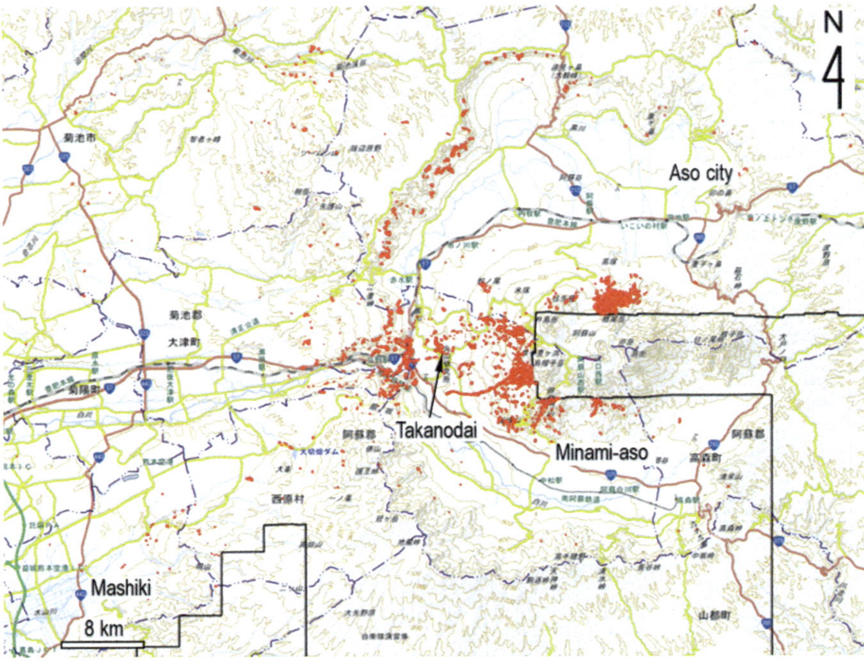

Fig. 26 Distribution of coseismic landslides related to the Kumamoto earthquake and the location of the Takanodai landslide (revised after NIED 2016)

Fig. 27 Landslide on the inner slope of the Aso caldera (taken on 16 April 2016)

brown color. The apparent friction angles for Part R, M, and L are estimated to be 6.1, 6.5, and 7.2 degrees, respectively. Through digging survey pits and trenches in the locations immediately below the scarp, it was identified that this whitish soil layer was mantled by a yellowish pumice layer, which was deposited about 26 ka ago (Fig. 31). Therefore, it is inferred that the sliding surface was formed along the boundary between the volcanic soil and this yellowish pumice layer. More details on the distribution of the landslides and their geologic features can be obtained in Sato (2017).

To better understand the initiation and movement mechanisms of the fluidized landslide occurring in the Hanokidaira area, we examined the geological features of this landslide and performed in-situ direct shear tests on the intact samples taken from three different layers (as shown in Fig. 31). We also took samples from these three layers and conducted drained/undrained ring shear tests on them that were prepared at different initial densities with different water contents. The samples were sheared by applying static or cyclic shear loadings.

The clay minerals of the weathered whitish pumice layer were first analyzed by using an X-ray diffractometer (XRD). The results are presented in Fig. 32, which revealed that the clay minerals of the sample consist mainly of halloysite. We further examined the shape of the halloysite minerals by means of a transmission electron microscope (TEM) and the results are presented in Fig. 33. Because we did not

Fig. 28 Google Earth image of Takanodai landslide (taken on 16 April 2016). R, M, L: Landslide materials slid along the right, central and left sides of the slope downstream, respectively

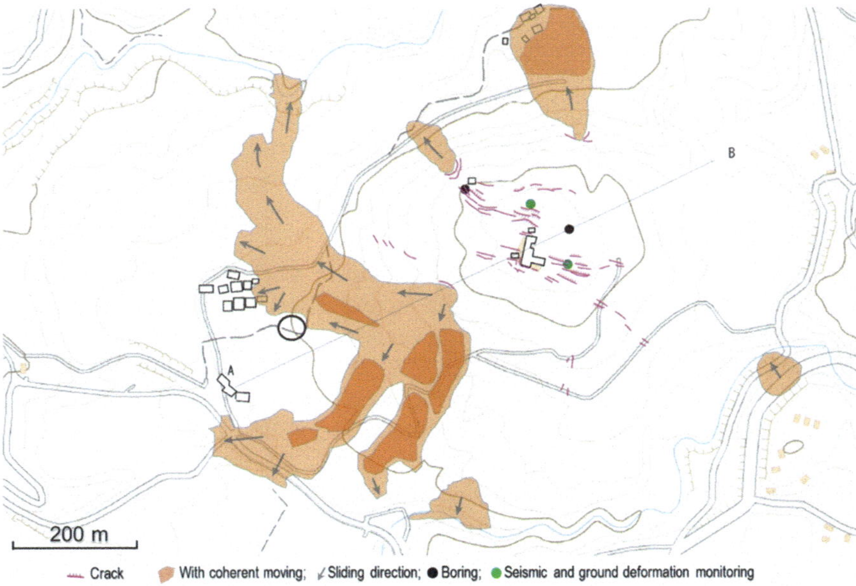

Fig. 29 Plan of Takanodai area after the earthquake. Addition to the landslides, many cracks also occurred on the hill slope. As the emergency measures, two borings were drilled, and systems for monitoring the seismic and ground deformation were installed on the hillslope immediately after the earthquake

Fig. 30 **a** Outcrop of sliding surface at location; **b** muddy striations on the sliding surface along the right-side margin

Fig. 31 A profile on the main scarp, showing the layers of pumice and underlying soil layer with a thin white layer sandwiched between them. The thin white layer is rich in halloysite

use dispersive agents to treat the sample before observation, the halloysites show a different topography. From Fig. 33, it can be noticed that the aggregate presents the shape of crumpled lamellae, while single halloysite tubes can also be identified.

For the Takanodai landslide, the samples from the yellowish pumice layer had high natural moisture contents (ranging from 133 to 178%), but small dry densities (ranging from 0.44 to 0.56 g/cm^3). The results obtained by the in-situ direct shear box tests on the intact samples showed that the whitish pumice layer developed on the bottom of the yellowish pumice has a very small peak shear strength (Fig. 34). Drained and undrained shear tests on these samples show that when the moisture content is high enough (close to a fully saturated state), the sample is very prone

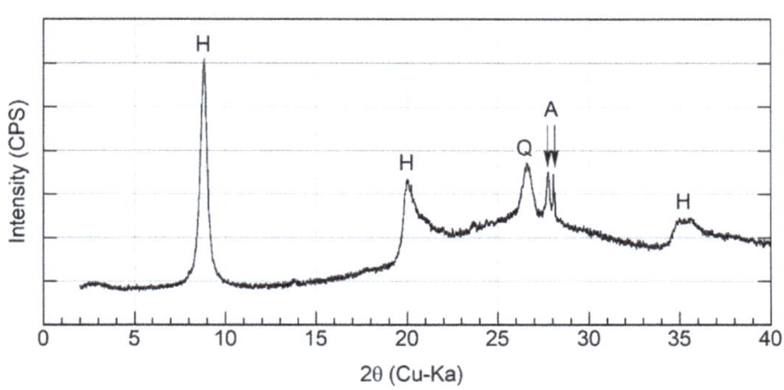

Fig. 32 XRD pattern for the sample taken from the thin layer of weathered whitish pumice (shown in Fig. 31), Takanodai area, Minamiaso, Kumamoto. H: Halloysite; Q: Quartz; A: Albite

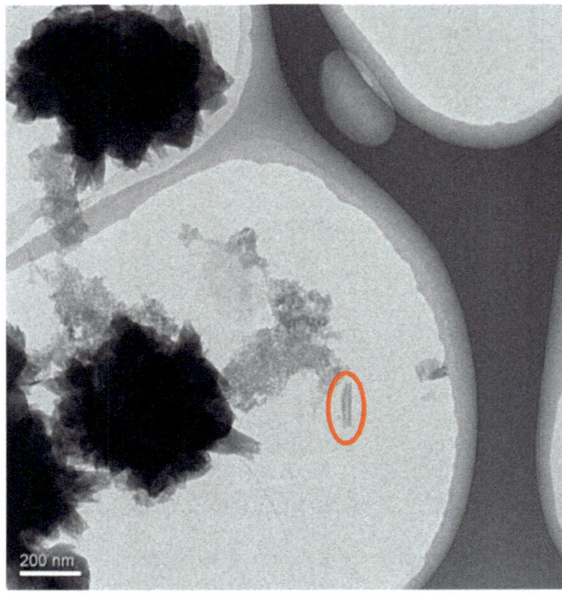

Fig. 33 TEM image of the sample from Minami Aso, Kumamoto (the red circle shows the tubular halloysite)

to suffer from liquefaction under the cyclic loading, and high pore-water pressure can also be built up within those samples with a lower saturation degree given the shear failure being triggered. Therefore, it is inferred that the whitish pumice layer enhanced the instability of the slope during the earthquake, and with the progress of sliding, high water pressure was further built up within the whitish shear failure and elevated the landsliding mobility. However, during the downslope movement, the high pore-water pressure might also be built up with the yellowish pumice layer with high water content and then further elevate the mobility of landslide materials.

By using the seismic records obtained at a nearby earthquake observation station, the possible coseismic loading resulting from the earthquake motion had been estimated (shown in Fig. 35), and applied to the samples with different initial densities and moisture contents. The results suggest that the coseismic response of the tephra slopes with different moisture content could be very complicated. Further study of this issue will be needed for us to clarify the initiation mechanism of this kind of earthquake-induced landslides of pyroclastic fall deposits occurring on gentle slopes.

5 2018 Hokkaido Eastern Iburi Earthquake

At 3:08 on September 6, 2018, local time, a magnitude 6.7 earthquake occurred in the eastern part of Hokkaido, with a focal depth of 37 km. Figure 36 shows the peak ground acceleration map for the 2018 Hokkaido Eastern Iburi Earthquake that was made by NIED (2019). The earthquake triggered a huge number of landslides (Asia

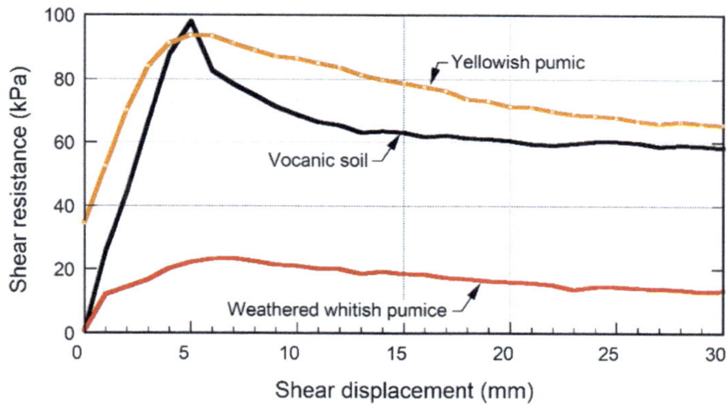

Fig. 34 In-situ direct shear box tests on the samples taken from the volcanic soil layer, yellowish pumice layer, and weathered whitish pumice layer, respectively. All the tests were conducted under normal stress of 81.3 kPa. The water contents were 178%, 167% and 198% for samples taken from the yellowish pumice, volcanic soil, and weathered whitish pumice layers, respectively

Air Survey Co., LTD 2018; Kameda et al. 2019). The high density of landslides triggered by this earthquake is evident on the distribution map shown in Fig. 37. According to the MLIT, the collapse area was about 13.4 square kilometers. The landslides caused 36 deaths in the Atsuma area. The most catastrophic ones were those occurring in the Yoshino area (Fig. 38), where 19 people were killed by the landsliding materials. In addition, the earthquake destroyed many houses, and the roads were blocked in the Atsuma and Yoshino areas. Figure 39 presents a view of the Takaoka area, Atsuma, showing how densely the landslides on slopes of pyroclastic fall deposits were triggered. These landslides prompted great concern from all over the world, and further indicate the importance and urgency of landslide hazard mitigation in areas of pyroclastic fall deposits, while our understanding of the coseismic response of this kind of slope and the basic geotechnical and hydrological properties of pyroclastic fall deposits are still limited. It is noted that great effort is now being paid to unraveling the initiation and movement of these coseismic landslides occurring in pyroclastic fall deposit areas, and it is expected that great progress may be obtained soon.

6 Summary

Earthquakes occur frequently in Japan, normally resulting in great damage to properties and large numbers of casualties. Landslides triggered by earthquakes are also a great threat to local properties and the lives of residents. In many cases, most of the casualties during an earthquake result from the coseismic landslides. In this report, we introduced those landslides triggered by some recent earthquakes in Japan. The information on these landslides can be summarized as follows.

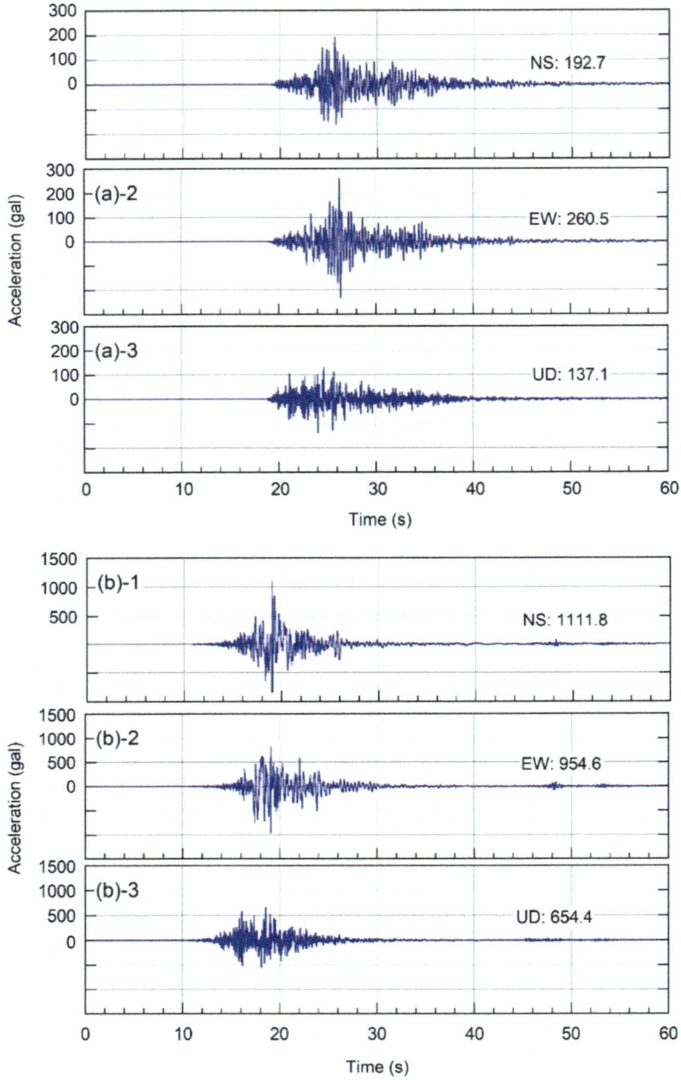

Fig. 35 Ground motion recorded at a seismic station in Minami Aso on April 14 (**a**) and April 16 (**b**), respectively

The 2004 Mid-Niigata Prefecture earthquake triggered a huge number of landslides including many reactivated ones. Among these reactivated landslides, the one that occurred in the Higashi-Takezawa area was introduced as an example. This landslide is a reactivated ancient landslide, and its sliding surface is developed above a smooth surface of siltstone. The rainfalls prior to the earthquake might have elevated the saturation degree of the soil layers and/or increased the groundwater table and

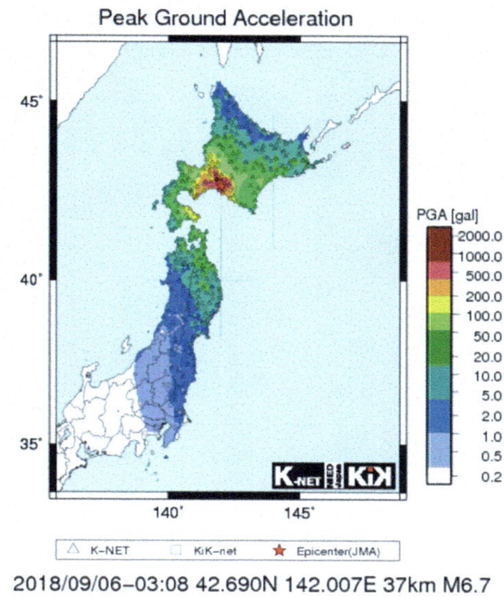

Fig. 36 The peak ground acceleration map for the 2018 Hokkaido Eastern Iburi Earthquake (NIED 2019)

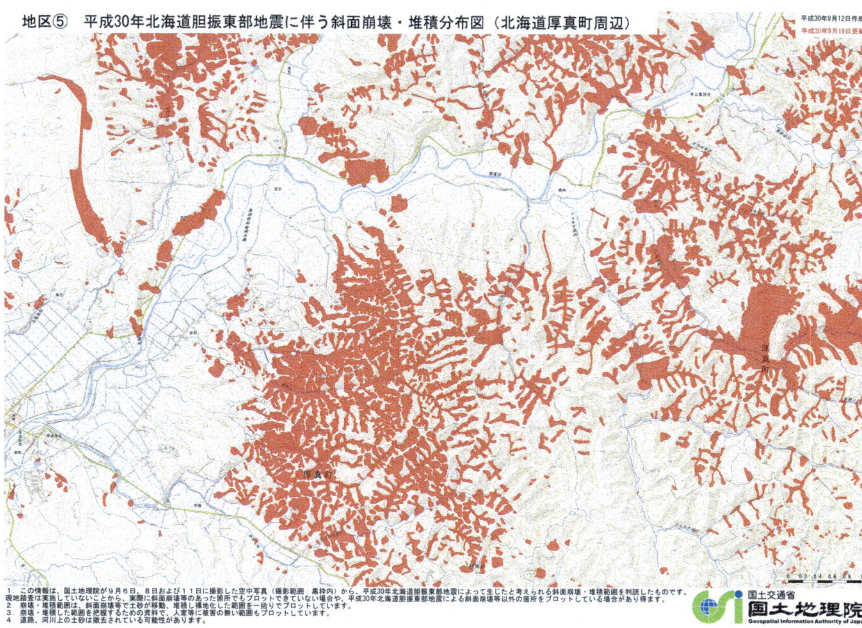

Fig. 37 Distribution map of landslides triggered by the Hokkaido Iburi Eastern Earthquake (from GIA: https://www.gsi.go.jp/BOUSAI/H30-hokkaidoiburi-east-earthquake-index.html#10)

Fig. 38 View of Yoshino area, Atsuma, Hokkaido after the earthquake (photo on October 20, 2018). The landslides directly resulted in 19 casualties

Fig. 39 Cluster of landslides on tephra slopes triggered by the 2018 Hokkaido Eastern Iburi Earthquake (photo on October 21, 2018)

favored the initiation of slope instability. The generation of high pore-water pressure within the saturated soil layer along the sliding surface after the failure elevated the mobility of the displaced landslide materials. The landslide materials dammed the river at the toe of the landslide, causing great damage to the houses on the upstream reach of the river due to backwater flooding, and bringing a great threat to a wide area downstream. The successful implementation of countermeasures for stabilizing the landslide dam and lowering the flood risk provides a good experience for the emergency response to future risks from landslide dams.

The landslides triggered by the 2011 Tohoku Earthquake in pyroclastic fall deposits showed very high mobility, with the occurrence of liquefaction. The landslide occurring in the Hanokidaira area was investigated and presented in detail as an example of some landslides that were triggered in Shirakawa City of Fukushima Prefecture and neighboring areas. Before the earthquake, distinct valley incision at the toe of these landslides might have lowered the stability of the upper slope of the landslide source area. Although a nearby permanent seismic station gave records that were relatively weak than those records obtained at other stations, the post-event seismic monitoring on the landslide area indicates that the seismic motion at the Hanokidaira landslide area might have been very strong because of the local site effects. The sliding surface was developed along the boundary between a palaeosol and the overlying pumice and scoria. Results of undrained shear tests indicate that these soil layers are highly liquefiable even in the state of natural water content. Therefore, it is inferred that shear failure might have been triggered first on the palaeosol layer, but liquefaction might have been triggered in both the palaeosol layer and the overlaying pumice layer, such that the displaced materials showed high mobility.

The 2016 Kumamoto earthquake triggered numerous landslides in and around Minamiaso Village, located at the western part of Aso caldera, Kumamoto Prefecture. Although the biggest landslide originated on the inner slope of the caldera in front of Aso bridge, most of the landslides were triggered within the pyroclastic deposits on gentle slopes inside the caldera, with high mobility and long travel displacement. Detailed field investigation and experimental shear tests on the samples from the landslide area revealed that the tephra slopes are rich in water content, which can favor the generation of high pore-water pressure during the progress of shearing after the shear failure occurred. We estimated the possible dynamic normal and shear stresses due to the introduction of the earthquake using the records from a nearby seismic station and then applied them to the samples with different degrees of saturation. The results showed that shear failure could be initiated, but the movement stopped when the seismic loadings terminated, indicating that the landslide would not have suffered from long runout movement under the inferred seismic loadings. These results suggest that the initiation and movement mechanisms of these coseismic landslides of pyroclastic fall deposits occurring on gentle slopes are very complicated, probably since the coseismic response of these slopes has not been fully understood. The cluster of landslides in Atsuma Town triggered by the 2018 Iburi earthquake in Hokkaido provides another example, indicating our poor understanding of coseismic landslides on slopes covered by pyroclastic fall deposits, and addressing the importance and urgency in promoting research on this issue.

Acknowledgements This study used the seismic recordings of the K-NET operated by the National Research Institute for Earth Science and Disaster Resilience (NIED), Japan, and of the Japan Meteorological Agency (JMA). The author thanks Emeritus Professor Kyoji Sassa of Kyoto University, Emeritus Professor Masahiro Chigira of Kyoto University, Emeritus Professor Toshitaka Kamai of Kyoto University, Assistant Professor Issei Doi at Kyoto University, Professor Gen Furuya at Toyama Prefectural University, Professor Fawu Wang at Tongji University, China, and many colleagues from other different institutes. Their great support and discussion are greatly appreciated.

References

Asia Air Survey Co., LTD (2018) Reports on the disaster triggered by 2018 Hokkaido Eastern Iburi Earthquake. http://www.ajiko.co.jp/article/detail/ID5G5WZ38R5. Accessed on 18 Sept 2018

Bishop AW (1973) The stability of tips and spoil heaps. Q J Eng GeolHydrogeol 6:335–376

Chigira M, Yagi H (2006) Geological and geomorphological characteristics of landslides triggered by the 2004 Mid Niigata prefecture earthquake in Japan. Eng Geol 82:202–221

Cruden DM, Varnes DJ (1996) Landslide types and processes. In: Turner and Schuster (ed) Landslides: investigation and mitigation. Press, National Academy, pp 36–71

Doi I, Kamai T, Azuma R, Wang G (2019) A landslide induced by the 2016 Kumamoto Earthquake adjacent to tectonic displacement—generation mechanism and long-term monitoring. Eng Geol 248:80–88

Doshida S, Uchiyama S (2012) Features and distribution of landslides in the 2011 off the Pacific coast of Tohoku earthquake. Nat Disaster Res Report Nat Res Inst Earth Sci Disaster Prevent 48:111–120

Geological Survey of Japan (2003) Geological map of Japan at scale 1:200 Million (5th version, CD-ROM)

Kameda J, Kamiya H, Masumoto H, Morisaki T, Hiratsuka T, Inaoi C (2019) Fluidized landslides triggered by the liquefaction of subsurface volcanic deposits during the 2018 Iburi-Tobu earthquake. Hokkaido Sci Rep 9:13119

Ministry of Land, Infrastructure and Transport (2005) On the landslides triggered by the 2004 Niigata Chuetsu earthquake. http://www.mlit.go.jp/kisha/kisha05/05/050113_.html. Accessed on 9 Nov 2021

NIED (2016) Distribution map of sediment disaster triggered by the 2016 Kumamoto Earthquake. https://map03.ecom-plat.jp/map/map/?cid=20&gid=587&mid=2908. Accessed on 3 June 2022

NIED (2019) Strong motion during the 2018 Hokkaido Eastern Iburi Earthquake. https://www.kyoshin.bosai.go.jp/kyoshin/topics/html20180906030750/main_20180906030750.html. Accessed on 8 Sept 2021

Research Group for Active Faults of Japan (ed) (1991) Active faults in Japan: sheet maps and inventories (revised edition). Tokyo University Press, Tokyo. p 437 (in Japanese with English Summary)

Sassa K (2005) Landslide disasters triggered by the 2004 Mid-Niigata prefecture earthquake in Japan. Landslides 2(2):135–142

Sassa K, Fukuoka H, Wang FW, Wang G (2007) Sliding mechanism of the 2004 mid-Niigata prefecture earthquake-triggered-rapid landslides occurred within the past landslide masses. J Japan Landslide Soc 44(2):71–78 (in Japanese with English abstract)

Sassa K, Fukuoka H, Wang FW, Wang G, Okada K, Marui H (2006) Landslide disasters triggered by the 2004 Mid-Niigata prefecture earthquake in Japan. Annuals Disas Prev Res Inst Kyoto Univ 49 C

Sato T (2017) Tephrostratigraphy and formative process of halloysite that control the potential sites of earthquake-induced landslides: case study of the landslides induced by the 2016 Kumamoto earthquake. Master Thesis, Graduate School of Science, Kyoto University

Suzuki T (1992) Tephrochronological study on Nasu Volcano. Bulletin of Volcanological Society of Japan (kazan). 37:251–263 (in Japanese with English abstract)

Takeshi T (2011) Landslide hazards immediately after the 2011 off the Pacific coast of Tohoku Earthquake and landslides in Shirakawa city, Fukushima. Report on the debriefing session of landslide hazards triggered by the 2011 off the Pacific coast of Tohoku Earthquake, Japan Landslide Society

Wang G, Suemine A, Zhang FY, Hata Y, Fukuoka H, Kamai T (2014) Some fluidized landslides triggered by the 2011 Tohoku Earthquake (Mw 9.0), Japan. Geomorphology 208:11–21

Wang G, Furuya G, Zhang FY, Doi I, Watanabe N, Wakai A, Marui H (2016) Layered internal structure and breaching risk assessment of Higashi-Takezawa landslide dam in Niigata, Japan. Geomorphology 267:48–58

Yamabe K, Marui H, Yoshimatsu H, Yamamoto S (2010) Landslide dam in Higashitakezawa and Terano area formed by the Mid Niigata Prefecture Earthquake. J Japan Landslide Soc 47(6):41–52 (in Japanese with English abstract)

Co-Seismic Landslides in Greece, a Review

Charalampos Saroglou, Vasilis Kallimogiannis, Kyriazis Pitilakis, and George Papathanassiou

Abstract This report presents a review of recorded co-seismic landslides in Greece since 1840. Various researchers have reported landslides triggered by historical and more recent earthquakes. The most significant co-seismic landslides in Greece since 1840 are discussed here briefly. A magnitude–distance relation chart is prepared, based on the recorded data. From this it is concluded that the magnitude of earthquakes that triggered landslides is between $M_w = 5.5$ and 7.9, while the minimum distance from the epicentre to a reported landslide was 5.5 km and the maximum distance was 131 km. Most of the landslides in Greece occur in the post-alpine deposits and weak flysch, as well as in weathered rock formations. Additionally, landslides are more pronounced in areas affected by tectonic activity. Recent earthquakes in Greece, that have triggered landslides, are presented in more detail. These case studies are encountered in the Ionian Sea in western Greece, which is considered as the region most prone to earthquakes, not only in Greece, but in all of Europe. It is characterized by very high seismicity, with earthquake magnitudes up to Mw=7.4. The recent earthquakes of Cephalonia Island in 2014 and Lefkada Island in 2015, which both have triggered landslides and caused a significant number of failures in road networks, buildings and ports, are presented in more detail. Finally, a reference to the most significant co-seismic rockfalls in Greece is given.

C. Saroglou (✉) · V. Kallimogiannis
School of Civil Engineering, National Technical University Athens, Athens, Greece
e-mail: saroglou@central.ntua.gr

V. Kallimogiannis
e-mail: vkallim@central.ntua.gr

K. Pitilakis
Department of Civil Engineering, Aristotle University of Thessaloniki, Thessaloniki, Greece
e-mail: kpitilak@civil.auth.gr

G. Papathanassiou
Department of Geology, Aristotle University of Thessaloniki, Thessaloniki, Greece
e-mail: gpapatha@geo.auth.gr

I. Towhata et al. (eds.), *Coseismic Landslides*, Springer Natural Hazards,
https://doi.org/10.1007/978-981-19-6597-5_8

1 Introduction

The geological setting of Greece is characterized by the presence of vulnerable rock formations, the existence of faults and significantly fractured rock masses in association with the steep topography and high seismicity. Consequently, Greece is prone to high likelihood of co-seismic landslides. In response to this situation, Koukis et al. (2005) proposed a landslide hazard zonation for Greece and Sabatakakis et al. (2013) developed a landslide susceptibility zonation using a landslide inventory derived from historical archives. The magnitude-distance relationship for Greek co-seismic landslides have been proposed by Papadopoulos and Plessa (2000). Furthermore, Chousianitis et al. (2016) investigated the aerial intensity and the spatial distribution of slope resistance demand in order to assess the hazards of earthquake-induced landslides in Greece.

Papazachos and Papazachou (1997), Pavlides and Caputo (2004), and Ambraseys and Jackson (1990) reported many co-seismic landslides in Greece in both historical and recent times. Papathanassiou et al. (2013, 2017), Zekkos et al. (2017) and Kallimogianis et al. (2022) investigated the earthquake-induced instabilities in Lefkada Island, which is in the region of the highest seismicity in Greece. The distribution of earthquake-induced geological effects in Lefkada and Cephalonia islands was presented by Papathanassiou et al. (2016, 2017). In the northern Peloponnese, the evolution of co-seismic rockfalls have been noted by Zygouri and Koukouvelas (2015). In line with this, Saroglou et al. (2017) discussed the earthquake-induced rockfalls during the 2014 Cephalonia ($M_w = 6.0$) and the 2015 Lefkada ($M_w = 6.4$) earthquakes. Accordingly, Saroglou (2019) made a rockfall susceptibility map with emphasis on co-seismic rockfalls. With these in mind, the present report reviews the history of earthquake-induced landslides in Greece. Note that the term 'landslide' is based on the principles and terminology of Varnes (1978) as adopted by Keefer (1984) in his study of landslides triggered by earthquakes.

2 Tectonic Setting and Seismicity of Greece

Greece is in the convergence boundary of two big lithospheric plates (Eurasian and African) and this is the main reason for the high seismicity in the area (Papazachos and Papazachou 1997). This convergence is taking place in a North–South direction with rate of 1 cm/yr, whilst the relevant value of motion of the Aegean area in respect to Africa is 5.5 cm/yr (Kastens et al. 1996; Papazachos and Papazachou 1997). In particular, a Benioff seismic zone has been detected by Papazachos and Comninakis (1970, 1971), and is considered as the strongest evidence for subduction of the eastern Mediterranean lithosphere under the Aegean lithosphere (Fig. 1a). The latter one marks the front part of Eurasian plate. Regarding the Anatolian plate, a westward motion of 2.5 cm/yr velocity is identified at the North Anatolian Fault (Oral et al. 1995).

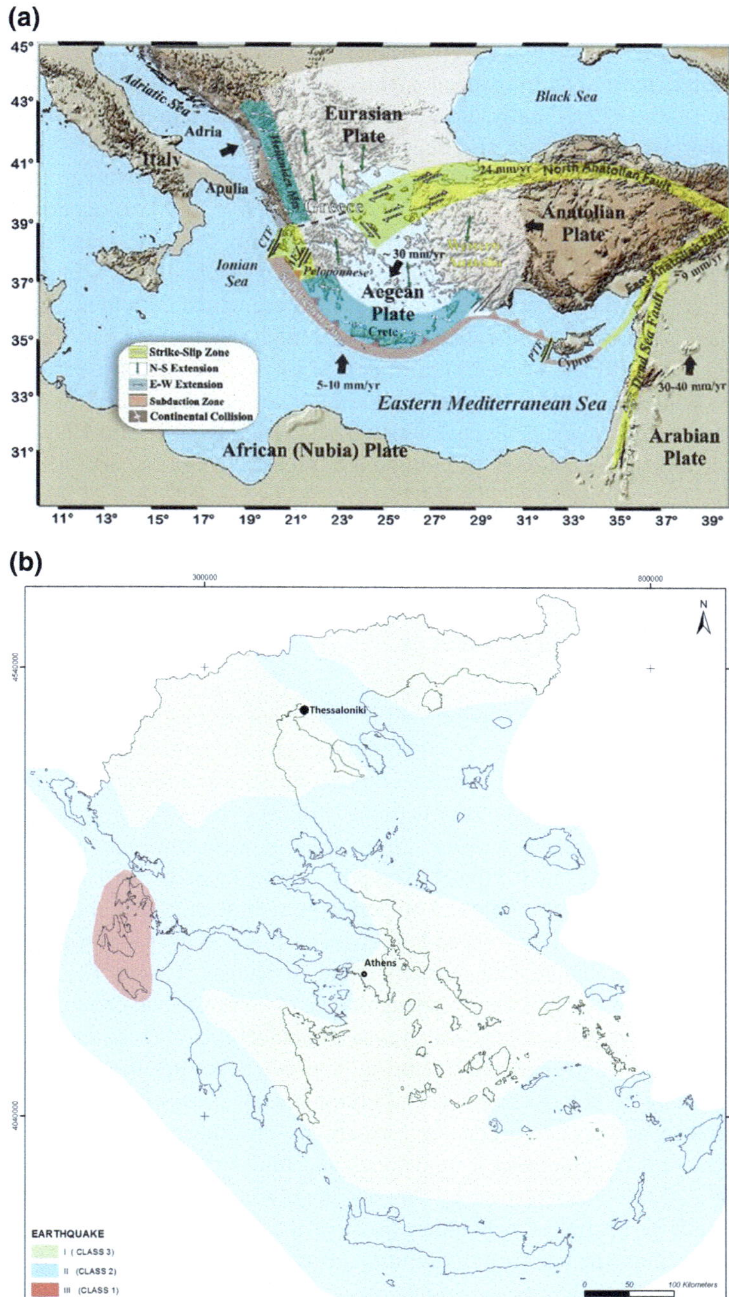

Fig. 1 **a** Active tectonics in the Aegean and surrounding area (reprinted from Kiratzi 2014, which was modified from Papazachos et al. 1998). CTF—cephalonia transform fault, PTF—pafos transform fault. **b** Earthquake intensity zonation in Greece (Reprinted from Saroglou 2019)

In addition, the outcome indicated by research on fault plane solutions is that thrust faulting is identified in the convex side of Hellenic arc and an extensional field and accordingly normal faulting mechanism, in the Aegean Sea and surrounding area (McKenzie 1970, 1972). Finally, along the 140 km long Cephalonia Transform Fault (CTF), a microplate motion exists (Scordilis et al. 1985).

The outcome indicated by quantitative investigation of the seismicity in Greece shows that intermediate depth earthquakes (60 km < h < 300 km) occurred in the southern part along the concave portion of the Hellenic arc, whilst no deep focus earthquakes were recorded in Greece (Papazachos and Papazachou 1997). Furthermore, shallow seismic activity is scattered all over the area, whilst certain seismic zones (Ionian Islands, Gulf of Corinth) are considered as high seismicity areas (Papazachos and Papazachou 1997). The mean return period for a M5.5 shallow earthquake is 0.16 years, whilst the relevant values for earthquakes of magnitude M6 and M6.5 are 180 days and 1.8 years, respectively (Papazachos and Papazachou 1997). Greece is characterized by three zones of earthquake intensity (Fig. 1b) according to the current Greek regulations (EPPO 2003). Zone I has a coefficient of acceleration equal to 0.12 g, Zone II is equal to 0.24 g and Zone III is equal to 0.36 g.

3 Geological Setting of Greece

The geological map of Greece published by the Institute of Geology and Mineral Exploration (scale 1:500.000) makes it possible to group the main geological formations encountered in Greece into eleven units comprising a number of lithologies of sedimentary, igneous and metamorphic rocks (Fig. 2).

The geological formations with the largest areas of occurrence are (a) post-alpine deposits, (b) limestones and (c) flysch. The post-alpine deposits consist of mainly of Quaternary alluvial deposits of fluvial or lacustrine sedimentary origin characterized by clayey, sandy and gravelly soils and mixtures of those, as well as Neogene formations, mainly marls and mudstones usually related to tectonic basins. The formations have a large range of soil types and varying degree of cohesion and cementation. The limestone formations have a very wide range of geotechnical behaviour, depending on their age, strength and degree of fracturing. Very often they are affected by tectonic activity, resulting in tectonic breccias and mylonites covering significant areas. The flysch in Greece is a very heterogeneous formation, consisting of alternations of sandstones, siltstones, conglomerates, mudstones and limestones. It is usually affected by tectonic activity characterized by thrusts, shear zones and faults, resulting in a generally poor geotechnical behaviour. Most of the landslides in Greece occur in the post-alpine deposits and flysch formation, as well as weathered rock formations, which in many areas have significant thickness and also form residual soils. In addition, in areas affected by tectonic activity, where mylonites and breccias are present, landslides are more pronounced.

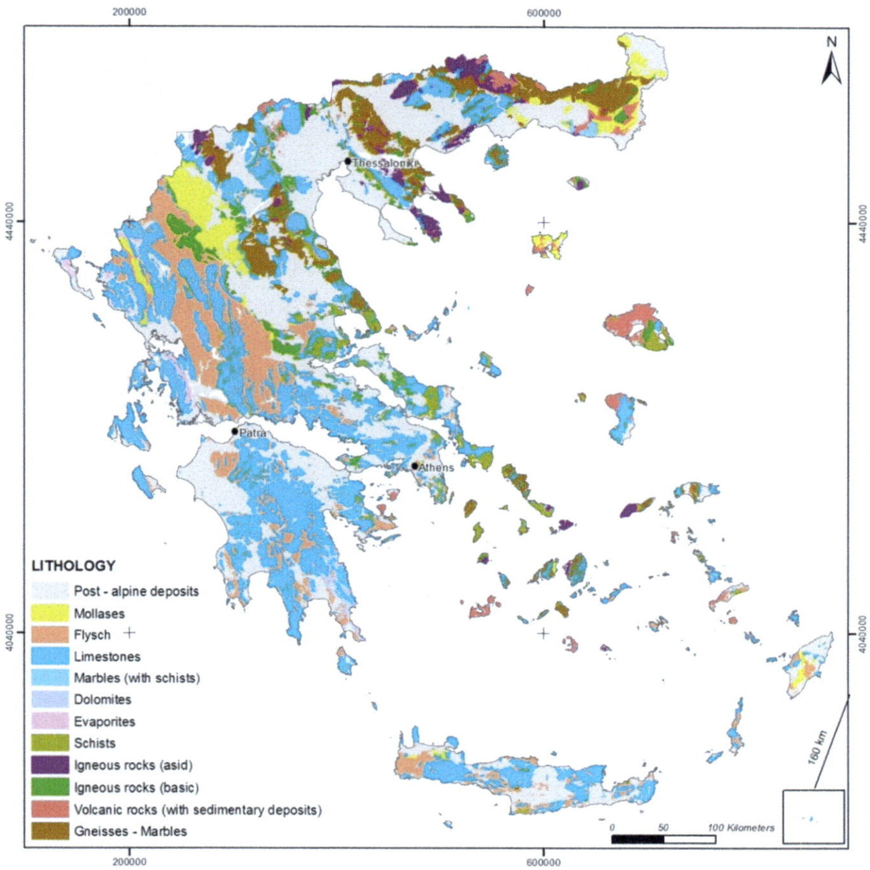

Fig. 2 Main geological units encountered in Greece (Reprinted from Saroglou 2019)

4 Past Records of Co-Seismic Landslides

Table 1 presents the most significant co-seismic landslides in Greece since 1840. Herein are listed the location, the date, the magnitude (M_w) of the earthquake and the distance of the landslide site from the earthquake epicentre based on data published by Papadopoulos and Plessa (2000). The magnitude for the earthquakes before 1968 is estimated (not recorded).

Papadopoulos and Plessa (2000) developed a magnitude–distance relations for earthquake-induced landslides in Greece. More recently, Chousianitis et al. (2016) investigated the earthquake-induced landslide hazard in Greece, paying attention to the aerial intensity and spatial distribution of slope strength demand, which is expressed by the critical acceleration required for slopes to keep within 10% the probability of the occurrence of events causing the exceedance of Newmark displacement thresholds. In this report, the magnitude–distance chart (Fig. 3) was plotted following

Table 1 Main co-seismic landslides in Greece

No	Location	Date	Magnitude (M_w)	φ° N	λ° E	Distance to landslide site
1	Zakynthos	30/10/1840	6.7	37.9	20.9	10 (±) 5
2	Chalki Isl	18/10/1843	6.5	36.3	27.7	18 (±) 5
3	Eretria	18/3/1874	6.0	38.5	23.5	24 (±) 5
4	Chios Isl	3/4/1881	6.4	38.2	26.1	26 (±) 5
5	Thiva	23/5/1893	6.0	38.3	23.2	5.7 (±) 1
6	Malesina, Martino	20/4/1894	6.4 (±) 0.3	38.65	23.16	8.7 (±) 1
7	Atalanti	27/4/1894	6.9 (±) 0.3	38.66	23.03	26.3 (±) 1
8	Kiparissia	22/1/1899	6.1 (±) 0.16	37.20	21.69	5 ± 1
9	Mitata, Kythira Isl	11/8/1903	7.9	36.0	23.0	131 ± 1
10	Skite, Magnesia	20/1/1905	6.3	39.6	23.0	14 ± 5
11	Fokis	30/5/1909	6.2	38.4	22.12	27 ± 5
12	Puri, Magnesia	31/3/1930	6.1	39.5	23.0	12 ± 5
13	Messinia	6/10/1947	6.92 (±) 0.24	36.8	22.0	47.3 ± 1
14	Velestino	8/3/1957	6.8	39.3	22.6	15.3 ± 1
15	Pitsidia, Crete Isl	14/5/1959	6.3	35.1	24.6	26 ± 5
16	Megalopolis	5/4/1965	6.1	37.7	22.0	40 ± 5
17	Eratini	6/7/1965	6.38 (±) 0.27	38.27	22.37	23.3 ± 1
18	Kremasta	2/5/1966	6.2	39.1	21.7	28 ± 1
19	Drosopighi, Arta	1/5/1967	6.4	39.5	21.2	7 ± 1
20	Ag. Efstratios Isl	19/2/1968	7.1	39.4	24.9	13 ± 1
21	Antikyra, Boeotia	8/4/1970	6.18 ± 0.29	38.23	22.6	3.6 ± 1
22	Trichonis lake	31/12/1975	5.7	38.5	21.7	3.9 ± 1
23	Alkyonides	24/2/1981	6.69 ± 0.21	38.1	22.84	31.6 ± 1
24	Pyrgos	26/3/1993	5.5	37.68	21.62	15 ± 1
25	Kozani-Grevena	13/5/1995	6.6	40.18	21.71	23.6 ± 1
26	Cephalonia	26/01/2014	5.9	38.11	20.24	16.4
27	Lefkada	11/2015	6.4	38.66	20.60	5.5

Keefer (1984), based on the records shown in Table 1. It is concluded that landslides are triggered by earthquakes with magnitudes greater than $M_w = 5.5$, whilst the maximum distance from the epicentre to a reported landslide was 131 km.

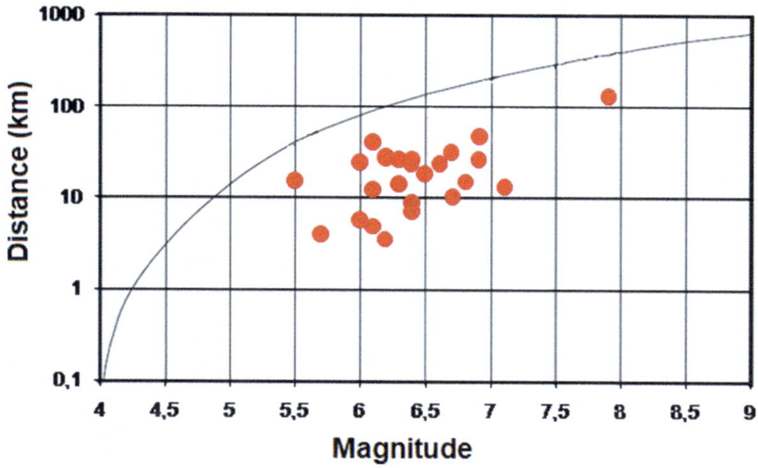

Fig. 3 Magnitude–source distance diagram of co-seismic landslides in Greece

5 Recent Co-Seismic Landlides

5.1 Earthquake-Induced Slope Failures in the Area of the Ionian Islands

Seismicity of the area

The area of the Ionian Islands is one of the regions that are most vulnerable to earthquakes not only in Greece but in the whole Europe. The high seismicity there is associated with the earthquake magnitude up to 7.4 (Louvari et al. 1999). As a result, detailed record is available for the characteristic seismic events that occurred within this area since the seventeenth century. In particular, ground failures triggered by at least 23 seismic events are known for the island of Lefkada. In more details, the western part of this island has been affected by co-seismic landslides and rockfalls because of the prevalence of steep slopes made of materials with low cohesion. On the island of Cephalonia, more than 10 damaging earthquakes occurred, and the relevant information regarding the triggered failures are provided by seismic catalogues and from reports published by local authorities, agencies and researchers.

The high seismicity of Lefkada and Cephalonia mainly stems from the complex deformation in the earth crust that is undergoing the effects of the subduction of the African Plate towards the northeast and the Apulian platform continental collision further to the northwest (Hatzfeld et al. 1995; Clement et al. 2000; Ganas et al. 2013). Both islands are located close to the Cephalonia Transform Fault and have experienced strong ground shaking many times (Fig. 1; Scordilis et al. 1985; Louvari et al. 1999; Sachpazi et al. 2000; Kokinou et al. 2006). During the most recent earthquake in Cephalonia in 2014, the ever-recorded highest PGA, equal to 0.75 G, occurred. This active tectonic structure is 140 km long with dextral (right-lateral) strike-slip movement and has a GPS slip rate between 10 and 25 mm/yr (Pérouse et al. 2012).

5.2 The Case of Lefkada Island

In the twenty-first century, two moderate earthquakes occurred on the island of Lefkada (Fig. 4). The most recent earthquake, with $M_w = 6.4$, (hypocentral depth = 11 km) occurred on 17th November 2015, at 07:10 GMT (09:10 local time) (Ganas et al. 2016; Papathanassiou et al. 2017), whilst the previous event (M_w = 6.2) occurred on 14th August 2003, offshore from the north-western coast of Lefkada Island (Papathanassiou et al. 2005). Both events were studied in detail by several researchers and a brief description of the earthquake-induced slope failures is presented in the following sections.

Regarding the geological units that are mapped in the island of Lefkada, they are comprised by Pleistocene and Holocene sediments deposited at the northern edge of the island, carbonate units of the Ionian zone and limestones of Apulia one, and Miocene marls and sandstones of Ionian zone (Bornovas 1964; Cushing 1985; Rondoyanni-Tsiambaou 1997) (Fig. 5). Onshore Lefkada Island, active or possible active thrust and normal faults with right-lateral component have been mapped.

5.3 Earthquake-Induced Landslides in Lefkada

2003 earthquake, M_w = 6.3

On 14th August 2003, a strong earthquake occurred close to the town of Lefkas. National Observatory of Athens, Institute of Geodynamics (NOAGI), identified its hypocentre at 38.810° N, 20.560° E (Fig. 3) together with the depth h = 12 km. The maximum intensity was I_o = VIII (EMS) at Lefkas municipality, where 550 buildings out of 800 were damaged. The focal mechanism of the 2003 event and smaller earthquakes were studied in association with the sea floor topography, the neotectonic pattern on land and the GPS measurements. It was thus found that the seismic fault was a right-lateral strike-slip fault with a small dip-slip component. Papathanassiou et al. (2005) revealed that this fault strikes NNE-SSW (10–20°) and dips to the ESE at a high angle. This earthquake-induced such typical ground failures as rock falls, soil liquefaction, ground cracks and slope failures (Margaris et al. 2003; Papathanassiou et al. 2005; Anastasiadis et al. 2006).

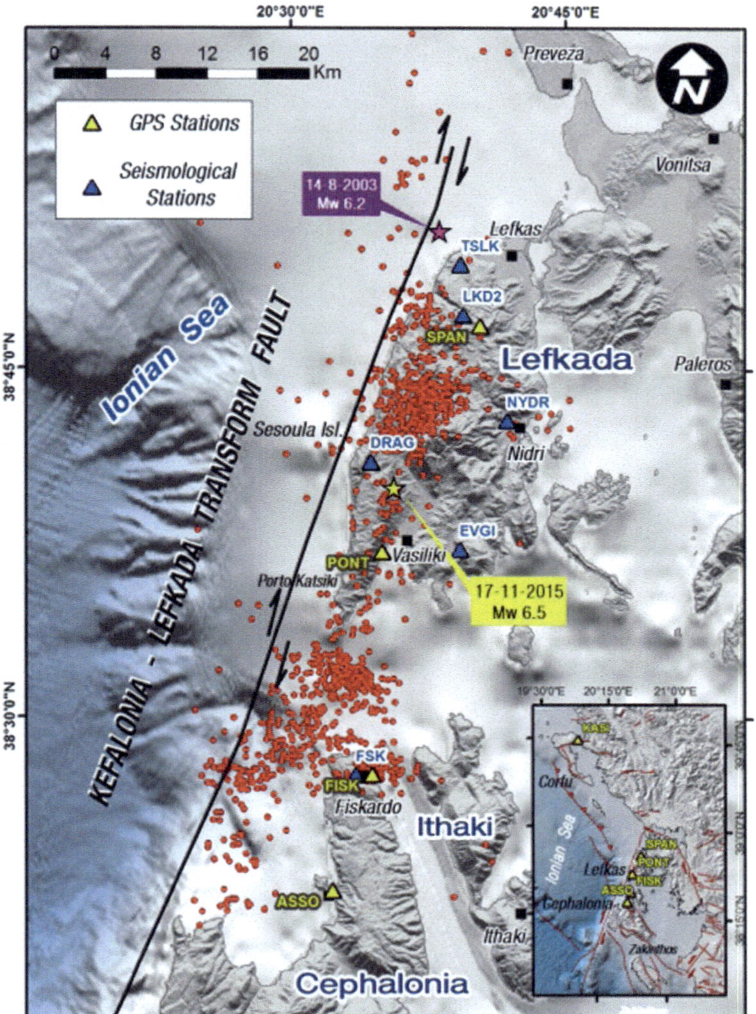

Fig. 4 The 17th November 2015 M6.5 Lefkada earthquake (yellow star) and aftershock epicentres recorded by National Observatory of Athens (NOA) until 17th December 2015. The black lines indicate the dextral faults along the Cephalonia-Lefkada Transform boundary, whereas the inset shows a regional map of the central and north Ionian Sea, with GPS stations shown by yellow triangles (Ganas et al. 2016)

2015 earthquake, $M_w = 6.4$

At 07:10 GMT (09:10 local time) on 17th November 2015, Lefkada was affected by a strong, shallow earthquake (NOA magnitudes ML = 6.0-$M_w = 6.4$, respectively; depth = 10.7 km). The National Observatory of Athens, Institute of Geodynamics

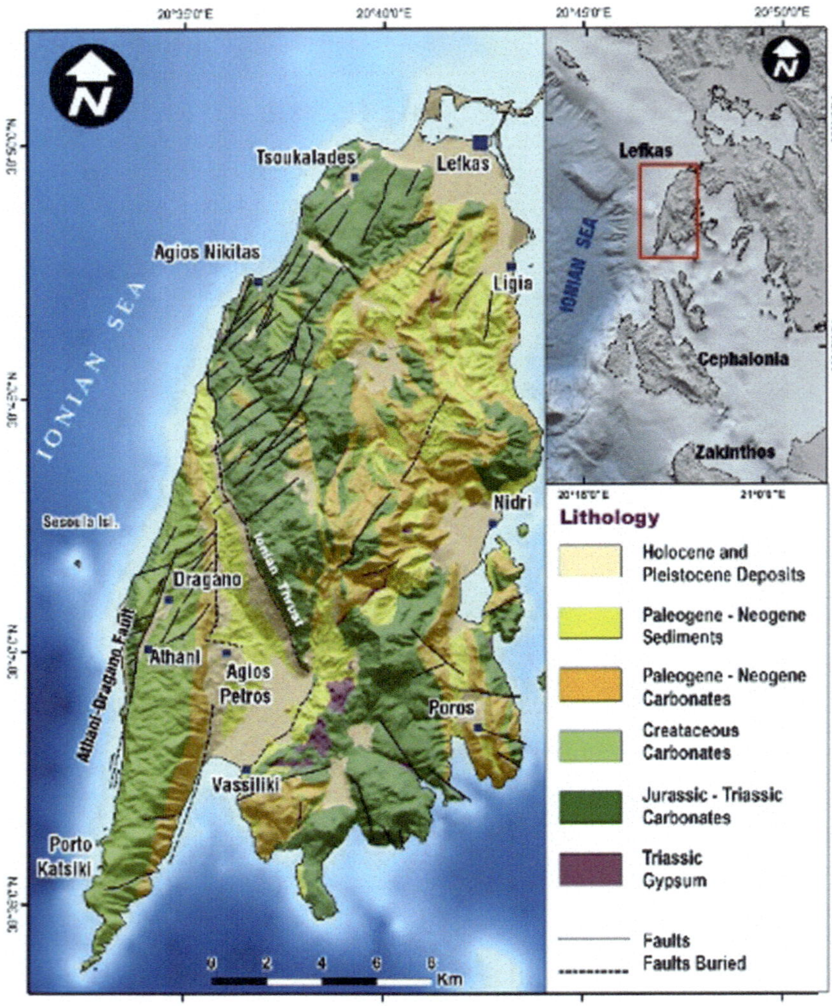

Fig. 5 Geological map and tectonic setting of the Lefkada Island (based on Cushing 1985 and Rondoyanni et al. 2007) (Reprinted from Papathanassiou et al. 2017, with author rights in Elsevier)

(NOA; www.gein.noa.gr) located the epicentre of the earthquake (38.6655° N, 20.6002° E) in the southwestern part of Lefkada Island. The range of the recorded PGA varies from 0.1 G to 0.36 G at the Vassiliki and the city of Lefkada instruments, respectively (EPPO-ITSAK 2015).

Most of the earthquake-induced landslides reportedly occurred in the western part of the island; particularly in the area called Egremnoi Beach and along the road connecting the Athani Village and the coastal area in Egremnoi. In this area, road network and residential houses were severely damaged by deep-seated landslides and

Fig. 6 Earthquake-induced slope failures, (left) Rock slide along the road between Tsoukalades and Agios Nikitas, (right) structural failure in the area of Okeanos

a large amount of debris movement. The debris consisted of coarse-grained materials mixed with a substantial amount of large gravels and boulders (Papathanassiou et al. 2017). In addition, structural failures occurred in an area of Okeanos, where recent luxury villas were situated upon an embankment that moved downslope and created a gap below the overlying structures.

The induced slope failures were dominated by rock and debris slides in association with a few shallow landslides (Fig. 6). What deserves attention is that most of the sliding material in Egremnoi and Gialos moved downwards along pre-existing tectonic discontinuities. It means that fault planes acted as sliding surfaces (Papathanassiou et al. 2017).

Furthermore, along the road connecting the villages of Tsoukalades and Agios Nikitas, large rock slides and rock falls occurred. The geology in this area is composed chiefly of a heavily fragmented limestone in association with occasional deposits of cohesionless materials (scree), which occur either as historical landslides or the consequence of rock weathering (Papathanassiou et al. 2017). This fragmentation affects the structure and the mechanical properties of the rock mass in this region and it is classified as a rock mass of low strength (Papathanassiou et al. 2014).

Taking into account pre- and post-earthquake very high resolution (0.5 m) Pleiades-1B satellite imagery, the landslide density was computed using a moving window of 500 m radius. As shown in Fig. 7, most of the slope failures are concentrated in a narrow zone that follows the western coast of Lefkada Island. According to Papathanassiou et al. (2017), this pattern could be preliminarily accounted for by the fact that (i) the geological formations in the western part of the island are fragmented profoundly by Holocene tectonic activity and that (ii) the fault rupture in 2015 started with a small slip in the vicinity of the hypocentre and propagated up dip for 3–4 s, followed by the further rupture propagation towards the SSW that created a major slip mostly to the south of the hypocentre (Sokos et al. 2016; Chousianitis et al. 2016).

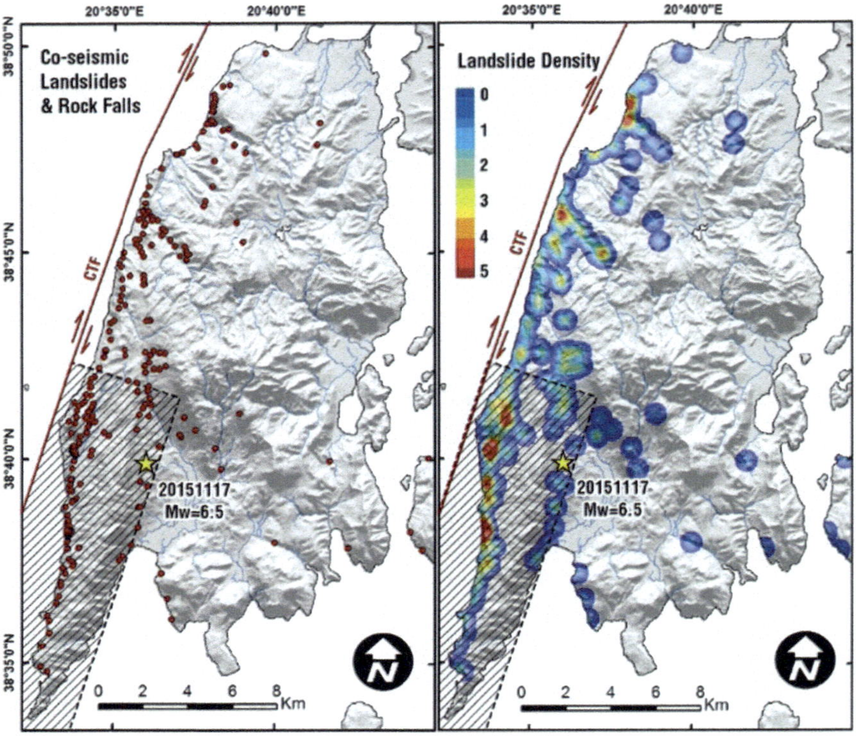

Fig. 7 Map showing the reported co-seismic landslides and rock falls triggered by the 17th November 2015 earthquake (left image). The right image illustrates the density map of the co-seismic landslides and rock falls. The colour scale indicates the number of landslides per 500 m × 500 m square (revised from Papathanassiou et al. 2017, with author rights in Elsevier), whilst the epicentre of the 2015 event is shown by a yellow star and the fault plane is provided by Ganas et al. (2016)

5.4 *Back Analysis of a Co-Seismic Landslide*

Landslide characteristics

The stability of a major landslide that occurred in Egremnoi beach (location shown in Fig. 5), was investigated in order to back-calculate the shear strength of the failed slope, which is characterized as a disintegrated and highly tectonized limestone. The geometric characteristics of the landslide in the Egremnoi area was evaluated using the pre- and post-earthquake geometries derived using the National cadastre surface model and imagery collected from an Unmanned Aerial Vehicle (UAV), respectively (Kallimogiannis 2018). Figure 8 illustrates the pre and post-earthquake surfaces, as well as the volume loss and gain using a scale coloured to illustrate height differences between the two 2.5D surfaces. A 2D longitudinal cross-section through the landslide is also shown in Fig. 8d. A height loss of up to 18.5 m is observed near the landslide

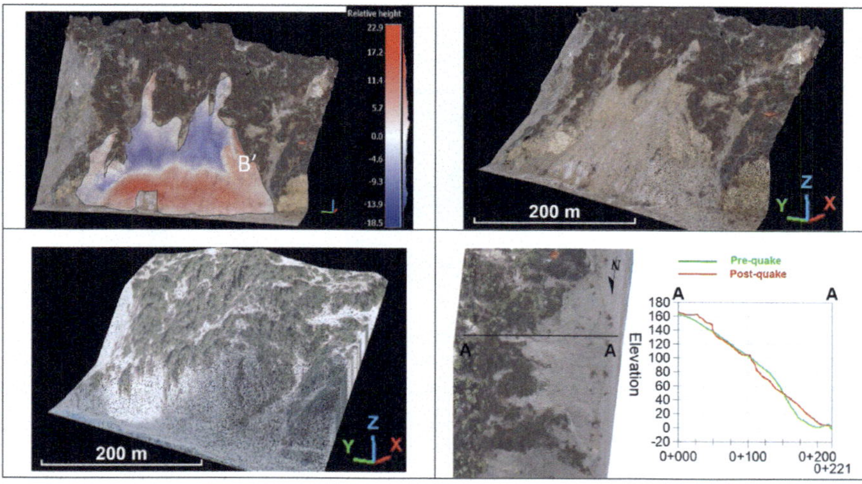

Fig. 8 **a** Pre-earthquake (2009) 3D Point Cloud interpolated from 5 m Hellenic Cadastre DEM, **b** Post-earthquake (April 2017) 3D Point Cloud generated by SfM processing based on UAV survey, **c** pre-post volume calculations presented as relative heights from reference surface (pre-quake DEM) with a 145-m section (BB') is shown for scale, **d** longitudinal cross-section (AA') through the landslide showing pre-earthquake and post-earthquake geometry (in m)

crest and a gain in height of 22.9 m due to deposition of the debris at the toe of the landslide is also measured. The volume loss due to landsliding is calculated equal to 62,000 m^3, whilst the debris volume at the toe of the slope was calculated at about 113,000 m^3. The average depth of the slope is 5.8 m, indicating relatively shallow sliding given the slope's height, which exceeds 200 m.

The failure surface is illustrated in software TSLOPE (Tagasoft Ltd) and compared with the real geometry in Fig. 9. During the UAV survey in April 2016, debris has been removed and the slip surface was revealed. The geometry of the landslide is complex. Its width remains relatively constant whilst its length (top–bottom distance) varies significantly. The modelled surface is practically identical to the actual one.

Analyses were conducted for a number of 2D cross-sections in an effort to identify the most critical ones for the complex geometry of the landslide. Figure 9b, c shows the location of three selected ones that had amongst the lowest factors of safety. The 2D surfaces of failure for all the cross-sections (AA', CC' and DD') are shown in Fig. 10.

Back Analysis Results

In back-calculating the shear strength of the material using the Mohr–Coulomb envelope, pairs of the two parameters, cohesion and friction angle, were derived since none of the two are known. Thus, the results are presented in terms of c-φ envelopes, and are shown in Fig. 11.

The strength parameters derived from the 2D analyses are higher than the strength parameters from 3D analysis, which is consistent with previous studies assuming that

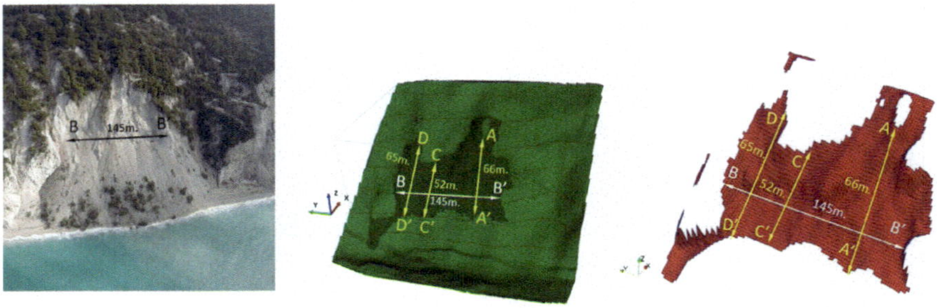

Fig. 9 Comparison of the Egremnoi site failure geometry from **a** photographic data after the earthquake; **b** 3D method of columns illustration in TSLOPE measured both in TSLOPE; and **c** the failure surface projected in TSLOPE. A 145-m. BB' length measured both in TSLOPE and CloudCompare along with the two cross-sections (AA', CC' and DD') used for the 2D analyses are also shown

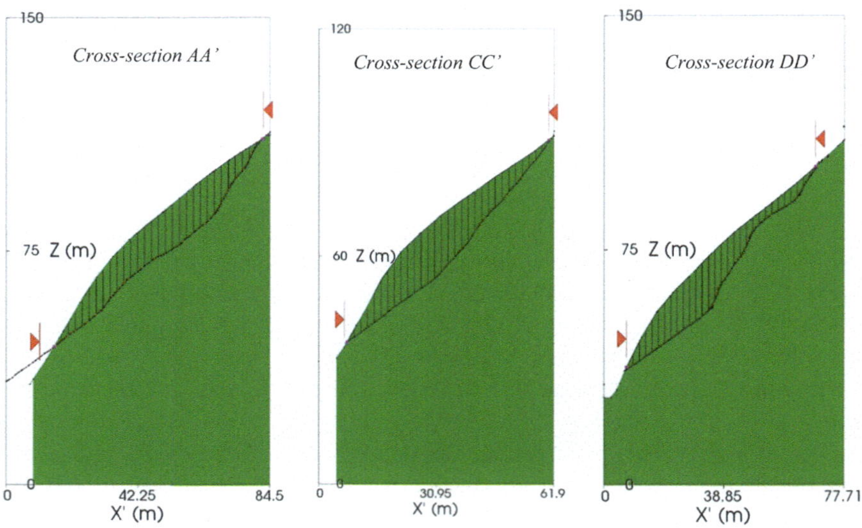

Fig. 10 2D stability analyses along AA', CC' and DD' cross-sections (see Fig. 9)

a critical cross-section is examined. Moreover, 2D analysis of cross-section CC' is more conservative than the 2D analysis of cross-section AA' and DD', showing that it is the most critical amongst all the cross-sections examined. The 2D shear strength for cross-sections AA', CC' and DD' was found to be 15.4%, 6.8% and 9.5%, respectively, higher compared to that calculated in 3D.

A factor of safety (FS) equal to 1 was assumed in the back-analysis. However, the actual value of the FS during the earthquake-induced failure may have been less than 1. Therefore, it is pointed out that the results derived from the back-analysis represent an upper limit of the material's shear strength.

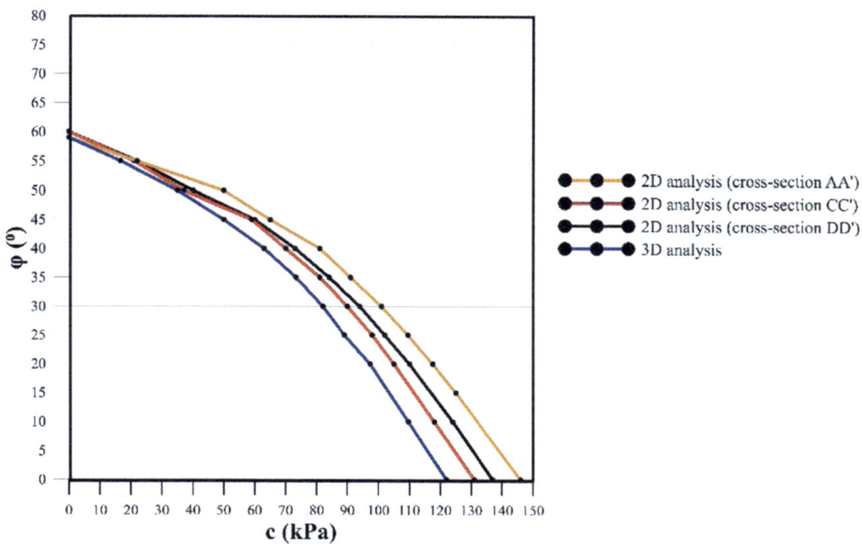

Fig. 11 C-φ envelopes of 2D and 3D analyses for the landslide at the Egremnoi site

Based on these results, the mechanical properties of the tectonized material are poor, as was anticipated. Disintegration due to tectonic stresses reduced the material's strength to that of a hard soil/weak rock. According to geological investigations conducted on the site (Kallimogiannis et al. 2019), the friction angle of the tectonized limestone may range from 35° to 45°, limiting the possible cohesion-friction angle combinations. On average, a uniaxial compressive strength of 0.27 MPa was calculated. The estimated strength was that of the material engaged in the landslide, which is near the surface of the ground. It was assumed that the limestone will exhibit better mechanical characteristics at greater depths, thus limiting the landslide to shallower depths.

5.5 *Impact of Seismic Coefficient* k_{hE}

Despite the existence of strong motion recordings in the vicinity of the site, there was some uncertainty associated with the derivation of an equivalent seismic coefficient. To investigate the sensitivity of the results to such a variation in the seismic coefficient, additional analyses were conducted for a range of k_{hE} values (Kallimogiannis et al. 2019). The results of back-analysis for four different seismic coefficient values are presented in Fig. 12. Small changes from the assumed value of 0.35 g would slightly affect the results.

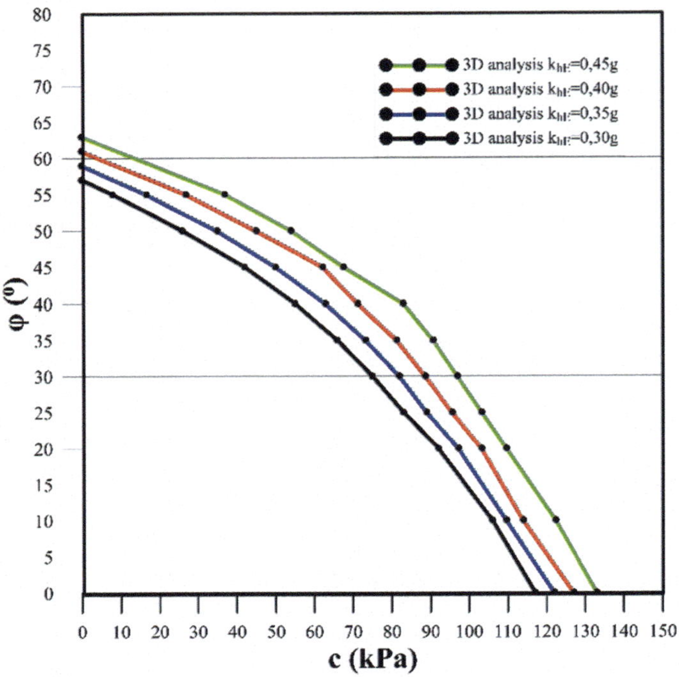

Fig. 12 C-φ envelopes for 4 different seismic coefficients k_{hE}

5.6 The Case of Cephalonia Island

Two earthquakes occurred within a period of 10 days triggering extensive secondary effects, i.e. landslides and liquefaction. The first event of M6.1 occurred on 26th January 2014, at 13:55 GMT (15:55 local time) whilst the second one of M6.0 occurred on 3rd February 2014, GMT 03:07 (05:07 local time) (Theodoulidis et al. 2015). The epicentre of the first event was offshore, at a few kilometres northwest of Argostoli Town, whilst the epicentre of the second one was located at Paliki peninsula, close to the village of Livadi (Karastathis et al (2015). The type of the causative faults of both events were high angle, strike-slip faults with a dextral sense of motion, (Ganas et al. 2013) (Fig. 13). The second event triggered the highest strong ground motion (e.g. PGA) ever recorded in Greece, equal to 752 cm/sec^2 (EPPO-ITSAK 2014; Kalogeras 2014).

The geology of Cephalonia Island is consisted of marls and sandstones of Oligocene–Middle Miocene age, limestones of Mesozoic–Oligocene belonging to the Paxos–Pre-Apulian zone (Underhill 1988; Lekkas et al. 2001), whilst Pliocene age marine deposits, marls, sands, sandstones and conglomerates are mapped in Paliki and south of Argostoli (BP Co. Ltd. 1971, Sorel 1976). Finally, lagoonal deposits are mapped in the coastal plains of Livadi and south of Argostoli (Hadler et al. 2011; Willershauser et al. 2011; Papathanassiou et al. 2016).

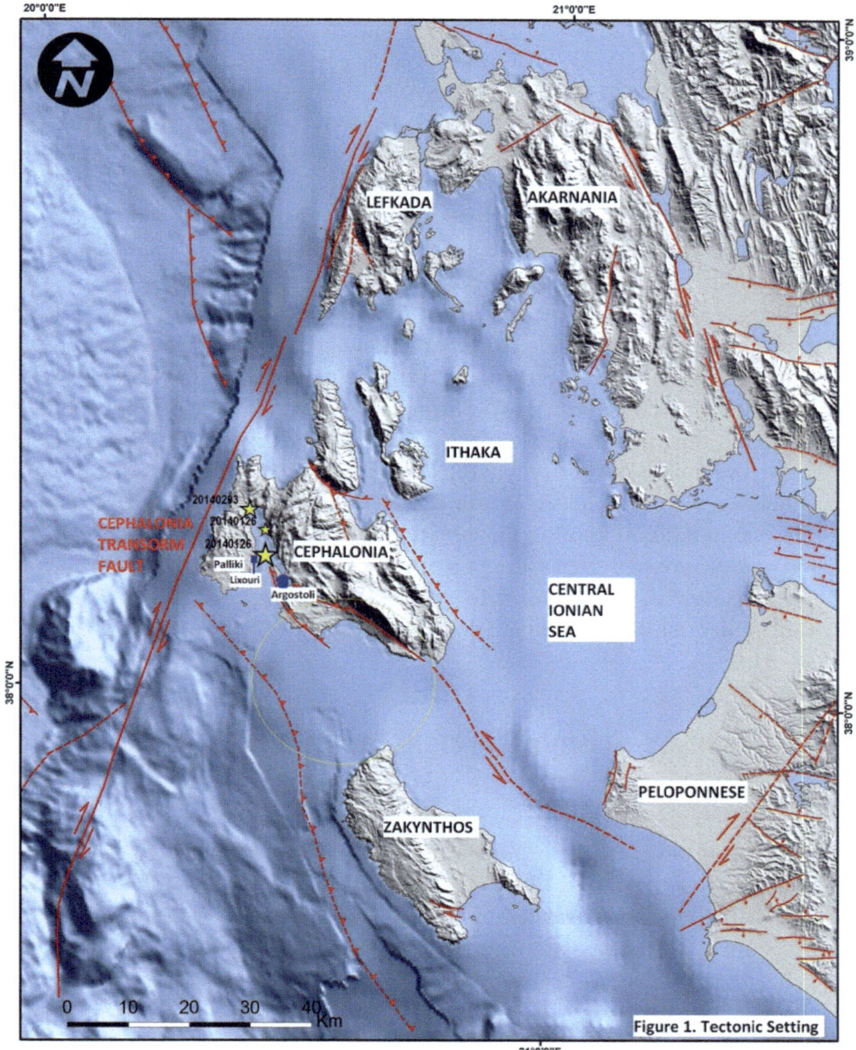

Fig. 13 Map showing as yellow stars the epicentres of the two mainshocks and the one of the strongest aftershock that occurred on 26th January 2014. The active faults are shown as red lines whilst the sense of motion is indicated by arrows (reprinted from Valkaniotis et al. 2014 with author rights in Elsevier)

5.7 *Earthquake-Induced Landslides*

The landslides triggered by the earthquakes were mainly documented at Paliki Peninsula west of Argostoli Bay, and the coastal zone from Argostoli area to Myrtos Bay, whilst no failures were reported in the northern and eastern part of Cephalonia (GEER 2014; Papathanassiou et al. 2016). The earthquake-induced slope failures destroyed the road network in Cephalonia Island and sometimes resulted in days-long closure of those roads for traffic. On the south Paliki, damage to the road network was observed due to the failure of artificial fills beneath the pavement.

The landslides of small to medium size were not only prevalent in Neogene and Quaternary soft sediments but also locally reported in clastic formations of Miocene flysch deposits. One of the largest mass movements in this area was the Soullari large landslide which was located to NE from the Soullari Village in Paliki Peninsula (Fig. 15). According to Valkaniotis et al. (2014), the landslide started during the first earthquake on January 26 on silt/clay Pliocene–Pleistocene sediments that occupies a small field on the top of a hill. Measuring 20 m in length and 30 m in width, the landslide initiated the rupture on the crown along pre-existing joints sets (NE-SW direction) and was re-activated during the second event on Feb. 3. It should be noted that an intense rainfall in the week following the first earthquake probably triggered further movement of the block in the downslope direction.

6 Records of Co-Seismic Rockfalls in Greece

Saroglou (2019) collected and reviewed data on the occurrence of rockfalls triggered by earthquakes in Greece since 1980. The most significant earthquakes have triggered rockfalls, in some cases, some of which occurred along re-activated fault scarps. A list of the major events is presented in Table 2, where information such as location, date of occurrence, earthquake magnitude and distance between the epicentre and

Table 2 Main co-seismic rockfalls in Greece

No	Location	Date	Magnitude (M_w)	Landslide site	Distance (km)
1	Gerania, Korinthos	24/2/1981	6.7	Alkyonides	27
2	Kalamata	13/9/1986	6	Kalamata-Sparti road	6.5
3	Konitsa	8/1998	5.7	Eptachori	36.8
4	Skyros	24/2/1981	6.5	Skyros castle	20.5
5	Lefkada	14/8/2003	6.3	Ag. Nikitas	7
6	Achaia	8/6/2008	6.4	Santomeri	5.9
7	Cephalonia	26/01/2014	5.9	Myrtos	16.4
8	Lefkada	11/2015	6.4	Ag. Petros	2.8

rockfall location are given. Earthquakes with magnitude greater than 5.7 Richter have triggered rockfalls and the maximum distance to a rockfall location was 36.8 km.

Rock falls and rock slides were observed along the western coast between the Myrtos beach and the south-west end of Paliki Peninsula and also along the steep east coast of Argostoli Bay. Especially in Myrtos Bay (Fig. 16), rock falls triggered along the steep cliffs caused damage to the roads and even reached the beach (GEER 2014; Valkaniotis et al. 2014).

Along the road cuts, numerous rock falls ranging from simple block falls to more extended but not very important ones, all within various types of limestones (blocky, fractured and/or weathered) were observed (GEER 2014; Valkaniotis et al. 2014). Construction and road service teams worked non-stop for two weeks after the January 26 and February 2 earthquakes in order to re-open the road to civil traffics and restore damage. Rock falls of volumes ranging between 0.5 to 2.0 m^3 were reported in Atheras Village at the northern part of Paliki peninsula (GEER 2014).

Many rock falls occurred in the entire island during the Lefkada earthquake in 2003 and especially in the north-western and central areas. The falls occurred mostly in natural and cut slopes as well as in downstream road embankment slopes. One of the most noteworthy rock falls, with rock diameters up to 4 m, occurred along the 6 km road of Tsoukalades-Agios Nikitas, accompanied by gravel, small rock and soil slides (Fig. 17). The falls occurred within a morphological scarp that was 300 m high from the sea level, and especially in a 10–40 m artificial slope. Although many slope failures and flows were reported in this area, only a few of them can be considered as typical landslides such as those along the road to Kalamitsi and Agios Petros Villages (Margaris et al. 2003; Papathanassiou et al. 2005).

The 2015 Lefkada earthquake caused detachment of a 2 m^3 limestone block from a rock slope near Ponti Village. This rock block struck a residential house, penetrated through two brick walls and killed an elderly lady. It was interesting herein that the rock block travelled over a very long distance which was about 800 m in plan view, from the point of detachment to the end of its run-out. This limestone slope has a height of 600 m and gradient of 35° to 40° (Fig. 18).

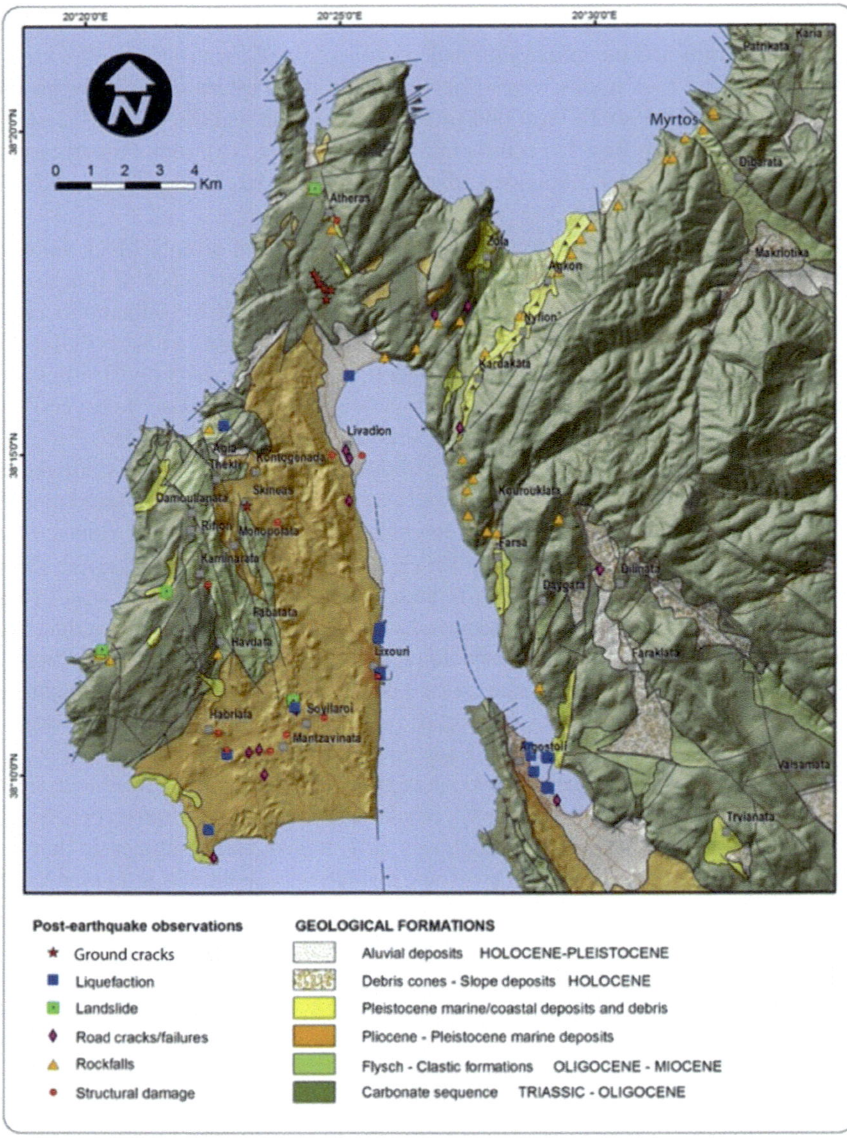

Fig. 14 Geological map of Cephalonia showing the distribution of earthquake-induced geological effects compiled by Papathanassiou et al. (2016). The geological formations and faults are modified from Lekkas et al. (2001) (From Valkaniotis et al. 2014 with author rights in Elsevier)

Fig. 15 Landslides reported in the area of Soullari (left) and Petani (right)

Fig. 16 Roll paths of the rock blocks at Myrtos Bay (GEER 2014)

Fig. 17 Earthquake-induced slope failures along the road Tsoukalades–Agios Nikitas (left) and at the entrance of Agios Nikitas (right), photos taken by G. Papathanassiou on 14th August 2003

Fig. 18 Top view orthophoto denoting the rolling section, bouncing positions and indicative close-ups of impact points of the Ponti village co-seismic rockfall. The impacted house is also shown (reproduction from Saroglou et al. 2018 with author rights)

7 Conclusions

This short report presented a review of recorded co-seismic landslides in Greece the last two centuries and in particular the last 20 years and reference was made to previous studies. In its long history, Greece has suffered from many earthquakes that induced landslides, as reported by a number of records and documents. The most significant co-seismic landslides in Greece, since 1840, were discussed briefly.

Papadopoulos and Plessa (2000) proposed a magnitude-distance relationship for co-seismic landslides in Greece. In this report, a magnitude—distance relation chart was prepared based on the records of Table 1 (Fig. 3). It was evident that landslides were triggered by earthquakes with a magnitude greater than $M_w = 5.5$, whilst the minimum distance from the epicentre to a reported landslide was 5.5 km and the maximum distance was 131 km.

Most of the landslides in Greece occur in the post-alpine deposits and weak flysch, as well as in weathered rock formations. Additionally, landslides are more pronounced in areas affected by tectonic activity, where mylonites and breccias are present.

The landslides triggered by recent earthquakes in Cephalonia and Lefkada were reported in more detail. These case studies are encountered in the Ionian Sea in western Greece, which is considered as the most earthquake-prone region not only in Greece but in all of Europe, and is characterized by a very high seismicity with potential maximum earthquake magnitudes up to Mw=7.4.

The subduction of the African Plate towards northeast as well as the Apulian platform continental collision further to the northeast produced a complex crustal deformation and this situation caused the high seismicity of Lefkada and Cephalonia. Both islands have been repeatedly affected by strong ground shaking because they are close to the Cephalonia Transform Fault, which is an active tectonic structure, being140 km long, with a GPS slip rate between 10 and 25 mm/yr.

The most recent earthquakes occurred in Cephalonia and Lefkada Islands in 2014 and 2015, both of which triggered landslides and caused failures of the road network and port facilities. The distribution of earthquake-induced geological effects in both islands was presented in Figs. 7 and 14.

The most pronounced effects are known in the western part of the island that was affected by the earthquake-induced landslides and rockfalls. In this area, steep slopes of low strength material (mainly tectonic breccias) are encountered.

In this report, a table of the most significant co-seismic rockfalls in Greece since 1980 is given. Those rockfalls were triggered by earthquakes whose magnitude was between $M_w = 5.7$ and 6.7, whilst the maximum distance from the epicentre to reported rockfalls was 36.8 km.

It can be concluded that Greece presents an area of increased interest for the study of co-seismic landslides due to the high seismicity of the region and the complexity of the geological and geotechnical conditions. It is obvious that more investigation is necessary for the understanding of the mechanisms of slope failure during earthquakes and prediction of landslide occurrence.

References

Ambraseys NN, Jackson JA (1990) Seismicity and associated strain of central Greece between 1890 and 1988. Geophys J Int 101:663–708

Anastasiadis A, Margaris V, Klimis N, Makra K, Pitilakis K (2006) The Lefkas Earthquake (M = 6.2, Aug 14, 2003): strong ground motion and evaluation of the subsoil's impact. In: 5th Hellenic conference of geotechnical and geoenvironmental engineering, Xanthi

Bornovas J (1964) Géologie de l'île de Lefkade. Geol. Geophys. Res. (Special Publication by Greek Geological Survey), V.10 (1) (Athens)

B.P. Co ltd. (1971) The geological results of petroleum exploration in western Greece. Institute of Geology and Subsurface Research, Athens, Greece, Special report 10

Chousianitis K, Konca AO, Tselentis G, Papadopoulos G, Gianniou M (2016) Slip model of the 17 November 2015 Mw =6.5 Lefkada earthquake from the joint inversion of geodetic and seismic data. Geophys Res Lett 43:7973–7981

Clement C, Hirn A, Charvis P (2000) Seismic structure and the active Hellenic subduction in the Ionian islands. Tectonophysics 329(1–4):141–156

Cushing M (1985) Evolution structurale de la marge nord ouest hellenique dans l'ıle de Levkas et ses environs (Grece nord occidentale). These 3me cycle, Univ. de Paris-Sud

EPPO (2003) Greek seismic code. Earthquake Planning and protection Organization, Athens

EPPO-ITSAK (2014) Strong ground motion of the February 3, 2014 Cephalonia Earthquake: effect on soil and built environment in combination with the January 26, 2014 Event. (http://www.slideshare.net/itsak-eppo/20140203-kefaloniaeq-report-en)

EPPO-ITSAK (2015) Preliminary report of the Lefkada 2015 earthquake (http://www.itsak.gr/uploads/news/earthquake_reports/Lefkas_M6.0_17-11-2015.pdf)

Ganas A, Marinou A, Anastasiou D, Paradissis D, Papazissi K, Tzavaras P, Drakatos G (2013) GPS-derived estimates of crustal deformation in the central and north Ionian Sea, Greece: 3-yr results from NOANET continuous network data. J Geodyn 67:62–71

Ganas A, Elias P, Bozionelos G, Papathanassiou G, Avallone A, Papastergios A, Valkaniotis S, Parcharidi I, Briole P (2016) Coseismic deformation, field observations and seismic fault of the 17 November 2015 M=6.5. Lefkada Island, Greece Earthquake, Tectonophysics 687:210–222

GEER (2014) Earthquake reconnaissance January 26th/February 2nd 2014 Cephalonia, Greece events, version 1. In: Nikolaou S, Zekkos D, Assimaki D, Gilsanz R (eds) GEER/EERI/ATC, p 492

Hadler H, Vott A, Bruckner B, Bareth G, Ntageretzis K, Warnecke H, Willershauser T (2011) The harbour of ancient Krane, Kutavos Bay (Cefalonia, Greece)—an excellent geo-archive for palaeo-tsunami research. In: Karius V, Hadler H, Deicke M, von Eynatten H, Bruckner H, Vott A (eds) Dynamische Kusten—Prozesse, Zusammenhange und Auswirkungen, Coastline Reports 17, pp 111–122

Hatzfeld D, Kassaras I, Panagiotopoulos D, Amorese D, Makropoulos K, Karakaisis G, Coutant O (1995) Microseismicity and strain pattern in northwestern Greece. Tectonics 14:773–785

Kallimogiannis V (2018) Determination of mechanical properties of Tectonized Limestones using back-analysis in landslides. The Case of the Earthquake (2015) in Lefkada Island. MSc Thesis, NTUA, p 124

Kallimogiannis V, Saroglou C, Zekkos D, Manousakis J (2019) 2D and 3D Back-analysis of a landslide in Egremnoi caused by the November 17 2015 Lefkada earthquake. In: Proceedings of ICONHIC, 2nd International Conference on Natural Hazards & Infrastructure

Kallimogiannis V, Saroglou C, Zekkos D, Papadimitriou A (2022) Back-analyses of co-seismic landslides in carbonate fault rocks due to the Mw 6.5 earthquake (2015). In: Lefkada island, Greece. Eng Geol 309(106850):1–18

Kalogeras I (2014) Preliminary analysis of the recorded strong ground motion triggered by the January 26, 2014 Cephalonia Earthquake (in Greek)

Karastathis VK, Mouzakiotis E, Ganas A, Papadopoulos GA (2015) High-precision relocation of seismic sequences above a dipping Moho: the case of the January–February 2014 seismic sequence on Cephalonia island (Greece). Solid Earth 6:173–184. https://doi.org/10.5194/se-6-173-2015

Kastens K, Gilbert L, Hurst K, Veis G, Paradissis D, Billiris H, Schlueter W, Seeger H (1996) GPS Evidence for Arc—Parallel Extension Along the Hellenic Arc, Greece. Tectonophysics

Keefer DK (1984) Landslides caused by earthquake. Bul Geol Soc Am 95(4):406–421

Kiratzi AA (2014) Mechanisms of Earthquakes in Aegean. In: Beer M, Kougioumtzoglou IA, Patelli E, Au IS-K Encyclopedia of Earthquake Engineering, vol 4, p 3966, Springer, ISBN 978-3-642-35343-7

Kokinou E, Papadimitriou E, Karakostas V, Kamberis E, Vallianatos F (2006) The Kefalonia transform zone (offshore western Greece) with special emphasis to its prolongation towards the Ionian abyssal plain. Mar Geophys Res 27(4):241–252. https://doi.org/10.1007/s11001-006-9005-2

Koukis G, Sabatakakis N, Nikolaou N, Loupasakis C (2005) Landslide hazard zonation in Greece. In: Sassa K, Fukuoka H, Wang F, Wang G (eds) Proceedings of 1st general assembly and the 4th session of board of representatives of the international consortium on landslides, pp 291–296

Lekkas E, Danamos G, Lozios S (2001) Neotectonic structure and neotectonic evolution of Lefkada Island. Bull Geol Soc Greece XXXIV(1):157–163

Louvari E, Kiratzi AA, Papazachos BC (1999) The CTF and its extension to western Lefkada Island. Tectonophysics 308:223–236

McKenzie DP (1970) Plate tectonics of the Mediterranean region. Nature 226:239–243

McKenzie D (1972) Active tectonics of the Mediterranean region. Geophys JR Astron Sot 30:1099185

Margaris B, Papaioannou Ch, Theodulidis N, Savaidis A, Anastasiadis Anastasios, Klimis N, Makra K, Demosthenous M, Karakostas Ch, Lekidis V, Makarios Triantafyllos, et al (2003) Preliminary observations on the August 14, 2003, Lefkada Island (Western Greece) Earthquake. EERI J Special Earthquake Report 1–12

Oral MB, Reilinger RE, Toksoz MN, King RW, Barka AA, Kinik I (1995) Coherent plate motions in the eastern Mediterranean continental collision zone, Eos. Trans AGU 76(2):9–11

Papadopoulos G, Plessa UV (2000) Magnitude–distance relations for earthquake-induced landslides in Greece. Eng Geol 58:377–386

Papathanassiou G, Pavlides S, Ganas A (2005) The 2003 Lefkada earthquake: field observations and preliminary microzonation map based on liquefaction potential index for the town of Lefkada. Eng Geol 82(1):12–31

Papathanassiou G, Valkaniotis S, Ganas A, Sp P (2013) GIS-based statistical analysis of the spatial distribution of earthquake-induced landslides in the island of Lefkada, Ionian Islands, Greece. Landslides 10(6):771–783

Papathanassiou G, Valkaniotis S, Dimaras K (2014) Validating the classification of earthquake-induced landslide hazard levels based on data provided by large scale mapping of failures induced by 2003 Lefkada, Greece Earthquake. Eng Geol Soc Territory 2:737–741

Papathanassiou G, Ganas A, Valkaniotis S (2016) Recurrent liquefaction-induced failures triggered by 2014 Cephalonia, Greece earthquakes: spatial distribution and quantitative analysis of liquefaction potential. Eng Geol 200:18–30

Papathanassiou G, Valkaniotis S, Ganas A, Grendas N, Kollia E (2017) The November 17th, 2015 Lefkada (Greece) strike-slip earthquake: field mapping of generated failures and assessment of macroseismic intensity ESI-07. Eng Geol 220:13–30

Papazachos BC, Comninakis PE (1969/70) Geophysical features of the Greek Island Arc and Eastern Mediterranean Ridge. Com. Ren. Des Seances de la Conference Reunie a Madrid, Sept. 1969 16:74–75

Papazachos BC, Comninakis PE (1971) Geophysical and tectonic features of the Aegean arc. J Geophys Res 76:8517–8533

Papazachos B, Papazachou C (1997) The Earthquakes of Greece. P. Ziti and Co, Thessaloniki, Greece, p 304

Papazachos BC, Papadimitriou EE, Kiratzi AA, Papazachos CB, Louvari E (1998) Fault plane solutions in the Aegean Sea and the surrounding area and their tectonic implication. Boll Geof Teor App 39:199–218

Pavlides SB, Caputo R (2004) Magnitude versus fault's surface parameters: quantitative relationships from the Aegean Region. Tectonophysics 380(3–4):159–188

Pérouse E, Chamot-Rooke N, Rabaute A, Briole P, Jouanne F, Georgiev I, Dimitrov D (2012) Bridging onshore and offshore present-day kinematics of central and eastern Mediterranean: implications for crustal dynamics and mantle flow. Geochem Geophys Geosyst 13:1525–2027. https://doi.org/10.1029/2012GC004289

Rondoyanni-Tsiambaou T (1997) Les seismes et l'environnement géologique de l'île de Lefkade, Grèce: Passe et Futur. In: Marinos et al. (eds) Engineering geology and the environment, Balkema, pp 1469–1474

Rondoyanni T, Mettos A, Paschos P, Georgiou C (2007) Neotectonic map of Greece, scale1:100.000, Lefkada sheet. I.G.M.E., Athens

Sabatakakis N, Koukis G, Vassiliades E, Lainas S (2013) Landslide susceptibility zonation in Greece. Nat Hazards 65:523–543

Sachpazi M, Hirn A, Clement C, Haslinger F, Laigle M, Kissling E, Charvis P, Hello Y, Lepine J-C, Sapin M, Ansorge J (2000) Western Hellenic subduction and Cephalonia transform: local earthquakes and plate transport and strain. Tectonophysics 319(4):301–319

Scordilis EM, Karakaisis GF, Karacostas BG, Panagiotopoulos DG, Comninakis PE, Papazachos BC (1985) Evidence from tranform faulting in the Ionian Sea: the Cephalonia island earthquake sequence of 1983. PAGEOPH 123:387–397

Sokos E, Zahradník J, Gallovič F, Serpetsidaki A, Plicka V, Kiratzi A (2016) Asperity break after 12 years: the M_W 6.4 2015 Lefkada (Greece) earthquake. Geophys Res Lett 42. https://doi.org/10.1002/2016GL069427

Sorel D (1976) Etude neotectonique des isles ioniennes de Cephalonie et Zanthe et de L'Elide occidentale (Grece) These du 3e cycle Universite de Paris-Sud, Centre d'Orsay

Saroglou H, Asteriou P, Tsiambaos G, Manousakis J, Zekkos D (2017) Study of co-seismic rockfalls during Lefkada and Cephallonia Earthquakes, Greece. In: Proceedings of 3rd North American symposium on landslides, Virginia, USA. Association of environmental and engineering geologists (AEG), pp 521–528

Saroglou C, Asteriou P, Zekkos D, Tsiambaos G, Clark M, Manousakis J (2018) UAV-enabled reconnaissance and trajectory modeling of a co-seismic rockfall in Lefkada. Nat Hazards Earth Syst Sci. https://doi.org/10.5194/nhess-2017-29

Saroglou C (2019) GIS-based rockfall susceptibility zoning in Greece. Geosciences 9:163. https://doi.org/10.3390/geosciences9040163

Theodoulidis N, Karakostas C, Lekidis V, Makra K, Margaris B, Morfidis K, Papaioannou C, Rovithis M, Salonikios T, Savvaidis A (2015) The Cephalonia, Greece, January 26 (M6.1) and February 3, 2014 (M6.0) earthquakes: near-fault ground motion and effects on soil and structures. Bull Earthq Eng. https://doi.org/10.1007/s10518-015-9807-1

Underhill JR (1988) Triassic evaporates and Plio-Quaternary diapirism in western Greece. J Geol Soc 145:269–282

Valkaniotis S, Ganas A, Papathanassiou G, Papanikolaou M (2014) Field observations of geological effects triggered by the January–February 2014 Cephalonia (Ionian Sea, Greece) earthquakes. Tectonophysics 630:150–157

Varnes DJ (1978) Slope movement types and processes. In: Scuster RL, Krizek RJ (eds) Landslides, analysis and control—special report 176. National Academy of Sciences, Washington DC, pp 11–33

Willershauser T, Vott A, Bruckner H, Bareth G, Hadler H, Ntageretzis K (2011_ New insights in the Holocene evolution of the Livadi coastal plain, Gulf of Argostoli (Cefalonia, Greece). In: Karius V, Hadler H, Deicke M, von Eynatten H, Bruckner H, Vott A (eds) Dynamische Kusten—Prozesse, Zusammenhange und Auswirkungen, Coastline Reports 17, pp 99–110

Zekkos D, Clark M, Cowell K, Manousakis J, Tsiambaos G, Saroglou H (2017) Characterization of landslides during the November 17th Mw 6.5 Lefkada Earthquake using satellite and UAV-based remote sensing approaches. In: Proceedings of ICSMGE, September 2017, Seoul, pp 2235–2238

Zygouri V, Koukouvelas IK (2015) Evolution of rock falls in the Northern part of the Peloponnese, Greece. IOP Conf Series Earth Environ 26:012043, Warwick UK, 10–11 September. https://doi.org/10.1088/1755-1315/26/1/012043

Characteristics of Landslides and Cases of Severe Landslides on Gentle Slopes Triggered by the Chi-Chi Earthquake, Taiwan, 1999

Meei-Ling Lin, Kuo-Lung Wang, and Jia-Jyun Dong

Abstract The Chi-Chi earthquake struck Central Taiwan on September 21, 1999, causing extensive landslide hazards. The distribution and characteristics of the landslides are presented in this paper, and the effects of the ground motion of the Chi-Chi earthquake are examined. In addition, three landslides on gentle slopes are discussed.

1 Introduction

Taiwan is situated at the juncture of the Euro-Asia continental plate and the Philippine Sea plate with the Philippine Sea plate pushing toward the northwest direction. The Central Mountain Range and the Coastal Mountain Range have been created due to the orogenic movement, and the prevailing geological formations and structural patterns align approximately parallel to the longitudinal axis of the island. The mountain area composes about 77% of the overall area and has geological conditions vulnerable to landslides. As a result of the active tectonic motion, Taiwan is among one of the most seismically active areas in the world. The Chi-Chi earthquake struck the central region of Taiwan on September 21, 1999, at 1:47 AM local time (GMT + 8), with a local magnitude of 7.3 (moment magnitude of 7.6). The epicenter was placed at 120.82° east longitude and 23.85° north latitude by the Central Weather Bureau (CWB) near the town of Chi-Chi, from which the earthquake derived its name. The earthquake caused a significant loss of lives and property and severe ground failures. Among different types of ground failure, slope failure was the most widely distributed and caused severe secondary hazards lasting for years after the

M.-L. Lin (✉)
National Taiwan University, Taipei, Taiwan
e-mail: linml@ntu.edu.tw

K.-L. Wang
National Chi-Nan University, Nantou, Taiwan
e-mail: klwang@ncnu.edu.tw

J.-J. Dong
National Central University, Chunli, Taiwan
e-mail: jjdong@geo.ncu.edu.tw

I. Towhata et al. (eds.), *Coseismic Landslides*, Springer Natural Hazards,
https://doi.org/10.1007/978-981-19-6597-5_9

earthquake. The effects of the earthquake on the landslides and three large landslides on gentle slopes are discussed in this paper.

2 The Chi-Chi Earthquake

The Chi-Chi earthquake occurred on 01:47:19 local time, September 21, 1999, and was triggered by movement along the Chelungpu fault, with a fault rupture length of about 105 km in an approximately north–south direction. The Chelungpu fault is a shallow thrust striking nearly north–south and is an east-dipping fault that moved westward. The focal depth of the earthquake was only 8 km, which suggested a tremendous amount of energy release near the ground surface. Due to the large westward thrusting motion, the recorded horizontal peak ground accelerations were very large and mainly in the east–west (E–W) direction, coinciding with the thrusting direction of the fault. Two strong motion stations recorded peak horizontal ground accelerations larger than 1 g; they were Sun Moon Lake Station (TCU084) with 989.21 gal in the E–W direction and Mingjein Station (TCU129) with 983.02 gal in the E–W direction (Shin 1999). The maximum peak vertical ground acceleration recorded was 0.7 g. Three events with magnitudes larger than six were triggered within 1 min after arrival of the main shock (Shin 1999). The epicenters of the main shock and three triggered events, ground rupture of the Chelungpu fault, and the distribution of horizontal peak ground acceleration are presented in Fig. 1. The isoseismal map in Fig. 1 was derived from the ground motion records of all strong motion stations of the Central Weather Bureau, Taiwan (CWB) (http://www.cwb.gov.tw/V5e/seismic/chichi.htm). Station TCU084 was close to the epicenter, and the TCU129 strong motion station was close to the fault rupture (see Fig. 1). During the Chi-Chi earthquake, the ground motion was strong for areas near the fault rupture and decreased rapidly with increasing distance away from the fault rupture. In general, the ground motions in the E–W direction are higher than motions in the N–S direction, which is consistent with the thrusting characteristics of the faulting action. A horizontal profile was taken across the Chelungpu fault near the maximum recorded ground motion to illustrate the variations of seismicity (Fig. 1). Distribution of the ground motion on the hanging wall side was high and decreased with distance from the fault rupture slowly compared to that on the foot wall side (Fig. 1). The triggered events led to higher seismic intensity in their neighboring areas at some distance away from the Chelungpu fault. Ground amplification effects showing high ground accelerations were observed in the Puli Basin series, starting from Puli Basin to the northeast of the epicenter (Fig. 1), with the basin series extending in the northeast to southwest direction. Figure 2 shows the distribution of the final ground displacements caused by the thrusting action of the Chelungpu fault, surveyed using a global positioning system (GPS) (Huang et al. 1999). The maximum horizontal displacement was about 9 m toward the northwest direction, and the maximum vertical displacement was about 6 m upward; both were measured on the hanging wall near the north tip of the fault rupture where the ground rupture propagated eastward. High ground accelerations were observed associated with the maximum displacements and propagation of ground ruptures. With such strong ground motions and large displacements, various types of ground failure occurred and caused tremendous amounts of damage and losses. Several stages of reconnaissance and mitigation were carried out as geotechnical and hazard mitigation aspects after the earthquake. The National Science Council (NCREE 2000) called upon researchers in related fields to participate in the reconnaissance mission the first week after the earthquake to collect and document important field data. From the ground-based geotechnical hazard reconnaissance report, landslides were reported as the most widely distributed hazard among all different types of ground failure, which resulted in tremendous losses and a large amount of landslide mass deposited in the field, causing concerns of secondary hazards (Lin et al. 2000a; NCREE et al. 1999).

3 Landslides Triggered by the Chi-Chi Earthquake

A landslide field reconnaissance was conducted where traffic allowed, and a total number of 436 cases were reported (Lin et al. 2000a). A follow-up investigation was conducted by the Council of Agriculture using aerial photographs and SPOT satellite images taken before and after the earthquake, and more than 21,900 landslides were mapped, covering a total area of more than 8600 ha. Further investigations were performed using available aerial photographs and SPOT satellite images, and more than 30,000 landslides were mapped. To ensure the accuracy of the mapped landslides, each mapped ground surface variation was checked for consistency in the elevation variations and slope angle (Lin and Tung 2004). After the screening process, a total of 31,978 landslides were mapped. The distribution of the mapped landslides and locations of field reconnaissance points are plotted in Fig. 3, along with the surface rupture of the Chelungpu fault and epicenters. The mapped landslides and field recorded points distributed widely from Miao-Li County to Jai-Yi County, as shown in Fig. 3, and almost all cases were located to the east or on the hanging wall of the thrust fault where the terrain was mountainous, as indicated by the shaded relief map (Fig. 3). Comparing the distribution of mapped landslides to the isoseismal map in Fig. 1, the distribution of mapped landslides appeared to decrease with decreasing seismicity. In addition, high concentrations of mapped landslides were observed where the maximum ground displacements occurred near the north tip of the fault rupture while it propagated eastward and in areas near both ends of the fault rupture. Lin et al. (2016) reported a general tendency of decreasing landslide area ratio (the ratio of landslide area in the designated area) with decreasing horizontal peak ground acceleration of the mapped landslides. The mapped landslides and field reconnaissance points are plotted in Fig. 4 versus zonation of the geological formations. More landslides are distributed in the sedimentary rock formation area than in the metamorphic rock formation area (Fig. 4). However, the sedimentary rock formation area was subjected to larger ground motions in general compared to the metamorphic rock formation area, as suggested in Fig. 1, and there was no

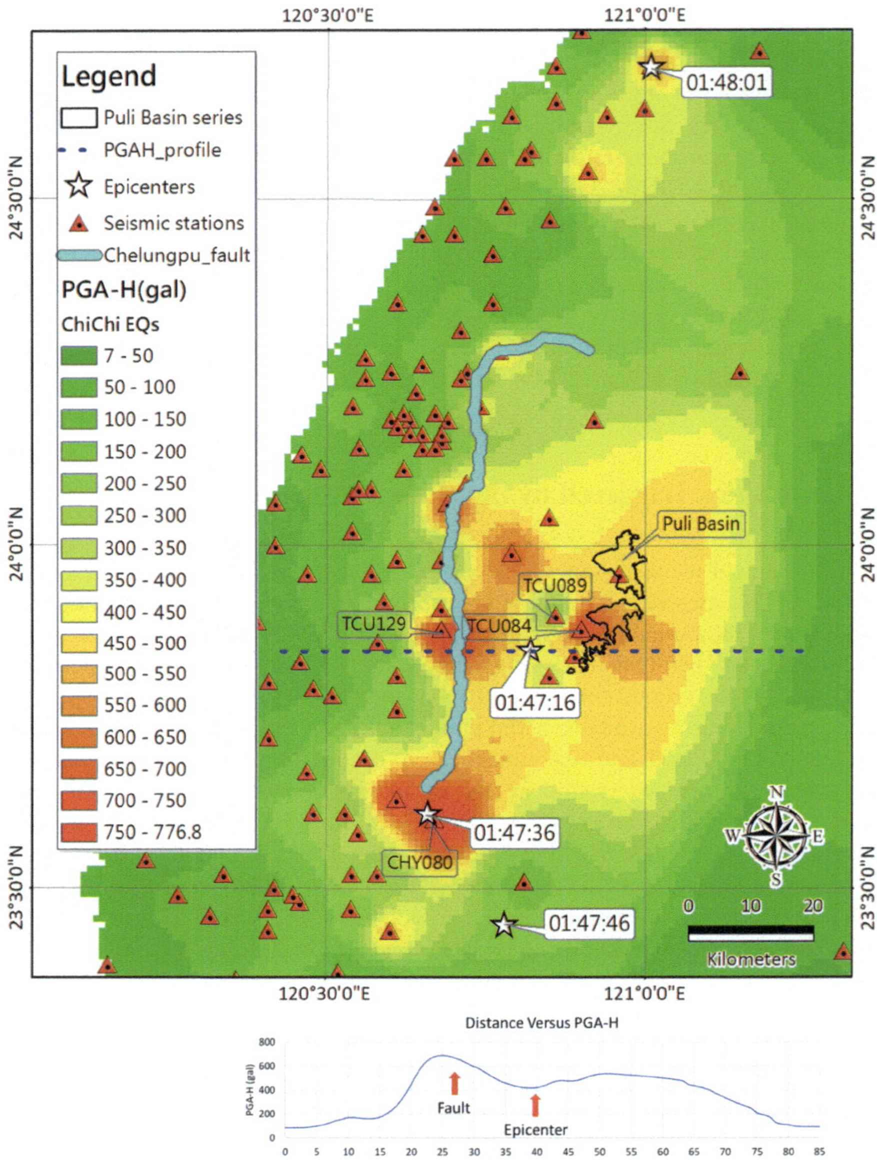

Fig. 1 Locations of epicenters and ground rupture of the Chelungpu fault and distribution of horizontal peak ground accelerations of the Chi-Chi earthquake

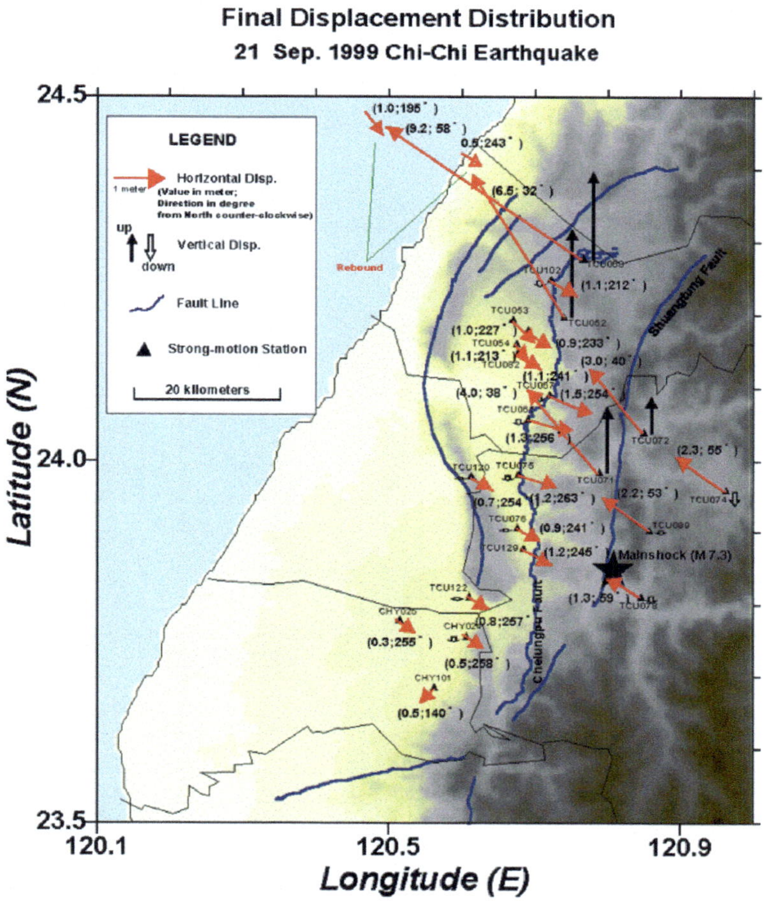

Fig. 2 Ground displacements caused by the thrusting fault surveyed using a global positioning system (GPS) (Huang et al. 1999, Central Weather Bureau (CWB) (http://www.cwb.gov.tw/V5e/seismic/chichi.htm))

clear evidence that different geological formations could have affected the landslide distribution. However, a high landslide distribution density was observed along the juncture area of the two different geological zones, which might suggest geological conditions vulnerable to landslide at the juncture area. The Shuangtung fault to the east of the Chelungpu fault (Fig. 2) situated in the juncture area of two geological zones also displaced during the Chi-Chi earthquake. It has been speculated that the Shuangtung fault was reactivated by the Chi-Chi earthquake (Central Geological Survey 2017), which might have caused a higher landslide density near the fault.

The Tai-Chung County and Nan-Tou County were most severely affected by earthquake-induced landslides due to the strong ground motions. In Tai-Chung County, the most severe slope failures occurred on slopes along the Central Trans-Island Highway, especially the section from Guguan to Techi, as shown in Fig. 5, and

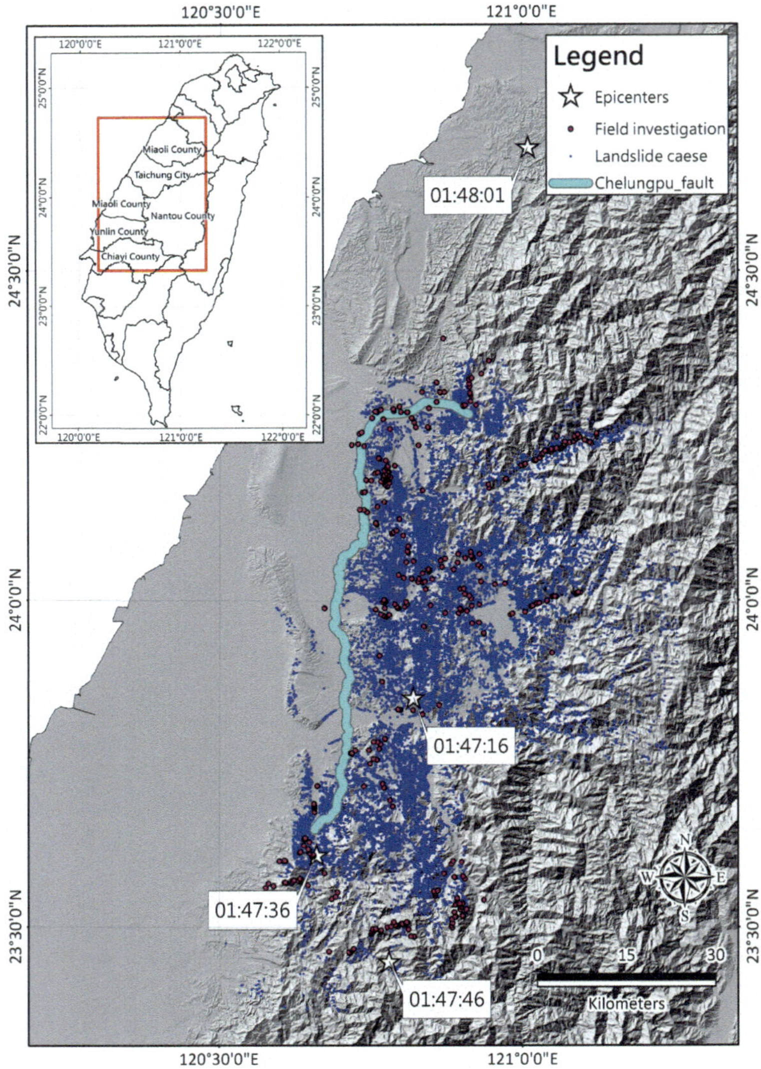

Fig. 3 Distribution of the mapped landslides and locations of field reconnaissance points along with the fault rupture and epicenters on the shaded relief map

almost all the slopes along the highway failed and damaged the highway severely after the Chi-Chi earthquake (Fig. 5). A slope failure affecting an area of approximately 950 ha. occurred at Juo-Juo Peaks in Nan-Tou County (Fig. 6). The landslides of Juo-Juo Peaks were mostly shallow landslides of top soil spalling due to the nature of the gravelly formation. In the Chiufengerhshan area, a dip-slope landslide occurred with an affected area of approximately 200 ha. (Fig. 7). The transported landslide mass formed two dammed-up lakes, and 41 people were buried (Lin et al.

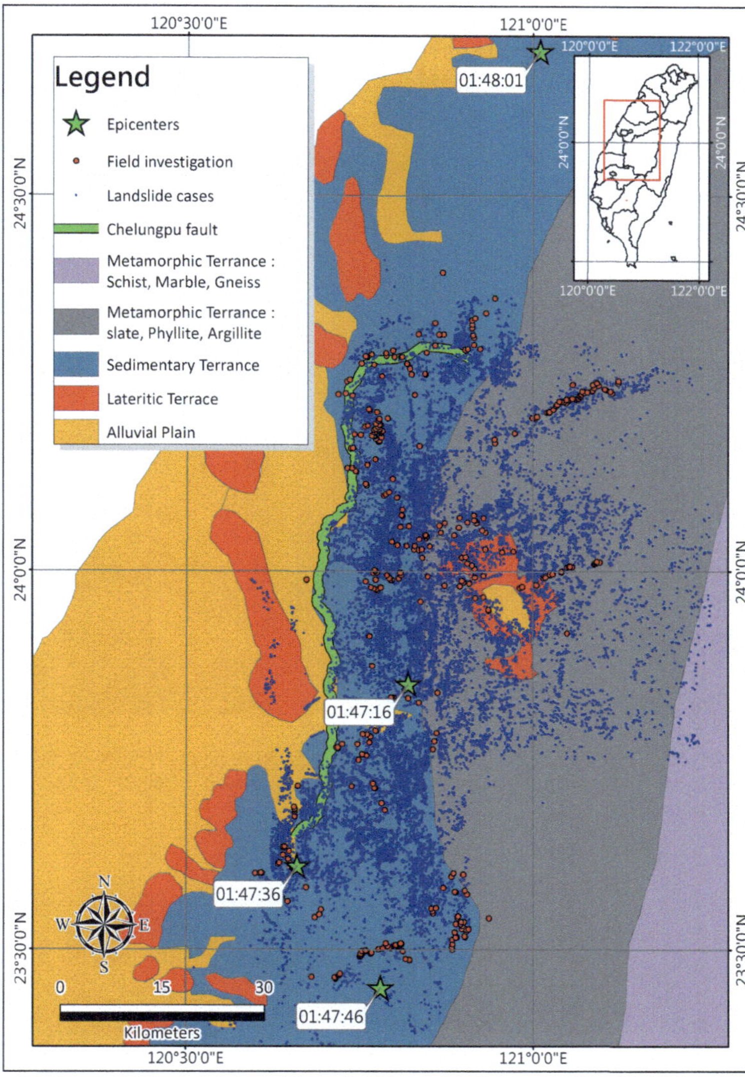

Fig. 4 Distribution of mapped landslides and locations of field reconnaissance points versus different geological formation zones

2000a). The Huntsaiping landslide, categorized as a modest displacement (~20 m on average), involved an area of approximately 100 ha. (Tseng et al. 2009; Dong et al. 2009) and was located west of the Chiufengerhshan landslide. The closest distance between these two landslides was only 2 km. In Yun-Lin County, a massive dip-slope landslide occurred in the Tsaoling area, covering an area of 400 ha. and with 120 million m^3 of deposited material; it formed a high mound across the Ching-Sui River (Fig. 8). Twenty-nine people were killed by the landslide, and a large amount of materials were deposited in the valley of the Ching-Sui River, damming the river channel for 5131 m upstream. Highways leading to the Tsaoling area were badly damaged due to landslides along the highways which blocked the road.

Fig. 5 Landslides along the Central Trans-Island Highway, from Guguan to Techi, the highway was severely damaged by slope failures (courtesy: Highway Bureau, Ministry of Transportation and Communication)

Fig. 6 Juo-Juo Peaks in Nan-Tou County after the Chi-Chi earthquake (courtesy: J. J. Hung)

Fig. 7 Chiufengerhshan dip-slope landslide, looking up to the sliding plane from the deposition area at the toe of the landslide (by M. L. Lin)

Fig. 8 Looking up to the sliding plane from a deposition mound across the Ching-Sui River near the toe of the dip-slope landslide in the Tsaoling area (courtesy: J. J. Hung)

To investigate whether weather conditions would have affected the landslide conditions, the rainfall record in September 1999 of the Sun Moon Lake rain gauge station of the Central Weather Bureau, which was close to the epicenter, was inspected (Fig. 9). Not much precipitation was recorded before the Chi-Chi earthquake, as shown in the rainfall figures, and it is postulated that the rainfall and weather condition were unlikely to have had significant effects on the landslide occurrences.

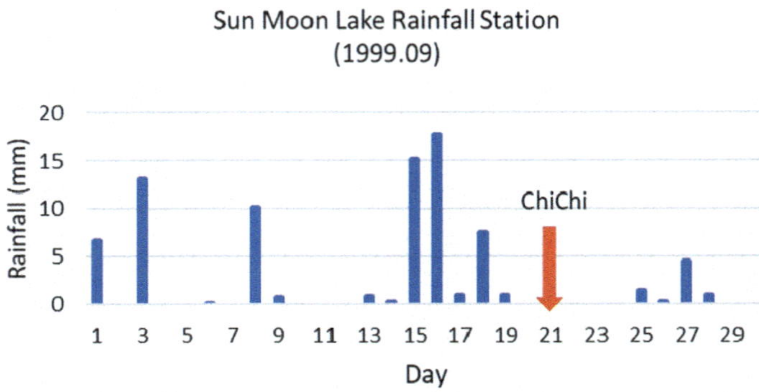

Fig. 9 Rainfall record of the Sun Moon Lake rain gauge station of the Central Weather Bureau, Taiwan in September 1999

4 Characteristics of the Earthquake-Induced Landslides

Statistical analysis of the field reconnaissance data revealed that most of the earthquake-induced landslides involved small to medium areas, and about 70% of the landslides had areas smaller than 4000 m^2 (Lin et al. 2000a, 2008). Figure 10 shows the distribution of types of slope failure based on field reconnaissance, and among them debris slides appear to be the most encountered slope failure type, accounting for 63% of the documented cases, while toppling and rock fall are the second most common type with 22% (Lin et al. 2000b). The large proportion of small-scale landslides consisted of the major slope failure types of debris slides and rock falls, in which the debris slides were usually shallow landslides with a small area. The field investigations confirmed that the most common type of landslide was shallow landslides near the crest of slopes. Similar results were also reported by Miles and Keefer (2000). Data from the 31,978 mapped landslides was also used to identify the characteristics of the earthquake-induced landslides. The distribution of the mapped landslide area is shown in Fig. 11, and about 85% of the mapped landslides have an area smaller than 5000 m^2. This result is consistent with the field reconnaissance data that small-scale landslides with a shallow depth are the major slope failure type. Only about 2.7% of the landslides had an area larger than 25,000 m^2. Due to the dry weather conditions, there were unlikely to be significant cases of deep-seated landslides. The distributions of the slope angles of the mapped landslides are presented in Fig. 12. It is noted that about 23% of landslides have a slope angle smaller than 20°, which might be caused by the large ground motions of the Chi-Chi earthquake. The landslide area data are plotted versus the slope angle in Fig. 12. The distribution of the landslide area is quite uniform and appears to be relatively independent of the slope angle. However, three distinctive cases were observed with a large area and gentle slope angles. The three cases were Tsaoling, Chiufengerhshan, and Huntsaiping landslides, and all three landslides had the characteristics of a dip slope. The locations of the three landslides are plotted in Fig. 13.

The Chiufengerhshan and Huntsaiping landslides were located near the epicenter of the Chi-Chi earthquake, and the Tsaoling landslide was near the epicenter of one of the triggered earthquake events. Thus, it is speculated that all three large gentle slope landslides were triggered due to large ground accelerations. The three cases are discussed in the following section.

Comparing the distribution of landslides in Fig. 3 to the distribution of horizontal peak ground acceleration in Fig. 1, it appeared that the landslides distribution is concentrated in high peak acceleration areas and decreased with decreasing peak ground acceleration. A similar distribution trend was observed with respect to the vertical peak ground acceleration. The threshold peak ground accelerations in horizontal and vertical directions for the mapped landslides were about 100 and 70 gal, respectively (Lin et al. 2008, 2009). The mapped scars of landslide were further

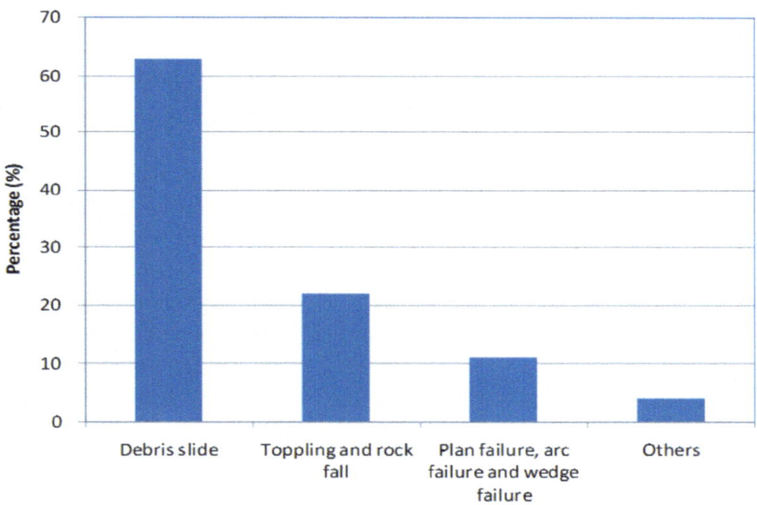

Fig. 10 Distribution of the different types of slope failure from field data after the Chi-Chi earthquake

Fig. 11 Distribution of the mapped landslide areas of the Chi-Chi earthquake

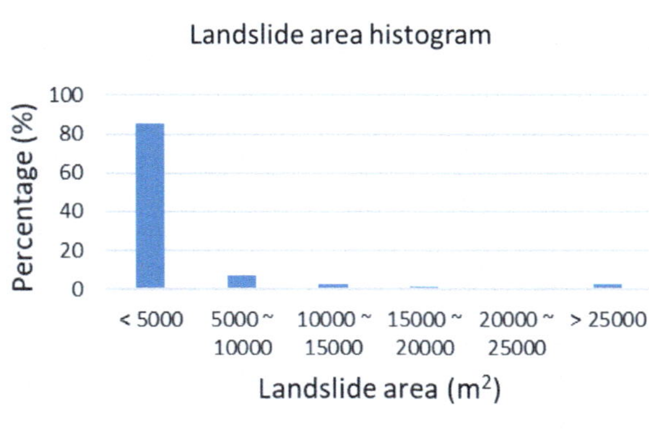

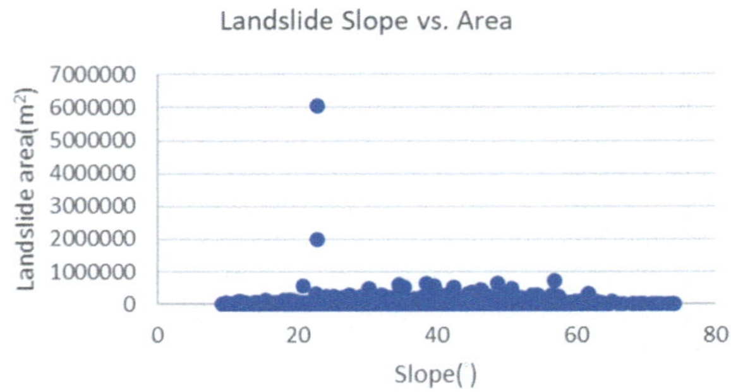

Fig. 12 Distribution of the landslide area versus slope angle of the mapped landslides of the Chi-Chi earthquake

edited to remove the scars on flat land, river channels, and on slopes with angles smaller than 10°. By overlaying distributions of landslides and peak ground accelerations, the variations of landslide ratios versus peak ground accelerations were calculated. Figure 14 presents the variation of landslide ratio with respect to the vertical peak ground acceleration, and Fig. 15 presents the variations of the landslide ratio with respect to the horizontal peak ground acceleration. The landslide ratio in both figures appears to decrease with decreasing peak ground acceleration and is consistent with the expected trend of variation. It appeared that the intensity of ground motion might be the most important factor for causing landslides. The mapped landslides are plotted versus the distribution of the vertical peak ground acceleration in Fig. 16, and mapped landslides are distributed densely near the high vertical peak ground acceleration area, even though the topography of these areas was not very steep. It was found that the vertical ground acceleration affected the landslide significantly with a lower threshold acceleration for triggering landslide, and the effect decreases rapidly with decreasing peak ground acceleration (Lin et al. 2008). In addition, the three landslides on gentle slopes are situated in high vertical peak ground acceleration areas, and it is likely that a high vertical ground acceleration contributes significantly to causing landslides on slopes with a gentle slope angle.

5 Three Large Landslides on Gentle Slopes Triggered by the Earthquake

During the Chi-Chi earthquake, three co-seismic landslides on gentle slopes involving large amount of mass transport were triggered and caused a severe hazard. The three cases were the Tsaoling, Chiufengerhshan, and Huntsaiping landslides. The geological map of the three cases is presented in Fig. 17, and it shows that all three landslides are situated in sedimentary rock formation zones. The three landslides all had characteristic dip-slope topography and geological formations, and a gentle slope angle, and could have been caused by strong ground motions in both vertical and horizontal directions. The three cases are presented in more detail as follows.

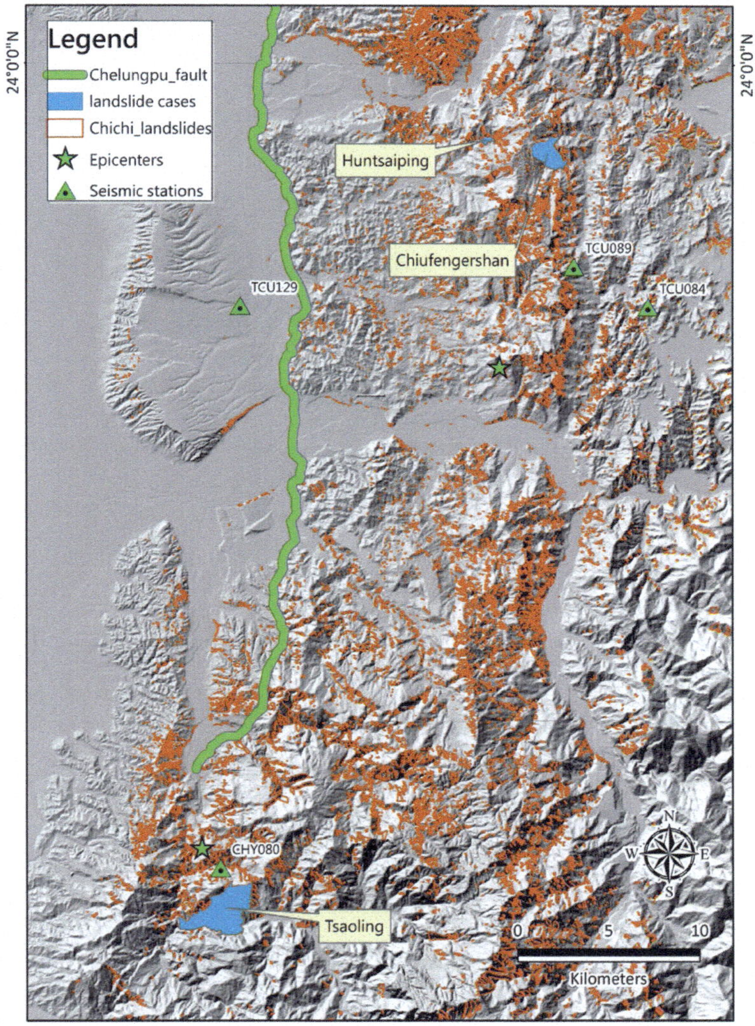

Fig. 13 Locations of Tsaoling, Chiufengerhshan, and Huntsaiping landslides on the shaded relief map

5.1 Tsaoling Landslide

During the Chi-Chi earthquake, a large dip-slope landslide of the Ku-Long Mountain, known as the Tsaoling landslide, was triggered. The landslide was located 6.6 km from the southern tip of Chelungpu fault at 120°40′24″ east longitude and 23°34′44″ north latitude in the Tsaoling village, Kou-Kern Township, Yun-Lin County, in Central Taiwan (Fig. 13). The elevation of the main landslide slope ranged from

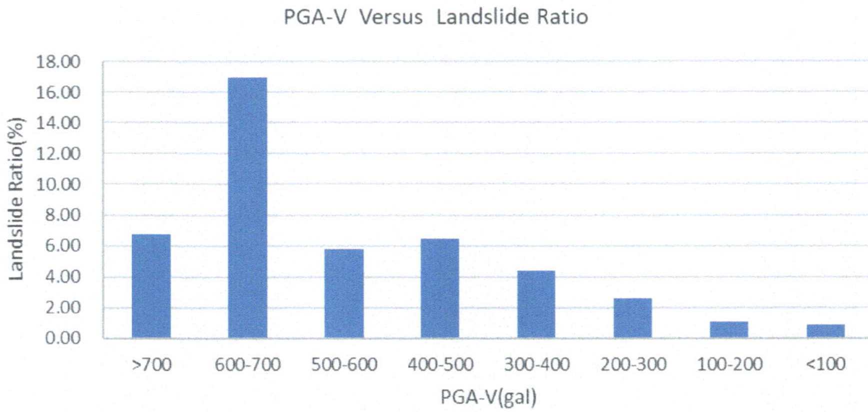

Fig. 14 Variations of landslide ratio with respect to the vertical peak ground acceleration

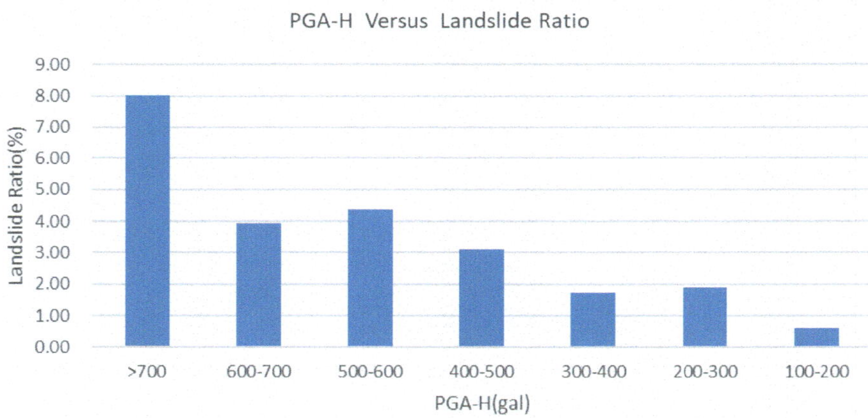

Fig. 15 Variations of the landslide ratio with respect to the horizontal peak ground acceleration

400 m to 1234 m above sea level (asl), and the landslide body slid from northeast to southwest during the earthquake, involving an area of approximately 440 ha. The average width of the landslide body was approximately 3800 m, and its length was 2800 m. The landslide mass transported by Tsaoling landslide was as high as 120 million m^3; however, only 25 million m^3 (approximately 20%) of the landslide mass dropped into the Ching-Shui River valley at the toe of the slope. The remaining landslide mass of about 100 million m^3 slid across the Ching-Shui River and landed on the remains of an old landslide dam (named "Daogiaoshan" by local people). Residents and their houses sitting on the landslide mass near the crest of the slope before the earthquake were translated with the landslide mass and landed on top of the old landslide dam (Fig. 8). The landslide mass translated downslope 3100 m in a horizontal direction and 800 m in a vertical direction. A total of 32 people were killed,

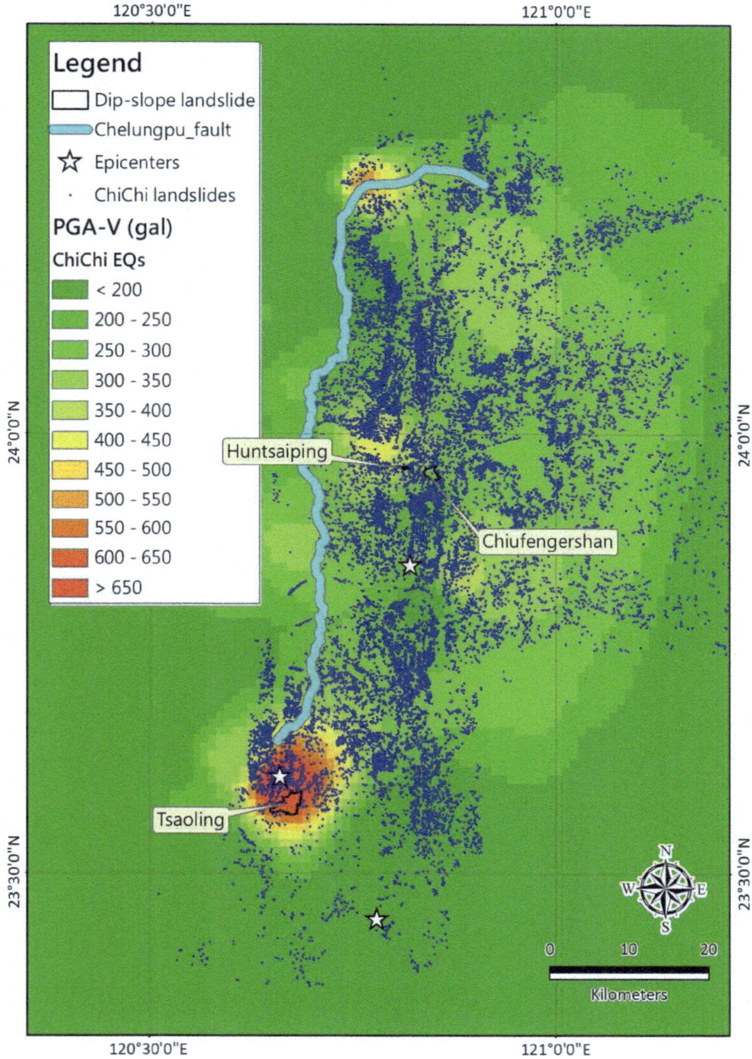

Fig. 16 Distribution of mapped landslides versus the distribution of the vertical peak ground acceleration

and seven people survived after the "sliding-landing" process (Fig. 8). The vegetation on a few hillslopes downstream of the landslide dam was stripped away, probably due to the air blast caused by the high-speed movement of the landslide mass. The Ching-Shui River was blocked by the huge landslide mass, and the upstream dammed-up lake held a total of 45 million m^3 of water. An emergency spillway was constructed on the debris dam to mitigate the potential hazard of dam breach, and overflow of the impounded water began on December 22, 1999, without causing damage to the

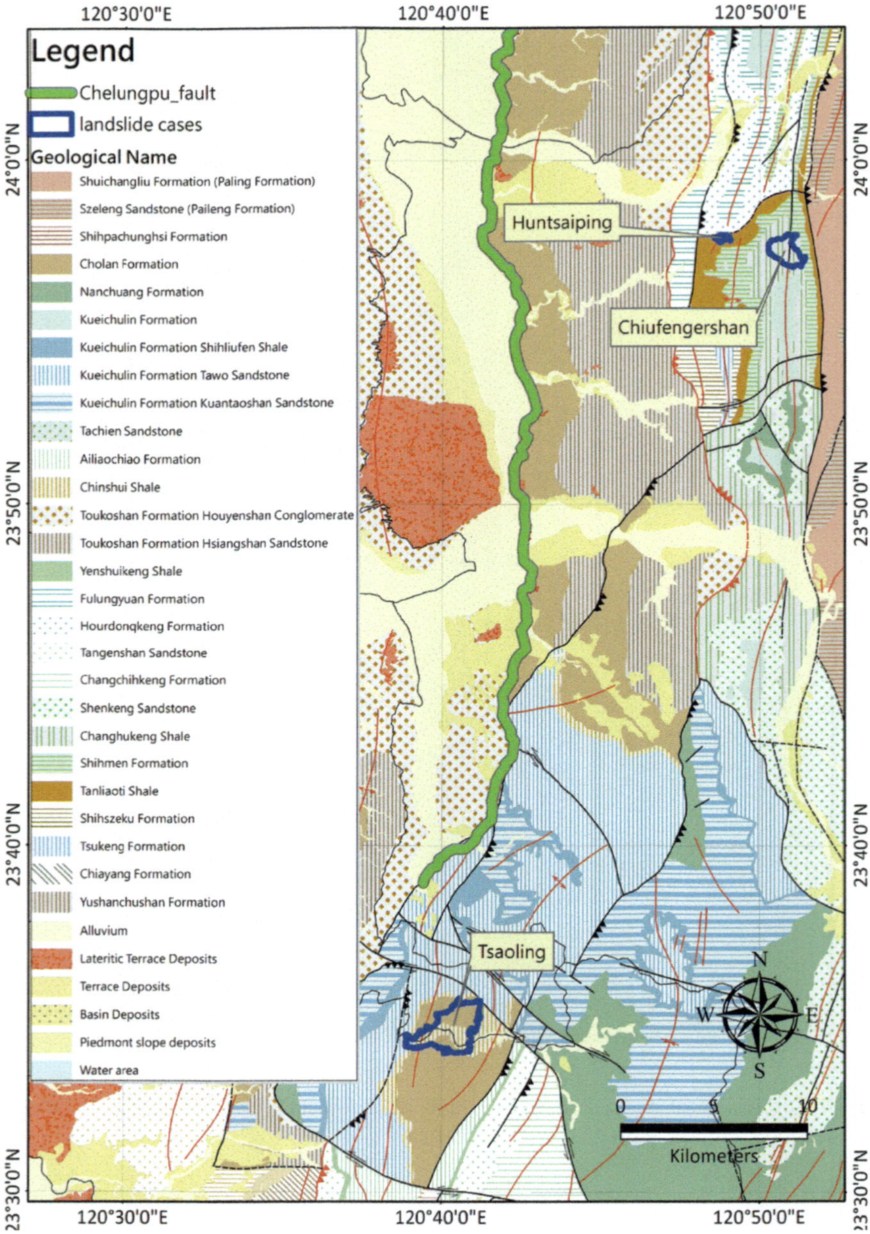

Fig. 17 Geological map for the three landslides on gentle slopes (modified from Central Geological Survey (CGS) 2013)

dam. Figure 18 is an aerial photograph of Tsaoling landslide taken on June 29, 2000, in which the landslide area, the debris mound of Daogiaoshan, the debris dam, and dammed-up lake can be observed clearly.

The Tsaoling dip-slope landslide has a long record of severe landslide incidents. Current records show that catastrophic landslides occurred in 1862, 1941, 1942, 1979, and 1999 (triggered by the Chi-Chi earthquake). A landslide dam was formed in each incident which blocked the Ching-Shui River, and breaching of the landslide dam took place in 1898, 1951, and 1979, respectively. Several studies were conducted on the case history of the Tsaoling landslide as early as 1942 (Taipei Observatory 1942; Chang 1951; Hsu 1951; Hsu and Leung 1977; Hung 1980; Hung and Lin 1981; Hung et al. 1982, 1994, 2000; Huang et al. 1983; Chang and Lee 1989; Lee et al. 1994; NCREE, NAPHM and Taiwan Geotechnical Society 1999; Lin et al. 2000a; Chen et al. 2004; Chen et al. 2014). A time history of the landslide and dam breach events is summarized in Table 1. Three major landslide events occurred prior to the landslide event of the Chi-Chi earthquake, namely landslides in 1862, 1941, and 1979. Note that the 1862 and 1941 events were both triggered by earthquakes. The event in 1942 was considered as a landslide of the residual slope of the 1941 event. In addition, the 1979 event was also suggested as a landslide of the residual slope of the 1941 event in several research studies (Hung 1980; Hung et al. 1994). Thus, the major triggering factor of the Tsaoling dip-slope landslide events appears to be seismic activity.

Fig. 18 Aerial view photograph of the landslide area, debris mound of Daogiaoshan, the dammed-up lake, and Tsaoling village, looking to the northwest direction. (photo taken on June 29, 2000, courtesy: Yung-Wang Lien)

Table 1 Time history of the dip-slope landslide events of the Tsaoling area (revised from Wu et al. 2001)

	Landslide event		Dam breach
	Date	Triggering factor	Date
(1)	1862	Earthquake	1875, 1898
(2)	1941.12.17	Earthquake	–
	1942.08.10	Rainfall	1951.05.18
(3)	1979.08.15	Rainfall	1979.08.24
(4)	1999.09.21	Earthquake	–

Tsaoling landslide is in the foothill region of Central Taiwan, and the geology formations are dominantly sedimentary rocks that dip in approximately the same direction as the slope aspect. The site geology is composed of Tawo sandstone (Pliocene epoch), Chinshui shale, and Cholan formation (Pliocene to Pleistocene epoch) from the bottom to the top, as illustrated in the profile cross section in Fig. 19. The statue of rock formations ranges from N20° W to N40° W dipping toward 14° SW. Two sets of orthogonal open joint were observed on the remaining dip slope. The main sliding occurred in the Cholan formation. The Ching-Shui River flows pass the toe of the Ku-Long Mountain, which causes the bedding plane to daylight due to erosion. Four day-lighted steps of scarps with heights ranging from 20 to 30 m were found on the remaining slope during field investigations. The dip angles measured on the remaining slope ranged from 12 to 14 degree, and the strike of the slope was N35° W, dipping south. Tension cracks and a graben near the crest of the remaining slope were also identified (Fig. 19). The topographic variations based on the digital terrain model (DTM) pairs before and after the landslide are illustrated in Fig. 20. The blue color in the figure indicates ground erosion or subsidence, and the red color indicates material deposition. The landslide mass estimated from the topographic variation was about 150 million m^3.

The free-field strong ground motion station CHY080 was located about 32 km south of the main shock and just north of the crest of Tsaoling (Figs. 13 and 19), and its records could be taken as the ground motions at the landslide site. The recorded peak horizontal ground accelerations were 841.5 gal in the north–south direction and

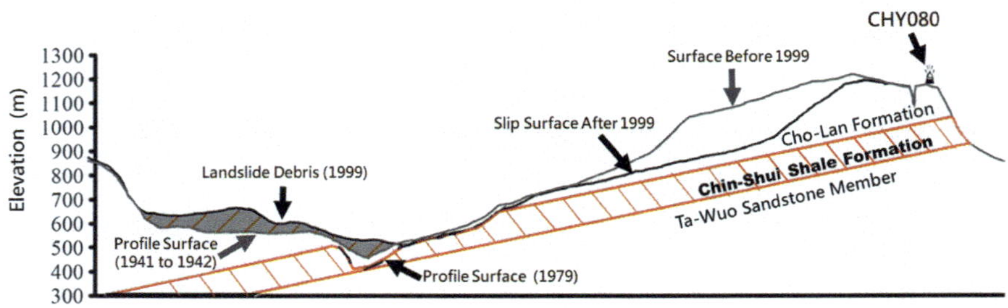

Fig. 19 Slope profile and sliding surface of the Tsaoling landslide caused by the Chi-Chi earthquake, 1999 (modified from Chen et al. 2004)

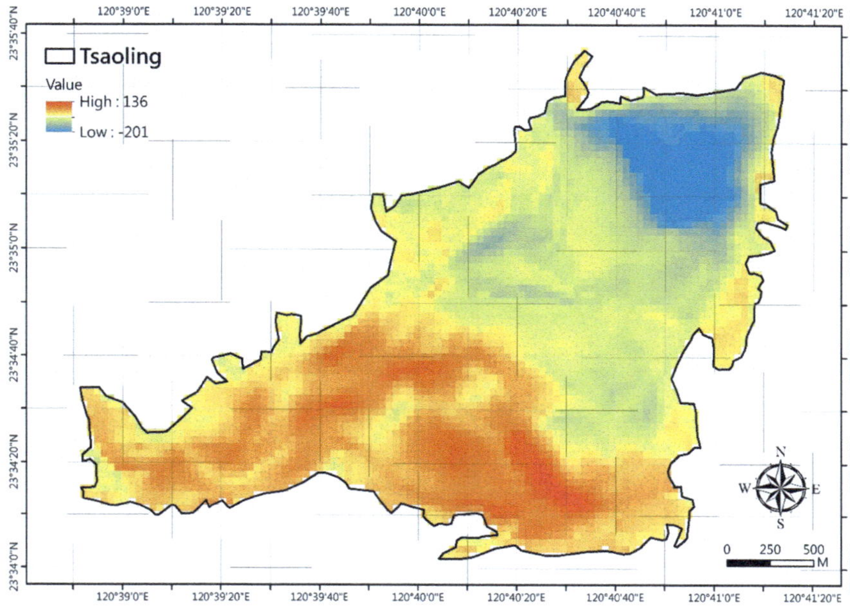

Fig. 20 Topographic variations caused by the Tsaoling landslide using DTMs before and after the Chi-Chi earthquake

792.4 gal in the east–west direction, and the peak vertical ground acceleration was 715.8 gal (as listed in Table 2). The peak values of all three components of ground acceleration were of a similar magnitude. Both the peak horizontal accelerations in the N–S and in the E–W directions were high, with the N–S direction being higher than that of the E–W direction. This was different from records of most other stations and the characteristics of fault thrusting. In addition, the high peak vertical ground acceleration suggested near-epicenter conditions. It was reported by Shin (1999) that among the three triggered earthquake events, the epicenter of the first triggered event was located approximately at 120.66° E, 23.61° N just to the north of the Tsaoling area and occurred at 01:47:36, with two distinctive sets of P- and S-wave arrivals being recorded. The CHY080 station was located very close to the epicenter of the first triggered event, and the arrival of the second set of waves was recorded about 20 s after the first set of wave arrivals, with maximum peak amplitudes. Ground motion records of CHY080 are shown in Fig. 21. Chen et al. (2004) conducted a pseudo-static analysis of the Tsaoling landslide triggered by the Chi-Chi earthquake and suggested that the high vertical acceleration was the key factor triggering the landslide. Further study by Chen et al. (2014) suggested that the ground motions caused by landslide triggering, landslide mass translation, and landing of the landslide mass all contributed to the seismic signals recorded by the strong motion station due to the huge landslide mass involved.

Table 2 Peak ground accelerations of station CHY080 for the Chi-Chi earthquake

Parameter\direction	Vertical (Z)	Horizontal (N–S)	Horizontal (E–W)	Horizontal (dip direction)
Peak ground acceleration (gal)	715.8	841.5	792.4	646.4

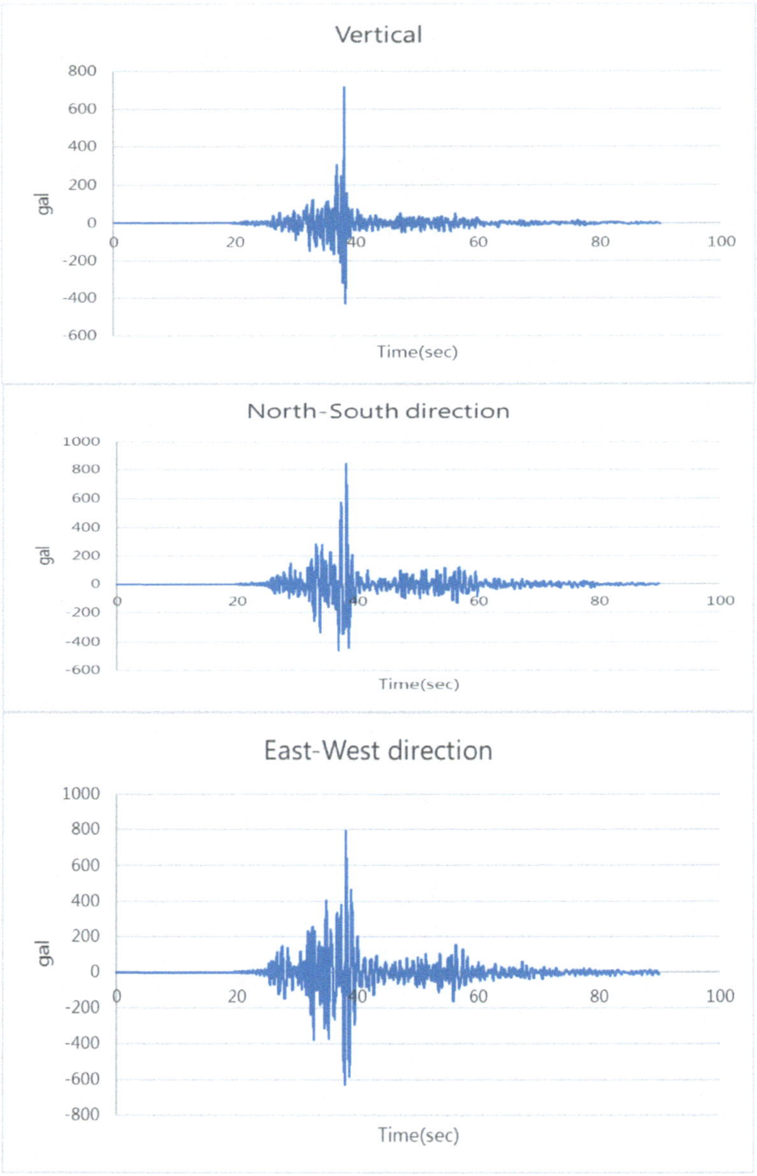

Fig. 21 Ground motion records of the CHY080 strong motion station for the Chi-Chi earthquake

5.2 Chiufengerhshan Landslide

A large-scale dip-slope landslide of the Ken-Dou Mountain in Nan-Tou County was triggered by the Chi-Chi earthquake; it is known as the Chiufengerhshan landslide. The Chiufengerhshan dip-slope landslide is located at 120°50′43″ E, 23°57′31″ N in the Changshi Alley, Nan-Kung Village, and Kuo-Shing Township of Nan-Tou County in Central Taiwan. The landslide was on the hanging wall side of the Chelungpu fault and was at a distance of 12 km from the epicenter and 14 km from the Chelungpu fault (Fig. 13). The landslide mass slid approximately from the northwest near the crest toward the southeast direction and has an area of approximately 195 ha. The elevation of the landslide slope ranged from 550 m to 1075 m asl; the width of the landslide was approximately 1100 m, and the length was 2000 m. The landslide mass transported by the Chiufengerhshan landslide was approximately 30–42 million m^3 (Lin et al. 2000a; Chang et al. 2005; Wu 2008). The transported landslide mass buried 19 houses and killed 41 people. Figure 7 shows the landslide looking up from deposition area to the crest of the slope. During the earthquake, it was found that the dip-slope landslide caused an abrupt rupture in the northeast part near the crest of landslide, which then induced a small landslide sliding in a direction approximately perpendicular to the direction of the main dip-slope landslide and toward the Bei-Keng area. Figure 22 shows the ground rupture near the crest of Ken-Dou Mountain. Three small rivers flowed through the toe area of the Ken-Dou Mountain, i.e., the Giou-Tsei-Hou River, the She-Tsu-Ken River, and the Gin-Gua-Liao River. The transported landslide mass formed a debris dam which dammed up the three rivers, forming lakes. Figure 23 shows the dammed-up lake of the She-Tsu-Ken River formed by the landslide. The total area of the three dammed-up lakes was approximately 14 ha, holding approximately 1.9 million m^3 of water. Emergency spillway and drainage channels were installed to mitigate the potential hazard that might be caused by a dam breach.

The Chiufengerhshan landslide area was located in the foothill region of Central Taiwan and with sedimentary rock formations dipping in approximately the same direction as the slope aspect. The geological map of the Chiufengerhshan landslide area is shown in Fig. 17. From the field investigation, the measured strike of the slope was N35° E, dipping to the southeast direction. The profile of the Chiufengerhshan landslide area was taken along the main sliding direction from northwest to southeast, as shown in Fig. 24. The geological formations from bottom to top are Tanliaoti formation (TL), Shihmen formation (SM), Changhukeng formation (divided into bottom, middle, and top sublayers, CHb, CHm, and CHt), and Kueichulin formation (KC) (Chen 2001) (Fig. 24). The strike of the formation is approximately the same as the slope—N35° E, and dipping to the southeast at about 24°, forming a dip-slope condition. Figure 24 shows the topographic variations of the slope profiles before and after the landslide. The slope angle before the landslide determined from the profile was approximately 21°. The dip angle measured in the field at the daylight of the slope surface near the toe was about 20°, which coincided with the profile analysis well. From the profile variations after the landslide in Fig. 24, the landslide plane was largely located in the Changhukeng formation, which was a thick sandstone layer

Fig. 22 Ground rupture near the crest of Ken-Dou Mountain caused by the Chiufengerhshan landslide (by M. L. Lin)

Fig. 23 Dammed-up lake of the She-Tsu-Ken River formed by the debris dam of the Chiufengerhshan landslide (by M. L. Lin)

with thin layers of shale. The topographic variations based on the digital terrain model pairs before and after the landslide are illustrated in Fig. 25. The blue color in the figure indicates ground erosion or subsidence, and the red color indicates material deposition. The landslide mass estimated from the topographic variations was about 36 million m^3. It was noted that an abrupt ground rupture near the crest can be observed in the northeast part of the mapped scar, which suggests a landslide in a direction approximately perpendicular to the direction of the main slide.

The ground motion at the Chiufengerhshan landslide caused by the Chi-Chi earthquake could be determined by interpolation of records of the free-field strong motion stations of the Central Weather Bureau (http://www.cwb.gov.tw/V5e/seismic/chichi.htm) in the neighboring area. There were twelve strong motion stations within distances ranging from 5 to 15 km. Among the twelve strong motion stations, seven were located on the hanging wall side and the other five on the foot wall side of the fault. A comparison study showed that the characteristics of ground motion records of the hanging wall side stations differed from those of the foot wall side (Lin et al. 2000a). The distribution of seismicity across the Chelungpu fault (Fig. 1) also suggested different ground motion properties on the foot wall and hanging wall. The differences in terrain across the fault might also contribute to the differences in ground motion characteristics, as shown by the shaded relief map in Fig. 3. Lin et al. (2000a) conducted a pseudo-static analysis of the Chiufengerhshan landslide, using the profile in Fig. 24 with the ground motion interpolated from the records of the strong motion stations on the hanging wall side. The horizontal ground motion was obtained by combining records in the east–west and north–south components and transforming them to the dip direction of the Chiufengerhshan landslide. The resulting peak vertical acceleration was 270 gal, and the peak horizontal ground acceleration in the dip direction was 450 gal. No prior dip-slope landslide event was recorded, which might be due to the gentle slope angle comparing to the strength of the sandstone/shale. The critical accelerations determined were 180 gal in the

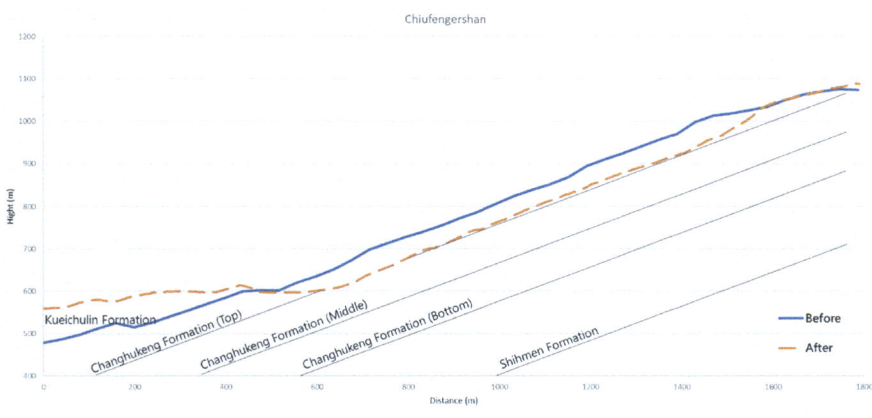

Fig. 24 Slope profile and sliding surface of the Chiufengerhshan landslide caused by the Chi-Chi earthquake, 1999

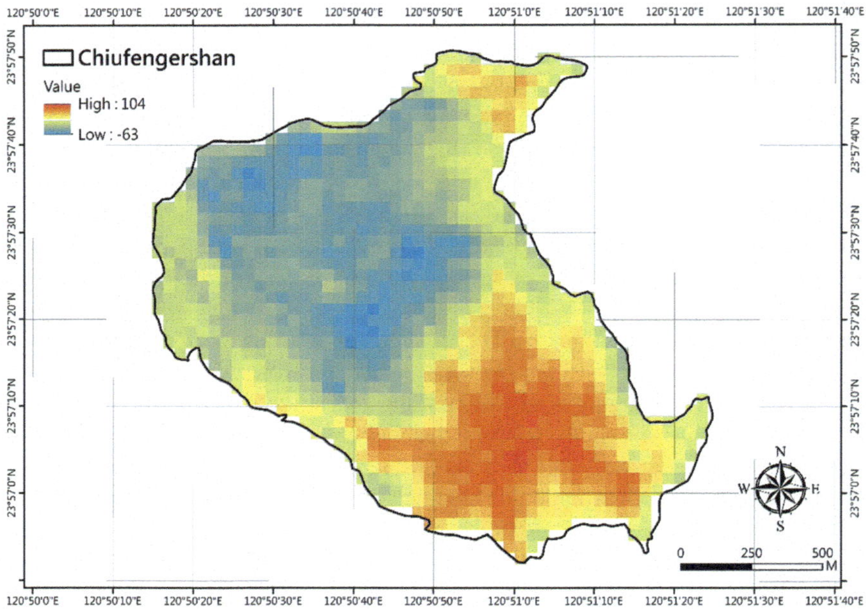

Fig. 25 Topographic variations caused by the Chiufengerhshan landslide using DTMs before and after the Chi-Chi earthquake

vertical direction and 280 gal in the horizontal direction by assigning a safety factor equal to 1. The critical accelerations thus determined were significantly smaller than the peak ground accelerations from the ground motion records. Because there are no prior records of landslide events caused by rainfall, it is likely that the most important factor causing the Chiufengerhshan landslide is the strong ground motion induced by the Chi-Chi earthquake, especially in the vertical direction.

5.3 Huntsaiping Landslide

The Huntsaiping landslide, a less noticeable landslide triggered by the Chi-Chi earthquake, is located west of Chiufengerhshan landslide (Figs. 13 and 17). The elevation of the landslide ranged from 500 and 1000 m asl. Most of the displaced area was only slightly disturbed. The morphological features at Huntsaiping landslide included a crown, tension cracks, a main scarp, steps of scarp, depressions, a hummocky ground surface, and a deposition zone at the toe. The displaced area and surface displacement vectors of the Huntsaiping landslide were identified using digital aerial photogrammetry (Lee et al. 2004) and particle image velocimetry technology (Tseng et al. 2009), and the estimated maximum horizontal surface displacements were 32 m and

24 m, respectively. The length and width of the Huntsaiping landslide were approximately 1500 m and 600 m, with an estimated area of 90 ha. (Dong et al. 2009). No documentation of the landslide volume was found in previous studies due to the difficulty in estimating sliding depth.

The engineering geological map of the Huntsaiping area is shown in Fig. 26. Miocene sedimentary rocks (interbeds of shales and sandstones) outcropped in the Huntsaiping area were overlain by thick colluvium (yellow polygon in Fig. 26), which might be the deposits of an ancient landslide (Dong et al. 2009; Chang et al. 2011). It is believed that the ancient landslide was a dip-slope slide along the bedding plane of the sedimentary rocks, with the source area located southeast of the Huntsaiping landslide. Lo et al. (2016) inferred that this landslide might have occurred during the 1916–1917 Taichung–Nantou series of earthquakes (four earthquakes with magnitudes of an average of 6 on the Richter scale, at a shallow focal depth within 1–45 km). These ancient displaced landslide deposits blocked the Yonglu Creek's paleochannel and resulted in a 90° turn of the channel, i.e., the movement of the slope during Chi-Chi earthquake is a reactivation of the ancient landslide deposits. An outcrop of the Tingshuiku fault was discovered in Yonglu Creek. The axis of Syncline A was constrained by the measured orientation of the bedding planes and M symmetry minor folds of thin-bedded shale, and sandstone layers crop out on the river bank of Yonglu Creek (Fig. 26). The A fault is a speculated structure and inferred from the discontinuous sandstone layer. Syncline B is a minor structure which is not significantly relevant to the landside.

The boundaries of this slightly disturbed landslide cannot be determined precisely, and different researchers referred to different landslide areas ranging from 90 to 120 ha. (Lee et al. 2004; Tseng et al. 2009; Dong et al. 2009). Dong et al. (2009) recognized the scarp of the landslide as the southern boundary. The maximum vertical displacement across the scarp during the Chi-Chi earthquake is roughly 20 m. The northern and the eastern boundaries of the Huntsaiping landslide follow the Yonglu Creek and Creek A. The western boundary of the Huntsaiping landslide is mainly along a ridge of massive sandstone. The sliding area estimated by Dong et al. (2009) was approximately 90 ha. Assuming an average sliding depth of 30 m, the estimated landslide volume was as high as 27 million m^3. This number is slightly less than the ancient landslide volume of 33 million m^3 estimated by Lo et al. (2016) and was considered reasonable.

There were three distinct sliding masses—A, B, and C (SA, SB, SC). The directions of surface movement of SA, SB, and SC shown by displacement vectors in blue in Fig. 26 were determined by Lee et al. (2004) via digital aerial photogrammetry, and they trend 330°, 315°, and 300°, respectively. Eight boreholes (Fig. 26) were drilled in the landslide mass between Creek A and Creek B (SA). Colluvial deposits overlying thick fractured shale were observed in the cores. The maximum thickness of colluvial deposits based on the borehole investigation was around 82 m. Slickensides observed in the shale cores (plunge of striations observed on the slickenside plane was 21°) and a bedding plane (N24°E/27°NW) outcrop in Creek A suggested that the interface between the shale and the colluvium was the sliding surface. The orientation of the interface between the colluvium and shale was determined as N19°E/23°NW

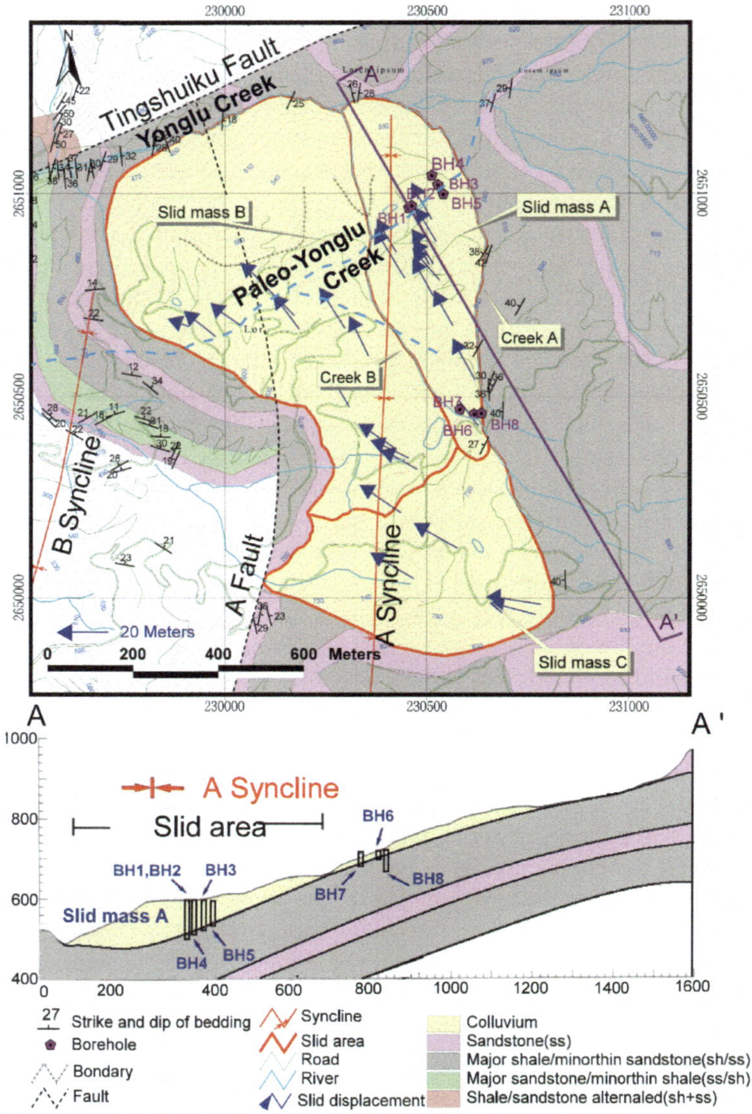

Fig. 26 Engineering geological map of the Huntsaiping area. Miocene sedimentary rocks (interbeds of shales and sandstones) overlain by thick colluvium. The cross section A–A' is along the direction of mass movement between Creek A and Creek B (330°). The interface between the shale and the colluvium was mainly a bedding plane (from Dong et al. 2009 with author rights)

using a three-point method (BH-1, BH-4, and BH-5), which was about parallel to the measured bedding plane. The observed striations in rock cores and in outcrops marked the footprints of colluvium sliding along the interface plane of this landslide triggered by the Chi-Chi earthquake. The trend of the striations observed on the slickenside plane outcrop in creek A was 341°, and the measured plunge was 15°. The trend of striations measured from the slickenside plane is almost identical with that of surface movement (330°) between Creek A and Creek B (SA) after the Chi-Chi earthquake.

It is interesting to note that the sliding direction did not coincide with the dip direction of the bedding plane. The apparent dip along the direction of surface movement (330°) of the displaced mass SA is 22° on the sliding plane measured in Creek A. Figure 26 shows the A–A' cross section along the direction of surface movement (330°) of displaced mass SA. The representative apparent dip is 21° along the direction of surface movement (330°) of SA. Near the toe of the slope, the sliding surface (mainly along the bedding plane) is inferred to become curved as the sliding surface passes through the axis of the Syncline A (Fig. 26).

By excluding particles with grain sizes larger than a #4 sieve, the grain size percentage of the colluvium sampled from Huntsaiping landslide site was 12% sand, 54% silt, and 34% clay. The liquid limit and plastic limit of the sample were 32 and 12, respectively. The plasticity index was 19. Illite (41.3%) was the dominant clay mineral of the Huntsaiping colluvium, based on X-ray diffraction test data of a glycolate-oriented sample and data of a slow scan at 24–26° 2θ position (Chen 2018). Kaolinite (11.3%) and chlorite (10.5%) were less abundant. Notably, the percentage of swelling minerals (such as smectite) was quite high (36.9%) and might account for the unstable nature of the landslide.

Chen (2018) conducted a pseudo-static analysis using STABL 5 M (modified Bishop method) and the critical accelerations (A_c, in g) required to trigger the sliding of the A–A' profile along the interface between colluvium and bed rock under different friction coefficients μ. The results show a linear relation as follows:

$$A_c = 0.63\mu - 0.23 \tag{1}$$

The equations for calculating the angular acceleration, angular displacement, and permanent displacement of a rigid block on a circular sliding plane can be found in Wan (2003). The representative acceleration histogram of strong motion station TCU089 (about 8.5 km southeast of the landslide) was used to perform Newmark displacement analysis (Fig. 27). A friction coefficient of 0.366 (friction angle of 20 degree) was used to calculate the time history of the down-dip acceleration S (black curve), velocity v (blue curve), and permanent displacement d (green curve). The accumulated displacement was 11.5 m, which was close to the measured average horizontal displacement 24.7 m during the Chi-Chi earthquake (Lee et al. 2004). The maximum sliding velocity during the earthquake was 0.5 m/s and decreased to zero after the ground shaking ceased. Thus, the run out of the slow-moving landslide was far less catastrophic than the Chiufengerhshan landslide.

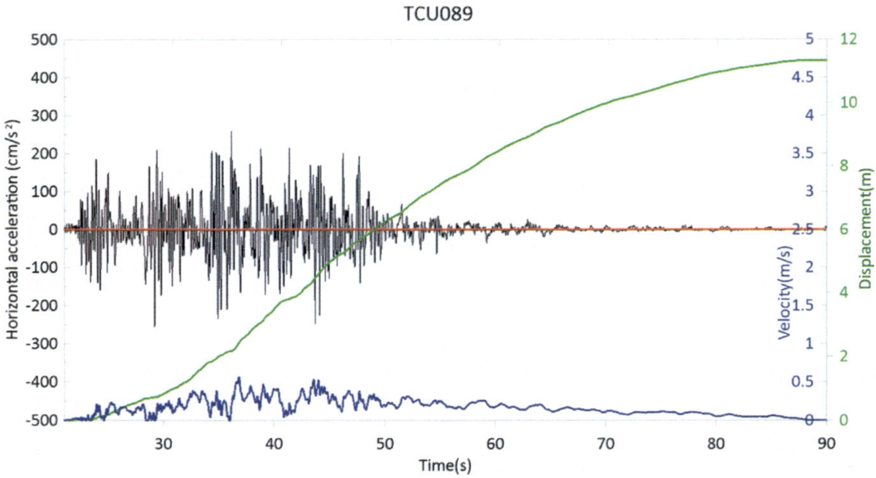

Fig. 27 Results of Newmark displacement analysis with the seismic data from TCU089 station and constant friction coefficient ($\mu = 0.366$). The down-dip acceleration S (black curve), velocity v (blue curve), and cumulative displacement d (green curve) are plotted. The accumulated displacement is 11.5 m, and the maximum sliding velocity is 0.5 m/s

6 Conclusions

In 1999, the Chi-Chi earthquake struck Central Taiwan, causing extensive landslides. Cross comparisons of the distribution of the landslides and ground motion data suggested that the strong ground motion was the most important factor for causing landslides, and the effect of the vertical peak ground acceleration was significant. The landslide ratios generally decrease with decreasing peak ground accelerations in both vertical and horizontal directions. No clear evidence was found for whether different geological formations would have affected the landslide distribution. However, an intensive distribution of landslides was observed along the juncture area of the two different geological zones, suggesting geological conditions vulnerable to landslide at the junction of different geological zones.

Three cases of large-scale landslides on gentle slopes were presented. All three cases had the specific geomorphological characteristics of a dip slope and being affected by high peak vertical and horizontal ground acceleration. Results of pseudo-static analysis suggested that the high vertical ground acceleration might be the key factor for triggering these landslides on gentle slopes involving large landslide areas.

Acknowledgements The authors appreciate the field photographs provided by Professor Ju-Jiang Hung, Mr. Yung-Wang Lien, and Highway Bureau, Ministry of Transportation and Communication, Taiwan.

References

Central Geological Survey (2013) Geological investigation and database for upstream watershed of flood-prone area. Central Geological Survey, Ministry of Economic Affairs, Taiwan

Central Geological Survey (2017) Geological map of active fault zone—Damaopu–Shuangtung fault. Central Geological Survey, Ministry of Economic Affairs, Taiwan, https://fault.moeacgs.gov.tw/MgFault/index.php/2017-06-02-08-00-54/85

Chang KJ, Taboada A, Chan YC (2005) Geological and morphological study of the Jiufengershan landslide triggered by the Chi-Chi Taiwan earthquake. Geomorphology 71:293–309

Chang KT, Lin ML, Dong JJ, Chien CH (2011) The Hungtsaiping landslides: from ancient to recent. Landslides 9(2):205–214

Chang LS (1951) Topographic features and geology in the vicinity of the nature reservoir near Tsao-Ling. (in Chinese). Taiwan Reconstruction Monthly 1(6):22–27

Chang SY, Lee CT (1989) The geology and landslides in Tsao-Ling area (in Chinese). Si-Notech Eng 19:27–43

Chen JH (2001) The study of the engineering geological characteristics of rock material in Juo-Feng-Err-Shan of the Nantou area. Master thesis, National Taiwan University, Taiwan

Chen YC (2018) Frictional and kinematical characteristics of the Hungtsaiping landslide, Taiwan. Master thesis, National Central University, Taiwan (in Chinese)

Chen TC, Lin ML, Hung JJ (2004) Pseudostatic analysis of Tsao-Ling rockslide caused by Chi-Chi Earthquake. Eng Geol 71:31–47

Chen TC, Lin ML, Wang KL (2014) Landslide seismic signal recognition and mobility for an earthquake-induced rockslide in Tsaoling, Taiwan. Eng Geol 171:31–44

Dong JJ, Lee WR, Lin ML, Huang AB, Lee YL (2009) Effects of seismic anisotropy and geological characteristics on the kinematics of the neighboring Jiufengershan and Hungtsaiping landslides during Chi-Chi earthquake. Tectonophysics 466(3–4):438–457

Hsu ST (1951) The Tsao-Ling dammed up lake, (in Chinese), vol 1, Department of Civil Engineering, National Taiwan University, Taiwan, pp 3–4

Hsu TL, Leung HP (1977) Mass movement in the Tsao-Ling area, Yunlin-Hsien, Taiwan (in Chinese). Proc Geol Soc China 20:114–118

Huang CS, Ho HC, Liu HC (1983) The geology and landslide of Tsao-Ling area, Yun-Lin, Taiwan (in Chinese). Bull Central Geol Surv 2:95–112

Huang CY, Lin CW, Chen WS, Chen YG, Yu SB, Chia IP, Lu MD, Hou CS, Wang YS (1999) Seismic geology of the Chi-Chi earthquake. In: Proceedings, international workshop on the September 21,1999 Chi-Chi Earthquake, pp 25–42

Hung JJ (1980) A study on Tsao-Ling rockslides, Taiwan (in Chinese). J Eng Environ 1:29–39

Hung JJ, Lin ML (1981) Preliminary study on Tsao-Ling landslide, material testing and stability analysis (in Chinese). Report for Council of Agriculture Development, Taiwan

Hung JJ, Lin ML, Liu ML, Lee CD, Jan KH, Hsieh PC (1982) Engineering geological investigation on Tsao-Ling landslide. Rock material testing report (in Chinese). Council of Agriculture Development, Taiwan

Hung JJ, Lin ML, Lee CT (1994) A stability analysis of the Tsao-Ling landslide area (in Chinese). In: Proceedings of 94 rock engineering symposium in Taiwan, Chung-Li, Taiwan, December, 15th-16th, pp 459–467

Hung JJ, Lee CT, Lin ML, Lin ML, Jeng FS, Chen CH (2000) A flying mountain and dam-up lake (Tsao-Ling rockslides) (in Chinese). Sino-Geotechnics 77:5–18

Lee CT, Lin ML, Wu LH, Cheng JS (1994) Geological investigations and the determination of sliding planes of rockslide events in the Tsao-Ling landslide area (in Chinese). In: Proceedings of 94 rock engineering symposium in Taiwan, Chung-Li, Taiwan, December, 15th-16th, pp 459–467

Lee JF, Wei CY, Huang CC (2004) The study of Hungtsaiping landslide using digital photogrametric technique. In: Proceeding of international symposium on landslide and debris flow hazard assessment, 7–8 October 2004. Taiwan, pp 51–59

Lin ML, Tung CC (2004) A GIS-based potential analysis of the landslides induced by the Chi-Chi earthquake. Eng Geol 71:63–77

Lin ML, Chen TC, Wang KL (2000a) The characteristic and preliminary study of landslides induced by Chi-Chi earthquake. Report of National Center for Research in Earthquake Engineering, Taiwan

Lin ML, Wang KL, Chen TC (2000b) Characteristics of the slope failure caused by Chi-Chi earthquake. In: Proceedings of international workshop on annual commemoration of Chi-Chi earthquake, III-Geotechnical Aspect, pp 199–209

Lin ML, Wang KL, Kao JC (2008) The effects of earthquake on landslides—a case study of Chi-Chi earthquake, 1999, invited lecture. In: Proceedings, the 10th international symposium on landslides and engineered slopes, pp 193–201

Lin ML, Huang GJ, Kao TC, Wang KL (2009) The threshold conditions of landslides induced by Chi-Chi earthquake. Sino-Geotechnics 121:57–66

Lin ML, Huang CY, Kao TC (2016) Threshold conditions and run-out displacements of the landslides induced by the Chi-Chi earthquake, Taiwan, the state of art lecture. In: Proceedings, the 12th international symposium on landslides, Napoli, Italy, pp 81–90

Lo CM, Huang WK, Lin ML (2016) Earthquake-induced deep-seated landslide and landscape evolution process at Hungtsaiping, Nantou County, Taiwan. Environ Earth Sci 75(8):645

Miles SB, Keefer DK (2000) Evaluation of seismic slope-performance models using a regional case study. Environ Eng Geosci 6(1):25–39

NCREE, NAPHM, and Taiwan Geotechnical Society (1999) Reconnaissance report of the geotechnical hazard caused by Chi-Chi earthquake. National Research Center on Earthquake Engineering, Taiwan, p 111

Shin TC (1999) Chi-Chi earthquake-seismology. In: Proceedings, international workshop on the September 21,1999 Chi-Chi earthquake, pp 1–14

Tai-Pei Observatory (1942) Report on Chia-Yi earthquake on 17th December 1941 (in Japanese). Taiwan Governor's Office, p 227

Tseng CH, Hu JC, Chan YC, Chu HT, Lee JF, Wei JY, Lu CY, Lin ML (2009) Non-catastrophic landslides induced by the Mw 7.6 Chi-Chi earthquake in central Taiwan as revealed by PIV analysis. Tectonophysics 466(3–4):427–437

Wan YC (2003) The displacement of rotational failure of the natural slopes under seismic condition. Master thesis, National Cheng Kung University, Taiwan (in Chinese)

Wu CY (2008) The structural and geomorphic characteristics before and after the Chiufen-erhshan landslide and possible mechanisms of the slope failure. Master thesis, National Taiwan University, Taiwan

Wu YS, Lai JS, Lin CH (2001) Summary of the emergency repairing of the Tsao-Ling dam-up lake. Q J Water Resour Manag (in Chinese) 3(1):18–23

Landslides Triggered by Recent Earthquakes in Italy

Ernesto Ausilio, Francesco Silvestri, Giuseppe Tropeano, and Paolo Zimmaro

Abstract This document provides a detailed description of seismically-induced landslides and rockfalls that occurred in Italy during the 2016 Central Italy earthquake sequence. Relevant ground motion characteristics for all mainshocks of the sequence are analyzed. The document also provides a quantitative overview of spatial and temporal characteristics of the landslides and rockfalls that occurred during the earthquake sequence which was characterized by three mainshocks: (1) **M**6.1 on 24 August, (2) **M**5.9 on 26 October, and (3) **M**6.5 on 30 October. A relevant fraction of landslides has been caused by the first **M**6.1 mainshock; however, both the following events, especially the third **M**6.5 earthquake, exacerbated the slope instability in many locations and triggered several new and more serious ground failures. The overwhelming majority of the instabilities induced by the sequence are rockfalls in fractured rocks. Details of three selected case histories are described: (1) the Nera rockslide, (2) the Pescara del Tronto landslide, and (3) the Accumoli landslide. Rainfall data during 2016 are also shown and described for selected locations in the epicentral area.

E. Ausilio · P. Zimmaro (✉)
University of Calabria, Rende, Italy
e-mail: pzimmaro@ucla.edu

E. Ausilio
e-mail: ernesto.ausilio@unical.it

F. Silvestri
University of Napoli Federico II, Napoli, Italy
e-mail: francesco.silvestri@unina.it

G. Tropeano
University of Cagliari, Cagliari, Italy
e-mail: giuseppe.tropeano@unica.it

P. Zimmaro
University of California, Los Angeles, USA

I. Towhata et al. (eds.), *Coseismic Landslides*, Springer Natural Hazards,
https://doi.org/10.1007/978-981-19-6597-5_10

1 Introduction

Landslides have been considered as secondary effects induced by earthquakes. These local phenomena affect the areas of interest and can substantially aggravate the seismic impact on human, social, and economic resilience during the aftermath. Recently, several studies have focused on the roles played by these secondary effects in the extents of damage and loss of human life (e.g., Bird and Bommer 2004).

Italy is characterized by very frequent seismic activity and high geomorphological and hydrological hazards. As a result, in recent years, numerous earthquakes were recorded in Italy, causing ground failures including landslides and rockfalls. Before 1783, historical accounts of the occurrence of earthquake-induced landslides were typically incomplete and vague (Keefer 2002). The first scientific post-earthquake reconnaissance including systematic documentation of landslides induced by earthquakes was undertaken in the Calabria region (Southern Italy) following the 1783 Calabrian earthquake sequence (Sarconi 1784, quoted by Cotecchia et al. 1986).

In the last century, Ambraseys (1976) and Govi (1977) documented ground failures triggered by the 1976 Friuli (Northern Italy) earthquake sequence, in which landslides were almost entirely rockfalls. Instability phenomena associated with the 1980 Irpinia earthquake (Southern Italy) were characterized by delayed initiation or reactivation. Cotecchia and Del Prete (1984) and D'Elia et al. (1985) observed coherent slides and earth flows that began moving between a few hours to a few days after the mainshock. These phenomena were associated with increased spring flow and pore-water pressures. This has resulted in an anomalous distribution of landslides where landslides were more numerous at greater distances from the epicenter. More recently, ground failures triggered by the 1997 Umbria-Marche (Central Italy) earthquake were mainly disrupted landslides. The areal distribution of landslides in Umbria-Marche was characterized by a large number of slope instabilities in the area adjacent to the epicentral area, with a decreasing trend with distance (Bozzano et al. 1998; Esposito et al. 2000).

In the period between August 24, 2016, and January 20, 2017, a wide area in Central Italy underwent a seismic sequence with three major earthquakes. The first event that had a moment magnitude $\mathbf{M} = 6.1$ occurred on August 24, 2016, the second ($\mathbf{M} = 5.9$) on October 26, and the third ($\mathbf{M} = 6.5$) on October 30, 2016. Each event was followed by numerous aftershocks.

The Mt. Vettore-Mt. Bove normal fault system caused all mainshocks of the sequence. The first mainshock (occurred on August 24, 2016) caused extensive damage especially to the villages of Amatrice, Accumoli, and Arquata del Tronto. The earthquake sequence caused 299 fatalities, generally from total collapses of unreinforced masonry buildings. The two mainshocks occurred in October caused massive new damage in the villages of Visso, Ussita, and Norcia. After the 24 August event, the epicentral area of the October earthquakes had largely been evacuated. As a result, these additional events did not produce fatalities.

Several landslides and rockfalls were triggered by the seismic sequence (Ausilio et al. 2019). The volume and area of the landslides caused by the August mainshock were limited despite that the regional morphology was steep and that a significant seasonal rainfall had occurred in the week preceding the August earthquake. For the instabilities triggered by the August event, this document focuses on the results (and subsequent analysis) of reconnaissance missions performed by the following agencies:

(1) The Geotechnical Extreme Event Reconnaissance Association (GEER 2016, 2017),
(2) The National Institute for Environmental Protection and Research (ISPRA 2016),
(3) The Research Center for Prediction, Prevention and Monitoring of Geological Risks of Sapienza University (CERI 2016),
(4) The National Institute for Geophysics and Volcanology (INGV 2016).

Information about landslide features triggered and/or exacerbated by the October events are mainly based on observations made by the GEER team and CERI. The authors of this document participated in the GEER reconnaissance mission and subsequent analysis. The GEER reconnaissance mission was conducted during two phases:

(1) Following the 24 August event, from early September to early October 2016,
(2) Following the October events, between the end of November and the beginning of December 2016.

For the 26 October event, limited field observations are described from the short time window between this event and the subsequent 30 October event and do not include detailed mapping. As a result, more emphasis is placed on features triggered by the 24 August event, and exacerbated and/or triggered by the October events (i.e., observed during the second reconnaissance trip following the 30 October event).

The large majority of the observed instability features are rockfalls in fractured rocks. Many rockfalls observed after the October events are likely due to breakage of joints that were likely weakened by the 24 August event. The document describes the following three major landslides that were judged as valuable case histories:

- The slope failure on the south-western flank of the Nera River valley, south-west of the village of Visso, where a large rockslide took place at the time of the **M**6.5 earthquake on 30 October 2016;
- Pescara del Tronto, where the 2016 events induced significant instabilities in a soil/rock slopes overlooking the SS4 motorway which connects the west to the east coast of Central Italy and hence is a very important transportation infrastructure;
- Accumoli, where the southern spur of a hill failed in spite of a village located atop.

For the Accumoli and Pescara del Tronto landslide case histories, particular attention was paid to the incremental damage (or lack thereof) caused by multiple events. This idea was particularly relevant as these areas were inspected repeatedly after both the 24 August and 26–30 October earthquakes. These multiple technical visits made it possible to document the damage level following each earthquake. This area of Central Italy has been affected by a large number of historical coseismic rockfalls. For example, previous studies by Antonini et al. (2002), Carro et al. (2003), and Gigli et al. (2014) documented and analyzed rockfall hazard and risk in the Umbria-Marche Apennine area of Central Italy following the 1997–1998 Umbria-Marche earthquake.

2 Seismic Sources and Ground Motion

The Central Apennine chain can be seen as the backbone of the Italian peninsula. The whole region is characterized by extensional tectonic activity. This regional extension is accommodated, in the inner Central Apennines, by a series of normal faults striking northwest-southeast (NW–SE) and dipping south-west (SW). Seismological, geometric, and kinematic characteristics of the main seismogenic sources of the area are summarized in GEER (2016, 2017) and Galadini et al. (2018).

This is a region with a long history of destructive earthquakes. The locations of faults have been well studied, and the effects of past earthquakes on villages and towns in the region is well documented. The 2016 earthquake sequence in Central Italy occurred in a seismic gap between two earlier earthquakes, which are the 1997 **M**6.1 Umbria-Marche earthquake to the northwest and the 2009 **M**6.1 L'Aquila earthquake to the southeast (Fig. 1a). Figure 1b shows the main active fault segments in the epicentral area of the 2016 Central Italy earthquake sequence. The **M**6.1 event on 24 August was a peculiar two-fault event caused by the rupture in the southernmost section of the Mt. Vettore-Mt. Bove fault system (MVBF in Fig. 1b) in association with the rupture of the northernmost section of the Laga Mts. fault system (Amatrice segment, AF in Fig. 1b). On the other hand, both the 26 and 30 October events occurred solely on the Mt. Vettore-Mt. Bove fault system. Table 1 shows parameters and locations of the mainshocks and three main aftershocks (after Galadini et al. 2018).

2.1 August 24, 2016 Event

The mainshock occurred on the August 24, 2016, at 01:36:32 (UTC) and was recorded by Italian National Seismic Networks (Rete Sismica Nazionale, RSN; www.gm.ingv.it/index.php/retesismica-nazionale/, last accessed November 21, 2016) owned by the Italian Institute of Geophysics and Volcanology (Istituto Nazionale di Geofisica e Vulcanologia, INGV). The August 24, 2016, mainshock event occurred along segments of normal fault systems trending NW–SE (strike ~165 degrees) with a dip angle SW of ~45 degrees. The strike of the fault from the moment tensor is generally consistent with the orientation of the Mt. Vettore fault to the north and the Laga Mountain fault to the south. As mentioned earlier, the event occurred as a result of

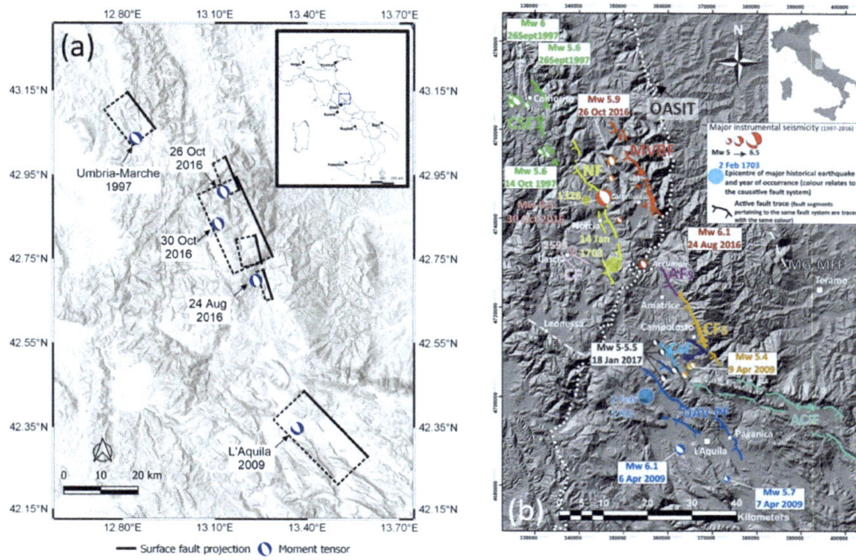

Fig. 1 **a** Map of Central Italy showing moment tensors of major earthquakes since 1997. Finite fault models from Chiaraluce et al. (2004; 1997 Umbria-Marche event), Piatanesi and Cirella (2009; 2009 L'Aquila event), and Galadini et al. (2018; 24 August, 26, and October 30, 2016, events) are also shown (from Lanzo et al. 2018); **b** Map showing active fault systems and locations of large seismic events in the region since 2009. Faults: Colfiorito-Sellano fault, CSF; Mt. Vettore-Mt. Bove fault, MVBF; Norcia fault, NF; Cascia fault, CF; Amatrice fault segment, AFs; Campotosto fault segment, CFs; Capitignano fault, CaF; Upper Aterno Valley-Paganica fault system, UAV-PF; Leonessa fault, LF; Monti Gemelli-Montagna dei Fiori fault, MGMFF; Olevano-Antrodoco-Sibillini inactive thrust system, OASIT (modified from Galadini et al. 2018)

Table 1 Summary of the six selected major seismic events in Central Italy between August 24, and October 30, 2016. Mainshocks are reported using bold-face fonts

Date	Hour (UTC)	Latitude (N)	Longitude (E)	Depth (km)	M	Strike (deg)	Dip (deg)
08/24/2016	**01:36:32**	**42.70**	**13.23**	**8**	**6.1**	**168**	**41**
						Mt. Vettore-Mt. Bove	
						163	**52**
						Amatrice segment	
08/24/2016	02:33:28	42.79	13.15	8	5.3	134	56
08/26/2016	04:28:25	42.60	13.29	9	4.8	165	36
10/26/2016	17:10:36	42.88	13.13	9	5.4	160	38
10/26/2016	**19:18:05**	**42.92**	**13.13**	**8**	**5.9**	**158**	**43**
10/30/2016	**06:40:17**	**42.84**	**13.11**	**5**	**6.5**	**156**	**43**

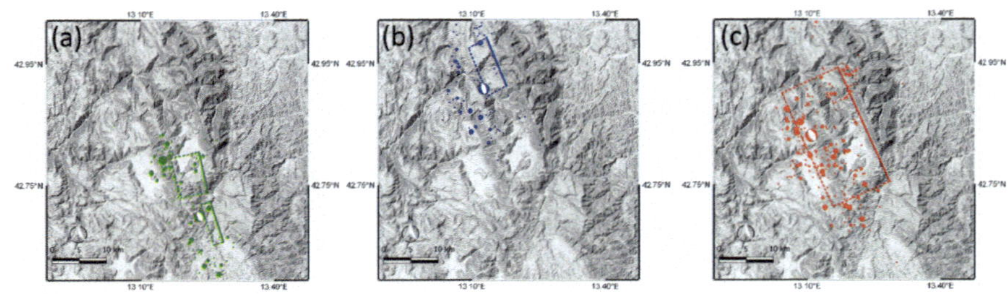

Fig. 2 Trimmed finite fault model, moment tensor, and epicenter location for the **a** M6.1 on 24 August, **b** M5.9 on 26 October, and **c** 30 October event, along with aftershocks recorded within 24-h periods following each mainshock

a multi-fault rupture between two-fault systems, the Mt. Vettore-Mt. Bove and the Laga.

Figure 2a shows the surface projection of the trimmed finite fault model for this event. The two-fault systems involved in this rupture are: the Laga Mountains fault system (Amatrice segment) and the Mt. Vettore-Mt. Bove fault system. The number of aftershocks within the 24-h periods following the 24 August mainshock was 121. Most of the aftershock epicenters are not within the surface projection of the hanging wall above the finite fault model, with many south and west of the rupture (Fig. 2a).

2.2 October 26 and 30, 2016 Events

The location of hypocenters, slip directions and surface rupture suggest that the events on 26 and 30 October were accompanied by ruptures in the segments of the Mt. Vettore-Mt. Bove Fault. As described in GEER (2016), this fault and the neighboring Laga Mts. fault to the south experienced rupture during the 24 August event.

The 26 October event has an along-strike length of 8 km and down-dip width of 4 km. The 30 October event has an along-strike length of 21 km and down-dip width of 16 km. Figures 2b, c shows the epicenter location and surface fault projection for the 26 and 30 October events, respectively, along with aftershocks recorded within 24 h of each mainshock.

Figure 3 shows the cumulative fault offset measurement (70 cm) on the lower or "western" fault trace (Lat = 42.812901°, Long = 13.24626°). A comprehensive overview of the fault surface rupture observed during the 2016 Central Italy earthquake sequence is provided by GEER (2017) and Gori et al. (2018).

The number of aftershocks within the 24-h period following each mainshock was 75 (26 October), and 258 (30 October). Aftershocks following these mainshocks have clear spatial patterns (Fig. 2b, c). For the 26 October event, very few aftershocks actually occur within the surface projection of the rupture. The aftershocks pattern

Fig. 3 Incremental surface fault rupture on the western slope of Mt. Vettore-Mt. Bove fault

for the largest event (30 October) suggests, as expected, that most epicenters are situated within the surface projection of the rupture.

Figure 4 shows the spatial distribution of peak ground acceleration (PGA) in the epicentral area for all three mainshocks. Also shown in Fig. 4 is the location of selected case histories. The measures of the ground motion intensity presented in this figure were estimated using kriging of within-event residuals (i.e., the difference between the recorded ground motions and those assessed by ground motion models for a specific earthquake). After calculating within-event residuals at recording station sites, the spatial distribution of the target intensity measure (e.g., PGA) can be estimated using the Jayaram and Baker (2009) correlation model based on global data. Residuals are equal to zero at recording station sites. The trimmed finite fault models presented in Galadini et al. (2018) were used to calculate source-to-site distances. In this document, within-event residuals were calculated using the average of the Italy-adjusted global ground motion models by Campbell and Bozorgnia (2014), Chiou and Youngs (2014), and Boore et al. (2014). This regional adjustment for Italy is needed to capture a relatively fast attenuation with distance of the ground motion observed in Italian earthquakes (e.g., Stewart et al. 2012 and Zimmaro et al. 2018). Zimmaro and Stewart (2017) recently illustrated the effectiveness of the adoption of global ground motion models with regional adjustments for applications in Italy. More details about the approach used to estimate ground motions used in this study are provided in GEER (2017) and Zimmaro et al. (2018). An estimate of PGA amplitudes at the three selected landslide sites during the 24 August, 26 October, and October 30, 2016, mainshocks is shown in Table 2.

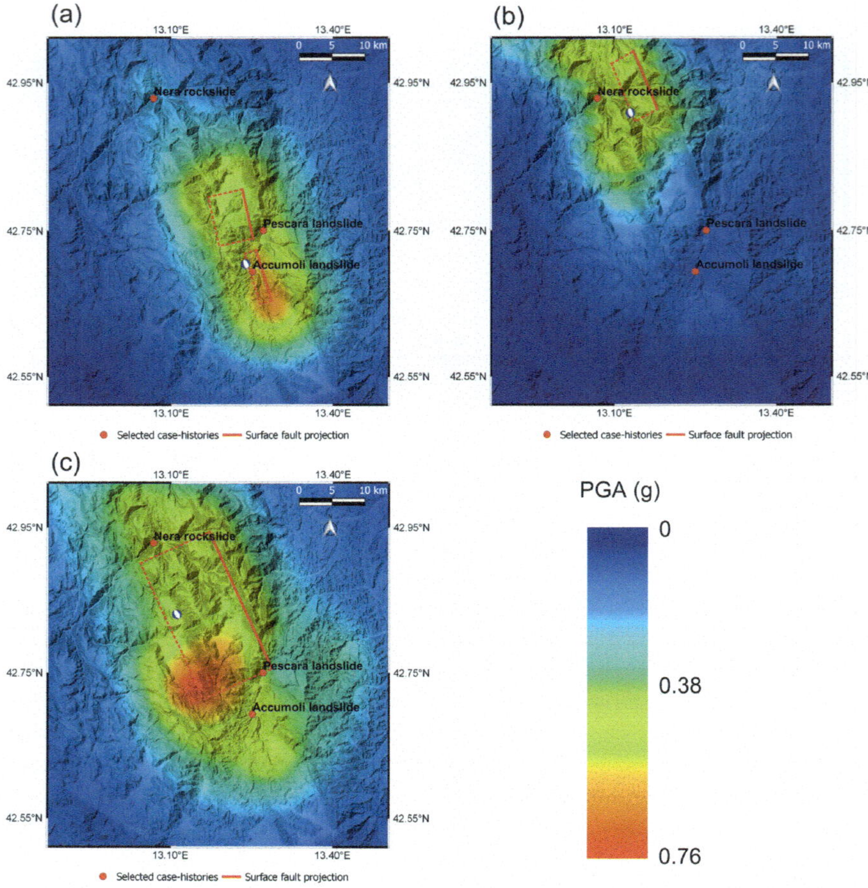

Fig. 4 Spatial distribution of PGA (in units of g) estimated in the epicentral area for the **a** M6.1 24 August, **b** M5.9 26 October, and **c** M6.5 30 October events

Table 2 PGA (in units of g) estimated at the selected mass-movement sites

Lat	Lon	PGA (g)			Location summary	Triggering event
		M6.1 24 August	M5.9 26 October	M6.5 30 October		
42.92900	13.06800	0.22	0.36	0.38	Nera Rockslide	**M6.5**, 10/30/2016
42.75057	13.27010	0.48	0.10	0.34	Pescara del Tronto	**M6.1**, 08/24/2016
42.69442	13.25029	0.55	0.07	0.40	Accumoli	**M6.1**, 08/24/2016

3 Rainfall Before and During the Earthquake Sequence

Three rain gauges are operated in the area affected by landslides/rockfalls: Endesa (Visso) and Ponte Tavola (Castelsantangelo Sul Nera) operated by Servizio Idrografico—Regione Marche and Nerito-Crognaleto operated by Servizio Idrografico—Regione Abruzzo. Figure 5 shows the precipitation recordings in these gauges over the time-period of the event sequence (August to December 2016). The analysis of the rain gauge recordings shows that the precipitation that occurred before the 24 August event was not significant. The largest amount of rain was recorded between August and October. However, as further discussed in the remainder of the document, none of the landslide features observed after the October events is related to rain effects.

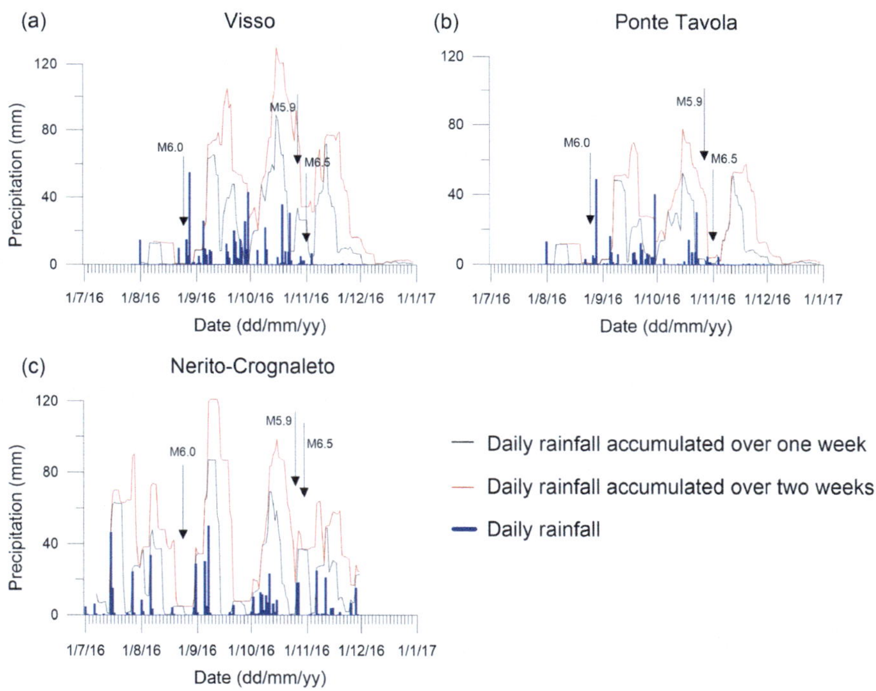

Fig. 5 Daily rainfall (blue bars) and daily rainfall accumulated over one (black line) and two weeks (red line) recorded by rainfall gauges at **a** Visso, **b** Ponte Tavola (courtesy of Servizio Idrografico—Regione Marche), and **c** Nerito-Crognaleto (courtesy of Servizio Idrografico—Regione Abruzzo) (modified from GEER 2016)

4 Landslide Data

Many landslide sites are located in areas with rugged and/or steep terrain, heavy vegetation, and/or limited site access. As a result, for the GEER (2016, 2017) reports, small Uninhabited Aerial Systems (sUAS) have been extensively used to digitally image several significant landslides (Franke et al. 2018). Traditional on-site field reconnaissance investigations have also been conducted at several locations when road access was possible. Following Hungr et al. (2014), which updated the well-known Cruden and Varnes (1996) classification, landslides in the area can be classified as (Franke et al. 2018):

(1) Rockfall and rock slides (rock planar- and rock wedge-slides involving the Miocene flysch and carbonatic units of the Umbria-Marche Succession),
(2) Shallow translational and rotational soil slides in natural slopes such as cliffs, gulley banks, and steep natural slopes,
(3) Shallow soil slides in anthropogenic slopes (e.g., steep roadway cuts and fills).

The information published in this document follows consistent description convention and categories. The landslides inventory was divided into three main categories as defined by Keefer (1984):

(1) Category I: falls and slides,
(2) Category II: coherent slides,
(3) Category III: lateral spreads or flow slides

No features belonging to Category III were observed during the 2016 Central Italy earthquake sequence. When observed landslides do not fit any of the standard categories or are not adequately described, they are labeled as "*other*" category. Consequently, the landslides during the three main events of the seismic sequence are described as follows:

4.1 Landslides Following the August 24, 2016 Event

Several landslides were observed following the 24 August mainshock. Figure 6 shows mapped slope instability locations. Table 3 and Fig. 7 summarize the percentage of landslides falling into each category and the different types of landslides for each category, following the 24 August event. Most of the observed phenomena were accompanied mainly by small rockfalls along roadway cuts throughout the earthquake zone. Therefore, they fall in Category I.

The area is characterized by the presence of the following geologic units of the Umbria-Marche Succession: the Miocene flysch units and the Carbonatic units. The former comprises the so-called Laga formation, which is a flysch unit consisting of alternating sandstone (which is the prevailing component) and marl layers. Such flysch units, similarly to other turbidite formations, are characterized by variations

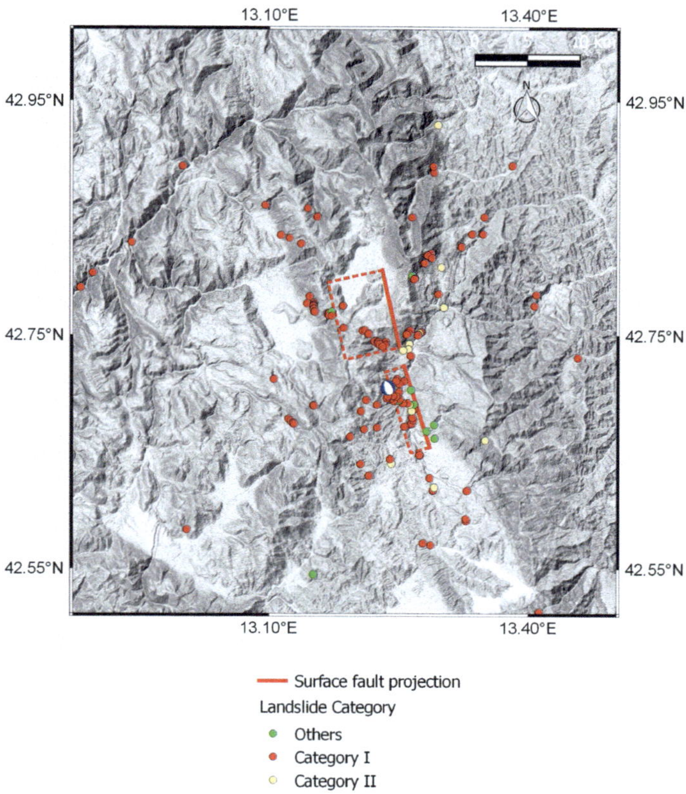

Fig. 6 Observed landslides following the M6.1 24 August event

Table 3 Categories of landslide occurred following the 24 August event

Category I	Category II	Other
83.1%	9.6%	7.3%
Rockfall 95% Rock slide 5% (7/148)	Debris slide 71% Rock slide 29% (5/17)	

in the sandstone/marl ratio and layer thickness. This spatial variability is related to the distance from the source area in the depositional basin. The latter (marl layers) are subjected to weathering but only at shallow depths. As a result, these marls appear weathered when they are exposed (e.g., in highway cuts and excavations). This weathering undercuts overlying sandstone slabs, which can break free if shaken by earthquake ground motions.

The most impressive rockfall features were observed in the area along state provincial routes SP64 (Tufo–Castelluccio) and SP477 (Castelluccio-Norcia). Almost all of the rockfalls occurred when isolated blocks of limestone detached from outcropping bedrock above the highway. Many of these blocks came to rest on the shoulder

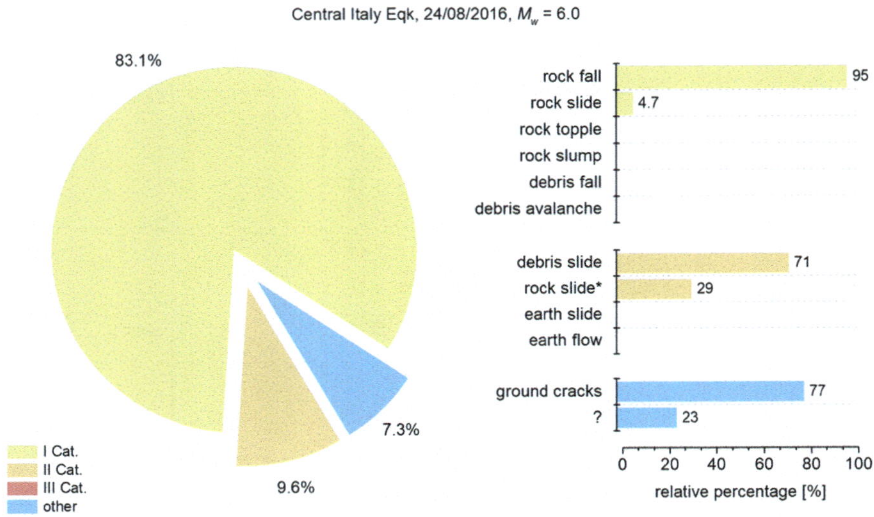

Fig. 7 Distribution of recognized landslides after the 24 August event

or pavement of the road, whereas other blocks maintained enough velocity to cross the road and continue their descent downslope. An example of isolated blocks on highway SP477 is shown in Fig. 8a, b. Figure 8c shows an image of the 3D model developed for the SP477 rockfall using sUAS-based imagery (GEER 2016). The data collected about Category II mainly concern rock slides in gentle slopes (labeled as "rock slide") and other phenomena referred to as "debris slide". Many ground cracks were reported along road embankments and landfills, mainly behind retaining structures.

Figure 9 illustrates the information concerning landslides triggered by the 24 August event compared with the empirical upper-bound limits defined by Keefer (1985). Herein, the epicentral distance (Fig. 9a) and Joyner and Boore distance (the closest distance to the surface projection of an extended fault) (Fig. 9b) were plotted against the Richter surface-wave magnitude (equal to six for the 24 August event). Furthermore, Fig. 9c compares the envelope of the area affected by landslides against the upper bounds proposed by Keefer (1984) as well as those by Rodriguez et al. (1999).

The same information is illustrated in a planar manner in Fig. 10 where the areal distribution of landslide sites is compared with the empirical upper-bound loci as defined by Keefer (1985) by using two source-to-site distance metrics: epicentral distance (Fig. 10a) and the Joyner and Boore distance (Fig. 10b).

When comparing the data against empirical limits (Fig. 9), it is evident that the zone characterized by the Category I landslides is limited by the radius of 40 km around the epicenter. This extent of radius corresponds to about 50% of the upper-bound distance proposed by Keefer (1985) for this category. The majority of the

Fig. 8 Example of rockfalls on SP477 (42.76729° N, 13.16983° E): **a** A large 2-m block of limestone that crossed the road and came to rest on the downslope side of the roadbed below a prominent outcrop of limestone (42.76729° N, 13.16983° E); **b** several blocks with sizes up to 3–4 m; **c** 3D model of the area (modified from GEER 2016)

observed landslides were located at a distance of 10 km or less from the earthquake surface ruptures. By comparing mapped landslides with the Italian landslide inventory (Inventario dei Fenomeni Franosi in Italia, IFFI project, ISPRA—Dipartimento Difesa del Suolo Servizio Geologico d' Italia, available at http://www.progettoiffi.isprambiente.it), it is found that the new landslides occurred mostly within the near-source zone (<10 km), while most of the far-field landslides come from the pre-existing inventory (Pavlides et al. 2017).

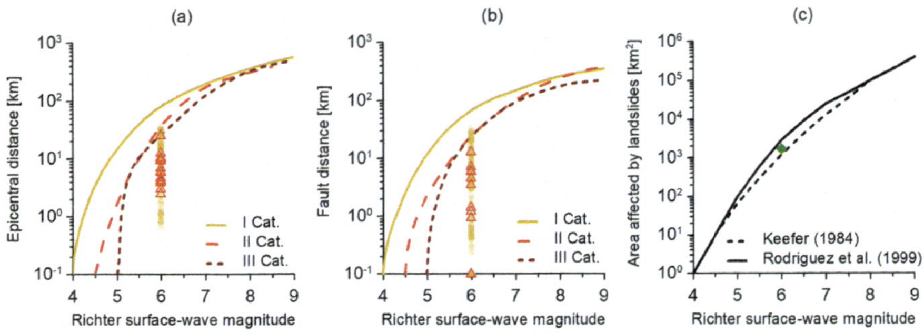

Fig. 9 Comparison between source-to-site distances for landslides that occurred during or after the 24 August event and the empirical upper-bound curves proposed by Keefer (1985) in terms of **a** epicentral distance and **b** Joyner and Boore distance; **c** comparison between the envelope area affected by landslides and the upper bound proposed by Keefer (1984) and Rodriguez et al. (1999)

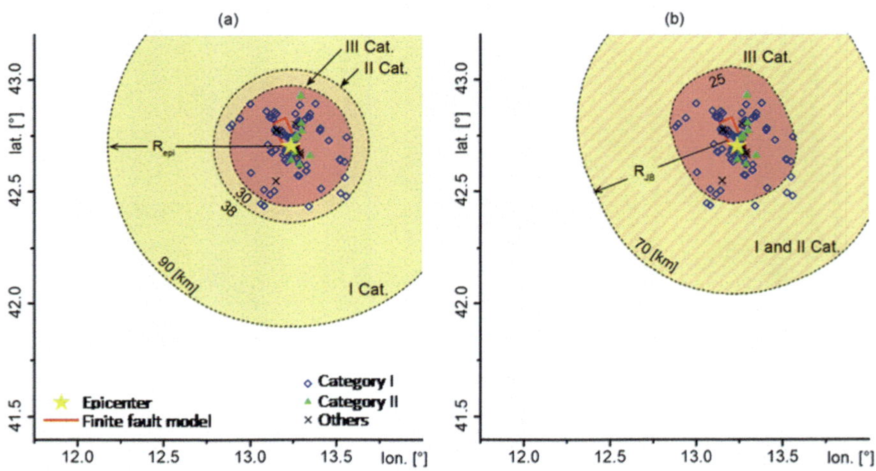

Fig. 10 Areal distribution of landslide sites during or after the 24 August event compared with the empirical upper-bound loci defined from Keefer (1985) for two source-to-site distance metrics: **a** epicentral distance and **b** Joyner and Boore distance

4.2 Landslides Following the 26 and October 30, 2016 Events

The October mainshocks triggered a larger number of landslide phenomena than the 24 August mainshock. These events also exacerbated the amount of displacement caused by the 24 August earthquake event in many areas. The types of instabilities are the same as those observed following the 24 August event. Figures 11a, b shows all slope instabilities following the 26 and 30 October event, respectively. Tables 4, 5 summarize the percentage of landslides falling into each category and the different types of landslides for each category following the October events. Figure 12 shows a graphical representation of the same information for the 30 October event. Similarly to what observed after the 24 August event, most of the observed phenomena fall in category I and involve rockfalls.

Several rockfalls observed following the October events are characterized by the large volumes of material involved. The most impressive features observed following the 26 October event are probably the series of rockfalls in the area of Mt. Bove, 4 km east-southeast of Ussita. The northern flank of Mt. Bove is characterized by massive or coarsely bedded Jurassic limestones of the Calcare Massiccio Formation.

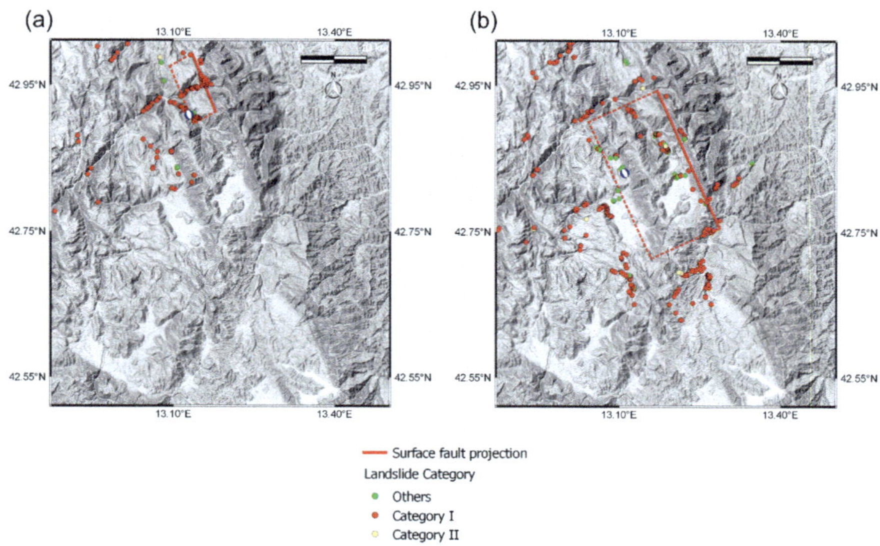

Fig. 11 Observed landslides following the: **a** M5.9 26 October and **b** M6.5 30 October events

Table 4 Categories of landslides that occurred following the 26 October event

Category I	Category II	Other
83%	10%	7%
Rockfall 95% Rock slide 5%	Debris slide 71% Rock slide 29%	

Table 5 Categories of landslides that occurred following the 30 October event

Category I	Category II	Other
83%	4%	13%
Rockfall 86% Rock slide 10% Rock topple >1% Rock slump 1% Debris fall 1% Debris avalanche >1%	Debris slide 66% Rock slide 13% Earth slide 13% Earth flow 8%	

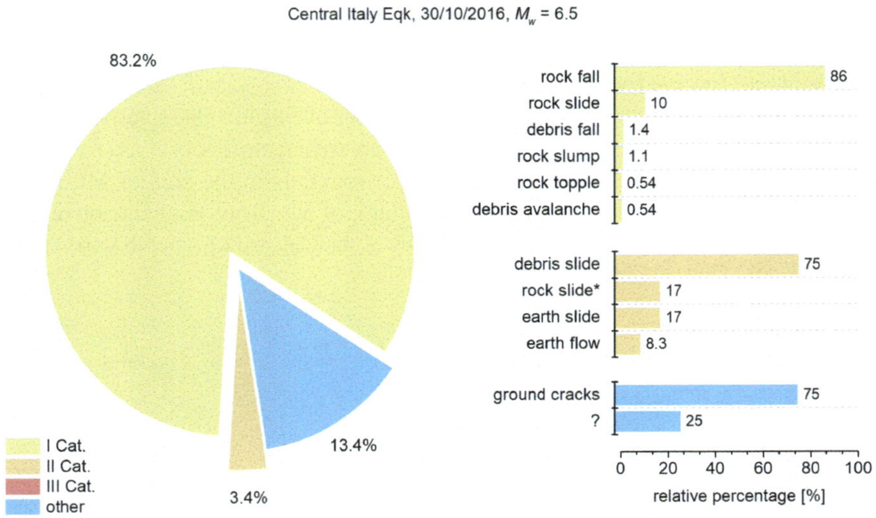

Fig. 12 Distribution of recognized landslides after the October 30, 2016, earthquake

At the top of the mountain, the Calcare Massiccio Formation is overlaid by limestones (Bugarone Formation) with more regular bedding (medium to thick beds). The Mt. Bove massif has been involved in intense Quaternary and ancient tectonics, which has produced faults trending in a strike range from ESE-WNW to N–S. The rock mass has closely spaced joints and persistent joints belonging to the major joint sets. Sever loosening of thick outer parts of the intensely fractured mountain flank could have favored the detachment of large irregular slices under seismic loading (GEER 2017). Figure 13 shows a sUAS-based image of a talus fields visible at the base of Mt. Bove. The same figure also highlights suspected source areas for these talus fields characterized by a lighter coloration of the freshly-exposed unweathered limestone (yellow circled areas in Fig. 13).

Another interesting rockfall feature is that observed in the Valle di Panico. This instability phenomenon has been likely triggered by the 30 October event. However, this information is only inferred, since prior information about it was not available at the time of reconnaissance (Lanzo et al. 2018). Mt. Bove represents the southern flank of the narrow valley called Valle di Panico. The northern flank of this valley is formed by a succession of limestone and marly units from Jurassic to Cretaceous in age. There are two landslide locations along the mountain road that winds along the northern flank of Valle di Panico. The road is cut into thinly-bedded marly limestones (Scaglia Bianca Formation) whose bedding joints frequently have a clay/bituminous infilling. This structural/lithologic feature, together with an intense fracturing, gives the rock mass a low quality. The source area of the rockfall at the second site is located in a marly-limestone unit featured by thicker layers. In this area, the bedrock is extensively covered by slope debris that are some meters thick (GEER 2017). The GEER reconnaissance team visited two main sites in the Valle di Panico area. The first

Fig. 13 sUAS image of Mt. Bove. Suspected rockfall sources are circled in yellow (modified from GEER 2017)

site was a landslide in the soil slope through and beneath the road. Figure 14 shows an aerial photograph of this 55–60 m wide landslide. The GEER team measured a vertical offset on the road pavement ranging between 10 and 70 cm, and horizontal deformation ranging from 2 to 40 cm. The GEER team also created sUAS-based 3D models of this landslide. As shown in Fig. 15a, the eastern portion of the landslide headscarp shows distinct cracks and associated displacements. The western portion of the headscarp appears more gradual and shows pavement cracking (Fig. 15b).

The second site in the Valle di Panico area is a 420 m-long rockfall along a mountain road. Figure 16 shows a 3D model of the entire rockfall produced by

Fig. 14 Aerial photograph of the Valle di Panico landslide (modified from GEER 2017)

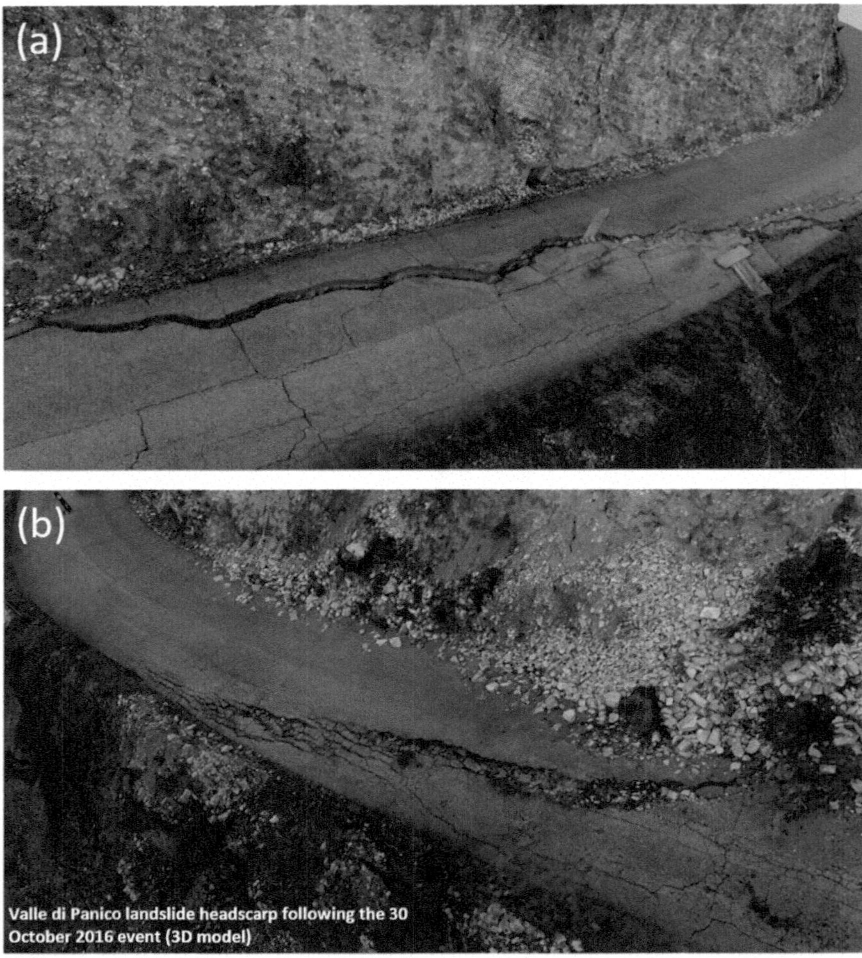

Fig. 15 **a** 3D model of the eastern half of the Valle di Panico landslide headscarp; **b** 3D model of the western half of the landslide headscarp (modified from GEER 2017)

the GEER team (GEER 2017). The total change in elevation from the source of the boulder to the final resting place on the bottom of the valley is approximately 235 m. The side slope of the valley rests at an angle of 34 degrees from the horizontal direction (1.5 H:1 V). Numerous boulder fragments ranging in diameter from gravel-size to over 3 m were observed and photographed along the entire rockfall length. The largest boulder measures 3 m in diameter. The source boulder is 11.2 m in length and belongs to a formation of heavily weathered and fractured limestone located 91 m above the mountain road (Fig. 17a). Much of the limestone boulder exploded into gravel-sized fragments in the first 90 m of the rockfall. The remaining parts of the boulder broke in larger fragments while tumbling to the bottom of the valley

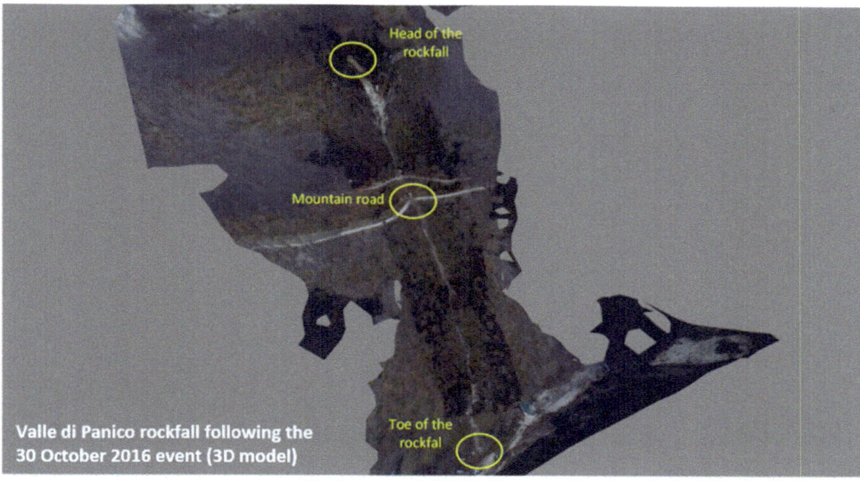

Fig. 16 sUAS-based 3D model of the Valle di Panico rockfall (modified from GEER 2017)

(GEER 2017). At the bottom of the valley, over ten large boulder fragments ranging in diameter from 0.8 to 2.9 m were observed to rest at the valley floor, just 13 m from what appeared to be a small pump or power house (Fig. 17b). Information reported by the GEER team shows that none of the boulders appear to have damaged this small structure (GEER 2017).

Figure 18 depicts the data of landslides triggered by the 30 October event and compare it with the empirical upper-bound limits defined by Keefer (1985). For this diagram, Richter surface-wave magnitude was employed in association with the epicentral distance (Fig. 18a) and Joyner and Boore distance (Fig. 18b). Figure 18c compares the envelope area affected by landslides and the upper bounds proposed by Keefer (1984) and Rodriguez et al. (1999). The same information is also presented in a planar style in Fig. 19. It draws the areal distribution of landslide sites and compare it with the empirical upper-bound loci as defined by Keefer (1985) for two source-to-site distance metrics: epicentral distance (Fig. 19a) and Joyner and Boore distance (Fig. 19b). The comparison with the empirical limits indicates that the zone involved in the activation of Category I landslides is limited to an area with a radius of about 40 km around the epicenter, with the majority being at 10 km or less from the surface ruptures.

5 Landslide Detection During the 2016 Central Italy Earthquake Sequence Using Geodetic Methods

Several geodetic methods use synthetic aperture radar (SAR) images to generate digital elevation models (DEM) for monitoring ground and structural deformation.

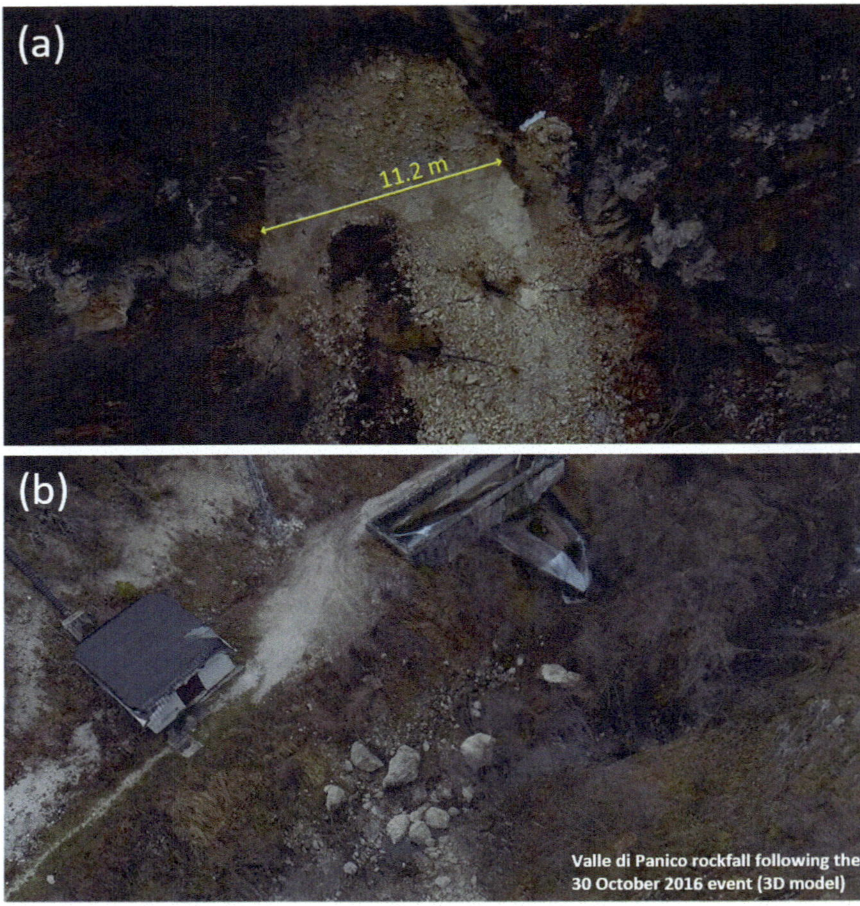

Fig. 17 **a** 3D model of the source of the Valle di Panico rockfall, located 90 m above the mountain road; **b** 3D model of the boulders and a small power or pump house at the toe of the Valle di Panico rockfall (modified from GEER 2017)

These methods are typically based on differences in the phase of waves returning to a moving platform (e.g., aircraft or satellites). SAR-based techniques are often utilized to identify deformation phenomena such as (1) earthquake-related surface deformations and ruptures (e.g., Jo et al. 2010), (2) volcanic eruptions (e.g., Jung et al. 2011; Lee et al. 2013), (3) subsidence (e.g. Choi et al. 2011; Zhang et al. 2012), and (4) massive landslides (e.g., Ausilio and Zimmaro 2017). Such techniques have been also recently used to perform rapid detection and regional mapping of landslides following earthquakes (e.g., Rathje and Franke 2016 following the 2004 Niigata-ken Chuetsu earthquake). It is worth mentioning that satellite-based detection of ground elevation often underestimates the size of the landslides (Rathje and Carr 2010; Rathje and Franke 2016). SAR-based methods are also employed to rapidly

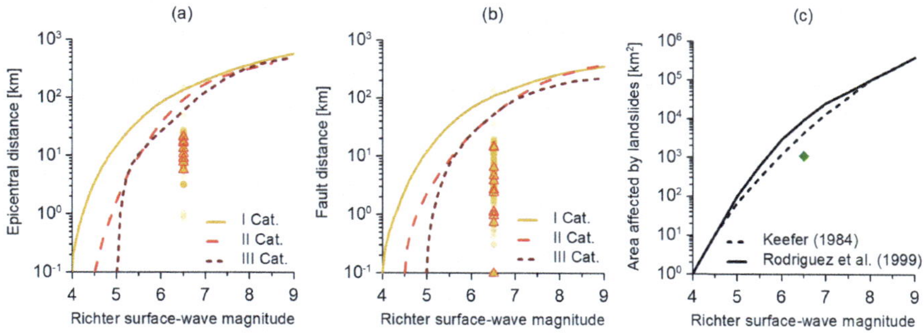

Fig. 18 Comparison between source-to-site distances for landslides that occurred following the 30 October event and the empirical upper-bound curves proposed by Keefer (1985) in terms of **a** epicentral distance and **b** Joyner and Boore distance; **c** comparison between the envelope area affected by landslides and the upper bound proposed by Keefer (1984) and Rodriguez et al. (1999)

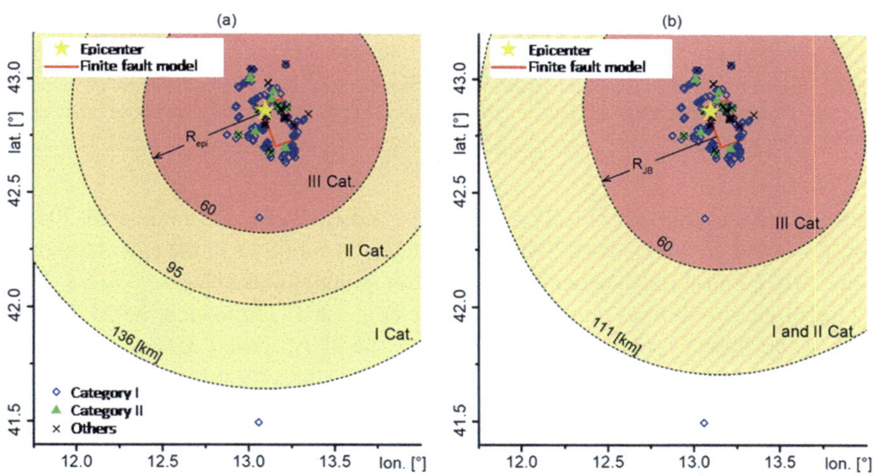

Fig. 19 Areal distribution of landslide sites following the 30 October event compared with empirical upper-bound loci defined from Keefer (1985) for two source-to-site distance metrics: **a** epicentral distance and **b** Joyner and Boore distance

produce post-disaster deformation maps. This is one of the goals to be achieved by the Advanced Rapid Imaging and Analysis (ARIA 2016) project. The ARIA project team releases coseismic interferograms and damage proxy maps following major natural disasters globally. The damage proxy maps are produced by comparing interferometric SAR coherence maps before and after an extreme event (e.g., Fielding et al. 2005; Yun et al. 2011). Yun et al. (2015) showed that the extents of coseismic slope instability caused by selected earthquakes were well captured by damage proxy maps. Such maps have been also used to identify possible landslide and rockfall locations following the 2016 Central Italy earthquake sequence, as shown in Fig. 20, which shows clusters of coherence changes in damage proxy maps versus detected

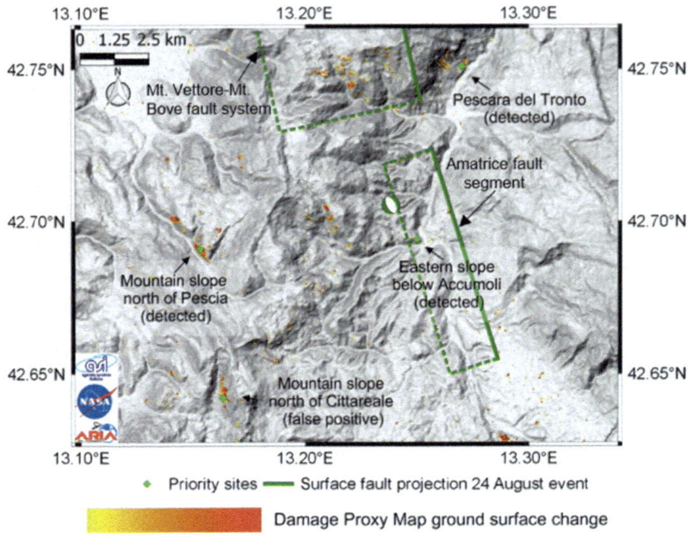

Fig. 20 Epicentral area of the M6.1 August 24 earthquake together with damage proxy maps of the area produced by the ARIA project and detected landslide features (modified from Franke et al. 2018)

landslide features following the 21 August event (GEER 2016 and Franke et al. 2018). Also Polcari et al. (2017) used SAR data to detect instabilities in the epicentral area of the 2016 Central Italy sequence, focusing on mountainous areas.

One of the instabilities detected by Franke et al. (2018) following the 24 August event is the Pescia rockfall. Figure 21 shows the sUAS-based 3D models of the Pescia rockfall. At this location, the GEER team produced multi-epoch 3D models showing that after the detection of rockfalls following the 24 August event, additional boulders fell as a result of the 30 October event. The rock face reported in Fig. 21 is approximately 90 m wide and 60 m high. The rock material that composes the face is a combination of limestone, argillaceous limestone, and marlstone. Boulder sizes reach 2.5 m. Next to this location, the GEER team detected additional rockfalls produced by the 30 October event that were not present after the 24 August event. Figure 22 shows a 3D model of these features.

6 Relevant Case Histories

This section focuses on three major landslide features, judged as valuable case histories:

(1) The south-western flank of the Nera River valley, located south-west of Visso Village, where a large rockslide was induced by the **M**6.5 earthquake on October 30, 2016, and is known as the Nera rockslide or Mount Sasso Pizzuto Rockfall;

Fig. 21 sUAS-based 3D models of the Pescia rockfall following the **a** 24 August and **b** 30 October event (modified from GEER 2016, 2017)

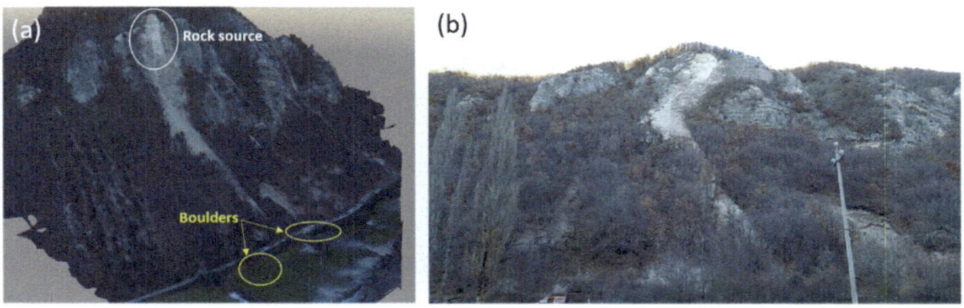

Fig. 22 **a** sUAS-based 3D models of additional rockfall features observed in the area of Pescia following the 30 October event; **b** source area of the rockfalls with runout damage (modified from GEER 2016, 2017)

(2) Pescara del Tronto, where the 2016 events caused significant cumulative damage in a soil/rock slope overlooking the SS4 motorway, connecting the west to the east coast of Central Italy;
(3) Accumoli, where slope failures occurred on the southern spur of the hill on top of which the village is built.

GEER reports (2016, 2017) showed that incremental accumulation of the structural damage was especially intense in Pescara del Tronto and Accumoli. The damage accumulation at these and other locations is related with the multiple cycles of mainshock-aftershock sequences observed in the area between August and October 2016. Additional information about these selected case histories are presented by Lanzo et al. (2018). Figure 23 shows the location of these case histories, along with epicenters of the three mainshocks of the sequence.

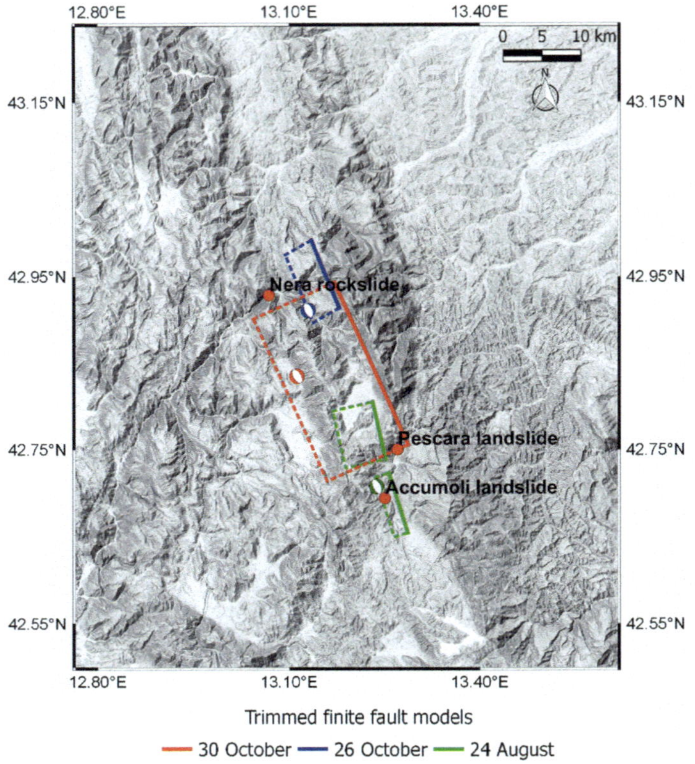

Fig. 23 Location of selected case histories

6.1 Nera Rockslide or Mount Sasso Pizzuto Rockfall

The Nera River Valley is a huge canyon formed by the erosion of the Nera River, a tributary of the Tiber River, in limestone formations between the Umbria and Marche regional border (Fig. 24). In some stretches, the regional road (SP 209) runs along the Nera River. The Italian landslide inventory (Inventario dei movimenti franosi in Italia, IFFI project, ISPRA Dipartimento Difesa del Suolo Servizio Geologico d'Italia) classifies this area as being subject to rockfalls and slumps. The geology in the Nera River valley area is characterized by sedimentary rocks of the carbonate Umbria-Marche stratigraphic sequence, which is an early Jurassic to Eocene age formation. The geology of the area (Romeo et al. 2017) is also associated with the outcropping of bedrock of the Umbria-Marche sequence between Calcare Massiccio (a massive platform limestone) and Scaglia Variegata-Cinerea (marly-calcareous rocks). The Maiolica Formation (stratified pelagic limestone) is present in the left bank of the Nera River. In some parts, the formations are pervasively folded (crumpled) and intensely fractured (Romeo et al. 2017).

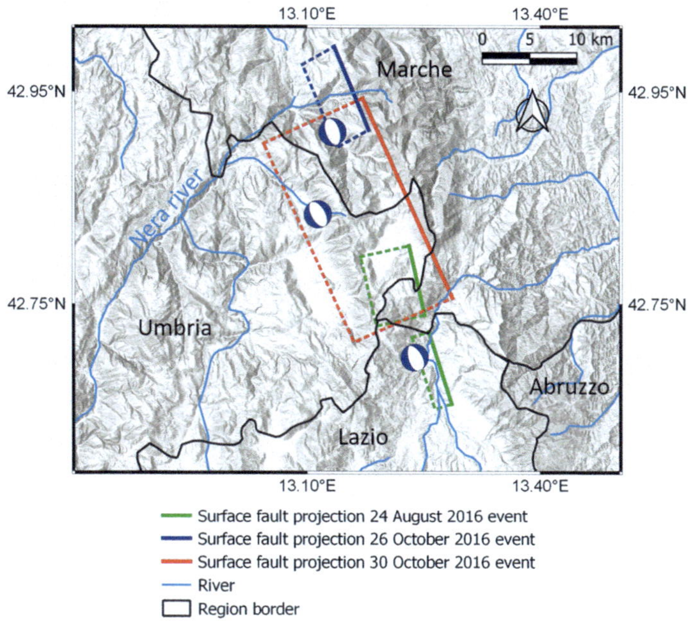

Fig. 24 Location of the Nera River area between the Umbria and Marche regional border

Several rockfalls and diffuse instabilities occurred along this valley during the 2016 Central Italy earthquake sequence. Very few and small landslides were triggered by the 24 August event, while severe rockfalls occurred following the October 2016 events. Boulders, with a volume ranging from a dozen cubic centimeters to a few cubic meters, have been observed on the shoulders and/or across the road. Figure 25 shows cases in which the installed rockfall protection measures were passed over, severely damaged or destroyed. These passive protection measures (i.e., rockfall nets, dynamic rockfall barriers, rigid barriers, or catch fences) were built along the SP 209 stretch after the 1997 Umbria-Marche earthquake sequence, which triggered about 200 rockfalls of various sizes along the Nera River Valley, and damaged man-made infrastructure such as roads, bridges, and tunnels (Carro et al. 2003; Gigli et al. 2014).

The largest and most important rockfall occurred following the 30 October earthquake, 1 km south-west of Visso, along the steep-faced north slope of Mount Sasso Pizzuto, which is characterized by highly deformed, stratified, and fractured rocks (Maiolica Formation). The avalanche generated by this rockfall interrupted the regional road (SP 209) and dammed the Nera River, forming a small lake of about 2000 m^2. This feature is known as the Nera rockslide or Mount Sasso Pizzuto rockfall. This rock avalanche affected the road transportation between Visso and the town of Terni (center of the administrative and industrial activities in the Umbria region) and the transportation network within the whole Central Apennine. Fish farming activities, which is a source of the significant financial income of this area, were damaged by the avalanche deposit as well. The presence of such fish farming activities, along

Fig. 25 Slope protection damaged by boulders and rock-blocks along the regional road SP 290 in the Nera River valley (modified from GEER 2017)

with obvious environmental concerns, imposed a need for extra care in the removal of the landslide dam, making this operation very lengthy. Detailed geometric information can be found in the report by GEER (2017) and in Romeo et al. (2017). In both documents, for safety reasons, remote-sensing measurement techniques have been used without direct site inspections in the landslide area. Table 2 showed that the estimated PGA value for the 30 October event at this location is 0.38 g (GEER 2017; Zimmaro et al. 2018; Franke et al. 2018).

The geometry of the rockslide scar and the avalanche deposit were quantitatively measured by photogrammetric reconstruction of images captured by sUAS flights at the beginning of December, 2016, by GEER (2017). The Structure-from-Motion (SfM) computer vision was employed for post-processing and geo-referencing of the acquired sUASdata in order to reconstruct a 3D image (more details in GEER 2017). The resulting 3D model is shown in Fig. 26. Figure 27 shows details of the talus damming the Nera River (Franke et al. 2018). Romeo et al. (2017) have used a TruPulse™ 200 laser rangefinder. The laser sensor was placed on a tripod in order to carry out the measurement of slope distance, horizontal and vertical distance, and inclination, or the calculation of the elevation of any target. Thirty points were installed on the landslide debris that covers an area of about 16,500 m^2. Furthermore, GIS analysis was defined in ArcGIS 10.1 (more details in Romeo et al. 2017).

Fig. 26 Nera rockslide photographed by the GEER sUASat an elevation of 400 m above Route SP 209. The primary rockfall was a wedge that detached on the right side of the image and disintegrated into talus debris as it fell 330 m to the river valley floor. The limestone bedding is visible in the scar of the wedge detachment (modified from GEER 2017)

The maximum elevation difference between the rockfall source area, close to the ridge of the slope, and the SP209 road is about 300 m. The slope angle at the detachment area was originally greater than 58°, while after the detachment is between 75° and 80°. The rockslide scar can be approximated by those left by a rock wedge delimited by two persistent joints and the cliff surface, as can be seen in the SfM imagery of Fig. 27. The bedding exhibits an apparent dip of 20°–30° toward the east and appears to be orthogonal to the joints delimiting the failed wedge, thus it did not play a major role in the sliding mechanism. The apron of debris accumulation has a slope angle of about 27° and is not exactly in the perpendicular direction from the source. The accumulation area covers the foot of the slope and has an extent of about 20,000 m^2. It is difficult to determine the volumetric ratio of the materials derived from the source area and that from the pre-existing debris along the slope. Romeo et al. (2017) have estimated the total volume (debris originated by the rockfall plus the mobilized pre-existing talus) at about 70,000 $\pm$ 8000 m^3, of which about 50,000 $\pm$ 60,000 m^3 is derived from the new event. This estimate is comparable to the volume of rock detached from the source area (about 40,000 m^3) determined by laser rangefinder, taking into account the considerable uncertainties in evaluating the pre-existing material. Similar conclusions on pre- and post-event talus deposits

Fig. 27 **a** 3D model of the Nera rockslide; **b** photograph of the dammed Nera River blocking the road (modified from Franke et al. 2018)

have been reached by Franke et al. (2018) using sUAS-based 3D models. It is not a particularly large slope failure, but probably is the biggest rockfall in the area during the last two centuries.

6.2 Slope Displacements in Pescara Del Tronto

Significant incremental ground deformations were observed in the village of Pescara del Tronto following the 24 August event (GEER 2016) and the October events (GEER 2017). For safety reasons, measurement techniques have been used without a direct approach to the landslides areas after the 24 August and 30 October events. A series of sUAS flights have been made to collect aerial imagery over the areas of interest due to landslides. sUAS-based photographs were subsequently processed with a SfM computer vision algorithm using Bentley ContextCapture software and Pix4D software to develop orthophotos and 3D point's clouds and meshed models. Additional details about these indirect measurement techniques are provided by Franke et al. (2018).

The village suffered very heavy damage, with many masonry building collapses and 48 fatalities on 122 inhabitants during the 24 August event. This village together with all the downtown of Amatrice and Petrana was characterized by the highest intensity 10–11 MCS (Mercalli-Cancani-Sieberg scale; Sieberg 1930) (Galli et al.

2016) during the sequence. A structure-by-structure damage assessment for Pescara del Tronto following the 24 August and 30 October events is given in Sextos et al. (2018).

The village of Pescara del Tronto is located at the foot of the southern escarpment of Mt. Vettore, downslope from the so-called Sibillini thrust. In this zone, the Umbro-Marchigiana carbonate sequence overlaps the Laga formation. Pescara del Tronto is built on a slope comprising pre-existing ancient landslide deposits. These deposits consist of angular carbonate clasts ranging in size from sands to cobbles, with interspersed pervasively fractured limestone blocks. The deposits thickness varies spatially, thinning when proceeding toward the slope foot (the Tronto River left bank). In this zone, they overlie fluvio-lacustrine sediments, including travertines. The bedrock is the turbiditic Laga Formation. In the cataclastic carbonate rocks, distributed along the thrust line, which are located a few hundred meters upslope from the Pescara del Tronto village, it is possible to observe a large landslide crown (at elevations ranging between 900 and 1150 m above the sea level). This complex configuration was likely generated by the coalescence of various rotational-translational and complex landslides. Some evidence of this complexity are the benches, counter-slopes, and high scarps still present along the slope. These landslides are probably the source of thick debris deposits that the village of Pescara del Tronto is built on (Aringoli et al. 2010 and Lanzo et al. 2018).

After the August events investigation of the 3D models of Pescara del Tronto produced by the GEER team (GEER 2016, 2017) revealed numerous shallow earthquake-induced landslides and retaining wall failures. The remainder of this section summarizes incremental slope instability features observed as a result of the 2016 Central Italy earthquake sequence. This analysis is largely based on multi-epoch observations mainly derived from sUAS-based 3D models. Seven locations of interest within Pescara del Tronto have been identified to carry out cumulative damage analysis (Fig. 28; GEER 2016, 2017).

The largest landslide occurred on the east slope below the city (42.75057° N, 13.27223° E), directly above Highway SS4 (marked as #1 in Fig. 28). The landslide after the 24 August event was approximately 75 m wide and 30 m high, and was shallow, with only the upper one meter or less of soil sliding downslope. The landslide damaged retaining structures around the city. Significant incremental movements occurred on the landslide above highway SS4 after the two October events as shown in Fig. 29a, b. Limestone boulders greater than 6 m in diameter were detached from the slope and rolled onto the highway.

A second, smaller shallow landslide was located above Highway SS4 (marked as #2 in Fig. 28), just south of the larger landslide #1 (42.7501° N, 13.2719° E). This landslide was approximately 16 m wide and occurred on a slope 26 m high. Figure 30 shows a comparison between 3D models produced for this feature following the 24 August and 30 October events. This smaller landslide was not observed to change significantly after the 24 August event (Fig. 30a) and the 30 October earthquake (Fig. 30b). In this section, the retaining wall supporting the road appeared to maintain its stability. However, significantly more structural debris from collapsed residences was observed at this location following the 30 October earthquake.

Fig. 28 3D model of Pescara del Tronto following the 24 August event. Yellow circles highlight the seven locations discussed in this document (modified from GEER 2017)

A few of the observed slope failures in Pescara del Tronto appear to be related to potential retaining wall failures. The lower portion of Pescara del Tronto was partially supported by a 24 m-tall masonry retaining wall that surrounded the foot of the village. This wall suffered serious damage following the 24 August event (Fig. 31a). Substantial damage accumulation was observed following the October events as shown in Fig. 31b. In Pescara del Tronto, structure debris was so ubiquitous that it was challenging to investigate ground deformation in this zone. Any attempt to directly investigate this zone was not realistic because of the dangerous and unstable nature of the debris field. As a result, sUAS-based imagery constituted the main damage detection and analysis tool, especially after the October events. Moderate to severe damage was found along most of the roads in proximity from the slopes and/or above the retaining walls throughout Pescara del Tronto.

Figure 32a shows a minor sloughing of the gulley wall observed following the 24 August event from sUAS-based imagery (feature #5 in Fig. 28). In the same area, sUAS-based imagery captured following the October earthquake events shown that a section of the slope nearly 9 m thick and 20 m wide failed into the gulley (Fig. 32b). The remaining scarp is nearly 12 m in height and stands at an angle of 52 degrees, remarkably similar to the same scarp angle that remains behind the failed lower retaining wall (i.e., Areas 3 and 4). A small residential structure that was observed

Fig. 29 D model of landslide #1 in Pescara del Tronto following: **a** the 24 August and **b** the 30 October event (modified from GEER 2017)

to rest on top of the slope after August event had plummeted into the gulley after the October event.

Another shallow landslide occurred on the northern side of Pescara del Tronto (marked as #6 in Fig. 28) following the 24 August event (Fig. 33a). This landslide is approximately 17 m wide by 20 m high. A significant amount of structural rubble from collapsed residences is also visible in the landslide debris. Additional observations carried out following the October events revealed that the depth of the landslide remained relatively unchanged, but its width expanded from 17 m to more than 36 m (Fig. 33b). In both cases (following the 24 August and the October events), landslide sizes have been measured using sUAS-based imagery (GEER 2017).

The last landslide feature present in the area of Pescara del Tronto is located on the slope bounding the gravel pit on the southern upper portion of the village adjacent to the SP 129 highway (marked as #7 in Fig. 28). Observations made following the 24 August event show that the gravel pit is approximately 70 m by 90 m, and has 2.4 V:1H side slopes comprised of slightly cemented cataclasized limestone blocks interspersed in a gravelly-sandy matrix comprised of angular limestone clasts (Fig. 34a). A dirt haul road extending up and along the edge of the slope is located on

Fig. 30 3D model of the landslide #2 in Pescara del Tronto following: **a** the 24 August and **b** the 30 October event (modified from GEER 2017)

the south side of the gravel pit. The landslide exposed about 15 m of a 50 cm-diameter pipeline that was shallowly buried beneath the dirt haul road. The shape and size of the landslide remained almost unchanged following the October earthquake events as shown in Fig. 34b. However, the slope beneath the haul road degraded slightly more, causing nearly half of the haul road above the landslide to disappear. Approximately 8 more meters of the pipeline became exposed from the additional slope movements, causing the pipeline to apparently sag and bow slightly. Another smaller pipe of a few centimeters' diameter was also exposed and sagging substantially along the scarp.

Fig. 31 3D model of the failed retaining wall area in Pescara del Tronto (areas #3–4 in Fig. 28) following: **a** the 24 August and **b** the 30 October event (modified from GEER 2017)

6.3 Slope Displacements in Accumoli

Accumoli is a small village in the Rieti province, located atop the elongated WNW-ESE ridge with its elevation ranging from 860 to 890 m a.s.l. (Fig. 35). The geology in this area is underlain by the Laga Formation as the bed rock and is mainly composed of arenaceous materials. This formation, whose shallow parts are often loosened and weathered, is locally covered by colluvium and landslide deposits. The village is located in the epicentral area of the August 24, 2016, mainshock. Accumoli suffered damage during the January 14, 1703, Valnerina earthquake (**M**6.9). Zimmaro et al. (2018) estimated a PGA = 0.40 g during the 30 October event and a PGA = 0.55 g during the 24 August event in Accumoli (Table 2). Structural and landslide-related damage observed following the October events (GEER 2017) was significantly more serious than that observed after the August event and was documented by GEER (2016).

We analyze landslide-related damage that occurred in two zones: (1) Point A in Fig. 35: incremental damage to a retaining wall, and (2) Point B in Fig. 35, a shallow slope failure on the eastern side of the village. Following the 24 August event, a

Fig. 32 3D model of the failure observed in area #5 in Fig. 28 following: **a** the 24 August and **b** the 30 October event (modified from GEER 2017)

4.8 m-tall retaining wall located at the eastern tip of the spur was observed to have rotated outwards 3.6 degrees, with horizontal movements of 57 cm and downward vertical movements of nearly 18 cm. A soil graben nearly 2.7 m wide was observed behind the rotated wall, with soil settlements of 45–50 cm. Much more damage was observed following the October events. It is thus supposed that a shallow landslide developed beneath the retaining wall and caused the entire structure to slide down over several meters along the slope.

Figure 36 presents an aerial screenshot of the 3D model developed from sUASimagery of the site, showing evident incremental damage as a result of the earthquake sequence. The shallow landslide appeared to be limited to the upper corner of the slope, near the crown. The slide, therefore, likely occurred in the man-made fill slopes that was placed during the construction of the village. The scarp of the landslide appeared to follow the scarp of the soil graben observed behind the wall following the 24 August event, suggesting that the graben may have been caused not necessarily by the rotation of the retaining wall alone. The landslide displaced the top of the wall by 5–6 m in the horizontal direction and between 3 and 4 m in the downward vertical direction. In total, approximately 50 m of the retaining wall failed and slid 5–6 m

Fig. 33 3D model of landslide #6 following: **a** the 24 August and **b** the 30 October event (modified from GEER 2017)

downslope. Further, south at Point A, a series of shallow cracks (orthogonal to the dip of the slope) approximately 5–14 cm in width were observed following the 24 August event. These cracks extended after the October events up to a width ranging between 80–190 cm. The depth of the cracks was approximately 55 cm.

Fig. 34 3D model of landslide #7 and exposed pipeline following: **a** the 24 August and **b** the 30 October event (modified from GEER 2017)

Fig. 35 Accumoli map with locations of the landslide-related damage observed during the August and October events (modified from Lanzo et al. 2018)

Fig. 36 3D model screenshots following: **a** the 24 August and **b** 30 October event (modified from Lanzo et al. 2018)

References

Advanced Rapid Imaging and Analysis (ARIA)—Center for Natural Hazards (2016) ARIA Data Share Available at: https://aria-share.jpl.nasa.gov/. Last Accessed 26 September 2017

Ambraseys NN (1976) The Gemona di Friuli earthquake of 6 May 1976. UNESCO Restricted Technical Report, RP/1975–76/2.222.3, Part II

Antonini G, Ardizzone F, Cardinali M, Galli M, Guzzetti F, Reichenbach P (2002) Surfaces deposits and landslide inventory map of the area affected by 1997 Umbria-Marche earthquakes. Boll Soc Geol It 1:843–853

Aringoli D, Gentili B, Materazzi M, Pambianchi G (2010) Mass movements in Adriatic Central Italy: activation and evolutive control factors. In: Werner, Friedman (eds) Landslides: causes, types and effects. Nova Science Publishers, Inc. New York, NY, USA, pp 1–71

Ausilio E, Zimmaro P (2017) Landslide characterization using a multidisciplinary approach. Measurement 104:294–301

Ausilio E, Silvestri F, Tropeano G, Zimmaro P (2019) Spatial distribution of landslides triggered by the 2016 Central Italy seismic sequence. In: Silvestri, Moraci (eds) Earthquake geotechnical engineering for protection and development of environment and constructions. Associazione Geotecnica Italiana, Rome, Italy. CRC Press, pp 1228–1235

Bird JF, Bommer JJ (2004) Earthquake losses due to ground failure. Eng Geol 75(2):147–179

Boore DM, Stewart JP, Seyhan E, Atkinson GM (2014) NGA-West 2 equations for predicting PGA, PGV, and 5%-damped PSA for shallow crustal earthquakes. Earthq Spectra 30:1057–1085

Bozzano F, Gambino P, Prestininzi A, Scarascia Mugnozza G, Valentini G (1998) Ground effects induced by the Umbria-Marche earthquakes of September–October 1997, Central Italy. In: Balkema AA (eds) Proceedings of the 8th international congress of the international association for engineering geology and the environment, Vancouver, Canada, 21–25 September, 1998, Rotterdam, The Netherlands, pp 825–830

Campbell KW, Bozorgnia Y (2014) NGA-West2 ground motion model for the average horizontal components of PGA, PGV, and 5% damped linear acceleration response spectra. Earthq Spectra 30:1087–1115

Carro M, De Amicis M, Luzi L, Marzorati S (2003) The application of predictive modeling techniques to landslides induced by earthquakes: the case study of the 26 September 1997 Umbria–Marche earthquake (Italy). Eng Geol 69:139–159

CERI working group, Martino S, Caporossi P, Della Seta M, Esposito C, Fantini A, Fiorucci M, Iannucci R, Marmoni GM, Mazzanti P, Missoni C, Moretto S, Rivellino S, Romeo RW, Sarandrea P, Troiani F, Varone C (2016). Sisma Centro Italia: Amatrice, Visso, Norcia Effetti-Interazione rete infrastrutturale. Available at http://www.ceri.uniroma1.it/. Last Accessed 19 April 2018

Chiaraluce L, Amato A, Cocco M, Chiarabba C, Selvaggi G, Di Bona M, Piccinini D, Deschamps A, Margheriti L, Courboulex F, Ripepe M (2004) Complex normal faulting in the apennines thrust-and-fold belt: the 1997 seismic sequence in central Italy. Bull Seismol Soc Am 94:99–116

Chiou BS-J, Youngs RR (2014) Update of the Chiou and Youngs NGA model for the average horizontal component of peak ground motion and response spectra. Earthq Spectra 30:1117–1153

Choi J-K, Won J-S, Lee S, Kim S-W, Kim K-D, Jung H-S (2011) Integration of GIS and SAR interferometry for a coal mine subsidence hazard map in Taebaek, Korea. Int J Remote Sens 32:8161–8181

Cotecchia V, Del Prete M (1984) The reactivation of large flows in the parts of Southern Italy affected by the earthquake of November 1980, with reference to the evolutive mechanism. Proceedings of the fourth international symposium on landslides, Toronto, Canada 2:33–37

Cotecchia V, Guerricchio A, Melidoro G (1986) The Geomorphogenetic Crisis Triggered by the 1783 earthquake in Calabria (Southern Italy). Proceedings of the international symposium on engineering geology problems in seismic areas, Bari, Italy 6:245–304

Cruden DM, Varnes DJ (1996) Landslide types and processes. In: Turner, Shuster (eds) Landslides: investigation and mitigation. Transportation Research Board, Special Reports 247:36–75

D'Elia B, Esu F, Pellegrino A, Pescatore T (1985) Some effects on natural slope stability induced by the 1980 Italian earthquake. In: Proceedings of the eleventh international society of soil mechanics and foundation engineering conference, San Francisco, USA

Esposito E, Porfido S, Simonelli AL, Mastrolorenzo G, Iaccarino G (2000) Landslides and other surface effects induced by the 1997 Umbria-Marche seismic sequence. Eng Geol 58:353–376

Fielding EJ, Talebian M, Rosen PA, Nazari H, Jackson A, Ghorashi M, Walker R (2005) Surface ruptures and building damage of the 2003 Bam, Iran, earthquake mapped by satellite synthetic aperture radar interferometric correlation. J Geophys Res 110:B03302

Franke KW, Lingwall BN, Zimmaro P, Kayen RE, Tommasi P, Chiabrando F, Santo A (2018) Phased reconnaissance approach to documenting landslides following the 2016 Central Italy earthquakes. Earthq Spectra 34:1693–1719

Galadini F, Falcucci E, Gori S, Zimmaro P, Cheloni D, Stewart JP (2018) Active faulting in source region of 2016–2017 central Italy earthquake sequence. Earthq Spectra 34:1557–1583

Galli P, Peronace E, Bramerini F, Castenetto S, Naso G, Cassone F, Pallone F (2016) The MCS intensity distribution of the devastating 24 August 2016 earthquake in central Italy (MW 6.2). Ann Geophys 59, FAST TRACK 5. https://doi.org/10.4401/ag-7287

GEER (2017) Engineering reconnaissance following the October 2016 Central Italy earthquakes. In: Version 2, Zimmaro, Stewart (eds) GEER Association Report No. GEER-050D. https://doi.org/10.18118/G6HS39

GEER (2016) Engineering reconnaissance of the 24 August 2016 Central Italy Earthquake. In: Version 2, Zimmaro, Stewart (eds). GEER Association Report No. GEER-050B. https://doi.org/10.18118/G61S3Z

Gigli G, Morelli S, Fornera S, Casagli N (2014) Terrestrial laser scanner and geomechanical surveys for the rapid evaluation of rockfall susceptibility scenarios. Landslides 11:1–14

Gori S, Falcucci E, Galadini F, Zimmaro P, Stewart JP, Kayen RE, Lingwall B, Moro M, Saroli M, Pizzi A, Di Domenica A (2018) Surface faulting caused by the 2016 Central Italy seismic sequence: field mapping and Lidar/UAV imaging. Earthq Spectra 34:1585–1610

Hungr O, Leroueil S, Picarelli L (2014) The Varnes classification of landslide types, an update. Landslides 11:167–194

INGV (2016) Gruppo di Lavoro INGV sul terremoto di Amatrice. Secondo rapporto di sintesi sul Terremoto di Amatrice Ml 6.0 del 24 Agosto 2016 (Italia Centrale). https://doi.org/10.5281/zenodo.154400

ISPRA (2016) Gruppo di Lavoro ISPRA. Report attività svolta da ISPRA in data 25–26/08/2016. Centro per la microzonazione sismica e le sue applicazioni. Available at: https://www.isprambiente.gov.it/files2017/notizie/sisma/ReportsinteticoISPRAagosto16marzo17.pdf. Last Accessed 18 Nov 2021

Jayaram N, Baker JW (2009) Correlation model for spatially distributed ground-motion intensities. Earthq Eng Struct Dynam 38:1687–1708

Jo M-J, Won J-S, Kim S-W, Jung H-S (2010) A time-series SAR observation of surface deformation at the southern end of the San Andreas fault zone. Geosci J 14:277–287

Jung H-S, Lu Z, Won JS, Poland MP, Miklius A (2011) Mapping three-dimensional surface deformation by combining multiple-aperture interferometry and conventional interferometry: application to the June 2007 eruption of Kilauea volcano, Hawaii. IEEE Geosci Remote Sens Lett 8:34–38

Keefer DK (1984) Landslides caused by earthquakes. Geol Soc Am Bull 95:406–421

Keefer DK (1985) Landslides caused by earthquakes: reply. Geol Soc Am Bull 96:1093–1094

Keefer DK (2002) Investigating landslides caused by earthquakes—a historical review. Surv Geohys 23:473–510

Lanzo G, Tommasi P, Ausilio E, Aversa S, Bozzoni F, Cairo R, d'Onofrio A, Durante MG, Foti S, Giallini S, Mucciacciaro M, Pagliaroli A, Sica S, Silvestri F, Vessia G, Zimmaro P (2018) Reconnaissance of geotechnical aspects of the 2016 Central Italy earthquakes. Bull Earthq Eng 17:5495–5532

Lee W, Lu Z, Won J-S, Jung H-S, Dzurisin D (2013) Dynamic deformation of Seguam Island, Alaska, 1992–2008, from multi-interferogram InSAR processing. J Volcanol Geotherm Res 260:43–51

Pavlides S, Chatzipetros A, Papathanasiou G, Georgiadis G, Sboras S, Valkaniotis S (2017) Ground deformation and fault modeling of the 2016 sequence (24 Aug–30 Oct) in central Apennines (Central Italy). Bull Geol Soc Greece 51:76–112

Piatanesi A, Cirella A (2009) Rupture process of the 2009 Mw6.3 L'Aquila (Central Italy) earthquake from nonlinear inversion of strong motion and GPS data. Istituto Nazionale di Geofisica e Vulcanologia (INGV), Rome, Italy

Polcari M, Montuori A, Bignami C, Moro M, Stramondo S, Tolomei C (2017) Using multi-band InSAR data for detecting local deformation phenomena induced by the 2016–2017 Central Italy seismic sequence. Remote Sens Environ 201:234–242

Rathje EM, Carr LP (2010) Satellite observations of landslides caused by the 2008 Wenchuan earthquake in China. In: Proceedings, 9th US national and 10th Canadian conference on earthquake engineering: reaching beyond borders, July 2010, Toronto, Canada

Rathje EM, Franke KW (2016) Remote sensing for geotechnical earthquake reconnaissance. Soil Dyn Earthq Eng 91:304–316

Rodríguez CE, Bommer JJ, Chandler RJ (1999) Earthquake-induced landslides: 1980–1997. Soil Dyn Earthq Eng 18:325–346

Romeo S, Di Matteo L, Melelli L, Cencetti C, Dragoni W, Fredduzzi A (2017) Seismic-induced rockfalls and landslide dam following the October 30, 2016 earthquake in Central Italy. Landslides 14:1457–1465

Sarconi M (1784) Osservazioni fatte nelle Calabrie e nella frontiera di Valdemone sui fenomeni del tramoto del 1783 e sulla geografia fisica di quelle regioni. Reale Accademia delle Scienze e Belle Lettere. Naples, Italy

Sextos A, De Risi R, Pagliaroli A, Foti S, Passeri F, Ausilio E, Cairo R, Capatti MC, Chiabrando F, Chiaradonna A, Dashti S, De Silva F, Dezi F, Durante MG, Giallini S, Lanzo G, Sica S, Simonelli AL, Zimmaro P (2018) Local site effects and incremental damage of buildings during the 2016 Central Italy earthquake sequence. Earthq Spectra 34:1639–1669

Sieberg A (1930) Geologie der Erdbeben. Handboch der Geophysic 2(4):552–554 [Tabb. 100, 101, 102, 103], Berlin, Germany

Stewart JP, Lanzo G, Pagliaroli A, Scasserra G, Di Capua G, Peppoloni S, Darragh R, Gregor N (2012) Ground motion recordings from the Mw 6.3 2009 L'Aquila earthquake in Italy and their engineering implications. Earthq Spectra 28:317–345

Yun S-H, Fielding EJ, Simons M, Rosen P, Owen S, Webb F (2011) Damage proxy map of M 6.3 Christchurch earthquake using InSAR coherence. Fringe Workshop. Advances in the science and applications of SAR interferometry from ESA and 3rd party missions. Frascati, Italy. Available at: https://earth.esa.int/documents/10174/1567329/Yun_FRINGE2011.pdf

Yun S-H, Hudnut K, Owen S, Webb F, Simons M, Sacco P, Gurrola E, Manipon G, Liang C, Fielding EJ, Milillo P, Hua H, Coletta A (2015) Rapid damage mapping for the 2015 Mw 7.8 Gorkha earthquake using synthetic aperture radar data from COSMO–SkyMed and ALOS-2 satellites. Seismol Res Lett 86:1549–1556

Zhang L, Lu Z, Ding X, Jung H-S, Feng G, Lee C-W (2012) Mapping ground surface deformation using temporarily coherent point SAR interferometry: application to Los Angeles Basin. Remote Sens Environ 117:429–439

Zimmaro P, Scasserra G, Stewart JP, Kishida T, Tropeano G, Castiglia M, Pelekis P (2018) Strong ground motion characteristics from 2016 central Italy earthquake sequence. Earthq Spectra 34:1611–1637

Zimmaro P, Stewart JP (2017) Site-specific seismic hazard analysis for Calabrian dam site using regionally customized seismic source and ground motion models. Soil Dyn Earthq Eng 94:179–192

Earthquake-Induced Landslides and Related Problems

Ikuo Towhata

Abstract The world has faced serious landslide problems since the beginning of the twenty-first century when gigantic earthquakes occurred in Pakistan, China, and many other parts of the world. The damage was substantial not only because of the burial of houses under earth but also because of river blockage, the breaching of landslide dams, induced tsunami, compound effects with rainfall during the aftermath, and long-term slope instability. The susceptibility of slopes is affected by rock weathering, hydration, the dip of geological strata, antecedent rainfall, melting glacier, and many other processes in nature. It is thus reasonable to state that earthquake-induced landslides have to be investigated by those in numerous fields of science and technology such as engineering geology, geomorphology, rock/soil mechanics, sedimentology, geohydrology, and many others. Nevertheless, landslide experts have not undertook comprehensive study to date, although research and practice in their respective fields have achieved tremendously detailed developments. With this viewpoint, this report attempts to review a number of landslide studies in a variety of fields both in the past and in recent times, as well as the experience and knowledge gained from recent disasters. On the basis of those, a holistic and comprehensive perspective is presented. Certainly, the review of real disasters is the key to understanding the facts of landslides.

Keywords Case history · Compound effect · Earthquake · Fault · Holistic view · Landslide · Mechanism · Problem soil · Weathering

1 Introduction

An important and difficult field in landslide studies is earthquake-induced landslides. This topic is important now because human habitation is expanding into hilly and mountainous regions where the threat of slope instability is high. Moreover, seismic damage to transportation and energy infrastructure directly affects the economic

I. Towhata (✉)
Kanto Gakuin University, 1-50-1 Mutsuura-Higashi Kanazawa-Ku, Yokohama 2368501, Japan
e-mail: towhata.ikuo.ikuo@gmail.com

I. Towhata et al. (eds.), *Coseismic Landslides*, Springer Natural Hazards,
https://doi.org/10.1007/978-981-19-6597-5_11

and social activities of entire nation. Nevertheless, the engineering communities are not well prepared to cope with this situation. Accordingly, several mega earthquakes toward the end of the twentieth century and at the beginning of the twenty-first century caused significant problems due to coseismic landslides. At the same time, people today have a strong desire for safety even under severe conditions. It is essential, therefore, that we understand the problems that we face now so that the extent of safety and resilience of our community may be improved in the upcoming decades.

Many memorable studies were made in the twentieth century of earthquake-induced landslides. Among them, the one by Keefer (1994) collected a great amount of data universally and interpreted them. His research emphasized the relationship between the earthquake magnitude and the maximum distance of landslides from epicenters or fault ruptures. An alternative of the maximum distance was the size of the affected area. Whatever parameter may be used, his discussion on distance helped develop improved risk assessment in earthquake-prone mountainous regions. Similarly, Yasuda (1993) presented a correlation between magnitude and maximum distance to earthquake-induced landslides. Rodríguez et al. (1999) collected more data and discussed the magnitude-distance (size of affected area) relationship, as well as analyzing the Modified Mercalli Intensity scale of seismic motion needed to trigger landslides.

There are two reasons why the studies of earthquake-induced landslides in the twenty-first century have to be different from those in the previous century. First, in the recent times, we have encountered gigantic earthquakes of magnitude 7.5, 8.0 or greater that trigger landslides over a huge area. Those earthquakes include the 1999 Chi-Chi earthquake of Mw (moment magnitude) = 7.6 in Taiwan, the event in Northern Pakistan of Mw = 7.6, the 2008 Wenchuan earthquake of Mw = 7.9 in China, and the 2016 Kaikoura earthquake of Mw = 7.8 in New Zealand together with many other quakes of slightly smaller magnitudes. Those gigantic events demonstrated more serious consequences of landslides than the events in the previous century. Second, the induced slope instability has caused variable types of damage to human communities with which we are not very familiar. Those new damage types related to slope instability comprise long-distance flow of debris, seismic behavior of volcanic slopes, seismic deterioration of slopes, natural dams, the effects of antecedent rainfall, tsunamigenic landslides, the compound effects of earthquake shaking and rainfall, and the seismic performance of artificial slopes. Thus, in conjunction with the increased probability of heavy precipitation induced probably by global warming, the world is now prone more to unprecedented types of slope disasters. With these in mind, this paper starts with reviewing the recent earthquakes from the viewpoint of landslides and related problems. The individual problems will be discussed in more detail so that their essence and their possible mitigation may be understood. The size of earthquakes is expressed by moment magnitude (Mw) as long as information is available. It is indispensable to note that the Japan Landslide Society (2012) carried out a comprehensive research project on earthquake-induced landslides and published a very good report. I deeply appreciate information and perspectives that were presented by this society. The earlier achievement of the present study (Towhata 2021a) was published by the 8th International Symposium on Landslides (2021). This paper is its continuation and presents substantially more information.

Studies of earthquake-induced landslides have achieved a tremendous development in recent decades. As soon as a disaster occurs, reconnaissance is carried out promptly and detailed reports are published. The problem is the difficulty in building-up a holistic scope. Landslide is a phenomenon that is related to geology, geomorphology, geohydrology, climate, weather conditions, the mechanics of soil and rock, fracture mechanics, human influence through farming, irrigation and construction, and, in the case of earthquake-induced landslides, tectonics. Because each field of study is now deep and detailed with ample precious knowledge, it is difficult to assemble them all in the limited time and within the capacity of an individual researcher. In this paper, I attempt to develop a holistic view, with an emphasis on case histories. Further efforts were made as well to pick emerging problems such as natural dams and long-term slope instability that were not taken into account in previous times. I agree that my effort was not sufficient, because the target is so huge. It is important to state here that this research was conducted as one of the activities of the Joint Technical Committee No. 1 (Prof. L. Picarelli was the Chair) on natural slopes and landslides under the auspices of Federation of International Geo-engineering Societies (FedIGS).

2 Review of Recent Landslide Disasters Induced by Earthquakes

In recent natural disaster studies, there are two major topics, which are extreme events and compound disasters. The extreme events exert unexpectedly significant influence on human community, and some engineered structures that can stay stable under design load suddenly collapse under load in excess of the design level. Those gigantic earthquakes in China (2008) and Japan (2011) were examples of such events but the 2009 Morakot Typhoon associated with heavy rainfall in Taiwan was another example. The compound disaster is induced by a chain of natural actions that occur consecutively within a short term. For example, an earthquake triggers landslide, the landslide mass falls into the sea, and the sea water produces tsunami disaster in the coastal area. For both extreme and compound disasters, the damage is characterized by a sequence of events such as.

Structural damage immediately upon the onset of disaster,

- Loss of service of infrastructures,
- No supply of food, water, energy, etc.
- Suspension of normal life, social activities, and operation of industries, and
- Economic damage.

Moreover, the affected mountain slopes may undergo long-term instability and generate sediment disasters for years or decades. To avoid the catastrophic situation after disasters, human has to be able to foresee what would happen and prepare necessary social systems (NILIM 2020). In this context, it is important to review and know what happened during the past disasters.

Whenever a natural disaster occurs, it is essential to conduct a reconnaissance and learn lessons that can be obtained only through site visits. In this respect, I start this report with the findings obtained by my own reconnaissance. Note that more specific and detailed information is available on respective regions in other chapters of this book.

2.1 Kobe Earthquake of Mw = 6.9 in 1995 in Japan

The review starts with the 1995 Kobe earthquake (Fig. 1). This earthquake is well known as the first event that hit a modern big city and demonstrated many problems occurring in urban environments. Those problems include the weakness of modern buildings and private houses, subsoil liquefaction in man-made islands, and the resilience of lifelines. In contrast, natural ground remained intact in spite of the strong acceleration of 304.9 cm/s^2, recorded on the campus of Kobe University on a hilly terrain. There were few big landslides (Sassa et al. 1996), probably because the earthquake occurred on January 17, 1995, which was in the middle of a dry winter (see Fig. 148). Fukuoka et al. (1997) showed that the precipitation in Kobe in 1994 was substantially less than the normal level. This earthquake attracted public concern with the seismic stability of residential land in hilly terrain because of the disasters such the one shown in Fig. 2. The Kobe earthquake was a rare (a previous rupture of the same fault system probably happened in 1597; Tsutsumi et al. 2004) but extremely strong event (the maximum acceleration at Takatori site on soft alluvium in Kobe reached 615.5 cm/s^2). After the earthquake, the philosophy of earthquake-resistant design adopted two-level design earthquakes; level 1 that may occur once during the life time of a structure and level 2 that is very rare; the return period being 500 or 1000 years or longer according to my idea. In design, no damage should occur for a level 1 earthquake, whereas no fatalities should occur under level 2, although significant deformation is allowed in designed structures. This philosophy was adopted by many other design codes and proved useful. However, the 2016 Kumamoto earthquake sequence of Mw = 6.2 and 7.0 exhibited its limitation, as will be touched upon later.

2.2 Chi-Chi Earthquake of Mw = 7.6 in 1999 in Taiwan

The 1999 Chi-Chi earthquake triggered many landslides in the mountain region of Taiwan (Fig. 3) where materials are geologically young and not fully lithified yet.

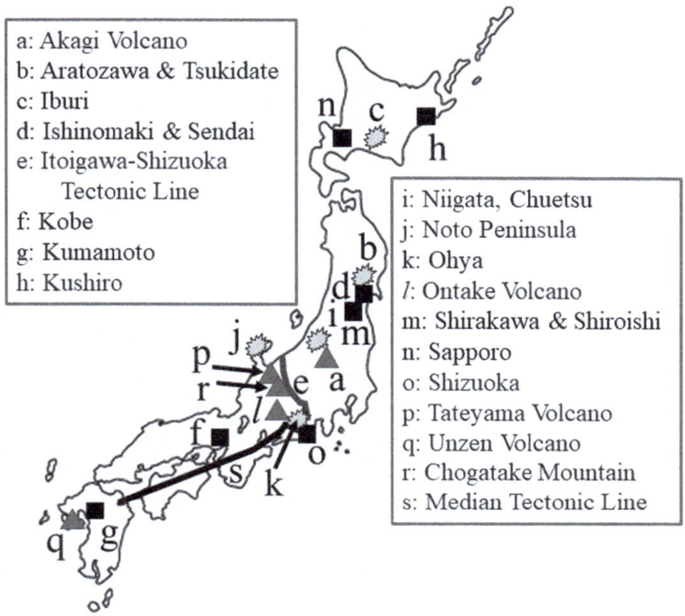

Fig. 1 Location of sites in Japan that are referred to in this paper

Fig. 2 Failure of residential fill in Kobe, Japan

Tectonic stress is high in this island because it is located between the Eurasian and the Philippine Sea plates. Figure 4 illustrates the location of Chelungpu Fault that triggered the earthquake. The length of the rupture was 95 km and most landslide occurred on the eastern or the hanging-wall side of the fault. Liao and Lee (2000) conducted a remote-sensing study and reported 9272 major landslides within the area with PGA > 250 cm/s^2 (*PGA* stands for the peak ground acceleration during the earthquakes).

Steep slopes and heavy precipitation, as well as the progress of rock weathering, promoted the coseismic landslides. Furthermore, the number of sediment disasters increased after this earthquake (Lin et al. 2004). Two gigantic landslides occurred at Tsaoling (120 million m^3; Hung 2000) and Jiu-Feng-Er-shan (36

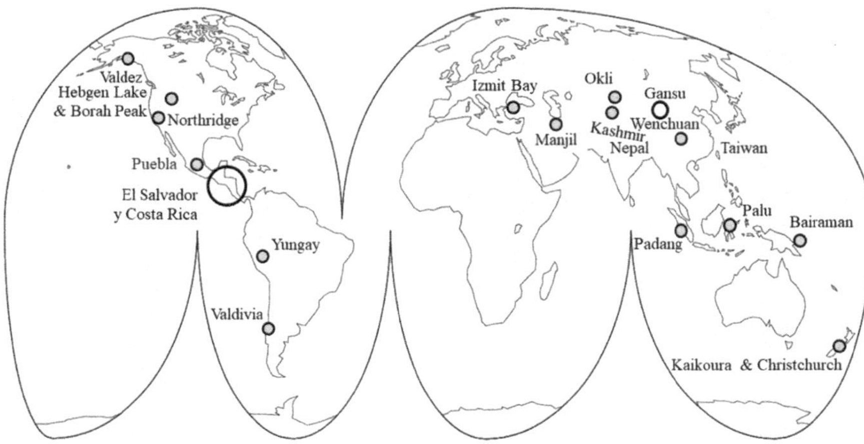

Fig. 3 Location of some of the sites referred to in this paper

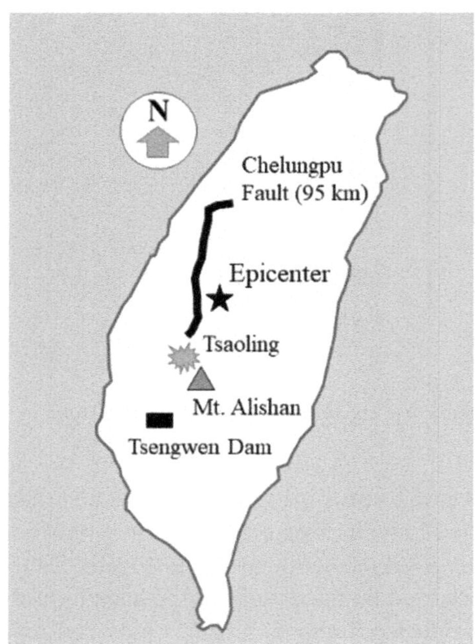

Fig. 4 Chelungpu fault as the source of the Chi-Chi earthquake (location of the fault was drawn after Shishikura et al. 2002)

Fig. 5 Jiu-Feng-Er-shan coseismic landslide in 1999, Taiwan

million m^3 according to Hung 2000) (Fig. 5) in central Taiwan. The failed slope at Tsaoling had repeated landslides since 1862 during earthquakes or rainfall because of its weak geological structure (layers parallel to the slope) (Ishihara 1985; Hung 2000). The previous seismic failure in 1941 produced a big landslide dam (Kawada 1943a, b) and the same process occurred in 1999 (Figs. 6 and 7). The new lake was filled with sediments quickly and did not exist anymore when I revisited the site in 2013 (Fig. 183). The Jiu-Feng-Er-shan landslide produced a small reservoir at its foot as well (Wang et al. 2006), and it still remains at the site. Chigira (1992, 1998) called attention to gravity-induced creep deformation of the mountain body prior to failure as a precursor of the disaster.

The coseismic landslide of an artificial fill at the Chungliao electric power supply substation damaged transformers and other equipment in the substation and profoundly affected the cities in northern Taiwan. The problem was serious, as half of the country's electric power passed through this substation, going from power stations in the south to the north (Ueshima et al. 1999; Charlwood et al. 2000).

2.3 *Izmit Earthquake of Mw = 7.6 in 1999 in Turkey*

Otherwise called the Kocaeli earthquake, the Izmit earthquake triggered instability of subaqueous slopes and subsidence in coastal areas. Figure 8 illustrates the subsidence of a coastal structure on the southern side of Izmit Bay (Fig. 3). Of particular interest was the disappearance of the shore line in Değirmendere (Kanatani et al. 2003; Aydan et al. 2008); see Fig. 9. Similar subsidence occurred at many places along the southern coast of Izmit Bay. Note that it was not a manifestation of fault rupture because the fault was located onshore (Fig. 10). Sapanca Lake to the east had similar subsidence

Fig. 6 Coseismic landslide at Tsaoling in 1999

Fig. 7 Coseismic landslide at Tsaoling in 1941 (Kawada 1943a, b)

as well (Fig. 11). Presumably, the reported subsidence was induced by a landslide or pull-apart movement of a strike-slip fault (Mann et al. 1983; Öztürk et al. 2000; Kuşçu et al. 2002) or their combination.

Fig. 8 Coseismic subsidence and inundation at Değirmendere during the 1999 Kocaeli earthquake, Turkey

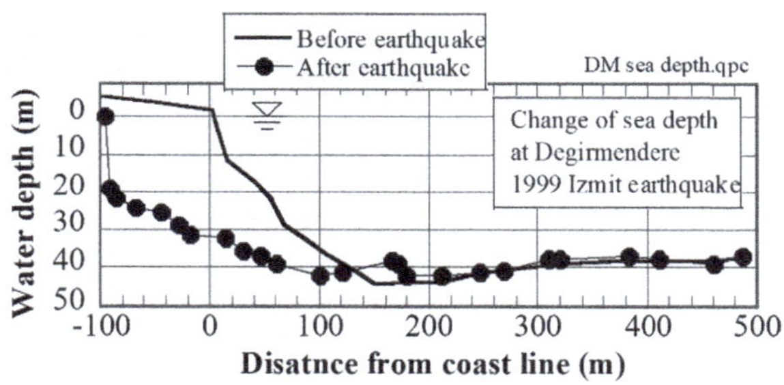

Fig. 9 Change of sea depth in front of the shore line at Değirmendere, Turkey (drawn after Kanatani et al. 2003)

Fig. 10 Onshore manifestation of fault rupture and depression shown by arrows; in Gölcük, Turkey

Fig. 11 Coseismic subsidence at the shore of Sapanca Lake, Turkey

2.4 *Niigata-Chuetsu Earthquake of Mw = 6.6 in 2004 in Japan*

This earthquake affected a hilly area in central Japan (area "*i*" in Fig. 1) where soft mudstone prevails. Landslides occurred at more than 3700 sites, affected local transportation and formed natural dams. Figure 12 illustrates one dam in Terano. The landslide at this site had a travel length (L) = 370 m, a fall height (H) = 60 m (see Fig. 40 for definition), hence $H/L = 0.162$, and the equivalent friction angle (otherwise called fahrböschung angle) = $\arctan(H/L) = 9.2°$ (Heim 1932; Scheidegger 1973; Hsü 1975), while the landslide volume was 1.04×10^6 m^3 (Yagi et al. 2007). It seems that moisture content in soils was high when the earthquake occurred on October 23rd, which was three days after antecedent rainfall due to a typhoon. Bommer and Rodríguez (2002) pointed out the importance of local climate and rainfall on the basis of their experiences in Central America. Another important source of water was the fish ponds that were prevalent in this area to support the local major carp hatchery industry (Fig. 13). Figure 14 illustrates two minor landslides. One of them was greater than the other, traveled laterally after falling and blocked the river flow, while the smaller one simply fell down and stopped at the valley bottom. This suggests the effects of landslide mass in which a bigger mass is able to travel over longer distances. Note that the embankment of a local road in this hilly region was easily destroyed at many places by this earthquake, making emergency rescue very difficult (Fig. 15). The embankments were made of locally available low-cost crushed mudstones that were susceptible to slaking and disintegration. Figure 16 shows a different type of seismic landslide at Yokowatashi where a rock mass slipped along a weak sandy seam (Deng et al. 2011). There is a rice paddy on the other side of the top of this landslide. Hence, it is possible that irrigation water affected the slope stability here.

Fig. 12 Natural dam at Terano during the 2004 Niigata-Chuetsu earthquake, Japan (photograph taken in 2009)

Fig. 13 Dainichi-Yama landslide moving along a circular slip plane; see fish ponds atop the sliding soil mass

Fig. 14 Effects of landslide volume on runout of lateral flow (Niigata-Chuetsu earthquake)

Fig. 15 Seismic damage to a local road situated on a mudstone slope (Niigata-Chuetsu earthquake)

2.5 Kashmir Earthquake of Mw = 7.6 in 2005

This earthquake occurred in Pakistan-administrated Kashmir (Fig. 3) and claimed more than 80,000 casualties. While many slopes fell down during the strong shaking (Figs. 17 and 18), instability continued during the post-earthquake period as well (Bulmer et al. 2007). This suggests deterioration of material due to shaking and will be discussed later (Fig. 222). Many landslides occurred along the Balakot-Bagh Fault and Sato et al. (2007) interpreted their distribution using satellite photographs. They stated that 47% of the landslides occurred within 2 km of the fault. Petley et al. (2010) pointed out that most landslides occurred on the hanging-wall side of the fault. Figure 19 illustrates the significant effect on local major transportation. Based on the interpretation of satellite photos by Pathier et al. (2006), I infer that the causative fault was 65 km long.

Fig. 16 Yokowatashi rock slide along a thin weak layer (slope angle = 19°) (Niigata-Chuetsu earthquake)

Fig. 17 Landslide to the east of Muzaffarabad in Pakistan

Fig. 18 Concentration of landslides along the Muzaffarabad-Balakot section of the causative fault (Fig. 4)

Fig. 19 Earthquake-induced landslides along a local major transportation route along Neelam River to the north of Muzaffarabad

Figure 20 illustrates the view of Balakot before and after the earthquake. A hill in this town had many buildings both atop and around it. All the houses suffered from total damage caused by the earthquake (Figs. 20b and 21). However, there seems to be a difference in the extent of "total damage" between the top and the bottom of the slope. As illustrated in Figs. 20b, 22 and the front view of Fig. 23, all the buildings at the top were totally flattened and, in contrast, those around the hill still maintained walls and columns after the disaster (Figs. 21 and 23). The different extents of damage suggest the topographic amplification of ground motion by which the motion at the crest is stronger than that at the bottom of the hill slope. Figure 24 depicts sliding of the slope surface around the hill. This movement was fatal to the foundations of the buildings as well.

(a) Before the earthquake (b) After the earthquake

Fig. 20 Small hill in Balakot before and after the earthquake in 2005 (photograph provided by Mr. K. Sajjaad)

Fig. 21 Seismically destroyed Balakot (after the 2005 earthquake)

Fig. 22 Heavy damage of buildings at the top of hill in Balakot

Fig. 23 Lesser extent of building damage at the foot of the hill in Balakot (seen from hill top)

2.6 Wenchuan Earthquake of Mw = 7.9 in 2008 in Sichuan Province of China

The Wenchuan earthquake (Fig. 3) was caused by the movement along the Longmenshan Fault, where rupture started at 10-km depth and propagated in the northeast

Fig. 24 Displacement of the slope surface around the heavily damaged hill in Balakot (Fig. 22); slope distortion is indicated by white dashed lines, and tilting of trees is noted by arrows

direction over the length of 300 km (China Earthquake Administration 2008). Gorum et al. (2010) interpreted satellite images to identify more than 60,000 landslides and showed that most of them occurred within a short distance (15 km on the west and 4 km on the east) of the coseismic fault. This finding suggests that distance from the fault plane is more meaningful than the epicentral distance for assessment of the coseismic landslide hazard. The biggest failure occurred at Daguangbao (Fig. 25), which was situated at a mountain top and measured 742 million cubic meters in volume (Huang et al. 2008, 2009). Part of the Daguangbao debris came down into a valley and was deposited at the valley bottom (Fig. 26). This has been the source of delayed disasters at the time of heavy rain.

Long-distance flows of landslide masses produced natural dams at many places. According to Cui et al. (2009, 2011), in total, 257 natural dams were formed and, among them, 33 were called major. To avoid their breaching and flood disaster in the

Fig. 25 Daguangbao landslide with 742 million cubic meters in volume (elevations above sea level were written on a photograph by Prof. G.H. Wang, Kyoto University)

Fig. 26 Huge soil deposit formed in the downstream valley of Daguangbao landslide (photographs by Mr. S.H. Cui, Chengdu University of Technology)

downstream areas, natural dams were destroyed by explosion and water was drained while people living in the communities in the downstream areas were evacuated. Figure 27 indicates the site of Hongshihe landslide dam at Donghekou, which was produced by 18 million m^3 of landslide mass and stored 4 million m^3 of water. After this landslide, fumaroles started with temperature exceeding 60 °C, which may imply potential instability of this slope, although Wang et al. (2014a) concluded that the fumarole was a consequence of accelerated weathering of rocks after the landslide. When the lake was filled with water, a small spillway was formed naturally and it was enlarged under control in order to drain the water and protect the downstream communities from breaching and flood (Chang et al. 2009). There are also a few cases in which a landslide dam did not breach (Fig. 28). Figure 29 shows a valley where a huge debris flow occurred during the earthquake (Wang et al. 2012).

In addition to the distance to the fault rupture, local geological conditions influenced the slope behavior to a certain extent. Tang et al. (2011) reported that regions of Silurian slate and phyllite, as well as Cambrian sandstone and siltstone, developed coseismic landslides, closely followed by limestone of different ages. Chigira et al. (2009a) stated that more slopes failed on the hanging-wall side of the seismic fault. Further, Chigira et al. (2009b) pointed out creep deformation of the mountain body at Daguangbao that was a precursor of a future landslide. Figures 30 and 31 compare the consequence of the earthquake in two different geological structures. Figure 30 shows significant failure because the strata were parallel to the slope and sliding could easily occur. Then, Fig. 31 indicates a different situation due probably to the opposite direction of bedding planes; see also Qi et al. (2010). Note that the distance between these two different locations was merely hundreds of meters.

A clear difference between rainfall-induced and earthquake-induced landslides is found at the location of the head scarp where sliding starts. Generally, rainfall-induced landslides start at knick points where a gentle slope at a mountain top becomes steeper on the flank at a lower elevation. This is obviously because of the natural erosion. In contrast, the initiation point of earthquake-induced landslides is affected by geometric amplification of seismic motion. A major landslide near Beichuan

Fig. 27 Site of Hongshihe landslide dam at Donghekou after controlled drainage

Fig. 28 Stabilized landslide dam near Yingxiu

(Fig. 32a) and near the Zipingpu Reservoir (Fig. 32b) clearly shows an example for this; the landslide started from the mountain top. This amplification occurs as the earthquake wave energy that comes in from the mountain bottom is focused as the wave propagates toward the mountain top, which has a smaller width. Analytical/numerical reproduction of this phenomenon will be presented later (Fig. 74 and Eq. 1).

Fig. 29 Appearance of a valley after a debris flow

Fig. 30 Significant seismic damage in a dip slope where the strata are parallel to the slope surface

Fig. 31 Intact cliff slope where strata are normal to the slope surface; see sliding on the opposite side of the valley

(a) Landslide near Beichuan; one village on this mountain slope disappeared

(b) Seismic landslide initiated at the mountain top where shaking energy was focused (near the Zipingpu Reservoir)

Fig. 32 Seismic slope instability starting from the mountain top (2008 Wenchuan earthquake)

Many slopes became unstable after the earthquake and triggered further disasters (Qiao and Yang 2010). In particular, heavy rain in September, 2008, brought about many post-earthquake landslides (Tang et al. 2011). This is an example of the compound effect by which two or more natural disasters together affect human communities and aggravate damage. This topic will be discussed in detail later.

Fig. 33 Seismic landslide in Costa Rica where sliding started from the top of a ridge

2.7 Cinchona Earthquake of Mw = 6.1 in 2009 in Costa Rica

Costa Rica in Central America was affected by earthquakes and seismically induced landslides many times in the past (Table 1 by Mora and Mora 1994). The earthquake in 2009 hit a mountainous region of this country (Fig. 3), where the ground surface is covered by volcanic ash (Barrantes-Castillo et al. 2013). Figures 33 and 34 indicate examples of coseismic landslides. The landslide in Fig. 33 started from the mountain top because of topographic amplification. The important lesson was the vulnerability of volcanic slopes (Fig. 34). Unlike ordinary mountain slopes, those of volcanoes consist of ash, pumice and layers of lava that are discrete and not bonded tightly together. Further, the volcanic slopes are porous and can contain much water, thus reducing the slope stability during earthquakes. It was still fortunate in Costa Rica that the earthquake occurred in a dry season (Fig. 35), which probably mitigated the extent of damage. In addition, Quesada-Román et al. (2019) reported a good example of compound disaster in Costa Rica. The 2016 Bijagua earthquake (Mw = 5.4) affected the slope stability, and then the Hurricane Otto brought heavy rainfall of 200–300 mm. Consequently, many debris flows occurred in volcanic slopes.

2.8 Earthquakes in 2003 and 2008 in Tohoku District of Japan

The 2003 Miyagiken-oki earthquake of Mw = 7.0 induced flow failure of fill that was composed of volcanic ash in Tsukidate Town of Miyagi Prefecture (Fig. 36).

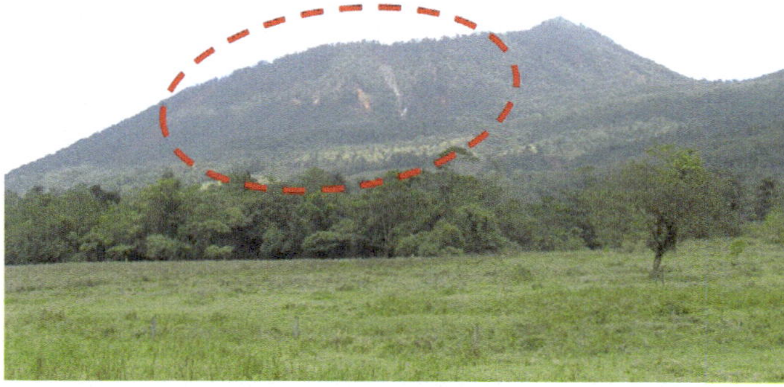

Fig. 34 Seismic landslides in Volcan Cerro Congo, Costa Rica

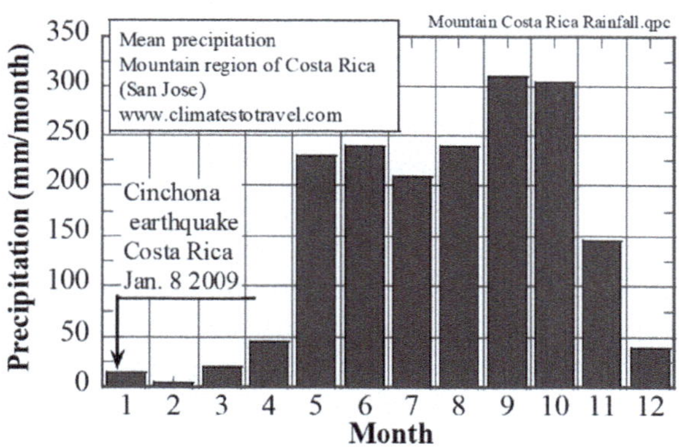

Fig. 35 Precipitation data for San José, Costa Rica (data retrieved from www.climatestotravel.com)

From an interview with a local resident who was an eyewitness to this flow failure, I supposed that the flow velocity at the foot of the slope was several m/s. This flow further moved into a rice paddy where friction with the flowing mass was small and its smooth movement was possible (Fig. 37). This site used to be a small valley and was filled with local materials around 1970. It seems that ground water used to flow into the valley before 1970 and seeped into the fill afterward, affecting its stability. In their report on this flow failure, Konagai and Kazama (2003) stated that the specific gravity of the volcanic material was small (Gs = 2.31–2.48) and that the fines content (Fc = 30%) was porous and held high water content under normal condition. This confined water was released upon cyclic loading, reducing shear strength, and this softening was further promoted by the crushing of fragile porous volcanic ash grains. Kazama et al. (2006) studied the cyclic behavior of volcanic ash and showed that water is held in ash grains and is released during cyclic shear deformation. Constant-volume triaxial compression tests on undisturbed soil specimens, collected from this landslide site, exhibited significant softening and, hence, unstable behavior prone to landslide in spite of their low degree of saturation (Ishihara and Ishizawa 2018). The importance of volcanic material was demonstrated by the 2016 Kumamoto earthquake again as well be described later.

Fig. 36 Flow failure of volcanic ash fill in Tsukidate, Japan, in 2003

Fig. 37 Sliding of a soil mass and bamboo trees on a rice paddy in Tsukidate, 2003 (thickness of sliding soil mass = 58 cm)

Fig. 38 Head scarp of Aratozawa coseismic landslide in Japan

The 2008 Iwate-Miyagi Inland earthquake of Mw = 6.9 affected mountainous areas covered by volcanic materials and triggered many landslides. The most remarkable event during this earthquake was the gigantic landslide in the Aratozawa area (Fig. 38) that is situated in the water shed of the dam of the same name. The volume of the failed soil mass was 67 million cubic meters, while the size of the landslide was 1300 m in length, 900 m in width and 150 m in thickness (Miyagi et al. 2011). The local geology consisted of tuff and pumice of volcanic origin together with sedimentary rocks. The landslide mass was broken into blocks (Fig. 39) and translated laterally along layer of weak volcanic materials (Gratchev and Towhata 2010) at the base. Fujisawa (2008) presented a cross section of this landslide from which I interpreted the H/L ratio (Scheidegger 1973; Hsü 1975; see Fig. 40) to be $0.175 = \tan(10°)$. In contrast, the inclination of the slip plane at the bottom of the landslide mass was 0°–2° and the runout distance was 300 m (Ohno et al. 2010). A small part of the landslide mass flowed into the Aratozawa dam reservoir and produced a small tsunami. This event is one of the cases that verify the risk of landslide in areas with volcanic geology.

2.9 Padang Earthquake of Mw = 7.5 in 2009 in Sumatra, Indonesia

Sumatra, Indonesia, has been affected by many big earthquakes since the gigantic event of Mw = 9.3 in 2004. In spite of its huge energy, this event in 2004 occurred in the ocean bed and did not trigger significant landslides on Sumatra Island. In

Fig. 39 Situation inside the failed mass of Aratozawa landslide (Gratchev and Towhata 2010)

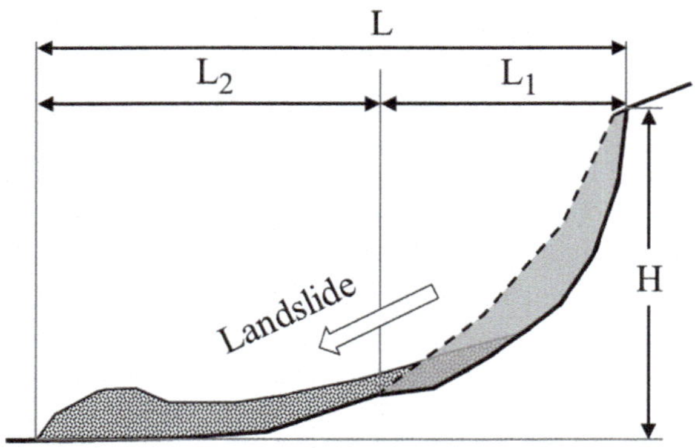

Fig. 40 Parameters on geometry of landslide event (L_1 and L_2 stand for the horizontal lengths of the landslide source and deposit areas, respectively)

contrast, the event in 2009 caused landslides and subsoil liquefaction as well as building damages in and around Padang City (Fig. 3) (Bothara et al. 2010). The affected area is covered widely with volcanic materials such as pumice and scoria that induced long-distance flows (Fig. 41). The earthquake in 2009 took place on September 30th. Because the Padang area has precipitation all year (Fig. 42), the slope stability was substantially affected by water. Similar to the events illustrated in Figs. 36 and 38, the volcanic material in regions with a wet climate contains substantial water and is prone to flow failure (Nakano et al. 2015). When I visited the site on October 31st, 2009, for reconnaissance, a shower started and the mountain slope shown in Fig. 43 fell down. This was an additional event after the coseismic landslide in the foreground of this photo. Figure 42 illustrates the mean distribution of rainfall in year. It is evident that the earthquakes during the rainy season are associated with a high risk of landslides.

Fig. 41 Long-distance flow of scoria induced by an earthquake (at Malakak, Sumatra Barat, Indonesia; 0.3–0.5 million m^3 and H/L angle = 11°)

Fig. 42 Mean rainfall distribution in the year in Padang, Sumatra, Indonesia

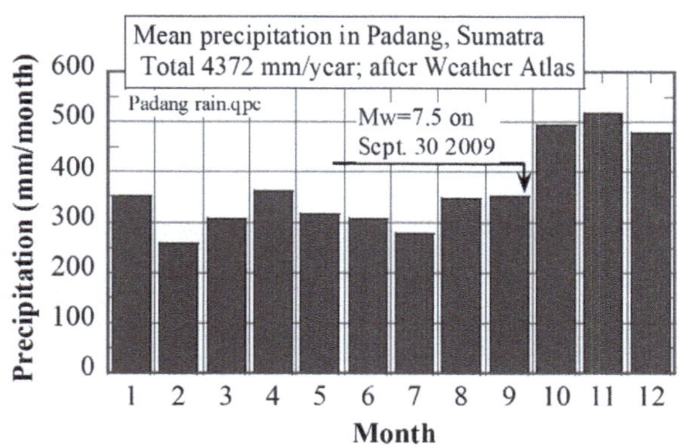

Fig. 43 Post-seismic landslide on a mountain slope during a shower (near Cumanak)

2.10 2011 Christchurch Earthquake of Mw = 6.2 in New Zealand

Christchurch and its suburbs were affected by a sequence of earthquakes in 2010 and 2011. The one in February, 2011, was the most devastating among others and induced building collapse and subsoil liquefaction in the center of Christchurch and its residential areas, where, in the nineteenth century, former swampy areas had been filled with sandy soil. Slope problems also occurred to the south of the city (such as in Lyttleton). Figure 44 illustrates a rock cliff near Lyttleton in which the 2011 earthquake caused a substantial failure of the cliff, and its extent was enlarged during the following years. The lava at the top forms a cap layer that is underlain by softer ash deposits. This is one of the examples of the earthquake-induced long-term instability of slopes which I am interested in. Moreover, rock falls on the slopes occurred and damaged a house (Fig. 45).

2.11 2011 Tohoku Earthquake of Mw = 9.0 in Japan

This earthquake is widely known for its devastating tsunami disaster and subsoil liquefaction. Very strong acceleration records were obtained at many places. For example, the record at Ishinomaki (Fig. 46) is characterized by strong amplitude as well as its long duration time because this time history suggests two huge fault ruptures. Thus, many number of cycles occurred in the soil and efficiently induced subsoil liquefaction. In contrasts, the number of sizeable landslides was small as reported by Nakamura et al. (2014). Figure 47 illustrates an example of a landslide at Hanoki-daira, which was made of volcanic materials that softened under cyclic loading (Wakai et al. 2016) with a total landslide volume of 30 thousand m^3 (Chigira

Fig. 44 Seismically destabilized rock cliff near Lyttleton, New Zealand

Fig. 45 Site of andesite rock falls, Port Hills, in a southern suburb of Christchurch, New Zealand

et al. 2012). I observed $H/L = \tan(5°)$. Figure 48 illustrates successful slope protection in Shiroishi City, where the slope failed drastically during the 1978 Miyagi-ken-oki earthquake of Mw = 7.5 (landslide volume = 60,000–80,000 m^3; Kobayashi 1980). After the disaster, drainage pipes and wells were installed. In 2011, no distortion occurred in this slope in contrast to nearby unreinforced slopes where many cracks opened and downslope movement occurred.

The reason for the small number of coseismic landslides in spite of the huge earthquake energy released is probably because the earthquake happened on March

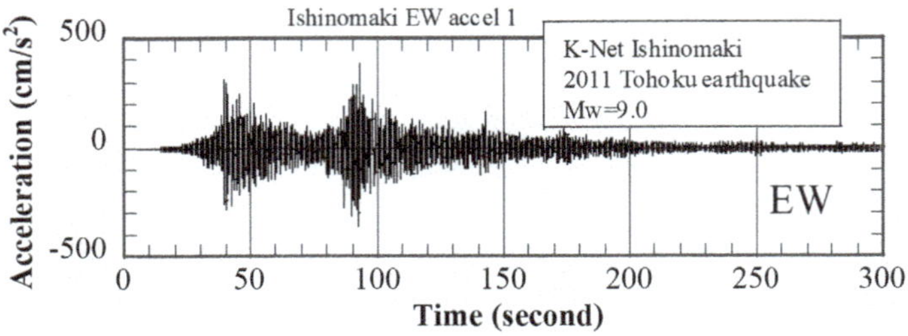

Fig. 46 Acceleration record at K-Net Ishinomaki Station during the 2011 Tohoku earthquake

Fig. 47 Coseismic landslide at Hanoki-daira in Shirakawa, Japan, consisting of volcanic materials (friction angle = 5° as inferred from H/L)

Fig. 48 Intact state of slope in Kotobuki-yama that had been installed with drainage facilities after its seismic failure in 1978

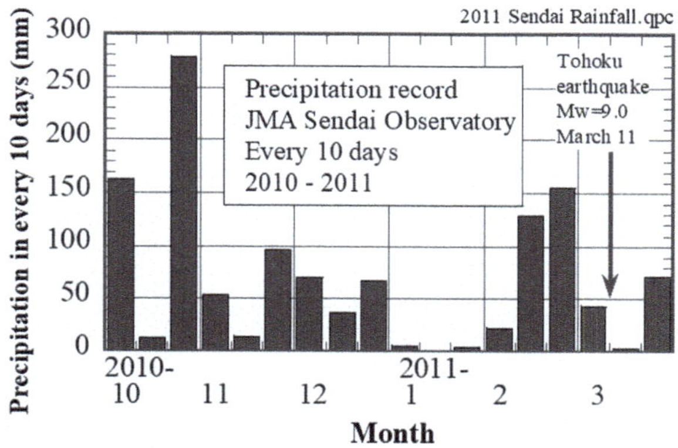

Fig. 49 Annual rainfall distribution in Sendai, Japan (drawn after JMA)

11th, within a dry winter (Fig. 49) (Nakamura et al. 2014). The climatic conditions were, thus, similar to what happened in Kobe in 1995.

Another point of concern was the poor seismic performance of residential land (Wakai et al. 2012). This problem came either from landslides in hilly areas or from subsoil liquefaction in young reclaimed land (Towhata et al. 2016). Figure 50 illustrates one of the examples of the first type; fill that was a part of a residential development collapsed, while, in contrast, a cut part within 200 m of this site was intact (Fig. 51). The problem is that neither land developers nor clients paid enough attention to seismic safety of the residential land. To cope with this problem of people who have no geotechnical background, the Japanese Geotechnical Society, in collaboration with several other institutions, established a new engineering qualification of "Qualified evaluator of residential land" whose mission is to provide people with necessary advice on disaster proneness of residential land (Towhata and Nakamura 2015).

2.12 *Gorkha Earthquake of Mw = 7.9 in 2015 in Nepal*

This earthquake triggered substantial damage both in the Kathmandu Basin by strong shaking and in mountain regions by landslides (Fig. 3). Valleys were blocked by landslide masses (Collins and Jibson 2015). Figure 52 illustrates a site of a gigantic landslide in Langtang Valley (Suwa 2015; Lacroix 2016) that claimed more than 500 lives. Many landslides began above glacier areas at high altitude (Lacroix 2016). The Langtang landslide mass fell down from a cliff (Fig. 52c) that is as high as 1000 m (Jones et al. 2020). It was fortunate that most natural dams released water without catastrophic breaching. The epicenter of this earthquake was situated in Barpak, Central Nepal, while great landslides occurred to the east in Sindhupalchok District, presumably because of the directivity that was induced by rupture propagation from west to east and the seismic wave propagation that was intensified to the east (Koketsu et al. 2016).

Fig. 50 Seismic failure of residential fill in a hilly suburb of Sendai City, Japan

Fig. 51 Intact shape of residential land near the failed site in Fig. 50

It was interesting to me that many human habitations in the mountainous regions avoided the damage from nearby landslides (Fig. 53), probably because people knew which place was safe to live. A noteworthy disaster was caused by snow avalanches that were triggered by this earthquake (Wilkinson 2015). It is shown in the literature that many snow avalanches have been triggered by earthquakes (Podolskiy et al.

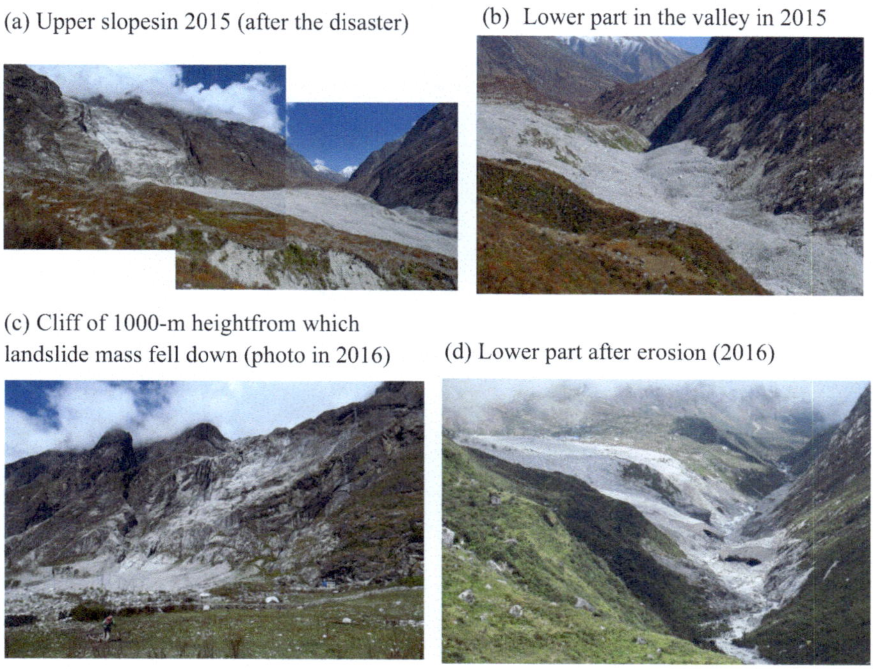

Fig. 52 Gigantic landslide in Langtang Valley, Nepal (Courtesy of Mr. S. Suwa 2015)

2010) during shaking or in minutes to hours after the earthquakes (Puzrin et al. 2019). The extent of damage was probably aggravated by snow (Fujita et al. 2017). The disaster in Langtang is reminiscent of what happened in Yungay, Peru, in 1970 (Fig. 115). Because of global warming, disasters related to snow and ice avalanches are becoming more likely.

2.13 *Kumamoto Earthquake Sequence of Mw = 6.2 and 7.0 in 2016 in Japan*

This earthquake sequence was associated with two major events on April 14th and 16th, 2016, both of which registered on the JMA (Japanese Meteorological Agency) seismic intensity scale of 7 in the epicentral Mashiki Town in Kumamoto Prefecture, Japan (Fig. 1). This scale of 7 is the highest and made these two events equivalent to the 1995 Kobe earthquake. After them, many aftershocks occurred for months. After the Kobe earthquake, the seismic resistant design principles adopted two-level design earthquakes, and structures have been required to withstand the level 2 strong and rare events with some plastic deformation but to avoid fatal collapse. The two events in Kumamoto registered an intensity scale of 7 and were certainly

Fig. 53 Farm land and human habitation on a steep slope that avoided damage from a landslide (near Barpak, Nepal)

real level 2 earthquakes. Many structures suffered plastic deformation to a certain extent during the first quake on April 14 but avoided collapse. Unfortunately, the plastic deformation during this first event reduced the strength of structures and they could not withstand the second stronger event 28 h later. Consequently, 50 people died under the debris of collapsed structures and slopes. Thus, the principle of two-level design earthquakes was found insufficient but experts could not propose a new practical design principle to mitigate against repeated level 2 earthquakes.

The Kumamoto earthquake sequence demonstrated again the seismic instability of volcanic soils. For example, Fig. 54 illustrates a big landslide in Aso volcanic caldera. This landslide measured 540 thousand m^3 in volume (Ishikawa et al. 2016a) and destroyed a railway and highways as well as a big bridge at the bottom. The cross section of this landslide (Ishikawa et al. 2016a) suggests $H/L = 0.385$ which is equivalent to a friction angle of 21°. Figure 55 shows a flow failure of a slope that was made of volcanic ash (Fukuoka and Sakai 2016). Figure 56 demonstrates the damage in a residential area in Minami-Aso Village where the entire fill moved downwards and the ground surface distorted (Yokota and Tori 2019). Figure 57 shows a surface depression at one hundred meters behind the slope shoulder. The slope distortion exhibits a clear contrast with the intact houses, which had reasonable earthquake resistance. This is one example of the evidence that recent seismic design is biased toward the safety of superstructures, and that the safety of foundation is not well considered. As of December, 2020, the reconstruction of damaged residential lands is not progressing well. Public assistance is unfortunately difficult to obtain for personal real estate. Furthermore, two months after the earthquake, the affected area in Kumamoto suffered from rainfall disasters and the seismically damaged slopes and sediments suffered from secondary disasters from April to July (Ishikawa et al.

2016b). This is a typical example of compound disaster induced by the combined effects of an earthquake and a heavy rainfall.

The geology in Kumamoto is characterized by the prevalence of volcanic materials. Tanimoto et al. (2018) collected soil samples from affected slopes and took X-ray photographs. Those photographs exhibited halloysite content in the specimens and it was concluded that water was released from halloysite during shaking and triggered slope instability. Midorikawa et al. (2020) carried out cyclic undrained shear tests on partially saturated volcanic ash specimens and demonstrated that the excess pore water pressure can reach 90% and induces softening (shear strain = 10%), even though the degree of saturation was only 64.8%. Thus, a volcanic ash slope is prone to coseismic instability. This situation becomes serious if ample ground water is available in slopes.

There is a certain resemblance between landslide disasters in Kumamoto and El Salvador, where an earthquake of Mw = 7.1 triggered many landslides (Bommer and Rodríguez 2002). The most well-known disaster happened at Las Colinas where the landslide mass measured 130 thousand m^3 in volume (Jibson et al. 2004) and the slope was made of volcanic deposits such as ash, tephra, and pumice with moisture. This landslide flowed a long distance and its H/L was 175/750 = 0.23, as determined from a field survey by Konagai et al. (2002).

The H/L ratio of volcanic materials may be smaller than that of non-volcanic materials. Ui (1983) defined the volcanic dry avalanche deposit as "a deposit formed as a result of large-scale sector collapse of a volcanic cone, associated with some form of volcanic activity." He investigated the topography of such avalanche deposits of both volcanic and non-volcanic origins and indicated that H/L = around 0.1 for the volcanic materials, while H/L = 0.2–0.62 for non-volcanic materials; thus, the volcanic avalanches with smaller H/L can travel more easily, associated with longer runout distances. The big volume of the collapsed volcanic cones, which was studied

Fig. 54 Coseismic landslide on the inner side of the caldera ring of Aso Volcano

Fig. 55 Coseismic flow failure of a slope in Aso comprising volcanic ash

Fig. 56 Slope distortion that made an intact house valueless during the Kumamoto earthquake

by him, possibly resulted in small values of H/L, whether volcanic or non-volcanic, due to the effects of landslide volume on this ratio (Fig. 114).

After two strong shakes in April, 2016, in Kumamoto, many aftershocks continued (Fig. 58). This slowed down the efforts toward recovery. Figure 59 shows the rainfall record in Kumamoto City. The earthquake occurred in the rainy season. It seems that the high water content in the subsoil aggravated the slope disasters. Moreover, in June of the same year, the Kumamoto and surrounding area had disastrous heavy rains. Many landslides occurred and suggested the deleterious effects of the earthquake shaking on slope stability (Inagaki 2017). This is one of the examples of a compound

Fig. 57 Surface depression in moving residential fill (the slope shoulder is on the far side)

disaster in which two different types of natural disaster occur at the same time or in sequence.

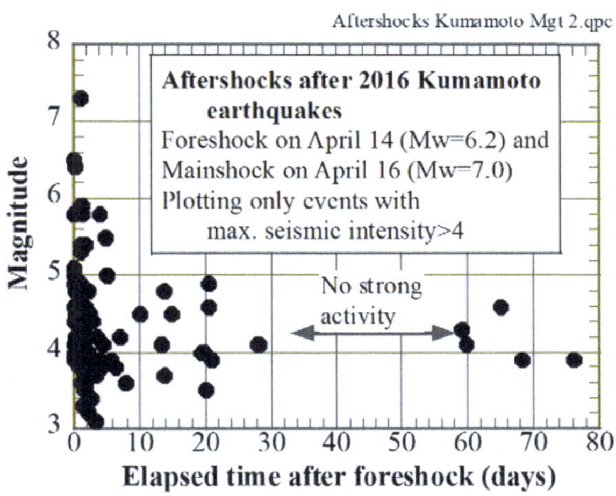

Fig. 58 Prolonged period of aftershocks in Kumamoto Region

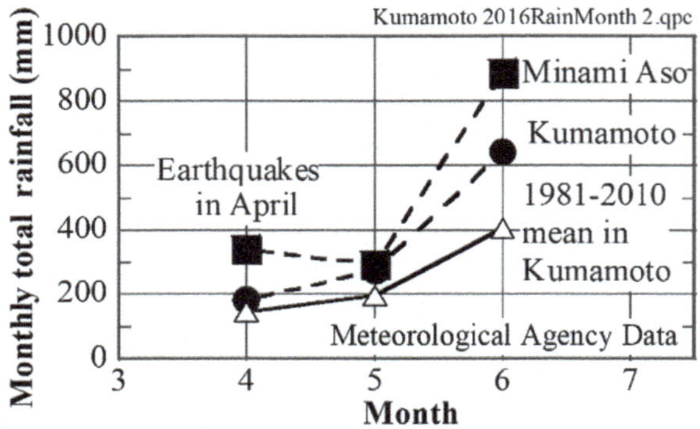

Fig. 59 Precipitation record in Kumamoto and Minami-Aso (JMA data)

2.14 *Kaikoura Earthquake of Mw = 7.8 in 2016 in South Island of New Zealand*

This earthquake induced many landslides in the mountain areas and coastal cliffs in the Kaikoura Region (Massey et al. 2018). Landslides were located along the fault rupture and more than 200 landslide dams were formed during this earthquake. Figure 60 exhibits one landslide that occurred in a limestone mountain in the coastal area; refer to Dellow et al. (2017). Some of those dams breached afterward.

Fig. 60 Landslide on a limestone mountain ridge in the coastal area (2016 Kaikoura earthquake in New Zealand)

Fig. 61 Seismic rock fall that hit a house at the slope bottom (near Oaxtepec, Mexico)

2.15 Earthquakes in 2017 in Mexico

Mexico was affected by two earthquakes in September, 2017. The first one occurred on 8th of September in Chiapas and Oaxaca on the Pacific Coast. Its magnitude was Mw = 8.2 and it destroyed many buildings. Although several landslides happened, they were not very serious and the resulting damage was limited to buildings (Hakhamaneshi et al. year not shown; Pozos-Estrada et al. 2019).

The second earthquake occurred on September 19th in the Puebla Province (Fig. 3) near Mexico City and its magnitude was Mw = 7.1. Although the magnitude was smaller than that of the preceding earthquake, it affected the mega-city of Mexico City. In the city, building damage was concentrated in the hill-basin boundary, similar to what happened during the 1985 catastrophic earthquake (Alberto et al. 2018). In contrast, the extent of induced slope disasters (Fig. 61) and other types of geotechnical damage (Fig. 62) were not serious (Alberto et al. 2018), which is consistent with Hakhamaneshi (year not shown) and Montgomery et al. (2020). The ground distortion originated due to consolidation settlement induced by excessive pumping of ground water (Auvinet et al. 2013, 2017) was enlarged by the earthquake.

2.16 Eastern Iburi Earthquake of Mw = 6.6 in 2018 in Japan

The hill slopes in the Iburi region of Hokkaido Island (Fig. 1) are widely covered by volcanic ash materials. As has been noted earlier, volcanic ash deposits with high water content are unstable during earthquakes and the same problem occurred during this earthquake at a greater scale. After the earthquake, a comprehensive study was made of slope instability by related academic societies and a report was published

Fig. 62 Aggravated crack and subsidence that had been induced by excessive ground water lowering (south of Mexico City)

by a joint "publication committee (2020)" under the title of "Landslides induced by the 2018 Hokkaido eastern Iburi earthquake."

Figure 63 is an aerial photo of the affected slopes. In total, 6117 landslides happened. One failed soil mass blocked a valley and formed a natural dam. While the earthquake occurred on Sept. 6th, the antecedent rainfall in August was ample and made the volcanic ash deposit wet and prone to slope instability (Hosoya et al. 2021). The total rainfall between August 1 and September 5 was 231 mm in 2018, which is greater than the mean annual rainfall of 197 mm between 1981 and 2010. Figure 64 shows a closer view of a failed slope. Many people lived near the foot of the slopes, reserving the open space in the valley center for rice agriculture. Consequently, their houses were easily destroyed by the landslide mass. Figure 65 shows that the failed soil mass traveled over a substantial distance because of the wet and soft soil in the rice field. Figure 66 illustrates a typical cross section of a volcanic ash deposit in which the bottom ash deposit exhibits lower material strength than the overlying materials (Hosoya et al. 2021) due possibly to chemical weathering (Chigira 2015). Afterward, Hosoya et al. (2021) studied the geometry of the landslides using geometric parameters, L and L_1 (see Fig. 40). The results are illustrated in Fig. 67, where the flow distance is very long in comparison with the size of the source.

Another important problem was the substantial damage to residential land in Sapporo City, which is 50 km from the epicenter (Fig. 1). This area was constructed in the 1960s and 1970s by filling valleys with local volcanic ash materials. Seismic subsoil damage has been repeated in its vicinity during past earthquakes in 1968 (Tokachi-oki earthquake of Mw = 8.2) (Kitago and Toki 1968), 2003 (Tokachi-oki earthquake of Mw = 8.0; Japanese Geotechnical Society—Reconnaissance Team 2003) and 2010 (magnitude = 4.6; Tajika et al. 2012). The disaster in 2018 (Yamashita

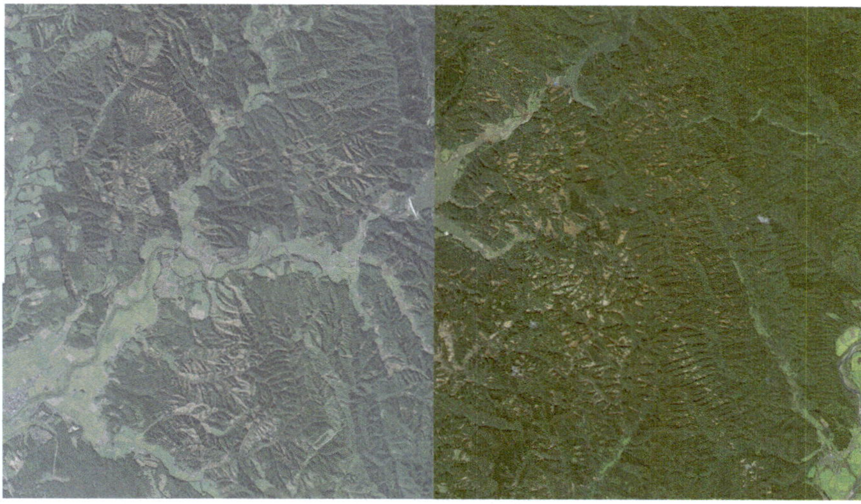

Fig. 63 Numerous failures in slopes covered by volcanic ash deposits (2018 Eastern Iburi earthquake, Hokkaido, Japan) (Combination of four aerial photographs provided by the Cabinet Satellite Intelligence Center in 2018; https://www.cas.go.jp/jp/houdou/180912csice.html)

Fig. 64 Close-up view of one of the failed slopes in Atsuma Town, Iburi District

et al. 2019) was the most significant. Figure 68 shows substantial depression of a house foundation that suggests liquefaction of the volcanic materials. This area was rapidly reconstructed using governmental funding. Note that there is a different possibility that the damage was aggravated by deterioration and collapse of embedded drainage pipes (Yasuda 2020).

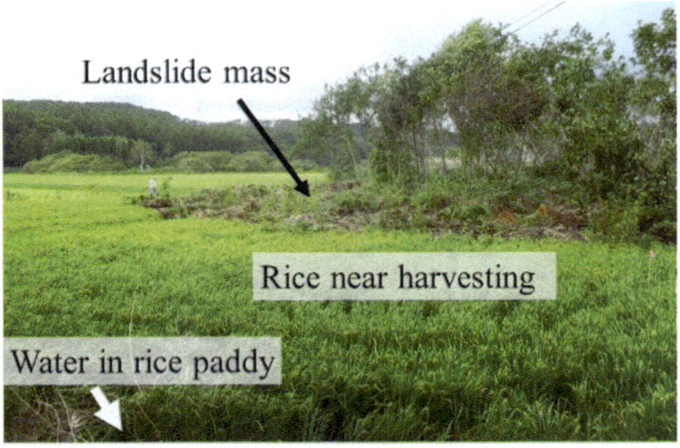

Fig. 65 Sliding of a failed soil mass on the wet and soft surface of a rice field

Fig. 66 Cross section of the volcanic ash layers in the failed slope in Atsuma Town (Hosoya et al. 2021)

2.17 Palu Earthquake of Mw = 7.5 in 2018 in Sulawesi, Indonesia

Palu City (Fig. 3) is located in the central part of Sulawesi Island and is situated on the very active Palu-Koro Fault. The damage caused by the earthquake on September 28th of 2018 was characterized by two phenomena; a tsunami in the coastal area and significant flow failure of inland gentle slopes (which was named Nalodo afterward). Tsunami is widely recognized as a consequence of significant bathymetric change due to coseismic tectonic movement. In contrast, the tsunami in Palu Bay arrived at Palu City very shortly (within tens of seconds) after shaking, in spite of the area being at 80 km from the epicenter and is considered to have originated "mainly" from nearby submarine landslides in the Palu Bay (e.g., Carvajal et al. 2019; Takagi

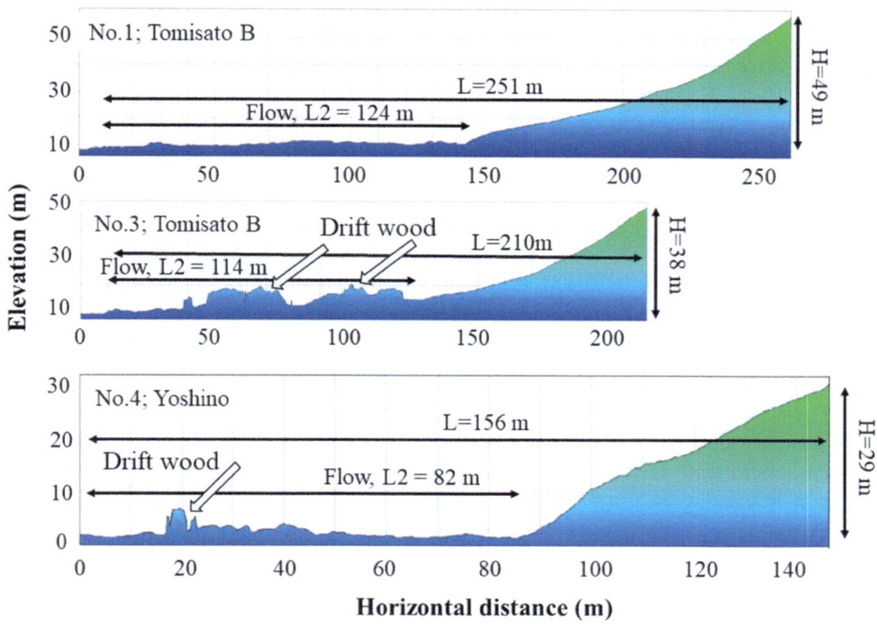

Fig. 67 Geometry of the landslide mass that flowed over the rice paddy such as shown in Fig. 65 (Hosoya et al. 2021)

Fig. 68 Depression in filled ground in Satozuka District, Kiyota Ward, Sapporo, Japan

et al. 2019a, b). The idea of the submarine landslide is further supported by the fact that the causative dislocation of the Palu-Koro Fault was a horizontal strike slip that had no vertical displacement. This situation is similar to the tsunamigenic Mindro earthquake of Mw = 7.1 in 1994 in the Philippines. After the Mindro event, Imamura et al. (1995) pointed out the need to think about additional seabed movement such as that of nearby islands. Tanioka and Satake (1996) calculated the vertical movement of seabed associating strike-slip fault rupture and discussed that the movement is too small to reproduce the observed tsunami in Mindro. In contrast, Elbanna et al. (2021) calculated the vertical displacement of seabed triggered by supershear rupture of a strike-slip fault in which the rupture propagates faster than S-wave. Because the 2018 rupture of the Palu-Koro Fault was associated with supershear (Fang et al. 2019; Natawidjaja et al. 2021), Elbanna et al. (2021) inferred that the tsunami in Palu Bay was produced by the strike-slip fault movement.

Nalodo (Irsyam et al. 2018; Okamura et al. 2019; Kiyota et al. 2020; Hazarika et al. 2021) is a local Kaili language term that means a depression or falling (Kamus Kaili-Ledo Indonesia Inggris 2003). It is a type of liquefaction disaster but involves a long-distance flow over 500 hundred meters in gentle slopes that had not previously been experienced by other liquefaction disasters (Fig. 69). It was a mysterious phenomenon of ground flow in gentle slopes that is possibly related to subsoil liquefaction but cannot be fully accounted for by liquefaction alone. Approximately, 1500 people were killed by Nalodo, and this number of victims overwhelmed the toll of seismic liquefaction in the past that had hardly claimed a human life. Details of Nalodo will be discussed later.

Fig. 69 Aerial view of one of the Nalodo sites (residential area of Balaroa to the west of Palu City) (photograph by study team of the Japan International Cooperation Agency abbreviated JICA)

3 Effects of Distance on the Distribution of Earthquake-Induced Landslides

The following parts of this report addresses a variety of natural and damage mechanisms of earthquake-induced landslides. At the beginning, it may be noted that there is no landslide that is 100% induced by earthquakes. Every landslide during or immediately after an earthquake is a consequence of the combined effects of geological and/or geomorphological weakness (basic factor), together with seismic action (cause). In this regard, Crozier (1992) presented the following characteristics of earthquake-induced landslides;

1. Association of strong seismic action,
2. Coincidence of the landslide distribution with an active fault, including the presence of some of the largest landslides on the fault trace,
3. Stability analysis exhibiting the need for some extra force to trigger a landslide,
4. Large size of all the landslides within the same-age population,
5. Presence of liquefied sand as a matrix within the displaced mass, and
6. Distribution of landslides appearing not to be related to geological or geomorphic conditions.

With the above remarks in mind, this section addresses the effects of distance from the epicenter or fault plane to mountain slopes. It seems that research interest in the twentieth century lay in the distance of coseismic landslides from the source of earthquakes. Keefer (1984) interpreted 40 worldwide cases of landslides and presented the maximum distance of landslides from the epicenter or fault rupture zones. Certainly, this distance increases with the seismic magnitude or energy. It was noted then that the influence of geological and soil mechanic conditions was very important. In line with this, Yasuda (1993) indicated that the maximum epicentral distance of landslides increases with the earthquake magnitude. Further, an Italian database uses the epicentral distance (Prestininzi and Romeo 2000; Martino et al. 2014). Bommer and Rodríguez (2002) interpreted case histories in Central America and obtained a similar relationship between the magnitude and the size of affected area. Rodríguez and Poran (1993) interpreted landslides in Colombia and presented their epicentral distances or areas of seismic effects. Furthermore, Keefer (1984) classified the earthquake-induced landslides into many types in accordance with the material types and landslide configurations. This study was further developed by adding additional case histories (Keefer 2002). There, remarks were made of the effects of intensity of shaking, landslide concentration, and related events during the 1995 Kobe earthquake located along the fault. Another important point was made of the delayed landslide that occurred five days after the 1959 Hebgen Lake ($Mw = 7.2$) and after the 1983 Borah Peak ($Mw = 6.9$) earthquakes.

The likelihood of earthquake-induced landslides depends on the intensity of shaking on a site. Keefer and Wilson (1989) attempted to use Arias Intensity of earthquake motion together with attenuation of motion in order to reproduce the size of the affected area. Harp and Wilson (1995) proposed an empirical formula to assess the threshold Arias Intensity using the distance from the earthquake source, magnitude and a probability parameter. Noro et al. (2011) summarized landslide information after the 2004 Niigata-Chuetsu earthquake (Mw = 6.6), the 2007 Noto Hanto earthquake (Mw = 6.7), the 2007 Niigata-Chuetsu-oki earthquake (Mw = 6.6) and the 2008 Iwate-Miyagi Inland earthquake of Mw = 6.9, all of which occurred in Japan. Their finding was that 87% of landslides occurred on the hanging-wall side of faults and that 93.3% occurred in the areas of acceleration (3-dimensional acceleration) exceeding 500 cm/s^2. Figure 70 illustrates one of the major reasons accounting for the higher vulnerability on the hanging-wall side of a fault. Due to the net short distance, the hanging-wall side is susceptible to stronger shaking (Abrahamson and Somerville 1996). Tiwari and Beena (2017) collected more information and showed that the Minimum *PGA* (Peak Ground Acceleration) to trigger landslide is 0.18G (G = 980 cm/s^2) and that the landslide likelihood profoundly increases above 200 cm/s^2. Wang et al. (2010a, b) examined 3000 landslides induced by the 2008 Wenchuan earthquake and suggested that PGA = 0.2G is the threshold to induce serious landslide disasters, while 0.05–0.07G is the minimum to trigger landslides. Note that acceleration in this and the following chapters are expressed either by cm/s^2 or G (1G = 980 cm/s^2), depending on the practice in the cited literature.

Khazai and Sitar (2003) interpreted a landslide database for the 1999 Chi-Chi earthquake in Taiwan and stated that 74% of all slope instabilities occurred in the region with vertical *PGA* greater than 0.2G, and that 81% was with horizontal *PGA* exceeding 0.15G. They also indicated that most of the landslides occurred to the east of the causative fault that means the hanging-wall side. Noteworthy to me is that as many as 22% of the failure was toppling, suggesting that the steep slope and weak rock mass of the Taiwan Mountains resulted in vulnerability to falls and toppling under strong seismic shaking. Another important lesson from Taiwan is that the area of slope instability was large and its outer edge was far from the fault rupture probably due to the weak geology, high precipitation, and erosion-prone steep slopes in the mountainous region of Taiwan. With these recent findings from earthquake case histories, it seems reasonable to state that shear failure is not necessarily the case

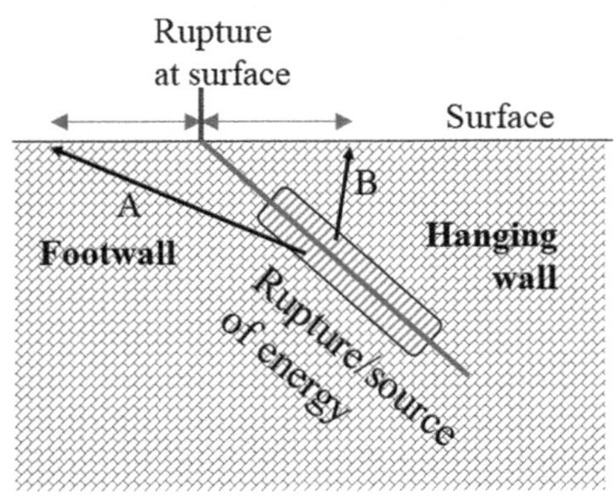

Fig. 70 Difference from the source of earthquake energy to the surface on the sides of hanging and foot wall; distance A > distance B, even though both are at the same distance from the surface rupture

in coseismic landslides and that *PGA* of 150–200 cm/s^2 is the threshold to trigger slope instability. Strictly speaking, this threshold *PGA* depends on the local material properties, topographic amplification, and slope angle and should be considered to be an approximate value.

Those attempts to define the range of landslide likelihood are useful in the preliminary stage of coseismic landslide risk assessment. The recent situation, however, demands technology to conduct a more detailed risk assessment by considering the local intensity of motion affected by topography, geological/soil mechanic conditions of the slope, the weathering of surface materials and many other factors. The experiences during the 2005 Kashmir earthquake and the 2008 Wenchuan earthquake clearly demonstrated that landslides are concentrated along the fault rupture and mostly on its hanging-wall side. Figure 71 for the Wenchuan earthquake depicts that the density of landslides, both in number and in area, is greater on the hanging-wall side of the fault rupture. The 1989 Loma Prieta earthquake (Mw = 6.9) triggered landslides mostly within 7 km of the surface projection or within 9 km of the updip edge of the fault plane, which was 34 km long (Keefer 2000). Therefore, the use of the maximum epicentral distance to landslide sites can be replaced by the distance to the fault rupture. With these current situations in mind, the following sections will discuss more details of the real landslide phenomena.

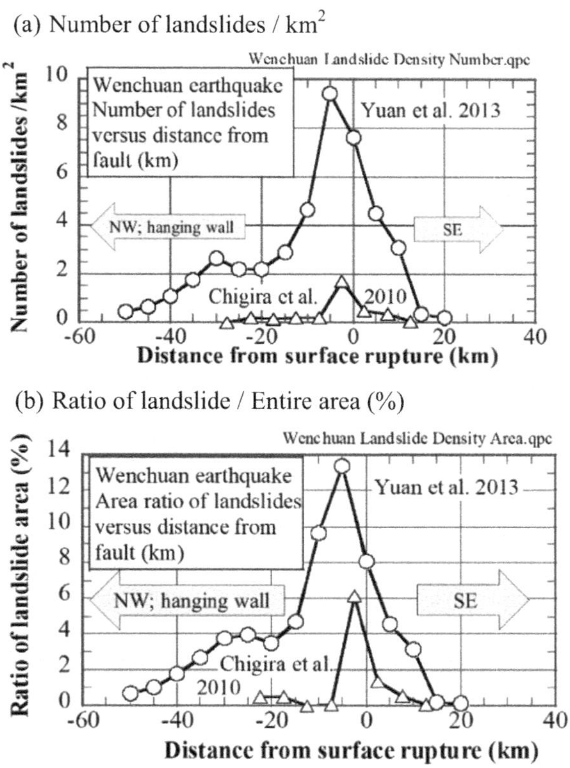

Fig. 71 Landslide density versus distance from the surface manifestation of the fault rupture (Wenchuan earthquake) (drawn after Chigira et al. 2010; Yuan et al. 2013)

4 Effects of Topography and Properties of Local Geomaterials

The topographic amplification of earthquake motion is well known for the intense acceleration records at the top of a small hill in Tarzana, California, during the 1994 Northridge earthquake of Mw = 6.7 (Huang et al. 1994). Afterward, Spudich et al. (1996) installed sensors to monitor ground motion during smaller earthquakes and clearly demonstrated this amplification. This chapter attempts to further study this issue.

Figures 22 and 23 compared the extent of building damage in Balakot. Pakistan, between the hill top and the foot. Because the hill top was subject to more serious damage than the foot, it was suggested that the earthquake motion was intensified by the hill topography. The 2008 Wenchuan earthquake provided good evidence of the topographic amplification of ground motion at a mountain top (Fig. 32). Meunier et al. (2008) investigated landslides caused by the 1994 Northridge (Mw = 6.7), the 1999 Chi-Chi (Mw = 7.7) and the Finisterre Mountains earthquakes in 1993 with Mw = 6.7 and 6.9 in Papua New Guinea. The studied coseismic landslides revealed concentration toward the mountain crests. This in particular was the case with bigger landslides. Meunier et al. (2008) also noted that topographic amplification is not symmetric across the mountain crests because the seismic wave propagates in an oblique direction from the hypocenter, and the slope away from the epicenter is subject to more significant amplification. The other reason is that the rate of weathering and deterioration of rock is affected by temperature change (sunlight) and underground stratification.

Kurita et al. (2005) verified the topographic amplification by comparing the observed intensity of ground motion at the top and foot of a mountain. Figure 72b illustrates the ratio of the amplitudes of two motions as a function of frequency. It is shown that the ratio or amplification increases with some fluctuation as the frequency gets higher. Field observations have been carried out by many other researchers as well (Davis and West 1973; Çelebi 1987; Pedersen et al. 1994; Bouchon and Barker 1996; Spudich et al. 1996; Le Brun et al. 1999; Fujinami et al. 2017). However, they did not necessarily focus on amplification of acceleration that is important from the design viewpoint. It seems that those studies suggested that approximately five times amplification is possible between the crest and the foot. In contrast, Nishimura and Mori (1983) reported a ten times greater velocity spectrum value at the crest. Conversely, Paolucci (2002) showed the amplification to be just 1.5–2 times. Thus, the range of the recorded topographic effect is quite variable. However, all the researchers have indicated that the extent of topographic amplification is a function of frequency. Although scientifically reasonable, this frequency dependency may not be convenient to practitioners who do not want to run advanced dynamic numerical analyses but still want a typical value of amplification.

Aki (1988) referred to a simple formula by Sanchez-Sesma (1985) who actually developed a differential equation of seismic wave propagation in a hill topography and solved it by referring to a mathematical solution of electric field by Macdonald

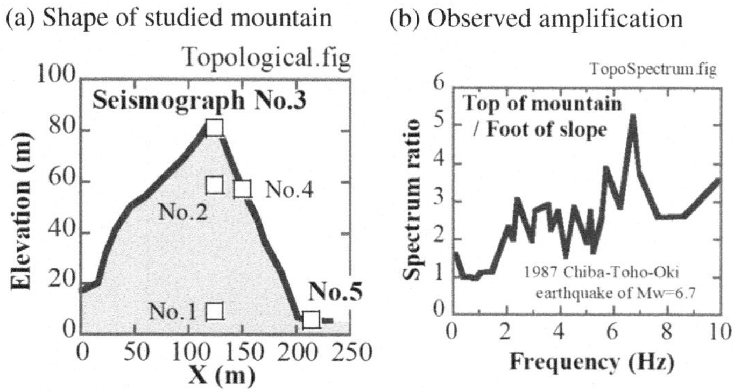

Fig. 72 Topographic amplification of earthquake motion observed between top and foot of a mountain (No. 3/No. 5) in Yokosuka, Japan, undergoing the 1987 Chibaken Toho-oki earthquake of Mw = 6.7 (drawn after Kurita et al. 2005)

(1902). Honestly, the author could not find the particular solution in Macdonald's book. In this simple solution, the apex angle of hill topography is expressed as $\nu\pi$ (Fig. 73) and Sanchez-Sesma (1985) as well as Aki (1988), who was interested in the ratio of the hill top motion over the incident wave, gave the amplification factor as $2/\nu$. Because the practitioners are rather interested in the ratio of the hill top motion over the free-field motion upon rock outcrop, Ohmachi and Midorikawa (1999) modified this formula to $1/\nu$. The wave length is assumed shorter than several times of the mountain width (Fig. 73). In the case of Fig. 72, S-wave velocity in the main body of the hill is 700 m/s, and the wave length of the component of 5 and 10 Hz is 140 and 70 m, respectively. They are obviously shorter than "several time of the mountain width (200 m)" and this simple assessment of amplification is feasible. Figure 74 compares this simple assessment and the observed value in the case of Fig. 72. Herein, the simple assessment shown by a straight line underestimates the observation. All in all, it seems that the amplification factor of 5 (rule of thumb) is not too bad, although care must be taken and a further study is desired on this topic.

Figure 75 plots the analytical (elasticity) solution of topographic amplification by which the seismic energy is focused at the mountain top as implied by Fig. 32

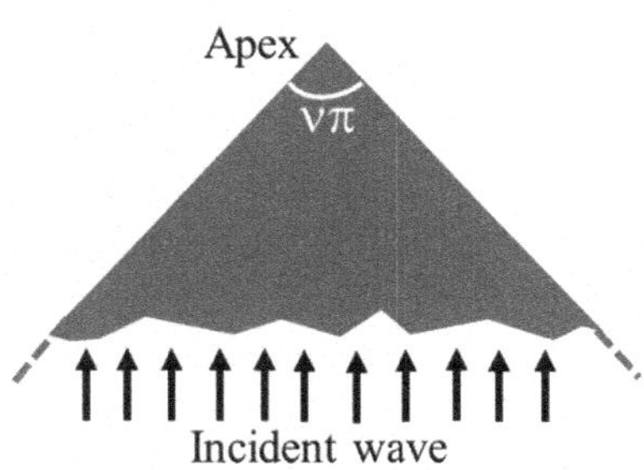

Fig. 73 Triangular idealization of hilly topography

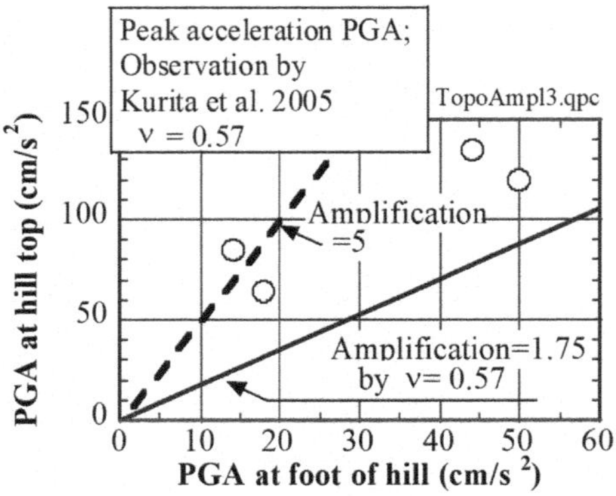

Fig. 74 Comparison of simple assessment of topographic amplification and the measured PGAs by Kurita et al. (2005)

$$\frac{\text{Top}}{\text{Free field}} = \frac{2V_s / \pi \omega z_o}{\sqrt{\left[J_0\left\{\frac{\omega(z_o+H)}{V_s}\right\}Y_1\left(\frac{\omega z_o}{V_s}\right) - J_1\left(\frac{\omega z_o}{V_s}\right)Y_0\left\{\frac{\omega(z_o+H)}{V_s}\right\}\right]^2 + \left[J_1\left\{\frac{\omega(z_o+H)}{V_s}\right\}Y_1\left(\frac{\omega z_o}{V_s}\right) - J_1\left(\frac{\omega z_o}{V_s}\right)Y_1\left\{\frac{\omega(z_o+H)}{V_s}\right\}\right]^2}} \quad (1)$$

where "J" and "Y" designate Bessel functions of the first and second kinds, while V_s stands for the S-wave velocity in the mountain body and the underlying medium (Fig. 75a), H is the height of mountain, z_o is the shape of the mountain (triangular or trapezoidal), f is the frequency (cycles/second), and $\omega_o = 2\pi f$. For derivation of this formula, refer to Sect. 6.12 in Towhata (2008).

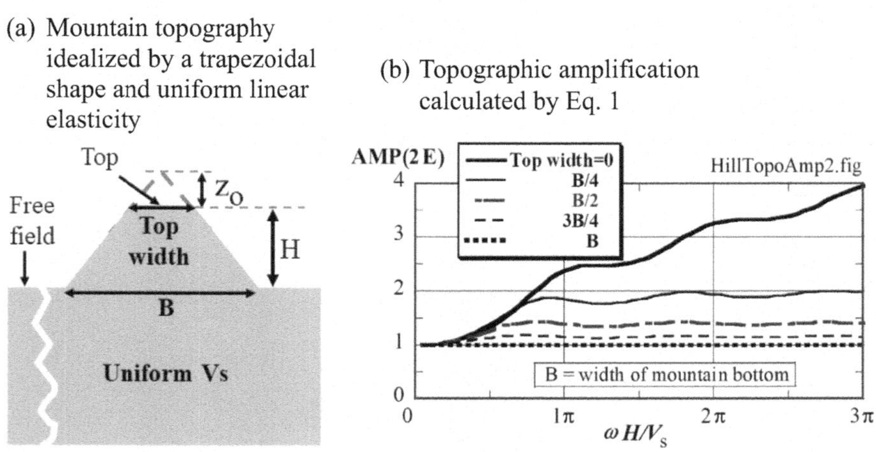

Fig. 75 Analytical solution of topographic amplification in an elastically uniform hill

The monotonic increase of the amplification shown in Fig. 61 (with a certain fluctuation) was reproduced by this solution. The monotonous increase of amplification with frequency is in a clear contrast with another kind of amplification of ground motion in soft surface soil in which an amplification factor $= 1/|cos[2\pi f \times (\text{soil thickness})/V_s]|$, where the maximum amplification occurs periodically when the frequency satisfies the condition of resonance. In general, the extent of topographic amplification in Figs. 72 and 75 remains 2–3 except for a very sharp mountain (top width = 0). This is consistent with observations by Pedersen et al. (1994) and Havenith et al. (2003). The difference is that the amplification observed by these two groups hits a maximum in the frequency range of 2–5 Hz, followed by less amplification under higher frequency. The maximum occurred in reality because the mountain top has some width, as illustrated in Fig. 75.

The significant amplification of ground motion does not necessarily imply a higher risk of landslide because there are two distinct material properties as illustrated in Fig. 76. The brittle material is prone to sudden collapse if the external load exceeds the peak strength. Since the peak strength is achieved after small deformation, a high frequency load can destroy a brittle slope if it is strong enough. In contrast, the ductile structure requires large deformation prior to the peak strength, and is thus not susceptible to sudden collapse. Strong cyclic loading has to occur many times, with a low frequency and long period. In general, cliffs and steep slopes made of fractured rock are brittle, while slopes made of soft soil are ductile. Physical weathering produces fissures and fractures in a rock mass. Consequently, an intact rock mass becomes brittle and prone to coseismic landslide. In contrast, soft clayey slopes made by chemical weathering may behave in a ductile manner.

The 2008 Wenchuan earthquake provided many important lessons. The effects of geological dip angle (see Figs. 31 and 32) were one of them. Alignment of landslides along the fault rupture was another one indicating that the distance from the fault is more important than the epicentral distance (Fig. 71). The third point is the earthquake-induced instability that lasted for years after the earthquake. For example, a heavy rain event in September, 2008, which was four months after the earthquake, induced many additional landslides. This point will be discussed later.

Yuan et al. (2013) reported that landslide concentrations along the surface rupture were characterized by $PGA > 0.4G$, the slope angle greater than 30°–40° and the

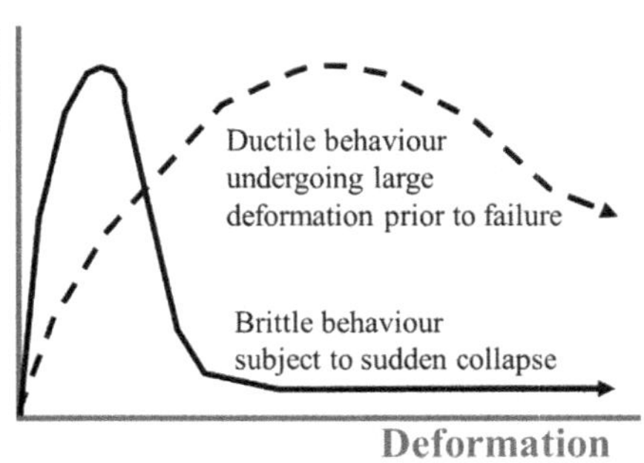

Fig. 76 Schematic illustration of brittle and ductile material properties

distance from the rupture shorter than 15 km on the hanging-wall side. Tang et al. (2011) chose a study area of 414 km^2, including seismically damaged Beichuan, and conducted field investigation as well as inspection of aerial and satellite images. In their study area, there were 40 landslides (2.68 km^2) prior to the Wenchuan earthquake. The earthquake induced 2221 coseismic landslides with a total area of 30.81 km^2. It was suggested that more landslides were triggered by stronger acceleration exceeding 500 cm/s^2. Four months after the earthquake, a heavy rainfall event produced 969 new landslides with an area of 6.9 km^2, together with 169 enlarged previous landslides, with an area of 2.48 km^2. Most coseismic events occurred on the hanging-wall side (northwest side) of the causative Longmenshan fault, which is consistent with experiences during other earthquakes.

Minami (1997) surveyed the literature on historical earthquake-induced landslides in Japan. Figure 77 presents his data, indicating that major historical landslides had slope angles at their source being between 20° and 40°, while there is very weak correlation for the angle increasing with the landslide volume. Data by Tang et al. (2011) was plotted in Fig. 78 to examine the effects of slope angle on the likelihood of landslides in the pre-earthquake, coseismic and post-earthquake periods. In this figure, the studied data were classified into seven categories based on the slope angle (horizontal axis). See that the area of landslides is maximum at 20°–30° and 30°–40°. Then, the landslide area density was calculated, that is the area of landslides in each category divided by the entire area of each category (shown by %). It is interesting that pre-earthquake slope instability exhibits the maximum density for 30°–40°, whereas the coseismic and post-seismic instabilities exhibit the greatest density at 40°–50°. This implies that slopes are affected by shaking in a different manner from static slope instability. This point is consistent with Keefer (2000), who demonstrated greater landslide concentrations with increasing slope angle. Parise and Jibson (2000) investigated landslides induced by the 1994 Northridge earthquake. They determined that landslide concentration (number of landslides per km^2) is maximum at slope angles of 40°–50° (Fig. 79). Thus, it is likely that earthquake-induced landslides occur at greater slope angles than gravitational landslides. In other words, shear failure may not be the unique failure process for coseismic landslides. This point will be discussed again later.

Sato et al. (2007) studied satellite photographs of landslides in Kashmir and stated that 47% (1139) out of 2424 landslides occurred within 2 km of the fault rupture and that the slope angle at the majority of landslide sites were 30°–40° or greater. Similarly, Ray et al. (2009) studied satellite photos and stated that most landslides occurred on slopes of 30°–40° and that 42% of the landslides occurred on the hanging-wall side, within 5 km of the fault. According to them, the fault length was 140 km. Collins et al. (2012) studied landslides induced by the 2007 Niigata-Chuetsu-oki earthquake of Mw = 6.6. They showed that most landslides occurred on slopes of 40°–45° and still many more were associated with angles greater than 50°. Owen et al. (2008), together with Khattak et al. (2010), noted that landslides were not common on the footwall side and that steep slopes (exceeding 50°) developed many failures. Further, a possible effect of slope angles on landslide volume is of interest. Inokuchi et al. (2012) prepared a comprehensive list of historical earthquake-induced

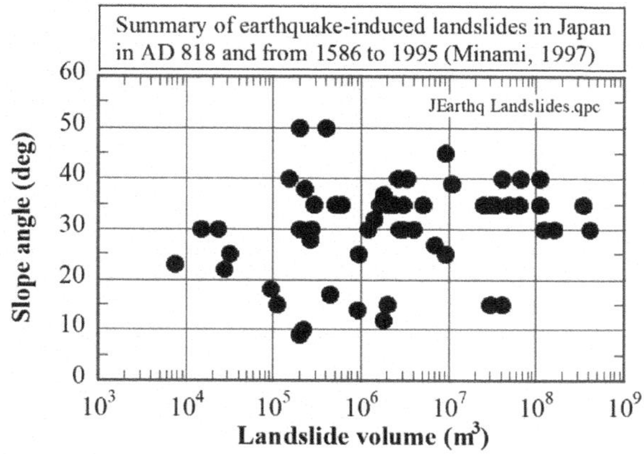

Fig. 77 Relation between the volume of historical earthquake-induced landslides in Japan and the slope gradient in the source area (data by Minami 1997)

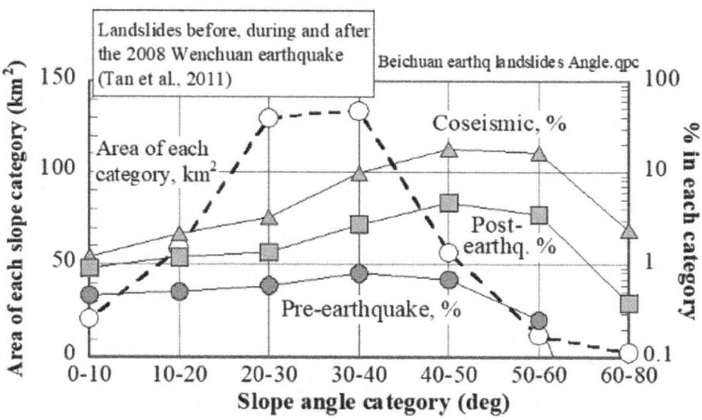

Fig. 78 Effects of slope angle on the likelihood of landslides (drawn after Tang et al. 2011)

landslides in Japan during the period ranging from prehistoric time and AD 715 to 1995. Figure 80 plots his data on the landslide volume and the angle of the slope where the sliding occurred. The data were classified into two groups before and after the 1891 Nobi earthquake of magnitude = about 8, which means that the earlier data is historical and the later data is more scientific. Although the data after 1891 may imply a slightly decreasing trend, it is not easy to draw any definite conclusions. Thus, the landslide volume is not clearly affected by the slope angle.

The material property of soils and rocks plays an important role in the earthquake proneness of slopes. While types of rock are of interest to geologists, of equal importance is discontinuity in rock masses. The slope at Tsaoling, Taiwan, (Fig. 6) repeatedly failed during earthquakes and heavy rainfalls in the past along a weak interface between the geological strata of Chinshui slate and the overlying

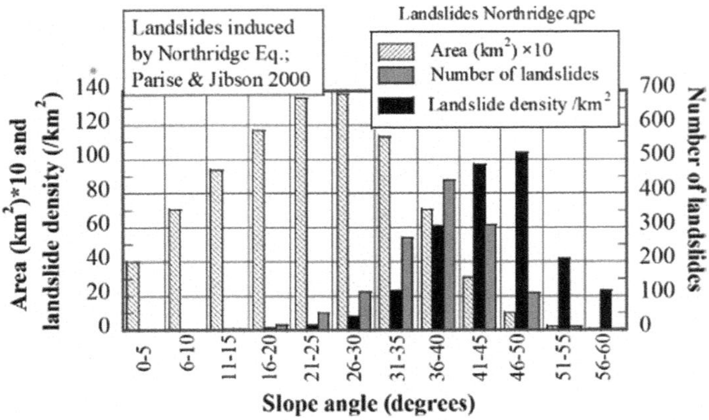

Fig. 79 Effects of slope gradient on the occurrence of landslides induced by the Northridge earthquake (drawn after Parise and Jibson 2000)

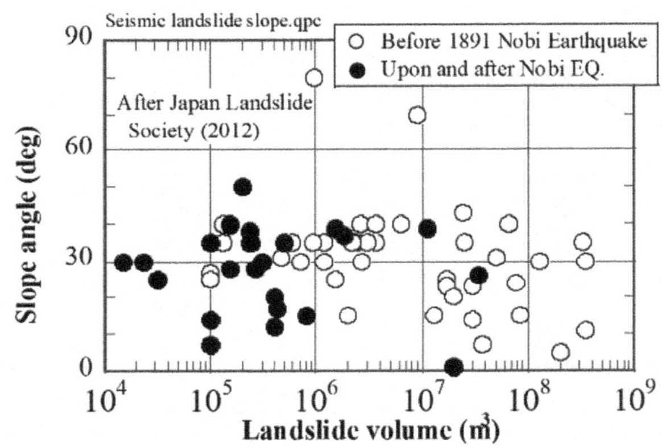

Fig. 80 Relationship between landslide volume and slope gradient for earthquake-induced landslides in Japan (data by Inokuchi et al. 2012)

Cholan Formation (Hung 2000), which is rich with slate fragments (Nagel et al. 2013). The important role played by inclination of geological strata was illustrated by the Tsaoling landside in 1999, where the strata was parallel to the slope surface (Fig. 81). In line with this, fractures and fissures produced by physical weathering affects the seismic stability. The weakening of the surface material well accounts for the facts that the majority of the earthquake-induced landslides are shallow events (Fig. 82). Fractures and fissures, which are produced by seismic shaking, allow rain water to infiltrate into slopes and trigger rainfall-induced landslides. This is one of the mechanisms of the compound effect of earthquake and subsequent rainfalls.

The problem of volcanic ash was addressed in the previous sections (Fig. 50). Another interesting material is called loess, an aeolian material abundant in Central Asia including Gansu Province, Northwestern China (Fig. 3). It has been known that water infiltration due to "over-irrigation" induces landslides in loess slopes (Wang et al. 2004; Peng et al. 2019). Loess has cementation among grains and, when dry, can

(a) Overall view with landslide lake (b) Exposed slip plane (by Dr. K .Watanabe)

Fig. 81 Tsaoling landslide site after the 1999 Chi-Chi earthquake in Taiwan

Fig. 82 Gravelly surface of Ninety-Nine Peaks after the 1999 Chi-Chi earthquake in Taiwan

be stable even with a nearly vertical configuration (Fig. 83). It is, however, weak when wet because the cementation material dissolves into water and cohesion disappears (Ueno et al. 2005). Figure 84 demonstrates one site of landslide in loess slope that was triggered by the 1920 Haiyuan or Gansu earthquake of Mw = 8.3 (Close and McCormick 1922; Wang et al. 2014b). The flow of loess at this site blocked a river and formed a dam that has been stable since then without breaching. The gentle slope here implies that the shear strength of the flowing loess was very low as is the case of liquefied sand. Consequently, many studies have been made of loess liquefaction (Wang et al. 2002; Tian et al. 2015). In the past, the 1718 Tongwei earthquake of Mw = 7.5 triggered slope instability in loess slopes as well (Sun et al. 2017).

Wang (2020) discussed the mechanism of loess liquefaction, saying that collapse of cementation by cyclic shear and the development of pore water pressure contribute together to the onset of liquefaction. Loess liquefaction is associated with high water content (near saturation) (He et al. 2020) in spite of the arid climate of the region (Fig. 83). The source of water is rain water, irrigation water, and dam impoundment (Zhang and Li 2011), as well as a mysterious rise of ground water from a deep elevation shortly before the 1920 Haiyuan earthquake (Wang 2020). Poor maintenance

Fig. 83 Landscape of dry loess plateau in Ningxia, Northwest of China

Fig. 84 Seismic failure of loess slope triggered by the 1920 Haiyuan earthquake

of irrigation channels leads to water leakage and dangerous saturation of the loess slopes.

Ishihara et al. (1990) reported long-distance flow failures in Tajikistan that were triggered by the 1989 Gissar earthquake (Negmatullaev et al. 1992) of body wave magnitude = 5.3. The biggest landslide occurred at Okuli. The source area of this landslide was 1.5 km in length along the slope inclination with a slope gradient of 2.6%. The total volume of the failed mass was 20 million cubic meters (Fig. 85). The runout distance was as long as 2 km. The second biggest one occurred at Sharara. Its source area was 150 m in length with a slope gradient of 15% or less. The failed soil mass comprised loess soil and traveled over a nearly horizontal plane. Mud ejection after failure implied the onset of liquefaction in the loess. It appears that water

Fig. 85 Okuli landslide in gentle loess slope in Tajikistan induced by Gissar earthquake in 1989 (Ishihara et al. 1990)

had been leaking from canals at the head of slopes and made the slopes unstable. Thus, profound flow failure occurred even though the seismic acceleration was just 50–90 cm/s^2.

5 Effects of Weathering and Deterioration of Materials

Weathering is a powerful process that alters the mechanical and hydrological properties of intact rock. Physical weathering produces fissures and cracks that reduce the shear strength of a rock mass and, in extreme cases, disassemble a rock mass into pieces (Qureshi et al. 2019). Chemical weathering, including interaction with water or hydration (Aziz et al. 2015), is able to reduce the strength. Both mechanisms convert the surface of a rock slope to a deposit of soil and increase the risk of failure during either seismic shaking or heavy rainfall. Figure 86 shows the surface of Ninety-Nine Peaks in Taiwan, where the surface material slid down during the 1999 Chi-Chi earthquake (Fig. 82). After the earthquake, the slopes in this area became less stable and started to produce more debris flows (Lu and Hsu 2004). Figure 87 shows a distortion of a small hill slope in Balakot during the 2005 Kashmir earthquake. It appears that the slope moved down by a few meters, disintegrating the foundation of buildings resting on it. These events imply the unstable nature of the surface weathered/disrupted materials.

Fissures, fractures, and disintegration of rock masses are consequences of not only physical weathering but also long-term tectonic actions. Once developed, a discontinuity is unlikely repaired and the slope instability progresses with time. Accordingly, if other conditions are unchanged, the older slope and rock masses are presumably more prone to failure. Figure 88 shows that the number of non-seismic landslides (ratio of landslides area/total area of the same geological unit) was very small for the rock in the age group of 800 million years, while it became much greater for coseismic and post-earthquake landslides. Similarly, Fig. 89 shows that the number of landslides per area of the same geological unit was smallest for the

Fig. 86 Surface of Ninety-Nine Peaks that was disturbed by the Chi-Chi earthquake, Taiwan

Fig. 87 Downward distortion of a hill slope in Balakot, Pakistan, where local buildings were totally destroyed; see tilting of a tree

youngest Quaternary materials. In contrast, slopes made of older Cretaceous—Paleogene rocks are more vulnerable and the maximum number of landslides occurs for the slightly younger Neogene rocks. This suggests the importance of discontinuities in the onset of earthquake-induced landslides in "rock."

After falling down, the surface detritus material is deposited at the valley bottom and washed by stream water immediately or during the following rainfalls. More materials that remain on the slope during the earthquake are finally washed down by rain water or by weaker aftershocks, and join the bottom deposits. This procedure is one of the important causes of sediment-related disasters in the post-earthquake period and will be discussed later. On the other hand, weathering continues and produces a new unstable layer on the slope surface. This process is promoted by the recovery of vegetation that adds mechanically weak organic materials to the surface.

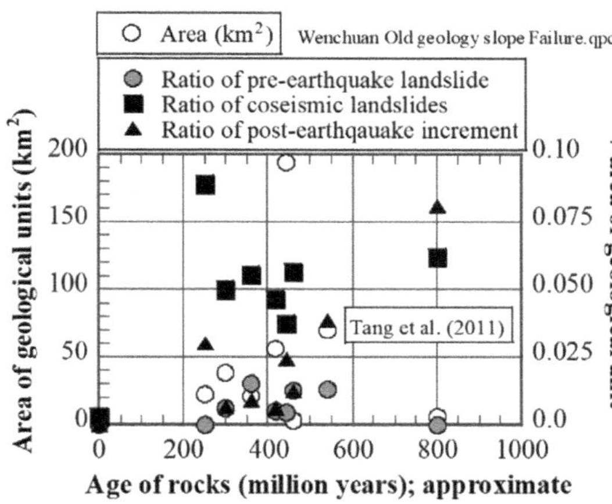

Fig. 88 Change of coseismic landslide numbers with rock age (2008 Wenchuan earthquake, China) (drawn after Tang et al. 2011)

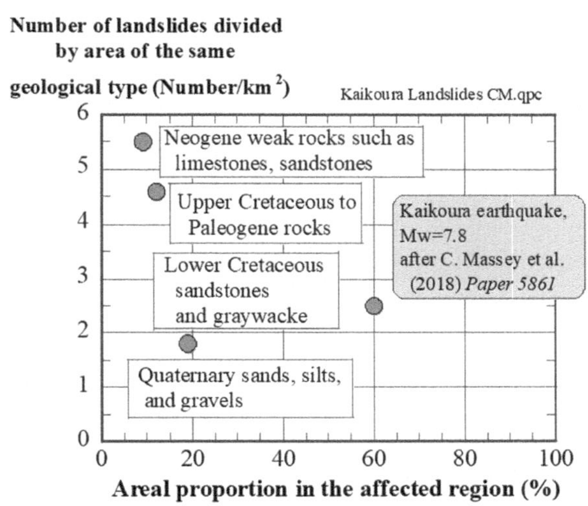

Fig. 89 Change of coseismic landslide density with age (2016 Kaikoura earthquake, New Zealand) (drawn after Massey et al. 2018)

After some time, this surface layer becomes thick enough and makes the slope again susceptible to landslides. The key issue is the time needed for recovery of the surface layer. Although this time is variable, depending on rock types, climate, temperature change, and many other factors, several studies suggested the time to be of the order of some hundred years. Table 1 shows various intervals of landslides. Figures 90 and 91 show the historical and current appearance of Mt. Tanzawa, Japan, which was devastated by coseismic landslides during the 1923 Kanto earthquake of Mw = 7.9–8.2 (on September 1st). The antecedent rainfalls were 5–10 mm on August 5th, a few mm/day from August 16th to 28th, and more than 80 mm on August 31st. After the quake, 250 mm of rainfall was brought by a typhoon on September 15th and many slopes failed (Nakamura et al. 2000). In Tanzawa, more landslides happened in steeper slopes and infiltration of rain water into seismically generated cracks triggered landslides (Moroto 1925). Inoue (1995) stated that the heavy rainfall two weeks after the earthquake triggered many landslides. Furthermore, the likelihood of landslides was high until 1929, which was six years after the earthquake, or possibly until the 1960s. This is an example of long-term slope instability after big earthquakes.

Obviously, the landslide risk decreases with time after seismic events as unstable soil masses fall down and vegetation recovers. Koide (1955) called this situation immunity to landslides. Soil formation resumes after a landslide as topsoil forms again on the slope surface. Chigira (2014) stated that earthquake-induced landslides occur on slopes affected by antecedent rainfall and chemical weathering of carbonate rocks and mudstone.

The idea of the rate of topsoil production and the time till the next slope disaster is suggested in Table 1. Note that a long recurrence period of more than 10,000 years does not imply the intact condition of a slope, because another part of the same

Table 1 Studies on time needed to recover a soil layer on a slope surface

Author(s) (year)	Studied rock/soil type or its location	Time for soil recovery and recurrence of landslides (years)
Kelsey (1982)	Northwest California	16,000–50,000
Takeshita (1985)	Kyushu Island, Japan	> 100,000 for chemical weathering
Trustrum and De Rose (1988)	Taranaki, New Zealand	Topsoil formation at 0.35 cm/year for 40 years; 0.12 cm/year for next 50 years
Shimokawa et al. (1989)	Shirasu = pyroclastic flow deposit	Top soil formation at 0.45 cm/year, > 500 years till next failure
Yanai and Igarashi (1990)	Marine terrace, Hokkaido, Japan	150–200
Shimizu et al. (1995)	Volcanic ash, Hokkaido, Japan	650
Matsumoto et al. (1996)	Granitic slope	60 for tree growth

Note the studied time interval is strongly affected by local climate and human action

Fig. 90 Debris flow in Tanzawa Mountain triggered by the 1923 Kanto earthquake, Japan (Office of Isehara City Council)

Fig. 91 Slope of Mt. Tanzawa, Japan, in 2019 that is mostly covered by vegetation but still prone to minor instability

slope may fall down during the next strong earthquake or heavy rainfall. Figure 92 illustrates weathered granite (locally called "Masa") in Rokko Mountains near Kobe, which easily fails during strong shaking or heavy rain. Matsukura (1997) reported that the rate of weathering of granite is around 10 mm per one thousand years. However, the rate of weathering in this photograph is much faster, as shown by the fact that this place is part of a wild park and the route here was constructed in the latest decade. Nevertheless, the granite surface has been already weathered and disintegrated. In contrast, Fig. 93 shows a Buddhism image carved in granite in AD 1594. Even though 400 years have elapsed, the image is still clearly recognizable. Thus, the rate of weathering varies with many conditions.

Creeping rock masses are prone to instability on a large scale. By describing several examples, Zischinsky (1969) addressed the importance of creep deformation in mountain slopes as a precursor of landslides. What he called "sackung" (=sagging

Fig. 92 Weathered granite slope near Kabutoyama, Nishinomiya, Japan

Fig. 93 Buddhism statue carved in granite with inscription of year of carving being AD 1594 (behind the graveyard, Sumiyoshi, Kobe, Japan)

or subsidence) implies depression or downward movement in a slope surface. Otherwise called gravitational deformation, the destructive effect of creep deformation in rocky slopes was discussed by Mahr (1977) together with Ter-Stepanian (1977). Voight and Kennedy (1979) referred to creep deformation and ultimate landsliding in the wall of Chuquicamata Open Pit Mine, which was substantially affected by an earthquake and blasting. They noted that microtremors increased with time toward the ultimate landslide. Hutchinson (1988) addressed a variety of creeping geometry in slopes prior to failure. Watari (2005) presented the meaning of "sackung" and "gleitung" (= sliding) as the precursory creep deformation over a large range of rock masses and its development/concentration in the weakest part of heterogeneous rock mass (Fig. 94). Pánek and Klimeš (2016) discussed the temporal evolution of creep toward the final landslide. Obviously, a rock slope with significant creep deformation is susceptible to failure under seismic action.

Chigira (1992, 1998) presented details of subsurface structure produced by the gravitational creep of rock masses and discussed that creeping rock mass may be prone to deep-seated landslides in the future, as exemplified by gigantic coseismic landslides at Tsaoling (Fig. 81), which had small ridges and scarps as precursors (Chigira et al. 2003) and Jiu-Feng-Er-shan (Fig. 95), which had buckling of the slope (Chigira 2009). Shou and Wang (2003) calculated the possibility of flexural buckling of a surface rock slab by using material properties of collected rock specimens. They used buckling formula obtained based on many hypotheses proposed by Cavers (1981) who attempted to consider the effects of gravity force acting on the rock slab. In other words, it is necessary to develop more advanced buckling theory for rock slabs. Tommasi et al. (2009) reported buckling distortion of a rock slope in Italy and discussed its failure in the future. Further studies on rock slope buckling have been conducted by Qin et al. (2001), Qi et al. (2015), Zhang et al. (2019) and Kawabe (2019). They infer the importance of pore water pressure that uplifts the surface rock slab and triggers buckling.

Furthermore, Chigira et al. (2009b) stated that the site of the gigantic Daguangbao landslide (750 million m^3; Figs. 25, 26 and 96) underwent creep depression prior to the Wenchuan earthquake. Figures 97 and 98 illustrate the creep distortion of unstable mountain bodies such as a double ridge near the mountain top at Ohya (Yagi and Inokuchi 2010) and Shichimenzan landslide sites in Japan, together with a schematic illustration (see Fig. 99a). Furthermore, Fig. 99b shows typical creep deformation of Mt. Chogatake in central Japan (Fig. 1) (Hohashi and Otsuka 2009; Yagi et al. 2016). This deformation started not later than the last glacial maximum (12.4 thousand years ago) (Nagata and Kojima 2020). Similar evidence of gravitational creep deformation was exhibited by Agliardi et al. (2013) in the European Alps

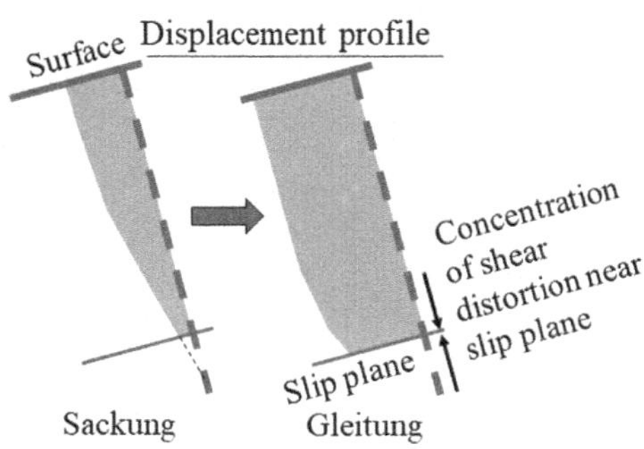

Fig. 94 Distributed creep deformation at an early stage and its concentration in the weakest part of a rock mass (drawn after Watari 2005)

Fig. 95 Jiu-Feng-Er-shan landslide triggered by the 1999 Chi-Chi earthquake, Taiwan

as well. Using an experimental perspective, Towhata and Gunji (2018) conducted creep tests on artificial rock materials in an unconfined compression manner, wherein the stress level was held constant for a long time (Fig. 100). It was found that the material strength is significantly reduced once the creep strain exceeds the strain at peak strength in conventional unconfined compression tests. Tajika and Ohtsu (2000) reported on their site observation of toppling failure in a shale slope after creep deformation. Note that the change of slope topography, such as over-steepening, is another damaging effect of creep deformation. Earthquake shaking can trigger failure of a slope that has reached critical condition after creep.

Hydration of mudstone is another kind of weathering and deterioration of material properties, leading to slope instability (Aziz et al. 2015), whereby unstable rock absorbs water and disintegrates. The 2004 Niigata-Chuetsu earthquake affected many

Fig. 96 Daguangbao coseismic landslide that was triggered by the 2008 Wenchuan earthquake; I visited this place twice but the weather was always foggy (see also Figs. 25 and 26)

Fig. 97 Double ridge being formed by gravity deformation of the mountain body at Ohya where a gigantic landslide was triggered by the 1707 Hoei, Japan, earthquake of Mw = 8.7–9.3 and has been followed by many landslides since then

Fig. 98 Depression at the mountain top immediately behind an ongoing slope instability, Shichimenzan landslide site, Japan

slopes made of mudstone and siltstone; see Figs. 13 and 14. Moreover, many man-made fills had been composed of locally available mudstone and were affected by the earthquake (Fig. 15).

Figure 101 conceptually illustrates the process of material deterioration of man-made fills that were constructed for residential development and were possibly affected by water action. This figure classifies the types into valley filling by cut-and-fill procedures and widening on a slope (see Fig. 3). Both are subject to water infiltration and possible hydration, together with an increase in weight.

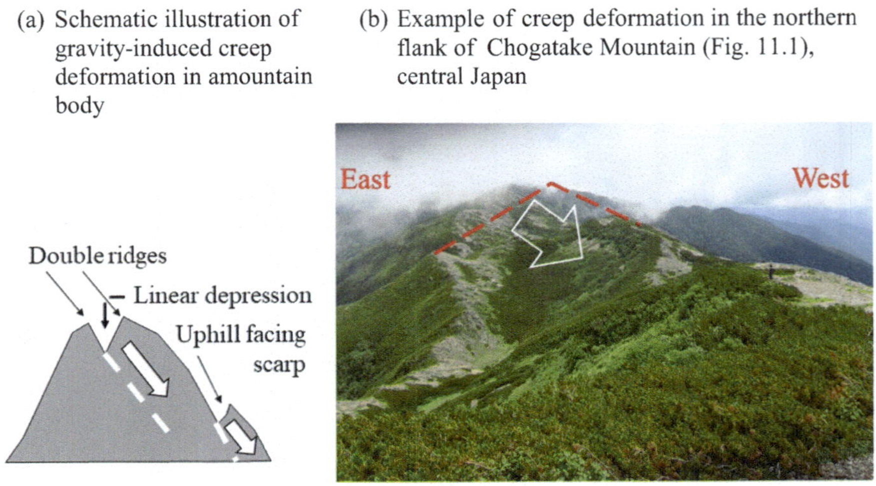

Fig. 99 Mechanism and example of gravity-induced creep deformation of a mountain body

Fig. 100 Comparison of strength in a conventional unconfined compression test and creep tests (Towhata and Gunji 2018)

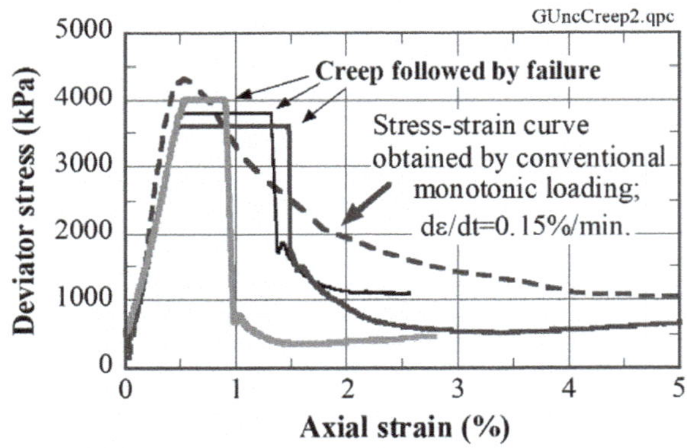

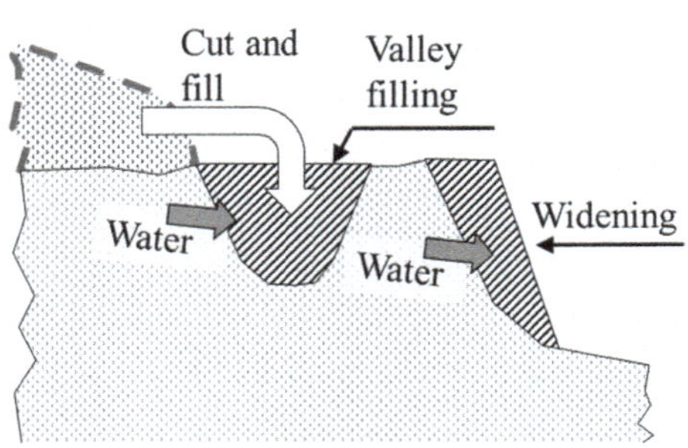

Fig. 101 Types of embankments for residential development in hilly terrain

Figure 102 shows the collapse of a local highway embankment caused by the Niigata-Chuetsu earthquake. This section of the highway crossed a water stream (valley filling) and was susceptible to hydration. Because this kind of damage occurred in many sections of the road network during this earthquake, emergency action and rescue of people had to be carried out by helicopters. An example of widening fill is exhibited in Fig. 103. The soil mass slipped down together with a concrete retaining wall. The lesson was that we have to be careful of using local materials for filling. Although inexpensive, poor materials will cause problems after some time.

Fig. 102 Collapse of a road embankment made of mudstone and crossing water channel; 2004 Niigata-Chuetsu earthquake, Japan

Fig. 103 Failure of widening fill around the Takamachi residential development; 2004 Niigata-Chuetsu earthquake, Japan

More serious damage occurred in valley filling during the 1993 Kushiro-oki earthquake of Mw = 7.6. Being located in a small valley, this residential fill (Fig. 104) had significant water infiltration coming in from the surrounding hills. Thus, the soil weight increased. Even worse, the bottom of the valley was made of peaty soft soil with low shear strength (Fig. 105). Peat was formed at the bottom of swamps because of a wet and cold climate. Consequently, very low shear strength was available to resist the seismic inertial force and large deformation occurred (Fig. 104). Then the problem was the reconstruction philosophy of reproducing the original shape, which was common in those days. Hence, the situation in Fig. 105 was reproduced and, when the 1994 Hokkaido-Toho-oki earthquake of Mw = 8.3 occurred next year, the same damage occurred again (Fig. 106). Accordingly, the philosophy was changed when the Kobe earthquake occurred in 1995 so that seismic resistance and performance would be improved after reconstruction. The fill in Kushiro was improved accordingly, and it was fortunate that no more damage occurred here when the 2003 Tokachi-oki earthquake of Mw = 8.0 occurred.

Fig. 104 Seismic deformation of residential valley filling resting on very soft peat deposit (1992 Kushiro-oki earthquake, Japan)

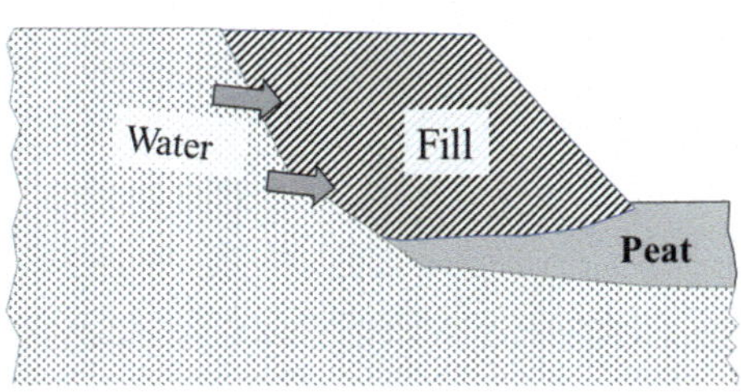

Fig. 105 Conceptual sketch of damage mechanism of valley filling resting on peat

Fig. 106 Repeated damage in the same fill as in Fig. 104 (1994 Hokkaido-Toho-oki earthquake, Japan)

Another lesson was learned when the 2009 Suruga-Bay earthquake of Mw = 6.3 hit Shizuoka, Japan. Figure 107 indicates the induced damage of the nation's No. 1 expressway (Yokota et al. 2011). This site was located in a former valley between two hills (Fig. 108). When the expressway was constructed, mudstone in the hill sections was cut and filled in the valley section. Thereinafter, ground water continuously flowed from the hills into the valley section, leading to hydration and deterioration whereby the well-compacted mudstone became weak clay. This clay had not only low shear strength but also low permeability that held more water inside the embankment and worsened the situation (Ishida and Iwabuki 2010; Takagi et al. 2010). The damage shown in Fig. 107 was the consequence of this process.

(a) Aerial view (Courtesy of NEXCO Central: Central Nippon Expressway Company)

(b) Proximity of cut-and-fill sections of the seismically affected expressway (by Prof. S. Yasuda, Tokyo Denki University)

Fig. 107 Seismic collapse of highway embankment, Tomei Express Way, Japan, in 2009

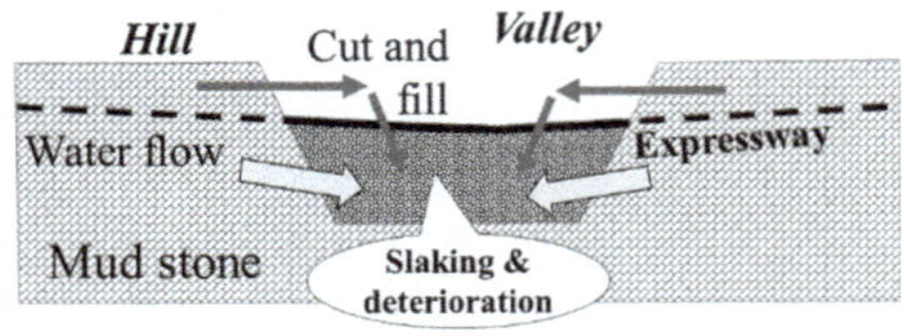

Fig. 108 Situation leading to the seismic damage of a motorway embankment in Shizuoka

Figure 109 illustrates another type of seismic damage in motorway embankment during the 2007 Noto Hanto-oki earthquake with Mw = 6.7 (Fig. 1). The local main motorway suffered embankment damage at 11 sites out of its total length of 42.85 km. Most of the damaged embankment had been constructed on mountain slopes without necessary seismic safety measures and slip planes developed during the earthquake along the interface between the fill and the surface of the original slope (Hashimoto 2009). This implies that no stabilization measures had been made there and the original soft surface soils remained between the slope surface and the road embankment. In earlier times, it was not uncommon that weathered soil on a slope surface was not removed prior to soil filling. Sometimes, it was even the case that trees and grasses at the slope surface were simply cut and buried under fill. This old practice formed a weak plane with low shear strength. Furthermore, the embankment was made of locally available tuff that was prone to slaking or water action. Nevertheless, this motorway was constructed in a rural region at a low cost and has served the community since 1982. In my opinion, its low construction cost made the cost–benefit ratio reasonable under low traffic demand.

Fig. 109 Seismic damage of motorway in Noto Peninsula, Japan, caused by earthquake in 2007

6 Long-Distance Flow of Landslide Mass

The long runout of landslide debris is one of the factors that govern the damage extent of a landslide. The prehistoric landslide at Seidmarreh or Saidmarreh, southwest Iran, measured 25 billion m^3 (Harrison and Falcon 1938) and is the world's second biggest onshore landside that is seen today. This huge event occurred more or less 10,000 years ago (Watson and Wright 1969) and was triggered by river erosion, possibly combined with an earthquake (Oberlander 1965; Shoaei 2014). The width of the slip plane is 16 km, while the length is 4 km (Fig. 110). Figure 111 depicts the slip plane, which is still smooth today because of little precipitation. Figure 112 demonstrates the high fluidity of this huge debris flow that overtopped a nearby hill and also formed two lakes (Shoaei 2014). Finally, the flow stopped at 26 km from the source. In Fig. 113, the entire hill comprises the debris deposit. It deserves attention that the world biggest onshore landslide is the one at Baga Bogd in Mongolia, which is of 50 billion m^3 in volume and $H/L = 0.4\ \text{km}/14\ \text{km} = \tan(1.6°)$. It is supposed that this event was triggered by a prehistoric earthquake as suggested by its proximity to an active fault and gigantic size (Philip and Ritz 1999).

Fig. 110 Overall view of Seidmarreh landslide, Iran, as seen from Pol Dokhtar (photograph by Dr. H. Shahnazari)

Fig. 111 Slip plane of Seidmarreh landslide (height of cliff on the right = 300 m)

Fig. 112 Seidmarreh debris flow overtopping a nearby hill (no debris deposit on the front side, which is out of the debris flow path; photograph by Mr. T. Shimomura)

Fig. 113 Terminal hill of debris flow of prehistoric Seidmarreh landslide in Lorestān, Iran (photograph taken in 2004)

Heim (1932) studied the height of landslide fall (H) and the distance from the head scarp to the end of flow (L) in order to indicate the importance of H/L ratio in the dynamism of long-distance flows. He presented a mathematical verification that this ratio is equal to tan(friction angle or fahrböschung angle). Figure 114a plots data of H/L and landslide volume in Heim (1932) which were obtained from both recent and prehistoric landslides in European Alps. It is important that the H/L ratio decreases as the landslide volume increases. In line with this and the past studies by Scheidegger (1973) and Hsü (1975), I collected H/L data of both seismic and non-seismic onshore landslides from many literature sources and plotted them against the landslide volume (Fig. 114b). H/L decreases with increasing landslide volume. Evidently, the bigger the landslide is, the longer is its runout; see Fig. 14 as well. It appears that the earthquake-induced landslides have a slightly smaller value of mean H/L than that of non-seismic landslides. However, the difference is not significant because movement of landslide mass is unanimously driven only by gravity, although coseismic landslides are triggered by dynamic inertial force. It deserves attention that Zhang and Wang (2007) studied the coseismic landslide in loess during the 1920 Haiyuan earthquake, China, and showed that the *H*/*L* value was substantially smaller than those in other coseismic landslides. Further note that submarine landslides have a profoundly smaller *H*/*L* than onshore events, as exhibited by data by Edgers and Kalsrud (1982), together with those that were collected and presented in Towhata (2008).

The city of Yungay in Peru was totally destroyed by a debris flow when the 1970 Ancash earthquake of Mw = 7.9 triggered collapse of a glacier and the body of Mt. Huascaran (Plafker et al. 1971). Although Yungay had survived another avalanche in 1962, the event in 1970 was greater in size and overtopped a hill behind the city (Fig. 115). Thus, the lesson was that empiricism is not always reliable and that an unexpectedly huge disaster is possible. Afterward, new Yungay was relocated to a safer place.

Breaching of natural dams may induce disasters by floods that travel hundreds of kilometers. The 1786 earthquake of magnitude = 7.6 blocked the Dadu River in Southwest of China. Although the detail is difficult to assess today, Dai et al.

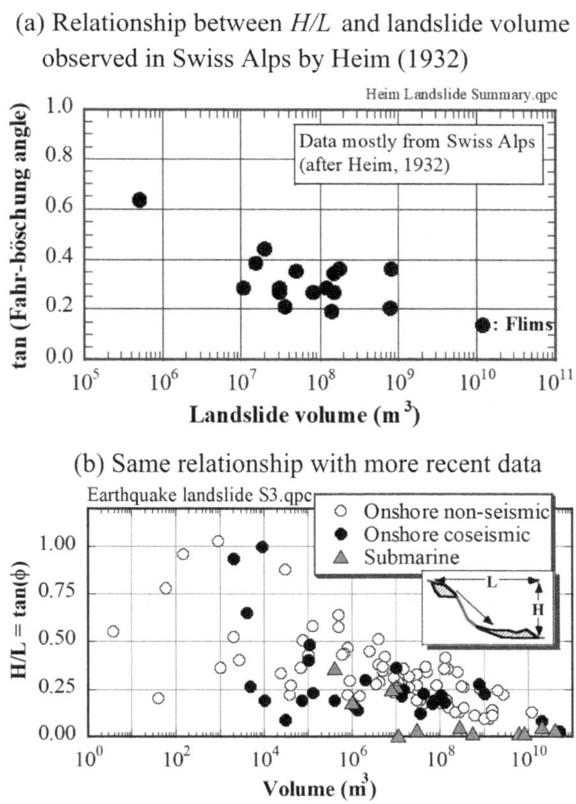

Fig. 114 Equivalent frictional coefficient of a landslide mass, *H*/*L*, decreasing with increasing landslide mass

Fig. 115 Ruin of Yungay, Peru, which was destroyed by an earthquake-induced debris flow in 1970

(2005), Wu et al. (2013) and Ling (2015) presented some information; landslide volume = 5, 6 or 45 million m^3, height of natural dam = 70–167 m and the stored water volume was 40–1150 million m^3. Consequently, the breaching of the dam caused a flood that induced disasters in Leshan (320 km downstream) and the water level rose by 1.54 m in Chongqing (1730 km). The 1985 Bairaman Landslide in New Britain Island of Papua New Guinea was triggered by an earthquake with its magnitude being 7.1. The landslide of 180 million m^3 in volume blocked the river, and dam breaching induced flood disasters at 39 km downstream distance (King et al. 1989). No fatalities, however, occurred due to evacuation. These examples imply the importance of monitoring and alerts in case that big natural dams are formed.

More recently, on April 9th, 2000, a non-seismic landslide of 300 million m^3 (Yin 2000; Zhou et al. 2016) blocked a river at Yigong in Tibet (Fig. 116). *H*/*L* of this landslide was 2.8 km/10.5 km according to a figure by Yin (2000). This natural dam breached on June 10th and initiated a debris flow that traveled over 500 km down along the Yarlung-Tsangpo/Brahmaputra River to Northeastern India (Delaney and Evans, 2015). For example, the river water level suddenly rose by 4 m on June 11th in Pasighat Town in Arunachal Pradesh Province of India (Tewari 2004). This area suffered from similar long-distance flows of debris in the past (Montgomery et al. 2004). Zhou et al. (2016) pointed out that the glacier played an important role in the initiation of the Yigong slope instability due to its own weight, cyclic loading upon thawing and freezing, and infiltration of melted water into the rock mass. The melting and freezing of ice in the rock disintegrates the rock mass and directly affects its mechanical properties as well (Qureshi et al. 2019). It should be further recalled that falls of glaciers caused disasters in the past, e.g., an ice fall at Mattmark Dam in 1965 (article in Securitas Magazine 2007) as well as a glacier fall and flood in Uttarakhand Province of India on February 7th, 2021. Given the current global warming and melting of ice, the instability of glacier (Röthlisberger 1978) will be more important in the upcoming decades in both seismic and non-seismic regions (Huggel et al. 2012).

A classical question is why a bigger landslide mass can travel over a longer distance. The first answer is the effects of water. If water content is high enough, undrained conditions are maintained in a large volume of the debris, excess pore water pressure is held high, and a flow is formed and can travel as fluid over a

Fig. 116 Site of natural dam at Yigong, Tibet (landslide came down from the right)

long distance. It has been known that liquefaction in a gentle slope (gradient of a few degrees) of loose water-saturated sand can trigger flow deformation in subsoil, with the flow distance being several meters (Hamada et al. 1986; Youd and Perkins 1987; Hamada 1992). It is then strange that Nalodo at four sites in Sulawesi in 2018 (Fig. 69) was able to travel 500 m or more over a small gradient of 2° or less. It is possibly important that a substantial amount of water was ejected out of the earth interior and maintained the state of null effective stress in the topsoil. This point will be discussed later.

There are many other ideas on the mechanisms of long-distance flow. Lubrication at the bottom of the soil mass (Figs. 37 and 65) is possible if very soft soil is available. This mechanism has something in common with what Hungr (2007) called "material entrainment" in which loose surface material is taken into landslide mass. Obviously, entrainment of water can promote a longer runout distance. The formation of an air cushion under the landslide mass is interesting. It is said that air is entrained under the mass when the mass jumps from a cliff into the air (similar to a ski jump) (Shreve 1966). This idea, however, cannot explain the huge landslides on the Moon (200 million m^3 and $H/L = 0.2$; Howard 1973) and Mars; 346, 376 and 458 km^3 in volume by Mazzanti et al. (2016) and five trillion m^3 by Quantin et al. (2004). Obviously, there is neither air nor water on the Moon and very little on the Mars. Melosh (1979) discussed acoustic fluidization in which stress in a landslide mass oscillates and allows grains to move rather freely when the grain-to-grain contact force is low. Likewise, Bartelt et al. (2006) argued the random motion of grains that generates expansion pressure in the landslide mass and reduces the effective stress and shear strength. Davies and McSaveney (1999) suggested that the high pressure in bigger landslide mass generates more grain crushing and promotes longer runout. Frictional heating promotes expansion of water volume and reduces the frictional resistance, as Voight and Faust (1982) together with Hendron and Patton (1985) suggested to explain the long-distance flow of the rock mass at the Vajont (Vaiont) Dam site. Hu et al. (2019) further pointed out the significance of heating because it generates CO_2 gas and the gas pressure lubricates the slip plane accordingly. Fusion of rock as found at the Köfels prehistoric landslide in Austria is the evidence of the high temperature generated by friction in the course of high-speed landslide (Erismann 1979). Masch et al. (1985) estimated the highest temperature at Köfels to have been 1520 °C during flow and also referred to the prehistoric Langtang landslide in Nepal (15 km^3) to be a similar example. There are thus many theories and ideas on the long runout of bigger landslide masses. It appears that further efforts are needed to understand this phenomenon.

The *H*/*L* analysis in Fig. 114 is well known in the landslide community. However, it is sometimes difficult in case history studies to determine the three-dimensional volume of a landslide mass. In such a case, analysis of just the length of the source area and the deposit area (L_1 and L_2 in Fig. 40) is convenient; see Fig. 117. Usuki et al. (2005) as well as Usuki and Mizuyama (2011) investigated 109 landslides in Japan, both coseismic and non-seismic, by using the runout coefficient defined as L_2/L_1 (Fig. 117). The greater value of this ratio, typically exceeding 0.9, means a long runout of the landslide. They then discussed the correlation between L_2/L_1 and the physical properties of soil such as grain size distribution, and liquid and plastic limits. It was then suggested that a longer runout (greater L_2/L_1) occurs for well-graded soil (similar amounts of sand, silt and clay mixed) and liquid limit ranging from 50 to 70%. Presumably, there is a roller-bearing effect of non-plastic fine particles among coarser grains.

Mizuhashi et al. (2006) studied soil properties at sites of landslides induced by the Niigata-Chuetsu earthquake, Japan. Conventionally, the possibility of an earthquake-induced landslide is often assessed by using geological and topographical information only and material properties are hardly taken into account. This is because detailed investigation within a vast area is difficult in practice. In their study, it was shown that the method to assess the coseismic landslide vulnerability by the Kanagawa Prefecture of Japan (1986) was improved by adding the effects of plasticity index (=liquid limit − plastic limit in %) and a better correlation was obtained between the assessed damage extent and the landslide volume. Thus, the risk of a coseismic landslide can be assessed more reasonably if information on the physical properties of local materials is available, although it is difficult to obtain necessary funds for such field work. Yagi et al. (2009), as well as Higaki et al. (2015), proposed to introduce the judgment of experienced experts into slope stability assessment when field

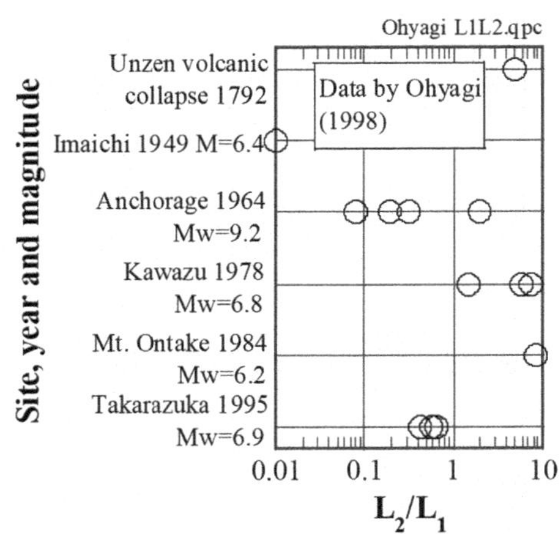

Fig. 117 Examples of L_2/L_1 ratio by Ohyagi (1998) (the original literature presented L/L_1 and I converted it to L_2/L_1)

investigation is limited to interpretation of aerial photographs because of financial requirements. They employed an analytic hierarchy process (AHP) scheme in which the importance of geological/topographical factors is discussed by experts. Those factors for assessment of gravitational slope instability were (i) distinctiveness of the main scarp, (ii) surface features of the landslide body, (iii) the position of the landslide body on the slope, (iv) the position of cracks in the landslide body, and (v) the stability conditions of the landslide toe (Yagi et al. 2009).

Thus, instability assessment still has to rely on a surface appearance of slopes that is hopefully affected by subsurface conditions. This is certainly the case with reactivation of past landslides, but primary landslides are still a difficult issue. Hopefully, gravity-induced creep deformation, if precisely detected, is a good clue to detect underground hazards.

7 Nalodo, a Long-Distance Flow of Earth in Sulawesi Island in 2018, and Roles Played by Ground Water

Nalodo occurred on September 28th, 2018, at four sites near Palu City, Sulawesi Island, Indonesia (Fig. 3) that consist of the Balaroa site to the west of the city, together with the sites of Petobo, Jono Oge and Sibalaya along the eastern edge of the Palu Valley where the seismic Palu-Koro Fault runs from north to south and has formed a pull-apart basin (Fig. 118). Because of the devastating effect of this disaster, Japan International Cooperation Agency (2019), that is abbreviated as JICA in what follows, carried out a project under the name of Project for Development of Regional Disaster Risk Resilience Plan in Central Sulawesi in the Republic of Indonesia. It was therein aimed to promote accurate and smooth reconstruction and realize a more resilient reconstruction (Build Back Better) by formulating and assisting the implementation of post-disaster reconstruction plans in the Central Sulawesi Province. The geological-geotechnical background of this disaster was studied comprehensively by a JICA Study Team based on the advice of the Academical Advisory Committee to which I belonged.

Figures 119, 120, 121 demonstrate what happened at Petobo. The original ground slope was more or less 2%. The top scarp (Fig. 119) implies that displacement was discontinuous across this place and that tensile stress was predominant. Beyond the head scarp in this photograph are many open cracks due to tension. There were still water springs below the scarp when I visited the site on November 2nd in 2018, which was about five weeks after the disaster. It was thus felt that high ground water pressure played some important role in development of Nalodo. Figure 120 shows a big open space after the surface soil crust had flowed away. The runout distance exceeded 500 m, which is a remarkable difference from the past experiences of liquefaction-induced soil flows. Previously, flow distances associated with liquefaction were merely 5–10 m in gentle slopes (Hamada et al. 1986, on gentle sloping ground; Towhata et al. 1996, on Port Island of Kobe). Accordingly, it was thought that

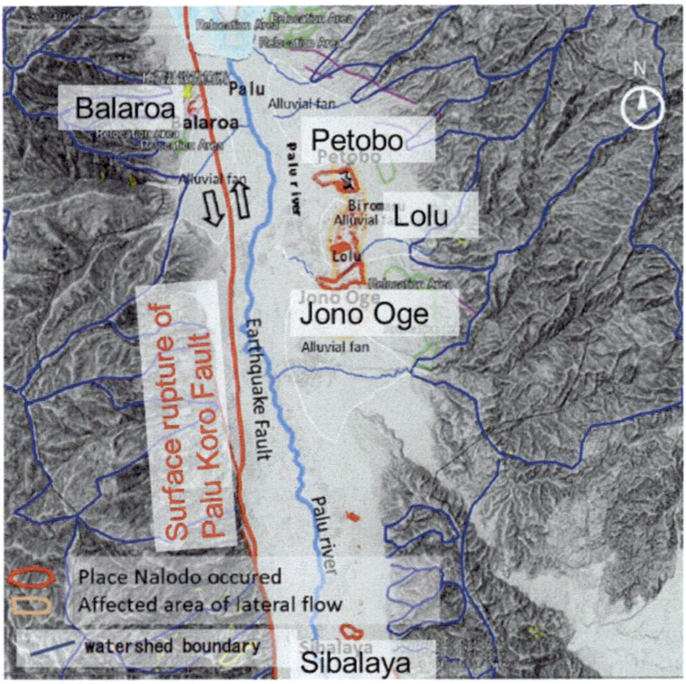

Fig. 118 Four locations of Nalodo seismic flow failure with a smaller event at Lolu during the 2018 Palu earthquake (drawn after JICA Study Team)

Nalodo is not a simple consequence of subsoil liquefaction but was affected by some special conditions. Figure 121 presents a huge deposit of soil and building debris at the downstream end of the Nalodo flow. Thus, the Nalodo area is divided into three parts; a tensile zone at the top, free flow in the middle and a bottom deposit. Figure 122 addresses another site at Balaroa (Fig. 69) and a huge amount of soil flowed away in spite of the gentle slope of the original topography. There were water springs at the bottom of this site as well even though five weeks had elapsed after the earthquake when this photograph was taken.

Figure 123 shows an aerial view of the Nalodo at Sibalaya. An irrigation canal breached at the head and affected the Nalodo flow. This irrigation channel goes along the ends of the alluvial fans on the eastern side of the valley and is located very close to the heads of three Nalodo (Petobo, Jono Oge and Sibalaya). Before the disaster, the water in the canal was supplied to rice fields on the gently sloping terrain. Moreover, the canal was not paved (Fig. 124) and allowed substantial water to leak and infiltrate into the Nalodo-prone slopes (Fig. 123). Breaching of the canal aggravated the disaster (Fig. 125). This situation impressed early authors of studies that Nalodo was triggered by water leakage from the unpaved canal, e.g., Nurdin et al. (2018), Bradley et al. (2019), Cummins (2019) and Watkinson and Hall (2019). However, the aforementioned huge Nalodo at Balaroa (Figs. 69 and 122) and another

Fig. 119 Head scarp of 2 m at Nalodo flow failure, Petobo, southeast of Palu

Fig. 120 Open space after Nalodo displaced the surface soil layer (at Petobo)

Fig. 121 Accumulation and deposition of soil and building debris at the downstream end of a Nalodo flow at Petobo

Fig. 122 Open space after Nalodo displaced the surface soil layer (at Balaroa southwest of Palu; see head scarp cliff and spring water)

small Nalodo at Lolu that developed limited ground displacement (<100 m) did not have a canal at their heads. Above the head of Balaroa site, the ground surface is occupied by houses and there is no canal (Fig. 126a), although Nalodo occurred here. When I visited Balaroa five weeks after the earthquake, the head scarp cliff was dry (Fig. 126b). This means leakage of sewage water was not occurring under houses in Balaroa. Therefore, leakage of canal water was not the main cause of Nalodo flow, although the extent of damage was aggravated by it. The same opinion was expressed by Natawidjaja et al. (2021) on the basis of their field reconnaissance.

In spite of the high activity of the causative Palu-Koro Fault (dislocation = as high as 29 ± 5 mm per year according to Bellier et al. 2001), little is known about the

Fig. 123 Nalodo at Sibalaya accompanied by a failed irrigation canal (drawn after JICA Study Team)

Fig. 124 Unpaved irrigation channel at the head of the Nalodo at Jono Oge that is empty after seismic breaching

Fig. 125 Breaching of irrigation channel at the head of the Nalodo at Sibalaya

(a) Ground surface without an irrigation channel

(b) Dry cliff at the head of the Nalodo

Fig. 126 Residential area above the head of the Nalodo in Balaroa

history of earthquakes and associating damage in the Palu region. Abendanon (1918) made a comprehensive technical trip in Sulawesi from 1909 to 1910. In pages 887–892 of his report, he mentioned an earthquake that hit the Palu (formerly called Paloe) basin in 1909 and was more devastating than the one in 1907. However, he referred only to house damage, cracks and new hot springs but did not mention landslides. It is interesting that he presented photographs of mountain slopes that were free of forest (Plates 275 and 276 therein). Trees had most probably been lost by cutting (shifting cultivation; slash and burn) and the photographed appearance of the mountain slopes is very similar to the present situation in Luzon Island and the Malay Peninsula. He paid attention to the substantial sediment yield and formation of huge alluvial fans.

Palu and north/central Sulawesi have been hit by damaging earthquakes in 1909 (Daryono and Natawidjaja, 2019), 1927 (Mw = 6.1; Thein et al. 2015), 1996 (M = 7.8 with tsunami, Pelinovsky et al. 1997; Gomez et al. 2000), 2005 (Mw = 6.2; Thein et al. 2015) and 2012 (Magnitude = 6.2; Thein et al. 2015). Nevertheless, no Nalodo is known. It is hence difficult to discuss the special cause of Nalodo long-distance flow on the basis of past experiences.

To date (September, 2021), there are several candidate causes of Nalodo long-distance flow in gentle slopes;

(1) Softening of very loose sand mixed possibly with cohesionless silt in which shear strength drastically decreases after peak strength in undrained conditions (see Fig. 127; refer to Ishihara 1993).
(2) Formation of thin layers of water under less pervious layers of soil, which is called the water film theory (see Fig. 128; Arulanandan et al. 1988; Boulanger and Truman 1996; Kokusho, 1999, 2000; Malvick et al. 2002),
(3) Ejection or bursting of artesian water pressure in an aquifer under alluvial fans, and
(4) Ejection or bursting of tectonic water that bears geopressure or over-pressure.

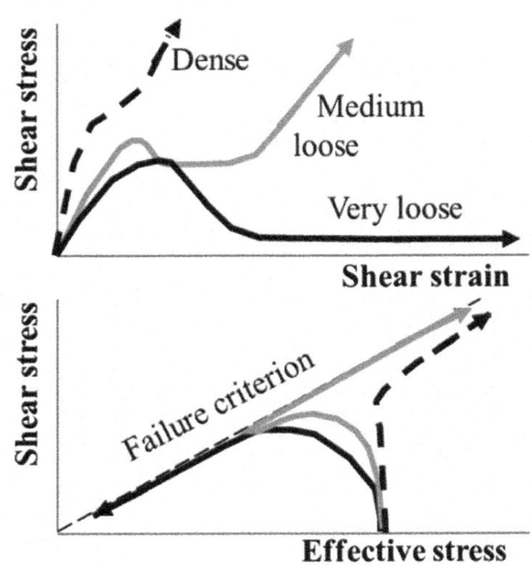

Fig. 127 Schematic illustration of softening behavior of loose liquefied sand undergoing undrained shear

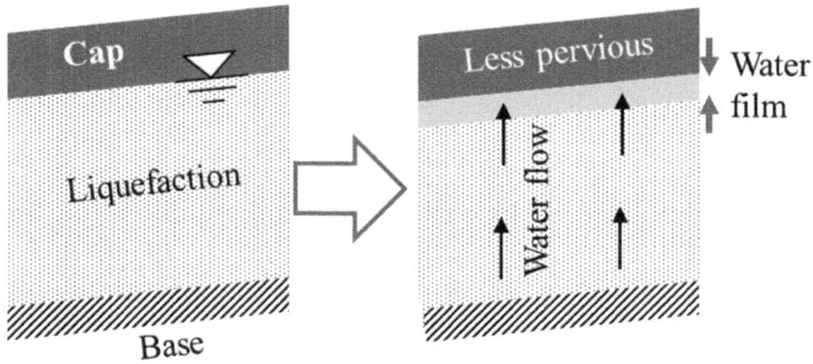

Fig. 128 Conceptual illustration of water film that promotes long-distance flow after liquefaction

The artesian pressure is referred to here because the Nalodo area had artesian springs before the earthquake.

The first idea of undrained softening was declined, however, because undisturbed soil specimens, that were collected at Nalodo sites and tested in Japan by the JICA Study Team, did not exhibit such a behavior. Since liquefaction occurred in Sulawesi, as verified by sand ejecta (Fig. 129), the formation of a water film between a liquefied sandy layer and a less pervious overlying cap layer is promising. This film is formed when pore water is expelled from the liquefied layer but cannot quickly flow through the less pervious cap layer. Accordingly, the water is retained between the cap and the liquefied sublayer (Fig. 128). Because this water has no shear resistance, the cap layer at the surface can translate freely under the action of gravity. Ironically, this idea is simple and likely to have occurred many times in the past. However, no liquefaction in sandy slopes resulted in a 500-m flow in the past. Therefore, the water film theory is not sufficient to account for Nalodo. An eyewitness report states that substantial water ejection started immediately (one minute or less) after the end of the earthquake shaking and that the ejected water was warm at some places (interview by JICA team). This is contradictory to the previous knowledge that water migration and drainage from the liquefied subsoil starts several minutes after shaking (Yoshida et al. 2007 on 1964 Niigata earthquake). The warm water suggests the effects of geothermal processes in the earth's crust. Thus, something more happened during the earthquake. Substantial water was provided to the ground surface and maintained the long-distance flow for a long time for several to 10 min according to eyewitness (interview by JICA team).

The third idea on artesian water aquifers comes from the geomorphology of the Nalodo area (Fig. 130). It is important here that Nalodo occurred in low places sandwiched by bigger and older alluvial fans. This finding suggests that the surface crust was disrupted upon liquefaction in a shallow sandy layer and allowed substantial water to be ejected from aquifers of artesian water, which is not uncommon in alluvial fans. Probably, many Nalodo have been repeated through geological history and the soil flow constructed the existing big fans. Now, the location of Nalodo has shifted to low places between fans, as observed in 2018.

Fig. 129 Ejection of liquefied sand around an active water spring (Jono Oge)

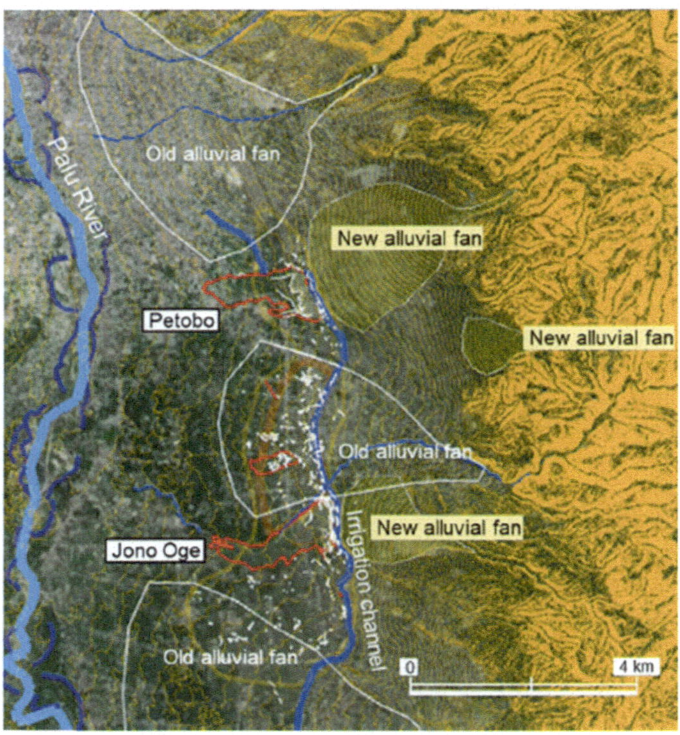

Fig. 130 Geomorphology of the Nalodo area (JICA Study Team)

The fourth idea is that substantial water was provided from the earth's crust, possibly through fault planes under the Nalodo sites. It should be recalled that Sulawesi is located on a triple junction of the Molucca, Banda, and Sunda tectonic plates and is extremely active in a tectonic sense, as evidenced by the high rate of offset along the Palu-Koro Fault (3–4 cm/year) (Bellier et al. 2001; Beaudouin et al. 2003; Socquet et al. 2006), as well as hot springs located along it (Jaya et al. 2019). Although the surface manifestation of the causative Palu-Koro Fault is located on the western side of the Palu Valley (Fig. 118), Patria and Putra (2020) suggested branch faults on the eastern side of the Palu Valley basin where three Nalodo (Petobo, Jono Oge and Sibalaya) happened.

To shed further light on water supply from deeper layers, which are artesian water or water from the earth's crust as candidate causes of Nalodo, historical studies were conducted to detect similar events. In Palu area, it is difficult to find any memory of Nalodo in the past, although earthquakes happened in 1907, 1909 and 1937 as well as in 1338 and 1468 (Daryono 2016). As an alternative, it was attempted to find a similar disaster elsewhere by using such keywords as earthquake, strike-slip fault, high rate of dislocation, pull-apart basin, alluvial fan, and ground water. Consequently, the candidate sites were the western part of the North Anatolian Fault (24 mm/year according to McHugh et al. 2006) that caused the 1999 Izmit earthquake and also many submarine landslides in the Marmara Sea and the Izmit Bay (Zitter et al. 2012), the Indawgyi Lake section (Tun and Watkinson 2017) of Sagaing Fault in Myanmar, and the Enriquillo–Plantain Garden Fault in Haiti (Mann et al. 1984, 1991) in association with the coseismic submarine landslides during the earthquake in 2010 ($M_w = 7.0$) (Hornbach et al. 2010; Rathje 2010). Nevertheless, no good information was detected.

A promising case was found in a volcanic mountain of Akagi (Fig. 1), Japan, whose height is 1827.6 m. Its slope collapsed most probably during the Konin earthquake of magnitude > 7.5 in AD 818 as shown by archeological and tephrochronological evidences (Noto 1990; Noto et al. 1990; Hayakawa et al. 2002). Because Mt. Akagi is a volcano, its slope is prone to seismic instability, as shown by other examples (e.g., Fig. 41 in Padang and Mt. Ontake in Fig. 158). No eruption is confirmed of Mt. Akagi during the historical era (Moriya 1971) but repeated sediment disasters have occurred on its slopes due to heavy rains (Koide 1950). I and colleagues searched a site of an ancient landslide and found the place in Maebashi City, Gunma Prefecture, central Japan (Figs. 131 and 132). In Fig. 131, a horseshoe-shaped depression is seen. Here, the cross section of the topography is as shown in Fig. 133, where there is a small head scarp at the top. Moreover, this area has been used for rice fields without houses, presumably because of an abundance of ground water (Fig. 134). This particular area has no archeological ruins in contrast to many of them in the surrounding area (Fig. 135). Since most archeological ruins in Maebashi City date back, at the youngest, to the Heian Era (AD 9th–12th Centuries), and few belong to the following Medieval Period or later (4193 sites until the Heian Period and 851 sites after the Medieval Period) (Board of Education of Maebashi City 2013), most sites in the Mt. Akagi area (Fig. 135) originates from Heian or older times and none has been found in the suspected location at the center of the area. Human activity in the lengthy Heian Period (lasting for four centuries) was declining due to the reliance on contract and its consequence in deterioration of governance of the central government (Sato 2020) in association with the regional unrest. The lack of archeological ruins in the center of the area in Fig. 135 is probably because of its suitability for rice agriculture and/or an ancient landslide. If the former is the case, the abundance of water for rice farming suggests the risk of landslides. If the latter is the case, the disaster occurred after the period of most of the archeological ruins; after AD eighth century or possibly earlier (Kumano 1992; Kasukawa Village 1994) but before the Medieval time. This perception is not contradictory to the idea of a huge landslide disaster at the beginning of the AD ninth century when the 818 Konin mega earthquake happened.

The elevation of this area is 270–300 m above sea level. Because of its flat topography, this area is not prone to common types of rainfall-induced landslides and erosion. The severest rainfall disaster in the past 80 years occurred in 1947 at a higher elevation on the Akagi Volcano (more than 600 m above sea level) where erosion is active (Koide and Noguchi 1950; Tsujimura 1950). In contrast, the lowest part of the volcano's flank is characterized by many avalanche hills that are products of ancient catastrophic eruption and collapse of the Akagi Volcano (Sawaguchi 2013). Thus, there is a good possibility that a big landslide occurred at the studied site in the middle flank of the mountain, but accurate information was not obtained on its age and the cause.

I found two examples of landslides induced by ejection of water coming from the deep earth crust. The first one is Makiuchi landslide (Mochizuki and Nakamura

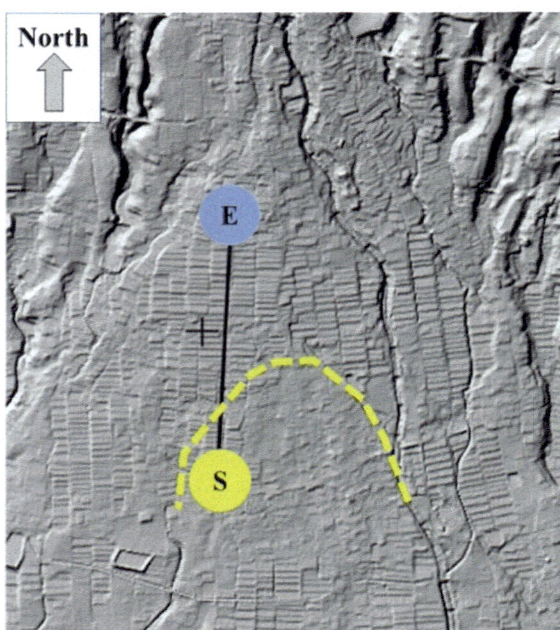

Fig. 131 Possible site of an ancient landslide in Miyagi and Naegashima of Maebashi City, south flank of Mt. Akagi (drawn on digital map, Geospatial Information Authority of Japan)

Fig. 132 Probable top of ancient landslide on the south flank of Mt. Akagi

Fig. 133 Cross section of slope of Mt. Akagi along the E-S line in Fig. 131 where an ancient landslide presumably occurred (data by field survey)

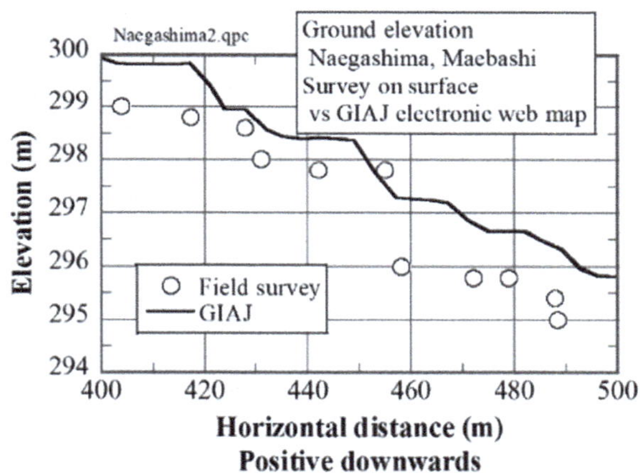

1982), which was triggered by water ejection during the Matsushiro earthquake swarm in 1960s in Japan (Fig. 136; Tsuneishi and Nakamura 1970). Interestingly, a water spring started from the head of the failed slope two hours before the onset of landsliding (Morimoto et al. 1967). This swarm was most probably caused by water ejection from the earth crust, wherein the effective stress and shear strength were reduced by excess pore water pressure (Hagiwara 1972; Nakamura 1981, 1984). Water ejection was ample during this swarm. Thus, there is a sequence of events starting from ground water ejection to earthquakes and then to landslides.

The other possibility is the Matsunoyama landslide, which occurred in 1962–1963 in Niigata, Japan (Taniguchi 1964a); see Fig. 137. The area of slope instability was 850 ha and the ground movement was associated with an ejection of salty water

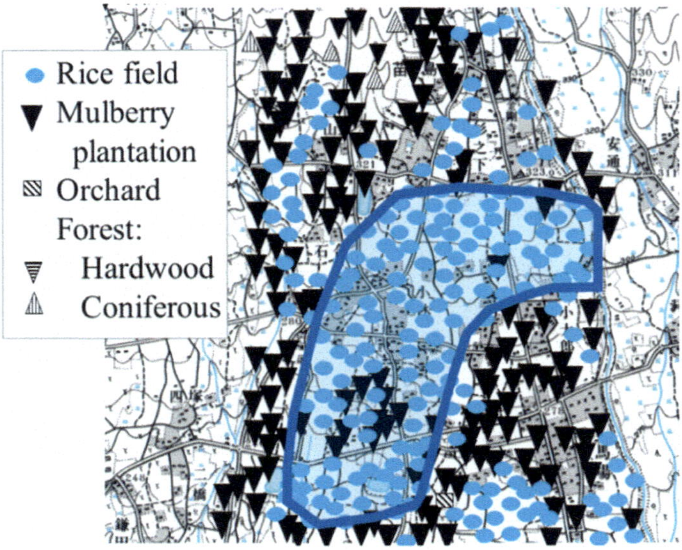

Fig. 134 Land use of the possible site of an ancient landslide on the south slope of Mt. Akagi (drawn on 1929 version of topographical map of Geospatial Information Authority of Japan)

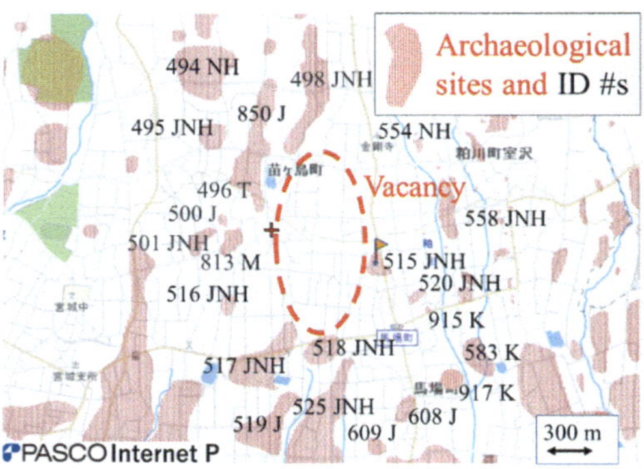

Name of archaeological or historical periods	Notation	Very approximate idea of centuries
"Jomon" (cord-marked pottery); Neolithic	J	Till BC 4th Century
"Kofun" tumulus period	K	AD 3rd or 4th - 7th Century
"Nara"	N	AD 8th Century
"Heian"; capital in Kyoto	H	AD 9th – 12th Century
Medieval	M	13th Century -

Board of Education, Maebashi City (2013)

Fig. 135 Distribution of archeological sites on the south flank of Mt. Akagi (drawn on Mapping Gunma after Board of Education, Maebashi City 2013)

(Fujita 1965). Hot springs in Matsunoyama have high salt concentrations (8800–9400 mg/l for Cl^-) and *ph* (acidity) = 7.3–7.7. In comparison, the salt concentration of sea water is 19,000 mg/l and ph is 7.4–8.0. Matsunoyama has been a site of slope instability for many years (Fig. 138) and a landslide occurred there in 2001 again (Yamagishi et al. 2001). Another interesting feature is mud volcanoes, which are good evidence of highly pressurized tectonic water, although their activity is low nowadays in Matsunoyama (Fig. 139). Bigger and longer activities of mud volcanoes have been known in Indonesia (LUSI; Mazzini et al. 2007) and Azerbaijan

信濃毎日新聞社

Fig. 136 Makiuchi landslide during the Matsushiro earthquake swarm (the Shinano Mainichi newspaper on September 17th, 1966)

(Gobustan; Babayev et al. 2014). There are many underground columns of mud in the area and they are under high pressure. Construction of the Nabetachiyama railway tunnel in Matsunoyama encountered one such column and heaving of clay made the tunneling extremely difficult: 22 years were spent on 9111 m tunneling (1973–1995; Tanaka and Ishihara 2009). Moreover, Matsunoyama is within tens of km of the area of the Matsushiro swarm and the Niigata-Chuetsu earthquake, and thus is located in a tectonically active region. With these in mind, Watanabe et al. (1997a, b) together with Oki et al. (1999) insisted that the fossil salt water is tectonically pressurized and ejected toward the surface, thus generating mud volcanoes and reducing effective stress, as well as shear strength in slopes. The pressurized water was called geopressured by them, and they considered it to be the cause of the Matsunoyama landslide and also earthquakes in the region (Oki et al. 1999).

Seismicity induced by high tectonic water pressure was discussed by Sibson (1995), Rice (1992), Bellot (2008) and Wei (2015) as well. High tectonic activity, faults, hot springs of salty water and slope instability are common features in the Matsunoyama region, Niigata in central Japan, and in the Palu Basin. Therefore, it deserves attention to discuss the effect of "geopressure" generated by tectonic and/or hydrothermal processes as one of the possible causes of Nalodo in Sulawesi. The

Fig. 137 Matsunoyama landslide in 1962–1963 associated with an eruption of ground water; photo by Mr. T. Takizawa (Taniguchi 1964b)

Fig. 138 One of the former landslide sites in Matsunoyama that has been stabilized by water drainage

Fig. 139 Tiny but active mud volcano at Gamoh near Matsunoyama

short interval of one minute or so between shaking and water ejection at the onset of Nalodo is understandable as the consequence of fault rupture triggered by elevated water pressure in the fault plane.

Sand liquefaction is not the only causative mechanism of ground water ejection. There are several earthquakes that were associated with ejection and other types of anomaly in ground water conditions. In some cases, the water ejection resulted in landslides shortly after the quakes, although not as immediately as was the case in Sulawesi. The event in Matsushiro was one of them (Fig. 136).

Muir-Wood and King (1993) called attention to the increased rates of ground water ejection and river water flow. For example, several hours or days after the 1983 Borah Peak Earthquake of Mw = 6.9 in the USA, water springs started (Wood and Steedman 1992). Waag (1985) mentioned substantial expulsion of ground water after the 1959 Hebgen Lake earthquake of Mw = 7.2 in the USA but did not refer to the elapsed time between shaking and water expulsion. This earthquake was associated with a landslide 5 days after the earthquake and increased the ground water flow (Witkind et al. 1962). Swenson (1964) said that the water was warmer than air temperature but did not provide details. Significant water ejection reported by Wood (1985) started shortly after the shaking of the Hebgen Lake earthquake. He mentioned that the mud flow of 200 thousand cubic meters at Lupine Lake, which is located 30 km from the epicenter, was triggered by increased spring water two days after the earthquake. Another water spring in the Clayton silver mine started 3 h after the earthquake. These water anomalies were attributed to the compressive strain of the earth crust, together with the increased permeability due to fissures generated by the earthquake.

Fujimori et al. (1995) reported a 10-time post-earthquake increase of spring water in Rokko Mountain, where the causative fault of the 1995 Kobe earthquake (onset at 5:47 a.m. on January 17th by local time) was located. This anomaly was attributed to the coseismic compressive strain of the mountain body. After this earthquake, the Kotaki landslide occurred in the epicentral Awaji Island in the midnight of January 18th or in the dawn of 19th (Fujiwara et al. 1996). This delayed landslide was interestingly associated with significant water ejection. Furthermore, Ohshima et al. (1996) reported ground water anomaly along the causative Nojima Fault in Awaji Island. Tokunaga (1999) found an increase in spring water in Awaji Island and attributed it to the change of permeability in the earth.

Similarly, the 1989 Loma Prieta earthquake was accompanied by increased spring water or stream flow (Rojstaczer and Wolf 1994). These were attributed to an increase in permeability in the earth. This means that fissures and fractures were created in the rock by strong seismic shaking. Curry et al. (1994) reported a similar phenomenon and referred to volumetric compression of the earth crust as another cause of the water ejection. A similar anomaly happened after the 1906 San Francisco earthquake of Mw = 7.9; Briggs (1994) stated that the increased flow in a stream took 50 days or half a year to drop down to the original level. Moreover, a new spring started 30 min after the Loma Prieta earthquake. Sato (2015) addressed a new spring that started after the 2011 Tohoku earthquake on March 11th and water ejection continued for 4 years. Interestingly, the region where well water rose was the region where tectonic strain was compressional during one of the strong aftershocks in April, 2011. Mori (1987)

Fig. 140 Sumiyoshi River with some water shortly "after" the 1995 Kobe earthquake on January 17, 1995; see local people collecting water for emergency use

found an official historical report that river water increased after the 1854 Ansei Tokai earthquake of Mw = 8.6. It is interesting, further, that the water level of the Sumiyoshi River (Fig. 140) suddenly increased three days "prior to" the 1995 Kobe earthquake in spite of the dry season (eyewitness report on earthquake precursors prior to the Kobe earthquake by Wadatsumi 1995). Montgomery and Manga (2003) collected data from case histories to plot the epicentral distance to sites of ground water anomalies against earthquake magnitude. It was shown that greater magnitude changes ground water conditions at a greater distance.

Big earthquakes are frequently associated with increased spring water and water ejection presumably because of tectonic compression of the earth's crust and/or fissures in a mountain body. Obviously, landslides are promoted under such circumstances. This is one of the reasons for the long-term instability of mountain slopes after big earthquakes.

In summary, the JICA Study Team on Nalodo concluded that this gigantic earth flow was triggered by liquefaction in the surface soil but inferred that its long runout was induced by something more. The possible mechanisms for this are (1) the effects of infiltration of leaked irrigation water (Petobo, Jono Oge and Sibalaya sites but the Balaroa and Lolu sites do not have irrigation channels), (2) drainage of water from liquefied sublayers and formation of water films under the cap layer, (3) the gigantic water flow from artesian aquifers under alluvial fans, and (4) dynamic ejection of water from the earth's crust or faults. Table 2 compares the candidate mechanisms and indicates that no single idea is best and that Nalodo long-distance flow was induced by the combined effects of several mechanisms. The idea of softening of sand during undrained shear is not included here because laboratory tests on undisturbed soil samples did not exhibit this behavior. Moreover, an irrigation channel as the water source is excluded because the Nalodo in Balaroa did not have it (Fig. 126).

Evaluation (A, B or C) is my personal idea.

Table 2 Comparison of three ideas on the cause of Nalodo

Candidate Mechanisms (2), (3) and (4)	Water film is formed by this mechanism?	Ground flow is resumed immediately after the end of shaking?	Water ejection is huge?	Flow runout > 500 m is possible?	Ejection of hot water is possible?
(2) Drainage from liquefied layer	Yes (A)	Drainage requires 10 min or longer (C)	3–4% of the volume of liquefied layer only (B)	No in case histories (B) possibly because liquefied slopes were not very long	It happened in the past; in the 1990 Luzon, 1999 Kocaeli, and 2001 Gujarat earthquakes. In the 1964 Niigata earthquake, the ejected water was not warm. (A)
(3) Water ejection from aquifer under alluvial fan	Yes (A)	Shorter time is enough if artesian pressure exists near the surface (B)	Possibly yes, if the aquifer is huge (A)	Yes (A)	No (C)
(4) Water ejection from earth crust	Yes (A)	Shorter time is enough if tectonic water pressure exists in the fault and triggers the earthquake (A)	Yes. See the examples of Matsushiro swarm. Also, Hebgen Lake and Borah Peak earthquakes; but at some time after the quake (B)	See Makiuchi landslide in Matsushiro (Fig. 136) (A)	Geologically possible. There are hot springs in the Palu Basin. (A)

8 Mechanisms of Earthquake-Induced Landslides

Shear is one of the most important keywords in soil mechanics. Slope stability analysis in soil mechanics always assumes a shear mechanism and calculates the factor of safety for both gravitational and seismic failures. This tradition, however, overlooks something important in reality. When the Kanagawa Prefecture of Japan (1986) conducted statistical analysis by referring to the landslides induced by the 1923 devastating Kanto earthquake (Fig. 90), it was found that hard rock slopes were more vulnerable than soil slopes. In other words, coseismic landslides occurred more frequently in rock slopes than in soil slopes during the earthquake. Since this was contradictory to soil mechanic assumptions, I wished to find the reason. The first idea was that many unstable soil slopes had slipped down during heavy rains before the earthquake and that only the more stable soil slopes experienced the Kanto earthquake. Conversely, rock slopes had survived heavy rains and then some of them were affected by the earthquake. Statistics interprets this situation that rock slopes are more likely to slide than soil slopes. Recently, I am thinking in a different way.

Landslides are classified into falls, topples, slides, spreads, and flows (Varnes 1958; Hungr et al. 2014). Although shear plays a major role in slides, spreads, and flows, and soil is more vulnerable to these types of landslides, falls and topples are controlled by different mechanisms. The rock shown in Fig. 141 is an example of a "fall." The Kashmir rock fall in Fig. 142 was initiated by a fall or exfoliation, followed by sliding downward on the slope. Figure 143 shows the gigantic Ohya landslide, Japan, which was triggered by the 1707 Hoei earthquake of Mw = 9.3 and involved 120 million m^3 of landslide mass (Takeuchi and Tsutsumi 1985). Mokudai and Chigira (2004) carried out a detailed study of the local topography and pointed out the toppling (inclination) behavior. Thus, it seems that a good number of earthquake-induced landslides have been associated with fall and/or toppling, not shear (Fig. 144). This is particularly the case for small but numerous events. Earthquake-induced landslides tend to occur on slopes at steeper angles than gravitational landslides (Figs. 78, 79, 80). The greater slope angle implies proximity to mountain crests where topographic amplification (Fig. 20) is significant. Thus, the increased seismic acceleration and inertial force can promote fall and exfoliation as well as toppling of an unstable (fractured) rock mass. See the significantly fractured rock at the Ohya site (Fig. 145), which Karasawa and Kano (1992) attributed to slaty cleavage that is a consequence of substantial compression due to tectonic actions; see folding in the Ohya site in Fig. 197. The important role played by compression, together with recrystallization and grain rotation, has been stressed by Sorby (1856) and Wood (1974), but further discussion is beyond the scope of this paper. All in all, the causative mechanism of earthquake-induced landslides consists of shear, fall and toppling. It is thus difficult and inappropriate to specify any single one of them in developing risk assessment methods.

The Shiratori-yama Mountain that is located near Ohya (Fig. 1) experienced the first coseismic landslide (1.8 million m^3) during the 1707 Hoei earthquake and this

Fig. 141 Big fallen rock in Yingxiu after the 2008 Wenchuan earthquake, China

Fig. 142 Fall or separation of surface materials during the 2005 Kashmir earthquake

Fig. 143 Ohya landslide with active creep toward toppling

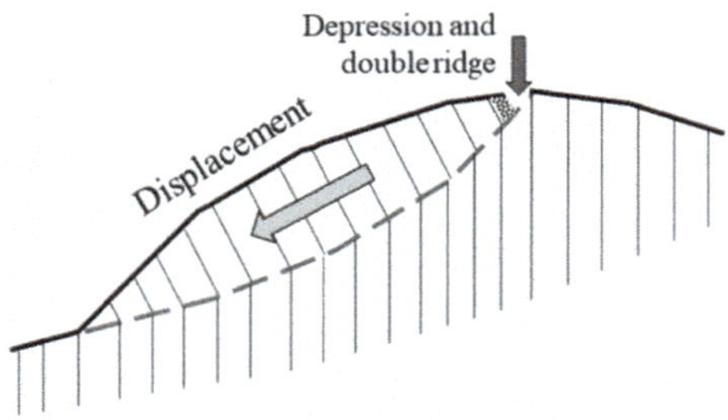

Fig. 144 Toppling and depression of a mountain ridge

Fig. 145 Fractured surface of a slope at Ohya exhibiting slaty cleavage in Setogawa Group according to Karasawa and Kano (1992)

landslide mass blocked the Fuji River (Fig. 146). Afterward, the remaining destabilized mass fell down again during the 1854 Ansei earthquake (Inokuchi and Yagi 2013; Watanabe et al. 2013). It is very interesting that the mountain slope above the failed part is still covered by fractured rock mass (Fig. 147), which is similar to the situation in Ohya (Fig. 145) (Towhata 2022). This situation suggests that the mountain body underneath has been affected by mechanical weathering and disintegration and the entire slope is subject to coseismic failure (Fujii 1977).

9 Effects of Antecedent Rainfall and Compound Disaster

Water is an important agent that can aggravate geotechnical problems and the earthquake-induced landslide is not an exception. Antecedent rainfall prior to earthquakes influence the extent of landslide. The couple of earthquakes (Mw = 6.7 and

Fig. 146 East flank of Shiratori-yama Mountain where the 1707 Hoei earthquake triggered a landslide and blockage of the Fuji River

Fig. 147 Upper slope of Shiratori-yama Mountain covered by fractured rock

7.1) on March 5th, 1987, in Ecuador induced several thousand landslides because of the antecedent rainfall of 600 mm in one month (Tibaldi et al. 1995). To facilitate a quantitative discussion on rainfall effects, Fig. 148 compares precipitations prior to the 1995 Kobe earthquake ($Mw = 6.9$) and the 2004 Niigata-Chuetsu earthquake ($Mw = 6.6$). Both earthquakes had similar magnitudes and had ruptures occur in landslide-prone mountains. Nevertheless, the consequences were profoundly different; in Kobe few landslides were triggered by the earthquake, while the Niigata-Chuetsu earthquake induced many. The most likely reason for this difference is the extents of the antecedent rainfalls. As shown in Fig. 148, the Kobe earthquake occurred on January 17th which was in the middle of a dry winter. The mountain slope surface was dry and developed few failures. Conversely, the Niigata-Chuetsu earthquake happened on October 23rd, at the end of the rainy season from June to October. A typhoon had brought 100 mm of precipitation three days before the earthquake. Consequently, many landslides occurred (Fig. 13).

The 1999 Chi-Chi earthquake in Taiwan happened on September 21st at the end of the rainy season (Fig. 149). This is one of the reasons why the triggered landslide disasters were very serious. The 2018 Eastern Iburi earthquake in Hokkaido, Japan,

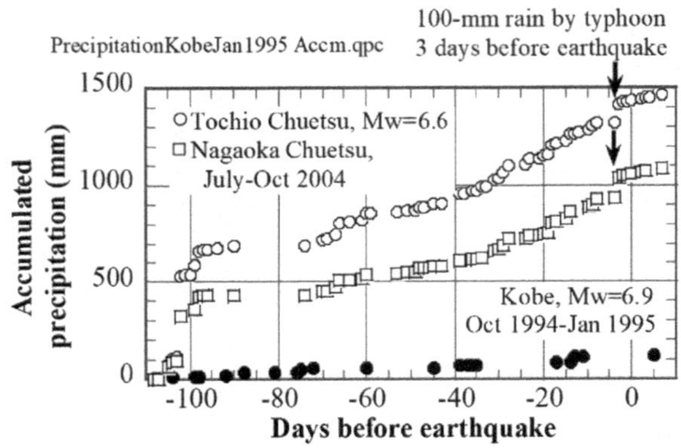

Fig. 148 Antecedent rainfalls before Kobe and Niigata-Chuetsu earthquakes (data by JMA)

triggered many landslides along wet tephra layers (Fig. 66). There was ample rainfall before the earthquake (Fig. 150). The total precipitation from July 1st to September 5th, the day before the earthquake, was 408 mm, and it is likely that this moisture condition induced the landslide disaster (Fig. 64).

An opposite example is the 2011 Tohoku earthquake of Mw = 9.0 that induced an unexpectedly small number of landslides in natural slopes, in spite of the substantial extent of subsoil liquefaction (Sasaki et al. 2012; Towhata et al. 2014). One earth-fill dam breached (Tanaka et al. 2013). Figure 151 depicts the rainfall record in Sendai City, which is near the middle of the affected region. The accumulated precipitation was only 360 mm in Sendai from the beginning of 2011 till the earthquake on March 11th. This number is substantially smaller than the accumulated precipitations before the Niigata-Chuetsu earthquake (Fig. 148) and the Chi-Chi earthquake (Fig. 149), and was less than the two-month rainfall before the Eastern Iburi earthquake (Fig. 148). It is therefore reasonable that the Tohoku earthquake occurred at the end of a dry winter of the affected region and did not trigger many landslides.

Water is not necessarily provided just by rainfall into seismically unstable slopes. The other source of water is inappropriate irrigation and leakage from water pipelines.

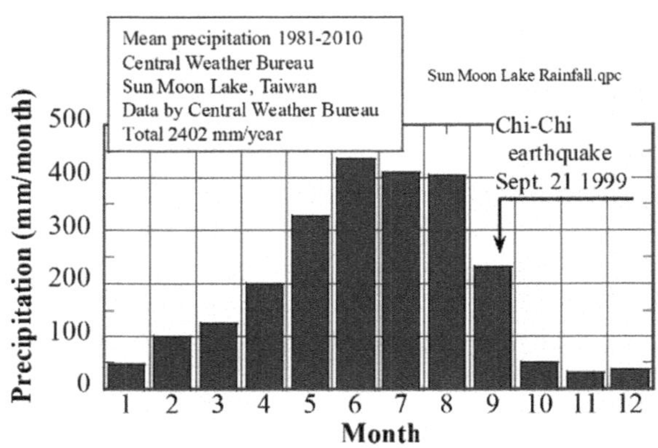

Fig. 149 Mean monthly precipitation at Sun Moon Lake Station, Taiwan

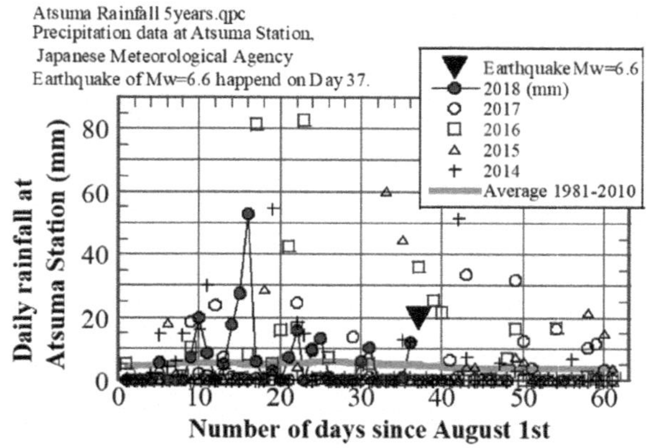

Fig. 150 Rainfall records at Atsuma near the landslide sites (Total rainfall from July 1 to September 5 = 408 mm in 2018)

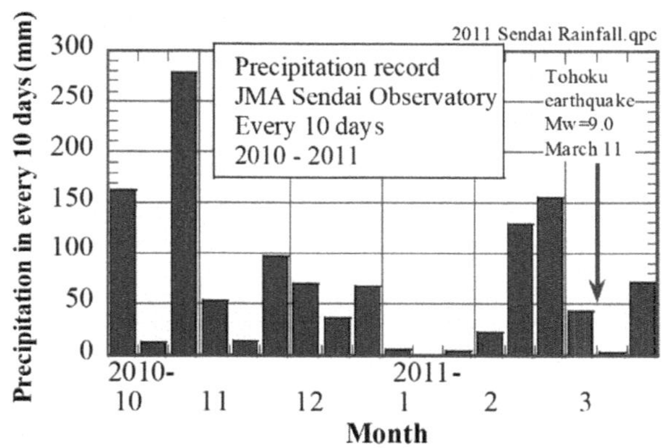

Fig. 151 Rainfall in Sendai prior to the 2011 Tohoku earthquake

This was the case at three Nalodo in Sulawesi (Fig. 125) and the event in Tajikistan (Fig. 85). Moreover, the Galdian landslide, which started one day after the 1990 Manjil earthquake of Mw = 7.4 in Iran (Fig. 152) and had 20 million cubic meters in volume, was most probably the consequence of irrigation (Ishihara et al. 1992). Careless water management in arid and semi-arid regions may lead to a catastrophic ground failure. Water submergence of partially saturated soil destroys the apparent cohesion among soil grains that is produced by surface tension of pore water. This phenomenon is known as collapse in soil mechanics.

10 Compound Effects of Landslide and Other Types of Disasters

In addition to any antecedent rainfall that may occur shortly after earthquakes, discussion is made here of the combined effects of earthquake-induced landslides and other

Fig. 152 Galdian landslide in Ira, triggered by the 1990 Manjil earthquake

disaster mechanisms. The 2016 Kumamoto earthquake in April was followed by a heavy rain-induced disaster in June. Hence, there was concern about further slope instability due to rain (Fire and Disaster Management Agency 2016). This situation is not exceptional and one can find similar situations at many other places. The 1923 Kanto earthquake was preceded by a typhoon and heavy rain one day before the earthquake and followed by heavy rain two weeks later, and slope stability was affected (Inoue et al. 2013). There are numerous combinations of different types of disaster in combination with each other. For example, earthquake-induced landslides are sometimes combined not only with rainfall but also with tsunami and floods. The 1707 Hoei earthquake of Mw = 9.3 in Japan was caused by the subsiding tectonic plate in the Pacific Ocean and was followed by the huge eruption (Volcanic Explosivity Index, VEI = 5) of Mt. Fuji seven weeks later; the activity of this volcano is governed by the subduction of the same tectonic plate. The following discussion concerns the short-term combination with rain and other issues will be treated later.

The 2008 Wenchuan earthquake occurred on May twelfth, and a huge amount of debris slid down into valleys and remained there. In the following years, heavy rainfalls in the damaged area washed the soil deposits downstream, and caused disasters. Heavy rains in July and September, 2008, triggered debris flows in Zoumaling (Yang 2019) (Fig. 153). Figure 154 shows the site of a debris flow on August 13, 2010, which had 3.1 million cubic meters in volume. As a consequence, seven people died and five were missing, while 479 houses were buried (Yu et al. 2010a). Qingping County, Sichuan Province (Fig. 155), is another site of debris flows on the same day; note that a huge amount of soil still remains after the wash-out. It was fortunate here that the local community had known that a devastating flash-flow disaster at Zhouqu had happened one week before (on August 7) and claimed 1744 lives

(Yu et al. 2010b; You et al. 2011). Hence, people had been on alert and evacuated, limiting the damage. Ma and Qi (1997) stated that the valleys of Zhouqu had been filled with sediments that were produced by earthquakes in 1879 and 1960. I also suppose that stream channels in Zhouqu Town did not have a sufficiently big cross section to accommodate the debris flow. Indeed, this is the case at many places in the world where water channels are designed to accommodate design "water" flows and do not consider debris flows.

A tsunamigenic landslide is another type of a compound disaster. Gutenberg (1939) pointed out that submarine landslides, whether coseismic or non-seismic, should be considered as one of the chief causes of tsunami. There are many cases in the past in which landslides induced tsunami. For example, the area of Messina Strait in Italy has been known for coincidence of earthquakes and tsunamis as discussed by Maramai et al. (2006) on the earthquake sequence in 1783 (the maximum magnitude = 7.0). After the earthquake (Mw = 7.1) and tsunami in 1908, Omori (1909)

Fig. 153 Site of 2008 Zoumaling post-earthquake debris flow of 30 million m^3 (the building in this photograph was constructed after the Wenchuan earthquake)

Fig. 154 Site of subsequent heavy rain and debris flows in Wenjia Gully

Fig. 155 Site of subsequent heavy rain and debris flows in Qingping County, Sichuan Province

carried out a reconnaissance study and showed the distribution of tsunami direction and run-up at many places along the coast. Ryan and Heezen (1965) stated that this distribution suggests a moving origin of the tsunami on the seabed, which was presumably submarine landslides. More recently, Favalli et al. (2009) back-calculated the tsunami propagation to identify its origin as a landslide in the seabed. Moreover, Tinti et al. (2004) provided a catalogue of historical tsunami through the history of Italy.

What was more striking was the event on May 22nd of 1792 in which the collapse of Mayu-yama Mountain (Fig. 156) in Unzen Volcano (Fig. 1), Kyushu, west Japan, triggered a volcanic avalanche (Figs. 157 and 158) that fell into the sea and triggered a huge tsunami disaster. Around the site of this collapse, 9745 were killed (Sato 1925). The volume of the landslide was 500 million m^3 (Omori 1908). This tsunami crossed the Ariake Bay and induced a catastrophic tsunami in Kumamoto (Fig. 159), where the run-up height was 55 m and 4000 people were killed (Sato 1925; Gusiakov 2009).

There are two opinions about the cause of the collapse of Mayu-yama Mountain; strong shaking (earthquake) of volcanic origin (Omori 1908, 1918) or a volcanic explosion (Sato 1925; Furuya 1974). The latter is based on the distribution of numerous avalanche hills (Fig. 158) as evidence of a volcanic eruption, because similar geomorphology is known around Mt. Bandai (eruption in 1888) and Mt. Chokai (the present Kisagata scenic point that used to have many small islands in a lagoon but was uplifted by the 1804 Kisagata earthquake of Mw = 7.5) volcanos, as well as many others in the world. However, there is no eyewitness report that volcanic ash fell down as evidence of an eruption (Omori 1918; Honma 1936). Mt. Bandai may not be a good example because its AD 1888 eruption was not a magma-induced eruption but a steam explosion in which the high pressure of heated vapor reduced the effective stress in the mountain body and finally destroyed the entire mountain (Mimura 1988). In this regard, the case of Mt. Bandai is similar to the idea

Fig. 156 Present view of collapsed Mayu-yama Mountain, Unzen, Japan

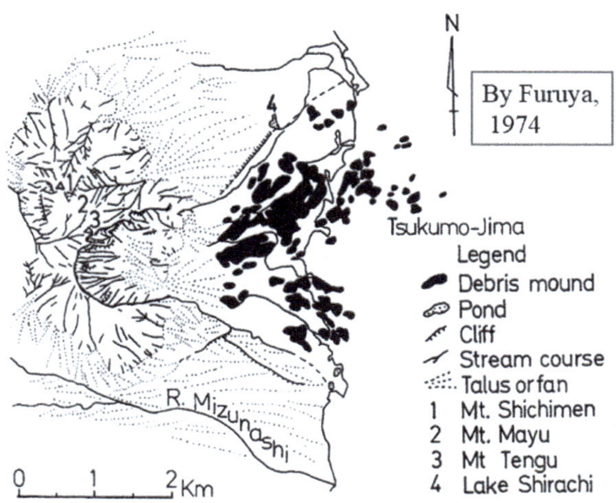

Fig. 157 Plan view of collapse of Mayu-yama, Unzen, Kyushu, Japan, and distribution of avalanche hills (Furuya 1974)

of shaking-induced collapse of the unstable Mayu-yama. Katayama (1974) carefully studied historical documents on the disaster and understood that hot water had been ejected before the disaster and that there was no ash fall during the mountain collapse. Summarizing all documents, he proposed that the disaster was triggered by an earthquake (of volcanic origin). By referring to this study, I briefly summarize the sequence of events as:

Fig. 158 One of the avalanche hills made by the Mayu-yama collapse that still remains today

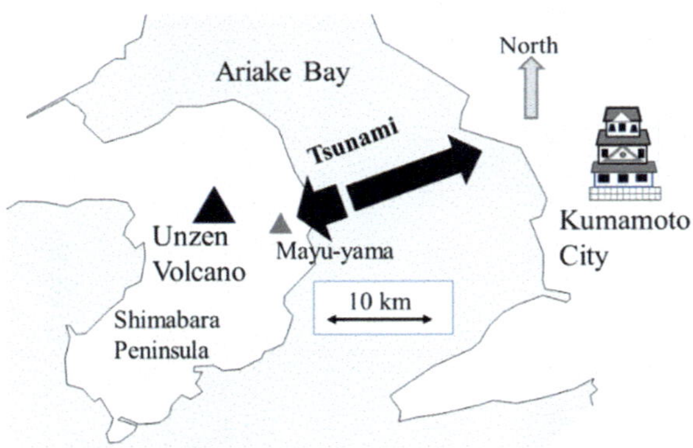

Fig. 159 Path of tsunami in 1792 that affected both sides of the Ariake Bay, Kyushu, Japan

- Initiation of volcanic seismicity by magma movement,
- Heating of ground water at depth by magma,
- Heated water rising toward surface,
- Deformation of the mountain by this water pressure associated with new water springs and cracks, leading to instability,
- An non-devastating volcanic earthquake on May 21st,
- Collapse of the mountain body, possibly with a gigantic landslide along a deep slip plane,
- An avalanche flowing into the sea and uplift of the seabed at the end of the slip plane,
- Tsunami.

The tsunami water was hot (Katayama 1974) possibly because ground water was ejected out of the seabed. After the disaster, a new artesian spring started in the former Mayu-yama site on May 30th and June 2nd (Katayama 1974). The body of Mayu-yama was still unstable and developed landslides, depressions, and new water springs during heavy rains and the earthquakes in 1922 (Ota 1969).

Ontake Volcano is situated in Nagano Prefecture of Japan (Fig. 1) and reaches 3067 m above sea level. A huge landslide was induced by the 1984 Naganoken-Seibu earthquake of $Mw = 6.2$ (Kitazawa et al. 1985; Momikura et al. 1985). Among several landslides, the one in Fig. 160 was the biggest and its volume was 34.5 million m^3. Voight and Sousa (1994) interpreted the flow path of this avalanche and obtained $H/L = 0.12$, which makes the Ontake debris flow highly fluidized. The length of the failed mass was 1 km, while width = 300 m, depth = 100 m and area = 44.18 ha. This huge mass fell down from the altitude of 2500 m and the drop height (H) was 1550 m. Accordingly, the avalanche velocity reached 80–100 km/h. Consequently, a landslide dam was formed at the valley bottom. It is believed that a layer of weathered pumice formed the slip plane. I suggest that the volcanic sliding mass of Ontake developed into avalanche because the steep mountain slope provided energy for the mass to be broken into pieces. In contrast, the mass of Mayu-yama was situated on a gentle slope and needed geothermal pressure to increase the pore pressure and reduce the material strength.

Collapse of Anak (=child) Krakatau Volcano, Indonesia, triggered a tsunami in 2018 (Ye et al. 2020). Fanetti et al. (2008) as well as Kremer et al. (2012) discussed ancient tsunami events in lakes triggered by landslides. Quick clay in Rissa, Norway, in 1978 flowed into Lake Botnen and triggered a small tsunami (L'Heureux et al. 2012). The island reclaimed for Nice Airport, France, partially flowed into the sea when it was affected by a tsunami (sudden drop of sea water level) generated by a submarine landslide of soft flood deposits on the sea bed (Auffret et al. 1982; Seed

Fig. 160 Coseismic landslide of Mt. Ontake, Japan, in 1984 (Photograph by Prof. S. Yasuda)

et al. 1988). The ancient gigantic Stregga subaqueous landslide off Norway generated a huge tsunami in the North Sea (Bondevik et al. 1997; Mountjoy and Micallef 2018).

From a seismic perspective, the 1929 Grand Banks earthquake of Mw = 7.2 triggered a gigantic submarine landslide or turbidity current (Heezen and Ewing 1952; Kuenen 1952) that produced a significant tsunami disaster (Fine et al. 2005) in the coastal area of Newfoundland, Canada. Hanrahan (2004) summarized the local situation starting from minor tremor and ending in the arrival of the tsunami. The most striking event occurred in Lituya Bay in Alaska in 1958, when an earthquake of Mw = 7.8 caused a huge mass of mountain slope to fall down into the fjord. The fallen material induced a tsunami, which overtopped another slope at a height of 524 m above sea level (Miller 1960).

The 1964 earthquake in Alaska (Mw = 9.2) induced submarine landslides and tsunami disasters. Valdez (Fig. 161) faces an inlet isolated from the outer ocean. The submarine slope in front of Valdez liquefied during the earthquake and flowed down, and the volume of this mass movement was 75 million cubic meters (Coulter and Migliaccio 1966, 1971). This movement triggered a tsunami that devastated the town and claimed 32 lives (Fig. 162). The submarine slope was susceptible to seismic liquefaction and flow failure for a number of reasons:

- The submarine slope consisted of rock flour of silty grain size that was produced by glacial erosion of mountain rocks and transported into the sea by river streams,
- Silt particles were deposited in water very slowly and softly because of its fine size and thus formed a very loose, soft, cohesionless, and water-saturated subsoil, which is prone to liquefaction,
- The inlet of Valdez was protected by peninsulas from ocean waves, and thus compaction by cyclic wave action was weak,

Fig. 161 Air view of the coastal area in former Valdez, Alaska, the USA (photo by K. Horikoshi)

Fig. 162 Ruin of Valdez Harbor

- Abundant sea water was mixed with the flow of soil and promoted a longer runout. Hence, the bathymetry in front of Valdez changed drastically and a tsunami was generated,
- The closed shape of the bay caused a seiche of the tsunami wave and aggravated the disaster (Coulter and Migliaccio 1966) and
- The tsunami run-up height was 66 m (=220 ft) (Houston 1979).

After the disaster, the Valdez community was relocated to a safer place.

The town of Seward suffered from a tsunami triggered by a submarine landslide during the same earthquake in 1964. Hampton et al. (1993) described that many cracks opened along the shore line and landslides occurred under water. Shortly after this, the tsunami came in. The short time that was elapsed between the landslide and the tsunami implies that the tsunami came from a nearby source, not from the offshore subduction zone. I suppose that the 1.5-m tectonic subsidence of ground level in the coastal region (Plafker 1969) affects its tsunami-proneness during gigantic earthquakes, e.g., the coseismic subsidence induced by the 1960 Valdivia earthquake of Mw = 9.6 in Chile (Weischet 1963) and by the 2011 Tohoku earthquake (Nishimura 2014).

The northern coast of Papua New Guinea was struck by a devastating tsunami in 1998 after an earthquake of Mw = 7.0. The cause of this tsunami is supposed to be a submarine landslide triggered by the earthquake (Kawata et al. 1999; Tappin et al. 1999; Synolakis et al. 2002). In 2008, a small part of the Aratozawa coseismic landslide mass (Figs. 38 and 39) fell into the dam reservoir (Fig. 163a) and triggered a tsunami. Fortunately, the over did not overtop the dam and the Aratozawa rockfill dam was maintained intact (Fig. 163b).

In summary, earthquake-induced landslides can trigger tsunami, as exemplified by many cases in the past. This mechanism of tsunami generation is hardly accounted

(a) Part of Aratozawa landslide moved into lake

(b) Intact shape of the Aratozawa rockfill dam

Fig. 163 Aratozawa rockfill dam after the 2008 Iwate-Miyagi Inland earthquake, Japan

for by current tsunami-hazard assessments. Furthermore, we have to be aware of a similar tsunamigenic mechanism in which glaciers fall into sea. It seems that the risk of glacier fall is currently increasing because of global warming and further that this type of problem is particularly important in seismic region such as Alaska (Dockrill 2020). For example, Mooney (2018) reported that an over-180-m tsunami occurred in 2015 in Alaska.

The remaining part of this chapter addresses the tsunami disaster that happened after the 2018 Palu earthquake, Sulawesi, Indonesia (Fig. 164). This disaster was peculiar in that the tsunami arrived shortly after the end of the earthquake tremors, in spite of 80-km epicentral distance to the Palu coast (Carvajal et al. 2019); see Fig. 165. Furthermore, it was considered unlikely that the left-lateral strike-slip dislocation of the causative Palu-Koro Fault, with no vertical displacement, moved sea water and generated the tsunami. Finally, Carvajal et al. (2019) as well as Takagi et al. (2019a, b) introduced aerial videos that exhibit tsunami waves coming not from the epicenter at a far distance, but from a nearby source area. It appears that the tsunamis started from the western coast of Palu Bay, where submarine landslides likely happened. Takagi et al. (2019a, b) further carried out a bathymetric survey to detect the loss of substantial masses from the submarine slopes. Overall, it is now thought that the tsunami in 2018 was generated by submarine landslides induced by the earthquake motion. The tsunami-affected area was limited to within one hundred meters or so behind the shore line (Fig. 164) because the volume of the mass displacement that triggered the tsunami was that of landslides, which were much smaller than displacement caused by the tectonic movement of the earth's crust.

The eastern part of Indonesia experienced many tsunami disasters in the past. Some of them were generated certainly by tectonic action of the earth (huge fault rupture) but others were possibly triggered by coseismic submarine landslides. It is not easy to judge which type of source triggered historical tsunamis. Heidarzadeh et al. (2014) discussed the difference between the two types of tsunami sources, noting that a tectonic source is associated with smaller vertical displacement at the sea bottom (10–20 m at maximum) and greater lateral size (hundreds of km at maximum),

Fig. 164 Tsunami disaster in the coastal area of Palu City, Indonesia; the building in the distant view was not destroyed

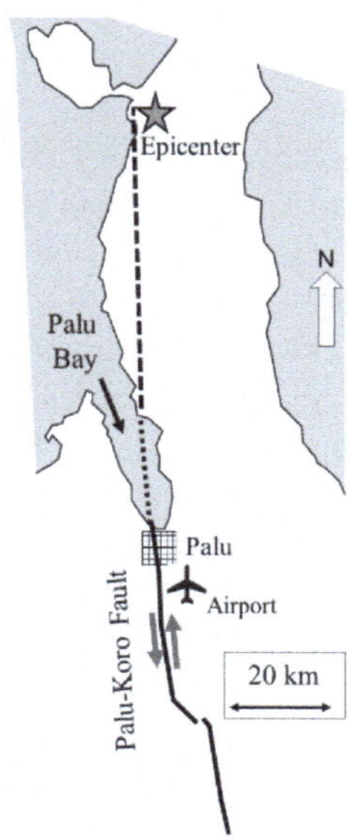

Fig. 165 Distance of Palu City from the epicenter of the 2018 earthquake

while submarine landslides are accompanied by as much as hundreds of meters of vertical displacement at the sea bottom, in contrast to their smaller lateral dimensions (of the order of 100 m). Okal and Synolakis (2004) proposed a new index, I_2, to discriminate these types of tsunami that are initiated at nearby sources;

$$I_2 = \frac{\text{(Height of tsunami run-up)}}{\text{Lateral size of tsunami source or Length of affected beach}} \quad (2)$$

It was stated that I_2 is greater than 0.0001 for tsunami generated by submarine landslides. On the basis of this idea, Pranantyo and Cummins (2020) judged that the historical tsunami that registered a 100-m run-up on Ambon Island in 1674 was caused by an earthquake-induced submarine landslide. In line with this, the tsunami in Palu in 2018 is likely to have been a landslide-generated event, because its run-up height was approximately 3 m and the beach length was more or less 3 km near the central part or Palu City, making $I_2 = 0.001$.

11 Landslide Dams Produced by Earthquake Action

The gigantic Valdivia earthquake of Mw = 9.6 in Chile in 1960 gave rise to landslide problems. Davis and Karzulovic (1963) stated that the drainage channel (San Pedro River) of Lake Riñihue was blocked by a landslide mass at three sites and the water level was raised by 26.5 m. To avoid a catastrophic breaching and flood, the natural dam was excavated to open a drainage channel (Hernandez 1960). The released water inundated Valdivia (Soto Melo), which had coseismically subsided by about 1.5 m during the earthquake and had sea water flow in (Weischet 1963) (Fig. 166). Many people had left Valdivia before the inundation. There is also a bigger and older landslide mass in the vicinity, which is presumably the landslide blockage of the river in AD 1575 that destroyed Valdivia after breaching (Araya et al. 2014). Thus, the history of natural disasters repeats.

The 2008 Wenchuan earthquake produced 256 landslide dams (Xu et al. 2009) with thousands of landslides induced by this event (Wang et al. 2010a, b). The dam at Donghekou of 10 million m^3 in volume was induced by a very long flow of debris (Fig. 167). $H/L = 0.364$ of this event suggests the equivalent friction angle = 20°. The blockage at Tangjiashan was the most famous and involved 20 million

Fig. 166 Recent Valdivia in Chile seen from downstream of San Pedro River

m^3 of landslide mass coming down from the right bank of the Jianjiang River. Qi et al. (2015) inferred that this landslide was a consequence of flexural buckling of the rock slab. The dam height exceeded 80 m (Xu et al. 2009). This dam lake was drained under control and the flash flood went downstream where the mega-city of Míanyáng had been evacuated. Figure 168 shows that some parts of the landslide dam still remained in October 2012. The problem here was that the remaining slope was susceptible to further failure and one disaster occurred in September, 2008 (Hu et al. 2009). Figure 169 illustrates the situation in the left-bank slope in 2012. Because the risk of repeated landslides and river blockage was high, a drainage tunnel was constructed on the opposite side of the valley (Fig. 170).

Many landslide dams were artificially breached after the Wenchuan earthquake, while a few of them overflowed naturally, as was the case for the Donghekou landslide (Fig. 167). Xu et al. (2009) examined 32 dams and presented geometric information. According to them, the natural dams pose risks for both upstream and downstream

Fig. 167 Donghekou landslide of 10 million m^3 volume and the equivalent friction angle of 20°

Fig. 168 Site of Tangjiashan landslide dam after drainage, Sichuan Province, China (October 2012)

Fig. 169 Unstable landslide slope at Tangjiashan in 2012

Fig. 170 Exit of emergency drainage tunnel on left bank of Jianjiang River at Tangjiashan

areas; inundation at the foot of the slopes induced a further landslide risk in the upstream reach and breaching/flooding in the downstream area. All of the studied dams were made of blocks and fragmented rocks. Eight of them did not breach, undergoing overflow of water and others were artificially breached for the safety of the locality. Figure 171 plots the dam height against the dam volume. Those dams that did not breach are shown by solid circles, and open circles designate those dams that were assessed to be dangerous and were artificially breached. There is no clear differences between the two groups. In contrast, for the Wenchuan Earthquake (Fig. 172), the lake volume (vertical axis) suggests that the dams without natural breaching are of a smaller volume. This point is in line with Casagli and Ermini (1999), who proposed that natural dams would not breach if the lake volume is smaller than the volume of dam.

Sichuan Province of China has an unstable mountain region that is affected by the tectonic movement of the Tibetan Plateau and is susceptible to earthquake-induced

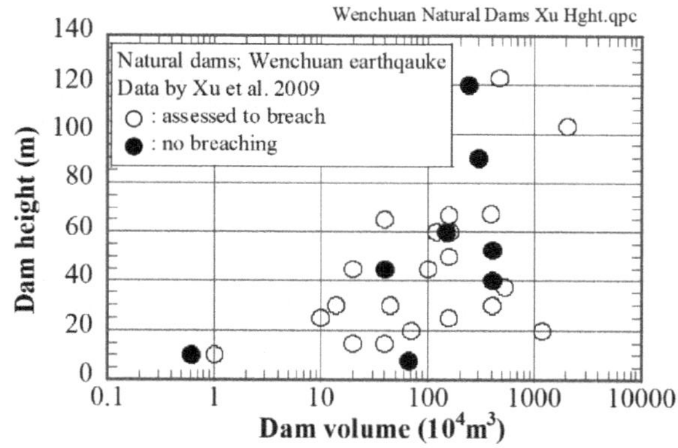

Fig. 171 Relationship between the volume and the height of natural dams, Wenchuan earthquake (drawn after Xu et al. 2009)

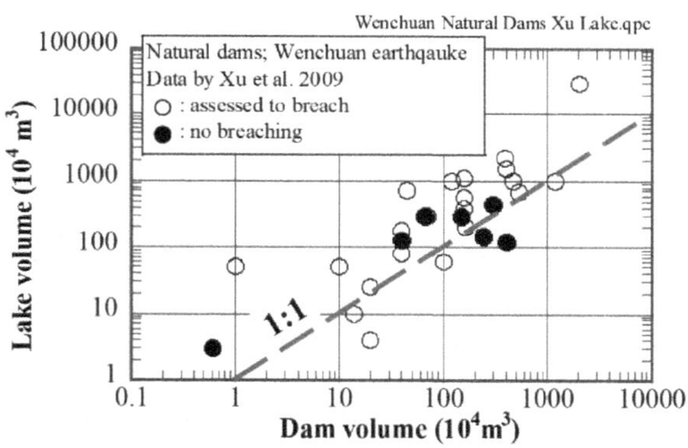

Fig. 172 Relationship between the volumes of dams and lakes, Wenchuan earthquake (drawn after Xu et al. 2009)

landslides. On August 25th, 1933, the Diexi earthquake of Mw = 7.3 produced a landslide dam; the dam height was 70 m and the lake volume exceeded 70 million m^3 (Wang et al. 2008). The lake was named Diexi Lake (Fig. 173). Although affected by increased inflow in 1993 and overtopping/flash floods over a distance of 250 km (Xu et al. 2009), this natural dam still remains until the present time. Chang (1938) conducted damage reconnaissance after this earthquake. He reported three landslide dams along the main stream of the Min Jiang River, among which the farthest downstream one breached on October 9th, 1983. Two more landslide dams were formed as well in the mountainous regions west of Diexi Town. This location is close to the site of the Xinmo landslide, which was triggered by gravity in 2017 (Fig. 218). Chang (1938) also referred to five historical earthquakes in this region since BC 12, three of which produced landslide dams. The photographs taken by Zhuang Xue-Ben during the aftermath of the Diexi earthquake illustrate that the mountain slopes became unstable and generated many landslides (Hong et al. 2011). Hong (2014) suggested that there was a bigger prehistoric earthquake in this region more than 20,000 years ago. Presumably, this earthquake-induced gigantic landslide formed a bigger lake at the same place. Ma et al. (2018) conducted detailed studies in the field and concluded that this prehistoric breaching of the dam occurred about 27 thousand years ago. A lacustrine deposit is still visible above the present lake water level (Fig. 174) and is an evidence of a bigger prehistoric landslide lake (Wang et al. 2014a, b, c). Thus, the history repeats.

The 2004 Niigata-Chuetsu earthquake formed many landslide dams in a hilly terrain (Yamabe et al. 2010). The Terano landslide site was one of them (Fig. 12). After a drainage channel was installed, the dam has survived to the present without breaching. The biggest one was the dam at Higashi-Takezawa, which blocked the Imo River (Figs. 175 and 176). This landslide had an length $(L) = 390$ m, an elevation difference $(H) = 55$ m, hence $H/L = 0.141$, and the equivalent friction angle $= 8.0°$, while the landslide volume was 1.3×10^6 m^3 (Yagi et al. 2007). Behind the head scarp (see Fig. 177) was a depression (Fig. 178) that means the 2004 landslide was a reactivation of an ancient landslide (Fig. 179). To prevent breaching, the lake water was drained urgently by pumping, the landslide dam was excavated for an emergency water channel, and the soil of the landslide mass was removed for stabilization. The drainage channel was protected by steel plates from erosion and uncontrolled water

Fig. 173 Diexi Lake produced by an earthquake-induced landslide in 1933 (the arrow indicates the direction of the coseismic landslide)

Fig. 174 Lacustrine deposit at an elevation higher than the present Diexi Lake (texture disturbed by rock fall)

discharge (Fig. 180). The stability of the natural dams was assessed by calculating the Dimensionless Blockage Index (DBI) by Ermini and Casagli (2002, 2003) (Marui 2007);

$$\mathrm{DBI} = \log_{10} \frac{A_\mathrm{b}}{V_\mathrm{d}/H_\mathrm{d}} \tag{3}$$

where A_b is the area of the catchment, H_d is the height of the dam and V_d is the volume of the landslide dam. In practice, DBI < 2.75 means stability and DBI > 3.08 implies instability; see Fig. 193.

Havenith and Bourdeau (2010) as well as Strom (2010) described earthquake-induced landslides and associated natural dams in the Tien Shan and Pamir regions of Central Asia where information is not abundant. Some of the events were the 1911 Kemin earthquake (surface-wave magnitude Ms = 7.7) that triggered the Ananevo

Fig. 175 Higashi-Takezawa landslide dam site after the Niigata-Chuetsu earthquake, Japan

Fig. 176 Inundation in Higashi-Takezawa

Fig. 177 Head scarp of the Higashi-Takezawa coseismic landslide

Fig. 178 Small depression behind the head scarp of the 2004 landslide

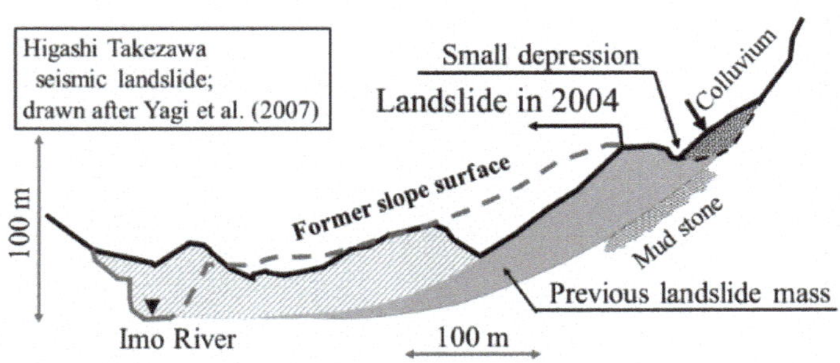

Fig. 179 Cross section of the 2004 Higashi-Takezawa landslide (drawn after Yagi et al. 2007)

Fig. 180 Protection of the drainage channel of Higashi-Takezawa landslide mass from erosion

landslide of 10–15 million m^3 in volume, the 1949 Khait earthquake of Mw = 7.5 that induced the gigantic rock-loess Khait landslide, and the 1992 Suusamyr earthquake, which had a surface-wave magnitude (Ms) = 7.5 and induced the Belaldy rock slide and natural dam breaching. Evans et al. (2009) assessed the volume of the Khait landslide to be 75 million m^3, while $H/L = \tan(10.8°) = 0.19$. Moreover, the Usoy (or Usoi) landslide of 2 billion m^3 in Tajikistan was induced in 1911 by the Sarez-Pamir earthquake with Ms = 7.4 (Alford et al. 2000). Ambraseys and Bilham (2012) stated that the volume of the landslide mass was 2.4×10^9 m^3, the length of the failed slope was 2500–3000 m and Mw of this earthquake was 7.7. This landslide dam measures 567 m in height (Alford et al. 2000). This dam still remains today which is 110 years after its formation. Strom (2010) made an excellent summary of landslide dams, both coseismic and non-seismic, in Central Asia.

The Hattian Bala landslide dam was formed by a huge landslide during the Kashmir earthquake in 2005 (Fig. 181); Fig. 182 shows the lake. This dam breached on February 9th, 2010. The landslide mass was 80.8 million m^3. Although DBI = 1.85, this dam breached. Because the dam-forming mudstone was prone to slaking and deterioration, it is likely that the dam breached by the weakening of the dam body (Kiyota et al. 2011).

Figure 183 shows the appearance of the Tsaoling landslide site in 2013. The landslide-induced lake had already been filled with sediments. The 2016 Kaikoura earthquake in New Zealand produced natural dams. Some dam breached (Fig. 184) while others still retain water (Fig. 185).

Although many landslide dams breach after a while, some remain stable for many years. Figure 186 shows a small example of a stable landslide dam that was produced during the 1923 Kanto earthquake in Japan (Terada and Miyabe 1932; Inoue et al. 2013). Chigira et al. (2017) stated that this landslip occurred along a layer of volcanic pumice. Because the catchment of this lake is tiny, this dam did not breach.

It has been shown that natural dams may or may not breach. Costa and Schuster (1988) and Chai et al. (2000), as well as Peng and Zhang (2012), studied natural

Fig. 181 Hattian Bala landslide in Pakistan, 2005

Fig. 182 Lake formed by the Hattian Bala coseismic landslide in Pakistan, 2005

dams that were both coseismic and non-seismic and showed that 13–15% survived for one year after their onset (Fig. 187). This issue is extremely important in a post-earthquake emergency situation. In this regard, efforts have been made to develop methods to quickly assess the stability and durability of natural dams.

Casagli et al. (2003) examined 70 Italian landslide dams and mentioned three possible mechanisms of landslide dam failure; overtopping (associated with erosion), piping (or internal erosion) and slope failure (due to reservoir water pressure). There is no seismically induced landslide dam in their database, but it does not affect their conclusion on the breaching mechanism. Figure 188 summarizes three literature sources on causes of breaching. They unanimously indicate that overtopping is the most important mechanism. Figures 189 and 190 answer the question why piping failure is uncommon. It is clear that, for the majority, the reservoir depth is "< 20 m" which is shallow, while most dams are longer than 100 m in the river stream direction. This length makes low hydraulic gradient in the stream direction and seepage (piping)

Fig. 183 Landscape of Tsaoling landslide site in Taiwan in 2013 when the lake had been filled with sediments

Fig. 184 Linton landslides site (2 million m^3) in New Zealand, where the natural dam breached

Fig. 185 Leaders landslides site in New Zealand, where the natural dam survived (as of January, 2020)

Fig. 186 Shinsei Lake that was produced in a small valley by landslide during the 1923 Kanto earthquake, Japan

Fig. 187 Development of dam breaching with time (Costa and Schuster 1988; Chai et al. 2000; Peng and Zhang 2012)

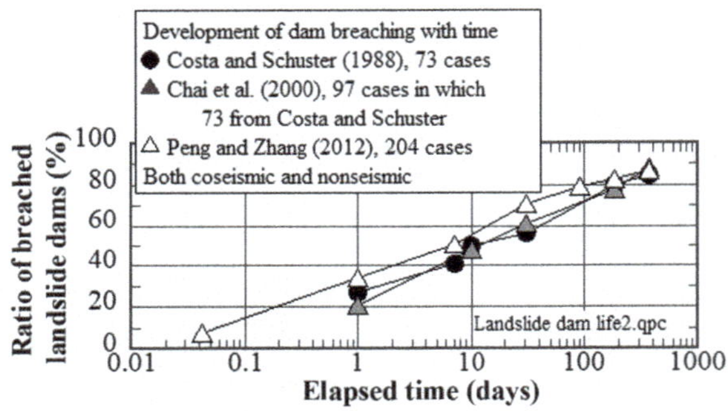

failure of the dam body unlikely. Moreover, because of the sufficient length in the stream direction, shear failure and breaching of dams under reservoir water pressure is unlikely, either. Figure 191 presents the data by Costa and Schuster (1988) on the geometry of natural dams. Although there is no discrimination in depth/length ratio between breached and stable dams, it is reasonable to say that the dams have sufficient length (hundreds meters) and likely resist erosion failure.

Fig. 188 Breaching types of natural dams (both coseismic and non-seismic)

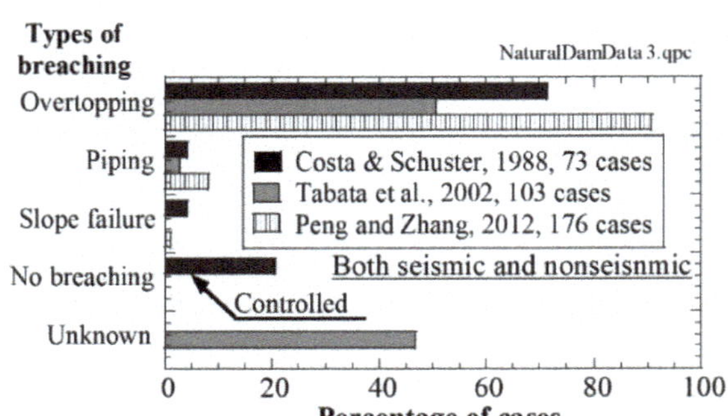

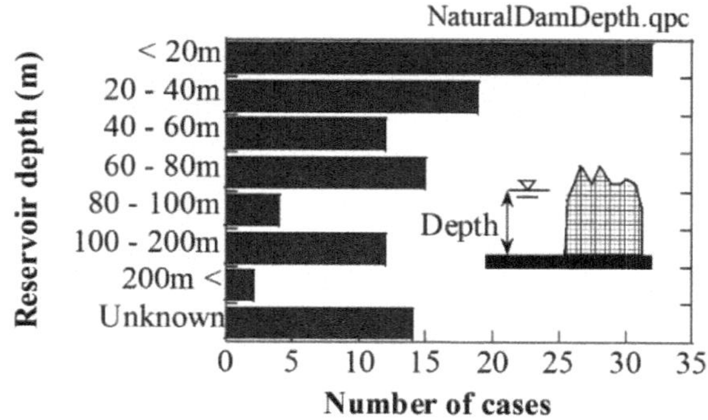

Fig. 189 Distribution of depth of natural dams; including both seismic and non-seismic cases (data by Tabata et al. 2002)

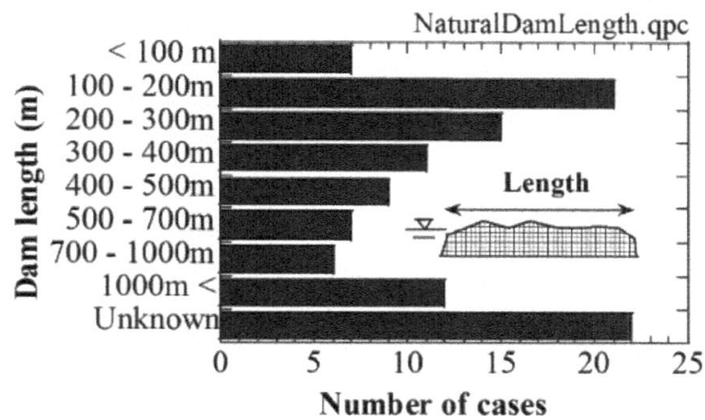

Fig. 190 Distribution of length of natural dams; including both seismic and non-seismic cases (data by Tabata et al. 2002)

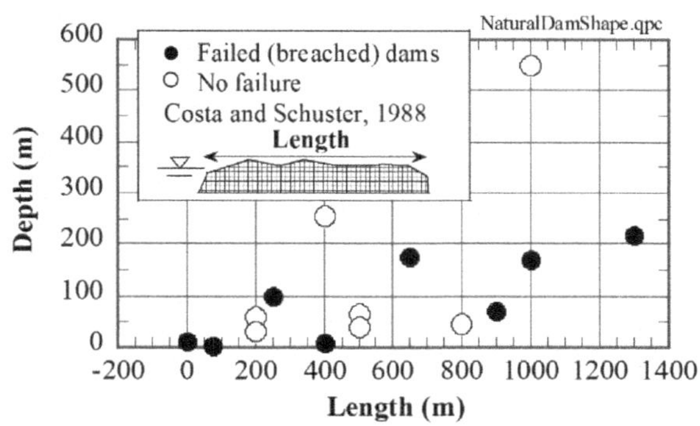

Fig. 191 Length and depth of natural dams (Costa and Schuster 1988)

The first idea on longevity (no breaching) of a natural dam is that a bigger landslide dam with a smaller lake has a better chance to survive. In this regard, Casagli and Ermini (1999) studied the ratio of the two volume parameters:

$$I_i = \log_{10}(V_d/V_l) \quad (4)$$

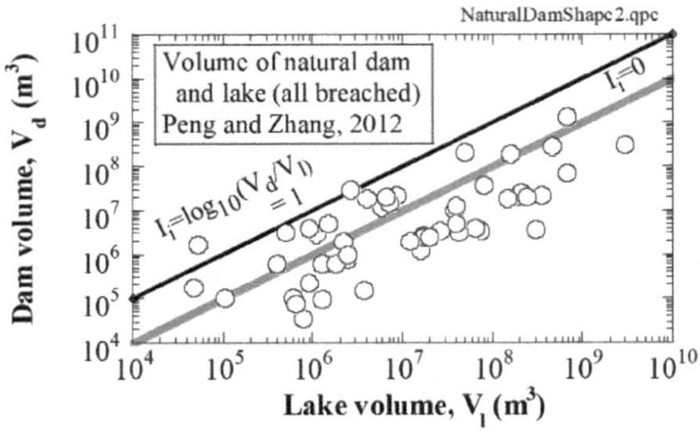

Fig. 192 Volumetric relationship between breached natural dams and lakes

in which V_d is the dam volume and V_l is the lake volume. Their data from non-seismic cases in the northern Apennine suggests that the dams are stable if I_i is greater than zero. In contrast, Korup (2005) examined the cases in New Zealand and suggested that $I_i > 1$ is a better criterion for safety. Figure 192 plots dam and lake data by Peng and Zhang (2012). All cases in this figure come from those of breaching and implies that $I_i > 1$ is a reasonable criterion for dam stability.

Dimensionless Blockage Index, DBI in Eq. 3, is another index for stability assessment of natural dams (see Fig. 193). This index increases as the watershed area (A_b) increases, the dam volume (V_d) decreases and the dam becomes higher (greater H_d), making the dam less stable. Ermini and Casagli (2003) suggested DBI < 2.75 for stability and DBI > 3.08 for the possibility of breaching. It seems that Fig. 193 supports this idea. However, care must be taken because a dam may breach for a smaller value of DBI (Figs. 181 and 182). Because precipitation in the water basin depends on climatic conditions, the thresholds of DBI (2.75 and 3.08) may be subject to regional variations. With this view, DBI data by Ermini and Casagli (2003) were replotted in Fig. 194 for different regions. The European and Japanese data may have different threshold values from North American and Australasian values. Because of uncertainty in details, however, further discussion of local differences is difficult. Further note that the data from Central Asia (Strom, 2010) are consistent with DBI = 2.75.

Shear strength, permeability and resistance against water effects are important factors to control the longevity of natural dams. It is, however, difficult to take all of them into account for stability assessment after big earthquakes. One reason for this is the heterogeneous properties in a huge landslide mass, while the other is the cost and time in emergency situations. Nevertheless, understanding soil mechanics aspects of dam behavior is important because judgment in an emergency is thereby improved. Casagli et al. (2003) investigated the physical properties of dam material. It was demonstrated that coarser dam material is more stable, while fine soil is prone to failure or requires human control.

My concern is about the seismic instability of glacier lakes, which are susceptible to melting of glaciers and rising of water levels. This situation may trigger collapse of moraine and GOLF (Glacier Lake Outburst Flood) (Komori et al. 2012). The Tasman

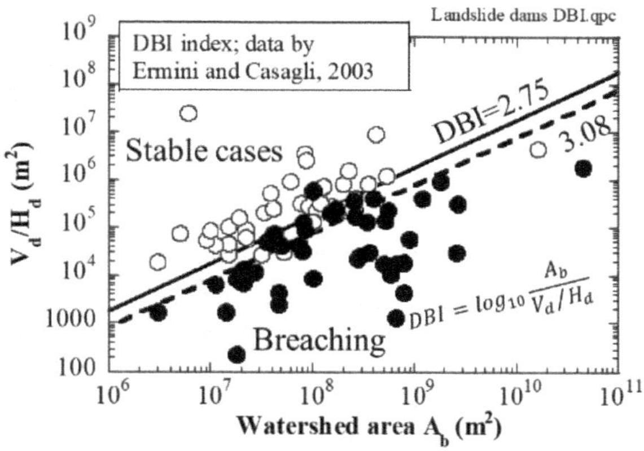

Fig. 193 DBI index for assessment of landslide dam stability (data by Ermini and Casagli 2003)

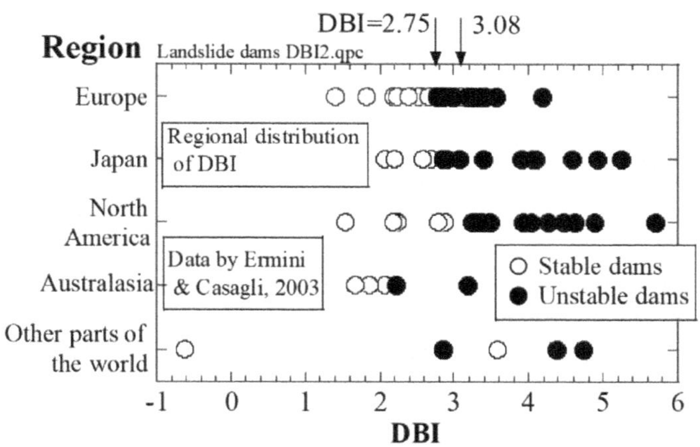

Fig. 194 Dimensionless blockage index (DBI) thresholds for natural dam breaching

Glacier Moraine (Fig. 195) illustrates an example of glacier lake moraine, which is stable now. However, some glacier lakes are situated in seismically active environments and their possible outburst would create a serious disaster in the downstream region (Röthlisberger 1978; Kuenza et al. 2004). Byers et al. (2017) investigated five glacier lakes in Nepal after the 2015 Gorkha earthquake. They reported fresh cracks and distortion of moraines and stated that one lake (Dig Tsho Lake) breached, triggering mud flows through the downstream villages. Moreover, the fall of moraine (mound that stores melt ice and water) into lake can generate tsunami that may destroy the damming part of the moraine (Klimeš et al. 2016). Strasser et al. (2008) reported outburst of Lake Zurich during the Pleistocene period and suggested that it was probably caused by an earthquake.

(a) View of lake

(b)Moraine hill made of rubbles (the arrow shows size of two people)

Fig. 195 Moraine lake made by the Tasman Glacier, South Island, New Zealand

12 Long-Term Instability of Mountain Slopes Affected by Strong Earthquakes

Seismically induced landslides may occur after the earthquakes as well. Keefer (1994) pointed out the importance of post-earthquake slope instability that lasts for tens of years or longer. Koi et al. (2008) analyzed the sediment in Miho dam reservoir in Japan to demonstrate that debris produced by the 1923 Kanto earthquake is still coming into the reservoir 80 years after the disaster. Further, Inoue (1995) counted the number of instabilities in the affected mountains after the Kanto earthquake and suggested that the long-term effect lasted until the 1960s. The coseismic gigantic landslide at Tsaoling in Taiwan in 1941 (Fig. 7) involved 150 million m^3 of falling mass. The mountains facing the dam probably developed cracks and became weaker (Kawada 1943c). Consequently, the heavy rainfall in August, 1942, triggered a second landslide of 200 million m^3.

The gigantic Ohya landslide (Fig. 143), with a volume of 120 million m^3, was triggered by the 1707 Hoei earthquake. This event was the beginning of a period of long-lasting sediment disasters in the downstream region. Ohya landslide (Fig. 145) exhibited highly fractured rock on the surface of the mountain slope, and heavy rainfalls have been washing it downstream many times until the present. It seems important that this site is near the Itoigawa-Shizuoka Tectonic Line (Figs. 1 and 196), accompanied by many short faults, and has been subject to significant distortion (Fig. 197). Figure 198 shows stone deposits and uplift of a river bed downstream of the landslide. In the past decades, substantial efforts have been made to stabilize the Ohya landslide by planting trees and constructing erosion control dams. Similarly, the gigantic coseismic Usoy landslide in Tajikistan occurred within a tectonically fractured rock mass (Schuster and Alford 2004; Havenith and Bourdeau 2010).

Shichimenzan landslide is located about 8 km northeast of Ohya landslide and has been unstable for centuries (Fig. 199). Because of its short distance to the tectonic

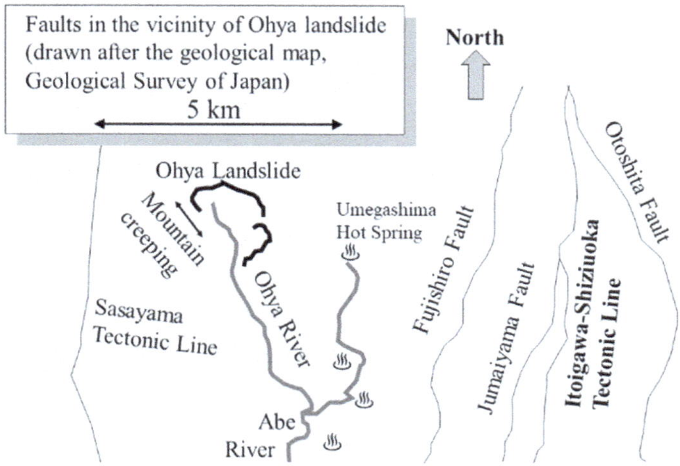

Fig. 196 Distribution of old and recent faults around Ohya landslide in central Japan (drawn after Geological Map NI-54–32-9 by Geological Survey of Japan)

Fig. 197 Folding of Ohya Mountain

subduction that caused the 1707 Hoei and other gigantic earthquakes, this site most likely has been affected by earthquakes. The effect of local faults seems important as well. According to Nagai and Nakamura (2000), the landslide morphology of Mt. Shichimenzan, which is locally called "Gare," is mentioned in historical documents in AD 1278, 1651 and 1666, with its picture drawn in the 1740s or 1847. They also stated that the rock at this site has been damaged by East–West compression (probably inducing slaty cleavage similar to the Ohya site), together with freeze–thaw cycles. The annual progress of weathering is 5–50 cm/year. Figure 200 illustrates the ongoing creep deformation near the shoulder of this landslide cliff.

Tombi landslide (Fig. 201) on Tateyama Volcano (Fig. 1) was induced by the February 1858 Hietsu earthquake, which had a magnitude = 7.0–7.1. The Joganji

Fig. 198 Stone deposit in the Abe River channel downstream of Ohya landslide

Fig. 199 Shichimenzan landslide

Fig. 200 Double ridge atop Shichimenzan landslide

Fig. 201 Present view of Tombi landslide site, Tateyama volcano, Japan

River, which originates from Mt. Tateyama, started to cause many debris flow and flood disasters since then (Ikeda 2011). In particular, the landslide mass formed natural dams and their breaching triggered substantial sediment disasters twice in the same year in the downstream region (Shimamoto et al. 2008). With beautiful illustrations, Inoue et al. (2010) stated that the volume of this earthquake-induced landslide was 114 million m^3. Figure 202 shows one of the avalanche hills that verifies the huge size of this earthquake-induced landslide. Two landslide dams were formed and breached later. Since then, the mountain has been unstable, the total volume of soil erosion has accumulated to 120 trillion m^3 and the mountain is still producing sediments during heavy rains. Huge efforts have been made to mitigate the sediment disaster by constructing erosion control dams at Shiraiwa in the upstream and at Hongu in the middle reach of the affected Joganji River (Figs. 203 and 204) together with railways for construction and maintenance of erosion control structures (Fig. 205) (Adachi et al. 1999; Ikeda 2011). The significant extent and longevity of the seismic effects on slope instability come from the weakness of the volcanic mountain (Mt. Tateyama), which is an accumulation of lava, pumice, and ash that have weak bonding among grains.

The 1995 Kobe earthquake occurred on January 17. Because it was a dry season, the mountain did not suffer much from coseismic landslides. However, the mountain slopes were affected to a certain extent. The Rokko Mountains are made of weathering-prone granite (Fig. 206) and suffered from rainfall-induced landslides many times in the past; in 1938, 1961 and 1967 in the recent past as well as in 1504, 1544, 1608, 1782, 1813, 1848, 1871, 1885, 1896, 1899, 1910, 1921, 1922, 1925 and 1935 (Sumiyoshi Village 1946) and in 1776, 1789 and 1799 (Kimura 2019). The seismic effect means fracturing and development of cracks in the mountain body, as suggested by water ejection. Tainosho and Fujita (1996) compared two mountain slopes changing during time intervals (April 28 vs. September 28, 1995; May 12 vs. October 2, 1995). The rainy season starts in late May and ends in September. They showed that the area of slope instability was enlarged with time after the earthquake. Furthermore, Ikeda (1996, 2007, 2015) as well as Ikeda and MLIT Rokko

Fig. 202 Avalanche hill produced by the Tombi landslide

Fig. 203 Shiraiwa check dams on Tateyama volcano (capacity = 1 million m^3)

Sabo Office (2016) undertook long-term observations of slopes in the Rokko Mountains for 20 years. It was found that the slopes continued to fall down for more than ten years, and that 20 years is needed for their stabilization. Okimura et al. (1998) also addressed a post-seismic progressive enlargement of slope instability behind the campus of Kobe University, which was particularly significant during the rainy

Fig. 204 Hongu dam for erosion control in midstream of the Joganji River; debris stored in the reservoir is seen from the upstream direction (capacity = 5 million m^3)

Fig. 205 Mountain railway in Tateyama for construction and maintenance of erosion control structures

season (May–September) of the same year as the earthquake. Tomita et al. (1996) attempted to quantitatively compare threshold rain fall intensities to trigger landslides before and after the earthquake. This was found to be difficult, however, because they could not identify which rainfall events triggered the studied landslides before the earthquake. Hirano and Ishii (1997) also showed that many landslides were triggered during the rainy season (From May to October) in the same year and in the same area of the Rokko Mountain (eastern part) as the seismic landslides in 1995.

Figure 207 shows the Horai Valley in Nishinomiya in the northeastern part of the Rokko Mountains. This valley is a typical badland, where rock is degraded and there is little soil or vegetation at the surface. Accordingly, erosion is very active therein during heavy rain (Nadal-Romero et al. 2011). During the 1995 Kobe earthquake, a substantial number of slope failures such as falling of rock masses, exfoliation of rock masses, stone falls and separation across vertical joints happened (Yokoyama and Kikuyama 1997). Ikeda (2015) found progress of slope deterioration in 10 years after the Kobe earthquake. The Horai Valley is situated in the Arima-Takatsuki fault zone. This fault zone most probably caused the 1596 Keicho-Fushimi earthquake

Fig. 206 Rokko Mountains to the north of Kobe City

of magnitude = 7 (Tsutsumi et al. 2004) that devastated Kyoto, present Osaka and Kobe. Ground deformation during past fault activities disrupted the rock masses, generated fissures, introduced water infiltration, promoted physical weathering, and created badland geomorphology here in conjunction with hydrothermal alteration in this fault zone (Utada 2003; Shimizu et al. 2017).

Landslides induced by the 1999 Chi-Chi earthquake were distributed over a huge area in Taiwan. In this sense, it is difficult to discuss the relation between landslides and faults. However, the long-term effect of a devastating earthquake on slope instability can be discussed in a quantitative manner. Meunier et al. (2008) investigated slope instability before, during and after the 1999 Chi-Chi earthquake; the pre- and post-earthquake events were triggered by heavy rainfall caused by typhoons. The pre-earthquake landslides were concentrated at the slope bottom, while the coseismic and the post-seismic events shifted toward the mountain tops. This suggests that the seismic shaking affected the slope stability at higher elevations by producing fissures there.

Fig. 207 Badland morphology in Horai Valley, Nishinomiya, Japan

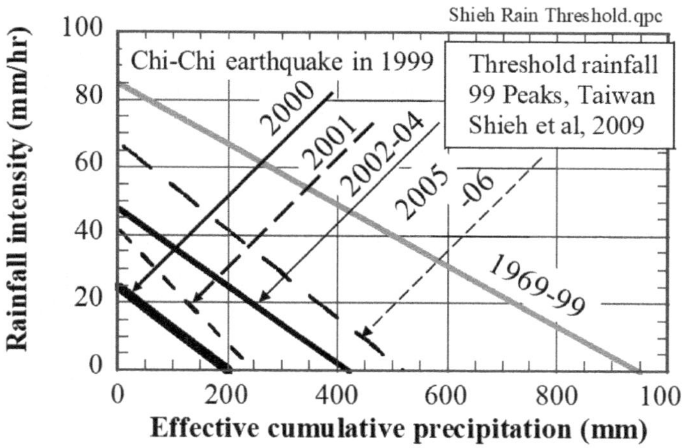

Fig. 208 Post-seismic change of threshold rainfall for triggering debris flows, Taiwan (Shieh et al. 2009)

Post-earthquake increase of landslide disasters in Taiwan has been reported by Lin et al. (2004, 2010), Lin and Lin (2013), and Lin and Chiang (2019). Sediment transport increased after the earthquake as well (Dadson et al. 2004). Shieh et al. (2009) examined the threshold rainfall intensity in Taiwan that could trigger debris flows. Immediately after the Chi-Chi earthquake, it was low and suggested the unstable state of the slopes and valley deposits. In 2005 and 2006, the threshold rainfall was getting greater because of the slope stabilization and termination of erosion at valley bottom (Fig. 208).

The Tsengwen Dam reservoir is situated to the south of Alishan (Fig. 4) whose slope was affected by the 1999 Chi-Chi earthquake and 2009 Morakot Typhoon that brought, at the maximum, 3000 mm precipitation in four days and caused devastating rainfall disasters (Chen and Wu 2009; Lee and Towhata 2010). Figures 209 and 210 show the Alishan Mountain slope and sediment deposited in the Lao Nong River channel in south Taiwan, verifying how this typhoon devastated this region. I presume that, after the Chi-Chi earthquake, the Tsengwen Dam Reservoir started to receive siltation more frequently from the disrupted upstream mountains (Fig. 211) (Ho et al. 2013), and this trend peaked at the time of the Morakot Typhoon.

Figure 212 shows disrupted mountains along the East–West Crossing Highway of Taiwan that were severely affected by the Chi-Chi earthquake. The extent of long-term instability is depicted in Fig. 213, where the area of newly added instability around existing ones is plotted against the years of typhoons. When the Chi-Chi earthquake affected the mountains, the area of new instability increased suddenly from the very low original level. Thereafter, the area has been decreasing continuously. Then a question arises, however: is this only because no big typhoon and rainfall occurred in recent times or because only small unstable slopes have been affected. In this respect, Fig. 214 plots the increment ratio of "area of new slope

Fig. 209 Alishan slope after the 2009 Morakot typhoon disaster (photograph on Feb 29 2012)

Fig. 210 Sediment deposit of Lao Nong River (a bridge was washed away here)

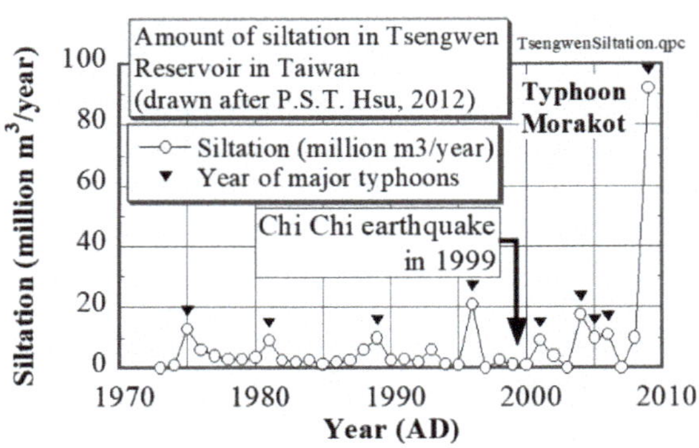

Fig. 211 Siltation in Tsengwen dam reservoir (drawn after Hsu 2012)

instability" over "existing instability area" against the total rainfall during respective typhoons. See a positive correlation. Then, the slope vulnerability index was defined by

$$\text{Vulnerability index} = {[\text{Areal increment ratio (\%) in Fig. 214}]}/{[\text{Total rainfall (mm)}]}/ \quad (5)$$

Figure 215 shows that the vulnerability has been decreasing with time and implies that the disrupted mountain slope has been stabilized gradually with time. Finally, the mountain slopes were stabilized 20 years after the 1999 Chi-Chi earthquake (Lin and Chiang 2019).

Fig. 212 Disrupted cliffs in central Taiwan after 1999 Chi-Chi earthquake

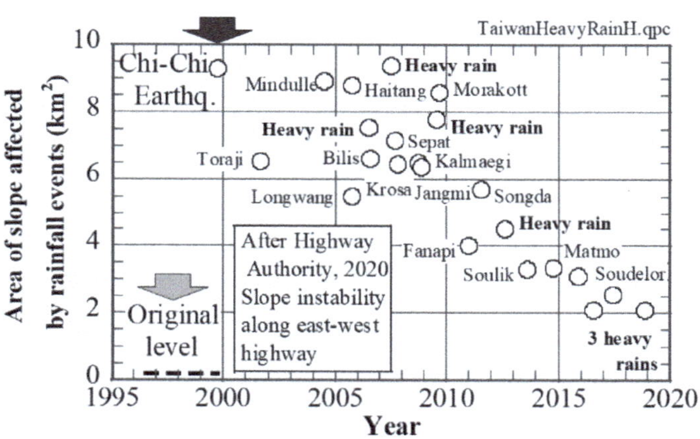

Fig. 213 Areal expansion in slope instability in the mountain area along the East–West crossing highway, Taiwan (drawn after Highway Office of Taiwan 2020)

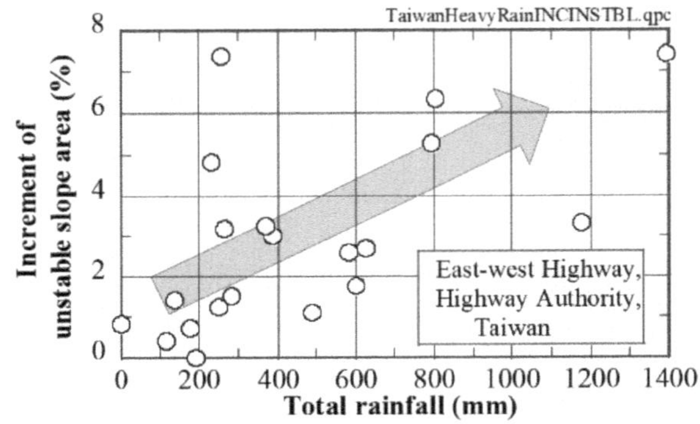

Fig. 214 Expansion of slope instability with total rainfall, expressed as percentage (Highway Office of Taiwan 2020)

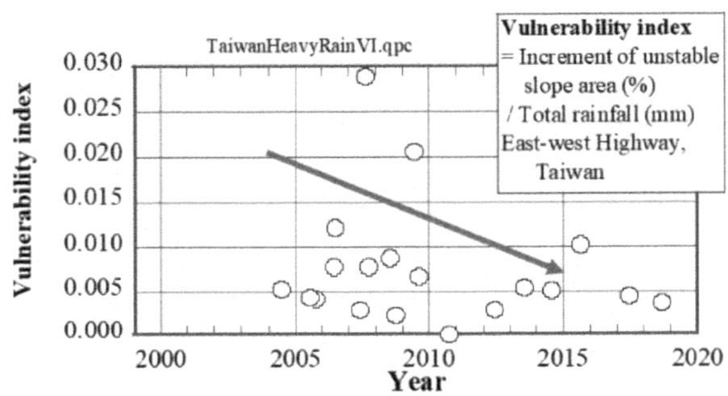

Fig. 215 Vulnerability of rainfall-induced landslides decreasing with time after the 1999 Chi-Chi earthquake (data by Highway Office of Taiwan 2020)

Marc et al. (2015) summarized the frequency of landslides in the epicentral area of the Chi-Chi earthquake. The obtained frequency in every four months or so was normalized by the background frequency (without seismic effects) and plotted against year. It was then found that the landslide frequency increased by approximately 20 times immediately after the Chi-Chi earthquake and then decreased gradually to the background level in 5 years. This rate of recovery appears faster than the reduction in debris flow disasters, possibly because sediments at the valley bottom can remain for a longer time. The same situation was reported by (Huang and Li 2014) in Sichuan Province, China, and will be addressed next.

The slope instabilities after the 2008 Wenchuan earthquake in China posed significant problems in Sichuan Province as well, in particular within 20 km from the fault (Huang 2011). The debris flows in Wenjia Gully (Fig. 154) was a disaster of this type, a consequence of sediment washed out during heavy rain. It seems that the delayed disasters are made worse when the mountain body has been deteriorated. Figure 216 illustrates Yinchang Gully where debris flows occurred in 2008, 2009 and 2010 (Ma et al. 2011). In particular, the effect of rainfall in September, 2008, was serious. As shown in Fig. 217, the mountain slope behind this site was extremely damaged and hazardous. Moreover, Huang and Li (2014) investigated the seismically loosened materials and showed that its amount (= 5.6×10^8 m^3 in total) is profoundly greater than that of the eroded sediments in 5 years (1.3×10^7 m^3). Because materials can thus remain in the valley, it was concluded that the difficult situation would last for a long time.

Since the area of the Wenchuan earthquake disaster undergoes the tectonic interaction caused by the Tibetan Plateau, the mountain body is distorted and damaged. This situation was illustrated by Cui et al (2019) on the geological cross section of the gigantic landslide at Daguangbao. I suppose that the long-term action of the Longmenshan and other faults in this region damaged the rock mass in a manner similar to the Ohya and Shichimenzan landslides (Figs. 197 and 199). Earthquake-induced landslides at Diexi Lake (Fig. 173) and the non-seismic Xinmo landslide in Maoxian in 2017 (Fig. 218) (4.3 million m^3; Fan et al. 2017) imply the overall instability of mountains in this region. Note that there was probably an older landslide next to the recent Xinmo event. There are many faults around this landslide (Meng et al. 2018).

Figure 219 demonstrates the appearance of a mountain slope near the Zipingpu Reservoir, Sichuan Province, that was severely affected by the earthquake in 2008. It appears that the vegetation has recovered to a satisfactory level in ten years after the earthquake. Fan et al. (2018) undertook long-term monitoring of the deposits and sediments and concluded that disaster risk decreased 7 years after the earthquake. I feel that their data indicate preliminary stabilization 3–4 years after the Wenchuan

Fig. 216 Tourist facility in Yinchang Gully that was destroyed as a consequence of post-earthquake debris flow

Fig. 217 Disrupted mountain slope behind Yinchang Gully resort and deposit of an induced debris flow

Fig. 218 Xinmo landslide in 2017, which is situated at around 4 km of the Diexi landslide dam

earthquake. The rate of recovery in Sichuan is faster than that in Taiwan (20 years needed for recovery as mentioned above) probably because Sichuan Province has less precipitation than Taiwan and therefore is less prone to new landslide disasters (Fig. 220). The information from the slope instability in Tanzawa Mountains in Japan (Fig. 90) should be recalled; slopes of therein were prone to rainfall-induced landslides until 1960s (Inoue 1995; Nakamura et al. 2000).

In a more quantitative manner, Shieh et al. (2009) showed that the threshold rainfall of landslide (rainfall that triggers slope failure) had decreased after the 1999 Chi-chi earthquake in Taiwan, increasing toward the original level after 2005 as mentioned above. Yu et al. (2014) made similar discussions on the situation after the Wenchuan earthquake, indicating that the greater coseismic acceleration (PGA) reduced the threshold rainfall more markedly and that several years had not been sufficient to make the threshold rainfall equal to the original level. Note that the recovery of the threshold rainfall is affected by two issues that are the increase of the slope resistance

Fig. 219 Recovery of surface vegetation in earthquake-affected slope near the Zipingpu Dam Reservoir (photograph on May 15, 2018; ten years after the Wenchuan earthquake)

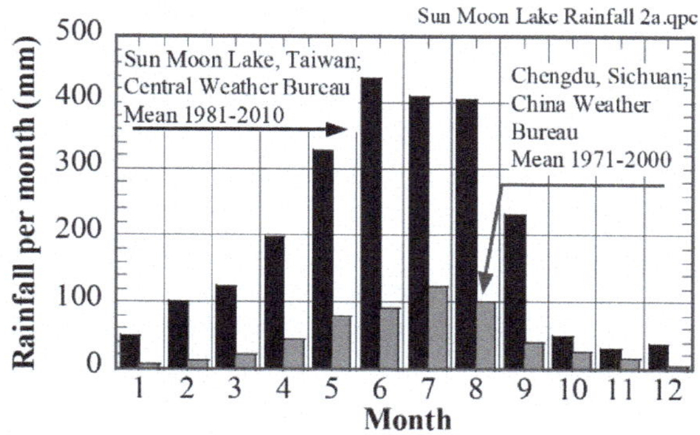

Fig. 220 Comparison of mean rainfalls in Central Taiwan and Chengdu of Sichuan Province

of the slope and the decrease of unstable soil deposit after many landslides during the aftermath.

In the Himalayan region, the Gorkha earthquake on April 25th, 2015, in Nepal was followed by increased landslide activities for three years (Tian et al. 2020). Figure 221 demonstrates a potentially unstable slope in the northeastern part of Sindhupalchok District which I visited on October 15th, 2015, which was six months after the earthquake. Detachment or sliding was in progress. Because of the tectonic interaction between India and Tibet, the bedrock is weak at many places in the Himalayan region. In particular, the Assam region of India has been prone to strong earthquakes associated with landslides. Oldham (1899) presented Plates III, XXVII and XXX of landslide disasters triggered by the seismic event in 1897 (Mw = 8.0). Many landslides were triggered again in Assam during the 1950 Assam-Tibet earthquake of Mw = 8.6 (Kingdon-Ward 1953). Floods occurred more frequently and river sediment increased after this gigantic earthquake, while the river bed became higher (Kingdon-Ward 1955). This is one of the examples of long-term seismic effects on slope instability.

Fig. 221 Potentially unstable slope that was visited in Nepal six months after the 2015 Gorkha earthquake

Kashmir belongs to the tectonically active Himalayan region. After the earthquake in 2005, landslides continued for some time. Figure 222 (Saba et al. 2010) illustrates that post-earthquake slope instability almost ceased in 2008, which was three years after the disaster. A similar point was made by Basharat et al. (2021) as well. This period for stabilization in Kashmir is shorter than those after the Wenchuan earthquake (10 years; see Fig. 219) and the Chi-Chi earthquake (20 years; Fig. 215) probably because the annual rainfall in Kashmir is less than that in Taiwan (Fig. 223). The reason for the difference between Kashmir and Sichuan is not known because the precipitations in the two regions are similar. Furthermore, mechanical damage of the mountain body triggers long-term slope instability. Figure 224 exhibits an open crack in a slope near the Hattian Bala landslide. In line with this, Petley et al. (2010) reported the occurrence of many cracks in the slope surface. Such cracks indicate the mechanical disruption of slopes and also introduce rain water into the slope. The slope in Fig. 225 started to fail after the earthquake probably because of seismically generated open cracks and fissures.

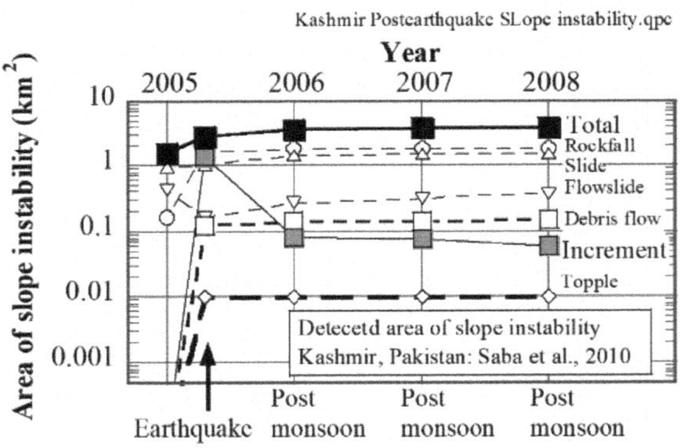

Fig. 222 Post-earthquake landslides around Muzaffarabad, Kashmir (Saba et al. 2010)

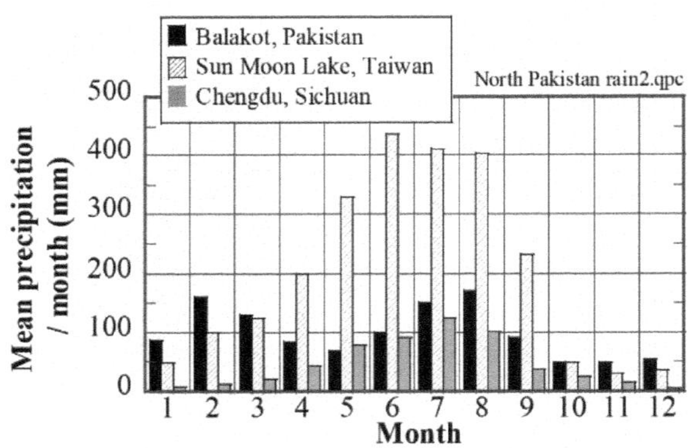

Fig. 223 Mean precipitation in Balakot, Pakistan

Fig. 224 Open cracks in a mountain slope near the Hattian Bala coseismic landslide; cracks are aligned between two triangles in this photograph

Fig. 225 Slope instability to the east of Muzaffarabad that started after the Kashmir earthquake

Overall, this chapter has addressed the post-seismic instability that may last for years or longer. The direct causes of landslides in the post-seismic period are aftershocks and heavy rainfalls. However, it has to be stressed that the potential of debris flows or landslides are increased by the earthquake by way of.

- Deposits of loose materials in valley bottom (see Fig. 154),
- Fissures in rock masses that cause deteriorations in the strength and integrity of slopes as suggested by increased water ejection after the Hebgen Lake and the Borah Peak earthquakes in the USA,
- Open cracks in mountain slope that allows infiltration of surface water (Fig. 224) and
- Tectonic effects and fault actions on the deterioration of rock masses (Ohya landslide in Fig. 197).

Note that the statements above employ terminologies for rock. From a soil mechanic viewpoint, the statements are replaced by "destruction of grain-to-grain bonding by cyclic shear" and "disappearance of suction during rain and water submergence" as discussed by Torii et al. (2007).

Delayed landslides that occur a few days after earthquakes are certainly related to the change of local geohydrology as Jibson et al. (1994) pointed out on the landslides triggered by the Racha earthquake of Mw = 7.0 in Georgia, Eastern Europe. The fault issue will be further discussed in the next chapter. There is an interesting case in which the 2011 Tohoku earthquake, Japan, stabilized the ongoing landslide activity (Suzuki et al. 2020). Most probably, the earthquake changed the local geohydrology, as suggested by lowered artesian water pressure. This case requires more detailed discussion.

13 Effects of Fault on Long-Term Instability of Slopes

The previous chapter illustrated that mechanical deterioration or cracks/fissures made by fault action is one of the important reasons for long-term slope instability. This effect includes both the consequences of recent fault rupture and what happened during past earthquakes. Korup (2004) pointed out that many slope instabilities occur along the Alpine Fault in New Zealand and called this situation tectonic weakening. Mochizuki (1976) investigated rock conditions around faults and showed that silty/clayey rock is highly fractured and susceptible to landslides. With these in mind, it is attempted in this chapter to shed light on fault action. Landslide occurring at the exact place of fault dislocation (Figs. 226 and 227) is an evidence that mountains are disrupted by long-term dislocation of a fault. In Fig. 227, the Guinsaugon landslide in the Philippines, which occurred on February 17, 2006, was triggered by heavy rain; its volume was 15 million m^3, while L/H = 810 m / 3800 m (Orense and Sapuay 2006; Evans et al. 2007). This landslide occurred between 10:30 and 10:45 AM and was preceded by local small earthquakes that occurred near the landslide site. Their magnitudes were $M = 2.3$ at 6:07 a.m. and $M = 2.7$ at 10:36 a.m. The weak shaking of the second earthquake may have been amplified to trigger the landslide after ample antecedent rainfall (Sassa et al. 2007). On the other hand, it is possible that the second tremor was the ground shaking induced by the landslide impact (Catane et al. 2008).

The earthquake-induced and other slope instabilities in the Ina Region of Japan are discussed in conjunction with the activities of the Median Tectonic Line (Naumann 1895), which has about 1000 km in length in Western Japan (Fig. 1). The Median Tectonic Line (MTL) is a strike-slip fault that was originally formed during the Cretaceous Period, as discussed by Taira (2001) and, since then, has been subject to complicated tectonic actions. As exemplified by the Great Glen Fault in Scotland (p. 89 of Wilson 1963) together with many other faults, fractured rocks around MTL have been eroded by water action and long valleys have been formed. The eastern part of MTL has been seismically inactive in the historical period except the section around Oshika and Tohyama Villages in Fig. 228 (Okada 1993). A typical outcrop of MTL is shown in Fig. 229 where felsite intrusion along MTL (shown by M; 15 ma = 15 million years ago) is sandwiched between A (100 ma) and B (70–80 ma). Takagi et al. (2019a, b) stated that the rate of dislocation of MTL in this area is only 0.01–0.02 m/1000 years, which is slow.

Fig. 226 Coseismic landslide to the north of Muzaffarabad, Pakistan, with fault dislocation in the foreground

Fig. 227 Rainfall-induced landslide in 2006 at Guinsaugon, Leyte Island, the Philippines, whose head is located upon the Philippine Fault (photograph taken by Prof. T. Uchimura)

At least three damaging earthquakes occurred along the Tohyama-Oshika section of MTL (Fig. 228) in AD 714 (Fuso Ryakki, a chronological revision of six official historical records of Japan in the ancient period and its completion around AD 1100) and in 1718, together with what Matsu'ura et al. (2006) inferred for the year of 1725. The first event (magnitude = 6.5–7.5) triggered the devastating Ikeguchi landslide (93 million m^3 according to Chigira 2015) in Tohyama Village (Fig. 228) and a natural dam was created. Buried trees (Fig. 230) in this landslide and lake deposit facilitated tree ring studies, which is called dendrochronology (Douglass 1941), that identified the year of this landslide to be AD 714 (Goto et al. 2008; Muramatsu and Teraoka 2015). The second event in AD 1718 is called the Tohyama earthquake with a magnitude of around 7 and it induced landslides as well. Figure 231 shows one of the sites of landslides in Tohyama Village where a new hill was created and a river was blocked (Ichimura 1933; Sakamoto 2014). Furthermore, the 1586 Tenshou earthquake (magnitude around 8) at a somewhat greater epicentral distance triggered

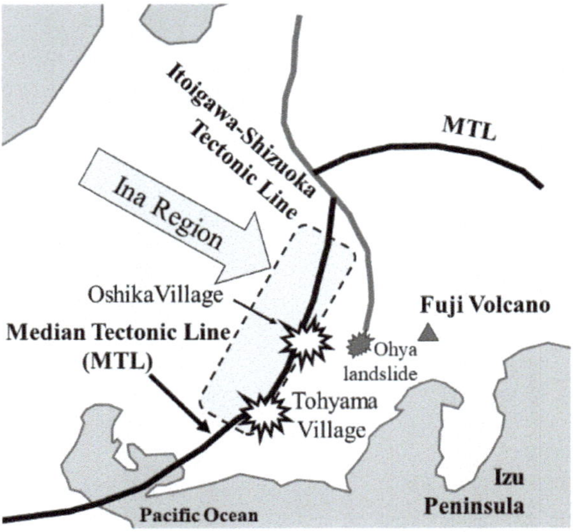

Fig. 228 Ina and eastern part of the Median Tectonic Line, Central Japan

Fig. 229 Outcrop MTL at Mizoguchi, to the north of Oshika Village

landslides in Ina, as verified by tree ring studies on buried cypress (Matsushima 2000a). The same earthquake produced a landslide dam in the vicinity as was verified by studies on grain size of silty deposits and ^{14}C dating (Suzuki et al. 2008).

In addition to the coseismic disasters, the Oshika-Tohyama area in Fig. 228 has many slope instabilities because of weak bedrock geology. Figure 232 shows the Tobigasu site, which had been in a critical state since the nineteenth century (Saito 1988). This site is within hundreds of meters of the underlying MTL. Figure 233 shows Ohnishi Yama Mountain, which had been unstable for a long time and fell down during the heavy rainfall in 1961. The volume of this failed mass was 2.4 million m^3 (Irazawa 1986). This slope is situated within one hundred meters of the underlying MTL.

Fig. 230 Tree in river bed at Kodohgi that was buried in a landslide lake after a coseismic event in AD 714

Fig. 231 Site of coseismic landslide in AD 1717

Fig. 232 Tobigasu slope exhibiting long-term instability in Oshika

Fig. 233 Ohnishi-yama landslide in Oshika triggered by heavy rainfall in 1961

In view of the many rainfall-induced disasters in the Ina Region, Matsushima (2000b) discussed a correlation between historical earthquakes and onset of rainfall-induced disasters during their aftermath. Because his idea was interesting, I attempted to investigate this correlation again by collecting information on damaging earthquakes in the Ina Region since the beginning of the seventeenth century from Kusakabe (1985), Osawa (1985), Matsushima (2000a), and Nagano Prefectural Government (2021). In contrast, there is virtually no historical summary of landslides in Ina. Therefore, similar to Matsushima (2000b), I used the historical summary of rainfall disasters by Murasawa (1983). It is, therefore, assumed that rainfall disasters are associated not only with flood but also with slope disasters. The extents of seismic intensity are assessed on the basis of described damage while those of slope disaster are evaluated based on the size of flood as quoted by Murasawa (1981) from historical documents: full, big, and ordinary floods. Consequently, a correlation between earthquakes and floods (Fig. 234) was drawn and it is not much different from what Matsushima (2000b) suggested. It may be said that three major seismic activities as illustrated by arrows are followed by many floods of scale factor = 3 and most likely slope disasters occurred.

A similar study was made for Kyoto on the possible correlation between strong earthquake shaking and floods of the Kamo River. Kyoto became the capital of Japan in AD 794 and, since then, the court people and priests have been concerned with the floods of this river. They left many documents on the occurrence of floods and Nakajima (1983) summarized those documents to illustrate the history of flooding over the centuries. He called some of those floods "severe" and I follow his idea. The history of earthquakes in Kyoto was summarized by Usami and Hisamoto (1971) who assessed the JMA seismic intensity scale in Kyoto for all the events. This scale is 7 at maximum, which implies serious damage. Figure 235 demonstrates the history of flood and earthquake history thus prepared. Although two seismic events of magnitude = 6.7 and 7.4, respectively, registered on the seismic intensity scale of 7 and were followed immediately by severe floods, the other one with magnitude = 7.0 was not accompanied by severe flooding. Floods are more closely related to population of Kyoto, which was very high three times in its history (indicated by

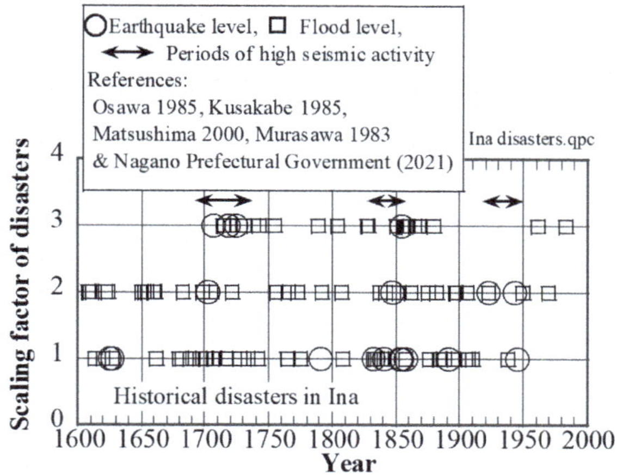

Fig. 234 Possible correlation between damaging earthquakes and flood disasters in Ina Region

two-way arrows). In those periods, Kyoto was the political and/or cultural as well as economic center of the nation with high population and people cut trees in the upstream region of the Kamo River in order to use them for construction and fuel needed by the big city. Moreover, the farmers in the suburbs of Kyoto preferred to keep grass lands in mountains in order to collect natural fertilizers, which was grass, and therefore remove trees. Thus, the mountains lost vegetation and became prone to erosion and flood. Due to lack of data, the seismic effect on slope instability in Kyoto was not verified.

The case histories indicated that there are two types of spatial distribution of earthquake-induced landslides (Fig. 236a). First, the landslide distribution is related to the epicentral distance and can be accounted for by regional geology and topography. This was the case of the Chi-Chi earthquake in Taiwan. Second, in contrast (Fig. 236b), the distance from the causative fault is the key issue. It is very likely that regional tectonics and past fault activities affected the material properties in the

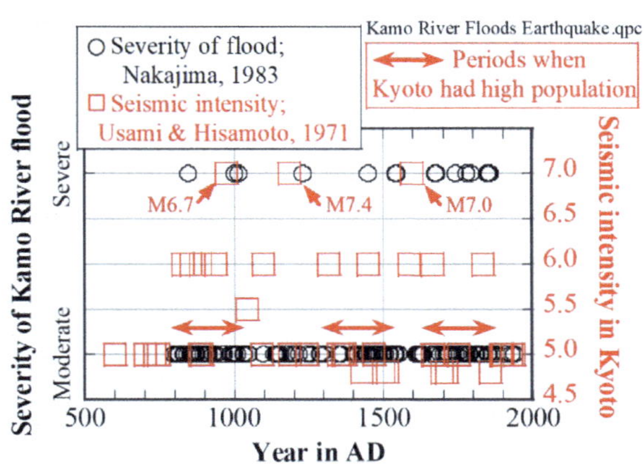

Fig. 235 Possibility of correlation in Kyoto between strong earthquakes and heavy floods of the Kamo River

affected region. This was likely the case with the Kashmir and Wenchuan earthquakes as well as the Ohya landslide (Fig. 145), Horai Valley (Fig. 178) and many others.

In mountain regions, slopes and valleys are formed through a variety of processes. In terms of the relationship with the fault, there are two types of valley formation, either displaced landform (fault cliff, pull-apart basin etc.) or denuded landform where materials are fissured or damaged and erosion is more active than in other places. In this chapter, I discuss fault-induced deterioration of materials along denuded landforms as a potential site of long-term slope instability. It is supposed that the band-shape distribution of landslides in Fig. 236b is the consequence of fault-induced deterioration.

From a macroscopic perspective, fault action can produce significant distortion in the adjacent rock mass such as the folding in Ohya Mountain as shown in Fig. 197. Bending deformation is associated with tension on one side of the distorted rock slab and tension cracks can develop there, leading to material weakening. From a microscopic viewpoint, Scholz et al. (1993) defined the process zone as the domain of ongoing crack generation near the tip of a growing fault rupture. Having both cracks and mechanical damage, the process zone is thought to induce long-term instability. Hence, the size of the process zone is important for assessment of the size of disaster-vulnerable areas. Figure 237 illustrates the definition of the width of process zone in accordance with Vermilye and Scholz (1998). In line with this, Anma (1987) discussed rock fracturing along fault planes.

Fig. 236 Two types of landslide distribution induced by earthquakes

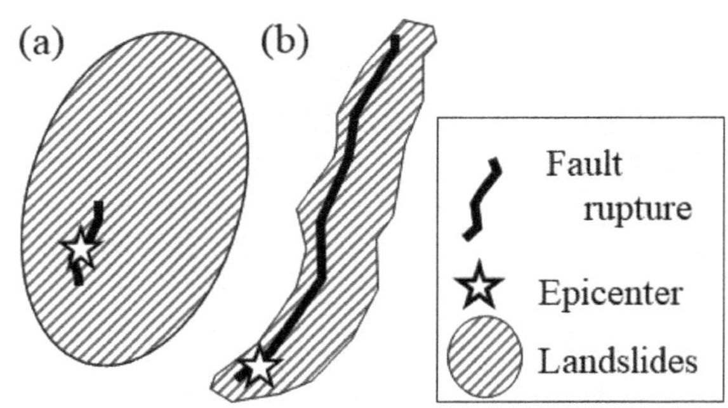

Fig. 237 Definition of width of process zone as the distance from fault rupture plane (drawn after Vermilye and Scholz 1998)

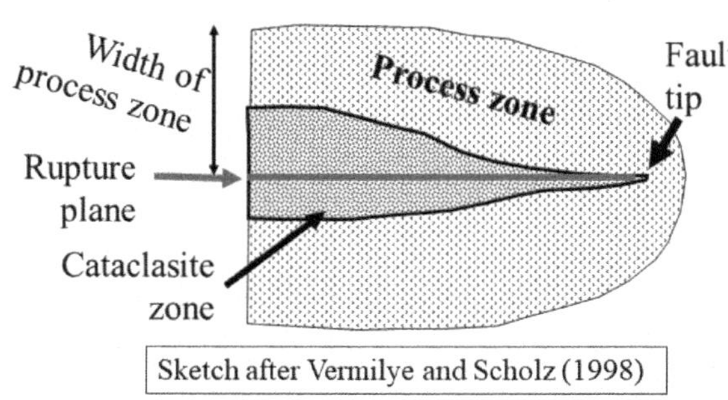

Vermilye and Scholz (1998) as well as Scholz (2002) undertook field investigations to collect data on the width of the process zone and the length of the fault, expecting a linear correlation between them. The original data came from their own as well as Anders and Wiltschko (1994), Brock and Engelder (1977), Chernyshev and Dearman (1991), Little (1995) and Yoshida et al. (2009), while I added more data obtained from the distribution of landslides induced by recent earthquakes. Of particular interest among them is Yoshida et al. (2009) whose investigation along the Atera Fault (70 km long or shorter) in Central Japan concerned not only the joint density but also the microscopic observation of the crystalline structure of affected rock specimens.

In my own data collection, Qi et al. (2010), for the Wenchuan earthquake, found that the range of significant landslide occurrence was limited to − 40 to + 10 km of distance. Therefore, I assume that the mean width of the landslide-prone area is (40 + 10)/2 = 25 km. The length of the causative fault is set equal to 300 km for this earthquake. For the 2016 Kaikoura earthquake in New Zealand, Massey et al. (2018) studied more than 10,000 landslides and showed that the majority of landslides are located within 2 km of the surface fault rupture. Although the Kaikoura earthquake was caused by a combined action of many faults, the present paper assumes 110 km, which is the total length of the rupture zone. By combining these data with the size of the process zone, the size of the area that is prone to long-term slope instability after big earthquakes in shown in Fig. 238. It is important that the proposal of 1% of the fault length by Vermilye and Scholz (1998) gives a reasonable assessment (or upper bound) for this type of problem.

Fig. 238 Assessed size of process zone or range of rock mass fracturing in which long-term slope instability is likely to occur after big earthquakes

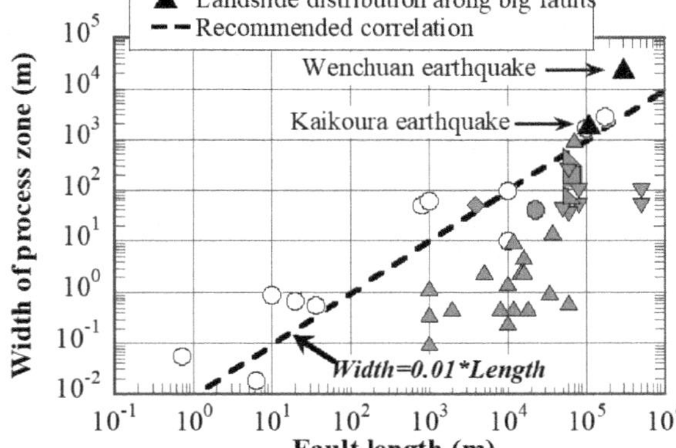

Note that this correlation between damage zone size and fault length is not universally accepted. Different authors pay attention to a correlation between damage size (W) and the geologically accumulated fault dislocation (D). Otsuki (1978) proposed an empirical equation;

$$\log_{10} D = 1.01 \times \log_{10} W + 1.78 \tag{6}$$

in which W was called the width of the shear zone and was probably equivalent with two-times the process zone.

14 Discussion

There have been profound developments in science and technology in the recent times. Nevertheless, disasters still occur as a consequence of earthquake-induced landslides. The limitations of the current studies are:

- We cannot predict earthquake-induced landslides because we cannot predict earthquakes. Therefore, the idea of emergency (early) warning and evacuation does not work.
- Underground conditions in vulnerable slopes are complicated and heterogeneous; they are not visible. Therefore, geotechnical subsurface investigation to assess the landslide hazard is difficult. In practice, budget limitations for investigation is a big problem as well.
- The value of a probabilistic approach has not been validated, although it is interesting.
- The earthquake disaster is a rare issue in the life of individual humans. People always have normalcy bias and do not want to worry about such a rare risk. Spending money on potential mitigation of rare disasters is not their choice.

Nevertheless, human in the past had a good way to reduce the risk without spending much time and money. This practice is called the wisdom of history in this section. Figure 239 is a view showing some of the many landslides caused by the 2004 Niigata-Chuetsu earthquake. It is noticeable that people's houses successfully avoided landslide damage. I suppose that the local community had learned lessons on disaster-prone places and safer sites after many past disasters during the one thousand year history of this village. I got a similar impression in Nepal as well (Fig. 53). Figures 240 and 241 illustrate shrines (religious places) that survived the tsunami disaster that claimed more than 22,000 lives during the 2011 Tohoku earthquake. At the site of Fig. 240, the tsunami run-up was very high but the shrine was located scarcely above it; no damage happened. The other shrine in Fig. 241 was situated at a very low elevation close to the beach but avoided tsunami as well. The bathymetry likely kept the tsunami low here but the shrine did not know it. In 2012, I traveled along the entire Pacific Ocean coast where tsunami disasters were severe and found many fortunate shrines. My interpretation was that shrines in the coastal area had been destroyed by tsunami many times in their history and every time the local community relocated them to higher and safer places. After repeating this relocation many times, the shrines were reconstructed at elevations higher than the maximum possible tsunami run-up. A similar finding was reported by Uda et al. (2012) and Endo and Mazereeuw (2015), while Yamazaki (2013) considers the situation just incidental. It seems to me that there are many things to be learned from history; see the houses on a steep slope near Barpak, Nepal, as well (Fig. 53). Unfortunately, the modern education and practice of science and technology miss the valuable wisdom of history.

Wasowski et al. (2011) presented the state of the art of studies on earthquake-induced landslides and suggested the future direction of research. Among their suggestions, I am interested in three issues and would like to add something. First, development of a landslide inventory is absolutely important. It will help develop holistic perspectives. My request is that presenting the number of landslides in a variety of categories (of geology, elevation, slope gradient etc.) is not sufficient. As practiced by Parise and Jibson (2000), landslide density (= number of landslides

Fig. 239 Residential houses that avoided coseismic landslides in 2004, Yamakoshi, Niigata, Japan

Fig. 240 Shrine at high elevation in Onagawa that avoided tsunami during the 2011 Tohoku earthquake, Japan

Fig. 241 Shrine at low elevation behind a beach in Ozuchi that avoided tsunami in 2011

per unit area) is more preferable and useful for general application. Furthermore, people suppose that GIS or satellite imagery can drastically improve the quality of inventories. This is true. However, the subsurface conditions such as shear strength, extent of weathering, fissures and cracks, ground water and many important other factors are overlooked. Reliance on "imagery" may tempt researchers to stay in their comfortable offices and work on computers, and the chance to visit the sites and "feel landslides" decreases. We should not forget the proper balance between digital work and field activities.

Second, the landslide hazard should be assessed in a quantitative manner by using a reasonably simple numerical method such as the Newmark rigid block analogy (Newmark 1965). In this analogy, it is assumed that landslides are induced by shear failure. On the contrary, landslides have been classified into many types (Varnes 1958) and shear failure is just one of them. Toppling and fall are other important triggering mechanisms. I tried to pay attention in this paper to the triggering mechanisms as well.

Third, the Newmark method cannot take into account the temporal changes in pore water pressure. However, in the recent geotechnical practice, this is considered to be an important issue. Antoine et al. (2016) conducted a step-by-step dynamic analysis in the time domain, using the Newmark analogy, to assess the earthquake-induced damage of earth dams. Also, the simplest approach is reducing the shear strength of soil in accordance with the factor of safety of soil against liquefaction. This reduction has been adopted by the Highway Bridge Design Code of Japan Road Association (JRA) for liquefaction risk assessment that is based on an early practical study by Iwasaki (1981).

Last but not least, attention must be paid to the effects of global climate change on landslide disasters. As is often discussed, climate change means global warming, which promotes water evaporation and the likelihood of heavy rain. Figure 242 shows the 24-h precipitation records (No. 1 ranking over the nation) during 316 rainfall-induced sediment disasters in Japan from 1945 to 2021 (Towhata 2021b).

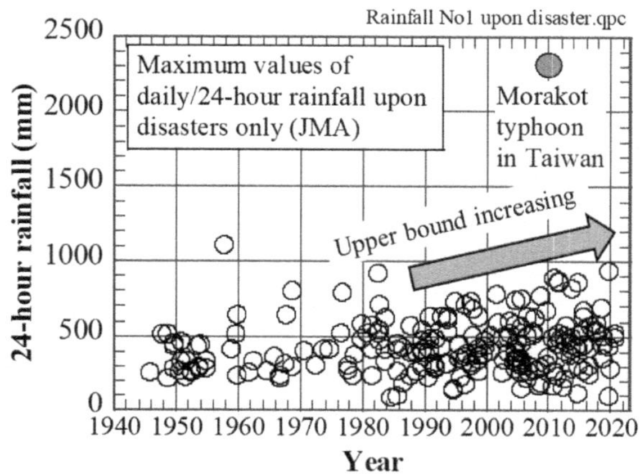

Fig. 242 Temporal variation of extreme rainfalls that have triggered sediment disasters in Japan (Towhata 2021b)

It can be seen herein that extreme rainfalls have been increasing since 1990. Most probably, many parts of the world are undergoing similar trends of rainfall. Figure 243 illustrates the magnitude of earthquakes (greater than or equal to 7.5) that occurred worldwide since 1950. It is interesting that the world seismic activity was low from 1970 to 2000, while the decades before and after this period experienced many earthquakes of magnitude greater than 8.5. Even the frequency of earthquakes with magnitudes = 7.5–8.0 was higher in these decades. This point is further illustrated by plotting in Fig. 244 the total annual earthquake energy that was simply calculated from the magnitude of 1153 earthquakes (both greater than and less than 7.5) by using the 1956 version of the Gutenberg equation (derived by substituting Eq. 1 of Gutenberg (1956) in his Eq. 3):

$$\log_{10} E = 4.8 + 1.5 \times \text{magnitude (energy in J unit)} \tag{7}$$

and then adding the energy together for each year. The magnitudes in Fig. 243 were substituted in Eq. 7, although Gutenberg intended to use his surface-wave magnitude (Ms), but this would not affect the conclusion. It is evident from the total annual earthquake energy (Fig. 244) that the world earthquake activity was high before 1970 and after 2000. While the effects of the gigantic earthquakes (Mw = 9.6 in Chile, 1960; 9.2 in Alaska, 1964; 9.3 in the Indian Ocean in 2004 and 9.0 in Tohoku, 2011) were profound, other smaller earthquakes made substantial contributions to the global seismic activity in the two active periods. In summary, the precipitation and seismic activities are greater in the twenty-first century than in the last 20 years of the previous century. When rainfalls and landslides are combined, they promote the occurrence of major compound disasters (rainfall + earthquake). Thus, the experience in late twentieth century (1970–2000) is not very helpful and earthquake-induced landslides become more likely today under the global climate change in the twenty-first century. Landslide fatalities, whether coseismic or non-seismic, have not decreased in the twenty-first century (Froude and Petley 2018).

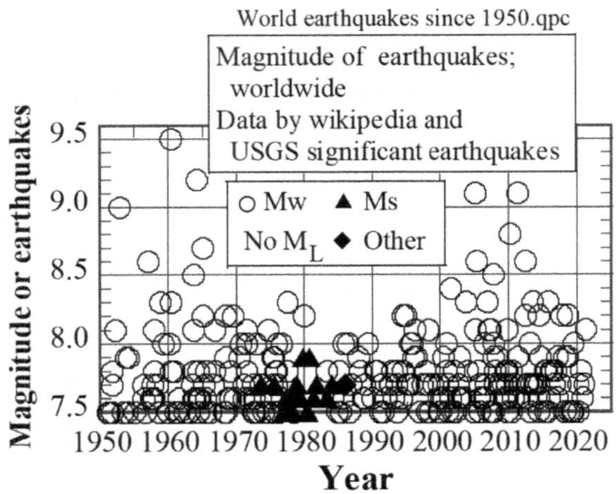

Fig. 243 Magnitude of earthquakes worldwide since 1950 (Mw: moment magnitude, Ms: surface-wave magnitude, and M_L: local magnitude)

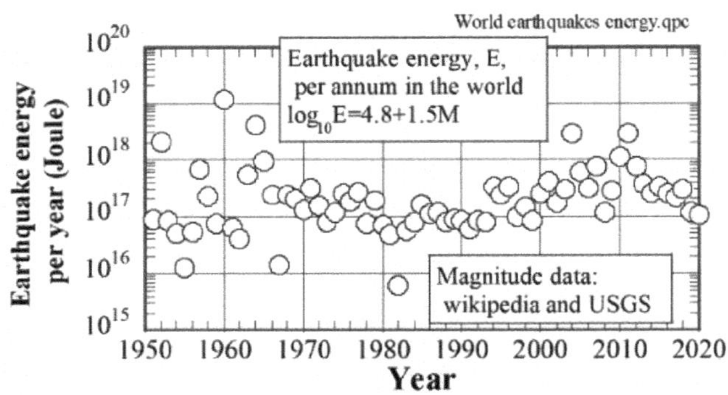

Fig. 244 Worldwide earthquake energy per annum in which the 1980s has the least seismic activity of the earth

The number of fatal earthquake-induced landslides and the number of victims have not decreased since the beginning of the twentieth century (Jessee et al. 2020) in spite of the development of science and technology.

15 Concluding Remarks

This review attempts to summarize the current state of the art of the studies on earthquake-induced landslides and to build up a holistic view. Because of the complex nature of this topic, I referred to the literatures in many disciplines. At the same time,

I visited many sites of landslide in the world. After all those efforts, the following conclusions can be drawn.

(1) It is essential nowadays to understand the nature of extreme and compound disasters in order to mitigate the induced social negative impact.
(2) It is certainly important to undertake elaborate and detailed research on many landslide topics. However, we should not forget the importance of a holistic view in order to remember what we have to achieve.
(3) The earthquake-induced landslides are a consequence of a complicated combination of geological and geomorphological environments, together with seismic action. Those slopes that are unstable under gravity are prone to earthquake-induced instability as well.
(4) Landslide inventories are a useful tool to develop a holistic view. It is important therein to pay more attention to landslide density (number of, or area of, landslides per unit area) than just the number of landslides.
(5) Shear is not the only mechanism of earthquake-induced landslides. Fall, toppling, detachment, and buckling of rock slabs can trigger landslides during strong shaking as well. This issue is often not reasonably considered in seismic hazard assessments.
(6) Creep deformation of slopes under gravitational action is an important signal for inferring the likelihood of failure during future earthquakes.
(7) The earthquakes in the twenty-first century have demonstrated the importance of new problems that were not seriously considered in the previous century. They are formation and breaching of landslide dams, the effects of antecedent rainfall, the long-term instability of affected mountains, and the nature of problem soils such as loess, volcanic ash and weathered rocks.
(8) The topic of compound disasters addresses the joint effects of different types of disasters that occur simultaneously or in sequence. For example, earthquakes trigger landslides and landslides trigger tsunami disasters. Heavy rainfall during the seismic aftermath is another example.
(9) Rock masses along major faults undergo high tectonic stress and are mechanically deteriorated. This is one of the reasons for long-term slope instability after devastating earthquakes.
(10) Earthquake-induced collapse of residential fills damages people's property. Because of insufficient funding, little effort has been made to address this issue.
(11) Mistakes in ground water management undermine the stability of slopes. Examples are found in over-irrigation in loess slopes, leakage from water supply channels and sewage water infiltrating into subsoil.
(12) Global warming promotes the melting of glaciers. This situation generates more slope instability and earthquakes will trigger landslides more easily than before. The instability of the glacier itself will be a problem as well. The tragedy of Yungay, Peru, was an important lesson.

(13) For hazard assessment, an empirical method based on topography, geomorphology and regional geology is the best choice. It is unfortunate that subsurface geotechnical investigation is not easy because of restraints due to cost and the nonuniformity under the ground. Consequently, an underground hazard has to be detected without knowing the underground condition. Gravity-induced creep deformation is a promising clue toward capturing the likelihood of future landslides.
(14) There are many mysteries in nature. Nalodo is one of them.
(15) Furthermore, there are cases in which mysterious behavior of ground water before, during and after earthquakes accompanied the coseismic landslides and likely promoted the slope failure.
(16) Global climate change and recent increase of stronger earthquakes affect the possibility of compound slope disasters.
(17) Earthquake-induced landslides cannot be predicted because earthquakes cannot be predicted by the current state of science. This is the biggest problem that we have.

Acknowledgements The content of this report relies extensively on reconnaissance trips. In this regard, I would like to express my sincere thanks to Dr. H. Shahnazari of Iran, Prof. K. Farooq in Lahore, Pakistan, Prof. L.M. Wang in Lanzhou, China, and Dr. R.P. Orense together with Dr. C. Massey in New Zealand who planned those trips and provided me valuable opportunities to learn from reality. Prof. A. Wakai of Gunma University helped me to study the ancient landslide in the flank of Mt. Akagi. The long collaboration with many people of the Taiwan Geotechnical Society has always stimulated me to proceed further. Another long collaboration through Dr. Wang Lin of Chuo Kaihatsu Co. with Prof. J.P. Qiao and Dr. Z.J. Yang as well as many members of the Institute of Mountain Hazards and Environment (CAS) in Chengdu, China, enabled me to visit numerous sites of landslides in southwest China. The reconnaissance after the 2018 earthquake in Sulawesi was conducted as a collaboration between Japan International Cooperation Agency (JICA) and Badan Meteorologi, Klimatologi, dan Geofisika (BMKG) (Indonesian Agency for Meteorology, Climatology and Geophysics) of the Indonesian Government. The Nalodo Study Team of JICA, including Mr. H. Furuichi of Yachiyo Engineering Company in particular, provided the author with invaluable opportunities to be involved in research on this mysterious disaster. Beautiful photographs of landslides were contributed to this publication by Mr. S. Suwa and Prof. S. Yasuda. Prof. M. Irsyam and HATTI (Indonesian Society for Geotechnical Engineering) provided fruitful joint activity on the Nalodo problem. The earthquake motion record in Fig. 46 was obtained by K-NET (doi: 10.17598/NIED.0004) that is operated by the National Research Institute for Earth Science and Disaster Resilience. I regret that I cannot mention all people who gave valuable support to me but wish them to achieve good works in future. The entire content of this paper is a part of the activities of JTC1 of FedIGS for which Prof. L. Picarelli was the Chair and gave me valuable opportunities to work on many interesting topics. Sincere thanks are extended to all of the institutions and individuals mentioned above.

References

Abendanon EC (1918) Expédition de la Célèbes centrale: Voyages géologiques et géographiques à travers la Célèbes centrale (1909–1910). Librairie et Imprimerie ci-devant E.J. Brill, Leyde vol. 2 (in French)

Abrahamson NA, Somerville PG (1996) Effects of the hanging wall and footwall on ground motions recorded during the Northridge earthquake. Bull Seismol Soc Am 86(1B):S93–S99

Adachi M, Otsuka Y, Kitaura M (1999) A study on flood control of the Joganji-River from feudal period to Meiji Era. Journal of Historical Studies in Civil Engineering, Japan Society of Civil Engineering 19:331–336 (in Japanese)

Agliardi F, Crosta GB, Frattini P, Malus MG (2013) Giant non-catastrophic landslides and the long-term exhumation of the European Alps. Earth Planet Sci Lett 365:263–274

Aki K (1988) Local site effects on strong ground motion. In: von Thun JL (ed) Earthquake engineering and soil dynamics II—recent advances in ground-motion evaluation, vol 20. Park City. Utah, ASCE Geotechnical special Publication, pp 103–155

Alberto Y, Otsubo M, Kyokawa H, Kiyota T, Towhata I (2018) Reconnaissance of the 2017 Puebla, Mexico earthquake. Soils Found 58(5):1073–1092

Alford D, Cunha SF, Ives JD (2000) Lake Sarez, Pamir Mountains, Tajikistan: mountain hazards and development assistance. Mt Res Dev 20(1):20–23

Ambraseys N, Bilham R (2012) The Sarez-Pamir earthquake and landslide of 18 February 1911. Seismol Res Lett 83(2):294–314

Anders MA, Wiltschko DV (1994) Microfracturing, paleostress and the growth of faults. J Struct Geol 16(6):795–815

Anma S (1987) Large scale landslides and its debris flows induced by earthquakes—a study on the hazard susceptibility & prevention scheme. Tokai University (in Japanese), Thesis for doctoral degree

Antoine D, Tatsuoka F, Shinbo T, Mohri Y (2016) A new simplified seismic stability analysis taking into account degradation of soil undrained stress-strain properties and effects of compaction: validation of dynamic analyses of dams and their equipment. International Symposium on the Qualification of Dynamic Analyses of Dams and Their Equipments, Saint-Malo, France, CRC Press, p 215

Araya C, Cisternas M, González F (2014) Evolución morfológica del principal deslizamiento del 'Riñihuazo', generado por el terremoto de 1960. VIII SLAGF, Simposio Latinoamericano de Geografía Física, IV SIAGF, Simposio Iberoamericano de Geografía Física: 1018–1026 (in Spanish)

Arulanandan K, Yogachandran C, Muraleetharan KK, Kutter B, Chang GS (1988) Seismically induced flow slide on centrifuge. ASCE Journal of Geotechnical Engineering 114(12):1442–1449

Auffret GA, Auzende JM, Gennesseaus M, Monti S, Pastouret L, Pautot G, Vanney JR (1982) Recent mass wasting processes on the Provencal margin (Western Mediterranean) marine slides and other mass movements. Plenum Press, pp 53–58

Auvinet G, Méndez E, Juárez M (2013) Soil fracturing induced by land subsidence in Mexico City. In: 18th international conference on soil mechanics and geotechnical engineering, Paris, pp 2921–2924

Auvinet G, Méndez E, Juárez, M (2017) Recent information on Mexico City subsidence. In: Proceedings of 19th international conference on soil mechanics and geotechnical engineering, Seoul, pp 3295–3298

Aydan Ö, Ulusay R, Atak VO (2008) Evaluation of ground deformations induced by the 1999 Kocaeli earthquake (Turkey) at selected sites on shorelines. Environ Geol 54(1):165–182

Aziz M, Towhata I, Irfan M (2015) Strength and deformation characteristics of degradable granular soils. Geotech Test J 39(3):452–461

Babayev G, Tibaldi A, Bonali FL, Kadirov F (2014) Evaluation of earthquake-induced strain in promoting mud eruptions: the case of Shamakhi–Gobustan–Absheron areas, Azerbaijan. Nat Hazards 72:789–808

Barrantes-Castillo G, Jiménez-Campos C, Ocón-García MJ (2013) Deslizamientos provocados por el terremoto de Cinchona de 2009, Costa Rica. Revista Geográfica De América Central 2(51):89–100 (in Spanish)

Bartelt P, Buser O, Platzer K (2006) Fluctuation-dissipation relations for granular snow avalanches. J Glaciol 52(179):631–643

Basharat M, Riaz MT, Jan MQ, Xu C, Riaz S (2021) A review of landslides related to the 2005 Kashmir Earthquake: implication and future challenges. Nat Hazards 108(1):1–30

Beaudouin T, Bellier O, Sébrier M (2003) Present-day stress and deformation fields within the Sulawesi Island area (Indonesia): geodynamic implications. Bull Soc Geol Fr 174(3):305–317 (in French)

Bellier O, Sébrier M, Beaudouin T, Villeneuve M, Braucher R, Bourles D, Siame L, Putranto E, Pratomo I (2001) High slip rate for a low seismicity along the Palu-Koro active fault in central Sulawesi (Indonesia). Terra Nova 13(6):463–470

Bellot JP (2008) Hydrothermal fluids assisted crustal-scale strike-slip on the Argentat fault zone. Tectonophysics 450(1–4):21–33

Board of Education, Maebashi City (2013) Map of archaeological sites in Maebashi (in Japanese)

Bommer JJ, Rodríguez CE (2002) Earthquake-induced landslides in Central America. Eng Geol 63:189–220

Bondevik S, Svendsen JI, Johnsen G, Mangerud J, Kaland PE (1997) The Storegga tsunami along the Norwegian coast, its age and run up. Boreas 26(1):29–53

Bothara J, Beetham D, Brunsdon D, Stannard M, Brown R, Hyland C, Lewis W, Miller S, Sanders R, Sulistio Y (2010) General observations of effects of the 30th September 2009 Padang earthquake, Indonesia. Bull N Z Soc Earthq Eng 43(3):143–172

Bouchon M, Barker JS (1996) Seismic response of a hill: the example of Tarzana, California. Bull Seismol Soc Am 86(1A):66–72

Boulanger RW, Truman SP (1996) Void redistribution in sand under post-earthquake loading. Can Geotech J 33(5):829–834

Bradley K, Mallick R, Andikagumi H, Hubbard J, Meilianda E, Switzer A, Du N, Brocard G, Alfia D, Benazir B, Feng G, Yun S-H, Majewski J, Wei S, Hill EM (2019) Earthquake-triggered 2018 Palu Valley landslides enabled by wet rice cultivation. Nat Geosci 12(11):932–939

Briggs RO (1994) Effects of the earthquake on surface waters in Waddell Valley, The Loma Prieta, California earthquake of October 17, 1989: Strong ground motion and ground failure. Hydrologic disturbances, USGS Professional Paper 1551-E:E21–E30

Brock WG, Engelder JT (1977) Deformation associated with the movement of the Muddy Mountains overthrust in the Buffington window, southeastern Nevada. Bulletin of Geological Society of America 88(11):1667–1677

Bulmer M, Farquhar T, Roshan M, Akhtar SS, Wahla SK (2007) Landslide hazards after the 2005 Kashmir earthquake. EOS Trans Am Geophys Union 88(5):53–68

Byers AC III, Byers EA, McKinney DC, Rounce DR (2017) A field-based study of impacts of the 2015 earthquake on potentially dangerous glacial lakes in Nepal. Himalaya Journal of the Association for Nepal and Himalayan Studies 37(2):26–41

Carvajal M, Araya-Cornejo C, Sepúlveda I, Melnick D, Haase JS (2019) Nearly instantaneous tsunamis following the Mw 7.5 2018 Palu earthquake. Geophysical Research Letters 46(10):5117–5126

Casagli N, Ermini L (1999) Geomorphic analysis of landslide dams in the Northern Apennine. Transactions of the Japanese Geomorphological Union 20(3):219–249

Casagli N, Ermini L, Rosati G (2003) Determining grain size distribution of the material composing landslide dams in the Northern Apennines: sampling and processing methods. Eng Geol 69(1–2):83–97

Catane SG, Cabria HB, Zarco MAH, Saturay RM, Mirasol-Robert AA (2008) The 17 February 2006 Guinsaugon rock slide-debris avalanche, Southern Leyte, Philippines: deposit characteristics and failure mechanism. Bull Eng Geol Env 67(3):305–320

Cavers DS (1981) Simple methods to analyze buckling of rock slopes. Rock Mech 14:87–104

Çelebi M (1987) Topographical and geological amplification determined from strong-motion and aftershock records of the 3 March 1985 Chile earthquake. Bull Seismol Soc Am 77(4):1147–1167
Chai HJ, Liu HC, Zhang ZY, Xu ZW (2000) The distribution, causes and effects of damming landslides in China. Journal of the Chengdu Institute of Technology 27(3):302–307
Chang LC (1938) Reconnaissance on Diexi earthquake in Sichuan Province. Geological Review, Geological Society of China 3(3):251–292 (in Chinese)
Chang DS, Zhang LM, Xu Y, Huang RQ (2009) Analysis of overtopping failure of Hongshhe landslide dam after Wenchuan earthquake. J Eng Geol 17(1):50–55 (in Chinese)
Charlwood RG, Little TE, Lou JK (2000) A review of the performance of two large sub-stations and eight large dams during the Chi-Chi Taiwan earthquake. ICLR Research Paper Series, No. 6, Institute for Catastrophic Loss Reduction
Chen SC, Wu CH (2009) The geomorphological variation triggered by catastrophic deep landslide in Siaolin Village. Journal of Chinese Soil and Water Conservation 40(4):359–376 (in Chinese)
Chernyshev SN, Dearman WR (1991) 2.3 Tectonic fractures. Rock fractures, publ. Butterworth-Heinemann, pp 41–56
Chigira M (1992) Long-term gravitational deformation of rocks by mass rock creep. Eng Geol 32(3):157–184
Chigira M (1998) Gravitational mass rock creep and landslides: from structural geology to hazard geology. Mem Geol Soc Jpn 50:241–250 (in Japanese)
Chigira M (2009) Mountain hazards induced by the 1999 Chi-Chi earthquake and their change after the earthquake. Journal of Japan Society for Natural Disaster Science 28(2):161–166 (in Japanese)
Chigira M (2014) Geological and geomorphological features of deep-seated catastrophic landslides in tectonically active regions of Asia and implications for hazard mapping. Episodes 37(4):284–294
Chigira M (2015) What we have learned from the landslides induced by recent earthquakes. Journal of Japan Landslide Society 52(2):93–98 (in Japanese)
Chigira M, Kasama T, Suzuki T, Furuki H (2017) Geological structure of the Shinseiko Landslide induced by the 1923 Kanto earthquake and its significance. Disaster Prevention Research Institute Annuals B, Kyoto University 60(B):417–430 (in Japanese)
Chigira M, Nakasuji A, Fujiwara S, Sakagami M (2012) Soil slide avalanches of pyroclastic fall deposits induced by the 2011 off the Pacific Coast of Tohoku earthquake. Journal of Japan Society of Engineering Geology 52(6):222–230 (in Japanese)
Chigira M, Wang WN, Furuya T, Kamai T (2003) Geological causes and geomorphological precursors of the Tsaoling landslide triggered by the 1999 Chi-Chi earthquake. Taiwan. Engineering Geology 68:259–273
Chigira M, Wu XY, Wang GH, Inokuchi T (2009a) Lessons we learned from the landslides induced by the Wenchuan earthquake and its application for the future. In: Proceedings of international conference on earthquake engineering—the first anniversary of Wenchuan Earthquake, Chengdu, China
Chigira M, Wu XY, Inokuchi T, Wang GH (2009b) Features and the distribution of landslides induced by the 2008 Wenchuan earthquake, Sichuan, China. Annuals of Disaster Prevention Institute, Kyoto University 52A:115–130 (in Japanese)
Chigira M, Wu X, Inokuchi T, Wang GH (2010) Landslides induced by the 2008 Wenchuan earthquake, Sichuan. China. Geomorphology 118(3–4):225–238
China Earthquake Administration (2008) Analysis of the cause of the M8.0 Wenchuan earthquake. http://www.cea.gov.cn/manage/html/8a8587881632fa5c0116674a018300cf/_content/08_05/30/1212119940937.html. Retrieved on 20210109 (in Chinese)
Close U, McCormick E (1922) "Where the mountains walked" an account of the recent earthquake in Kansu Province, China, which destroyed 100,000 lives. National Geographic Magazine 41(5):445–472
Collins BD, Jibson RW (2015) Assessment of existing and potential landslide hazards resulting from the April 25, 2015 Gorkha, Nepal Earthquake sequence. USGS Open File Report 2015–1142

Collins BD, Kayen R, Tanaka Y (2012) Spatial distribution of landslides triggered from the 2007 Niigata Chuetsu-Oki Japan Earthquake. Eng Geol 127:14–26

Costa JE, Schuster R (1988) The formation and failure of natural dams. Geol Soc Am Bull 100(7):1054–1068

Coulter HW, Migliaccio RR (1966) Effects of the earthquake of March 27, 1964 at Valdez, Alaska. US Geological Survey, Professional Report 542-C

Coulter HW, Migliaccio RR (1971) Effects at Valdez. The Great Alaska Earthquake of 1964, Chapter on Geology, National Academy of Sciences, pp 358–394

Crozier MJ (1992) Determination of palaeoseismicity from landslides. In: 6th international symposium on landslides, Christchurch, vol 2, pp 1173–1180

Cui P, Chen XQ, Zhu YY, Su FH, Wei FQ, Han YS, Liu HJ, Zhuang JQ (2011) The Wenchuan earthquake (May 12, 2008), Sichuan Province, China, and resulting geohazards. Nat Hazards 56(1):19–36

Cui SH, Pei XJ, Huang RQ, Zhang WF, Liang YF (2019) Discontinuities and engineering geological significances of strong earthquake-induced Daguangbao landslide. Journal of Southwest Jiaotong University 54(1):61–72 (in Chinese)

Cui P, Zhu YY, Han YS, Chen XQ, Zhuang JQ (2009) The 12 May Wenchuan earthquake-induced landslide lakes: distribution and preliminary risk evaluation. Landslides 6(3):209–223

Cummins PR (2019) Irrigation and the Palu landslides. Nat Geosci 12(11):881–882

Curry RR, Emery BA, Kidwell TG (1994) Sources and magnitudes of increased streamflow in the Santa Cruz Mountains for the 1990 water year after the earthquake. The Loma Prieta, California earthquake of October 17, 1989: strong ground motion and ground failure, hydrologic disturbances. USGS Professional Paper 1551-E:E31–E50

Dadson SJ, Hovius N, Chen H, Dade WB, Lin JC, Hsu ML, Lin CW, Horng MJ, Chen TC, Milliman J, Stark CP (2004) Earthquake triggered increase in sediment delivery from an active mountain belt. Geology 32(8):733–736

Dai FC, Lee CF, Deng JH, Tham LG (2005) The 1786 earthquake-triggered landslide dam and subsequent dam-break flood on the Dadu River, southwestern China. Geomorphology 65(3–4):205–221

Daryono MR (2016) Paleoseismology Tropis Indonesia (dengan Studi Kasus di Sesar Sumatra, Sesar Palukoro-Matano, dan Sesar Lembang). Ph.D. Thesis, Bandung Institute of Technology (in Indonesian)

Daryono MR, Natawidjaja DH (2019) Surface rupture of the 1909 event (M7) on Palukoro Fault in tropical forest of Central Sulawesi, Indonesia. In: Proceedings of joint convention Yokyakarta 2019, publ. HAGI–IAGI–IAFMI-IATMI (JCY 2019)

Davies TR, McSaveney MJ (1999) Runout of dry granular avalanches. Can Geotech J 36(2):313–320

Davis SN, Karzulovic J (1963) Landslides at Lago Riñihue. Bull Seismol Soc Am 53(6):1403–1414

Davis LL, West LR (1973) Observed effects of topography on ground motion. Bull Seismol Soc Am 63(1):283–298

Delaney KB, Evans SG (2015) The 2000 Yigong landslide (Tibetan Plateau), rockslide-dammed lake and outburst flood: review, remote sensing analysis, and process modelling. Geomorphology 246:377–393

Dellow S, Massey C, Cox S, Archibald G, Begg J, Bruce Z, Carey J, Davidson J, Della Pasqua F, Glassey P, Hill M, Jones K, Lyndsell B, Lukovic B, McColl S, Rattenbury M, Read S, Rosser B, Singeisen C, Townsend D, Villamor P, Villeneuve M, Wartman J, Rathje E, Sitar N, Athanasopoulos-Zekkos A, Manousakis J, Little M (2017) Landslides caused by the 14 November 2016 Mw7.8 Kaikōura earthquake and the immediate response. Bulletin of the New Zealand Society of Earthquake Engineering 50(2):106–116

Deng J, Kameya H, Miyashita Y, Kuwano J, Kuwano R, Koseki J (2011) Study on a failed dip slope with a thin sandy layer in 2004 Niigata-ken Chuetsu Earthquake. Eng Geol 123(4):302–314

Dockrill P (2020) Ice melt in Alaska threatens to unleash unprecedented 'Mega-Tsunami', Scientists warn. Science Alert, https://www.sciencealert.com/ice-melt-in-alaska-threatens-to-unleash-unprecedented-mega-tsunami-scientists-warn. Retrieved on 11 Jan 2021

Douglass AE (1941) Crossdating in dendrochronology. J Forest 39(10):825–831
Edgers L, Karlsrud K (1982) Soil flows generated by submarine slides—case studies and consequences. Norw Geotech Inst Bull 143:1–11
Elbanna A, Abdelmeguid M, Ma X, Amlani F, Bhat HS, Synolakis C, Rosakis AJ (2021) Anatomy of strike-slip fault tsunami genesis. Proceedings of the National Academy of Sciences 118(19)
Endo K, Mazereeuw M (2015) Historical study on geographical features of shrines in town of Minami Sanriku, Miyagi Prefecture. Journal of the Japanese Institute of Landscape Architecture 78(5):693–696 (in Japanese)
Erismann TH (1979) Mechanisms of Large Landslides. Rock Mechanics 12(1):15–46
Ermini L, Casagli M (2002) Criteria for a preliminary assessment of landslide dam evolution. Landslides. In: Rybar J, Stemberk J, Wagner P (ed) Proceedings of 1st European conference on landslides, pp 157–162
Ermini L, Casagli M (2003) Prediction of the behaviour of landslide dams using a geomorphological dimensionless index. Earth Surface Processes and Landform 28(1):31–47
Evans SG, Guthrie RH, Roberts NJ, Bishop NF (2007) The disastrous 17 February 2006 rockslide-debris avalanche on Leyte Island, Philippines: a catastrophic landslide in tropical mountain terrain. Nat Hazard 7(1):89–101
Evans SG, Roberts NJ, Ischuk A, Delaney KB, Morozova GS, Tutubalina O (2009) Landslides triggered by the 1949 Khait earthquake, Tajikistan, and associated loss of life. Eng Geol 109(3–4):195–212
Fan XM, Domènech G, Scaringi G, Huang RQ, Xu Q, Hales TC, Dai LX, Yang Q, Francis O (2018) Spatio-temporal evolution of mass wasting after the 2008 Mw 7.9 Wenchuan earthquake revealed by a detailed multi-temporal inventory. Landslides 15:2325–2341
Fan XM, Xu Q, Scaringi G, Dai LX, Li WL, Dong XJ, Zhu X, Pei XJ, Dai K, Havenith H-B (2017) Failure mechanism and kinematics of the deadly June 24th 2017 Sinmo landslide, Maoxian, Sichuan. China. Landslides 14(6):2129–2146
Fanetti D, Anselmetti FS, Chapron E, Sturm M, Vezzoli L (2008) Megaturbidite deposits in the Holocene basin fill of Lake Como (Southern Alps, Italy). Palaeogeogr Palaeoclimatol Palaeoecol 259(2–3):323–340
Fang J, Xu C, Wen Y, Wang S, Xu G, Zhao Yi L (2019) The 2018 Mw 7.5 Palu earthquake: a supershear rupture event constrained by InSAR and broadband regional seismograms. Remote Sensing 11(11):1330
Favalli M, Boschi E, Mazzarini F, Pareschi MT (2009) Seismic and landslide source of the 1908 Straits of Messina tsunami (Sicily, Italy). Geophys Res Lett 36(16):L16304
Fine IV, Rabinovich AB, Bornhold BD, Thomson RE, Kulikov EA (2005) The Grand Banks landslide-generated tsunami of November 18, 1929: preliminary analysis and numerical modeling. Mar Geol 215(1–2):45–57
Fire and Disaster Management Agency (2016) Damage caused by the 2016 Kumamoto earthquake and governmental response. https://www.fdma.go.jp/publication/hakusho/h28/items/special1.pdf. Retrieved on 1 Jan 2021 (in Japanese)
Froude MJ, Petley DN (2018) Global fatal landslide occurrence from 2004 to 2016. Natural Hazards and Earth System Science 18:2161–2181
Fujii T (1977) Coseismic landslide of Shiratori-yama Mountain triggered by Ansei Tokai earthquake. Shizuoka Chigaku 35:33–37 (in Japanese)
Fujimori K, Yamamoto T, Otsuka S, Ishii H (1995) Observations of crustal movements and discharge at Rokko-Takao station—changes associated with the 1995 Hyogo-ken-Nanbu earthquake. Annuals Disaster Prevention Research Institute, Kyoto University 38(B-1):287–296 (in Japanese)
Fujinami Y, Yoshimi M, Sakai H, Sakai K, Ikeda Y (2017) Consideration of ground amplification factor of the mountain terrain based on the earthquake observation. Proc JSCE A1 73(4):907–913 (in Japanese)

Fujisawa K (2008) Gigantic coseismic landslide in the vicinity of Aratozawa Dam. Meeting on reconnaissance after 2008 Iwate Miyagi Inland earthquake. https://www.pwri.go.jp/jpn/research/disaster/iwate-miyagi/04_jisuberi.pdf. Retrieved on 4 Dec 2020 (in Japanese)
Fujita T (1965) Matsunoyama landslide, Higashi-Kubiki, Niigata. Civil Engineering Journal 7:24–30 (in Japanese)
Fujita K, Inoue H, Izumi T, Yamaguchi S, Sadakane A, Sunako S, Nishimura K, Immerzeel WW, Shea JM, Kayastha RB, Sawagaki T (2017) Anomalous winter-snow-amplified earthquake-induced disaster of the 2015 Langtang avalanche in Nepal. Nat Hazard 17(5):749–764
Fujiwara J, Tohyama S, Suzuki S, Kokugan S (1996) On the mechanism of the landslide by Southern Hyogo earthquake and how to control it. The case of Kotaki landslide, Tsuna, Awaji Island. Monthly Magazine of the Japanese Geotechnical Society 44(2):79–82 (in Japanese)
Fukuoka H, Sakai N (2016) Reconnaissance of the landslides triggered by the 16 April 2016 Kumamoto earthquake in Minami-Aso. Japan. Journal of Japan Landslide Society 53(3):95–99 (in Japanese)
Fukuoka H, Sassa K, Scarascia-Mugnozza G (1997) Distribution of landslides triggered by the 1995 Hyogo-ken Nanbu earthquake and long runout mechanism of the Takarazuka Golf Course landslide. J Phys Earth 45(2):83–90
Furuya T (1974) A geomorphological consideration of the Mayu-Yama great landslide in 1792. Disaster Prevention Research Institute Annuals 17B:259–264 (in Japanese)
Gomez JM, Madariaga R, Walpersdorf A, Chalard E (2000) The 1996 earthquakes in Sulawesi, Indonesia. Bull Seismol Soc Am 90(3):739–751
Gorum T, Fan X, van Westen CJ, Huang RQ, Xu Q, Tang C, Wang G (2010) Distribution pattern of earthquake-induced landslides triggered by the 12 May Wenchuan earthquake. Geomorphology 133(3–4):152–167
Goto A, Muramatsu T, Teraoka Y (2008) Dates of buried trees formed by ancient landslide-dammed lake along the Toyama-river, Minami-Shinano Nagano Prefecture. In: Summaries of researches using AMS at Nagoya University and proceedings of the 20th symposium on chronological studies at the Nagoya University Center for Chronological Research XIX, pp 99–102 (in Japanese)
Gratchev I, Towhata I (2010) Geotechnical characteristics of volcanic soil from seismically induced Aratozawa landslide. Japan. Landslides 7(4):503–510
Gusiakov VK (2009) Tsunami History: Recorded. the Sea 15:23–53
Gutenberg B (1939) Tsunamis and earthquakes. Bull Seismol Soc Am 29:517–526
Gutenberg B (1956) The energy of earthquakes. Q J Geol Soc Lond 112(1–4):1–14
Hagiwara T (1972) Cause of shallow earthquakes occurring in the earth's crust as suggested by the accompanying land deformation. Phys Earth Planet Inter 6(4):250–255
Hakhamaneshi M, Alberto Y, Sharma K (Year not shown) EERI VERT Chiapas earthquake. https://www.eeri.org/wp-content/uploads/Landslide-rockfalls-landseparation.pdf. Retrieved on 7 Dec 2020
Hamada M, Yasuda S, Isoyama R, Emoto K (1986) Generation of permanent ground displacements induced by soil liquefaction. Proc JSCE 376/III-6:211–220 (in Japanese)
Hamada M (1992) Large ground deformations and their effects on lifelines: 1964 Niigata earthquake, case studies of liquefaction and lifeline performance during past earthquakes, Chap. 3, NCEER-92–0001, National Center for Earthquake Engineering Research
Hampton MA, Lemke RW, Coulter HW (1993) Submarine landslides that had a significant impact on man and his activities: Seward and Valdez, Alaska. Submarine landslides: selected studies in the US exclusive economic zone. USGS Bulletin 2002:123–134
Hanrahan M (2004) Tsunami the Newfoundland tidal wave disaster. Flanker Press Ltd, St. John's, Newfoundland. ISBN 1-894463-63-3
Harp EL, Wilson RC (1995) Shaking intensity thresholds for rock falls and slides: evidence from Whittier Narrows and Superstition Hills earthquake strong-motion records. Bull Seismol Soc Am 85(6):1739–1757
Harrison JV, Falcon NL (1938) An ancient landslip at Saidmarreh in southwestern Iran. J Geol 46:296–309

Hashimoto T (2009) Analysis of highway embankment failure during the 2007 Noto-Hanto earthquake, Japan. Proc JSCE A1 65(1):850–863 (in Japanese)
Havenith HB, Bourdeau C (2010) Earthquake-induced landslide hazards in mountain regions: a review of case histories from Central Asia. An inaugural lecture to the society. Geologica Belgica 13(3):137–152
Havenith HB, Venini M, Jongmans D, Faccioli E (2003) Initiation of earthquake-induced slope failure: influence of topographical and other site specific amplification effects. J Seismolog 7:397–412
Hayakawa Y, Morita T, Nakashimada E, Kabe N (2002) The AD 818 earthquake described in Ruiju Kokushi, and its geologic evidence printed on the southern foot of Akagi Volcano. Historical Earthquakes 18:34–41 (in Japanese)
Hazarika H, Rohit D, Pasha SMK, Maeda T, Masyhur I, Arsyad A, Nurdin S (2021) Large distance flow-slide at Jono-Oge due to the 2018 Sulawesi Earthquake. Indonesia. Soils and Foundations 61(1):239–255
He SY, Wang XL, Fan HB, Wang H, Ren R, Guo CX (2020) The study on loess liquefaction in China: a systematic review. Nat Hazards 103:1639–1669
Heezen BC, Ewing M (1952) Turbidity currents and submarine slumps, and the 1929 Grand Banks Earthquake. Am J Sci 250:849–873
Heidarzadeh M, Krastel S, Yalciner AC (2014) The state-of-the-art numerical tools for modeling landslide tsunamis: a short review. Submarine Mass Movements and Their Consequences 37:483–495; Krastel S et al (ed) Advances in Natural and Technological Hazards Research. Springer International
Heim A (1932) Bergstürz und Menschenleben, Fretz & Wamuth Verlag A.G. Zürich. 114–119 (in German)
Hendron AJ, Patton FD (1985) The Vaiont slide, a geotechnical analysis based on new geologic observations on the failure surface, 1. US Corps of Engineers, Technical Report GL-85-5: 87–88
Hernandez PL (1960) La Epopeya del Riñihue, Editorial Ercilla. http://www.memoriachilena.gob.cl/archivos2/pdfs/MC0000526.pdf. Retrieved on 12 Jan 2021 (in Spanish)
Higaki D, Hayashi K, Hamazaki, E, The project team of collaborative research for river and erosion control, Japan Landslide Society, Kanbara J (2015) A report on the collaborative research for river and erosion control by the Japan landslide society and ministry of land, infrastructure, transportation and tourism. Susceptibility mapping of earthquake-induced landslides. Journal of Japan Landslide Society 52(2):85–92 (in Japanese)
Highway Office of Taiwan (2020) Field survey for reconstruction of East-West No.8 highway. Transportation Authority, Taiwan (in Chinese)
Hirano M, Ishii T (1997) Characteristics of slope failures induced by the 1995 Hyogoken-Nanbu earthquake in the Rokko Mountain and their change by subsequent rainfall. Journal of the Japan Society of Erosion Control Engineering 50(1):23–32 (in Japanese)
Ho HC, Lin BS, Chang YL, Chi SY, Chien YD, Tsai MF, Feit LY (2013) Assessment of conservation program for mountain slopes in Tsengwen Reservoir catchment after Morakot Typhoon). Sinotech Engineering 118:21–30 (in Chinese)
Hohashi R, Otsuka T (2009) Topographical and geological studies of multiple ridges at Mt. Chogatake, Hida Mountains, central Japan. The Annals of Environmental Science, Shinshu University 31:90–95 (in Japanese)
Hong SZ, Liu SL, Xu JT, Zhu WJ (2011) Interpretation and research on photographs about Diexi earthquake shoot by Zhuang Xue-Ben in 1934. Seismology and Geology 33(1):208–224 (in Chinese)
Hong SZ (2014) Recent advances pf research on Diexi earthquake. Journal of Southwest Jiatong University 49(2):185–194 (in Chinese)
Honma F (1936) Mount Unzen. Bull Volc Soc Japan 3:75–124 (in Japanese)
Hornbach MJ, Braudy N, Briggs RW, Cormier MH, Davis MB, Diebold JB, Dieudonne N, Douilly R, Frohlich C, Gulick SP, Johnson HE III (2010) High tsunami frequency as a result of combined strike-slip faulting and coastal landslides. Nat Geosci 3(11):783–788

Hosoya T, Towhata I, Ikeda M (2021) Reconnaissance survey on earthquake-induced failures of ground composed of volcanic materials in Hokkaido, Japan, during the 2018 Hokkaido Eastern Iburi earthquake of Mw=6.6. In: 6th international conference on geotechnical and geophysical site characterization, Budapest: ISC2020–405

Houston JR (1979) State-of-the-art for assessing earthquake hazards in the United States. Miscellaneous Paper S-73-I, Report 15, Tsunamis, Seiches, and landslide-induced water waves, US army engineer, Waterways Experiment Station, Vicksburg

Howard KA (1973) Avalanche mode of motion: implication from lunar examples. Science 180:1052–1055

Hsu PST (2012) Tsengwen Reservoir siltation problem and remedial strategy. GT International, Taiwan (in Chinese)

Hsü KJ (1975) Catastrophic debris streams (Sturzstroms) generated by rockfalls. Geol Soc Am Bull 86:129–140

Hu W, Huang R, McSaveney M, Yao L, Xu Q, Feng M, Zhang X (2019) Superheated steam, hot CO_2 and dynamic recrystallization from frictional heat jointly lubricated a giant landslide: field and experimental evidence. Earth Planet Sci Lett 510:85–93

Hu XW, Lu XP, Huang RQ, Shi YB, Liu J, Li YB (2009) Developmental features and evaluation of blocking dangers of Dashui Ditch debris flow in Tangjiashan dammed lake. Chin J Rock Mechan Eng 28(4):850–858 (in Chinese)

Huang R (2011) After effect of geohazards induced by the Wenchuan Earthquake. J Eng Geol 19:145–151 (in Chinese)

Huang R, Li W (2014) Post-earthquake landsliding and long-term impacts in the Wenchuan earthquake area, China. Engineering Geology 182 (Part B):111–120

Huang RQ, Pei XJ, Li TB (2008) Basic characteristics and formation mechanism of the largest scale landslide at Daguangbao occurred during the Wenchuan earthquake. J Eng Geol 16(6):730–741 (in Chinese)

Huang RQ, Pei XJ, Zhang WF, Li SG, Li BL (2009) Further examination on characteristics and formation mechanism of Daguangbao landslide. J Eng Geol 17(6):725–736 (in Chinese)

Huang MJ, Shakal AF, Darragh RB (1994) Recorded strong-motion data from the Northridge, California earthquake of January 17, 1994. In: 9th Japan earthquake engineering symposium, Tokyo, 3:I-16–I-25

Huggel C, Clague JJ, Korup O (2012) Is climate change responsible for changing landslide activity in high mountains? Earth Surf Proc Land 37(1):77–91

Hung JJ (2000) Chi-Chi earthquake induced landslides in Taiwan. Earthquake Engineering and Engineering Seismology 2(2):25–33

Hungr O (2007) Dynamics of rapid landslides. Progress in landslide science. Springer, Berlin, pp 47–57

Hungr O, Leroueil S, Picarelli L (2014) The Varnes classification of landslide types, an update. Landslides 11(2):167–194

Hutchinson JN (1988) General report: morphological and geotechnical parameters of landslides in relation to geology and hydrogeology. In: Proceedings of 5th international symposium on landslides, Lausanne, pp 3–35

Ichimura M (1933) Documents on Kyoho earthquake that created a new mountain in Tohyama valley. Shinano Series 1 2(6):200–202 (in Japanese)

Ikeda H (1996) A strong earthquake and the changes it caused in granitic landscape: the case of Mt. Rokko due to the great Hanshin Earthquake of 1995. Memoirs of the Nara University 24:105–112 (in Japanese)

Ikeda H (2007) A strong earthquake and the changes it caused in granitic landscape—the case of Mt. Rokko ten years after the Great Hanshin earthquake. Memoirs of the Nara University 35:101–120 (in Japanese)

Ikeda A (2011) Sediment discharge from large-scale collapse and its countermeasure—a case history of the Sabo Works at Joganji River. Journal of the Japan Society of Erosion Control Engineering 64(3):57–63 (in Japanese)

Ikeda H (2015) A strong earthquake and the changes it caused in granitic landscape: the case of Mt. Rokko ten years after the Great Hanshin Earthquake. Memoirs of the Nara University 43:104–124 (in Japanese)
Ikeda H, MLIT Rokko Sabo Office (2016) Mt. Rokko and the Great Hanshin Earthquake of January 17, 1995: the photographic records of landform changes and conditions after 20 years. Report, https://www.kkr.mlit.go.jp/rokko/disaster/pdf/20y-from1995.pdf. Retrieved on 6 July 2020 (in Japanese)
Imamura F, Synolakis CE, Gica E, Titov V, Listanco E, Lee HJ (1995) Field survey of the 1994 Mindoro Island, Philippines tsunami. Pure Appl Geophys 144(3):875–890
Inagaki H (2017) Relationship between slope failure and the active fault based on Kumamoto Earthquake in 2016 and following heavy rainfall induced disaster. Journal of Japan Society of Engineering Geology 58(3):188–196 (in Japanese)
Inokuchi T, Inoue K, Komatsubara T, Doshida S (2012) Review of earthquake-induced landslides. Section 2.1, Earthquake-induced landslides. Japan Landslide Society pp 11–23 (in Japanese)
Inokuchi T, Yagi H (2013) Shiratori-yama landslide caused by the Hoei earthquake in 1707. Journal of the Japan Landslide Society 50(3):45–46 (in Japanese)
Inoue K (1995) 1923 Kanto earthquake and sediment disasters. Sabo-to-Chisui (erosion Control and Water Management) 28(2):14–20 (in Japanese)
Inoue K, Kawabe H, Aihara N, Chino M, Kurishima A (2013) Sedimentary disasters caused by 1923 Kanto earthquake. Kokon Shoin, Tokyo (in Japanese)
Inoue K, Mizuyama T, Sakatani Y (2010) The catastrophic Tombi landslide and accompanying landslide dams induced by the 1858 Hietsu Earthquake. J Disaster Res 5(3):245–256
Irazawa M (1986) Collapse of Ohnishi yama mountain. Journal of the Japan Society of Erosion Control Engineering 39(1):30–32 (in Japanese)
Irsyam M, Hanifa NR, Sahadewa A, Muntohar AS, Prakoso, WA, Sri Harninto D, Djarwadi D, Ridwan M, Natawjdjaja DH, Daryono M, Gunawan E, Meilano I, Pamumpuni A, Rudyanto A, Pramono S, Nazit R (2018) Damages associated with geotechnical problems in 2018 Palu earthquake, Indonesia. In: Proceedings of 20th Southeast Asian geotechnical conference, Jakarta, pp 5–14
Ishida M, Iwabuki K (2010) Seismic damage and its restoration in embankment of Tomei expressway that was caused by the Suruga-Bay earthquake (emergency restoration by using EPS). Geosynthetics Technical Information, Japan Chapter of International Geosynthetics Society, pp 9–13 (in Japanese)
Ishihara K (1985) Stability of natural deposits during earthquakes. Theme Lecture. In: 11th international conference on soil mechanics and foundation engineering, vol 1. San Francisco, pp 321–376
Ishihara K (1993) Liquefaction and flow failure during earthquakes. Géotechnique 43(3):351–415
Ishihara K, Haeri SM, Moinfar AA, Towhata I, Tsujino S (1992) Geotechnical aspects of the June 20, 1990 Manjil Earthquake in Iran. Soils Found 32(3):61–78
Ishihara K, Ishizawa T (2018) Steady-state residual strength of partly saturated soils to evaluate runout distance of landslides during earthquakes. In: ASCE innovations in geotechnical engineering, pp 277–298
Ishihara K, Okusa S, Oyagi N, Ischuk A (1990) Liquefaction-induced flow slide in the collapsible loess deposit in Soviet Tajik. Soils Found 30(4):73–89
Ishikawa Y, Kubota T, Aoto K, Iijima Y, Ikawa T, Ikegami T, Ikeda M, Ue H, Uehara Y, Uchimura Y, Egawa K, Ohishi H, Okano K, Kaibori M, Katsura S, Kato N, Kawahara S, Koga S, Sakashima T, Sagara W, Jitousono T, Shinohara Y, Shimizu O, Shimoda Y, Suzuki S, Suzuki M, Seto K, Tagata S, Terada H, Teramoto Y, Dounowaki M, Tobioka S, Torita E, Nakano K, Nishikawa T, Hanada R, Hirakawa Y, Fukuzuka K, Fujisawa Y, Fujita M, Masaki K, Miyama N, Yamaguchi K, Yamashita S, Yamane M, Yokoo K (2016a) Sediment-related disasters induced by the Kumamoto Earthquake in April 2016. Journal of the Japan Society of Erosion Control Engineering 69(3):55–66 (in Japanese)

Ishikawa Y, Akazawa F, Ue H, Ohno H, Osanai N, Kaibori M, Kubota T, Koga S, Gonda Y, Sakashima T, Jidousono T, Shimizu O, Takeshi T, Tarukado A, Torita E, Nakano K, Nishi M, Noro T, Hirakawa Y, Hiramatsu S, Fujita M, Matsu S, Yamada T (2016b) Secondary sediment movements and secondary sediment disasters caused by rainfalls after the Kumamoto Earthquake in April 2016. Journal of the Japan Society of Erosion Control Engineering 69(4):25–36 (in Japanese)

Iwasaki T (1981) Dynamic soil-structure interaction with emphasis on geotechnical engineering aspects. Monthly Magazine of Japanese Geotechnical Society 29(9):7–10 (in Japanese)

Japanese Geotechnical Society—Reconnaissance Team (2003) Quick damage report on geotechnical damage caused by the 2003 Tokachi-oki earthquake. Monthly Magazine of JGS 51(12):77–81 (in Japanese)

Japan Landslide Society (2012) Earthquake-induced landslides. Report of a Special Committee, ISBN978–4–902628–03–6 (in Japanese)

Jaya A, Nishikawa O, Jumadil S (2019) Distribution and morphology of the surface ruptures of the 2018 Donggala-Palu earthquake, Central Sulawesi, Indonesia. Earth, Planets and Space 71(1):1–13

Jessee MAN, Hamburger MW, Ferrara MR, McLean A, FitzGerald C (2020) A global dataset and model of earthquake-induced landslide fatalities. Landslides 17:1363–1376

Jibson RW, Crone AJ, Harp EL, Baum RL, Major JJ, Pullinger CR, Escobar CD, Martínez M, Smith ME (2004) Landslides triggered by the 13 January and 13 February 2001 earthquakes in El Salvador. Geological Society of America, Special papers, pp 69–88

Jibson RW, Prentice CS, Borissoff BA, Rogozhin EA, Langer CJ (1994) Some observations of landslides triggered by the 29 April 1991 Racha earthquake, Republic of Georgia. Bull Seismol Soc Am 84(4):963–973

JICA, Japan International Cooperation Agency (2019) Brief explanation of "Nalodo" assessment and mitigation. Presentation material in National Workshop on Joint Research, Assessment and Mitigation of Liquefaction Hazards (Lesson learned from the 2018 Palu earthquake)

Jones JN, Stokes M, Boulton SJ, Bennett GL, Whitworth MRZ (2020) Coseismic and monsoon-triggered landslide impacts on remote trekking infrastructure, Langtang Valley. Nepal. Quarterly Journal of Engineering Geology and Hydrogeology 53(2):159–166

Kamus Kaili-Ledo Indonesia Inggris (Dictionary of Kaili-Ledo language, Indonesian and English) (2003) Prime Edition. Evans D (ed) Service of culture and tourism, Government of Area of Central Sulawesi Province, p 109

Kanagawa Prefectural Government (1986) Prediction of seismic damage in Kanagawa Prefecture, pp 13–63 (in Japanese)

Kanatani M, Kiku H, Yasuda S, Yoshida N, Ishihara K, Kokusho T, Mimura M, Goto Y, Morimoto I (2003) Damages on waterfront ground during the 1999 Kocaeli earthquake in Turkey. Soils Found 43(5):29–40

Karasawa Y, Kano K (1992) Formation and deformation process of the slate belt in the Setogawa Group (the Shimanto Belt) of the eastern Akaishi Mountains, easternmost Southwest Japan. J Geol Soc Jpn 98(8):761–777 (in Japanese)

Kuenza K, Dorji Y, Wangda D (2004) Landslides in Bhutan. SAARC workshop on landslide risk management in South Asia, Thimpu, pp 80–87

Kasukawa Village (1994) One-hundred-year history of Kasukawa Village, pp 22–76 (in Japanese)

Katayama N (1974) Old records of natural phenomena concerning the Shimabara catastrophe. Science Reports of the Shimabara Volcano Observatory, the Faculty of Science, Kyushu University 9:1–45 (in Japanese)

Kawabe H (2019) On the conditions for occurrence of buckling of sliding surface exposed after large-scale landslide. Journal of Japan Landslide Society 56(4):162–164 (in Japanese)

Kawada S (1943a) Untersuchung des neuen sees gebildet nach erfolge des erdbebens vom jahre 1941 im Formosa (Taiwan). Jpn J Limnol 13(1):1–8 (in Japanese)

Kawada S (1943b) Study of new lake created by the earthquake in 1941 in Taiwan. Bulletin of the Earthquake Research Institute, University of Tokyo 21(2):317–325 (in Japanese)

Kawada S (1943c) Some record on the topography in Taiwan. Geographical Review of Japan 19(2):57–78 (in Japanese)
Kawata Y, Benson BC, Borrero JL, Davies HL, De Lange WP, Imamura F, Letz H, Nott J, Synolakis C (1999) Tsunami in Papua New Guinea was as intense as first thought. EOS (transactions, American Geophysical Union) 80:101–105
Kazama M, Takamura H, Unno T, Sento N, Uzuoka R (2006) Liquefaction mechanism of unsaturated volcanic sandy soils. Proc JSCE C 62(2):546–561 (in Japanese)
Keefer DK (1984) Landslides caused by earthquakes. Geol Soc Am Bull 95(4):406–421
Keefer DK (1994) The importance of earthquake-induced landslides to long-term slope erosion and slope-failure hazards in seismically active regions. Geomorphology 10(1–4):265–284
Keefer DK (2000) Statistical analysis of an earthquake-induced landslide distribution—the 1989 Loma Prieta, California event. Eng Geol 58:231–249
Keefer DK (2002) Investigating landslides caused by earthquakes–a historical review. Surv Geophys 23(6):473–510
Keefer DK, Wilson RC (1989) Predicting earthquake-induced landslides, with emphasis on arid and semi-add environments. In: Sadler PM, Morton DM (ed) Landslides in a semi-arid environment, vol 2. Inland Geological Society, pp 118–149
Kelsey H (1982) Hillslope evolution and sediment movement in a forested headwater basin, Van Duzen River, North Coastal California. General Technical Report PNW-141, Pacific Northwest Forest and Range Experiment Station, Forest Service, US Department of Agriculture, pp 86–96
Khattak GA, Owen LA, Kamp U, Harp EL (2010) Evolution of earthquake-triggered landslides in the Kashmir Himalaya, northern Pakistan. Geomorphology 115(1–2):102–108
Khazai B, Sitar N (2003) Evaluation of factors controlling earthquake-induced landslides caused by Chi-Chi earthquake and comparison with the Northridge and Loma Prieta events. Eng Geol 71:79–95
Kimura S (2019) Chronologies of villages and natural disasters: a case study of chronology of Mikage Village. Annual report of the Community Outreach Center, vol 11. Graduate School of Humanities, Kobe University, pp 109–132 (in Japanese)
King J, Loveday I, Schuster RL (1989) The 1985 Bairaman landslide dam and resulting debris flow, Papua New Guinea. Q J Eng GeolHydrogeol 22(4):257–270
Kingdon-Ward F (1953) The Assam earthquake of 1950. Geogr J 119:150–182
Kingdon-Ward F (1955) Aftermath of the great Assam earthquake of 1950. Geogr J 121:295–303
Kitago S, Toki S (1968) Geotechnical investigation on the settlements of volcanic ashes and sand fills caused by the 1968 Tokachi-oki earthquake. Report on the Tokachi-Oki earthquake 1968. Special Committee for the Investigation of the Tokachi-Oki Earthquake 1968:463–494 (in Japanese)
Kitazawa S, Miyazaki T, Horiuchi T (1985) Geological study of Ontake landslide triggered by the Western Nagano Earthquake, 1984. Journal of the Japan Society of Erosion Control Engineering 38(3):12–19 (in Japanese)
Kiyota T, Furuichi H, Hidayat RF, Tada N, Nawir H (2020) Overview of long-distance flow-slide caused by the 2018 Sulawesi earthquake. Indonesia. Soils and Foundations 60(3):722–735
Kiyota T, Sattar A, Konagai K, Kazmi ZA, Okuno D, Ikeda T (2011) Breaching failure of a huge landslide dam formed by the 2005 Kashmir earthquake. Soils Found 51(6):1179–1190
Klimeš J, Novotný J, Novotná I, Jordán de Urries B, Vilímek V, Emmer A, Strozzi T, Kusák M, Cochachin Rapre A, Hartvich F, Frey H (2016) Landslides in moraines as triggers of glacial lake outburst floods: example from Palcacocha Lake (Cordillera Blanca, Peru). Landslides 13(6):1461–1477
Koi T, Hotta N, Ishigaki I, Matuzaki N, Uchiyama Y, Suzuki M (2008) Prolonged impact of earthquake-induced landslides on sediment yield in a mountain watershed: the Tanzawa region. Japan. Geomorphology 101(4):692–702
Koide H (1950) On landslip and the mud stone flow at Akagi volcano, in Sept. 1947. Report of Geological Survey of Japan 133 (in Japanese)
Koide H, Noguchi Y (1950) Collapse of Akagi Volcano and debris flow. Study on Kathleen Typhoon. Gunma Prefecture, pp 199–219 (in Japanese)

Koide H (1955) Immunity to slope disasters. Engineering Geology 2 (landslides). Kokon Book Publisher, pp 15–23 (in Japanese)
Koketsu K, Miyake H, Guo Y, Kobayashi H, Masuda T, Davuluri S, Bhattarai M, Adhikari LB, Sapkota SN (2016) Widespread ground motion distribution caused by rupture directivity during the 2015 Gorkha. Nepal Earthquake. Scientific Reports 6(1):1–9
Kokusho T (1999) Water film in liquefied sand and its effect on lateral spread. ASCE Journal of Geotechnical and Geoenvironmental Engineering 125(10):817–826
Kokusho T (2000) Mechanism of water film generation and lateral flow in liquefied sand layer. Soils Found 40(5):99–111
Komori J, Koike T, Yamanokuchi T, Tshering P (2012) Glacial lake outburst events in the Bhutan Himalayas. Glob Environ Res 16:59–70
Konagai K, Johansson J, Mayorca P, Yamamoto T, Miyajima M, Uzuoka R, Pulido NE, Duran FC, Sassa K, Fukuoka H (2002) Las Colinas landslide caused by the January 13, 2001 off the coast of El Salvador earthquake. Journal of Japan Association for Earthquake Engineering 2(1):1–15
Konagai K, Kazama M (2003) Slope failure in Tsukidate. Sections 5.1 and 5.2, reconnaissance report on the May 26, 2003 Miyagiken no oki earthquake, Joint delegation team with Japan Society of Civil Engineering and the Japanese Geotechnical Society, pp 9–12 (in Japanese)
Korup O (2004) Geomorphic implications of fault zone weakening: slope instability along the Alpine Fault, South Westland to Fiordland, New Zealand. Journal of Geology and Geophysics 47:257–267
Korup O (2005) Geomorphic hazard assessment of landslide dams in South Westland, New Zealand: fundamental problems and approaches. Geomorphology 66:167–188
Kremer K, Simpson G, Girardclos S (2012) Giant Lake Geneva tsunami in AD 563. Nat Geosci 5(11):756–757
Kuenen PH (1952) Estimated size of the Grand Banks turbidity current. Am J Sci 250:874–884
Kusakabe A (1985) Earthquake in Ina recorded in historical documents. Ina 33(2):65–73 (in Japanese)
Kuşçu İ, Okamura M, Matsuoka H, Awata Y (2002) Active faults in the Gulf of İzmit on the North Anatolian Fault, NW Turkey: a high-resolution shallow seismic study. Mar Geol 190(1–2):421–443
Kurita T, Annaka T, Takahashi S, Shimada M, Suehiro T (2005) Effect of irregular topography on strong ground motion amplification. Proc Japan Association for Earthquake Engineering 5(3):1–11 (in Japanese)
Kuwahara T, Hirama K (1991) Study on engineering material evaluation of fault and sheared zone. Report of the Technical Research Institute, Ohbayashi-Gumi Ltd 43:99–106 (in Japanese)
Lacroix P (2016) Landslides triggered by the Gorkha earthquake in the Langtang valley, volumes and initiation processes. Earth, Planets and Space 68(1):1–10
Le Brun B, Hatzfeld D, Bard PY, Bouchon M (1999) Experimental study of the ground motion on a large scale topographic hill at Kitherion (Greece). J Seismolog 3(1):1–15
Lee WF, Towhata I (2010) Geotechnical structure damages during the 2009 Typhoon Morakot. Bulletin of ISSMGE 4(1):25–35
L'Heureux JS, Eilertsen SE, Glimsdal S, Issler D, Solberg I-L, Harbitz CB (2012) The 1978 quick clay landslide at Rissa, mid Norway: subaqueous morphology and tsunami simulations. Submarine mass movements and their consequences. Springer, Dordrecht, pp 507–516
Liao HW, Lee CT (2000) Landslides triggered by the Chi-Chi earthquake. In: 21st Asian conference on remote sensing, Taipei vol 1, pp 383–388
Lin A, Shimamoto T, Maruyama T, Sigetomi M, Miyata T, Takemura K, Tanaka H, Uda S, Murata A (2001) Comparative study of cataclastic rocks from a drill core and outcrops of the Nojima Fault zone on Awaji Island. Japan. Island Arc 10(3–4):368–380
Lin CW, Shieh CL, Yuan BD, Shieh YC, Liu SH, Lee SY (2004) Impact of Chi-Chi earthquake on the occurrence of landslides and debris flows: example from the Chenyulan River watershed, Nantou. Taiwan. Engineering Geology 71:49–61

Lin M, Chiang, T-K (2019) Long term effects of landslides induced by catastrophic events. In: 16th Asian regional conference on soil mechanics and geotechnical engineering, Taipei
Lin ML, Lin YS (2013) Estimation and prediction of debris flow potential using discrimination analysis. In: Proceedings of 19th ICSMGE, vol 3. Paris, pp 2209–2212
Lin ML, Wang KL, Chen TC (2010) Debris flow hazard and mitigation in Taiwan following Chi-Chi earthquake, 1999. In: Proceedings of 17th Southeast Asian geotechnical conference, Taipei, pp 123–126
Ling S (2015) Landslide damming in Western Sichuan Province, China, with special reference to the 1786 Dadu River and 1933 Diexi events. University of Waterloo, Thesis for master of earth science
Little TA (1995) Brittle deformation adjacent to the Awatere strike-slip fault in New Zealand: faulting patterns, scaling relationships, and displacement partitioning. Geol Soc Am Bull 107(11):1255–1271
Lu KK, Hsu ML (2004) The geomorphic characteristics of debris flows in the Ninety-Nine Peaks area, Nantou. Journal of Geographical Science, Taiwan National University 38:1–16 (in Chinese)
Ma Y, Ma DT, Li F, Wang XL (2011) Characteristics of earthquake debris flow in Wenchuan earthquake area—a case from Donglinsi—Haihui Bridge in Yinchang Gully. Journal of Guilin University of Technology 31(3):351–356 (in Chinese)
Ma DT, Qi L (1997) Study on comprehensive controlling of debris flow hazard in Sanyanyu Valley. Bull Soil Water Conserv 17(4):26–31 (in Chinese)
Ma JX, Chen J, Cui ZJ, Zhou W, Liu C, Guo P, Shi Q (2018) Sedimentary evidence of outburst deposits induced by the Diexi paleolandslide-dammed lake of the upper Minjiang River in China. Quatern Int 464:460–481
Macdonald HM (1902) Electric waves. Cambridge University Press, Cambridge
Mann P, Burke K, Matumoto T (1984) Neotectonics of Hispaniola: plate motion, sedimentation, and seismicity at a restraining bend. Earth Planet Sci Lett 70(2):311–324
Mann P, Draper G, Lewis JF (1991) An overview of the geologic and tectonic development of Hispaniola. Geologic and tectonic development of the North America-Caribbean plate boundary in Hispaniola. Geol Soc Am Spec Pap 262:1–28
Marc O, Hovius N, Meunier P, Uchida T, Hayashi S (2015) Transient changes of landslide rates after earthquakes. Geology 43(10):883–886
Mahr T (1977) Deep—Reaching gravitational deformations of high mountain slopes. Bull Int Assoc Eng Geol 16(1):121–127
Malvick EJ, Kulasingam R, Kutter BL, Boulanger RW (2002) Void redistribution and localized shear strains in slopes during liquefaction. Physical Modelling in Geotechnics ICPMG'02:495–500
Mann P, Hempton MR, Bradley DC, Burke K (1983) Development of pull-apart basins. J Geol 91(5):529–554
Maramai A, Graziani L, Tinti S (2006) A revision of the 1783–1784 Calabrian (southern Italy) tsunamis. Nat Hazard 6(6):1053–1060
Martino S, Prestininzi A, Romeo RW (2014) Earthquake-induced ground failures in Italy from a reviewed database. Nat Hazard 14(4):799–814
Marui H (2007) Large slope failure induced by earthquakes. Geotechnical Engineering Committee, Japan Society of Civil Engineers (in Japanese), Presented at Subcommittee on slope engineering
Masch L, Wenk HR, Preuss E (1985) Electron microscopy study of hyalomylonites—evidence for frictional melting in landslides. Tectonophysics 115(1–2):131–160
Massey C, Townsend D, Rathje E, Allstadt KE, Lukovic B, Kaneko Y, Bradley B, Wartman J, Jibson RW, Petley DN, Horspool N, Hamling I, Carey J, Cox S, Davidson J, Dellow S, Godt JW, Holden C, Jones K, Kaiser A, Little M, Lyndsell B, McColl S, Morgenstern R, Rengers, FK, Rhoades D, Rosser B, Strong D, Singeisen C, Villeneuve M (2018) Landslides triggered by the 14 November 2016 Mw 7.8 Kaikoura Earthquake, New Zealand. Bulletin of the Seismological Society of America 108(3B):1630–1648
Matsukura Y (1997) Weathering rates of bedrocks in artificial and natural slopes. Journal of Japan Society of Engineering Geology 38:224–231 (in Japanese)

Matsumoto M, Shimokawa E, Jidousono T (1996) A natural revegetation process on shallow landslide scars of deep weathering granite slopes. Journal of Kyushu Chapter, Japanese Forest Society 49:155–156 (in Japanese)
Matsushima N (2000a) Tensyou earthquakes in the Ina Valley. Historical Earthquakes 16:53–58 (in Japanese)
Matsushima N (2000b) History of debris flow disasters (called "Mansui") in the Ina Valley—in relation to periods of seismic activity. Natural History Reports of Inadani 1:11–15 (in Japanese)
Matsu'ura R, Nakamura M, Karakama I (2006) Summary of 7-year study on reconsideration of historical earthquakes in Edo Era. Historical Earthquakes 21:255–256 (in Japanese)
Mazzanti P, De Blasio FV, Di Bastiano C, Bozzano F (2016) Inferring the high velocity of landslides in Valles Marineris on Mars from morphological analysis. Earth, Planets and Space 68(1):1–16
Mazzini A, Svensen H, Akhmanov GG, Aloisi G, Planke S, Malthe-Sørenssen A, Istadi B (2007) Triggering and dynamic evolution of the LUSI mud volcano. Indonesia. Earth and Planetary Science Letters 261(3–4):375–388
McHugh CMG, Seeber L, Cormier M-H, Dutton J, Cagatay N, Alina P, William BFR, Gorur N (2006) Submarine earthquake geology along the North Anatolia Fault in the Marmara Sea, Turkey: a model for transform basin sedimentation. Earth Planet Sci Lett 248(3–4):661–684
Melosh HJ (1979) Acoustic fluidization: a new geologic process? Journal of Geophysical Research: Solid Earth 84(B13):7513–7520
Meng WL, Xu YH, Cheng WC, Arulrajah A (2018) Landslide event on 24 June in Sichuan Province, China: preliminary investigation and analysis. Geosciences 8(2):39
Meunier P, Hovius N, Haines JA (2008) Topographic site effects and the location of earthquake induced landslides. Earth Planet Sci Lett 275(3–4):221–232
Midorikawa Y, Unno T, Kamura A, Sento N, Kazama M (2020) Physical and mechanical properties of volcanic soils damaged in the 2018 Hokkaido Iburi Eastern earthquake. Proc JSCE A1 76(4):741–754 (in Japanese)
Miller DJ (1960) Giant waves in Lituya Bay Alaska. USGS Professional Paper 354-C
Mimura K (1988) Geology of Bandai volcano and its volcanic history. J Geogr 97(4):280–284 (in Japanese)
Minami T (1997) Study on the actual situation of the earthquake-induced large-scale landslide and it's disaster characteristic. Sabo planning laboratory, technical memorandum of public works research laboratory 3501 (in Japanese)
Miyagi T, Hamasaki E, Esaka F (2011) Massive landslide triggered by earthquake in the Aratozawa Dam area, Tohoku. Japan. Landslides 8(1):99–108
Mizuhashi M, Towhata I, Sato J, Tsujimura T (2006) Examination of slope hazard assessment by using case studies of earthquake- and rainfall-induced landslides. Soils Found 46(6):843–853
Mochizuki K (1976) Landslide behaviors along the faults. Journal of Japan Landslide Society 13(4):11–18 (in Japanese)
Mochizuki K, Nakamura S (1982) Roles of landslides to the landform development: a landslide survey in the northern mountain district of Nagano Prefecture. Journal of Japan Landslide Society 19(1):1–8 (in Japanese)
Mokudai K, Chigira M (2004) Geomorphic features and processes of gravitational mountain deformation in the area from Mt. Yambushi to Oya-Kuzure, Central Japan. Geographical Review of Japan, The Association of Japanese Geographers 77(2):55–76 (in Japanese)
Momikura Y, Yasuda S, Sakaki Y (1985) A prediction method for landslides by an earthquake and its appraisal based on the Naganokei-seibu earthquake. Monthly Magazine of Japanese Geotechnical Society 33(11):41–46 (in Japanese)
Montgomery J, Candia G, Lemnitzer A, Martinez A (2020) The September 19, 2017 Mw 7.1 Puebla-Mexico City earthquake: observed rockfall and landslide activity. Soil Dynamics and Earthquake Engineering 130:105972
Montgomery DR, Hallet B, Yuping L, Finnegan N, Anders A, Gillespie A, Greenberg HM (2004) Evidence for Holocene megafloods down the Tsangpo River gorge, southeastern Tibet. Quatern Res 62(2):201–207

Montgomery DR, Manga M (2003) Streamflow and water well responses to earthquakes. Science 300(5628):2047–2049

Mooney C (2018) Scientists have found the mystery cause of one of the biggest tsunamis ever recorded. Science Alert. https://www.sciencealert.com/alaskan-fjord-one-of-largest-tsunami-ever-recorded-600-feet-climate-change. Retrieved 7 Sept 2018

Mora S, Mora R (1994) Los deslizamientos causados por el terremoto de Limón: Factores de control y comparación con otros eventos en Costa Rica. Revista Geológica de América Central Vol. especial pare Terremoto de Limón: 139–152 (in Spanish)

Mori T (1987) Study on historical official documents on disasters in Umegashima Village, upstream of Abe River, caused by the 1854 Ansei Tokai earthquake. Historical Earthquakes 3:93–113 (in Japanese)

Morimoto R, Nakamura K, Tsuneishi Y, Osaka J, Tsunoda N (1967) Landslides in the epicentral area of the Matsushiro earthquake swarm—their relation to the earthquake fault. Bulletin of the Earthquake Research Institute, University of Tokyo 45(1):241–263

Moriya I (1971) History of Akagi volcano. Bulletin of the Volcanological Society of Japan 2 15(3):120–131 (in Japanese)

Moroto K (1925) On landslips caused by the great earthquake. Report of Earthquake Disaster Prevention Research Committee 100B:79 (in Japanese)

Mountjoy J, Micallef A (2018) Submarine landslides. In: Micallef A, Krastel S, Savin A (eds) Submarine geomorphology. Springer International Publishing, Cham, pp 235–251

Muir-Wood R, King GCP (1993) Hydrological signatures of earthquake strain. J Geophys Res 98(B12):22035–22068

Muramatsu T, Teraoka Y (2015) Blockage of Tohyama River and buried forest as a consequence of the 714 Tohtoumi earthquake. Shinano, Shinano Historical Society 67(4):241–257 (in Japanese)

Murasawa T (1983) Natural disasters and poor harvest in Ina, 2nd edn. Kitahara Technical Office (in Japanese)

Nadal-Romero E, Martínez-Murillo JF, Vanmaercke M, Poesen J (2011) Scale-dependency of sediment yield from badland areas in Mediterranean environments. Prog Phys Geogr 35(3):297–332

Nagai O, Nakamura H (2000) The large-scale landslide of Mt. Shichimen—the study on the history of the landslide and the enlargement process. Journal of Japan Landslide Society 37(2):20–29 (in Japanese)

Nagano Prefectural Government (2021) List of damaging historical earthquakes. Appendix 1 for Disaster Mitigation Plan (in Japanese)

Nagata H, Kojima S (2020) Gravitational deformation on mountain slope to the northeast of Tokugotoge Pass, Nagano Prefecture, Japan. JpGU-AGU Joint Meeting Paper HDS12-P03 (in Japanese)

Nagel S, Castelltort S, Wetzel A, Willett SD, Mouthereau F, Lin AT (2013) Sedimentology and foreland basin paleogeography during Taiwan arc continent collision. J Asian Earth Sci 62:180–204

Nakajima C (1983) History of flooding of Kamo River (1). DPRI Annuals, Kyoto University 26B(2):75–92 (in Japanese)

Nakamura H, Tsuchiya S, Inoue K, Ishikawa Y (2000) Sabo against Earthquakes. Kokon Shoin, Tokyo, ISBN 4–7722–4014–4 C3051, pp 68–70 (in Japanese)

Nakamura K (1981) A hypothesis on mechanism of the Matsushiro earthquake swarm. Kagaku Asahi October, pp 127–133 (in Japanese)

Nakamura K (1984) Three different mechanisms for elongate distribution of shallow earthquake epicenters. Workshop XIX on Active Tectonic and Magmatic Processes beneath Long Valley Caldera, Eastern California, USGS I:482–492

Nakamura S, Wakai A, Umemura J, Sugimoto H, Takeshi T (2014) Earthquake-induced landslides: Distribution, motion and mechanisms. Soils Found 54(4):544–559

Nakano M, Chigira M, Choun Sian L, Sumaryono G (2015) Geomorphological and geological features of the collapsing landslides induced by the 2009 Padang earthquake. In: 10th Asian regional conference of international association for engineering geology, Kyoto.

Natawidjaja DH, Daryono MR, Prasetya G, Liu PL, Hananto ND, Kongko W, Triyoso W, Puji AR, Meilano I, Gunawan E, Supendi P (2021) The 2018 Mw7.5 Palu 'supershear' earthquake ruptures geological fault's multisegment separated by large bends: results from integrating field measurements, LiDAR, swath bathymetry and seismic-reflection data. Geophysical Journal International 224(2):985–1002

Naumann E (1895) Neue Beiträge zur geologie und geographie Japans. Petermanns Mitteilungen, Ergänzungsheft (Supplementary booklet) No. 108, p 28 (in German)

Negmatullaev SK et al (1992) The Gissar earthquake (23 January 1989) seismotectonics, seismology and engineering geology, engineering aspects of earthquake risk mitigation; lessons from management of recent earthquakes, and consequential mudflow and landslides. In: Proceedings of UNDRO/USSR/UNDP training seminars, Dushanbe, 17–28 Oct 1988, pp 157–163. https://inis.iaea.org/collection/NCLCollectionStore/_Public/24/029/ 24029480.pdf?r=1. Retrieved on 14 Dec 2020

Newmark NM (1965) Effects of earthquakes on dams and embankments. Géotechnique 5(2):137–160

NILIM, National Institute for Land and Infrastructure Management (2020) Research on crisis management for excessive external forces and complex natural disasters. Project Research Report, 64, ISSN 1880–0114 (in Japanese)

Nishimura K, Mori W (1983) An observed effect of topography on seismic ground motions. Journal of the Seismological Society of Japan. 36(3):383–392 (2nd series)

Nishimura T (2014) Pre-, co-, and post-seismic deformation of the 2011 Tohoku-Oki earthquake and its implication to a paradox in short-term and long-term deformation. J Disaster Res 9(3):294–302

Noro T, Maruyama K, Baator Has, Nakamura, A (2011) Study on earthquake landslide occurrence estimate of risk evaluation method on landslide topography. Technical Note of PWRI, Public Works Research Institute 4204 (in Japanese)

Noto T, Uchida K, Soda T (1990) Paleoseismological study on sediment disaster in south slope of Mt. Akagi caused by Konin earthquake in AD 818. Shinano 42:755–772 (in Japanese)

Noto T (1990) Research note on earthquake-induced disaster in AD818. History of Gunma Prefecture 34:38–50 (in Japanese)

Nurdin S, Harianto T, Aswad S, Arsyad A, Alexsander S (2018) Liquefaction disaster mitigation and geohydrology conditions, lessons from the Palu earthquake magnitude 7.4 Mw 28 September 2018. In: Proceedings of 23rd annual national conference on geotechnical engineering, HATTI, Indonesian Society for Geotechnical Engineering

Oberlander T (1965) The Zagros streams. Syracuse Geographical Series, 1, Syracuse University Press

Ogata S, Honsho S (1981) Fault activity evaluation in the case of electric power plants. J Jpn Soc Eng Geol 22(1):67–81 (in Japanese)

Ohmachi T, Midorikawa S (1999) Earthquake ground motion. Geotech Notebook, 9, Japanese Geotechnical Society, p 61 (in Japanese)

Ohno R, Yamashina S, Yamasaki T, Koyama T, Esaka F, Kasai S (2010) Mechanisms of a large-scale landslide triggered by the earthquake in 2008—a study of Arato-sawa Landslide. Journal of Japan Landslide Society 47(2):84–90 (in Japanese)

Ohshima H, Tokunaga T, Miyajima F, Tanaka K, Ishibashi H (1996) Groundwater fluctuations caused by the earthquake. J Jpn Soc Eng Geol 37(4):351–358 (in Japanese)

Ohyagi N (1998) Appearance and inside of landslides. Fukadaken Library, p 14 (in Japanese)

Okada A (1993) Segmentation and Paleo−seismicity on the median tectonic line active fault system—a typical example of active faults in Japan. Monthly Magazine of the Japanese Geotechnical Society 41(3):7–12 (in Japanese)

Okal EA, Synolakis CE (2004) Source discriminants for near-field tsunamis. Geophys J Int 158(3):899–912

Okamura M, Ono K, Arsyad A, Minaka US, Nurdin S (2019) Large-scale flowslide in Sibalaya caused by the 2018 Sulawesi Earthquake. Soils Found 59(5):1148–1159

Oki Y, Xu HL, Ishizaka N, Kawauchi K (1999) Geopressured hydrothermal system associated with active faults and historical destructive earthquakes in the Shinano River seismic zone. Journal of Hot Spring Sciences, Balneological Society of Japan 48(4):163–180 (in Japanese)

Okimura T, Torii N, Nagai H (1998) An analysis of slope failures mechanism caused by rainfall after the earthquake. Memoirs of Construction Engineering Research Institute Foundation 40B:97–114 (in Japanese)

Oldham RD (1899) Report on the great earthquake of 12th June 1897. Memoir of Geological Survey India 29

Omori F (1908) Collapse of Unzen volcano in 1792. J Geol Soc Jpn 15(181):447–450 (in Japanese)

Omori F (1909) Preliminary report on the Messina-Reggio earthquake of December 28, 1908. Bulletin of Imperial Earthquake Investigation Committee 3:37–46

Omori F (1918) Discussion on cause of collapse of Mt. Mae-yama, Unzen, in 1792. Journal of the Geological Society of Japan 25(296):256–258 (in Japanese)

Osawa K (1985) Toyama earthquake in Ina. Ina 23(2):74–77 (in Japanese)

Orense RP, Sapuay SE (2006) Preliminary report on the 17 February 2006 Leyte. Philippines Landslide. Soils and Foundations 46(5):685–693

Öztürk H, Koral H, Geist EL (2000) Intra-basinal water movements induced by faulting: the August 17, 1999, Gölcük (İzmit Bay) earthquake (Mw=7.4). Marine Geology 170(3–4):263–270

Ota K (1969) Study on the collapses in the Mayu-yama I. On the mechanism of collapse. Science Reports of the Shimabara Institute of Volcanology and Balneology, Kyushu University 5:6–35 (in Japanese)

Otsuki K (1978) On the relationship between the width of shear zone and the displacement along fault. J Geol Soc Jpn 84(11):661–669 (in Japanese)

Owen L, Kamp U, Khattak GA, Harp E, Keefer D, Bauer MA (2008) Landslides triggered by the 8 October 2005 Kashmir earthquake. Geomorphology 94(1–2):1–9

Pánek T, Klimeš J (2016) Temporal behavior of deep-seated gravitational slope deformations: a review. Earth Sci Rev 156:14–38

Paolucci R (2002) Amplification of earthquake ground motion by steep topographic irregularities. Earthquake Eng Struct Dynam 31(10):1831–1853

Parise M, Jibson RW (2000) A seismic landslide susceptibility rating of geologic units based on analysis of characteristics of landslides triggered by the 17 January, 1994 Northridge. California Earthquake. Engineering Geology 58(3–4):251–270

Pathier E, Fielding EJ, Wright TJ, Walker R, Parsons BE, Hensley S (2006) Displacement field and slip distribution of the 2005 Kashmir earthquake from SAR imagery. Geophys Res Lett 33:L20310

Patria A, Putra PS (2020) Development of the Palu–Koro Fault in NW Palu Valley, Indonesia. Geoscience Letters 7(1)

Pedersen H, Le Brun B, Hatzfeld D, Campillo M, Bard PY (1994) Ground-motion amplitude across ridges. Bull Seismol Soc Am 84(6):1786–1800

Pelinovsky E, Yuliadi D, Prasetya G, Hidayat R (1997) The 1996 Sulawesi tsunami. Nat Hazards 16(1):29–38

Peng M, Zhang LM (2012) Breaching parameters of landslide dams. Landslides 9(1):13–31

Peng D, Xu Q, Zhang X, Xing H, Zhang S, Kang K, Qi X, Ju Y, Zhao K (2019) Hydrological response of loess slopes with reference to widespread landslide events in the Heifangtai terrace, NW China. J Asian Earth Sci 171:259–276

Petley D, Dunning S, Rosser N, Kausar AB (2010) Incipient landslides in the Jhelum Valley, Pakistan following the 8th October 2005 earthquake. SAARC Workshop on Landslide Risk Management in South Asia, Thimpu, pp 107–116

Philip H, Ritz J-F (1999) Gigantic paleolandslide associated with active faulting along the Bogd fault (Gobi-Altay, Mongolia). Geology 27:211–214

Plafker G (1969) Tectonics of the March 27, 1964, Alaska earthquake, The Alaska Earthquake, March 27, 1964: regional effects. USGS Professional Report 543-I

Plafker G, Ericksen GE, Fernandez Concha J (1971) Geological aspects of the May 31, 1970, Peru earthquake. Bull Seismol Soc Am 61(3):543–578

Podolskiy EA, Nishimura K, Abe O, Chernous PA (2010) Earthquake-induced snow avalanches: I. Historical case studies. J Glaciol 56(197):431–446

Pozos-Estrada A, Chávez MM, Jaimes MÁ, Arnau O, Guerrero H (2019) Damages observed in locations of Oaxaca due to the Tehuantepec Mw 8.2 earthquake, Mexico. Natural Hazards 97(2):623–641

Prestininzi A, Romeo R (2000) Earthquake-induced ground failures in Italy. Eng Geol 58(3–4):387–397

Pranantyo IR, Cummins PR (2020) The 1674 Ambon tsunami: extreme run-up caused by an earthquake-triggered landslide. Pure Appl Geophys 177(3):1639–1657

Publication Committee (2020) Landslides induced by the 2018 Hokkaido Eastern Iburi earthquake. Hokkaido University Press. ISBN978–4–8329–8235–2 (in Japanese)

Puzrin AM, Faug T, Einav I (2019) The mechanism of delayed release in earthquake-induced avalanches. Proc Royal Society A 475(2227):20190092

Quesada-Román A, Fallas-López B, Hernández-Espinoza K, Stoffel M, Ballesteros-Cánovas JA (2019) Relationships between earthquakes, hurricanes, and landslides in Costa Rica. Landslides 16(8):1539–1550

Qi SW, Xu Q, Lan HX, Zhang B, Liu JY (2010) Spatial distribution analysis of landslides triggered by 2008512 Wenchuan earthquake. China. Engineering Geology 116(1–2):95–108

Qi S, Lan H, Dong J (2015) An analytical solution to slip buckling slope failure triggered by earthquake. Eng Geol 194:4–11

Qiao JP, Yang ZJ (2010) Distribution and impacts of rainfall on seismic landslides following the 5.12 Wenchuan earthquake. In: Proceedings of 17th Southeast Asian geotechnical conference, Taipei, pp 103–108

Qin S, Jiao JJ, Wang S (2001) A cusp catastrophic model of instability of slip-buckling slope. Rock Mech Rock Eng 34(2):119–134

Quantin C, Allemand P, Delacourt C (2004) Morphology and geometry of Valles Marineris landslides. Planet Space Sci 52(11):1011–1022

Qureshi MU, Towhata I, Yamada S (2019) Experimental relation between shear strength under low pressure and S-wave velocity of rock subjected to mechanical weathering. Soils Found 59:1468–1480

Rathje EM (2010) The geotechnical aspects of the Haiti earthquake. ISSMGE Bulletin 4(3):8–20

Ray PC, Parvaiz I, Jayangondaperumal R, Thakur VC, Dadhwal VK, Bhat FA (2009) Analysis of seismicity-induced landslides due to the 8 October 2005 earthquake in Kashmir Himalaya. Curr Sci 97(12):1742–1751

Rice JR (1992) Fault stress states, pore pressure distributions, and the weakness of the San Andreas Fault. Fault mechanics and transport properties of rocks. International Geophysics 51:475–503

Rodríguez CE, Bommer JJ, Chandler RJ (1999) Earthquake-induced landslides: 1980–1997. Soil Dyn Earthq Eng 18(5):325–346

Rodríguez CE, Poran CJ (1993) Earthquake-induced landslides in Colombia. In: International engineering conference on case histories in geotechnical engineering, Missouri-Rolla, pp 337–344

Rojstaczer S, Wolf S (1994) Hydrologic changes associated with the earthquake in the San Lorenzo and Pescadero drainage basins, The Loma Prieta, California earthquake of October 17, 1989: strong ground motion and ground failure. Hydrologic disturbances, USGS Professional Paper 1551-E:E51–E64

Röthlisberger H (1978) Eislawinen und ausbrüche von gletscherseen. Jahrbuch der Schweizerischen Naturforschenden Gesellschaft, Wissenschaflicher Teil: 170–212 (in German)

Ryan WBF, Heezen BC (1965) Ionian Sea submarine canyons and the 1908 Messina turbidity current. Geol Soc Am Bull 76(8):915–932

Saba SB, Van der Meijde M, Van der Werff H (2010) Spatiotemporal landslide detection for the 2005 Kashmir earthquake region. Geomorphology 124(1–2):17–25
Saito N (1988) Slope instability at Tobigasu. Journal of the Japan Society of Erosion Control Engineering 41(2):38–39 (in Japanese)
Sakamoto M (2014) Disaster research in the Toyama Earthquake (1718), Iida city, Nagano Prefecture. Natural History Reports of Inadani 15:1–18 (in Japanese)
Sanchez-Sesma FJ (1985) Diffraction of elastic SH waves by wedges. Bull Seismol Soc Am 75:1435–1446
Sasaki Y, Towhata I, Miyamoto K, Shirato M, Narita A, Sasaki T, Sako S (2012) Reconnaissance report on damage in and around river levees caused by the 2011 off the Pacific coast of Tohoku Earthquake. Soils Found 52(5):1016–1032
Sassa K, Fukuoka H, Scarascia-Mugnozza G, Evans S (1996) Earthquake-induced-landslides: distribution, motion and mechanisms. Soils Found 36:53–64
Sassa K, Fukuoka H, Wang F, Wang G (2007) Landslides induced by a combined effect of earthquake and rainfall. Progress in landslide science. Springer, Berlin, pp 193–207
Sato D (1925) Discussion against mountain collapse theory on 1792 Unzen disaster. Earth 4(6):437–446 (in Japanese)
Sato S (2020) Dynastic government of dynastic nation (Chap. 1). In: Medieval Japan, N130–1, Iwanami books, ISBN978–4–00–381283–9, pp 25–78 (in Japanese)
Sato T (2015) Hot spring activity that continues for 4 years after the 2011 Tohoku earthquake. IEVG News Letter, Research Institute of Earthquake and Volcano Geology 2(3):1–4 (in Japanese)
Sato HP, Hasegawa H, Fujiwara S, Tobita M, Koarai M, Une H, Iwahashi J (2007) Interpretation of landslide distribution triggered by the 2005 Northern Pakistan earthquake using SPOT 5 imagery. Landslides 4(2):113–122
Sawaguchi H (2013) Mudflow hills in Kiryu/Isesaki/Maebashi area. Report on Good Natural Environment of Locality 38:79–94 (in Japanese)
Scheidegger AE (1973) On the prediction of the reach and velocity of catastrophic landslides. Rock Mech 5(4):231–236
Scholz CH (2002) The mechanism of earthquakes and faulting, 2nd edn. Cambridge University Press, Cambridge, pp 110–127
Scholz CH, Dawers NH, Yu JZ, Anders MH (1993) Fault growth and fault scaling laws: preliminary results. Journal of Geophysical Research: Solid Earth 98(B12):21951–21961
Schuster RL, Alford D (2004) Usoi landslide dam and Lake Sarez, Pamir Mountains, Tajikistan. Environ Eng Geosci 10(2):151–168
Securitas Magazine (2007) Switzerland's Allalin Glacier Avalanche-Mattmark Dam Construction Disaster, 1965. 6(4):12–15
Seed HB, Seed RB, Schlosser F, Blondeau F, Juran I (1988) The landslide at the Port of Nice on October 16 1979. EERC Report, University of California, Berkeley, pp 88–10
Shieh CL, Chen YS, Tsai YJ, Wu JH (2009) Variability in rainfall threshold for debris flow after the Chi-Chi earthquake in central Taiwan. China. International Journal of Sediment Research 24(2):177–188
Shimamoto R, Takano Y, Maeda I (2008) Disaster response of Kaga local government to earthquake and landslide disaster in Ansei Period, 19th Century. Bulletin of the Tateyama Caldera Sabo Museum 9:21–57 (in Japanese)
Shimizu T, Manaka M, Tsukamoto H, Kazahaya K (2017) Changes in long-term geological environment in slab-related upwelling region, part (1): hydrothermal alteration along the Arima-Takatsuki fault zone. 124th Annual Convention of the Geological Society of Japan, Ehime 544 (in Japanese).
Shimizu O, Nagayama T, Saito M (1995) Distribution and frequency of landsides for the last 8000 years in a small basin, Central Hokkaido. Transactions, Japanese Geomorphological Union 16(2):115–136 (in Japanese)
Shimokawa E, Jitousono T, Takano S (1989) Periodicity of shallow landslide on Shirasu (Ito pyroclastic flow deposits) steep slopes and prediction of potential landslide sites. Transactions of the Japanese Geomorphological Union 10(4):267–284 (in Japanese)

Shipton ZK, Cowie PA (2003) A conceptual model for the origin of fault damage zone structures in high-porosity sandstone. J Struct Geol 25(3):333–344
Shishikura M, Ota Y, Watanabe M, Sawa H, Yamaguchi M, Yanagida M, Tanaka T, Ichikawa K, Lee YH, Lu ST, Shih TS (2002) Paleoseismological study at Fengyuan site on the northern part of the Chelungpu Fault, central Taiwan. Annual Report on Active Fault and Paleoearthquake Researches 2:279–289 (in Japanese)
Shoaei Z (2014) Mechanism of the giant Seimareh landslide, Iran, and the longevity of its landslide dams. Environmental Earth Sciences 72(7):2411–2422
Shou KJ, Wang CF (2003) Analysis of the Chiufengershan landslide triggered by the 1999 Chi-Chi earthquake in Taiwan. Eng Geol 68(3–4):237–250
Shreve RL (1966) Sherman landslide. Alaska. Science 154(3757):1639–1643
Sibson H (1995) Implication of fault-valve behavior for rupture nucleation and recurrence. Tectonophysics 211:283–293
Socquet A, Simons W, Vigny C, McCaffrey R, Subarya C, Sarsito D, Ambrosius B, Spakman W (2006) Microblock rotations and fault coupling in SE Asia triple junction (Sulawesi, Indonesia) from GPS and earthquake slip vector data. Journal of Geophysical Research: Solid Earth 111:B08049
Sorby HC (1856) On the theory of the origin of slaty cleavage. The London, Edinburgh, and Dublin Philosophical Magazine and Journal of Science 12(77):127–129
Soto Melo G Valdivia 1960. Terremoto e inundacion de 1960 (in Spanish)
Spudich P, Hellweg M, Lee WHK (1996) Directional topographic site response at Tarzana observed in aftershocks of the 1994 Northridge, California, earthquake: implications for mainshock motions. Bull Seismol Soc Am 86:193–208
Strasser M, Schindler C, Anselmetti FS (2008) Late Pleistocene earthquake-triggered moraine dam failure and outburst of Lake Zurich, Switzerland. J Geophys Res Earth Surf 113(F2):F02003
Strom A (2010) Landslide dams in Central Asia region. Journal of the Japan Landslide Society 47(6):309–324
Sumiyoshi Village of Muko County (1946) History of Sumiyoshi Village, p 1105 (in Japanese)
Sun P, Li R, Jiang H, Igwe O, Shi J (2017) Earthquake-triggered landslides by the 1718 Tongwei earthquake in Gansu Province, northwest China. Bull Eng Geol Env 76(4):1281–1295
Suwa S (2015) Reconnaissance report on damage caused by the 2015 Nepal Gorkha earthquake. Report of Calamity Science Institute, Osaka (in Japanese)
Suzuki K, Nakamura T, Kato T, Ikeda A, Goto A, Oda M, Minami M, Kamikubo H, Kajizuka I, Adachi K, Tsuboi M, Tokiwa T, Oota T, Nishida M, Esaka N, Tanaka A, Mori S, Dunkley DJ, Kusiak MA, Suzuki S, Niu E, Nakazaki M, Senda R, Kanaagwa K, Kumazawa H (2008) The place name 'Umi' in Kamiyahagi of Ena City originated from the damming by a landslide triggered by Tensho Earthquake in 1586. Summaries of Researches using AMS at Nagoya University, XIX (in Japanese)
Suzuki S, Fujimoto R, Ishii Y (2020) Inactivation of landslide after the 2011 off the Pacific coast of Tohoku earthquake—case study of Shionokawa landslide, Fukushima Prefecture, Japan. Journal of the Japan Landslide Society 57(5):7–12 (in Japanese)
Swenson FA (1964) Ground-water phenomena associated with the Hebgen Lake earthquake. USGS Professional Paper 435-K, The Hebgen Lake, Montana, earthquake of August 17, vol 1959, pp 159–165
Synolakis CE, Bardet JP, Borrero JC, Davis HL, Okal EA, Silver EA, Sweet S, Tappin DR (2002) The slump origin of the 1998 Papua New Guinea Tsunami. Proc Royal Society of London A 458:763–789
Tabata S, Mizuyzma T, Inoue K (2002) Natural dam and disasters. Kokon Shoin, pp 50–53 (in Japanese)
Tainosho Y, Fujita T (1996) Mountain slope failures caused by the Hyogoken-Nanbu earthquake and the further failures by subsequent rainfalls. J Jpn Soc Eng Geol 37(3):174–184 (in Japanese)
Taira A (2001) Tectonic evolution of the Japanese island arc system. Annu Rev Earth Planet Sci 29(1):109–134

Tajika J, Ohtsu S (2000) Collapse of creeping mass of hard shale. Earth Sci 54(3):149–151 (in Japanese)
Tajika J, Ishimaru S, Kawakami G, Okazaki T, Yokohama S, Miura S (2012) Landslides in Kiyota-ku, Sapporo, induced by the shallow earthquake (M4.6), Dec. 2, 2010. Journal of Japan Landslide Society 49(2):75–79 (in Japanese)
Takagi H, Pratama MB, Kurobe S, Esteban M, Aránguiz R, Ke B (2019a) Analysis of generation and arrival time of landslide tsunami to Palu City due to the 2018 Sulawesi earthquake. Landslides 16(5):983–991
Takagi H, Sugiyama K, Tamura I, Mizuno K, Kitazawa N, Kawamoto K (2019b) Identification of the youngest activity of the Median Tectonic Line at Hiji outcrop, Ina City, Nagano Prefecture. Active Fault Research 50(1–12) (in Japanese)
Takeshita K (1985) Process of slope and soil formations on steep mountain slopes. Transactions, Japanese Geomorphological Union 6(4):317–332 (in Japanese)
Takagi M, Yokota S, Suga K, Yasuda S, Ota H (2010) The actual situation of the slope of earthfill that collapsed by an earthquake disaster in the Tomei Expressway Makinohara District. In: 55th geotechnical engineering symposium, JGS, pp 193–196 (in Japanese)
Takeuchi T, Tsutsumi H (1985) Ohya Kuzure landslide. Journal of the Japan Society of Erosion Control Engineering 38(3):20–22 (in Japanese)
Tanaka K, Ishihara T (2009) Mud volcanism near the Nabetachiyama Tunnel and the formation mechanism of swelling rock mass. J Geogr 118(3):499–510 (in Japanese)
Tanaka T, Tatsuoka F, Mori H (2013) Causative mechanism of Fujinuma reservoir during the 2011 Tohoku earthquake. Journal of Japan Society of Dam Engineers 23:99–114 (in Japanese)
Tanioka Y, Satake K (1996) Tsunami generation by horizontal displacement of ocean bottom. Geophys Res Lett 23(8):861–864
Tang C, Zhu J, Qi X, Ding J (2011) Landslides induced by the Wenchuan earthquake and the subsequent strong rainfall event: a case study in the Beichuan area of China. Eng Geol 122(1–2):22–33
Taniguchi T (1964a) Matsunoyama landslide in Niigata. Civil Engineering for Life 19:47–64 (in Japanese)
Taniguchi T (1964b) Brief on Matsunoyama landslide. Journal of Japan Landslide Society 1:53–54 (in Japanese)
Tanimoto S, Kaibori M, Takada K (2018) Study on distinctive landslides induced by an earthquake. Studies of Environmental Sciences 13:41–52 (in Japanese)
Tappin DR, Matsumoto T, Watts P, Satake K, McMurtry GM, Matsuyama M, Lafoy Y, Tsuji Y, Kanamatsu T, Lus W, Iwabuchi Y, Yeh H, Matsumoto Y, Nakamura M, Mahoi M, Hill P, Crook K, Anton L, Walsh JP (1999) Sediment slump likely caused 1998 Papua New Guinea tsunami. EOS Trans Am Geophys Union 80(30):329–340
Terada T, Miyabe N (1932) Landslide at Hatano. Bulletin of Earthquake Research Institute, University of Tokyo 10(1):192–199 (in Japanese)
Ter-Stepanian GI (1977) Deep-reaching gravitational deformation of mountain slopes. Bull Int Assoc Eng Geol 16(1):87–94
Tewari P (2004) A study on soil erosion in Pasighat Town (Arunachal Pradesh) India. Nat Hazards 32(2):257–275
Thein PS, Pramumijoyo S, Brotopuspito K, Kiyono J, Wilpo W, Furukawa A, Setianto A, Putra RR (2015) Estimation of S-wave velocity for sedimentary layered media using Microtremor array measurements in Palu City. Procedia Environ Sci 28:595–605
Tian Y, Owen LA, Xu C, Ma S, Li K, Xu X, Figueiredo PM, Kang W, Guo P, Wang S, Liang X (2020) Landslide development within 3 years after the 2015 Mw 7.8 Gorkha earthquake. Nepal. Landslides 17:1251–1267
Tian WT, Sun JJ, Wang LM, Xu SH, Liu K, Sun Y (2015) Research progress and frontier scientific problems in loess dynamics. Chinese Journal of Geotechnical Engineering 37(11):2119–2127 (in Chinese)

Tibaldi A, Ferrari L, Pasquarè G (1995) Landslides triggered by earthquakes and their relations with faults and mountain slope geometry: an example from Ecuador. Geomorphology 11(3):215–226
Tinti S, Maramai A, Graziani L (2004) The new catalogue of Italian tsunamis. Nat Hazards 33(3):439–465
Tiwari B, Beena A (2017) Landslides triggered by earthquakes from 1920 to 2015. In: 4th world landslide forum, Ljubljana, pp 5–15
Tokunaga T (1999) Estimating permeability change and specific yield from earthquake-induced hydrological changes: an example from Kobe earthquake. Journal of Japan Society of Engineering Geology 40(2):99–106 (in Japanese)
Tomita Y, Sakurai W, Naka Y (1996) Study on the extension of collapse caused by rainfall after the earthquake in Rokko Mountain Range. Journal of the Japan Society of Erosion Control Engineering 48(6):15–21 (in Japanese)
Tommasi P, Verrucci L, Campedel P, Veronese L, Pettinelli E, Ribacchi R (2009) Buckling of high natural slopes: the case of Lavini di Marco (Trento-Italy). Eng Geol 109(1–2):93–108
Torii N, Okimura T, Kato S (2007) Experimental study on slope failure mechanism induced by rainfall after earthquake. Proc JSCE C 63(1):140–149 (in Japanese)
Towhata I (2008) Geotechnical Earthquake Engineering. Springer, Berlin
Towhata I (2021a) Earthquake-induced landslides; complicated mechanisms and outputs. In: 8th international symposium on landslides, Cartagena, Colombia, Keynote lecture
Towhata I (2021b) Views on recent rainfall-induced slope disasters and floods. In: CREST 2021b, 1st international symposium on construction resources for environmentally sustainable technologies, March, Fukuoka, Keynote lecture
Towhata I (2022) Coseismic and post-seismic slope instability along existing faults. In: 4th international conference on performance based design in earthquake geotechnical engineering, Beijing, Keynote lecture
Towhata I, Ghalandarzadeh A, Sundarraj KP, Vargas-Monge W (1996) Dynamic failures of subsoils observed in water-front areas. Special issue on geotechnical aspects of the January 17 1995 Hyogoken-Nambu earthquake. Soils Found 1:149–160
Towhata I, Gunji K (2018) Experimental interpretation of seismically induced instability of mountain slopes. XVI Danube-European Conference on Geotechnical Engineering, Skopje 1:255–260
Towhata I, Maruyama S, Kasuda K, Koseki J, Wakamatsu K, Kiku H, Kiyota T, Yasuda S, Taguchi Y, Aoyama S, Hayashida T (2014) Liquefaction in Kanto region during the East-Japan gigantic earthquake on March 11. Soils Found 54(4):859–873
Towhata I, Nakamura H (2015) Qualified evaluator of subsoil quality for safety of residential land. In: International conference on soft ground engineering (ICSGE2015), Singapore, Pape No. 180
Towhata I, Taguchi Y, Hayashida T, Goto S, Shintaku Y, Hamada Y, Aoyama S (2016) Liquefaction perspective of soil ageing. Geotechnique 67(6):467–478
Trustrum NA, De Rose RC (1988) Soil depth-age relationship of landslides on deforested hillslopes, Taranaki, New Zealand. Geomorphology 1:143–160
Tsujimura T (1950) Reconnaissance of flood disaster in Gunma in September 1947. Study on Kathleen Typhoon. Gunma Prefecture, pp 107–130 (in Japanese)
Tsuneishi Y, Nakamura K (1970) Faulting associated with the Matsushiro swarm earthquakes. Bulletin of Earthquake Research Institute, University of Tokyo 48(1):29–51
Tsutsumi H, Goto H, Tani M (2004) Right-lateral offsets on the Arima-Takatsuki fault zone in southwest Japan associated with prehistoric and historic earthquakes. Active Fault Research 24:157–165 (in Japanese)
Tun ST, Watkinson IM (2017) The Sagaing fault (Chap. 19). In: Myanmar: geology, resources and tectonics, vol 48. Geological Society, London, Memoir, pp 413–442
Uda T, San-nami T, Hoshigami Y, Sakai K (2012) Damage due to the 2011 great tsunami and shrines escaped from being damaged. Proc Japan Society of Civil Engineers B3 68(2):I_43–I_48 (in Japanese)

Ueno M, Nishimura T, Kato M, Nakamura H (2005) Variation of shearing characteristics of loess soil after irrigation. Northwestern Seismological Journal 27(2):128–134
Ueshima T, Shiba Y, Hataya R, Kawai T, Toyoda Y, Yajima H, Matsuda T (1999) Quick investigation report on the 1999 Taiwan Jiji earthquake. CRIEPI Report, Central Research Institute of Electric Power Industry U99023 (in Japanese)
Ui T (1983) Volcanic dry avalanche deposits—identification and comparison with nonvolcanic debris stream deposits. J Volcanol Geoth Res 18:135–150
Usami T, Hisamoto S (1971) Future probability of a coming earthquake with intensity V or more in the Kyoto area. Bulletin of the Seismological Institute, University of Tokyo 4:115–125 (in Japanese)
Usuki N, Mizuyama T (2011) Soil properties and fluidity of long-travelling landslides, Italian Journal of Engineering Geology and Environment. In: Special issue for 5th international conference on debris-flow hazards mitigation: mechanics, prediction and assessment, topic, vol 7, pp 787–795.
Usuki N, Tanaka Y, Mizuyama T (2005) Investigation of the distribution of long-traveling landslides. Journal of the Japan Society of Erosion Control Engineering 57(5):47–52
Utada M (2003) Alteration of Rokko granites: mineralogical and magnetic susceptibility changes. J Geogr 112(3):360–371 (in Japanese)
Varnes DJ (1958) Landslide types and processes. Landslides and Engineering Practice 24:20–47
Vermilye JM, Scholz CH (1998) The process zone: a microstructural view of fault growth. J Geophys Res 103(B6):12223–12237
Voight B, Faust C (1982) Frictional heat and strength loss in some rapid landslides. Geotechnique 32(1):43–54
Voight B, Kennedy BA (1979) Slope failure of 1967–1969, Chuquicamata mine, Chile. Developments in Geotechnical Engineering, Elsevier, 14, Part B, pp 595–632
Voight B, Sousa J (1994) Lessons from Ontake-san: a comparative analysis of debris avalanche dynamics. Eng Geol 38(3–4):261–297
Waag CJ (1985) Groundwater eruptions and sediment boil formation in the Chilly Buttes area, MT. Borah, Idaho earthquake. Proceedings of Workshop XXVIII on the Borah Peak Earthquake, Idaho, USGS Open-File Report 85–290:593–611
Wadatsumi K (1995) 1519 earthquake precursors preceding the 1995 Kobe earthquake. Tokyo Shuppan, p133 of short version (in Japanese)
Wakai A, Cai F, Ugai K, Soda T (2016) Mechanism of fluidized landslides triggered by cyclic softening of cohesive soils during earthquakes. Journal of Japan Landslide Society 53(4):117–127 (in Japanese)
Wakai A, Sato S, Mitsuji K, Mori T, Kazama M, Koseki J (2012) Investigation of damage caused by the 2011 off the Pacific coast of Tohoku earthquake—brief report on the residential lands in Sendai City. Geotechnical Engineering Journal, Japanese Geotechnical Society 7(1):79–90 (in Japanese)
Wang CL, Yin CY, Wang WN (2006) Observation of the Chiu-fen-er-shan Landslide. J Soil Water Conserv 38(4):303–316 (in Chinese)
Wang G, Zhang D, Furuya G, Yang J (2014a) Pore-pressure generation and fluidization in a loess landslide triggered by the 1920 Haiyuan earthquake, China: a case study. Eng Geol 174:36–45
Wang LM (2020) Mechanism and risk evaluation of sliding flow triggered by liquefaction of loess deposit during earthquakes. Chinese Journal of Geotechnical Engineering 42:11–19 (in Chinese)
Wang LM, He KM, Shi YC, Wang J (2002) The study on liquefaction of saturated loess by in-situ explosive test. In: Liu H-L (ed) Soil dynamics and geotechnical earthquake engineering. China Architecture & Building Press, pp 560–565 (in Chinese)
Wang ZR, Wu WJ, Zhou ZQ (2004) Landslide induced by over-irrigation in loess platform areas in Gansu Province. Chinese Journal of Geological Hazard and Control 15(3):43–46 (in Chinese)
Wang LM, Yuan ZX, Sun HM, Deng J (2010a) Criterions, prediction and prevention of loess liquefaction. In: 5th international conference on recent advances in geotechnical earthquake engineering and soil dynamics, San Diego, Paper No. OSP11

Wang LS, Yang LZ, Li TB, Xu XN, Wang XQ, Cui J (2008) Diexi earthquake-induced landslide, Min River, Sichuan Province (1933) (Chap. 3). In: Hao RQ, Xu Q et al (eds) Catastrophic landslides in China. Science Press, Beijing, pp 57–93 (in Chinese)
Wang XY, Nie GZ, Wang DW (2010b) Research on relationship between landslides and peak ground accelerations induced by Wenchuan earthquake. Chin J Rock Mechan Eng 29(1):82–89 (in Chinese)
Wang ZH, Guo ZC, Du ML, Cheng ZL (2012) Model study of predicting rainstorm induced landslide and debris flow at Niumian Gully. Earth Sci Front 19(1):228–238 (in Chinese)
Wang G, Huang R, Lourenço SD, Kamai T (2014b) A large landslide triggered by the 2008 Wenchuan (M8.0) earthquake in Donghekou area: phenomena and mechanisms. Eng Geol 182:148–157
Wang XQ, Li YR, Yuan Y, Zhou Z, Wang LS (2014c) Palaeoclimate and palaeoseismic events discovered in Diexi barrier lake on the Minjiang River, China. Nat Hazard 14:2069–2078
Wasowski J, Keefer DK, Lee CT (2011) Toward the next generation of research on earthquake-induced landslides: current issues and future challenges. Eng Geol 122(1–2):1–8
Watanabe N, Oki Y, Sato O, Kusakabe M (1997a) Origin of Na-Cl type groundwaters in the Matsunoyama landslides, Niigata Prefecture. Annual report of research institute for hazards in snowy areas, vol 18. Niigata University, pp 81–92 (in Japanese)
Watanabe N, Sato T, Oki Y, Shiraishi S, Sato O, Kusakabe M (1997b) Deep groundwaters in the tertiary type landslides. Monthly Magazine of Japanese Geotechnical Society 32(6):32–34 (in Japanese)
Watanabe S, Orui K, Tsuchiya S (2013) Measurement of the large-scale landslide on the east flank of Mt. Shiratori using airborne laser scanning. Chubu Forestry Research 61:41–44 (in Japanese)
Watari M (2005) Original birth of active landslide and "Sagging" by gravitational deformation on mountain slope. Journal of Japan Landslide Society 41(5):57–66 (in Japanese)
Watkinson IM, Hall R (2019) Impact of communal irrigation on the 2018 Palu earthquake-triggered landslides. Nat Geosci 12(11):940–945
Watson RA, Wright HE Jr (1969) The Saidmarreh landslide. Iran. Geological Society of America, Special Paper 123:115–139
Wei A (2015) Characteristics, origin and quantitative evaluation of overpressure in strike-slip and compression-shear booster zone of Tan-Lu Fault. Pet Geol Exp 37(1):47–52 (in Chinese)
Weischet W (1963) Further observation of geologic and geomorphic changes resulting from the catastrophic earthquake of May 1960, in Chile. Bull Seismol Soc Am 53(2):1237–1257
Wilkinson F (2015) Everest Base Camp a 'War Zone' after earthquake triggers avalanche. National Geographic Magazine, 25 April
Wilson JT (1963) Continental drift. Sci Am 208(4):86–103
Witkind IJ, Myers WB, Hadley JB, Hamilton W, Fraser GD (1962) Geologic features of the earthquake at Hebgen Lake, Montana, August 17. Bull Seismol Soc Am 52(2):163–180
Wood DS (1974) Current views of the development of slaty cleavage. Annu Rev Earth Planet Sci 2(1):369–401
Wood SH (1985) Regional increase in groundwater discharge after the 1983 Idaho earthquake: coseismic strain release, tectonic and natural-hydraulic fracturing. Proceedings of Workshop XXVIII on the Borah Peak Earthquake, Idaho, USGS Open-File Report 85–290:573–592
Wood RM, Steedman RS (1992) Earthquake induced groundwater changes and liquefaction. Proceedings of Tenth World Conference on Earthquake Engineering, Madrid 3:1337–1341
Wu J, Wang Y, Dong S, Chen Y, Wang L (2013) Genetic mechanism and failure process of the Mogangling seismic landslide. J Geol Soc India 82(3):277–282
Xu Q, Fan XM, Huang RQ, Van Westen C (2009) Landslide dams triggered by the Wenchuan earthquake, Sichuan Province, south west China. Bull Eng Geol Env 68(3):373–386
Yagi H, Inokuchi T (2010) Yambushi, and Oya-Kuzure; gravitational deformation of mountain body and subsequent landslides due to earthquake, Central Japan. Journal of Japan Landslide Society 47(1):66–68 (in Japanese)

Yagi H, Inokuchi T, Sato G (2016) Aerial watching of landslides in Japan—44 cradle-shaped depression, "Funakubo", as gravitational deformation occurring along gentle crest of Mt. Chogatake, Central Japan. Journal of the Japan Landslide Society 53(4):156–158 (in Japanese)

Yagi H, Yamasaki T, Atsumi M (2007) GIS analysis on geomorphological features and soil mechanical implication of landslides caused by 2004 Niigata Chuetsu earthquake. Journal of Japan Landslide Society 43:294–306 (in Japanese)

Yagi K, Higaki D, Japan Landslide Society Research Committee for Detection of Landslide Hazardous Sites in Tertiary Distributed Area (2009) Methodological study on landslide hazard assessment by interpretation of aerial photographs combined with AHP in the middle course area of Agano River, Central Japan. Journal of Japan Landslide Society 45(5):358–366 (in Japanese)

Yamabe K, Marui H, Yoshimatsu H, Yamamoto S (2010) Landslide dam in Higashitakezawa and Terano area formed by the Mid Niigata Prefecture earthquake. Journal of Japan Landslide Society 47(6):349–360 (in Japanese)

Yamagishi H, Miyaki H, Yoshizumi Y, Hotta T (2001) Higashikawa landslide taking place on April 28th, 2001, in the Matsunoyama Town, Niigata Prefecture. Journal of Japan Landslide Society 38(2):157 (in Japanese)

Yamashita K, Lin A (2009) Width of fault-fracture zone in the Rokko Fault, Arima-Takatsuki tectonic lines. In: 116th annual conference, Geological Society of Japan, Okayama, 195 (in Japanese)

Yamashita S, Ogawa K, Kawajiri S, Kawaguchi T, Watanabe T (2019) Liquefaction characteristics of residential volcanic soil ground suffered by 2018 Hokkaido Iburi Tobu earthquake—comparison with agricultural volcanic soil ground suffered by 2003 Tokachi-oki Earthquake. Geotechnical Engineering Journal, Japanese Geotechnical Society 14(4):353–361 (in Japanese)

Yamazaki K (2013) Disaster and shrine. In: Proceedings of the annual meeting of the association of Japanese geographers 100038 (in Japanese)

Yanai S, Igarashi Y (1990) History of the slope failure and paleoenvironment on the marine terrace of Hidaka District. Central Hokkaido. the Quaternary Research 29(4):319–336 (in Japanese)

Yang BJ (2019) Mitigation of debris flow disaster at Zoumaling. Building Technology Research 2(4):89–90 (in Chinese)

Yasuda S (1993) Zoning for slope instability, manual for zonation of seismic geotechnical hazards. Technical Committee 4, International Society for Soil Mechanics and Foundation Engineering, pp 48–71

Ye L, Kanamori H, Rivera L, Lay T, Zhou Y, Sianipar D, Satake K (2020) The 22 December 2018 tsunami from flank collapse of Anak Krakatau volcano during eruption. Science Advances 6(3):eaaz1377

Yin YP (2000) Characteristics of Bomi-Yigong huge high speed landslide in Tibet and the research on disaster prevention. Hydrogeology and Engineering Geology 27(4):8–11 (in Chinese)

Yokota S, Takagi M, Yasuda S (2011) Seismic failure of embankment of Tomei expressway at Makinohara. The Foundation Engineering & Equipment, Monthly May, pp 50–54 (in Japanese)

Yokota S, Tori M (2019) Surface cracks appeared in a residential area, Minamiaso Village, at the time of the 2016 Kumamoto earthquake. The Quaternary Research 58(2):109–119 (in Japanese)

Yokoyama S, Kikuyama H (1997) Movement types and mechanism of slope failures occurred in the Rokko granite area during the 1995 Hyogo-Ken Nanbu earthquake. Journal of Japan Landslide Society 34(3):17–24 (in Japanese)

Yoshida N, Tazo T, Wakamatsu K, Yasuda S, Towhata I, Nakazawa H, Kiku H (2007) Causes of Showa Bridge collapse in the 1964 Niigata earthquake based on eyewitness testimony. Soils Found 47(6):1075–1087

Yoshida E, Oshima A, Yoshimura K, Nagamoto A, Nishimoto S (2009) Fracture characterization along the fault—a case study of 'damaged zone' analysis on Atera Fault, Central Japan. Journal of Japan Society of Engineering Geology 50(1):16–28 (in Japanese)

You Y, Chen XZ, Liu JF (2011) "8.13" extra-large debris flow disaster in Wenjia gully of Qingping Township, Mianzhu, Sichuan Province. Journal of Catastrophology, Earthquake Bureau of Shaanxi Province 26(4):68–72 (in Chinese)

Youd TL, Perkins DM (1987) Mapping of liquefaction severity index. ASCE Journal of Geotechnical Engineering 113(11):1374–1392
Yu B, Ma Y, Yu YF (2010a) Investigation of severe debris flow hazards in Wenjia Gully of Sichuan province after the Wenchuan earthquake. J Eng Geol 18(6):827–836 (in Chinese)
Yu B, Yang YH, Su YC, Huang WJ, Wang GF (2010b) Research on the giant debris flow hazards in Zhouqu County, Gansu Province on August 7, 2010. J Eng Geol 18(4):437–444 (in Chinese)
Yu B, Wu Y, Chu S (2014) Preliminary study of the effect of earthquakes on the rainfall threshold of debris flows. Eng Geol 182:130–135
Yuan RM, Deng QH, Cunningham D, Xu C, Xu XW, Chang CP (2013) Density distribution of landslides triggered by the 2008 Wenchuan earthquake and their relationships to peak ground acceleration. Bull Seismol Soc Am 103(4):2344–2355
Zhang M-S, Li T-L (2011) Triggering factors and forming mechanism of loess landslides. J Eng Geol 19(4):530–540 (in Chinese)
Zhang D, Wang G (2007) Study of the 1920 Haiyuan earthquake-induced landslides in loess (China). Eng Geol 94(1–2):76–88
Zhang YJ, Nian TK, Guo XS, Chen GQ, Zheng L (2019) Modelling the flexural buckling failure of stratified rock slopes based on the multilayer beam model. J Mt Sci 16(5):1170–1183
Zhou JW, Cui P, Hao MH (2016) Comprehensive analyses of the initiation and entrainment processes of the 2000 Yigong catastrophic landslide in Tibet. China. Landslides 13(1):39–54
Zischinsky U (1969) Über sackungen. Rock Mech 1(1):30–52 (in German)
Zitter TAC, Grall C, Henry P, Özeren MS, Çağatay MN, Şengör AMC, Gasperini L, de Lépinay BM, Géli L (2012) Distribution, morphology and triggers of submarine mass wasting in the Sea of Marmara. Mar Geol 329:58–74

Characteristics of Sites Response of a Series of Monitored Slopes in Sichuan Mountain Area

Qiang Xu, Yonghong Luo, and Yunsheng Wang

Abstract Slope site seismic response is different from one of the flat sites, it has been known for many years; and this phenomenon has also attracted great attention in China since the "5.12" Wenchuan earthquake. From 2008, the State Key Laboratory of Geohazard prevention and Geoenvironment protection (SKLGP) has started to complete long term and systematic monitoring of the site amplification for different slopes. This chapter presents monitoring characteristics of slope site response of seven typical sites. A total of four main shocks were monitored by the Zigong (recorded Wenchuan earthquake), Luding (recorded Lushan and Kangding earthquakes) and Shimian (recorded Changning earthquake) monitoring points. Hundreds of aftershocks were observed by our monitoring stations. The Peak ground acceleration (PGA), horizontal to vertical spectral ratio (HVSR), normalized Arias intensity (Ia), standard spectral ratio (SSR) and azimuth information has been extracted from the monitoring data. The monitoring characteristics of the slope seismic response show that the soil site is always affected by stronger amplification than the bedrock site in a small amplitude range (Reference site < 0.076 g). This is due to the high wave impedance between soil and bedrock promoting site ground motion effects. Meanwhile, significant polarization effects have also been observed on the mountain top (hilltop), the normalized Ia and polarization diagram of the HVSR show the same directivity characteristics of ground motion near the mountain tops, marked by strongest amplitudes transverse to the ridge, where the extension direction of the ridge and orientation of geological structures influence the seismic energy redistribution. In addition, the PGA amplification shows nonlinear increasing with the topographic relief and changing geological conditions. Amplification of PGA of the slope in the upper part can be quite high (>10) with respect to a reference site (for weaker motion of an intensity less than V), but is less marked (<4.5) for strong motion conditions (Intensity $\geq$ VI). This means that the PGA amplification effect has significant variable characteristics under low intensity of earthquake. The HVSR analysis shows that the amplified resonance frequency of most monitoring sites is in the range of 2–4 Hz, and part are in the range of 6–8 Hz, and that multi peak frequency values can

Q. Xu (✉) · Y. Luo · Y. Wang
State Key Laboratory of Geohazard Prevention and Geoenvironment Protection, Chengdu University of Technology, Chengdu 610059, China
e-mail: xq@cdut.edu.cn

I. Towhata et al. (eds.), *Coseismic Landslides*, Springer Natural Hazards,
https://doi.org/10.1007/978-981-19-6597-5_12

be observed in most slope monitoring sites. The complexity of the slope morphology and of the geology conditions has an influence on related results. This paper presents some analyses of the seismic response of a series of monitored slope sites, which show that predicting the slope response is still very difficult, due to the variability of site effects. An even greater challenge is the prediction of the seismic slope failure potential as there is a great lack of recorded strong motion data on sites affected by permanent deformation after an earthquake.

1 Introduction

The slope site seismic response is a complex dynamic scientific question and have been known for decades (e.g., Boore 1972; Bouchon 1973; Davis and West 1973; Çelebi 1987; Ashford et al. 1997). To understand the dynamic response, topography, geological structure, lithology and other factors or combined effects etc. were used to analyze the influence. A series of representative views of the seismic amplification effect are: size and shape of topography (e.g., Boore 1972; Luo et al. 2013), high impedance contrast between surface materials and substratum (e.g., Del Gaudio and Wasowski 2011), surface morphology and the presence of a low-velocity layer (e.g., Havenith et al. 2002, 2003; Graizer 2009), and the combined effect of stratigraphy and topography (Hailemikael et al. 2016; Luo et al. 2020). Geology (rock type, fractures) also contributes to seismic amplification (e.g., Gischig et al. 2015). In China, the 2008 "5.12" Wenchuan earthquake promoted to further study the dynamic failure of slopes, thus the slope seismic monitoring, numerical simulations, etc. were applied to reveal the mechanism of slope seismic response. In longmenshan mountain areas Sichuan province, southwest of China, we installed more than twenty monitoring stations in the past ten years. Seven typical monitoring sites are shown in Fig. 1. Table 1 lists the site information, the monitoring earthquake and largest recorded magnitude. In this chapter, we want to show the site response though analysis of PGA, FTT, Arias Intensity, horizontal to vertical spectral ratios (HVSR) (Nakamura 2000). The reasons for the dynamic response characteristics are not analyzed in depth.

2 The PGA and HVSR Characteristics on Bedrock and Soil Sites

PGA is an important parameter to describe the ground vibration features, and the HVSR can reveal the peak frequency of site. In this section, both PGA and HVSR are used to analyze the site effects. At the Shimian, Shuanghe and Zigong monitoring points, stations were installed on bedrock and soil sites, main shock or aftershocks were recorded by those stations. Each slope seismic response characteristic is analysis as follows.

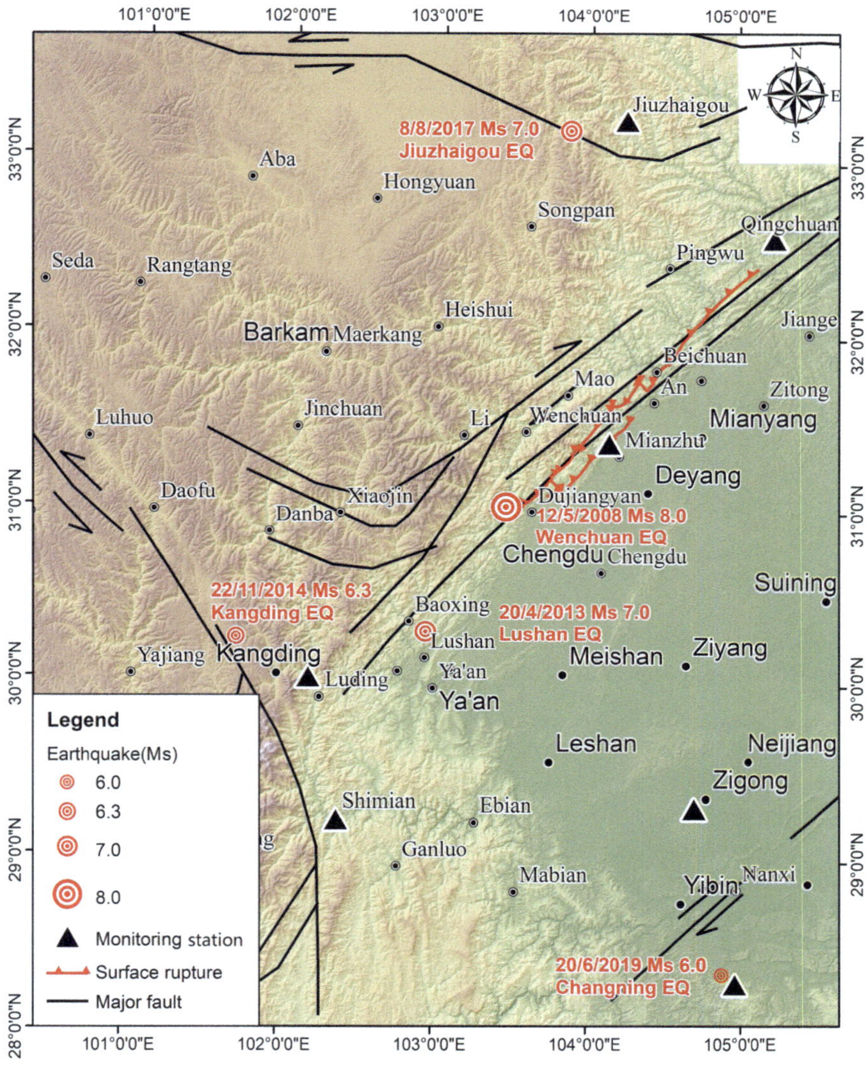

Fig. 1 Distribution of typical monitoring sites on slopes in the Sichuan Province

(1) Shimian monitoring sites

The Shimian monitoring sites locate in Shimian County, Sichuan province. We set three monitoring stations on the bank slopes of both Nanya river (Fig. 2). The #1 and #2 stations distribute on the right bank slope at the elevation of 1150 and 1060 m, where lithology is outcrop weak weathering granite. The #3 station distributes in the middle of the left bank slope, where the soil thickness is more than 30 m. On June 17th 2019, those stations monitored the Changning Ms 6.0

Table 1 The basic information of typical monitoring sites

Location	Year	Monitoring events	The largest magnitude	Site characteristics
Shimian County	2019	Changning earthquake	Ms 6.0	Bedrock and soil
Shuang He Town	2019	Changning aftershocks	Ms 5.4	Bedrock and soil
Jiuzhaigou	2017	Jiuzhaigou aftershocks	Ms 5.4	Bedrock
Luding	2014	Lushan and Kingding main shocks	Ms 7.0	Bedrock and soil
Zigong Xishan park	2008	Wenchuan mainshock	Ms 8.0	Bedrock
Qingchuan	2009–2014	Wenchuan aftershocks	Ms 5.2	Bedrock and soil
Mianzhu	2009	Wenchuan aftershocks	Ms 4.6	Bedrock and soil

earthquake, where the distance was about 265 km from the epicenter. Figure 2 presents the seismic waves of north–south component.

Figure 2 shows that the acceleration amplitude of #3 is much bigger than the two other stations. Table 2 presents the PGA, Arias intensity and dominant frequency value of the horizontal north–south (NS), east–west (EW) and up-down (UD) components in the monitoring sites. It can be observed that the highest PGA response is at the soil site #3 for those three sites, and that the horizontal PGA values of #3 are 2–4 times higher than the other two sites. The Arias intensity values of #3 are also much higher than the #1 and #2 sites. Additionally, the dominant frequency of Fast Fourier Transform (FFT) shows less than 3.2 Hz in all monitoring sites (Table 2).

Figure 3 presents the mean HVSR analysis of three sites. They have the similar resonance frequency at about 1.5 Hz, but the amplification factor of #3 soil site is bigger than the bedrock sites. According to the code for seismic design of buildings (GB 50011-2010), the #3 soil site is class III site, which shear wave velocity is in the range of 150–250 m/s. We can estimate the site resonance frequency (f = Vs/4H) is in the range from 1.25 to 2.08 Hz. This is close to the monitoring result. It means that the soil layer of #3 contributes to the amplification effects. On the other hand, the vibration azimuth of #3 is in the range of 60°–120°, different from 0°–40°and 140°–180° at #1 and #2 sites. We can infer that the slope topography of #3 affects the azimuth of the site seismic response.

(2) **Shuanghe town monitoring sites**

On June 17, 2019, a strong earthquake of Ms 6.0 occurred in Shuanghe town, Changning County, Sichuan province. After the main shock, three stations were installed on slope near the epicenter (Fig. 4). Among them, the #1 station as a

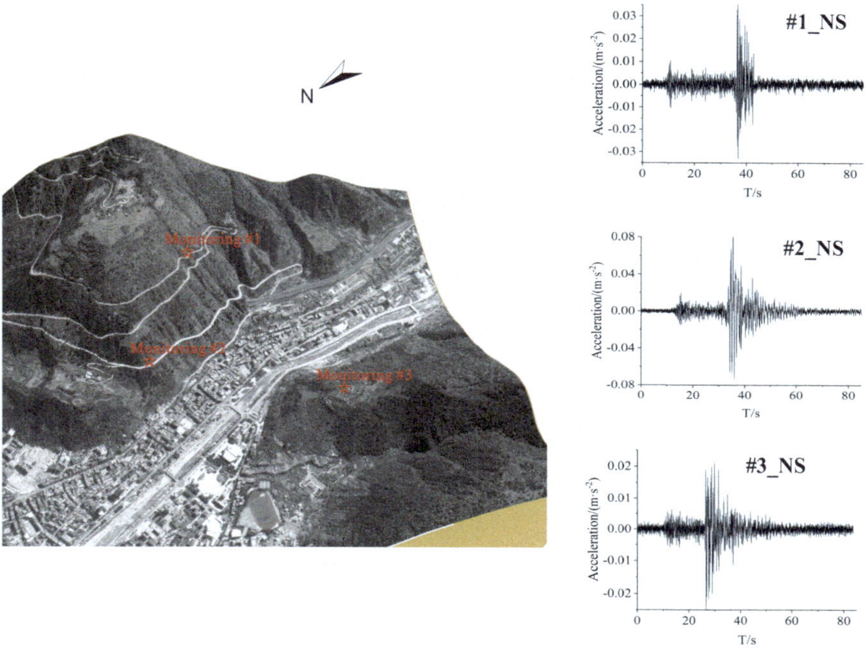

Fig. 2 Monitoring stations in Shimian County and the waveform in the north–south component of Changning Ms 6.0 earthquake (modified from Jin et al. 2019 with author rights)

Table 2 The PGA, arias intensity and dominant frequency at each monitoring site

Monitoring site	PGA ($cm{\cdot}s^{-2}$)			Arias intensity ($cm{\cdot}s^{-1}$)			Dominant frequency (Hz)		
	EW	NS	UD	EW	NS	UD	EW	NS	UD
#1	1.22	2.94	2.23	0.36	1.22	0.89	0.98	1.94	1.02
#2	2.13	3.64	2.35	0.55	1.80	0.81	3.16	3.20	3.16
#3	7.42	10.18	2.94	15.09	13.26	2.75	1.26	1.01	2.61

reference was at the valley bottom with the site outcrop of middle to strongly weathered bedrock and collapse deposit. The #2 site is in soil, with thickness of 4–5 m, and the #3 site is on the mountain top, where is middle to strong weathering bedrock crops out. At 22:29 on June 22th, 2019, a magnitude Ms 5.4 aftershock occurred near the monitoring sites. All stations recorded this event (Fig. 4). Tables 3 and 4 list the PGA and Arias intensity value of each station. Refer to the #1, the biggest amplification factor of the horizontal component PGA is about 4.30 at #2, about 3.51 at #3 (Table 3). The up-down component, the biggest amplification factor is in the range of 2.05–2.2 on both sites. In addition, the Arias intensity also shows strong amplification in the north–south

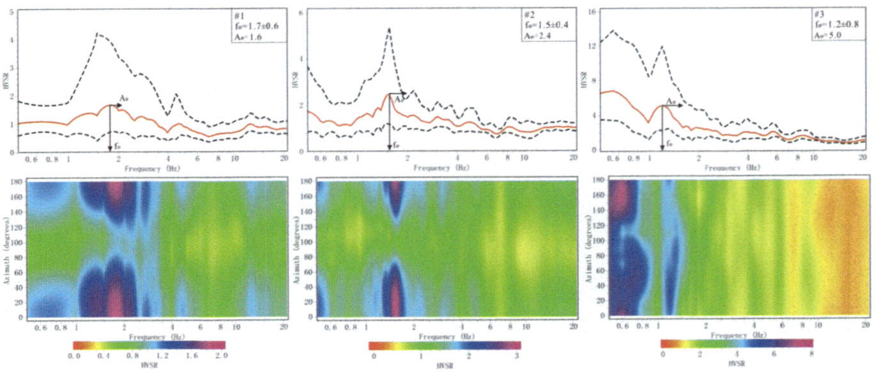

Fig. 3 The horizontal to vertical spectral ratio (HVSR) of three monitoring sites

component (Table 4), then the east–west component, the up-down component amplified in the range of 3.0–3.57.

Figure 5 presents the mean HVSR analysis results of each monitoring site. It can be observed that the 1# site has weak amplification in the frequency range of 2–3, 6–8 and 15–20 Hz. Regarding to the #2 site, the peak frequency has three obvious values, the 2.6, 4.6 and 10 Hz. According to the code for seismic design of buildings

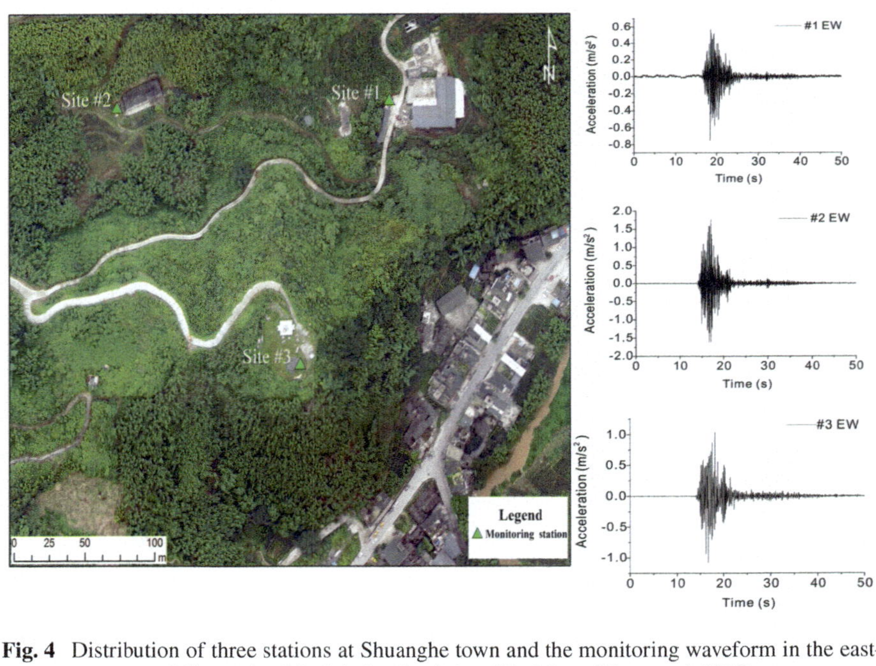

Fig. 4 Distribution of three stations at Shuanghe town and the monitoring waveform in the east–west component of Changning Ms 5.4 aftershock (modified from Wang et al. 2019)

Table 3 PGA value and amplification factors at each monitoring site

Monitoring site	11/08/2017 M_L3.5					
	PGA (m s^{-2})			Amplification factors		
	EW	NS	UD	EW	NS	UD
#1 (315 m)	0.76	0.64	0.46	/	/	/
#2 (346 m)	1.76	2.74	1.03	2.33	4.29	2.22
#3 (378 m)	1.08	2.24	0.95	1.43	3.51	2.04

Table 4 Arias intensity value and amplification factors at each monitoring site

Monitoring site	11/08/2017 M_L3.5					
	Arias intensity/m s^{-1}			Amplification factors		
	EW	NS	UD	EW	NS	UD
#1 (315 m)	0.03	0.02	0.01	/	/	/
#2 (346 m)	0.19	0.34	0.05	5.81	15.91	3.57
#3 (378 m)	0.10	0.33	0.04	2.91	15.37	3.04

(GB 50011-2010), the #2 site is class III as Vs in the range 150–250 m/s, and we can estimate that the site resonance frequency is in the range from 9.37 to 15.6 Hz. High peak frequency of 10 Hz is probably due to the soil resonance, while other peak frequencies of 2.6 and 4.6 Hz maybe influenced by other reasons, such as the slope or the combined effects of layer with topography. At the #3 site, the HVSR amplified effects have wider frequency range of 2–6 Hz than the #1 and #2 site, there is no soil contribution for this amplification, it means that the topography of #3 combined with the weathered bedrock contributed the site amplification effects. In addition, Fig. 5 also presents the dynamic azimuth characteristics. It can be observed that the #2 and #3 sites have obvious azimuth in the range of 0°–40° and 140°–180°, those probably affected by the micro-relief at both sites. The azimuth of #1 site is about 120°–160° at 2.5 Hz, 40°–120° at 7.5–8.0 Hz, this mainly affected by the stratigraphic structure.

(3) **Zigong monitoring sites**

The Zigong monitoring site locates at the Xishan Park, Zigong City, Sichuan Province. The outcrop lithology is moderately to weakly weathered sandstone, mudstone and sandy mudstone. Before the 2008, the Sichuan Seismological Bureau installed 7 strong earthquake monitoring stations for constructing a monitoring array to study the slope amplification effects (Figs. 6 and 7). Besides, the Z0 was on the infill (Fig. 6).

Table 5 presents the PGA values and the amplification factor in the three components of each monitoring stations, recorded for the "5.12" Wenchuan earthquake. The Z1 station was used as the reference site. It can be observed that the soil site of Z0 has the highest values, the maximum factor is 2.39 in the EW component. In addition, the amplification factor of other sites is in the range of 1.09–1.77. On the

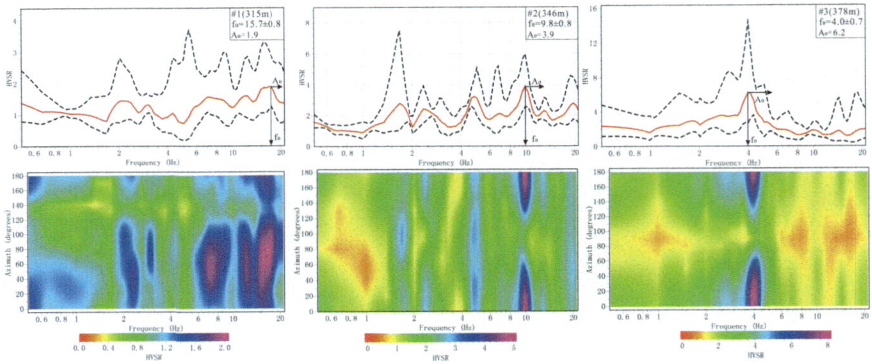

Fig. 5 The horizontal to vertical spectral ratio (HVSR) of three monitoring sites

Fig. 6 The distribution of monitoring station in the Xishan Park, Zigong City, Sichuan province (modified from Huang 2020)

hilltop, the Z6 and Z7 sites are amplify more than other sites (Z2–Z5). By comparing the horizontal and vertical components, the horizontal PGA shows higher values and amplified factors than the vertical PGA.

Figure 8 presents the HVSR and azimuth characteristics for each monitoring site. At the site Z0, the HVSR shows that the peak frequency is about 7 Hz**,** much bigger than the bedrock site of the slope. Besides, the seismic azimuth of the Z0 shows a significant dynamic azimuth (20°–100°) at the frequency of 7 Hz, this may be affected by the geological layers.

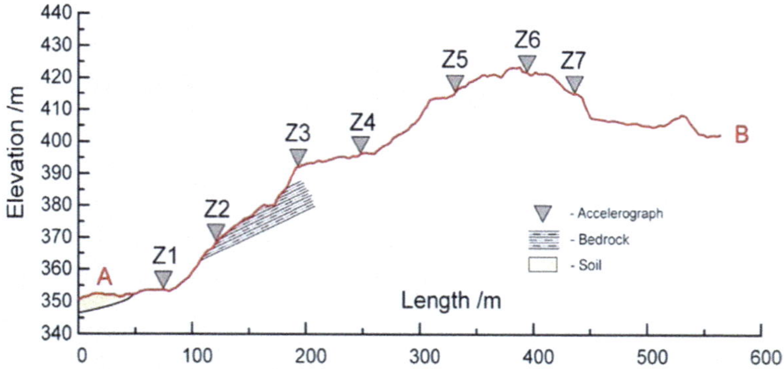

Fig. 7 The A-B monitoring profile in the Xishan Park (modified from Huang 2020)

Table 5 The PGA and amplification factors characteristics of each station at Zigong

Station no	Elevation (m)	Lithology	PGA (cm/s^2)			Amplification factor		
			EW	NS	UD	EW	NS	UD
Z1	345	Bedrock	23.04	26.49	14.73	Reference station		
Z0	345	Soil	55.03	47.72	19.97	**2.39**	**1.80**	**1.36**
Z2	367	Bedrock	28.08	29.95	15.95	1.22	1.13	1.08
Z3	385	Bedrock	34.86	32.53	18.34	1.51	1.23	1.25
Z4	390	Bedrock	32.53	32.23	19.72	1.41	1.22	1.34
Z5	392	Bedrock	33.03	42.17	17.24	1.43	1.59	1.17
Z6	417	Bedrock	40.74	42.24	19.74	**1.77**	**1.59**	**1.34**
Z7	397	Bedrock	39.30	45.49	16.05	1.71	1.72	1.09

From the Z1 to Z7, the site amplification effects appear in a wide frequency range of 2.0–3.0 Hz, and there is another peak in the high frequency range (>10 Hz). Comparison of the mean HVSR shows that both the Z6 and Z7 around the hilltop have stronger amplification than other sites. With regard to the azimuth, the site Z1 shows azimuth in the frequency range of 1.5–2.5 Hz, site Z2 in the frequency range of 1–4 Hz, Z4, Z5 and Z6 sites have azimuth in the frequency range of 2–4 Hz. On the other hand, the site Z0 shows azimuth in the frequency range 6–8 Hz. The azimuth directional amplification of Zigong monitoring sites is not only affected by the topography characteristics (such as the slope inclination), but also related to the geological features (such as the layers, rock structures).

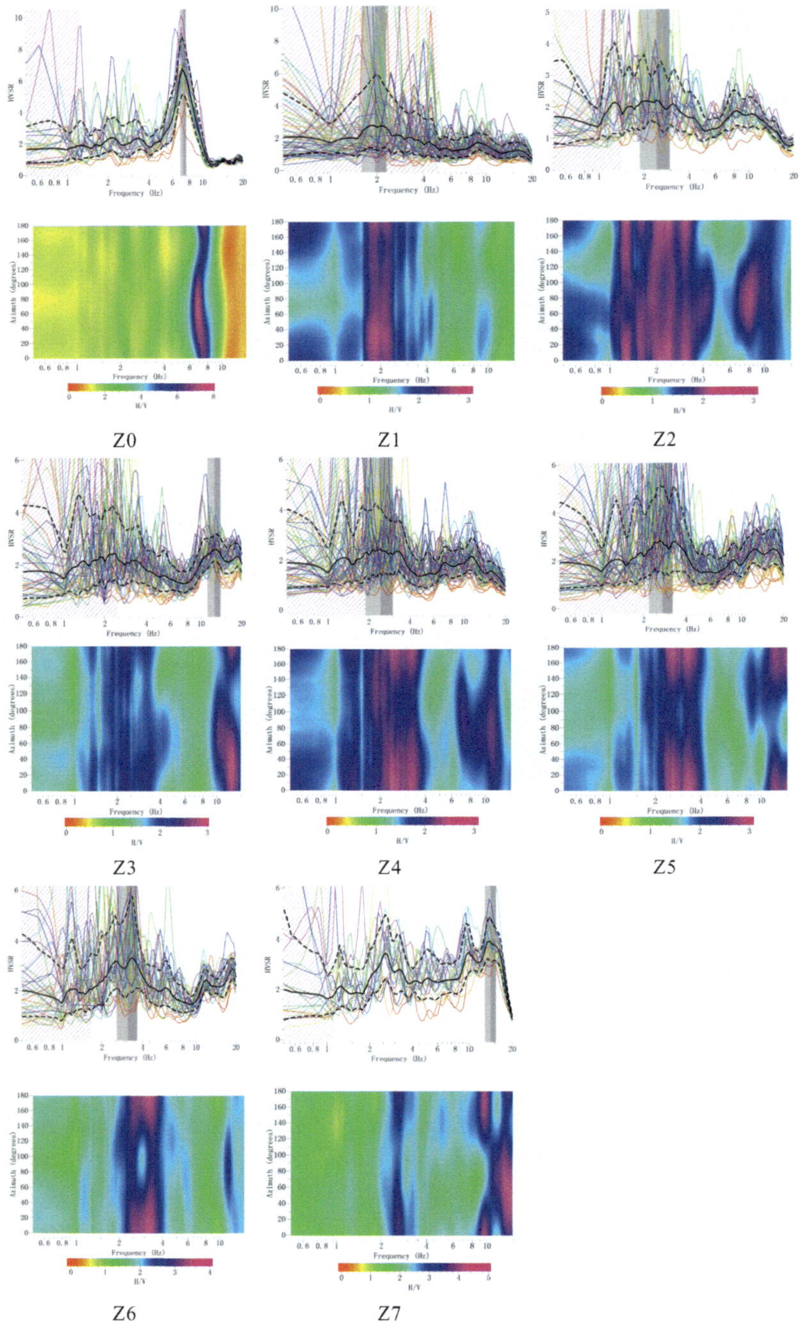

Fig. 8 The HVSR amplification and azimuth characteristics of Z0–Z7

3 Polarization Effects of Mountain Top (Hilltop)

(1) **Jiuzhaigou monitoring sites**

After the "8.8" Jiuzhai Ms 7.0 earthquake, 2017, we had installed two stations to observe the ground motion on the Jubao mountain close to the city, where the ground surface of the hilltop was affected by cracks after the earthquake (Fig. 9). The J1 is located at the slope foot, at an elevation of 1414 m, where the bedrock is made of outcropping weathered slate. The J2 station located on the hilltop (1551 m), where moderately to weakly weathered slate crops out (Fig. 10). Table 6 lists basic information of five typical aftershocks, those are of small magnitudes with an epicentral distances of 37–45 km. Figure 10 presents the monitoring profile and a typical recording.

Table 7 presents the PGA values and dominant frequencies at J1 and J2 for five recordings. With J1 as reference, the PGA amplification factor of J2 shows more than 10.0 in the EW component at 7.7 Hz, but the NS component is amplified less than 2.5 times. A significant shaking polarization in EW direction can be observed for each recording. Further analysis of the normalized Ia and the mean HVSR amplification for sites (Figs. 11 and 12), shows that the J1 presents no polarization regular, but the site J2 shows EW directivity amplification with Ia shape liking a peanut.

The directivity amplification effects in the Mt. Dong and Weigan hill slope analysis show that the reason of this phenomenon probably is the redistribution of the seismic energy by the influence of topography, geological structure and other factors (Luo et al. 2014). The J2 site of JuBao Mountain is a NS elongation ridge, where

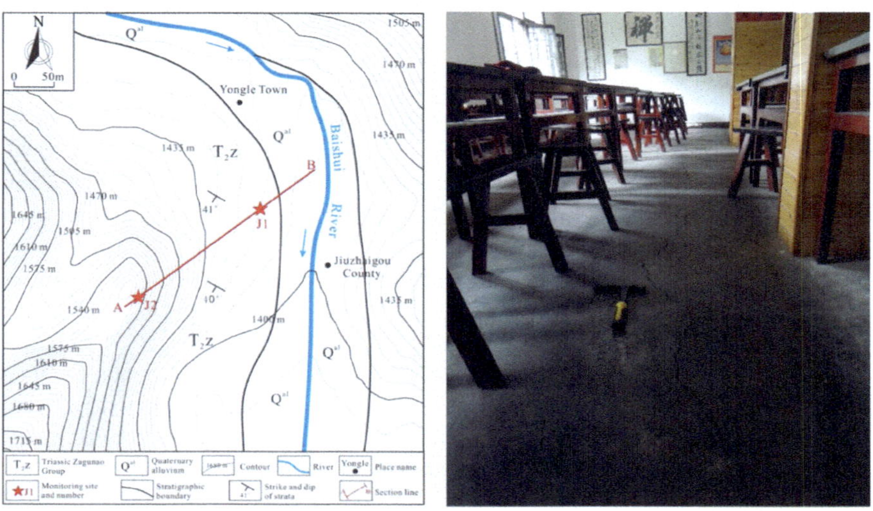

Fig. 9 Geological map (up left) and cracks causing by the earthquake at Jubao Mountain (up right) (modified from Shen et al. 2019)

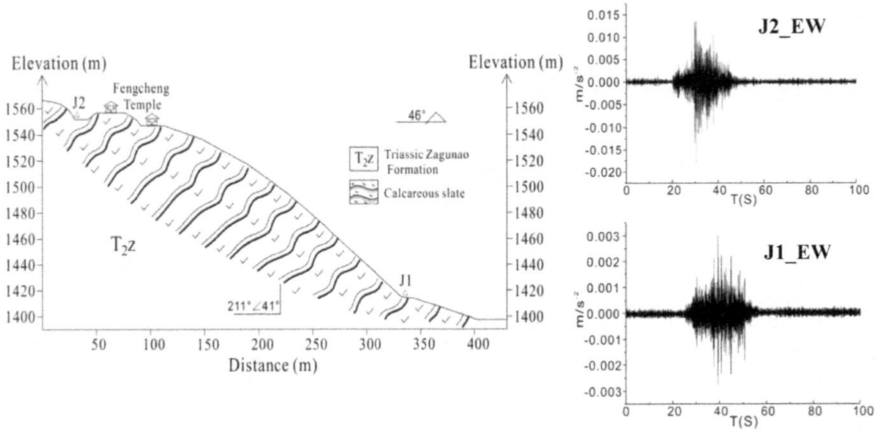

Fig. 10 The monitoring profile (up left) and typical acceleration waves (up right) of monitoring aftershock on August 14th 2017 (modified from Shen et al. 2019)

Table 6 The basic information of the aftershocks

Date	Time	Latitude (°)	Longitude (°)	Depth (km)	Epicentral distance (km)	Magnitude (M_L)
11/08/2017	19: 26: 21	103.86	33.12	16	38.0	3.5
12/08/2017	07: 56: 39	103.86	33.11	16	38.8	3.7
13/08/2017	22: 38: 31	103.89	33.09	16	37.4	3.4
14/08/2017	16: 16: 51	103.83	33.20	14	38.0	2.6
27/08/2017	08: 58: 13	103.75	33.22	15	45.0	3.3

the topography probably affects seismic energy redistribution to the EW component. Regarding the Polar diagram of HVSR (Shen et al. 2019), J1 shows a weak amplification effect in the frequency range of 8–12 Hz, but J2 shows a significant amplification at 3, 7 and 8–9 Hz (Fig. 12). And J2 has a directivity amplification consistent with the Ia analysis.

(2) **Luding monitoring sites**

The Luding monitoring site is situated in Lengzhuguan village, on the right bank of the Dadu River, approximately 17 km from Luding County, Sichuan Province, in Southwest China. The nearest active faults are about 15–17 km from the study area, and this area is one of the most seismically active regions of China (Allen et al. 1991). The main lithological units are characterized by plagioclase granite and quartz diorite (Jinning-Chengjiang period in Proterozoic time). Additionally, the bedrock is overlain by a layer of colluvium. After the "5.12" Wenchuan earthquake, we had installed two stations on the right bank of Lengzhuguan gully (Fig. 13). S1 site is on an alluvium layer with thickness of

Table 7 PGA and dominant frequency characteristics at each monitoring site of aftershocks

Monitoring site	11/08/2017 M_L 3.5					
	PGA ($m \cdot s^{-2}$)			Dominant frequency (Hz)		
	EW	NS	UD	EW	NS	UD
J1(1414 m)	0.003	0.017	0.014	12.04	12.06	12.55
J2(1551 m)	0.043	0.021	0.021	7.72	4.37	12.3
AF(J2/J1)	14.33	1.24	1.5			
Monitoring site	12/08/2017 M_L 3.7					
	PGA ($m \cdot s^{-2}$)			Dominant frequency (Hz)		
	EW	NS	UD	EW	NS	UD
J1(1414 m)	0.006	0.033	0.028	16.21	15.95	19.36
J2(1551 m)	0.13	0.073	0.08	7.37	10.28	11.8
AF(J2/J1)	21.67	2.21	2.86			
Monitoring site	13/08/2017 M_L 3.4					
	PGA ($m \cdot s^{-2}$)			Dominant frequency (Hz)		
	EW	NS	UD	EW	NS	UD
J1(1414 m)	0.004	0.009	0.008	17.47	14.71	14.71
J2(1551 m)	0.038	0.011	0.01	5.3	6.2	6.31
AF(J2/J1)	9.5	1.22	1.25			
Monitoring site	14/08/2017 M_L 2.6					
	PGA ($m \cdot s^{-2}$)			Dominant frequency (Hz)		
	EW	NS	UD	EW	NS	UD
J1(1414 m)	0.003	0.01	0.008	17.88	17.5	15.7
J2(1551 m)	0.018	0.012	0.013	8.21	5.86	8.23
AF(J2/J1)	6.0	1.2	1.63			
Monitoring site	27/08/2017 M_L 3.3					
	PGA ($m \cdot s^{-2}$)			Dominant frequency (Hz)		
	EW	NS	UD	EW	NS	UD
J1(1414 m)	0.0045	0.009	0.0126	14.73	14.61	15.01
J2(1551 m)	0.017	0.016	0.0127	5.15	4.85	8.67
AF(J2/J1)	3.78	1.78	1.01			

more than 3 m, and it locates at the hilltop with convex topography characteristics (Fig. 14). The S2 site locates in the middle of the slope, the lithology is of colluvium deposits with 1.5–2 m thickness (Fig. 14).

In the past 6 years, both stations recorded three strong earthquakes, those are Lushan Ms 7.0, Kangding Ms 6.3 and Kangding Ms 5.8, respectively. Table 8 lists the PGA values of each monitoring site. The national strong seismic monitoring station (The number is 51KDZ) near the study area at about 5 km, and is the reference station.

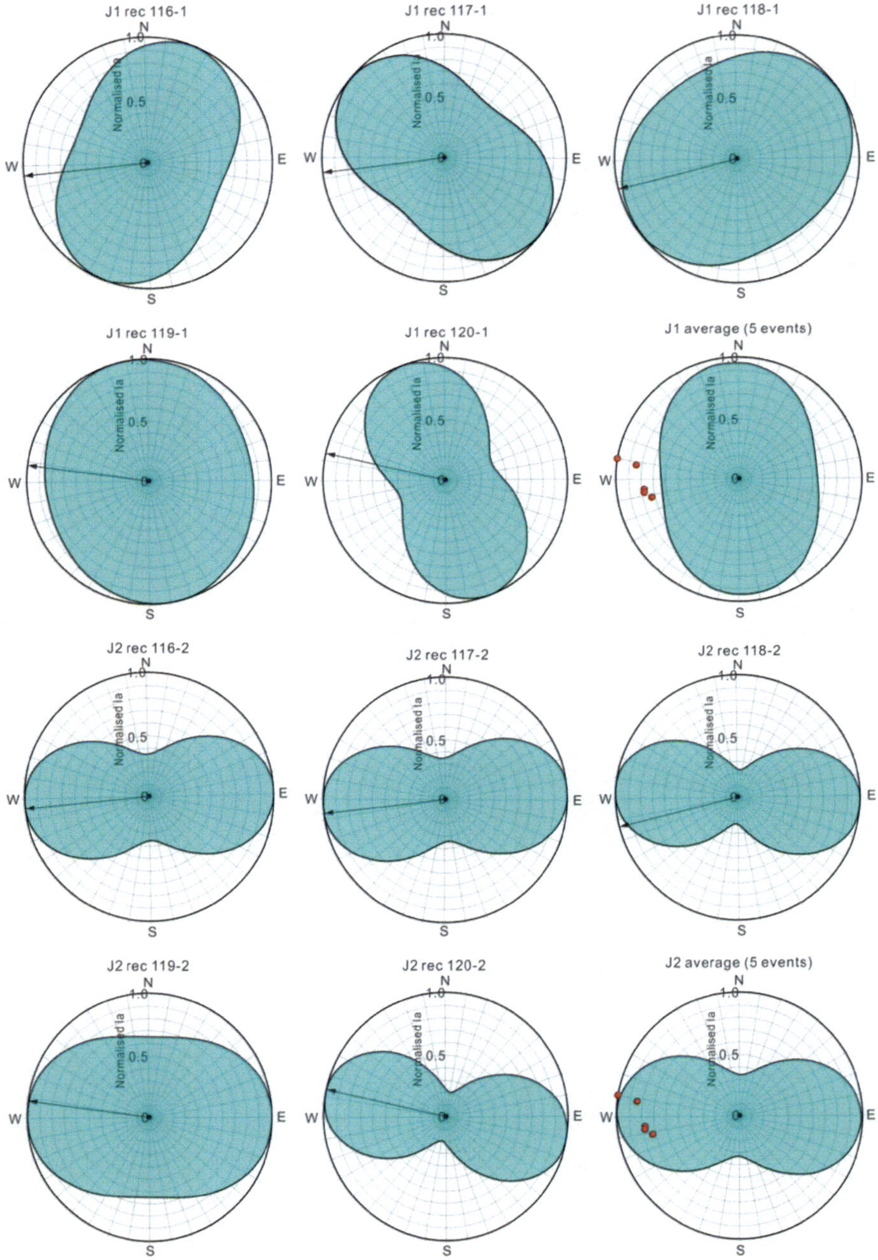

Fig. 11 Polar diagrams of the normalized Ia (the maximum is equal to 1) calculated for different monitoring sites during the aftershocks; the event date, magnitude and epicentral distance are reported in Table 6. Arrows mark the event source back-azimuth (station-epicenter azimuth). Diagrams in the right-most column show the mean normalized Arias intensities calculated along each direction from all the events recorded at each station: dots represent the relative positions of the epicenters of the events with respect to the site (back-azimuth and distance; the latter normalized by the maximum distance equated to the polar diagram radius) (modified from Shen et al. 2019)

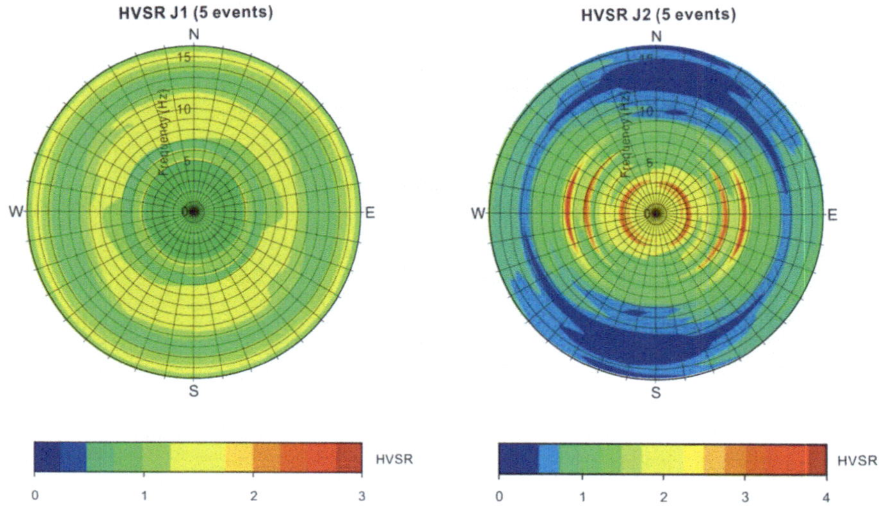

Fig. 12 Polar diagrams of the horizontal to vertical spectral ratio (HVSR) values for stations J1 and J2 (modified from Shen et al. 2019)

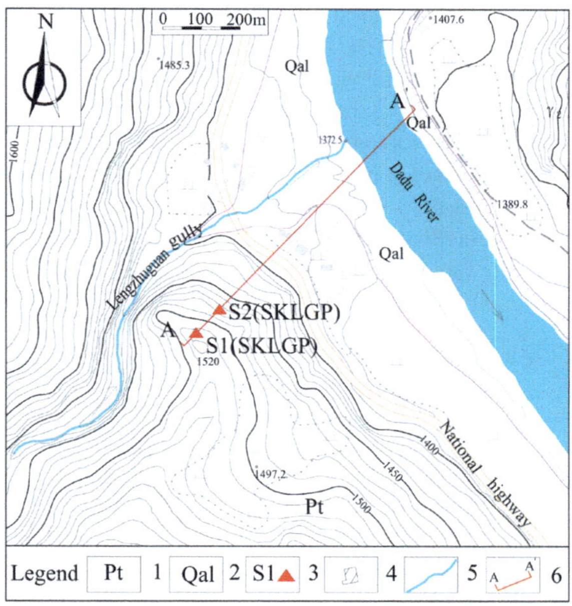

Fig. 13 Topographic and geological sketch map of the Lengzhuguan slope area: *1* Plagioclase granite and quartz diorite in Proterozoic time; *2* Alluvial deposits in Quaternary period; *3* Monitoring station; *4* Residential area; *5* River; *6* Geological section (modified from He et al. 2020 with author rights in Elsevier)

The amplified results show that the largest PGA amplification factor of S1 site is more than 10 times in the NS component, while the EW and UD (up-down) components also have high amplification factors (Table 8). In S2, the PGA amplification factor

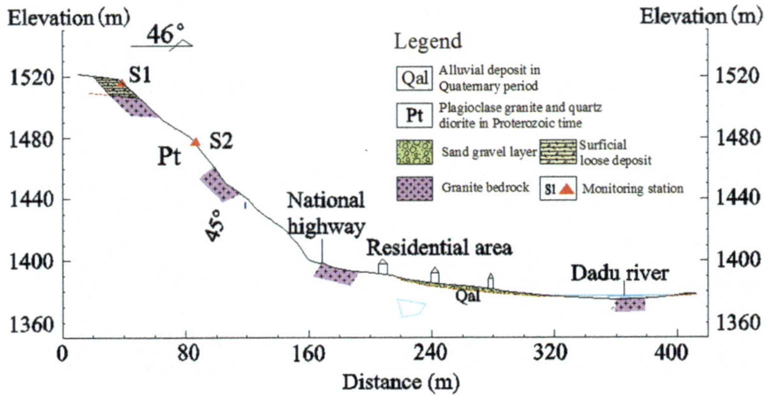

Fig. 14 Geological profile of monitoring stations installed along the slope (modified from He et al. 2020 with author rights in Elsevier)

is less than 6.0. Comparing both monitoring sites, S1 site had a higher amplification effect than the S2 site.

The mean horizontal to vertical Fourier spectra of S1, S2 and 51KDZ is shown in Fig. 15. The spectral ratio amplitude of S1 and S2 exhibit peak values of approximately 11 and 6, in the frequency range of 1.5–2.5 Hz, while station 51KDZ displays a steady state in the whole frequency range; The largest amplification of S1 and S2 occurs at the 50° azimuth clockwise from North, which is approximately perpendicular to the elongation direction of the ridge (see Fig. 13); in contrast, the seismic amplification of station 51KDZ is slightly affected by the directional resonance. The analysis results from the H/V also show that the S1 spectral ratio amplification is much higher than the S2, it agrees with the PGA amplification. The slope amplification only reproduced by a heterogeneous model containing with deposit layers (He

Table 8 Obtained PGA of stations from three strong earthquakes

Events	Stations	PGA ($cm{\cdot}s^{-2}$)			PGA amplification factor (refer to the 51KDZ)		
		EW	NS	Vertical	EW	NS	Vertical
Lushan Ms 7.0	S1	163.5	153.8	66.7	6.96	5.70	3.39
	S2	42.4	39.3	29.0	1.80	1.46	1.47
	51KDZ	23.5	27.0	19.7	Reference station		
Kangding Ms 6.3	S1	188.1	147.6	111.9	9.50	10.70	7.72
	S2	70.4	69.9	36.5	3.56	5.07	2.52
	51KDZ	19.8	13.8	14.5	Reference station		
Kangding Ms 5.8	S1	82.3	85.9	74.7	6.69	8.03	8.69
	S2	19.3	29.1	24.1	1.57	2.72	2.80
	51KDZ	12.3	10.7	8.6	Reference station		

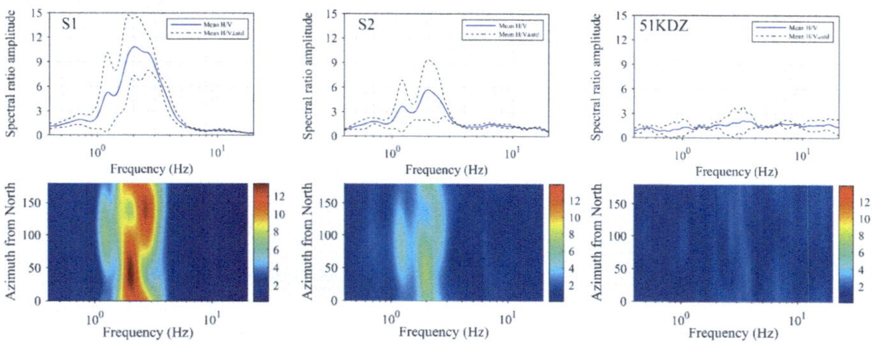

Fig. 15 Top left: mean Fourier spectra H/V and with ±1 std of S1; bottom left: average directional H/V of S1. Top middle: mean Fourier spectra H/V and with ±1 std of S2; bottom middle: average directional H/V of S2. Top right: mean Fourier spectra H/V and with ±1 std of 51KDZ; bottom right: average directional H/V of 51KDZ (modified from He et al. 2020)

et al. 2020). So, we inferred that the combination amplification effects of topography and deposits were the main reason for the strong amplification. But the deflection polarization effects are probably mostly influenced by the elongation direction of mountain ridge of S1 site.

At station of 51KDZ, which presents less amplified as a reference for the stand spectral ratio (SSR) analysis. In the Fig. 16, it can be observed that the spectral ratio amplitude of S1 is about 16.3 at 2 Hz, but this is only 5.4 for the S2. Additionally, the amplified frequency of S1 and S2 is in a wide range of 1.5–6 and 1.5–3 Hz. And the directivity amplification effect is significant at the S1 site, consistently with the H/V analysis.

(3) **Qingchuan monitoring sites**

The Qingchuan monitoring sites locate at the Qiaozhuang town, Qingchuan County, in the northeast of Sichuan province. It is about 250 km from the epicenter of Wenchuan earthquake, and the nearest surface rupture is about 25 km from the Qingchuan monitoring sites. Strong earthquake energy induced the slopes of Mt. Dong, Mt. Shizi and Weigan hill cracking or failure around the Qiaozhuang town. After the Wenchuan main shock, a large number of aftershocks occurred near the Qiaozhuan town, so we chose those slopes for monitoring the slope seismic response. 11 stations were successively set at those slopes (Fig. 17). The basic parameters of each monitoring site are in Table 9.

From 2009 to 2016, those stations monitored more than 200 aftershocks, the magnitude is between M_L 1.3–Ms 5.2 with the source depth less than 30 km. All the recordings are in the distance of 2.0–46.0 km.

(1) **Mt. Dong**

In Mt. Dong, six stations (Fig. 18) are distributed from the slope foot to the mountain top, three stations (Q3, Q4, Q6) were put in a short tunnel (<20 m),

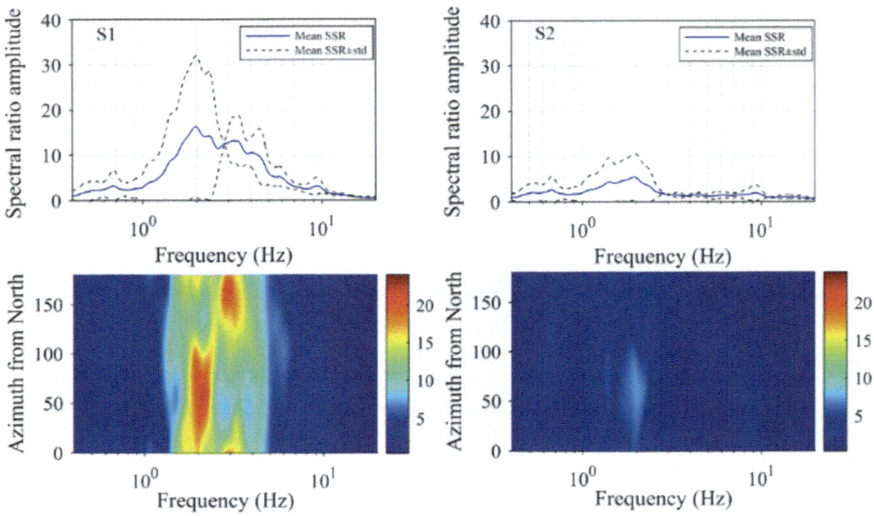

Fig. 16 Top left: mean horizontal SSR of S1; bottom left: mean directional SSR of S1. Top right: mean horizontal SSR of S2; bottom right: mean directional SSR of S2 (modified from He et al. 2020)

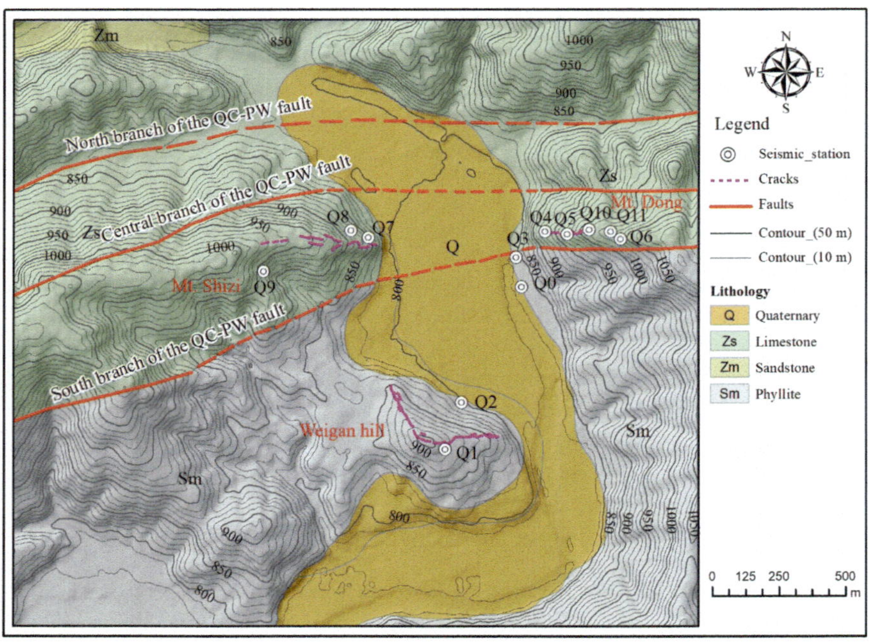

Fig. 17 The geological sketch map and the distribution of monitoring station at Qingchuan

Table 9 The parameters of the monitoring sites (modified from Luo et al. 2011)

Number	Elevation (m)	Geological condition	Location
Q0	780	Overlay (Sand and gravel)	Qiaozhuang town
Q1	875	Bedrock (Phyllite)	Weigan hill
Q2	805	Bedrock (Phyllite)	Weigan hill
Q3	788	Bedrock (Phyllite)	Mt. Dong
Q4	871	Bedrock (Phyllite)	Mt. Dong
Q5	921	Bedrock (Dolomite)	Mt. Dong
Q6	962	Bedrock (Dolomite)	Mt. Dong
Q10	960	Bedrock (Dolomite)	Mt. Dong
Q11	981	Bedrock (Dolomite)	Mt. Dong
Q7	902	Bedrock (Dolomite)	Mt. Shizi
Q8	893	Bedrock (Dolomite)	Mt. Shizi
Q9	962	Bedrock (Dolomite)	Mt. Shizi

and other stations are installed on the slope surface. Referred to the Q3 site, four typical aftershocks analysis indicate that the PGA amplification factor is vary nonlinearly along the slope, where the Q10 has the strongest amplification (Fig. 19). The north–south component shows higher amplification effects than the east–west and vertical components. The largest amplified factor of PGA is 4.64.

Regarding to the normalized Ia, the mean statistics analysis of Q3, Q4, Q5, Q6, Q10 and Q11 is from 4 to 100 events (Fig. 20). Results indicate that the ground motion is polarized in the direction of 60°–45°at Q3, Q4 and Q6 sites, 350°–5° at Q5, Q10 and Q11 sites. The former polarization direction is parallel to the slope inclination and the latter is transverse to the elongation of the ridge (Fig. 17). Figure 21 presents the polar diagrams of HVSR, it can be observed that Q3 has a small amplification in the high frequency (>13 Hz), Q4 is amplified in the frequency range of 6–14 Hz, Q5, Q10 and Q11 have the same amplified frequency of 2–5 and 6–10 Hz, Q6 has small amplification in the range of 2–9 Hz. On the other hand, the polar diagrams of HVSR analysis shows the same polarization direction effects as the Ia results (Fig. 21). The reasons of the polarization preliminary analysis by Luo et al (2014)

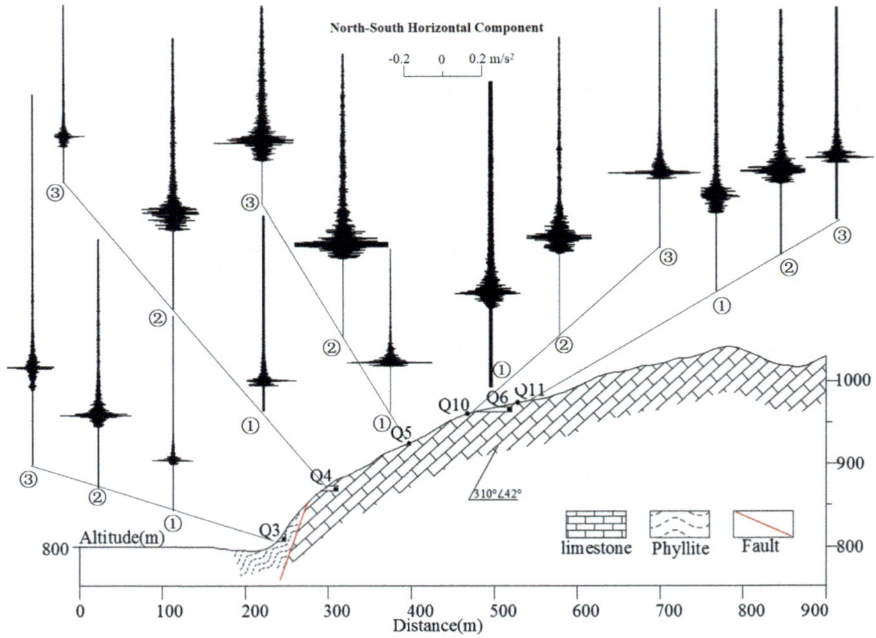

Fig. 18 The geological profile of Mt. Dong, six stations for aftershock recordings of M_L 2.7 (06–05-2014, ①), M_L 3.2 (08–11-2014, ②) and M_L 3.8 (08–22-2014, ③) (modified from Luo et al. 2020 with author rights in Elsevier)

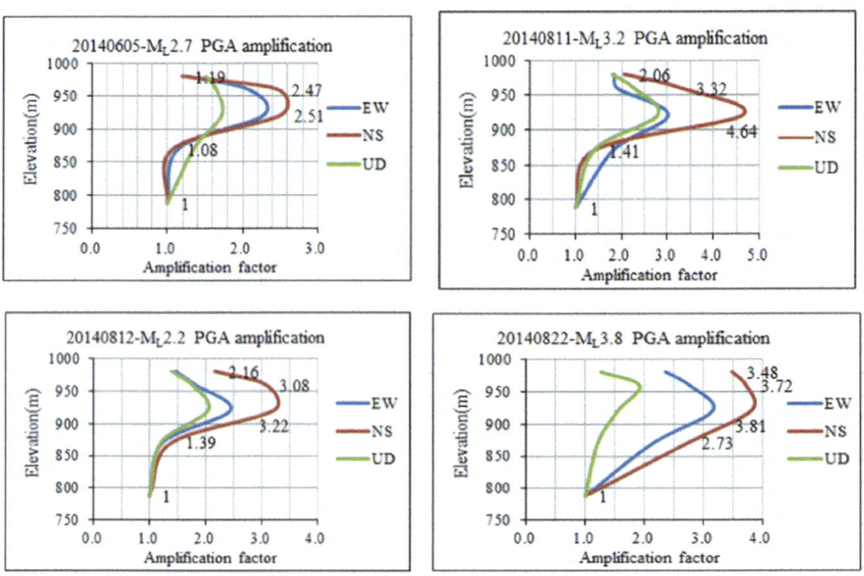

Fig. 19 PGA amplification factor of Q4, Q5, Q10 and Q11

were the terrain and geological features of local site. But at the top of Mt. Dong Mountain, the polarization transverse to the ridge elongation is mainly controlled by the topography characteristics. Regarding to the high amplification effects at the upper part of the slope, Luo et al (2020)., constructed 2D and 3D models analysis, which indicated that the combined topography with layer amplification are the main contributions, and in particular, the strong amplification occurs at the break position of the slope, convex mountain top and steep slopes. It means that the slope topography characteristics is the first reason for occurring high amplification, but these effects only combined with geological factors can induce dynamic failure or cracking.

(2) **Weigan hill**

In Weigan hill, two stations were installed on the hill top (Q1) and slope foot (Q2), on outcropping relatively hard phyllite, another station of Q0 was installed on alluvium deposit (Fig. 22). From 2009 to 2010, both stations recorded more than 40 aftershocks of Wenchuan earthquake, Fig. 22 presents the geological profile and typical recordings for those monitoring stations. Luo et al (2010) analysis revealed that the PGA amplification factor of the Q1 (Q2 is the reference site) is about 7–8 in horizontal NS component, and this polarization amplification of NS shows the dominant frequency value of FFT only in the range of 2–4 Hz. For further analysis site effects, the normalized Ia and polarization diagrams of HVSR are shown in the Figs. 23 and 24. It can be observed that the Ia presents different types, such as Q0 is ellipse, Q2 is a small change of ellipse, but the Q2 is a standard peanuts polarization in the direction of 0° (Fig. 23).

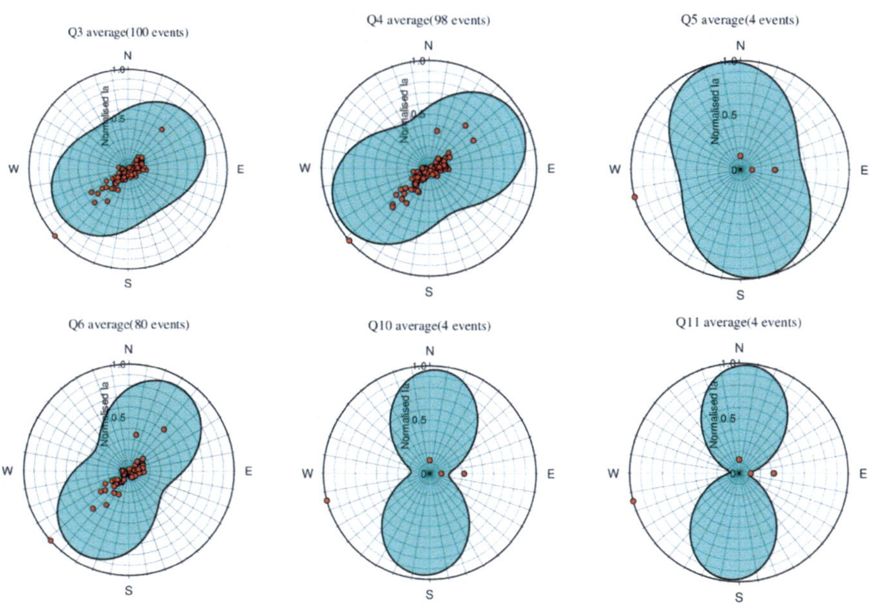

Fig. 20 The normalized Ia of Q3, Q4, Q5, Q10 and Q11 (Mt. Dong)

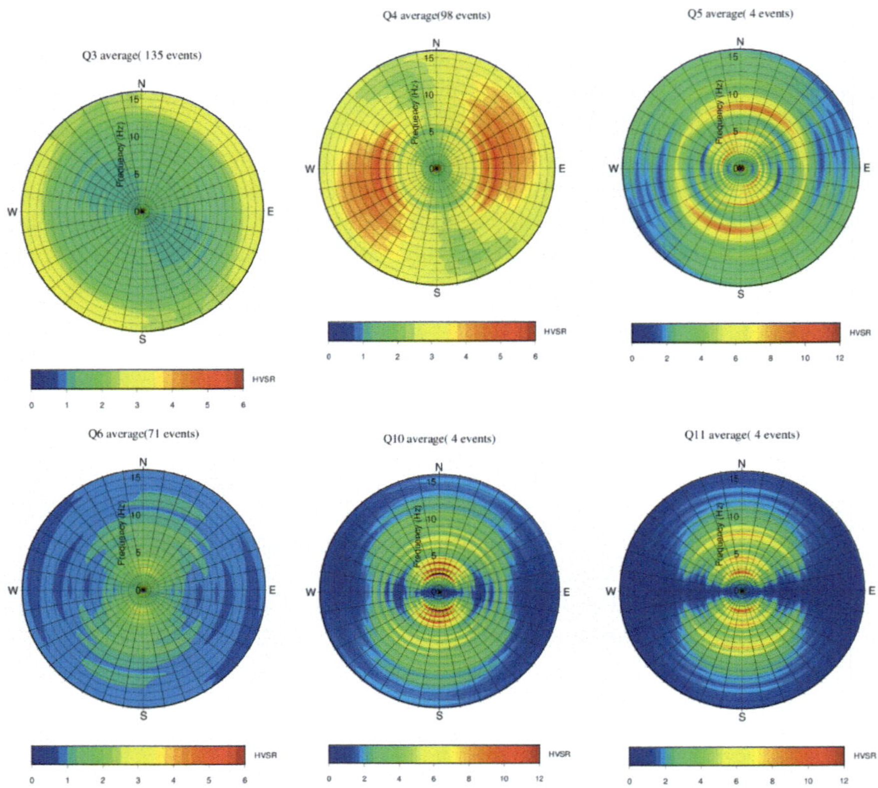

Fig. 21 Polarization diagrams of HVSR analysis of Q3, Q4, Q5, Q10 and Q11 (Mt. Dong)

Polarization diagrams of HVSR of three sites present the amplified frequency in different range (Fig. 24). In Q0, the amplification effects occur in the high frequency range of 11–13 Hz, and this occurs between 8 and 12 Hz in Q2, but the amplification effects of Q1 is in a range from 2 to 4 Hz consistent with the FFT analysis results. On the other hand, the polarization diagrams of HVSR of Q1 shows significant polarization direction in the NS, transverse to the ridge extension, consistently with the Ia analysis results. This shows the same polarization phenomenon as the J2 at Jiuzhaigou monitoring site and the Q10 and Q11 of Mt. Dong mountain top monitoring sites. Therefore, we have enough monitoring evidence to prove that a ridge will produce polarization direction effects perpendicular to the extension of the ridge. This polarization effects are probably the main reason of cracking along the ridge during the earthquake.

(3) **Mt. Shizi monitoring sites**

In Mt. Shizi, the Wenchuan earthquake induced a nearly 1 km length crack along the mountain ridge (Luo et al. 2014). Three stations (Q7, Q8 and Q9) were set in middle of the slope and mountain top (Fig. 17). Both Q8 and Q9 stations

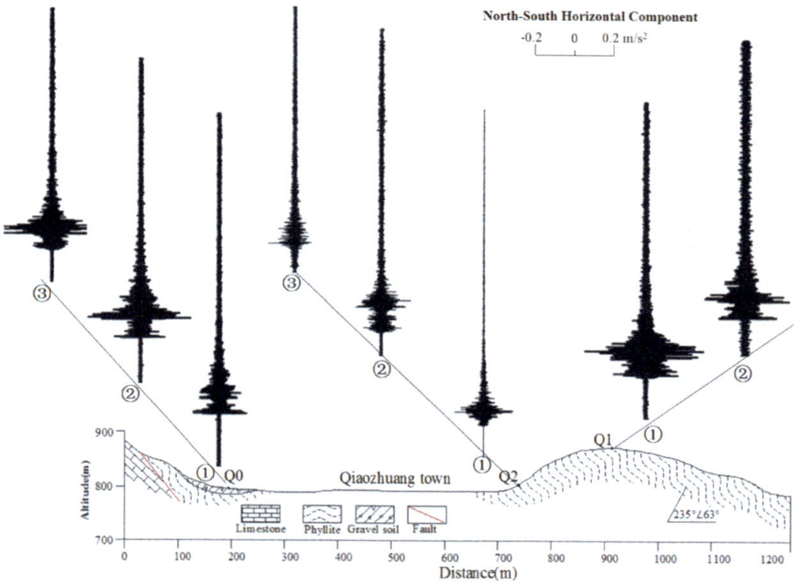

Fig.22 The geological profile of Weigan hill, three stations for aftershock recordings of M_L 2.7 (06–05-2014, ①), M_L 3.2 (08–11-2014, ②) and M_L 3.8 (08–22-2014, ③) (modified from Luo et al. 2020 with author rights in Elsevier)

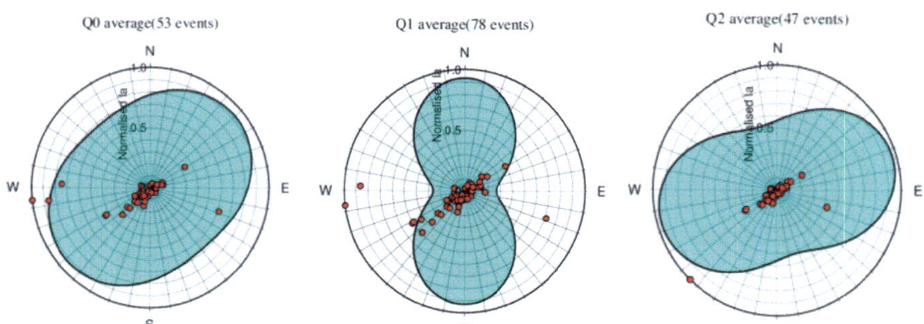

Fig. 23 The normalized Ia of Q0, Q1 and Q2 (Weigan hill)

were put in the short tunnel (<20 m), Q7 was set at the surface around 2014, but got few recordings. Figure 25 presents the normalized Ia statistics analysis the magnitudes between M_L 1.9 and Ms 5.2, it can be observed that in Q8, Ia in different magnitude intervals has a stable directivity in 60°, in Q9 the directivity is in a range of 60°–80°.

Regarding to the HVSR, Fig. 26 presents that the Q8 has a stronger amplification in the frequency range of 5–6 and 9–10 Hz, the Q9 has an amplification in the frequency range of 2–3, 7–8 and 10–12 Hz. Besides, between the M_L 1.4 to Ms

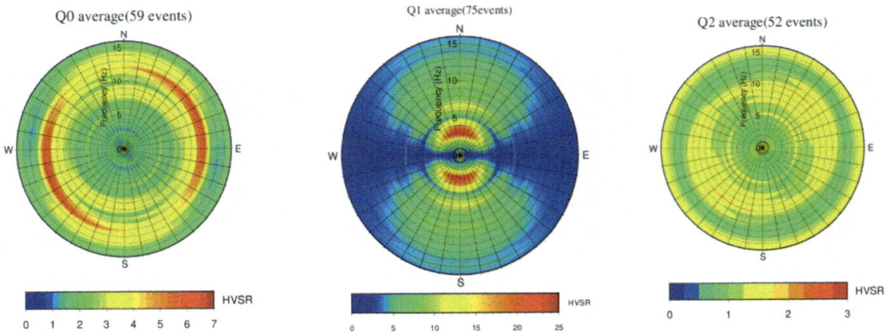

Fig. 24 Polarization diagrams of HVSR analysis of Q0, Q1 and Q2 (Weigan hill)

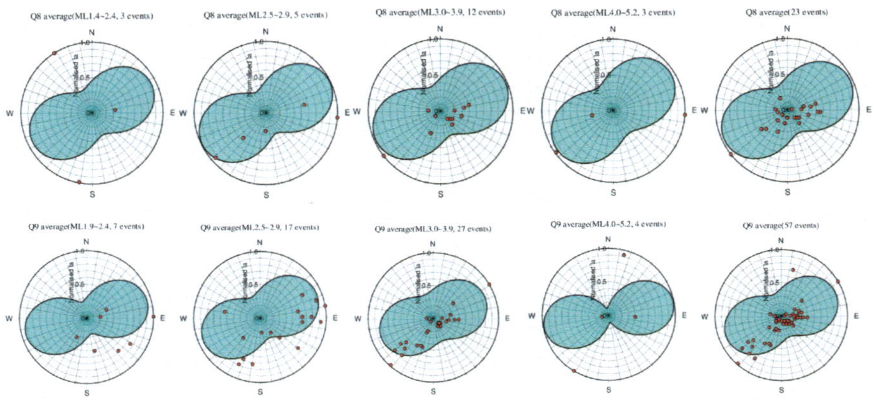

Fig. 25 The normalized Ia of Q8 and Q9 (Mt. Shizi)

5.2, the HVSR characteristics of different magnitude intervals were amplified higher with increasing seismic energy, in particular in the low frequency range (~5 Hz). This cannot be observed by the mean analysis of HVSR for all recordings. Certainly, we should pay more attention to strong energy recordings although difficult to record those events, which are always the main reason for triggering seismic failure.

4 Dynamic Response Including Fault Zone in Mianzhu Monitoring Sites

Mianzhu monitoring sites are located at Jiulong town, Mianzhu County, near the boundary between the Sichuan basin and the Longmenshan Mountains. The frontal thrust (Peng-Guan fault) was active during the Wenchuan earthquake, where the ground surface was uplifted up to 2–3 m, inducing a serious damage inside the Jiulong

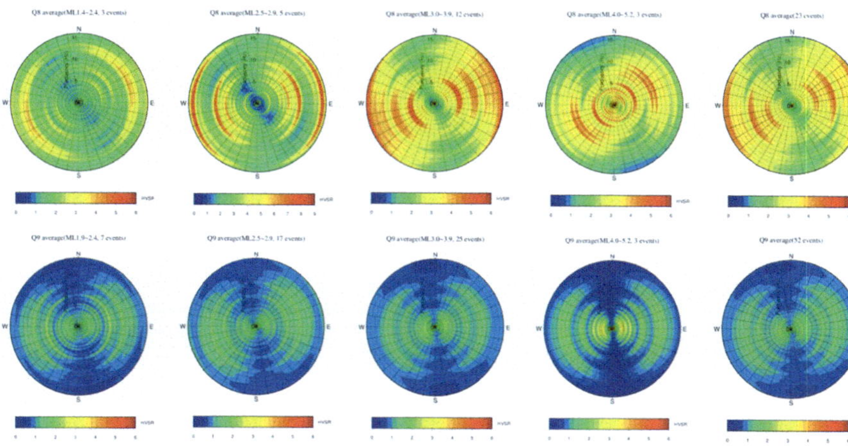

Fig. 26 Polar diagrams of HVSR calculation of different magnitude of Q8 and Q9 (Mt. Shizi)

town (Wang et al. 2012). After the main shock, three stations were installed to monitor the slope site seismic response (Fig. 28). Those were distributed at the sediments (M3), near the fracture (M2) and the hanging wall of the fault (M1) (Fig. 27).

Figure 28 presents the mean HVSR amplification of horizontal component in those three stations. It can be observed that the resonance frequency, in M1 and M2, is about 5 and 4 Hz, and there is no peak frequency for less than 6 Hz for M3. Regarding to the HVSR value, the EW component is larger than the NS component in the M1 and M2, this shows a certainly polarization dynamic slope response. The site of M1 is more amplified higher than the site of M2. In M3, the HVSR shows weak amplification in the frequency range of 6–10 Hz. Therefore, M3 is suitable as a reference station to analyze the amplification effects in low frequency. Figure 29

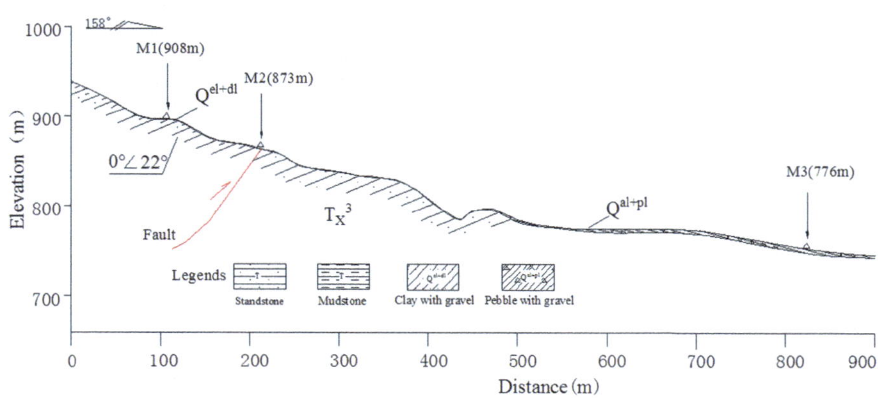

Fig. 27 Monitoring stations in Jiulong slope, Mianzhu County (modified from Luo 2011 with author rights)

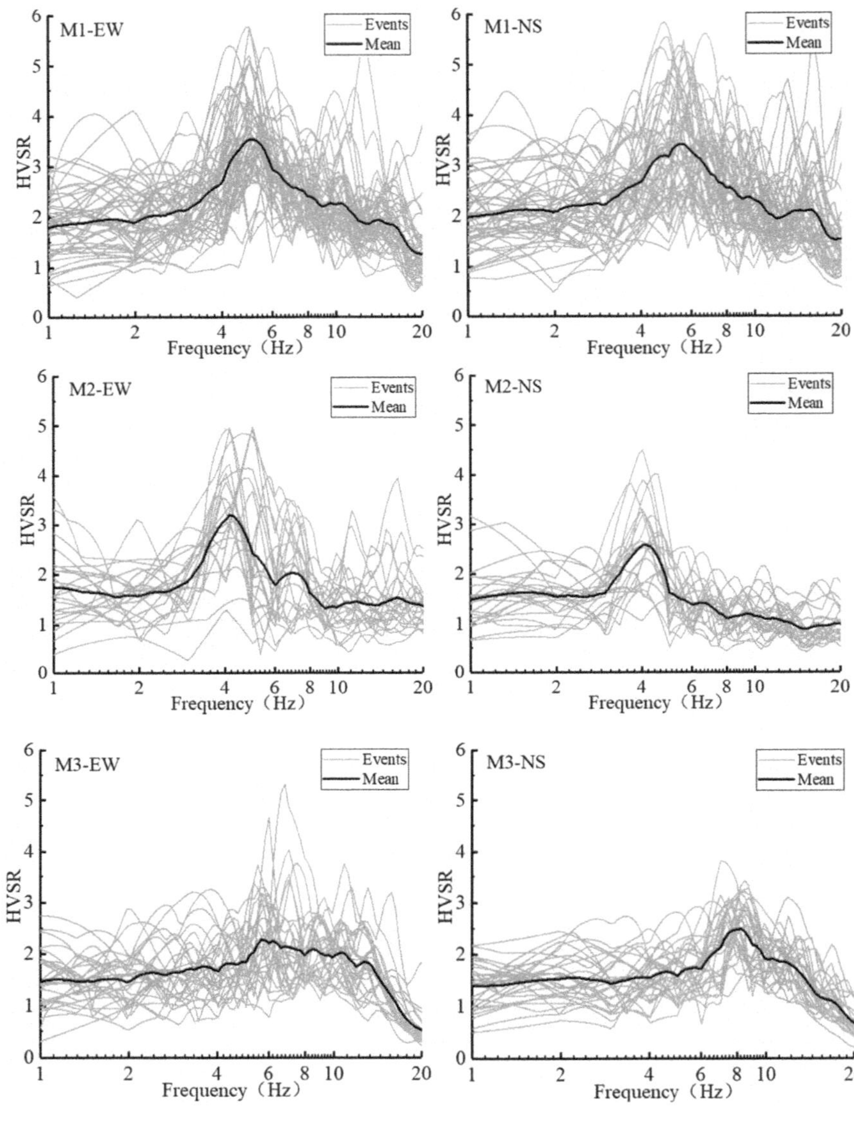

Fig. 28 HVSR amplification characteristics of M1, M2 and M3 (modified from Zhu 2020 with author rights)

presents the SSR amplification of M1 and M2. The peak value of SSR, in M1, is in the range of 3.5–3.8 at 5 Hz, which is amplified between 2 and 2.5 at 4.2 Hz in M2. On the other hand, the PGA is another parameter for analysis of the site amplification effects. Figure 30 presents the PGA amplified curve of recordings. It can be observed that the amplified curve shows a changing increasing type. Referred to the M3 site, the PGA amplified factor of M2 was between 0.42 and 1.31, but M1 is

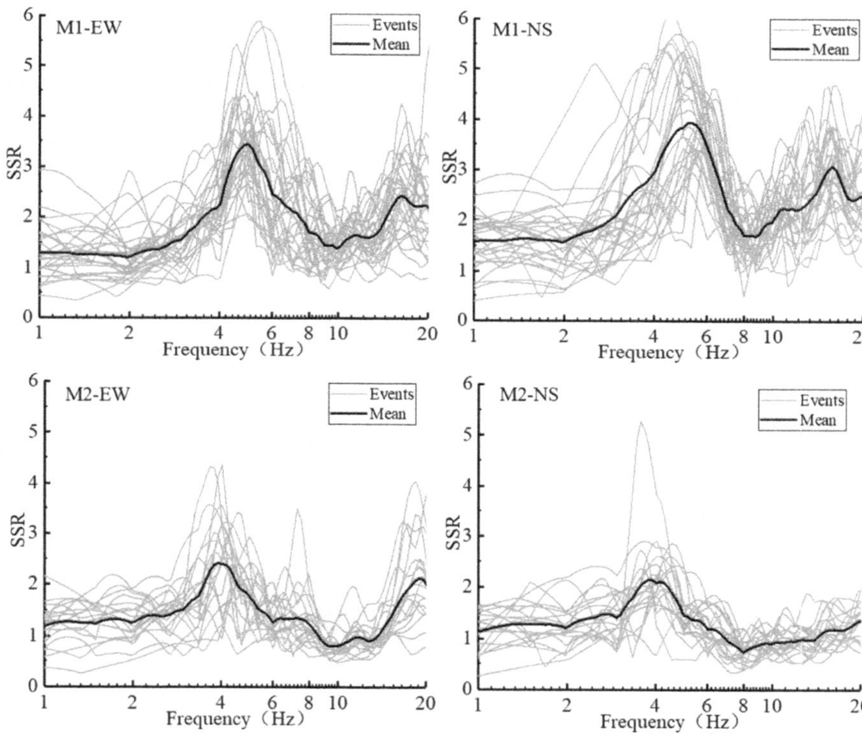

Fig. 29 SSR amplification characteristics of M1 and M2 (modified from Zhu 2020 with author rights)

in the range of 1.19–4.49 (Fig. 30). Comparing those site amplification observations, most recordings of M2 PGA amplification shows attenuation (the amplified factor less than 1.0, Fig. 30), while M1 shows high site amplification effects. Further analysis by numerical simulation revealed that the fault zone of M2 site has a significant attenuation effect on during the ground motion (Zhu 2020). A complex dynamic effect of wave propagation in the fault zone probably leads to this ground motion, which may be due to fault zone trapped seismic wave (Li and Leary 1990), the barrier effects (Zhou et al. 2006; Liu et al. 2008).

5 Discussion and Conclusions

In this report, several monitoring sites present the earthquake recordings analysis results of PGA, HVSR, SSR and Ia are presented for several monitoring sites; thereby, we can observe that ground motion can highly vary along a slope and also changes from one slope site to another. However, there are also similar patterns of slope

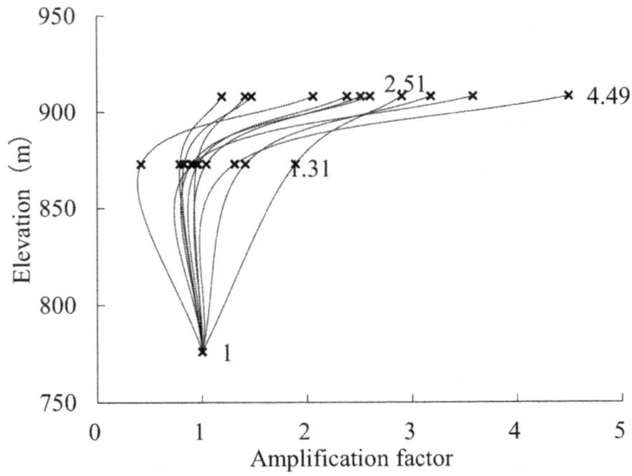

Fig. 30 The horizontal PGA amplification characteristics of M1 and M2 (modified from Luo 2011 with author rights)

amplified position, amplification effects in sediments (Shuanghe, Zigong), and slope combined amplification effects. The monitoring data revealed that the amplification factor of PGA is changing in a large range at the same monitoring site (e.g., Jiuzhaigou J2, Luding S1). When the amplitude of reference site is very low (such as less than 0.02 g), the amplification factor of slope upper part can reach more than 10, but when the amplitude of reference site was changing highly with the earthquake (such as the biggest PGA of reference site in this report is 0.076 at Shuanghe #1 site), the amplification factor was small (less than 4.0). This monitoring results indicate that the variability of slope site effects is more obvious in the small earthquake with low seismic intensity than the big earthquake with high seismic intensity. Along the intensity increasing of reference background site (such as the intensity is more than VII), the PGA amplification factor of slope site may be in a low range (such as less than 3.0), but the amplified effect of slope upper part is enough to induce failure for slope site. Otherwise, there are more complicate reasons for a slope dynamic failure in a low intensity of background site (intensity less than V). Thus, it is necessary to distinguish and discuss the slope dynamic effects of different intensity condition for ground motion results.

In this report, although a low intensity of reference site shows a complicated site amplification effect of PGA, the polarization direction and resonance frequency of slope site show little changing. Those characteristics of ground motion can be used to infer the slope site seismic response under strong earthquakes. On the other hand, part of the monitoring sites has been further analyzed by modeling for the influenced factors of site effects (such as the Mt. Dong Qingchuan, Mianzhu), other sites need combined monitoring and numerical simulation for deeper analysis of the factors influencing the site effects.

References

Allen CR, Luo ZL, Qian H et al (1991) Field study of a highly active fault zone: the Xianshuihe fault of southwestern China. Geol Soc Am Bull 103(9):1178–1199

Ashford SA, Sitar N, Lysmer J, Deng N (1997) Topographic effects on the seismic response of steep slopes. Bull Seismol Soc Am 87(3):701–709

Boore DM (1972) A note on the effect of simple topography on seismic SH waves. Bull Seismol Soc Am 62(1):275–284

Bouchon M (1973) Effect of topography on surface motion. Bull Seismol Soc Am 63(2):615–632

Çelebi M (1987) Topographical and geological amplifications determined from strong-motion and aftershock records of the 3 March 1985 Chile earthquake. Bull Seismol Soc Am 77(4):1147–1167

Davis LL, West LR (1973) Observed effects of topography on ground motion. Bull Seismol Soc Am 63(1):283–298

Del Gaudio V, Wasowski J (2011) Advances and problems in understanding the seismic response of potentially unstable slopes. Eng Geol 122:73–83

Gischig VS, Eberhardt E, Moore JR, Hungr O (2015) On the seismic response of deep-seated rock slope instabilities-Insights from numerical modeling. Eng Geol 193:1–18

Graizer V (2009) Low-velocity zone and topography as a source of site amplification effect on Tarzana hill, California. Soil Dyn Earthq Eng 29(2):324–332

Hailemikael S, Lenti L, Martino S, Paciello A, Mugnozza GS (2016) Ground-motion amplification at the Colle di Roio ridge, central Italy: a combined effect of stratigraphy and topography. Geophys J Int 206(1):1–18

Havenith HB, Jongmans D, Faccioli E, Abdrakhmatov K, Bard P-Y (2002) Site effect analysis around the seismically induced Ananevo rockslide, Kyrgyzstan. Bull Seismol Soc Am 92(8):3190–3209

Havenith HB, Vanini M, Jongmans D, Faccioli E (2003) Initiation of earthquake-induced slope failure: influence of topographical and other site specific amplification effects. J Seismolog 7(3):397–412

He JX, Qi SW, Wang YS, Saroglou C (2020) Seismic response of the Lengzhuguan slope caused by topographic and geological effects. Eng Geol. https://doi.org/10.1016/j.enggeo

Huang JG (2020) Research on seismic dynamic response characteristics of ridge topography slope of Xishan Park in Zigong. Chengdu University of Technology, Chengdu (in Chinese)

Jin G, Wang YS, Shi BX, Luo Y, Ming WT (2019) Characteristics of seismic response of the 6.0-magnitude earthquake, Changing county of Yibin in Southwest China's Sichuan province. J Mountain Sci 37(6):943–954 (in Chinese)

Li YG, Leary PC (1990) Fault zone trapped seismic waves. Bull Seismol Soc Am 80(5):1245–1271

Liu JB, Liu XQ, Zhao DD (2008) Effects of vertical fault fracture zones on the spreading of Rayleigh wave and ground motion. Explosion Shock Waves 28(6):507–514

Luo YH (2011) Study on complex slopes response law under earthquake action. Chengdu University of Technology, Chengdu (in Chinese)

Luo YH, Wang YS, Wang FH, Deng X (2010) Monitoring of slope seismic response during aftershocks of Wenchuan earthquake in Qingchuan County. J Eng Geol 18(1):27–34

Luo YH, Del Gaudio V, Huang RQ et al (2014) Evidence of hillslope directional amplification from accelerometer recordings at Qiaozhuang (Sichuan-China). Eng Geol 183:193–207

Luo YH, Fan XM, Huang RQ, Wang YS, Yunus Ali P, Havenith HB (2020) Topographic and near-surface stratigraphic amplification of the seismic response of a mountain slope revealed by field monitoring and numerical simulations. Eng Geol. https://doi.org/10.1016/j.enggeo.2020.105607

Luo YH, Xu Q, Zhan WW, Grelle G (2011) Seismic hazard prediction using multispectral amplification maps in a complex topographic area: a case study of Qiaozhuang town, Sichuan province, Southwest China. J Mountain Sci 18. https://doi.org/10.1007/10.1007/s11629-021-6837-9

Luo Y, Wang Y, Fu R, Liu J (2013) Study on mountain slopes topographic effect induced by Wenchuan earthquake. Global View Eng Geol Environ 73–80

Ministry of housing and urban rural development of the people's Republic of China general administration of quality supervision, inspection and Quarantine of the people's Republic of China (2010) Code for seismic design of buildings, GB 50011-2010, 19

Nakamura Y (2000) Clear identification of fundamental idea of Nakamura's technique and its applications. In: Proceedings of 12th WCEE, Auckland, paper no. 2656

Shen T, Wang YS, Luo YH et al (2019) Seismic response of cracking features in Jubao Mountain during the aftershocks of Jiuzhaigou Ms7.0 earthquake. J Mountain Sci 16(11):2532–2547

Wang YS, Luo YH, Wang FH, Wang D, Ma X, Li S, Deng X (2012) Slope seismic response monitoring on the aftershocks of the Wenchuan earthquake in Mianzhu section. J Mt Sci 9:523–528

Wang YS, Liu JW, Zhao B, Luo YH, Ming WT, Luo Y, Jing G, Zhou YH (2019) Response characteristics of slope seismic to Gongxian Ms 5.4 earthquake in Sichuan, China. J Earth Sci Environ 41(5):613–622 (in Chinese)

Zhou ZH, Wen RZ, Mao GB, Wang YS, Wang YH, Yang C (2006) Effect of fault on ground motion in the case of incident Rayleigh wave. Earthq Eng Eng Vib 26(5):1–6 (in Chinese)

Zhu XM (2020) Study on seismic dynamic response of cross fault slope in Jiulongzhen, Mianzhu, Sichuan Province. Chengdu University of Technology, Chengdu (in Chinese)

Post-earthquake Landscape Response

Qiang Xu, Xuanmei Fan, and Ali Pulpadan Yunus

Abstract Strong earthquakes shake mountains and trigger thousands of landslides, from small and shallow soil failures to large, fast, and deadly rock avalanches. Although a large body of literature in the past discussed the dynamics and surface processes of earthquake-induced landslides, less attention has been received on the post-earthquake surface geomorphic processes, primarily owing to lack of monitoring data. Based on a multi-temporal landslide inventory created post-2008 Wenchuan earthquake, a large interplay of surface geo-dynamic processes responsible to landslide remobilizations and debris flows in the epicentral area was discussed by the authors in the past several years. This report provides a brief overview on the development process of the multi-temporal landslide inventory, their applicability on understanding the spatiotemporal evolution of post-earthquake landslides, mechanism of post-earthquake debris flows, and the factors controlling the remobilized landslides. The long-term landscape response to earthquakes was also discussed in a risk perspective of delayed response of some seismic events.

1 Introduction

Knowing the patterns of earthquake-induced surface processes (and geological hazards) is fundamental to better understand their causal mechanisms, which depend on a complex interplay between seismological, morphological, lithological, hydrological, and geotechnical factors (Fan et al. 2018a, b; Gorum et al. 2011). Understanding causal mechanisms will improve our methods for landslide hazard assessment, landslide prediction, and reduction of the related risk for the

Q. Xu (✉) · X. Fan · A. P. Yunus
State Key Laboratory of Geohazard Prevention and Geoenvironment Protection, Chengdu University of Technology, Chengdu 610059, China
e-mail: xq@cdut.edu.cn

I. Towhata et al. (eds.), *Coseismic Landslides*, Springer Natural Hazards,
https://doi.org/10.1007/978-981-19-6597-5_13

population and infrastructure (Dai et al. 2002, 2011; Gorum et al. 2011). Furthermore, the investigation of the patterns of post-earthquake landsliding in space and in time will provide insight on the earthquake-induced processes controlling landscape evolution in the short and long terms (Hovius et al. 2011; Marc et al. 2015). This improved understanding of slope processes will facilitate the implementation of targeted and more effective risk mitigation countermeasures, which will contribute to safer post-emergency response efforts, more efficient reconstruction management, and increased resilience of communities and the built environment (Das et al. 2018; Flentje and Chowdhury 2018; Petley 2011). More broadly, understanding the processes that act, in particular, over long time scales, will reveal the way in which earthquakes shape mountain belts, providing us with better tools for identifying signatures of past earthquake histories in the current landforms, as well as foreseeing future landform evolutions.

A large body of literature on earthquake-induced surface processes has been produced over the past decades, especially after the magnitude 7.7 Chi-Chi, Taiwan, 1999 earthquake and the magnitude 7.9 Wenchuan, Sichuan, China, 2008 earthquake. Research on all aspects related to the latter, in particular, has been extremely prolific, with more than 1000 journal articles that focused on the earthquake-induced geohazards (Fan et al. 2018b). Many aspects have been investigated in detail (Table 1).

Despite the large quantity of research output, no comprehensive works have been published that focus on the post-earthquake landscape response, in particular the post-seismic remobilization, debris flows, and their controlling factors. In this work, we focus on some themes that we have experienced and worked after the Wenchuan earthquake of 2008 and their immediate implications for risk assessment and mitigation. We discuss the spatiotemporal evolution of post-earthquake landslides, mechanism of post-earthquake debris flows, controlling factors and evolution of remobilized landslides, and long-term landscape response to earthquakes in the following sections.

Table 1 Summary of literature and meta-analyses on aspects of earthquake-induced surface processes

Earthquake-induced landslide inventories: first experiences	Accounts from earthquakes until the late eighteenth century have been always incomplete concerning the landslide numbers and vague about landslide characteristics (Keefer 1999). Extensive ground-based studies initially, followed eventually by the use of aerial photography, progressively increased the availability of coseismic landslide inventories, at least for epicentral areas. This lent insight into the characteristics of these landslides and their associated hazards and how they relate to several seismic, geologic, and topographic parameters. Keefer (1984, 1994, 2002) summarized and presented important analyses of reports of earthquake-induced landslides in dozens of worldwide earthquakes

(continued)

Table 1 (continued)

Methods for generating earthquake-induced landslide inventories	Several methods have come into use (Guzzetti et al. 2012; Xu et al. 2010), but the lack of commonly accepted principles has led to differences in the outputs and thus usefulness of the inventories. Harp et al. (2011) and Xu (2015) suggest guidelines for the preparation of high-quality earthquake-induced landslide inventories. Xu (2015) recognized the following principles: (a) All the landslides should be mapped as long as they can be identified in the imagery; (b) both the boundary and the source area should be mapped; (c) the pattern of spatial distribution of earthquake-triggered landslides should be continuous; (d) complex landslides should be divided into distinct groups; (e) errors in the location and boundary of landslides, false positive errors, and false negative errors should be controlled and reduced; and (f) visual interpretation should be preferred to automatic extraction methods. This agrees with the earlier guidelines by Harp et al. (2011) who, in addition, had pointed out that the imagery from which the landslides are identified (g) must be continuous and span the entire landslide distribution; (h) must have a resolution that is good enough to identify individual landslides that are as small as a few meters across; (i) must have stereo coverage or be able to be draped over a digital elevation model to obtain a stereo-like perspective view; and (j) must be acquired as soon as possible after the earthquake to capture the initial aspects of the landslides and the terrain or infrastructure that they affected. The necessity of polygon-mapping (in opposition to dot mapping) emerged. This mapping has to be plotted in a georeferenced GIS environment and has to be complete in order to obtain robust hazard analyses
Factors controlling earthquake-induced landslide patterns	Several approaches to analyzing the factors contributing to seismic slope failure have been proposed. Jibson et al. (1998, 2000) used observations from the 1994 Northridge earthquake, California, to propose a physically based model for GIS-based landslide hazard prediction that accounts for the effects of seismic shaking as well as slope steepness, material strength, and the effects of groundwater on soil strength. Nowicki et al. (2014) and Nowicki Jessee et al. (2018) used statistical modeling to analyze regional landslide distributions with respect to ground shaking, topographic slope, lithology, land-cover type, and a topographic index related to soil moisture and ground water. Other studies have focused on far-field (Delgado et al. 2011) and near-field (Gorum et al. 2011) patterns of earthquake-induced landslides. Meunier et al. (2008) compared the effects of several earthquakes to determine how topographic amplification of seismic shaking affected where on slopes landslides tended to occur during earthquakes

(continued)

Table 1 (continued)

Initiation and displacements of earthquake-induced landslides	In the most recent review, Jibson (2011) elaborated a review of the physical methods for seismic slope-stability analysis, including methods of limit-equilibrium (pseudostatic), displacement, and deformation analysis based on Newmark's sliding block method (Newmark 1965). These methods provide a physical basis for understanding and predicting the initiation and movement of earthquake-induced landslides
From earthquake-induced landslide inventories to hazard and risk assessments	Keefer (2002) emphasized that meta-analyses of earthquake-induced landslides have led to greatly increased understanding of the related hazards and to the development of models and methods for hazard mapping and evaluation. These are aimed, ultimately, at being integrated into quantitative risk assessments (Corominas et al. (2013)), which are fundamental tools for risk management and mitigation. Statistical models proposed by Nowicki et al. (2014) and Nowicki Jessee et al. (2018) use coseismic landslide inventories and data layers of contributing physical characteristics to produce near-real-time hazard and risk maps following significant earthquakes. Godt et al. (2008) apply a displacement-based physical model for the same purpose. These models are currently in use by the U.S. Geological Survey to produce ground-failure estimates from earthquakes in near real time as part of the ShakeMap (Wald et al. 1999) program
Earthquake-induced mass wasting and long-term landscape evolution	Hovius et al. (1997) looked in detail at sediment removal from mountainous areas as a result of landsliding; although these studies looked at landsliding of all types, they both were in seismically active areas where earthquake-triggered landslides dominate. Malamud et al. (2004) analyzed landslide size distributions and how they relate to earthquakes and other forcing parameters
Research on earthquake-induced landslides (broad overview)	Wasowski et al. (2011) presented a general, simple overview of earthquake-induced landslides. The authors briefly highlighted some new developments in investigation techniques, modeling and data analyses, and the progresses in our understanding of "collateral seismic hazards." The authors formulated the following recommendations for research priorities: (a) compilation of many more complete seismic landslide inventories having adequate contextual information and retrospective inventories; (b) improvement of regional assessments of seismic landslide susceptibility and hazard; (c) development of new methods for regional hazard analysis concerning large catastrophic landslides; and (d) long-term monitoring of representative test slopes instrumented with an array of accelerometers

(continued)

Table 1 (continued)

Chains of geohazards	Multiple chains of geohazards can initiate from the same event and can mutually interact. Xu et al. (2014) reviewed case studies of natural hazards in China, pointing out that research on hazard chains is still at its infancy, and neither the studies on their mechanics nor the methodology are yet mature. The mechanism of transformation from one hazard to the other is often unclear because of the difficulties in assessing the interactions between hazards quantitatively. Currently, only some portions of the hazard chains are studied in detail (i.e., the two-hazards chains such as the earthquake–landslide chain, the storm-flood chain, and the rainfall–landslide chain), while comprehensive studies of (multiple and interacting) hazard chains are lacking Gill and Malamud (2014) recognized four types of multiple hazard interactions: (1) interactions where a hazard is triggered, which can be of the same type as that of the primary hazard or of a different type and can, in turn, trigger tertiary hazards and so on in a chain effect; (2) interactions where the probability of a hazard is increased, when one hazard may not trigger another hazard directly, but may change environmental parameters so as to alter (and, in particular, increase) the frequency or magnitude of another hazard; (3) interactions where the probability of a hazard is decreased; (4) events involving the spatial and temporal coincidence of natural hazards, because of which risks are generated that are not simply the sum of the risks deriving from the individual events. For the latter point, Gill and Malamud (2014), also referring to Alexander (1993), discuss the importance of the spatial and temporal observation scales for the identification and assessment of such risks that also relate to the magnitude of the events that are considered. For the spatial overlap, the type of hazard under investigation will influence the scale of interest, whether it be a town, a region, or larger. For the temporal aspect, the authors point out that there could be a range of possible definitions of before, during, and after in the occurrence of a hazard that will have obvious implications for our understanding of the interactions and the ways in which temporal overlap is considered and the mitigation strategies are devised and implemented Kappes et al. (2010) also warn about the complications when dealing with multi-hazard scenarios in the assessment of the hazard level, as well as in the vulnerability toward distinct processes, and hence in the resulting risk. The nonlinearity of the interactions between hazards and the complexity in the definition of risk from that generated by the individual hazards is such that a change of perspective is required, Kappes et al. (2012) write, in which multi-hazards should be assessed either by identification of coincidences (overlay) or by detailed scenario development, and this requires the account of newer concepts such as nonlinearity, complexity, self-organization, and self-organized criticality

2 Spatiotemporal Evolution of Post-earthquake Landslides

Landslide mapping and inventorying on various scales are essential tools that assist qualitative investigations on the spatial and temporal patterns of mass movements, produce more insight into the failure-run-out mechanisms and damage of the mass movements, and capture topographic, seismic, geological, hydrological, climatic, biological, and anthropogenic preconditions and causal factors that are indicated in their distribution and fate (Corominas et al. 2013; Gariano and Guzzetti 2016; Guzzetti et al. 2009, 2012; Nowicki Jessee et al. 2018; Parker et al. 2017). In this regard, many researchers compiled inventories of earthquake-induced landslides are in response to major events, and the quality and completeness of the inventories have been improved with time (Keefer 2002; Schmitt et al. 2017; Tanyaş et al. 2017). They play essential roles in assessment of the extent of the earthquake-affected areas and assist the post-earthquake emergency response because they are the basis for susceptibility, hazard, and risk analyses. Moreover, the earthquake mechanisms can be reconstructed more precisely by referring to the landslide inventories (Fan et al. 2018a; Gorum et al. 2011). For instance, Keefer (1994) and Rodrıguez et al. (1999) first related the spatial distribution of coseismic landslides around the seismogenic fault to the earthquake magnitude. It deserves attention that complete and detailed inventories are indispensable when the landscape response to earthquakes is evaluated quantitatively and the descriptive and predictive models are calibrated effectively (Marc et al. 2016, 2015).

A substantial number of coseismic landslide inventories have been released to date, and some of them can be collected from freely accessible repositories (Schmitt et al. 2017). We believe that the standardization of data collection and presentation will be promoted by assembling of inventories, and that meta-analyses and modeling efforts will be facilitated remarkably. Several inventories have been compiled for the Wenchuan earthquake-affected areas (Dai et al. 2011; Gorum et al. 2011; Tang et al. 2016, 2009). Therein, landslides were mapped either as points (identifying the landslide scar) or polygons with varying extents of accuracy (Tang et al. 2016) with or without distinction between landslide types (slides, flows, etc.). Both accuracy and presenting styles (by points or polygons) have direct implications for the analysis of landslide size-frequency distributions as well as hazard and risk assessments together with identification of influential parameters. In addition, more detailed studies of distinctive controlling factors and characteristics will be made possible by distinction among different types of landslide.

Few landslide inventories exhibit temporal scenes before and after earthquakes. Actually, the temporal evolution of landsliding after major earthquakes has been greatly more interested in during the past decades and is particularly so after the 1999 Mw 7.7 Chi-Chi and the 2008 Mw 7.9 Wenchuan earthquakes (Dadson et al. 2004; Fan et al. 2018a; c; Yang et al. 2017; Yunus et al. 2020; Zhang and Zhang 2017). This temporal perspective has a potential to answer many research questions concerning, for instance, the patterns of erosion and morphological signatures in seismically active mountain ranges. It is true that researchers have been facilitated by the increased availability of repeated and high-resolution remote-sensing imageries (Fan et al. 2018b) as well as by nearly real-time monitoring networks (Huang et al. 2013). Nevertheless, to the knowledge of the authors, none of the multi-temporal inventories of post-seismic landslides compiled so far have been made open access.

2.1 Pre-, Coseismic, and Post-seismic Landslide Inventory Mapping

The authors have created a multi-temporal inventory that addresses pre- and coseismic landslides, landslides remobilized after the earthquake, and new slope failures. A significant portion of the earthquake's epicentral area is covered by this inventory. It is stressed that this inventory is a detailed multi-temporal landslide inventory that covers larger landslide-affected area after the Wenchuan earthquake than any other so far developed (Domènech et al. 2018; Fan et al. 2018a, c). The covered area comprises 42 catchments with 471 km^2 in area, extending from the town of Yingxiu (the epicenter) to the town of Wenchuan (Fig. 1). The coseismic landslides in the covered area had a total volume of 0.8–1.5 billion m^3, as estimated through a set of first-order empirical area–volume correlation (Parker et al. 2017; Xu et al. 2016). The reliability and uncertainties involved in this empirical correlation were discussed in detail by Fan et al. (2018c).

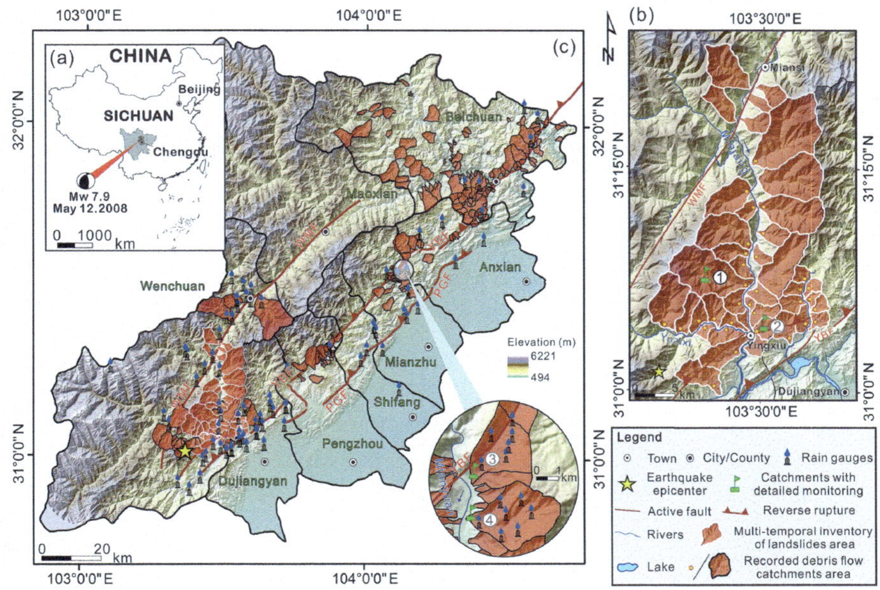

Fig. 1 Location map of the study area

The investigated region has rugged mountains that rise rapidly from 420 m a.s.l. of elevation in the main river valley to over 4000 m a.s.l. at some mountain ridges. The slopes are generally steep, exceeding 36° in more than half of the area. Two of the main faults of the Longmenshan, the Wenchuan–Maowen and the Yingxiu–Beichuan faults, delimit this study area. The dense vegetation and a variably thick soil cover upon the surface are underlain by weathered and highly fractured rocks, in which area is mostly igneous (granite, diorite) associated with metamorphic and sedimentary rocks (schist, shale, sandstone, limestone) as well as recent Quaternary deposits. The climate in this area is subtropical affected by the monsoonal circulation, while the mean annual temperature is 13 °C and most of the annual precipitation that exceeds 1250 mm/year occurs in summer. The Minjiang River, which is a tributary of the Yangtze River in its upper course, crosses this region through Wenchuan and Yingxiu and discharges 452 m^3/s of water, on average, throughout the year (Tang et al. 2011).

To facilitate its visual interpretation, the authors' inventory is provided with high-resolution aerial and satellite images (Spot 5, Spot 6, Worldview 2, Pleiades). In total, eight sets of images were acquired, covering the period from 2005 to 2018 (Table 2).

The scenes of imagery were selected by referring to their availability, date of acquisition, coverage, absence of clouds, and resolution. In the targeted study regions, the selected images have the areal coverage close to 99% in 2007, 2011, and 2015, 97% in 2008, 95% in 2013, and 93% in 2005, 2017, and 2018. Among them, the scene in 2011 was chosen as georeferencing base in the ArcGIS environment (Environmental Systems Research Institute, Inc., USA) and underwent the orthorectification by Pix4D software (Pix4D S.A., Switzerland). A digital elevation model of 25 m resolution that was acquired from the Sichuan Bureau of Surveying and Mapping was used to delineate the catchment boundaries. This inventory provides a polygon-based delineation of landslides that occurred before the earthquake (2005 and 2007 scenes), and this information was used to capture the pre-earthquake landslide activities and their patterns in studies with emphasis on a short-term time window.

Table 2 Reference images used to map the landslides

ID	Reference image	Month and year of acquisition
2005	Spot 5 (2.5 m)	July 2005
2007	Landsat 4 (30 m)	September 2007
2008	Aerial photos (1–2.5 m)	May–July 2008
2011	Aerial photos, World view 2 (0.5–1 m)	April 2011
2013	Aerial photos, Pleiades (0.5–2 m)	April 2013
2015	Spot 6 (1.5 m)	April 2015
2017	Spot 6 (3 m)	April 2017
2018	Spot 6 (3 m)	April 2018

The 2008 scene helped delineate the coseismic landslides, while those in 2011, 2013, 2015, 2017, and 2018 were used to identify the new post-earthquake landslide together with the post-seismic remobilization of coseismic landslides. All the landslide areas were classified into three categories, which are slides, debris flows, and channel deposits. Slides were mapped as if preferential movement paths could not be identified in their debris-covered deposition areas (slides are differentiated from debris flows by checking if there are apparent flow paths). This set of landslides comprises debris, rockslides, and debris with rockfalls; refer to Hungr et al. (2014) for the definitions of the landslide types. This simplification was inevitable because the remote-sensing images could not recognize the difference between these types of movement and their combinations. In contrast, the second classification (debris flows) exhibited a finer material texture along its preferred movement path. They occurred along the hillslopes (hillslope debris flows) and in small channels (channelized debris flows). Large amounts of accumulated debris were detected as well in the main channels. The source area of these materials cannot be clearly determined because they come from multiple debris flows and slides that are located upstream in which clear remobilization was hardly identified. Consequently, these deposits were mapped as channel deposits. They are very important source areas for subsequent debris flows initiated by run-off erosion during intense storms. Some examples of landslide types are demonstrated in Fig. 2.

The authors' inventory map defines four levels of landslide activity so that landslide remobilization may be captured (Fig. 3), following Tang et al. (2016);

(i) Activity level A1: less than one-third of the coseismic or post-seismic landslide area displays signs of remobilization,
(ii) A2: the remobilization ratio being between one-third and two-thirds of the area,
(iii) A3I: the remobilization exceeds two-thirds of the area, and
(iv) A0: no remobilizations in progress.

2.2 Landslide Statistics

The authors identified 133 and 71 landslides in the 2005 and 2007 (pre-earthquake) scenes, whereas 8917 coseismic landslides were detected in the 2008 scene, of which 8259 were classified as slides, 571 as debris flows, and 87 as channel deposits. Moreover, 832 new landslides were delineated in the 2011 scenes (589 slides, 193 debris flows, 50 channel deposits), 387 in the 2013 scene (254 slides, 106 debris flows, 27 channel deposits), 14 in the 2015 scene (7 slides, 1 debris flow, 6 channel deposits), 5 in the 2017 scenes (2 slides, 2 debris flows, 1 channel deposit), and 9 (all types) in the 2018 scene. The number of identified remobilizations was 4099 in 2011, 1152 in 2013, 273 in 2015, 22 in 2017, and 29 in 2018 in association with the activity levels of A1–A3. Table 3 provides more details. Note that the present study and Fan et al. (2018a) may present different numbers until 2015, despite that the two studies concern the same study area. This is simply because the data in the present

Fig. 2 Types of landslides and deposits mapped in the inventory: channel deposit (**a**), slide (**b**), debris flow in a channel (**c**), and debris flow on a hillslope (**d**)

inventory has been refined and improved in the recent years. It is important that the annual number of remobilization of the coseismic deposits, the area affected by the remobilization, and the number and the areas of new failure (i.e., the post-seismic landslides) have been decreasing significantly over time (Table 3). For example, the coseismic landslides in 2011 covered 124 and 37 km^2 had some activity (A1–A3), while only 14 km^2 was active in 2013, 6 km^2 in 2015, 0.27 km^2 in 2017, and eventually 0.34 km^2 in 2018.

The landslide activity of the coseismic deposits decreased quickly with time. The number and the areas of active slides decayed faster than those of debris flows. While a profound number of remobilizations were identified in the form of debris flows within the first 3 years after the earthquake (2008–2011), this number decreased considerably during the time period of 2013–2018 in spite of a large amount of coseismic materials still remaining on the hillslopes (Fan et al. 2018a). This suggests that the effects of the earthquake on landslide rates might be less significant than the early expectations.

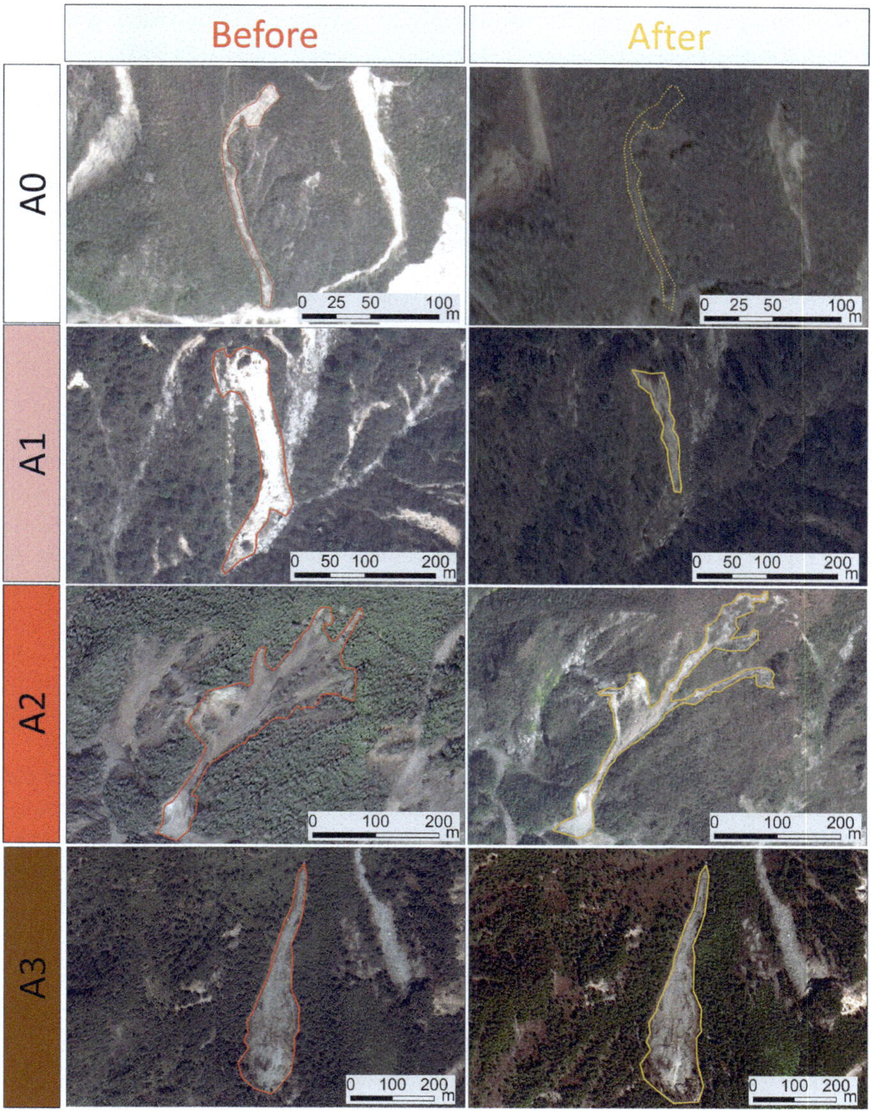

Fig. 3 Examples of four different types of landslide activity level

2.3 *Evolution of Landslide Activity*

9189 coseismic landslides were identified in the 2008 remote-sensing imageries in association with 5598 post-seismic remobilizations in 2011, 1539 in 2013, 285 in 2015, 27 in 2017, and 38 in 2018. The coverage of the coseismic landslides was 131.6 and 47.6 km^2 was remobilized until 2011, whereas only 6.5 km^2 was

Table 3 Landslide statistics (detailed statistics on corresponding number of landslides for each activity levels are referred to Fan et al. 2019)

Year	Channel deposits	Slides	Debris flows
2005		133	
2007		71	
2008	98	8825	364
2011	109	3544	1273
2013	59	1041	439
2015	72	108	105
2017	2	21	4
2018	2	29	7

remobilized between 2013 and 2015. Thus, the landslide inventory exhibited quick decay of the degree of activity in the coseismic deposits with time. In line with this, the hillslope slides became proportionally less common with respect to debris flows. Figure 4a highlights a clear decrease of the remobilized areas over time. More detailed examination of the major debris flow triggering rainfall events that occurred each year reveals that a large number of new landslides and remobilizations in 2011 were affected by the rainstorm on August 13, 2010, in Yingxiu that triggered an enormous number of landslides and debris flows (Tang et al. 2011). On the other hand, it is peculiar that the extreme rainfall event on July 10, 2013, which brought 328.5 mm of rainfall, did not result in a significant number of slope instabilities.

Care must be taken of the material transport during heavy rains. Shortly after the Wenchuan earthquake, the loose coseismic landslide deposits are readily to be remobilized by rainstorms. Therefore, some coseismic debris material was transported to most of the low-order drainage channels, while most still remained along the hillslopes or in channels within catchments. With time, the hillslope deposits became stabilized, while deposits stored in catchment channels can still be converted into

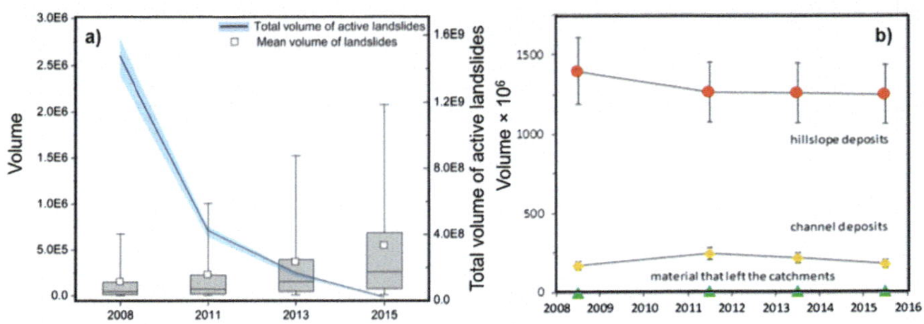

Fig. 4 **a** Volume of coseismic landslides triggered in 2008 and later remobilized in 2011, 2013, and 2015. **b** Evolution of hillslope deposits and channel deposits and volume of material that left the catchments with time

debris flows by heavy rain storms or transported by fluvial erosion down into the outlet of the catchments. These processes are quantified in Fig. 4b.

According to the estimation by the authors, just about 10% mass of the hillslope deposits was lost between 2008 and 2015. Most of this loss took place from 2008 to 2011. It is thus considered that, as of 2015, about 90% of the coseismic debris on hillslope still remains in place. On the other hand, only a very small proportion of the hillslope deposits (<2%) is still affected by landslides in the same year. As for the volume of channel deposits, it increased from 2008 to 2011 and then progressively decreased. In 2008, 89% of the coseismic debris was stored as hillslope deposits, while the remaining 11% had been already channelized within the catchments. Three years later in 2011, the relative proportion between the hillslope and channel materials was different slightly: 84 versus 16%, and became 87 versus 13% in 2015, as the mobilization and removal of the channelized deposits are relatively faster. As time passes on, the coseismic debris is progressively removed/evacuated from the catchment. Nevertheless, the volume of the removed material is still very small as compared with that of the remaining coseismic deposits. Between 2008 and 2015, only a very small fraction of the coseismic materials (about 0.7%) reached the higher-order fluvial system. The remaining 99.3% is still situated within the catchments (mostly along the hillslopes) and has been largely stabilized.

3 Mechanism of Post-earthquake Debris Flows

As mentioned in Sect. 2, the amount of removed/evacuated sediments in the Wenchuan catchments are much smaller than those resting in the hillslopes. In case that the weather is not severe, such deposits lie in a quasi-stable state. On the contrary in wet seasons, these deposits can easily evolve into debris flows once instability is triggered by profound precipitation. Much research on rainfall-induced debris flows has been carried out after the earthquakes, and some studies predicted that the debris flow hazards would increase with time. The rainfall threshold to trigger debris flows significantly decreased after the 1999 Chi-Chi earthquake in Taiwan (Shieh et al. 2009). Tang et al. (2009) drew a similar conclusion on the basis of field investigation and statistical analyses of debris flows that occurred in the Beichuan County on September 24, 2008. It is prerequisite to understand the various hydrological mechanisms that trigger these complex debris flows in order to accurately assess the hazard, to develop thresholds for warning, and to develop risk management (Berti and Simoni 2007).

The Wenchuan earthquake induced many coseismic landslides that generated several km^3 of debris, and this material deposited along the slopes and in low-order channels, forming a large and distributed source of successive post-seismic debris flows (Parker et al. 2011). It is common that debris flows start as a failure of shallow stratum and/or run-off erosion of loose deposits. Excess pore pressure is generated in soil and consequently the shear strength in soil decreases, possibly leading to liquefaction failure of granular deposits and rapid debris flows characterized by long runout distances (e.g., van Asch 2005). The mechanism known as the "temporal blockage of gullies by landslides or side slope failures" aggravated the extent of debris flows in several gullies (Xu et al. 2012). The sudden breach of these blockages may associate debris flows, thus increasing the mobility and damage potential of debris flows.

3.1 Processes and Effects of the Debris Flows—Some Insights from Wenchuan

3.1.1 The Wenjia Gully and Zoumaling Gully Debris Flows

At 1:00 a.m. on August 12, 2010, the Wenjia Gully and Zoumaling Gully developed two debris flows, and a huge amount of debris was removed from the Wenjia Gully. This debris traveled over a long distance even across the main river, forming a deposit fan that immediately blocked the river. This event was triggered by rainfall that brought 55 mm precipitation from its start until the occurrence of the debris flow, lasting for 11 h. The site reconnaissance and the interpretation of remote-sensing imageries revealed that a total of 11 debris flows occurred simultaneously in eleven valleys. Table 4 presents the basic information on the gullies where the debris flow occurred. The total volume of the deposited materials was as much as 5.677×10^6 m^3. Among many debris flows, the one in the Wenjia Gully was the largest, having the volume of 450.0×10^4 m^3, while the flow in the Zoumaling Gully had its volume equal to 0.82×10^6 m^3. Note that the loose materials distributed in the Wenjia Gully were the deposits as a consequence of a mass movement during the Wenchuan earthquake.

When the earthquake happened, approximately 5000×10^4 m^3 of limestone rock mass fell down from the elevation between 1780 and 2340 m. This huge volume of destabilized rock mass was transformed into a rock avalanche, translated into SW direction, and eventually entered the so-called 1300 m platform. Afterward, it turned into SEE direction along the downstream part of the gully. The high energy of the rock avalanche made it possible to entrain the shallow, loose, and soil materials along its flow path, thus increasing the total amount of the displaced and depositing materials to more than 70×10^6 m^3. One part of the massive deposits with a volume of 22.60×10^6 m^3 accumulated at the toe of the sliding source area in the Hanjia Plain. This loose and coarse material on the Hanjia Plain has been proved to have substantial resistance against erosion during water overtopping caused by several rainstorms in subsequent years. Therefore, it is reasonable the finer material that deposited in the lower part of the Wenjia catchment near the 1300 m platform is the main source of the debris flows. In addition to the huge amount of available materials, the gradient of the gullies in this section is relatively steep, ranging between 32.5 and 46%. Accordingly, the loose and relatively fine materials in the gully floor could be easily eroded and transformed into debris flows under heavy rainstorm conditions.

Table 4 Some basic information for the debris flow gully

Name of Gully	Catchment area (km^2)	Channel length (km)	Catchment relief (km)	Channel gradient (°)	Slope angle of imitation zone (°)	Debris supply (10^4 m^3)	Volume (10^4 m^3)
Zoumaling	5.76	3.39	1.1	18	21.3	432.7	82
Luojia	1.64	2.91	1.4	25.7	28	130.8	16
Dongzi	0.12	0.64	0.38	30.7	33.1	4.4	1.2
Wawa	0.51	1.43	0.7	26.1	31.4	40.2	9.8
Didong	0.2	0.76	0.42	28.9	35.6	3.2	0.7
Wenjia	7.65	4.33	1.5	19.1	25.2	5000	450
Linjia	1.27	2.16	0.76	19.4	22.7	7.8	1.5
Taiyang	0.48	1.13	0.68	31.1	39.8	8.6	2
Maliuwan	1.01	2.06	0.8	21.2	27.4	8.3	1.3
Pujia	0.85	1.75	0.8	24.6	33.5	10	0.8
Caodun	0.29	1.31	0.68	27.4	34	10.3	2.4

3.1.2 Hongchun Gully Debris Flow

Situated on the opposite side of the Yingxiu Town, the Hongchun Gully has a catchment area of 5.35 km^2. The highest part of the source of the debris flow is at 2168.4 m a.s.l. The elevation of the gully ranges from 880 to 1700 m. The average gully gradient is 35.8%, and the length of the main gully is approximately 3.6 km.

The Hongchun catchment forms a gully of an ancient debris flow. Before the Wenchuan earthquake, around 2.0×10^6 m^3 of loose deposits were distributed along both sides of the gully. A total of 1.5×10^6 m^3 new loose deposits were formed during Wenchuan earthquake. We interpreted six large landslides in the catchment from aerial photos. These landslides blocked the drainage channel and profoundly increased the risk of debris flow initiation. The rainstorm that occurred from 17:00 on August 12 to 02:00 on August 14, 2010, in the Yingxiu Town region almost instantaneously triggered many hill slope and channelized debris flows.

Among many disasters during this rainfall event, what deserved significant attention was the debris dam produced by the Hongchun Gully debris flow. As a consequence, the river stream changed the course of flow and flooded the newly reconstructed Yingxiu Town. On August 12, 2010, 19.9 mm of rain occurred between 17:00 and 24:00. Thereinafter, a total of 126.8 mm rainfall was recorded throughout the entire day of August 13. On August 14, the precipitation was 23.4 mm between 0:00 and 02:00. Thus, the total cumulative rainfall reached 162.1 mm in 33 h which was recorded from 17:00 on August 12 until 02:00 on August 14. Many eyewitnesses reported that the most damaging debris flow started at around 03:00 and ended at 04:30. As a consequence, approximately 40×10^4 m^3 of debris material moved into the plane of the Minjiang River, forming a 10-m-high natural debris dam, which measured 100 m in length across the river and 150 m in width along the river. Thereinafter, the aforementioned flood destroyed the new Yingxiu Town. The estimated discharge of the river flow through the town was 570 m^3/s shortly after the debris flow occurred and reached the value of 1400 m^3/s on August 15 (Tang et al. 2011).

3.2 *Triggering Rainfalls for Debris Flows*

Although it is well recognized that rainfall is the major trigger of debris flows (e.g., Crosta and Frattini 2001; Segoni et al. 2018), much is not known about the physical processes that relate the rainfall with the occurrence and magnitude of slope failures, landslide remobilization, acceleration, and/or fluidization of the debris flow (e.g., Crozier and Glade 1999; Montgomery and Dietrich 1994; Wieczorek et al. 2000). This understanding is partly hindered by the complexity, variability, and scale dependency of the factors controlling slope failure (Almeida et al. 2017; Beven and Germann 2013; Corominas et al. 2013; Crosta and Frattini 2003). Nonetheless, some types of landslides (shallow landslides and debris flows) are more straightforwardly related to the patterns of rainfall, i.e., to rainfall intensity, duration, and antecedent rainfall (e.g., Aleotti 2004; Gariano et al. 2017; Guzzetti et al. 2007). Empirical observations of the rainfall characteristics in relation to past landslide occurrence are thus the most widespread tool to forecast debris flow occurrence in a defined spatial scale.

Rainfall thresholds are the rainfall conditions that are likely to trigger debris flows when reached or exceeded. They represent the lower bounds of geo-hydrological processes (e.g., rainfall, infiltration, soil moisture) that are experienced during landslide disasters in the past, or the upper bounds that were not associated with landslides (Reichenbach et al. 1998). It is true that there is a spread of values between them due to many uncertainties in nature. In fact, the laws relating landslide occurrence, rainfall intensity, rainfall duration, and antecedent rainfall can be complex, and so are their definitions. Hence, the simplified empirical relations might not be fully accurate, primarily because they neglect other important controlling factors (and their spatial variability). Thus, they should better be characterized probabilistically for the implementation in early warning and risk mitigation systems.

3.2.1 Changing Rainfall Thresholds for Post-seismic Debris Flow Occurrence

Rainfall thresholds can evolve with time as a result of alterations in both hydrological and geotechnical conditions of the slopes, which can be driven by a variety of events and processes, including earthquake-induced slope weakening, weathering, wildfires, climate changes, and new land use. More factors that control the threshold are slope strengthening processes and soil cover depletion (e.g., Fan et al. 2018b; Gariano et al. 2017; Shieh et al. 2009). Statistical and physically based models are widely employed to calibrate the rainfall thresholds above which debris flows may occur. Statistical models informed by historical events have been commonly used to elucidate the form of rainfall intensity–duration (I-D) relationships and event rainfall–duration (E-D) thresholds.

After the 1999 Chi-Chi and the 2008 Wenchuan earthquakes, a profound availability of information on landslides and debris deposits on slopes promoted several studies to recognize the significantly reduced rainfall thresholds and the consequently increased frequency of rainfall-triggered landslide remobilizations and debris flows (Lin et al. 2006; Shieh et al. 2009). In a portion of the area affected by the 1999 Chi-Chi, Taiwan, earthquake, Chang et al. (2011) identified significant differences in pre-earthquake and post-earthquake levels of the rainfall intensity and cumulated precipitation that were required to trigger debris flows as well as the differences in the frequency of large debris flows and also in the size of the catchments in which debris flows occurred. Chen and Hawkins (2009) also identified relationships among earthquake disturbance, tropical rainstorms, and enhanced frequency of debris flows in Taiwan. After the 2008 Wenchuan earthquake, more frequent and destructive debris flows than before occurred in the earthquake-stricken region as well. At the same time, the rainfall threshold for debris flow was significantly lowered (Yu et al. 2014; Zhou and Tang 2014). Zhang and Zhang (2017) reported that the rainfall intensities that triggered debris flows after the Wenchuan earthquake were as much as 75% lower than that before the earthquake (decrease to 25% of the original level). Similarly, they reported reductions by as much as 60% after the Chi-Chi earthquake and as much as 50% after the 2005 Kashmir earthquake (moment magnitude Mw = 7.6).

Due to a variety of natural processes, such as depletion of the erodible material and slope strengthening processes, the rainfall thresholds tend to gradually recover to the pre-earthquake values. How long this process takes for a complete recovery is still an open question, with estimates ranging from less than a decade to as much as a century (Cui et al. 2009; Huang and Fan 2013). On the one hand, insufficient research has been conducted so far on the temporal patterns of debris flows over a long time span after a strong earthquake (Yu et al. 2014); on the other hand, the mechanistic reasons and the main controls of these patterns, as well as those that produce time-dependent changes in the prevailing initiation mechanisms of post-seismic landslides (e.g., Fan et al. 2018b; Zhang and Zhang 2017), need to be better understood.

4 Controlling Factors for Post-earthquake Landslide Evolution

Spatial patterns of landsliding within a region can be determined by local and regional controls, and by their evolution, at a scale larger than that of individual slopes or stream lines. From our multispectral inventory mapping (Domènech et al. 2018; Fan et al. 2019), we notice a good qualitative correlation between the locations of the coseismic landslides (2008) and those of the active landslides in 2011, which is not surprising since the coseismic debris is initially prone to be remobilized by rainfalls rather easily. The region with the highest density corresponds to the area along the seismogenic fault, moving from the epicenter toward the northeast, is the direction that the fault rupture propagated to. However, from 2013 onwards, the (relatively minor) hotspots of landslide activity do not seem as strongly correlated with the map of coseismic landslides. This allows for two considerations: that the seismic shaking does drive focused landsliding in specific areas, but also that this focused landsliding has a limited persistence in time if one looks at landslide activity (in fact, this does not imply that other hillslope erosion processes might not remain enhanced for a longer time). A few years after the earthquake, other controls on landsliding become dominant. For instance, the low year-over-year spatial correlation could be explained by the control exerted on landslide occurrence by local-scale patterns of rainfall. In fact, landslide-triggering rainstorms can occur with extremely inhomogeneous spatial patterns which, moreover, can differ significantly from year to year.

4.1 Quantitative Correlation of Landslide Locations

We correlated the volume of the active landslides based on the catchments in which they occurred across the years, using the subdivision in 42 catchments provided in the inventory release. This is shown (Fig. 5) by comparing the sum of volume density (SVD) (with volumes estimated using Larsen's empirical relationship) in each catchment in each year and quantifying the correlation from year to year. Again, it is apparent that the control exerted by the coseismic landslide volumes on the subsequent landslide activity is initially very high and then decreases (except for 2015). The correlation is close to 90% for the SVD in 2011 compared to that in 2008, which means that remobilizations and post-seismic landslides (especially the former, as they are in greater number) occurred almost proportionally to the volume density of the coseismic landslides in each catchment. This result is important, because it also implies that potential catchment-specific factors have initially a very minor role in controlling landsliding compared to that exerted by the coseismic landslides and by the seismic factors that triggered them and controlled the post-seismic failures. After 2011, the correlation with the SVD in 2008 decreases significantly: It is less than 60% in the subsequent years.

4.2 Controlling Factors

In order to get an insight into the factors controlling landslide activity and their evolution, we evaluated the importance of a number of parameters (indices of topography, hydrology, material, climate, and seismic forcing as independent variables) as potential factors controlling landslide occurrences. We performed the analysis for each year since 2005 by permuting out-of-bag (OOB) observations using a random forest

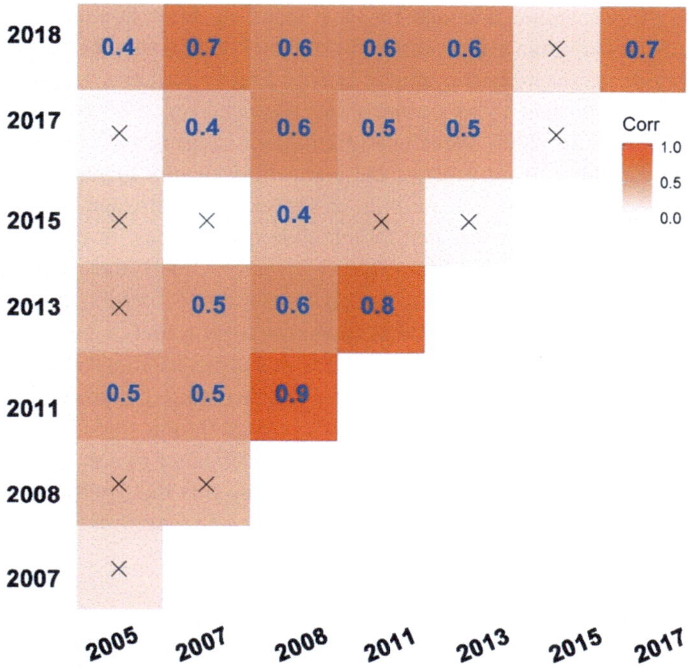

Fig. 5 Correlogram of catchment-level temporal active landslide locations. We correlated the landslide volume density (with volumes obtained through Larsen's empirical relationship (Larsen et al. 2010)) for each of the 42 catchments in which the study area has been divided. The correlation coefficients R are summarized ($p < 0.01$); the crossmark indicates cases in which the p-values are not significant

regression analysis (50,000 random pixels having 0 and 1 have dependent variable; see Methods). A greater value of estimates in OOB indicates more important predictors. The advantage of RF models is that they are robust and their performance is proved good on complex data sets, with little need for fine-tuning and in the presence of many noisy variables (Catani et al. 2013; Dou et al. 2019; Yunus et al. 2019). In order to perform the analysis, we firstly selected the following independent variables derived from the digital elevation model (DEM), published geological maps, seismic parameters during earthquake, and landslide inventory maps, namely ELE—elevation; SLO—slope; ASP—aspect; CUR—planar curvature; POS—positive openness; MRI—Melton ruggedness index; MPI—morphometric protection index; TPI—topographic position index; LITH—lithology; FACC—flow accumulation; DTS—distance to streams; RAIN—rain anomaly between 2007 and 2017; DAH—diurnal anisotropic heat; PGA—peak ground acceleration; PGV—peak ground velocity; MINT—maximum seismic intensity; COS—area of coseismic landslides; AAP—area of previous year active landslides; AAA—area of all landslides. The dependent variable, i.e., the presence or absence of landslide remobilization in each year, is then derived with the help of mapped landslide inventories.

According to the values obtained for each factor in different years, we normalized the predictor importance and assigned weights. Figure 6 shows that topographic factors (elevation, slope, aspect) as well as hydro-climatic variables (rainfall, distance to streams) are important factors controlling the occurrence of landslides in the study area prior to the 2008 Wenchuan earthquake. During the coseismic phase, the predictor significance for most of the controlling factors increased: Elevation, slope, positive openness, aspect, rainfall, distance to streams, and seismic factors (including the PGA) were correlated to landslide occurrences. In the years immediately following the coseismic phase (2011 inventory map), the most significant controlling factor was the rainfall, followed by the area of pre-existing landslides (here, the area of coseismic landslides), elevation, aspect, and distance to streams. After 2011, the OOB observations identified the area of pre-existing landslides (active landslide area identified in the previous remote-sensing image), rainfall and elevation as the strongest controlling factors, and to a lesser extent the slope and the diurnal anisotropic heat. It can also be observed that the significance of area of pre-existing landslides is decreasing with time (Fig. 6), suggesting that the significance of topographic and hydro-climatic variables tends to go back to that of the pre-earthquake conditions.

4.3 *Implications for Landslide Susceptibility Assessments*

The evolution of the controlling factors over time (Fig. 6) implies that there are serious limitations of relying upon landslide susceptibility assessments based on a trained model from a specific set of controlling factors. Because the predictor significance of certain factors decreases or increases after the earthquake, it is essential to understand how long a trained model can still be used for successful prediction of remobilization of landslide bodies or first-failure landslides. With the factors given in Fig. 6, we have trained a susceptibility model using random pixels (of remobilized landslides) from the 2011 images and a random forest (RF) machine learning model implanted in MATLAB (see Methods). RF models are ensembles of decision trees grown from boot-strapped training data. The advantage of RF models is that they reduce classification and regression errors and can be applied on a variety of data structures (Dou et al. 2019). The model performance is then tested and validated using the receiver operating characteristic (ROC) curve and area under curve (AUC) values, which are shown in Fig. 7. The AUC for 2011 is very high (0.89), indicating a very good performance of the training model, as expected. Validation of the model for predicting landslides in 2013 yields lower, yet still satisfactory AUC values (0.81). Validation of the same model on the 2015 and 2017–18 inventories yields AUC values of only 0.73 and 0.60, respectively, which are unsatisfactory (considering that 0.50 means no predictive ability) (Fig. 6a). Similarly, we have also trained new susceptibility models based on the 2013 and on the 2015 inventories and validated the models on the maps of the subsequent years. The AUC results (Fig. 7) indicate that the predicting capacity of a susceptibility model trained for a year is satisfactory

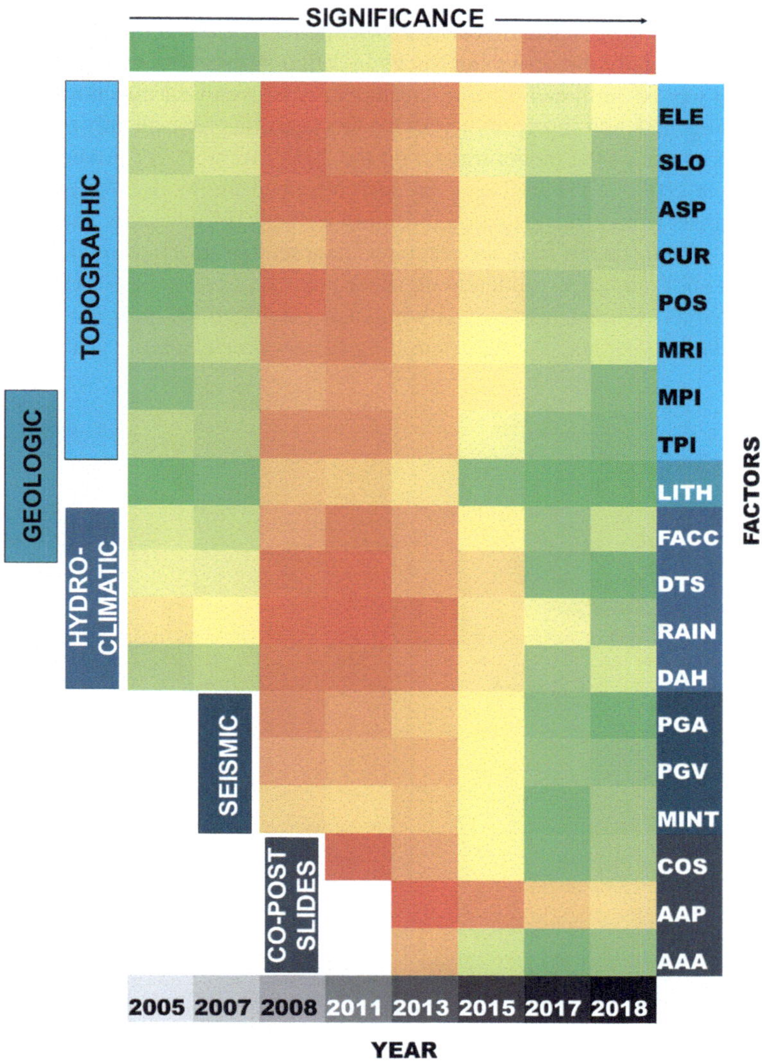

Fig. 6 Estimates of out-of-bag (OOB) predictor significance obtained from a random forest model for each of the possible controlling factors for different years. The level of significance of each factor over the years is indicated by the color bar (dark red indicates high significance; dark green indicates low significance). The estimates are normalized across the factors and across the years. ELE—elevation; SLO—slope; ASP—aspect; CUR—planar curvature; POS—positive openness; MRI—Melton ruggedness index; MPI—morphometric protection index; TPI—topographic position index; LITH—lithology; FACC—flow accumulation; DTS—distance to streams; RAIN—rain anomaly between 2007 and 2017; DAH—diurnal anisotropic heat; PGA—peak ground acceleration; PGV—peak ground velocity; MINT—maximum seismic intensity; COS—area of coseismic landslides; AAP—area of previous year active landslides; AAA—area of all landslides (see the supplementary information file for definitions)

only for the following mapped year (e.g., the model trained on the 2011 inventory can predict well only the active landslides in 2013). Moreover, a set of controlling factors that were determined from an inventory for a given year may not be suitable for predicting landslide susceptibility after a few years that have already passed (e.g., poor performance of the model trained on the 2015 inventory for predicting active landslides in 2017–18) (Fig. 7a–c). We also retrained the model, for example we mixed samples from 2013 landslide inventory to the 2011 inventory and randomly selected 50,000 instances from the combined data set to train a new model, test with 30 percent split, and then validate with subsequent years (Fig. 7d–f). However, no significant improvement is observed with such combinations, hence warranting a new model each year because the controlling factors keep changing year by year.

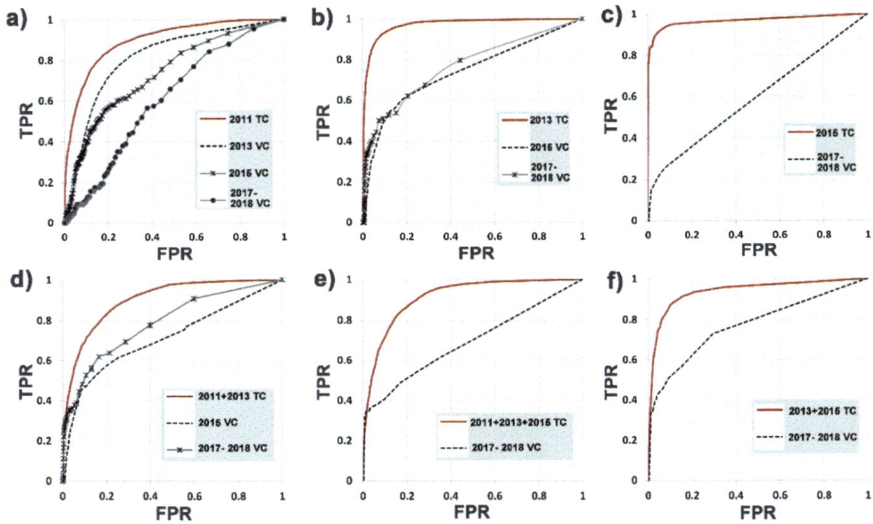

AUC VALUES			
	2011	2013	2015
2011	0.89		
2013	0.81	0.97	
2015	0.73	0.73	0.96
2017-18	0.60	0.76	0.59

AUC VALUES (combinations)			
	2011-13	2011-13-15	2013-15
2011-13	0.90		
2011-13-15		0.91	
2013-15			0.93
2015	0.71		
2017-18	0.76	0.69	0.77

Fig. 7 Performance of susceptibility models. Receiver operating characteristic (ROC) curves and area under the curve (AUC) values showing the performance of the susceptibility models trained on landslide inventories from different years (TC—testing curve; VC—validation curve). The red lines show the results of the testing curve from the trained model for **a** 2011 landslide inventory, **b** 2013 landslide inventory, **c** 2013 landslide inventory, **d** combined 2011 and 2013 landslide inventory, **e** combined 2011, 2013, and 2015 landslide inventory, and **f** combined 2013 and 2015 landslide inventory; dashed and marked lines in each graph represent the validation curves for subsequent years as shown in the legend. The controlling factors shown in Fig. 5 are used in the models as independent variables

5 Long-Term Effect After a Major Earthquake

The long-term effect of strong earthquakes on geological hazards is an underestimated issue (Fan et al. 2018a). Great effort is put in the study of coseismic landslides, and considerable attention is given to the effects of earthquakes on debris flow occurrence and sediment fluxes; on the other hand, the long-term effect of earthquakes on slope stability is often neglected.

Earthquakes can initiate, propagate, widen, or interconnect cracks on rock exposures and down to some depth, particularly near ridge or crests and high on the slopes where the shaking is amplified by topographic effects. The presence of open and/or interconnected cracks facilitates rock weathering by thermal stress and freeze–thaw cycles, oxidation, and hydrolysis by exposure to air, moisture, and infiltration of aqueous-electrolyte solutions. However, rock weathering and crack propagation are complex time-dependent processes. Thus, the patterns of post-seismic landslides might not follow clear trends, and destructive events might still happen "randomly" for several decades after the earthquake (see, e.g., Towhata 2013). Some rock slopes damaged by the Mw 7.7 Chi-Chi earthquake in 1999 in Taiwan exhibit rain-triggered collapses during rainy seasons and cause severe fatalities years after the earthquake (Lin et al. 2006). Post-seismic landslides and long-term rock degradation have been reported in highly seismic areas of Japan (Okamoto et al. 2013). A role of past earthquakes in enhanced landsliding triggered by a subsequent earthquake was identified in New Zealand (Parker et al. 2017). In Sichuan, China, Wei et al. (2018) suggested that the occurrence of some recent landslides can be associated with the Ms 7.5 Xichang earthquake that hit the region more than 160 years ago.

Five to ten thousand people die because of a non-coseismic landslide each year in the world (Froude and Petley 2018; Petley, 2012) with an increasing trend over time. Some (or a significant part) of these events occurring in seismic regions might actually be delayed post-seismic landslides. Fan et al. (2018a) argued that high-elevation mountain ridges in seismic inhabited regions should be systematically scanned for features (e.g., extensive cracks and slow deformations or creep deformation) that can foretell ongoing rock damage and approaching failures, so that real-time monitoring, warning systems, and other countermeasures can be enforced to reduce the risk associated with such failures. However, delayed landslides in the post-seismic period remain a poorly investigated research field. Much efforts are needed to constrain the role of past earthquakes in the long-term spatiotemporal patterns of landslides and prepare long-term predictive models.

The most recent example of a destructive landslide on a slope preconditioned by a past earthquake is that of the 2017 Xinmo landslide, again in Sichuan, that mobilized several millions of cubic meters of rock from a high-elevation mountain ridge, which rushed along a steep slope and destroyed an entire village causing many fatalities (Fan et al. 2017; Scaringi et al. 2018).

5.1 *Xinmo Landslide Event*

The village of Xinmo is situated in the eastern margin of the Tibetan plateau, Sichuan, China. It was on June 24, 2017, that a 13-million-m^3 rock and debris avalanche submerged this village with impressive energy, rushing further toward the river, and blocked the stream course over more than 1 km. The rescue operations were initiated promptly, and all possible efforts were made by local experts. Despite this, ten people were found dead, and 73 more were reported missing in one of the deadliest landslides in the recent history (Fan et al. 2017; Fig. 8).

Almost a century ago in 1933, Xinmo and its surrounding area were struck by the Diexi earthquake of Mw = 7.3. This seismic event triggered large-scale landslides that destroyed villages and choked rivers, producing dammed lakes that collapsed later and initiated devastating floods. Consequently, thousands of people died (Chang 1938). A large amount of loose material was left along the slopes after coseismic landslides, which, in turn, caused deadly debris flows and avalanches for decades during every rainy season. The same disaster was repeated in 2008, and it is still happening after 2008 within several kilometers from the previous sites, after the Ms = 8.0 (Mw = 7.9) Wenchuan earthquake (Fan and Huang 2013). Note that what did not collapse during past earthquakes is not exempt from damage. Strong seismic shaking can produce cracks and fractures in the rock that can be noticed paradoxically from the space by the eyes of satellites but are hidden from the human eye if covered under dense vegetation on high-elevation ridges (e.g., Fan et al. 2017). The damaged rock can remain in place for decades or even centuries, but rainwater can infiltrate into the fractures, dissolve minerals, fill the cracks, freeze and expand in volume, and push the rock blocks apart with its pressure, until, more or less suddenly, the proverbial last straw—a rainfall or a minor shake—makes the rock mass collapse.

Fig. 8 Panoramic view of the Xinmo rock avalanche as seen from the opposite slope. Notice the red trucks for the scale

6 Summary

Landscapes in tectonically active areas are shaped by interplay of various geodynamic processes such as earthquakes, coseismic landslides, erosion, debris flows, post-seismic landslides, flooding, and sediment evacuation. The details on post-seismic processes have long been considered in isolation, but recent earthquake events that happened in Chi-Chi, Taiwan in 1999, Wenchuan, China, in 2008 and Gorkha, Nepal, in 2015, and availability of archived data from remote-sensing imageries enhanced our understanding of spatiotemporal evolution of landslides. This report collectively briefed the processes and controls of post-seismic landslide mechanism and complexities based on the research insights and monitoring data that acquired post-Wenchuan earthquake. The research outlook presented in this work helps us in deeper understanding of post-earthquake landsliding, which can be used to robustly determine its role in landscape dynamics through models and observations.

References

Aleotti P (2004) A warning system for rainfall-induced shallow failures. Eng Geol 73:247–265. https://doi.org/10.1016/j.enggeo.2004.01.007

Alexander DC (1993) Natural disasters. Kluwer Academic Publishers, Dordrecht, Boston, London

Almeida S, Holcombe EA, Pianosi F, Wagener T (2017) Dealing with deep uncertainties in landslide modelling for disaster risk reduction under climate change. Nat Hazard 17:225–241. https://doi.org/10.5194/nhess-17-225-2017

Berti M, Simoni A (2007) Prediction of debris flow inundation areas using empirical mobility relationships. Geomorphology 90:144–161. https://doi.org/10.1016/j.geomorph.2007.01.014

Beven K, Germann P (2013) Macropores and water flow in soils revisited: review. Water Resour Res 49:3071–3092. https://doi.org/10.1002/wrcr.20156

Catani F, Lagomarsino D, Segoni S, Tofani V (2013) Landslide susceptibility estimation by random forests technique: sensitivity and scaling issues. Nat Hazard 13:2815–2831

Chang LC (1938) Reconnaissance on Diexi earthquake in Sichuan Province, Geological Review by Geological Society of China. Publ Geol Library Beijing 3(3):251–292 (in Chinese)

Chang C-W, Lin P-S, Tsai C-L (2011) Estimation of sediment volume of debris flow caused by extreme rainfall in Taiwan. Eng Geol 123:83–90. https://doi.org/10.1016/j.enggeo.2011.07.004

Chen H, Hawkins AB (2009) Relationship between earthquake disturbance, tropical rainstorms and debris movement: an overview from Taiwan. Bull Eng Geol Env 68:161–186. https://doi.org/10.1007/s10064-009-0209-y

Corominas J, van Westen C, Frattini P, Cascini L, Malet J-P, Fotopoulou S, Catani F, Van Den Eeckhaut M, Mavrouli O, Agliardi F, Pitilakis K, Winter MG, Pastor M, Ferlisi S, Tofani V, Hervás J, Smith JT (2013) Recommendations for the quantitative analysis of landslide risk. Bull Eng Geol Env. https://doi.org/10.1007/s10064-013-0538-8

Crosta GB, Frattini P (2001) Rainfall thresholds for triggering soil slips and debris flow. In: Proceedings of the 2nd EGS Plinius conference on mediterranean storms: publication CNR GNDCI, 2547, pp 463–487

Crosta GB, Frattini P (2003) Distributed modelling of shallow landslides triggered by intense rainfall. Nat Hazards Earth Syst Sci 3:81–93. https://doi.org/10.5194/nhess-3-81-2003

Crozier MJ, Glade T (1999) Frequency and magnitude of landsliding: fundamental research issues. Zeitschrift Für Geomorphologie Supplement Volumes 115:141–155. https://doi.org/10.1127/zfgsuppl/115/1999/141

Cui P, Zhu Y, Han Y, Chen X, Zhuang J (2009) The 12 May {Wenchuan} earthquake-induced landslide lakes: distribution and preliminary risk evaluation. Landslides 6:209–223

Dadson SJ, Hovius N, Chen H, Dade WB, Lin JC, Hsu ML, Lin CW, Horng MJ, Chen TC, Milliman J, Stark CP (2004) Earthquake-triggered increase in sediment delivery from an active mountain belt. Geology 32:733–736. https://doi.org/10.1130/G20639.1

Dai FC, Lee CF, Ngai YY (2002) Landslide risk assessment and management: an overview. Eng Geol 64:65–87. https://doi.org/10.1016/S0013-7952(01)00093-X

Dai FC, Xu C, Yao X, Xu L, Tu XB, Gong QM (2011) Spatial distribution of landslides triggered by the 2008 Ms 8.0 Wenchuan earthquake, China. J Asian Earth Sci 40:883–895. https://doi.org/10.1016/j.jseaes.2010.04.010

Das JT, Puppala AJ, Bheemasetti TV, Walshire LA, Corcoran MK (2018) Sustainability and resilience analyses in slope stabilisation. In: Proceedings of the institution of civil engineers—engineering sustainability, vol 171, pp 25–36. https://doi.org/10.1680/jensu.16.00054

Delgado J, Garrido J, López-Casado C, Martino S, Peláez JA (2011) On far field occurrence of seismically induced landslides. Eng Geol 123:204–213

Domènech G, Yang F, Guo X, Fan X, Scaringi G, Dai L, He C, Xu Q, Huang R (2018b) Two multi-temporal datasets to track the enhanced landsliding after the 2008 Wenchuan earthquake. https://doi.org/10.5281/zenodo.1405489

Dou J, Yunus AP, Bui DT, Merghadi A, Sahana M, Zhu Z, Chen C-W, Khosravi K, Yang Y, Pham BT (2019) Assessment of advanced random forest and decision tree algorithms for modeling rainfall-induced landslide susceptibility in the Izu-Oshima Volcanic Island, Japan. Sci Total Environ 662:332–346

Fan X, Domènech G, Scaringi G, Huang R, Xu Q, Hales TC, Dai L, Yang Q, Francis O (2018a) Spatio-temporal evolution of mass wasting after the 2008 Mw 7.9 Wenchuan Earthquake revealed by a detailed multi-temporal inventory. Landslides. https://doi.org/10.1007/s10346-018-1054-5

Fan X, Juang CH, Wasowski J, Huang R, Xu Q, Scaringi G, van Westen CJ, Havenith H-B (2018b) What we have learned from the 2008 Wenchuan earthquake and its aftermath: a decade of research and challenges. Eng Geol 241:25–32

Fan X, Scaringi G, Yang F, Domènech G, Guo X, Dai L, He C, Xu Q, Huang R (2018c) Two multi-temporal datasets to track the enhanced landsliding after the 2008 Wenchuan earthquake. Earth Syst Sci Data Discuss 1–29. https://doi.org/10.5194/essd-2018c-105

Fan X, Xu Q, Scaringi G, Dai L, Li W, Dong X, Zhu X, Pei X, Dai K, Havenith H-B (2017) Failure mechanism and kinematics of the deadly June 24th 2017 Xinmo landslide, Maoxian, Sichuan, China. Landslides 14:2129–2146. https://doi.org/10.1007/s10346-017-0907-7

Fan X, Scaringi G, Domènech G, Yang F, Guo X, Dai L, He C, Xu Q, Huang R (2019) Two multi-temporal datasets that track the enhanced landsliding after the 2008 Wenchuan earthquake. Earth Syst Sci Data 11:35–55. https://doi.org/10.5194/essd-11-35-2019

Flentje P, Chowdhury R (2018) Resilience and sustainability in the management of landslides. In: Proceedings of the institution of civil engineers—engineering sustainability, vol 171, pp 3–14. https://doi.org/10.1680/jensu.16.00045

Froude MJ, Petley D (2018) Global fatal landslide occurrence from 2004 to 2016. Nat Hazard 18:2161–2181

Gariano SL, Guzzetti F (2016) Landslides in a changing climate. Earth Sci Rev 162:227–252. https://doi.org/10.1016/j.earscirev.2016.08.011

Gariano SL, Rianna G, Petrucci O, Guzzetti F (2017) Assessing future changes in the occurrence of rainfall-induced landslides at a regional scale. Sci Total Environ 596–597:417–426. https://doi.org/10.1016/j.scitotenv.2017.03.103

Gill JC, Malamud BD (2014) Reviewing and visualizing the interactions of natural hazards: interactions of natural hazards. Rev Geophys 52:680–722. https://doi.org/10.1002/2013RG000445

Godt JW, Sener B, Verdin K, Wald DJ, Earle PS, Harp EL, Jibson RW (2008) Rapid assessment of earthquake-induced landsliding. In: Proceedings of the first world landslide forum, United Nations University, Tokyo, pp 3161–3166

Gorum T, Fan X, van Westen CJ, Huang R, Xu Q, Tang C, Wang G (2011) Distribution pattern of earthquake-induced landslides triggered by the 12 May 2008 Wenchuan earthquake. Geomorphology 133:152–167. https://doi.org/10.1016/j.geomorph.2010.12.030

Guzzetti F, Peruccacci S, Rossi M, Stark CP (2007) Rainfall thresholds for the initiation of landslides in central and southern Europe. Meteorol Atmos Phys 98:239–267

Guzzetti F, Ardizzone F, Cardinali M, Rossi M, Valigi D (2009) Landslide volumes and landslide mobilization rates in Umbria, central Italy. Earth Planet Sci Lett 279:222–229. https://doi.org/10.1016/j.epsl.2009.01.005

Guzzetti F, Mondini AC, Cardinali M, Fiorucci F, Santangelo M, Chang K-T (2012) Landslide inventory maps: new tools for an old problem. Earth Sci Rev 112:42–66

Harp EL, Keefer DK, Sato HP, Yagi H (2011) Landslide inventories: the essential part of seismic landslide hazard analyses. Eng Geol 122:9–21

Hovius N, Stark CP, Allen PA (1997) Sediment flux from a mountain belt derived by landslide mapping. Geology 25:231–234

Hovius N, Meunier P, Lin C-W, Chen H, Chen Y-G, Dadson S, Horng M-J, Lines M (2011) Prolonged seismically induced erosion and the mass balance of a large earthquake. Earth Planet Sci Lett 304:347–355. https://doi.org/10.1016/j.epsl.2011.02.005

Huang R, Fan X (2013) The landslide story. Nat Geosci 6:325–326. https://doi.org/10.1038/ngeo1806

Huang R, Zhao J, Ju N, Li G, Lee ML, Li Y (2013) Analysis of an anti-dip landslide triggered by the 2008 Wenchuan earthquake in China. Nat Hazards 68:1021–1039. https://doi.org/10.1007/s11069-013-0671-5

Hungr O, Leroueil S, Picarelli L (2014) The Varnes classification of landslide types, an update. Landslides 11:167–194. https://doi.org/10.1007/s10346-013-0436-y

Jibson RW (2011) Methods for assessing the stability of slopes during earthquakes—a retrospective. Eng Geol 122:43–50

Jibson RW, Harp EL, Michael JA (1998) A method for producing digital probabilistic seismic landslide hazard maps—an example from the Los Angeles, California, area. U.S. Geol Survey Open-File Rep 98–113, 17, 2 pl

Jibson RW, Harp EL, Michael JA (2000) A method for producing digital probabilistic seismic landslide hazard maps. Eng Geol 58:271–289

Kappes MS, Keiler M, Glade T (2010) From single-to multi-hazard risk analyses: a concept addressing emerging challenge

Kappes MS, Keiler M, von Elverfeldt K, Glade T (2012) Challenges of analyzing multi-hazard risk: a review. Nat Hazards 64:1925–1958. https://doi.org/10.1007/s11069-012-0294-2

Keefer DK (1984) Landslides caused by earthquakes. Geol Soc Am Bull 95:406. https://doi.org/10.1130/0016-7606(1984)95%3c406:LCBE%3e2.0.CO;2

Keefer DK (1994) The importance of earthquake-induced landslides to long-term slope erosion and slope-failure hazards in seismically active regions. In: Geomorphology and natural hazards. Elsevier, pp 265–284. https://doi.org/10.1016/B978-0-444-82012-9.50022-0

Keefer DK (1999) Earthquake-induced landslides and their effects on alluvial fans. J Sediment Res 69:84–104

Keefer DK (2002) Investigating landslides caused by earthquake—a historical review. Surv Geophys 23:473–510

Larsen IJ, Montgomery DR, Korup O (2010) Landslide erosion controlled by hillslope material. Nat Geosci 3:247–251. https://doi.org/10.1038/ngeo776

Lin W-T, Lin C-Y, Chou W-C (2006) Assessment of vegetation recovery and soil erosion at landslides caused by a catastrophic earthquake: a case study in Central Taiwan. Ecol Eng 28:79–89

Malamud BD, Turcotte DL, Guzzetti F, Reichenbach P (2004) Landslides, earthquakes, and erosion. Earth Planet Sci Lett 229:45–59

Marc O, Hovius N, Meunier P, Uchida T, Hayashi S (2015) Transient changes of landslide rates after earthquakes. Geology 43:883–886. https://doi.org/10.1130/G36961.1

Marc O, Hovius N, Meunier P (2016) The mass balance of earthquakes and earthquake sequences. Geophys Res Lett 43:3708–3716. https://doi.org/10.1002/2016GL068333

Meunier P, Hovius N, Haines JA (2008) Topographic site effects and the location of earthquake induced landslides. Earth Planet Sci Lett 275:221–232. https://doi.org/10.1016/j.epsl.2008.07.020

Montgomery DR, Dietrich WE (1994) A physically based model for the topographic control on shallow landsliding. Water Resour Res 30:1153–1171. https://doi.org/10.1029/93WR02979

Newmark NM (1965) Effects of earthquakes on dams and embankments. Geotechnique 15:139–160

Nowicki MA, Wald DJ, Hamburger MW, Hearne M, Thompson EM (2014) Development of a globally applicable model for near real-time prediction of seismically induced landslides. Eng Geol 173:54–65

Nowicki Jessee MA, Hamburger MW, Allstadt K, Wald DJ, Robeson SM, Tanyas H, Hearne M, Thompson EM (2018) A global empirical model for near real-time assessment of seismically induced landslides. J Geophys Res: Earth Surf

Okamoto T, Sakurai M, Tsuchiya S, Yoshimatsu H, Ogawa K, Wang G (2013) Secondary hazards associated with coseismic landslide. In: Ugai K, Yagi H, Wakai A (eds) Earthquake-induced landslides. Springer, Berlin, Heidelberg, pp 77–82. https://doi.org/10.1007/978-3-642-32238-9_8

Parker RN, Densmore AL, Rosser NJ, de Michele M, Li Y, Huang R, Whadcoat S, Petley DN (2011) Mass wasting triggered by the 2008 Wenchuan earthquake is greater than orogenic growth. Nat Geosci 4:449–452. https://doi.org/10.1038/ngeo1154

Parker RN, Rosser NJ, Hales TC (2017) Spatial prediction of earthquake-induced landslide probability. Nat Hazards Earth Syst Sci Discuss 1–29. https://doi.org/10.5194/nhess-2017-193

Petley DN (2011) Building resilience to landslides in mountain communities. https://blogs.agu.org/landslideblog/2011/10/12/building-resilience-to-landslides-in-mountain-communities/

Petley DN (2012) Global patterns of loss of life from landslides. Geology 40:927–930. https://doi.org/10.1130/G33217.1

Reichenbach P, Cardinali M, De Vita P, Guzzetti F (1998) Regional hydrological thresholds for landslides and floods in the Tiber River Basin (central Italy). Environ Geol 35:146–159. https://doi.org/10.1007/s002540050301

Rodrıguez CE, Bommer JJ, Chandler RJ (1999) Earthquake-induced landslides: 1980–1997. Soil Dyn Earthq Eng 18:325–346

Scaringi G, Fan X, Xu Q, Liu C, Ouyang C, Domènech G, Yang F, Dai L (2018) Some considerations on the use of numerical methods to simulate past landslides and possible new failures: the case of the recent Xinmo landslide (Sichuan, China). Landslides 15:1359–1375. https://doi.org/10.1007/s10346-018-0953-9

Schmitt R, Tanyas H, Jessee MAN, Zhu J, Biegel K, Allstadt KE, Jibson RW, van Westen C, Sato HP (2017) An open repository of earthquake-triggered ground-failure inventories: data release collection

Segoni S, Piciullo L, Gariano SL (2018) A review of the recent literature on rainfall thresholds for landslide occurrence. Landslides 15:1483–1501. https://doi.org/10.1007/s10346-018-0966-4

Shieh C-L, Chen YS, Tsai YJ, Wu JH (2009) Variability in rainfall threshold for debris flow after the Chi-Chi earthquake in central Taiwan, China. Int J Sedim Res 24:177–188

Tang C, Zhu J, Li WL, Liang JT (2009) Rainfall-triggered debris flows following the Wenchuan earthquake. Bull Eng Geol Env 68:187–194. https://doi.org/10.1007/s10064-009-0201-6

Tang C, Zhu J, Ding J, Cui XF, Chen L, Zhang JS (2011) Catastrophic debris flows triggered by a 14 August 2010 rainfall at the epicenter of the Wenchuan earthquake. Landslides 8:485–497. https://doi.org/10.1007/s10346-011-0269-5

Tang C, Van Westen CJ, Tanyas H, Jetten VG (2016) Analysing post-earthquake landslide activity using multi-temporal landslide inventories near the epicentral area of the 2008 Wenchuan earthquake. Nat Hazard 16:2641–2655. https://doi.org/10.5194/nhess-16-2641-2016

Tanyaş H, van Westen CJ, Allstadt KE, Anna Nowicki Jessee M, Görüm T, Jibson RW, Godt JW, Sato HP, Schmitt RG, Marc O, Hovius N (2017) Presentation and analysis of a worldwide database of earthquake-induced landslide inventories: earthquake-induced landslide inventories. J Geophys Res: Earth Surf 122:1991–2015. https://doi.org/10.1002/2017JF004236

Towhata I (2013) Long-term effects of earthquake-induced slope failures. In: Proceedings of the 7th international conference on case histories in geotechnical engineering, Chicago 10. http://scholarsmine.mst.edu/icchge/7icchge/session16/10

van Asch TWJ (2005) Modelling the hysteresis in the velocity pattern of slow-moving earth flows: the role of excess pore pressure. Earth Surf Proc Land 30:403–411. https://doi.org/10.1002/esp.1147

Wald DJ, Quitoriano V, Heaton TH, Kanamori H, Scrivner CW, Worden CB (1999) TriNet "ShakeMaps": rapid generation of peak ground motion and intensity maps for earthquakes in southern California. Earthq Spectra 15:537–555

Wasowski J, Keefer DK, Lee C-T (2011) Toward the next generation of research on earthquake-induced landslides: current issues and future challenges. Eng Geol 122:1–8. https://doi.org/10.1016/j.enggeo.2011.06.001

Wei R, Zeng Q, Davies T, Yuan G, Wang K, Xue X, Yin Q (2018) Geohazard cascade and mechanism of large debris flows in Tianmo gully, SE Tibetan Plateau and implications to hazard monitoring. Eng Geol 233:172–182

Wieczorek GF, Morgan BA, Campbell RH (2000) Debris-flow hazards in the Blue Ridge of central Virginia. Environ Eng Geosci 6:3–23. https://doi.org/10.2113/gseegeosci.6.1.3

Xu C (2015) Preparation of earthquake-triggered landslide inventory maps using remote sensing and GIS technologies: principles and case studies. Geosci Front 6:825–836. https://doi.org/10.1016/j.gsf.2014.03.004

Xu C, Shyu JBH, Xu X (2010) Landslides triggered by the 12 January 2010 Port-au-Prince, Haiti, Mw = 7.0 earthquake: visual interpretation, inventory compiling, and spatial distribution statistical analysis. Nat Hazards Earth Syst Sci 31

Xu L, Meng X, Xu X (2014) Natural hazard chain research in China: a review. Nat Hazards 70:1631–1659

Xu C, Xu X, Shen L, Yao Q, Tan X, Kang W, Ma S, Wu X, Cai J, Gao M, Li K (2016) Optimized volume models of earthquake-triggered landslides. Sci Rep 6. https://doi.org/10.1038/srep29797

Xu Q, Zhang S, Li WL, van Asch ThWJ (2012) The 13 August 2010 catastrophic debris flows after the 2008 Wenchuan earthquake, China. Nat Hazard 12:201–216. https://doi.org/10.5194/nhess-12-201-2012

Yang W, Qi W, Wang M, Zhang J, Zhang Y (2017) Spatial and temporal analyses of post-seismic landslide changes near the epicentre of the Wenchuan earthquake. Geomorphology 276:8–15. https://doi.org/10.1016/j.geomorph.2016.10.010

Yu B, Wu Y, Chu S (2014) Preliminary study of the effect of earthquakes on the rainfall threshold of debris flows. Eng Geol 182:130–135. https://doi.org/10.1016/j.enggeo.2014.04.007

Yunus AP, Dou J, Song X, Avtar R (2019) Improved bathymetric mapping of coastal and lake environments using sentinel-2 and landsat-8 images. Sensors 19:2788

Yunus AP, Fan X, Tang X, Jie D, Xu Q, Huang R (2020) Decadal vegetation succession from MODIS reveals the spatio-temporal evolution of post-seismic landsliding after the 2008 Wenchuan earthquake. Remote Sens Environ 236:111476

Zhang S, Zhang LM (2017) Impact of the 2008 Wenchuan earthquake in China on subsequent long-term debris flow activities in the epicentral area. Geomorphology 276:86–103. https://doi.org/10.1016/j.geomorph.2016.10.009

Zhou W, Tang C (2014) Rainfall thresholds for debris flow initiation in the Wenchuan earthquake-stricken area, southwestern China. Landslides 11:877–887

Earthquake-Induced Landslide Susceptibility and Hazard Assessment Approaches

Sergio A. Sepúlveda

Abstract The assessment of earthquake-induced landslide susceptibility and hazard is a complex problem, as it must consider elements from both the geological and geotechnical properties of the slope and strong ground motion parameters from expected earthquakes scenarios. In recent decades, quantitative and computing tools have quickly developed and are increasingly applied for the evaluation of susceptibility and hazard. While traditional semi-quantitative analyses, statistical approaches or geotechnical methods such as Newmark displacement analysis are still frequently used, soft computing methods like artificial neural networks and machine learning tools are increasingly being utilized. Improved regional seismic hazard models and tools to represent seismic scenarios in some regions such as ShakeMap have also contributed to develop new hazard mapping techniques. Nevertheless, large uncertainties related to seismic parameters as well as geotechnical and geological variability are still key aspects that must be dealt with. Many challenges remain in this field, including forward hazard assessment in areas with infrequent large earthquakes and little data on earthquake-induced landsliding, better site-specific models of seismic scenarios in different tectonic contexts and better distinction between susceptibility and hazard assessment, among others.

1 Introduction

Landslide susceptibility and hazard assessment and mapping are key elements for disaster risk management and prevention. In the last few decades, quantitative and computing tools for landslide hazard mapping have quickly developed and expanded, including earthquake-induced landslides. Recognition of seismic and geological slope parameters and their properties influencing the slope response during an earthquake and thus its likelihood of triggering a landslide is essential for susceptibility and hazard assessment and is the basis for the preparation of hazard maps. In this chapter, we review those seismic, geomorphological, geological, geotechnical

---

S. A. Sepúlveda (✉)
Department of Earth Sciences, Simon Fraser University, Burnaby, BC, Canada
e-mail: sergio_sepulveda@sfu.ca

I. Towhata et al. (eds.), *Coseismic Landslides*, Springer Natural Hazards,
https://doi.org/10.1007/978-981-19-6597-5_14

and other factors conditioning the slope coseismic response, and examine the most used methods for earthquake-induced landslide susceptibility and hazard mapping described in the literature in the last twenty years.

2 Seismic and Slope Parameters for Susceptibility and Hazard Assessment

The engineering geological characteristics of any region susceptible to being affected by earthquakes are of primary importance in order to assess the hazard of earthquake-triggered landslides. The geomorphological, geological and geotechnical characteristics of rocks and soils and their interaction with the ground motion define the susceptibility of the slopes to suffering failure. The intrinsic parameters of the slopes include the type of soil or rock, the slope gradient, height and geometry, structural and groundwater conditions, rock or soil strength, and increasingly, man-made alterations. Meanwhile, the level of intensity of shaking will directly influence the likelihood of slope failure during an earthquake. This intensity of shaking is in turn influenced by the local topographic and geological conditions. Therefore, for the study of susceptibility and hazards of earthquake-induced landslides, it is necessary to recognize, characterize and acquire data on several slope and strong ground motion parameters.

2.1 Strong Ground Motion Parameters and Site Effects

The amplitude of the strong motion is perhaps the most relevant parameter that may be directly retrieved from a seismogram or earthquake time history record. The strong motion amplitude parameter may be acceleration, velocity or displacement. Because of its direct relation with the seismic force acting on the slope, the acceleration is one of the most significant ground motion parameters for the study of earthquake-induced landslides. The most widely used measure of a particular ground motion is the peak ground acceleration (PGA). This is commonly defined as the largest (absolute) value of horizontal acceleration recorded in a time history for the different components. In some cases, the peak vertical acceleration may be larger than the horizontal, or at least it can be sufficiently large to also be considered, and it has been recognized to also correlate with landslide occurrence (Serey et al. 2019). While acceleration in the form of PGA, or most often a fraction of PGA (Kramer 1996), has been widely used for seismic slope stability analyses, the importance of the peak ground velocity (PGV) has been recognized in recent studies of earthquake-induced landslides (e.g. Jessee 2017; Nowicki Jessee et al. 2018; Allstadt et al. 2018). PGV has the advantage of not being as sensitive as PGA to transient high-amplitude ground accelerations and not saturating with magnitude as much as PGA, thus it may reflect better the ground motions for moderate to large earthquake events (Allstadt et al. 2018; Nowicki Jessee et al. 2018).

The availability of strong motion records is widely variable, with concentrations of dense seismic networks in some regions while other are barely covered. The use of new tools for assessment of seismic scenarios, such the ShakeMap, may be a good option for hazard studies. The USGS ShakeMap is a near-real-time product that provides spatial estimates of PGA, PGV, peak spectral accelerations and shaking intensities within minutes of any significant earthquake (widely felt and/or Mw > 5.5) worldwide (Allen et al. 2008; Worden and Wald 2016; Allstadt et al. 2018). For larger earthquakes, the ShakeMap can be updated several times after an event as the magnitude and location are refined, thus the resulting susceptibility or hazard analysis results may vary depending on the ShakeMap version used. While providing global information, particularly useful in places with poor seismic networks, ShakeMap has limitations in the variable quality and/or spatial resolution of ground motion parameters provided, which does not account for some site effects such as topographic amplification (Tanyas et al. 2019).

One important site-specific effect for the generation of landslides during earthquakes is topographic amplification, in which the interaction of ground motion with the topographic surface configuration results in increased ground accelerations, which may trigger failure on slopes that would be stable under regional shaking conditions (Sepúlveda et al. 2005a). The topographic amplification effect is characterized by the complex response of steep slopes and ridges to seismic ground motion. The lateral propagation of diffracted waves and their interference with primary *P* and *S* waves create rapid variations in amplitude and phase, giving rise to substantial differential motions along the slope (Aki 1988; Sepúlveda et al. 2005a). However, the amplification from topography is often combined with those site effects from impedance contrasts of geological materials and cannot be easily decoupled (Ashford et al. 1997; Sepúlveda et al. 2005a; Rizzitano et al. 2014; Massey et al. 2017), thus its prediction to be included in landslide hazard analysis is not straightforward. In any case, when using PGA from deterministic or probabilistic seismic hazard analysis as an input parameter for coseismic landslide analyses, combined site and topographic effects should be taken into account and a site-corrected PGA should be used, with an amplification factor obtained from empirical, numerical methods or code recommendations (e.g. Eurocode 8, CEN 2004).

Studies of earthquake-induced landslides have shown the influence of topographic amplification in locally higher coseismic landslide activity, either from field evidence and records of seismic instruments available at the top of slopes close to the landslide areas (e.g. Harp and Jibson 2002; Havenith et al. 2002; Sepúlveda et al. 2005b) or from numerical modelling of site response and slope stability of landslide sites (e.g. Havenith et al. 2003; Massey et al. 2017). Long-term monitoring of slopes in the region affected by the 2008 Wenchuan earthquake showed that, in addition to topography, the lithology and geological structure also influences the site amplification and ground motion directivity (Fan et al. 2018).

Topographic amplification has been empirically recorded and studied from seismic arrays deployed in some landslide sites and slopes (e.g. Poppeliers and

Pavlis 2002; Wasowski et al. 2013; Massey et al. 2017). For example, Wasowski et al. (2013) found directional amplifications by a factor of about 4 and 20 for PGA and Arias Intensity, respectively, in an old landslide site during the 2009 L'Aquila earthquake in Italy.

Uchida et al. (2004), from a study of the 1995 Kobe earthquake in Japan, developed a method to quantify the topographic effect on shallow earthquake-induced landslide susceptibility in terms of the slope, mean curvature of the ground surface and maximum seismic acceleration at the site, which allows estimation of the topographic amplification factor from DEM data. The method of Uchida et al. (2004) for estimating topographic amplification effects has been used by authors such as Nonomura and Hasegawa (2017). Topographic amplification can also be computed using 2D and 3D numerical modelling of slopes from different methods, including finite-difference, finite-element and distinct-element methods (e.g. Peng et al. 2009; Massey et al. 2017; Gischig et al. 2015; García et al. 2018) but this is more restrictive for single landslides and difficult to apply for hazard analysis of wider areas.

While the PGA is the most commonly used parameter to describe strong ground motion, in the landslide community, the use of Arias intensity proposed by Wilson and Keefer (1985) has become increasingly popular, as it better describes the whole ground motion shaking energy in a site, and includes characteristics of amplitude, frequency content and duration of the ground motion (Chousianitis et al. 2014). The Arias intensity (Ia, Arias 1970) is defined as

$$\mathrm{Ia} = \frac{\pi}{2g}\int_{0}^{tf} [a(t)]^2 dt \tag{1}$$

where g is the acceleration due to gravity, $a(t)$ is the recorded acceleration time history and tf is the duration of the ground motion. Earthquake-induced landslide studies from Keefer and Wilson (1989) and Wilson (1993) suggested Arias intensity threshold values around 0.10–0.15 m/s for disrupted landslides, 0.32–0.50 m/s for coherent slides and 0.50–0.55 m/s for lateral spreads and flows.

Similarly to PGA, the Arias intensity, apart from being computed directly from a ground motion record, can also be estimated from ground motion predictive equations from large ground motion data sets. These equations are usually as a function of earthquake magnitude and source distance, sometimes with specifications for site type to account for site effects and focal mechanisms (e.g. Wilson 1993; Jibson 1993; Stafford et al. 2009; Lee et al. 2012; Chousianitis et al. 2014). These the equations can in turn be used for regional to local scale prediction of Newmark displacements (see below) for susceptibility and hazard assessments (e.g. Katz and Crouvi 2007). Corrections for topographic amplification have also been proposed, such as an empirical formula by Lin and Lee (2003, in Lee et al. 2008), in which the amplification factor for Arias intensity is a function of the height relative to a riverbed.

2.2 Geological and Geomorphological Parameters

Soil and rock strength are usually difficult parameters to obtain for a widespread area, due to the little and normally very localized geotechnical data from in-situ or laboratory testing. Gallen et al. (2015) presented an inversion method to assess rock strength for earthquake-induced landslides, based on PGA, a coseismic landslide inventory and using the Newmark displacement method. By fixing the friction angle, an effective cohesion value, and thus the total rock strength, can be obtained by the combination of both for a pilot study area of coseismic landslides triggered by the Wenchuan earthquake (Gallen et al. 2015). Applying this method for the 2015 Gorkha earthquake in Nepal, Gallen et al. (2017) found that the number of landslides decreases exponentially with increasing rock strength. Similarly, Shinoda et al. (2019) assumed a certain friction angle for a geological unit and back-calculated a lower-limit cohesion from a pseudostatic analysis of slopes during a foreshock with no failure. Then, they assumed that during failure in the mainshock, cohesion decreased to zero in a residual state, which is then applied to landslide susceptibility analysis based on Newmark displacement calculations (Shinoda et al. 2019). If rock and soil strength is used in a forward stability analysis, a further complication is that seismic shaking may cause a material strength-weakening, thus if static strength parameters are used they should be corrected. For instance, Nian et al. (2019) use a strength-weakening factor formula for sensitive marine clays that is dependent on the seismic coefficient and soil sensitivity value. For fractured rock masses, Sepúlveda et al. (2005a) applied a reduction to 40% for the rock mass apparent cohesion in order to assess coseismic stability for rock slides triggered by the 1999 Chi-Chi earthquake. For those slides in slopes with discontinuity control, the reduced strength may be estimated as the residual strength of joint surfaces, which is a conservative approach (Hencher 1981).

Another factor that may influence rock slope failures such as toppling and rock avalanches is the "looseness" of fractured rock masses on the surface. This factor may be captured by airborne-electromagnetic resistivity (Nonomura and Hasegawa 2017).

The morphology of the slopes has a direct influence on their stability. The two key geomorphological controls are slope angle or gradient, and the slope height, while local changes in slope geometry and the slope orientation or aspect may also influence the occurrence of landslides. Uchida et al. (2004, 2006) proposed a method for coseismic landslide probability based only on slope parameters (slope gradient and curvature) and PGA, applying discriminant analysis to shallow landslide data from the 1995 Kobe earthquake, and later tested the analysis for other Japanese earthquakes. This method has been applied by Hasegawa et al. (2009) for only DEM-based susceptibility maps in Japan.

While slope angle and height are factors that favour slope instability, it is not always possible to generalize threshold values as an indication of failure. As the driving component or earthquake forces decrease for steeper slopes, gentler slopes are more sensitive to seismic shaking, and their susceptibility can significantly increase compared to static conditions (Das et al. 2013). There may be differences depending on the location of the slopes, as shown by Massey et al. (2018) for the Kaikoura earthquake in New Zealand, comparing the landslide activity and statistical regression models to assess landslide probability for coastal and inland slopes. They found that coastal slopes have consistently more landslides for a given slope angle than inland slopes, albeit of smaller size, and that landslide probability increases with decreasing elevation, corresponding to coastal settings (Massey et al. 2018). The coastal slopes have particular geometry, erosion and weathering processes that affect their properties and ground motion amplification, and thus must be treated carefully.

While pore-water pressure calculated from the water table is normally used in limit equilibrium slope stability analysis methods used for landslide susceptibility and hazard assessment, the dynamic variations of pore-water pressure due to seismic loading is often disregarded. This may be particularly important for seismic regions also affected by large amounts of rainfall and big storms, where the slope instability during the earthquake may be heavily conditioned by the groundwater conditions. Nguyen et al. (2020) propose a combined model that integrates a hydrological model with an infinite slope model in order to consider both the change in pore-water pressure due to rainfall and the generation of excess pore-water pressure during earthquakes for assessment of spatial probability of rainfall-earthquake-induced landslides, applied to case studies in Japan and South Korea. The model considers recharge of groundwater from rainfall, subsurface flow and calculation of factors of safety for different PGA scenarios. The model was tested for areas affected by a typhoon and then an earthquake, showing the big influence of excess pore-water pressure generation under seismic loading for slope instability (Nguyen et al. 2020).

The distance to mapped faults is also commonly used for susceptibility models. While the presence of a nearby fault may influence the geotechnical quality of rock masses and thus the susceptibility to landsliding, when used for active faults it may also be used with good results as an indicator of potential seismic intensity for shallow crustal earthquake scenarios (Massey et al. 2018).

2.3 Anthropic Parameters

Human influence on coseismic landslide susceptibility may result from different issues. Slopes on anthropogenic fill soil covers are one of them, as old, non-engineering fill materials usually are geotechnically weak and prone to seismic amplification and slope instability, as found in urban areas of ancient cities (e.g. Katz and Crouvi 2007).

Cao et al. (2019) included the anthropogenic influence of tourism in the landslide susceptibility assessment of the west Sichuan plateau in China. They included as a conditioning factor the distance to scenic spots and tourist attraction sites. Notably, when included, this factor has more impact than geomorphic or seismic parameters.

The authors suggest that the increase in build-up areas, road networks and construction at slope toes, with impacts on slope drainage, may explain this influence on landslide susceptibility (Cao et al. 2019).

All these seismic, geological or anthropic parameters are commonly expressed in thematic layer maps. For most statistical and soft computing methods, such those described in the next section, they are expressed in pixels as mapping units, as they are easily rasterized and useful in GIS platforms. The use of slope or terrain units with similar hydrological and geomorphological conditions may provide more meaningful analyses, but are more time-consuming and difficult to apply for some methods.

3 Landslide Susceptibility and Hazard Mapping Methods

Landslide susceptibility can be defined as the propensity of an area to undergo landsliding (Glade et al. 2005). A landslide susceptibility map depicts areas that have the potential for landsliding, determined by correlating some of the principal factors (typically geological and geomorphological parameters) that contribute to the generation of landslides (Highland and Bobrowsky 2008). In fact, many susceptibility maps are focused on the source areas and may not consider runout, thus not all hazard areas are necessarily covered. That difference with hazard maps may be particularly important for long-runout landslides such as debris flows and rock avalanches (Lara et al. 2018). Furthermore, hazard maps should consider the temporal occurrence or recurrence and magnitude of failure, while susceptibility maps just consider whether conditions would likely lead to failure (Hervás and Bobrowsky 2009; Hess et al. 2017), thus they may be "event-independent".

However, in earthquake-induced landslide research, the difference so far between susceptibility and hazard mapping is less clear. On the one hand, most landslide inventories, on which many methods rely as described below, do not differentiate between source and runout areas. On the other hand, most studies include seismic parameters based on event data, and are called susceptibility of hazard studies by different authors. But they may not be probabilistic or if they do not consider temporal recurrence, then those are not proper hazard assessments.

Therefore, in this review, we will not make strict distinction between susceptibility and hazard assessments, although the latter will be mostly used when some kind of temporal probabilistic analysis is included. In practice, earthquake-induced landslide probabilistic hazard mapping is mostly based on a combination of susceptibility analysis and seismic parameters (PGA or Arias intensity) obtained from a probabilistic seismic hazard assessment. The latter provide the temporal probability (scenarios with different return periods, e.g. Lee 2014; Caccavale et al. 2017; Grant et al. 2016) to the analysis, assuming a direct effect on landsliding over certain threshold values. They rarely consider either landslide runout or landslide temporal frequency, which should be included in a fully probabilistic hazard assessment.

The landslide susceptibility and hazard can be determined by different approaches. Aleotti and Chowdhury (1999) proposed a general classification of landslide hazard assessment methods, including qualitative and quantitative methods. Qualitative methods include expert field geomorphological analyses and the use of index or parameter maps (most often for susceptibility analyses), while quantitative methods include statistical analyses, geotechnical engineering approaches and neural network analyses. This classification may be updated and adapted for the current state of earthquake-induced landslide research (Fig. 1). Qualitative methods in earthquake-induced landslide research mainly rely on the overlay and subjective weighting of parameter maps, while quantitative methods include several statistical variants, deterministic analyses based on the geotechnical approach of Newmark displacement calculations and several computing methods including neural networks (Fig. 1).

3.1 Geotechnical Engineering Approaches

3.1.1 Deterministic Newmark Displacement Methods

To assess earthquake-induced landslide susceptibility deterministically, the most popular approach is the application of the Newmark sliding block method to estimate the displacement of the sliding mass. The final products are digital maps of expected permanent displacements (Newmark accumulated displacement) for pre-defined earthquake scenarios. These are generally represented by Arias intensity (e.g. Bojadjieva et al. 2018; Wang et al. 2016) or PGA as inputs for simplified Newmark methods. PGA is obtained from either probabilistic seismic hazard analysis (e.g. Caccavale et al. 2017; Kwag and Hahm 2018), deterministic analysis for specific faults and earthquake scenarios or strong ground motion seismic records (e.g. Peng et al. 2009; Shinoda et al. 2019) or from ShakeMaps (e.g. Godt et al. 2008; Allstadt et al. 2018; Gallen et al. 2017; Ma et al. 2020). Sometimes the seismic scenarios are combined with other variables such as water saturation levels (Bojadjieva et al. 2018).

The Newmark Sliding Block Method

Newmark (1965) proposed a method of estimating the displacements that a slope may suffer when subjected to seismic shaking. The method is based on an analogy between the sliding mass and a block resting on an inclined plane. The block is considered in static equilibrium, and then the seismic forces are applied. When the factor of safety drops below 1.0, the block is no longer in equilibrium and begins to move relative to the plane. That yield situation is represented by a critical acceleration (*Ac*). The displacement may be calculated by double integration of the part of acceleration time history that is above the critical acceleration (Wilson and Keefer 1985). Repeating the process over the whole duration of the earthquake, the total accumulated displacement—known as "Newmark displacement" (*DN*)—is computed (Jibson et al. 2000; Romeo 2000; Sepúlveda 2004; Fig. 2). This method, also known as Newmark rigorous analysis, can be implemented in software for use with real

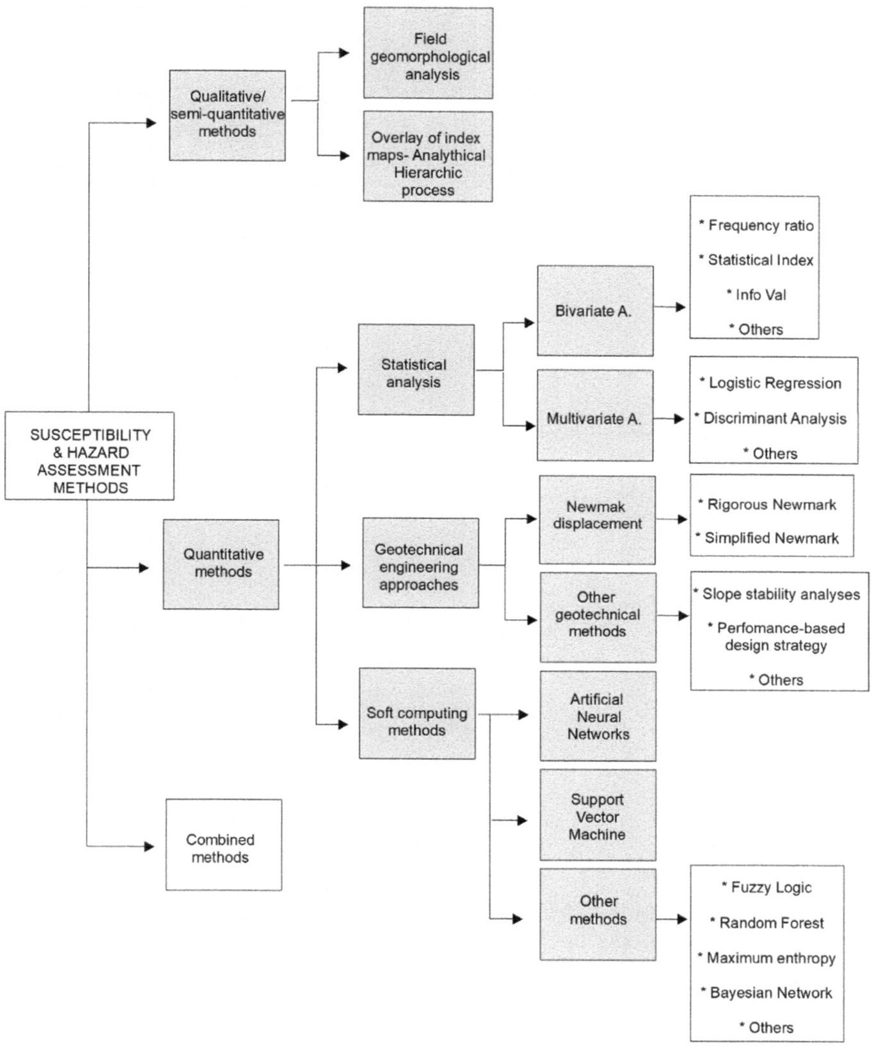

Fig. 1 Classification of methods for earthquake-induced landslides susceptibility and hazard mapping and assessment (modified after Aleotti and Chowdhury 1999 and Lara and Sepúlveda 2010)

or synthetic acceleration time histories representing an earthquake scenario for the study area.

Wasowski et al. (2013) applied rigorous Newmark displacement methods from artificial accelerograms accounting for site amplification for interpretation of a historical earthquake-induced landslide from 1627 in southern Italy. Song et al. (2018) studied the influence on displacements of ground motion directivity effects, which are common near source faults for shallow crustal earthquakes. They found that

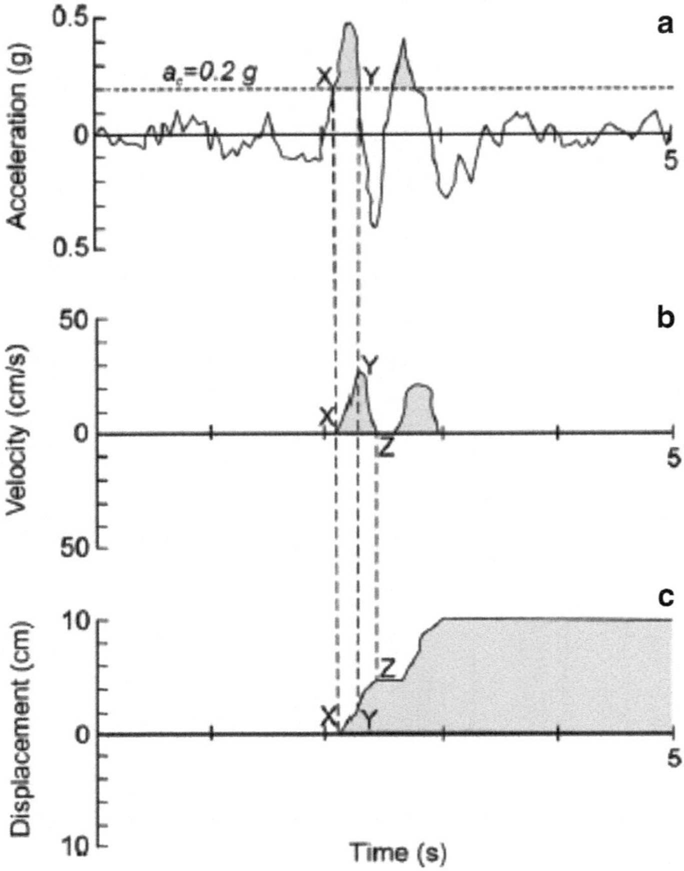

Fig. 2 Example of the Newmark rigorous method to compute sliding block displacement through double integration of an acceleration versus time record **a** over a critical acceleration of 0.2 g. **b** velocity of the landslide block versus time; **c** accumulated displacement versus time. The final value of accumulated displacement is known as the Newmark displacement (after Jibson 2007)

better correlations with "intensity measures" vary, being Arias intensity for ordinary ground motion, and PGV for forward-directivity pulse ground motions for Ac/PGA below about 0.5 and PGA for larger Ac/PGA, regardless of the ground motion type (Song et al. 2018). Shinoda and Miyata (2017) also applied the Newmark method, accounting for horizontal and vertical acceleration time histories at each slope azimuth estimated from local ground motion records using inverse distance weighting, assuming that phase characteristics remain the same as those of the closest observatory.

Simplified Newmark Methods

As the rigorous Newmark analysis can be difficult to implement for several slopes in a regional susceptibility or hazard analysis, several researchers have statistically obtained predictive relationships for Newmark displacement, an application known as Newmark simplified analysis. In these relationships, the Newmark displacement is expressed as a logarithmic function of critical acceleration and a ground motion parameter, such as Arias Intensity (e.g. Jibson 1993; Jibson et al. 2000) or PGA (e.g. Ambraseys and Menu 1988; Jibson 2007). A good summary and discussion of these regression models for Newmark displacement is found in Jibson (2007). Examples of Newmark simplified analysis equations are (Jibson et al. 2000; Jibson 2007)

$$\log DN = 1.52 \log Ia - 1.993 \log Ac - 1.546 \tag{2}$$

$$\log DN = 0.215 + \log\left[\left(1 - \frac{Ac}{Amax}\right)^{2.341}\left(\frac{Ac}{Amax}\right)^{-1.438}\right] \pm 0.51 \tag{3}$$

where log DN is the base 10 logarithm of Newmark displacement, Amax is the site-corrected PGA, and *Ac* is the critical acceleration. *DN* represents the final permanent displacement due to the seismic input. The corrected PGA values should include eventual amplification effects corrections (e.g. Peng et al. 2009; Rodríguez-Peces et al. 2011; Caccavale et al. 2017).

Wilson and Keefer (1985) suggested fixing a minimum critical acceleration of 0.05 g because, for very low *Ac* values, the slope is very unstable and thus, it may likely fail under non-seismic conditions from more frequent triggers.

Many authors compute *DN* based on simplified analyses based on PGA, Ac and/or Arias intensity using equations such as Eqs. 2 and 3. Other authors such as Chousianitis et al. (2014) in Greece have developed their own regression equations for *Ia* and then for *DN* from local seismic datasets.

In susceptibility and hazard assessment studies, critical accelerations for Newmark displacement computations are commonly obtained from pseudostatic analysis using the infinite slope model, which restricts the application of the method to shallow landslides. Some authors adopt multimodal approaches for different failure modes such as rock blocks, disrupted slides and coherent rotational slides (Grant et al. 2016). The use of such critical accelerations for regional or district scale analysis imposes uncertainties from the geotechnical properties such as cohesion, internal friction angle and unit weight of the materials, which at regional scale are assigned for different geologic units mostly from extrapolation of singular points data or from the literature (e.g. Bojadjieva et al. 2018; Shinoda et al. 2019).

The resulting Newmark displacements are often classified in susceptibility or hazard categories, which vary from different authors. For example, Bojadjieva et al. (2018) proposed three categories of relative hazard: low (*DN* = 0–20 cm), moderately low (20–50 cm) and high (>50 cm). Caccavale et al. (2017) chose *DN* = 2, 5 and 10 cm as threshold values for hazard zoning at different seismic return periods, while

Shinoda et al. (2019) used 5, 15 and 30 cm to define four susceptibility categories from very low (<5 cm) to high (>30 cm).

While application of the Newmark displacement method is usually deterministic, if probability distribution functions for geotechnical, slope and seismic factors are available, then a probabilistic approach can be used, as shown by Refice and Capolongo (2002), who applied Montecarlo techniques to obtain probability distribution functions, or by Khalaj et al. (2020) who used Bayesian network methods for probabilistic modelling of Newmark displacement. In a different approach, Caccavale et al. (2017) combined different probabilistic seismic scenarios with related Newmark displacement values over a certain threshold, to construct earthquake-induced landslide hazard maps for different return periods based on the estimated annual frequency of occurrence of the seismic triggering event.

3.1.2 Other Geotechnical Methods

Slope stability analyses have limited application for regional susceptibility assessment as they are thought to be applied to specific slopes with detailed geotechnical and topographic data rather than regionally, as used in susceptibility and hazard maps. Nevertheless, susceptibility maps for shallow landslides applying the simpler, 2D infinite slope limit equilibrium method has long been applied in landslide susceptibility for static conditions, assuming the soil strength data by extrapolation or from the literature and using slope gradient values for each cell from digital elevation models. The introduction of seismic forces adds even more uncertainty to the analyses. Hess et al. (2017) proposed an interesting expansion to 3D slope stability analysis for shallow landslides considering seismicity, lateral earth pressures and tree root reinforcement. While still based on the limit equilibrium theory, but by not considering dynamic response to pore-water pressure or soil strength, and using pseudostatic accelerations, the Hess et al. (2017) three-dimensional topographic limit equilibrium (**3DTLE**) model also includes back-calculation of soil strength from a local landslide inventory, which improves the results, and allow calculation of the probability of failure.

Sharifi-Mood et al. (2017) present a probabilistic method using a **performance-based design strategy**, approach in which how likely an engineering demand parameter would be exceeded is calculated. This method uses a DEM, a landslide inventory, probabilistic seismic hazard curves, engineering soil site classification and geological mapping to compute the exceedance probability of seismically-induced landslides at a location and the probability of exceeding certain displacement thresholds.

3.2 *Statistical and Computing Methods*

Statistical methods are very popular for earthquake-induced landslide susceptibility and hazard assessment. They can be implemented in GIS platforms for regional analysis based on a series of available geomorphological, geological and/or seismic parameters, and their accuracy can be tested and their results in many cases can be expressed in probabilistic terms. Other mathematical techniques, including machine learning or neural networks methods, have also been implemented for the same purpose. The main limitation of all of these methods is that they rely on the existence of landslide inventories in the study area, which are not always available, in particular if the aim is a specific analysis for earthquake-induced landslides. While there have been compilations of earthquake-induced landslide inventories (Tanyas et al. 2017), they are far from complete. Thus, these methods are usually applied in regions recently affected by large earthquakes, for which landslide inventories have been constructed. This generally results in an analysis that is performed with data from just one seismic event (e.g. García-Rodríguez et al. 2008; Zhang et al. 2012), which may result in a biased analysis. The general approach consists of the statistical comparison of landslide presence or absence with certain landslide controlling factors such as relief, slope gradient and curvature, lithology, distance to faults or streams, and often seismic parameters such PGA or PGV, among others. While most authors use the concept of susceptibility for these analyses, but oriented to earthquake-induced landslides, some consider that when parameters related with the seismic trigger (e.g. PGA) are part of the variables, then the resulting map is a hazard map (e.g. Budimir et al. 2015). This is consistent with the concept that for hazard analysis both intrinsic factors (e.g. geology, slope gradient) and extrinsic or causative factors (rainfall, shaking) must be included (Hervas and Bobrowsky 2009).

In the last 20 years, several bivariate and multivariate classic statistical methods and other sophisticated algorithmic techniques have been implemented for the analysis of landslide susceptibility in general, and for earthquake-induced landslide susceptibility and hazard in particular. These approaches include multivariate methods such as logistic regression and discriminant analysis, bivariate statistical methods such as frequency ratios, statistical index models and Info Val, and soft computing methods including artificial neural networks, random forest and support vector machine (SVM) models (Fig. 1).

The **accuracy assessment** for landslide susceptibility and hazard maps based on statistical methods is in most cases carried out using the area under the curve (AUC, Fig. 3) of a receiver-operating characteristic curve (ROC). The observed landslide areas should coincide with the areas of high susceptibility on the resulting landslide susceptibility map. In addition, high susceptibility values should cover small areas (Can et al. 2005; Bai et al. 2012). In model evaluation, AUC > 0.9 can be considered excellent, AUC > 0.8 as good and AUC > 0.7 as fair (Lee 2014).

3.2.1 Multivariate Methods

A simple multivariate approach is the use of **discriminant analysis**, consisting of a linear multiple regression in which the presence/absence or percentage of landslides in each terrain unit or pixel is a linear function of the instability factors. The final

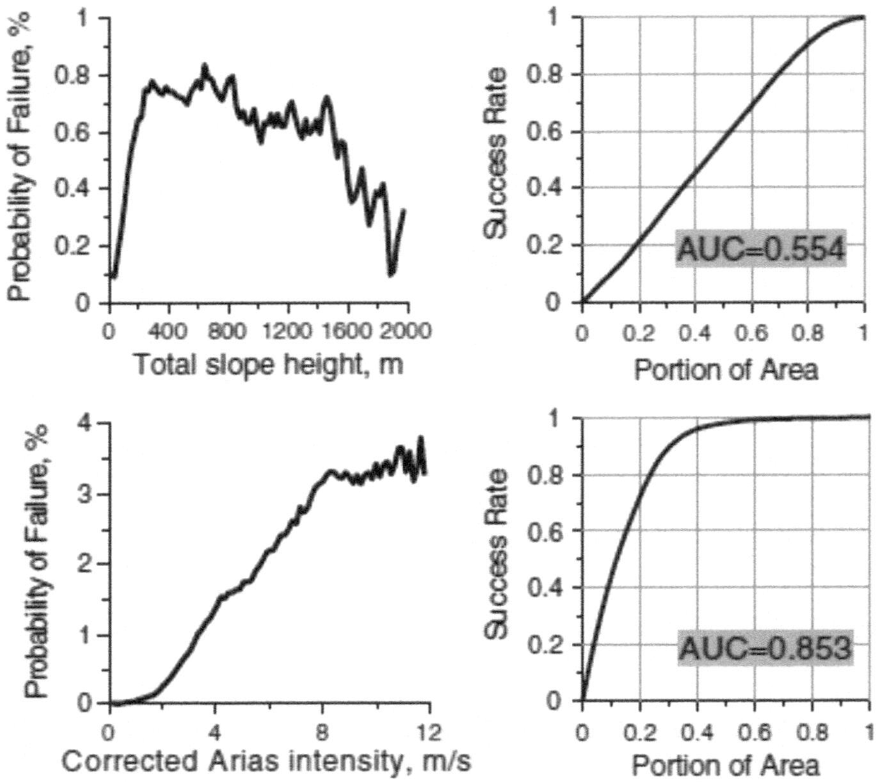

Fig. 3 Example of statistical analyses results for Taiwan, using a logistic regression multivariate method (Lee 2014). Probability of failure (left) and success rate curve (right) are shown for slope height and corrected Arias intensity parameters. The value of the area under the curve (AUC) is the area under the success rate curve, with higher values representing better accuracy of the hazard assessment (after Lee 2014)

susceptibility model can be obtained by applying a sigmoidal function to the discriminant function, to convert the discriminant values into probabilities (Valagussa et al. 2017). For example, discriminant analysis has been applied to coseismic landslides in Nepal by Valagussa et al. (2017), with reasonably good prediction power, and by Lee et al. (2008) in Taiwan, who include among the variables the Arias intensity corrected for topographic amplification, which substantially improves the correlation with the Arias intensity parameter.

Most popular among multivariate methods is **logistic regression,** a method that estimates the probability of a certain event dependent of a number of variables. In this case, the dependent variable is a binary variable representing the presence or absence of landslides (Bai et al. 2012). The probability of landslide occurrence P is expressed in the general form (Hosmer and Lemeshow 2000):

$$P = \frac{1}{1 + e^{-z}}, \quad z = \beta_0 + \beta_1 \beta x_1 + \beta_2 x_2 + \cdots + \beta_n x_n \tag{4}$$

where x_i ($i = 1, 2, 3, \ldots, n$) are the independent variables and β_i are the model coefficients for positive or negative correlations. The independent variables represent pre-failure conditions. P varies from 0 to 1 on an s-shaped curve. Logistic regression has some advantages over discriminant analysis, in that variables can be used for categorical data, for instance lithology and slope aspect, normality is less important, and the shape and concentration of data points in the probability of failure curve are better (Lee 2014).

Seismic parameters for the logistic regression analysis can be included in different ways. The most common parameter is PGA, although some may prefer PGV, which was found to work better than PGA by Nowicki et al. (2014). Budimir et al. (2015) made a hazard analysis of an area affected by the 1994 Northridge earthquake in California using PGA from seven shaking maps for different earthquake scenarios with a common fault source of the 1994 earthquake, using the open source OpenSHA application for seismic hazard analysis (opensha.org). Nowicki et al. (2014) proposed a global model based on logistic regression and using PGA from the USGS ShakeMap as the shaking factor, which was expanded by Jessee (2017) and Nowicki Jessee et al. (2018), which preferred the use of PGV, also from ShakeMap. Allstadt et al. (2018) tested the global logistic regression models of Nowicki et al. (2014) and Nowicki Jessee et al. (2018), together with a Newmark method model by Godt et al. (2008), for the 2016 Kaikoura earthquake in New Zealand. They used several versions of the USGS ShakeMap that evolved in time with increasing data, from a few minutes to three months after the earthquake. The results show that the ShakeMap details strongly control the spatial pattern of landslide hazard and that the models tend to overpredict the hazard, especially for far source distances, but improve when the ShakeMap is based on an extended rupture rather than a point source used in the quick initial ShakeMap outputs (Allstadt et al. 2018). Also, studying the Kaikoura earthquake, Massey et al. (2018) performed logistic regression analyses using both PGA and PGV from ShakeMap calibrated for New Zealand, plus several other susceptibility variables. Interestingly, they found that the distance to the surface expression of the fault that ruptured used as a proxy of ground motion intensity has more power as a landslide predictor than PGA or PGV. This may be due the higher frequency of ground motions close to the faults that favour amplification of slopes responding to such high frequencies (Massey et al. 2018). In a country-scale study for Taiwan, Lee (2014) used Arias intensity as the seismic factor, obtained from probabilistic seismic hazard analysis and corrected for topographic amplification, which was found to have better predictive results than PGA and PGV (Lee 2014, Fig. 4).

ShakeMap inputs have also been used for global susceptibility analysis, such as the work by Tanyas et al. (2019). They used logistic regression with software developed by Rossi and Reichenbach (2016) for its application on landslide susceptibility evaluation, and based on slope units instead of pixels, applying a software module by Alvioli et al. (2016) for their automatic delineation. They worked with a selection of earthquake-induced landslide inventories from Tanyas et al. (2017). Results show

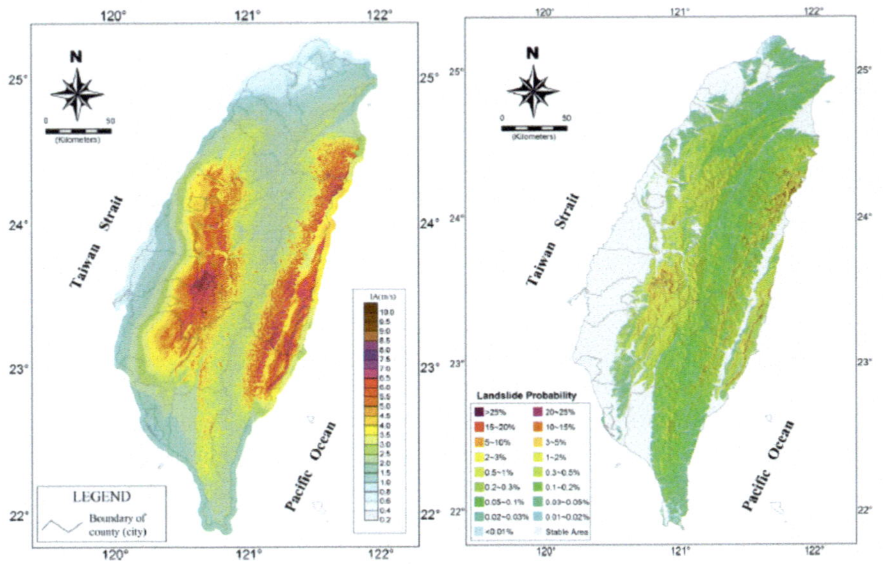

Fig. 4 Example of an earthquake-induced landslide hazard map at a regional scale for Taiwan, using Arias intensity corrected for topographic effects as the seismic parameter and a logistic regression method. Left: 475-year topographically corrected Arias intensity map; Right: 475-year seismic landslide hazard map (after Lee 2014)

that models trained by specific inventories give a low performance in most cases, while if the inventories are grouped and categorized, model performance improves (Tanyas et al. 2019). This shows the importance of the landslide inventory quality for the method and the difficulties of their use in other regions.

Some authors have made comparative studies of landslide susceptibility using logistic regression with different landslide inventories for the same area. Bai et al. (2012) applied logistic regression for two landslide inventories in the Zhouqu segment region of north-western China. The first was an inventory of historical landslides, and the second of earthquake-induced landslides triggered by the 2008 Wenchuan earthquake. Their results show that susceptibility only varies strongly for 6% of the area between maps, while 58% of the area has the same susceptibility class. Meanwhile, Bai et al. (2013) compared susceptibility maps made with rainfall-induced and earthquake-induced landslide inventories in Wudu County, China. Dependent variables were almost the same, with exception of those related to the triggering factor, as average annual distribution was used for the rainfall-induced susceptibility map while PGA was chosen for the earthquake-induced susceptibility one. As would be expected, they found in general a similar distribution, with main differences in the lower susceptibility areas. In both studies, Bai et al. (2012, 2013) used the combination of both susceptibility maps to describe the maximum likelihood of landslide occurrence in the area.

3.2.2 Bivariate Methods

Bivariate statistics are another approach, in which each individual factor is compared with the landslide inventory map. Each factor is categorized in classes, and the weighted value of each class is determined on the basis of landslide density for each individual class (Xu et al. 2012a, b). Different bivariate models have been used for earthquake-induced landslide studies, such as the **statistical index model** of Yalcin (2008) applied by Xu et al. (2012a) to the Wenchuan earthquake, in which a weight value for a factor class is defined as the normalized natural logarithm of the landslide density for each class. For the **frequency ratio model**, the ratio of the area with landslide occurrence in the total study area to the ratio of the probabilities of landslide occurrence to a non-occurrence for number of susceptibility factors, named the frequency ratio, is computed (e.g. Umar et al. 2014; Regmi et al. 2016; Tanoli et al. 2017).

Another bivariate method is the **Info Val method**, proposed by van Westen (1997). The method is based on the determination of the probability of landslide occurrence within each class of thematic data layer, in which the weight for such class is calculated from the ratio of landslide density within that particular class and landslide density within the entire layer. The Info Val method has been applied in the Himalayas by Pareek et al. (2010, 2013) for a series of conditioning factors with and without using PGA obtained from probabilistic seismic hazard assessment, finding that the inclusion of seismic factors improves the results of the susceptibility mapping.

3.2.3 Soft Computing and Machine Learning Methods

Advanced statistical methods mainly include soft computing technologies such as artificial neural networks and support vector machine as well as other methods. Some of these methods have been widely applied for non-seismic landslide susceptibility, and adopted for coseismic susceptibility in recent years (Xu et al. 2012a, 2016 and references therein). The **artificial neural network (ANN)** is an attempt to imitate the human brain neural system, applying a learning algorithm for prediction. There are different types of algorithm, such as the back-propagation training algorithm used by Xu et al. (2012a, 2016) and Das et al. (2013) for earthquake-induced landslides, using input layers based on the conditioning factors defined for the landslide susceptibility analysis. The ANN learns by adjusting the weights in response to errors between the target and actual output values (Pradhan and Lee 2010). The autoencoder is another type of unsupervised learning algorithm that can be trained to reconstruct input onto the output layer, as used by Nam and Wang (2019) in Japan.

Support vector machine modelling, or SVM, is another procedure based on statistical learning theory, with a training phase and then application of the trained model to separate data sets (Yao et al. 2008; Xu et al. 2012b). The method is used as a machine learning algorithm, based on nonlinear transformations of covariates into higher dimensional space, using kernel functions, and finding an optimum linear separating hyperplane that separate data patterns (Xu et al. 2012b; Huang and Zhao

2018). Kernel functions found to work well for this purpose are linear, polynomial, radial-basis and sigmoid functions (Xu et al. 2012b). A study by Xu et al. (2012b) on earthquake-induced landslides from the Wenchuan earthquake in China compared SVM modelling for different training samples grouped by landslide size, and a series of common input parameters such as elevation, slope distance to the fault and drainages and lithology. They found that radial-basis and polynomial kernel functions are suitable for modelling any training data, while best results are obtained with a radial-basis function for training samples ignoring landslide size, with success rates (AUC values) of the order of 80%.

Xu et al. (2012a) compared three SVM models with bivariate statistics, logistic regression and artificial neural network for the same area affected by the 2008 Wenchuan earthquake. The AUC values of the results showed that the logistic regression model got the best predictive results, followed by the SVM-radial-basis function model (both with AUC around 80%), while the SVM-linear model shows the lowest AUC values (Xu et al. 2012a). In another comparative study, Xu et al. (2016) compared SVM and ANN models for earthquake-induced landslides triggered by the 2013 Minxian-Zhangxian earthquake in China. After testing ten different models, they found the best results using the SVM polynomial kernel function, in contrast with previous studies pointing to the radial-basis function as most suitable SVM model type for landslide susceptibility assessment. Meanwhile, they also found that the sigmoid kernel function is inappropriate for this purpose (Xu et al. 2016). A comparative analysis, by Huang and Zhao (2018) of SVM performance against other methods such as ANN, logistic regression and random forest, proposes that SVM has some advantages for landslide susceptibility mapping, such as a good performance with a small group of samples and that it solves in an easier way nonlinear and high-dimensional classification problems. However, it is not user-friendly, may present overestimation issues and is not suitable for very large datasets due to lower performance in the speed of training and testing datasets (Huang and Zhao 2018).

Another approach is the use of **fuzzy logic** in GIS (Lee et al. 2008; Miles and Keefer 2009; Kritikos et al. 2015). Fuzzy logic is derived from fuzzy set theory, in which a relationship, or "membership", between 0 and 1 is obtained for each landslide causative factor based on a frequency ratio analysis, and then combined through a fuzzy gamma operator to obtain relative hazard values (Kritikos et al. 2015). Robinson et al. (2017) showed that the fuzzy logic method can be used for quick hazard assessment based on preliminary inventories after an earthquake, provided they have sufficiently wide areal coverage. Meanwhile, Kritikos et al. (2015) showed that using relationships obtained from applying the method to well documented cases and complete inventories, it can then be successfully applied to cases without landslide inventories. However, all their case studies are related to shallow crustal earthquakes and the results may not work for other earthquake types such as megathrust earthquakes in subduction zones, which tend to show different landslide patterns (Wartman et al. 2013; Serey et al. 2019), thus such results must be used with caution.

The **random forest** method is an ensemble-learning technique in which models of decision trees are built and then combined to obtain the final predicted results by a voting process. It has good stability and is included in common statistical software packages (Cao et al. 2019). It also reduces calculations compared with other computing intensive machine learning algorithms such as ANN (Huang and Zhao 2018). Cao et al. (2019) performed a comparative study of logistic regression and random forest methods for the Western Sichuan plateau region in China affected by the 2008 Wenchuan, 2013 Lushan and 2017 Jiuzhaigou earthquakes. Landslide susceptibility analysis using the landslide inventories for those three events show that the level of prediction using the ROC assessment was consistently higher for the random forest method, as shown in other studies (e.g. Youssef et al. 2016). The authors point out that the random forest method can perform better for nonlinear relationships between the landslides and influencing factors, while it can also handle a large number of variables including numerical and categorical without need to be rescaled or transformed (Cao et al. 2019).

Shrestha and Kang (2019) used the **maximum entropy model**, originally from the field of statistical physics and used in ecology, to assess earthquake-induced landslide susceptibility in Nepal. This is another type of machine learning model that calculates patterns and processes without requiring large amounts of data, to estimate a probability distribution of landslides over a set of sites in the study area (Shrestha and Kang 2019; Phillips et al. 2006).

Song et al. (2012) developed a **Bayesian network** method, based on Bayes' theorem, for landslide susceptibility. The Bayesian network is a graphical model for probabilistic relationships among a set of variables, with a first component corresponding to the landslide-causing factors, and a second that describes a conditional distribution for each node. The conditional probability distribution is determined by Bayesian network parameter learning. This method was tested for the Wenchuan earthquake by Song et al. (2012), using seven environmental factors and Arias intensity, finding major influence from the latter, and lithology, with high accuracy of the model (AUC of 93%).

3.3 *Qualitative or Semi-quantitative Methods*

Expert judgement and experience is always an option for susceptibility and hazard assessment, although it use should be restricted for specific cases at consultancy level, or as a first approach before using more quantitative and objective methods. In any case, given the inherent uncertainties in all the previously described methods expert engineering and geological judgement for critical revision of results is always recommended. A different qualitative approach for the assessment, but not mapping, of earthquake-induced landslide hazards and transmission of expert judgement to authorities and practitioners is the use of conceptual geomodels, three-dimensional computer drawings with indication of expected landslide type, size and effects for different seismic and geomorphological settings, as proposed by Serey et al. (2020) for the Chilean Andes.

Apart from a direct expert assessment based on field geomorphological analyses, semi-quantitative methods based on overlay of factor maps using GIS platforms are popular and low cost methods for landslide assessment, and earthquake-induced landslide research is not an exception. These kind of methods may be particularly appropriate for regions with data that is too scarce or low quality to apply more sophisticated methods. A **holistic approach** based on overlay of factor maps similar to statistical analyses, but in which the relative weights are mostly assigned based on field surveys, literature review and author experience, sometimes supported by partial statistical analysis of some factors, has been adopted for coseismic landslides by several authors. This method, in which the subjective judgement is decomposed into a layer structure to express results quantitatively, is also known as the **analytical hierarchic process** (Saaty 1980). Examples of such analysis for earthquake-induced landslide susceptibility and hazard analysis have been produced by Abuzied et al. (2016) in Egypt, Yoshimatsu and Abe (2006) in Japan, as well as by Kamp et al. (2008), Tang et al. (2009) in Pakistan, Lara and Sepúlveda (2010) in Chile and James and Sitharam (2017) in India. The method is a simple and easy tool compared with statistical or machine learning type methods, but it has the limitation of introducing more subjectivity to the analysis (Huang and Zhao 2018). However, it has the advantage of not needing landslide inventories, although their use to assign weighting factors should improve their predicting capacity.

3.4 Combined Methods

Ma et al. (2020), aiming at response after an earthquake, propose a combination of methods in three phases: for quick emergency response, a Newmark displacement method using PGA data from shaking maps is applied. Then, a second, temporary resettlement phase is implemented using logistic regression methods with limited landslide inventory data collected after the earthquake. A refined logistic regression model with the complete landslide inventory is then applied for a third, reconstruction phase.

Wang et al. (2016) present another example of combination of Newmark displacement and logistic regression methods, applied to the Wenchuan earthquake. They first assess the Newmark displacement, which is in turn used as a variable for logistic regression, together with morphological factors. Results show that the combined model improves the landslide prediction more than using the Newmark method alone (Wang et al. 2016).

Umar et al. (2014) presented an ensemble method of frequency ratio and logistic regression for landslide susceptibility applied to the 2009 West Sumatra earthquake. The combined method allows assessing the impact of conditioning factors, as well as the impact of classes of each conditioning factor, on landslide occurrence (Umar et al. 2014).

Yang et al. (2015) proposed a combination of analytical hierarchic process and bivariate analysis to obtain factor weights and a sensitivity coefficient, which were multiplied to obtain a landslide susceptibility index for quick susceptibility assessment after the 2013 Lushan earthquake in China. The combined results reduced the subjectivity and uncertainty of the single methods.

4 Landslide Related Hazards and Risk: Beyond the Susceptibility or Hazard Map

Landslide dams are an important hazard associated with large earthquakes. For example, for the 2008 Wenchuan earthquake, at least ten rivers were blocked by landslide debris, with landslide-dammed lakes closely related to proximity to the earthquake source fault (Yin et al. 2009). These authors proposed a qualitative risk assessment of landslide dam breaching based on dam height, capacity of the lake and the materials and materials forming the dam (Yin et al. 2009). Gallen et al. (2017) applied their landslide displacement analyses, based on ShakeMaps and the Newmark method, to the analyses of landslide dam's hazards in Nepal. Instead of modelling each landslide, they used the density of failure cells (over a 5 cm Newmark displacement threshold) as a proxy for landslide debris available to impound flow. Then, using GIS hydrology tools, they identified those streams with a potential for being blocked.

Van Westen et al. (2006) recommend that a **risk assessment** at a medium scale could only be a qualitative, or at best a semi-quantitative one. **The risk classes**, categorized with such terms as "very high", "high", "moderate", "low", and "very low" risk, should be defined (Bojadjieva et al. 2018).

Bojadjieva et al. (2018), based on their hazard maps computed using Newark displacement, performed simple qualitative risk assessment by combination with an **exposure map** of the housing and local roads of their study area in Macedonia (Fig. 5). Similarly, Budimir et al. (2015) combined their hazard map prepared with logistic regression with the exposure of assets such as population, housing and roads in Southern California. Exposure is a binary expression (exposed or not) applied to elements present and thus subject to damage or potential losses (Budimir et al. 2015), which can be identified for areas of high landslide probability as a simplified form of risk assessment. Hess et al. (2017) combined their susceptibility maps with the road network to create "infrastructure susceptibility maps" under different scenarios.

Zhang et al. (2012) considered not just the presence of assets, but took the area of houses and farmland and the length of roads as economic indices of **vulnerability** in areas of high earthquake-induced landslide susceptibility affected by the Wenchuan earthquake in China. These vulnerability proxy values were multiplied by the susceptibility index obtained by logistic regression to compute a risk value. The authors considered not only the susceptibility value at the location of the exposed unit, but also those adjacent and lower in altitude to high susceptibility terrain units, to account for burial potential (Zhang et al. 2012).

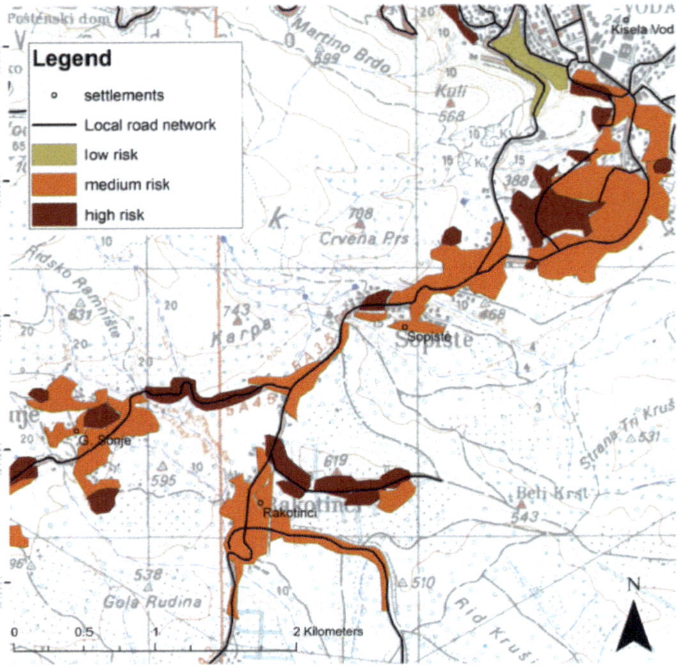

Fig. 5 Example of a landslide risk map in Macedonia, created by combination of Newmark displacement-based hazard maps and exposure maps of settlements and roads (after Bojadijeva et al. 2018)

A more comprehensive model to quantitatively estimate the human vulnerability to earthquake-induced landslides was proposed by Zhang et al. (2020), using Bayesian networks. The network considers as main nodes warning time, resistance capacity and evacuation, with several dependent parameters, and conditional probabilities are established between them to finally assess the probability of loss of life as a human vulnerability index. Applying the model to a case study for the Wenchuan earthquake, they got a good agreement in the fatality rate with recorded landslides and found that time of day, evacuation distance, warning time and evacuation are the four most important parameters governing the loss of life (Zhang et al. 2020).

Some specific risk studies for particular types of infrastructure can be found in the literature. For instance, Kwag and Hahm (2018) presented an approach for probabilistic safety assessment for nuclear power plants subject to seismic and earthquake-induced landslide hazards. In their work, the landslide hazard is modelled using Newmark analyses. A probability of slope failure is computed and combined with the probability of the landslide reaching the plant, based on travel distance calculations, and the seismic hazard from probabilistic analysis. Then, Kwag and Hahm (2018) obtain seismic fragility curves for different accident scenarios to obtain a core damage frequency as an indicator of safety for the nuclear power plant.

5 Concluding Remarks

Landslide susceptibility and hazard assessment are some of the great challenges for effective disaster risk management of earthquake-induced landslides. The choice of appropriate methodologies and the large uncertainties related to seismic parameters, as well as geotechnical and geological variability, are key aspects that must be dealt with. Despite this, important advances in the development and application of deterministic and probabilistic methods including novel machine learning and other computing methods have allowed an increased variety of options to assess landslide hazards during earthquakes. Some remaining challenges are the development of more accurate seismic scenarios models for forward hazard assessment, as most studies are still based in specific case studies of recent earthquakes, and of tools for quick hazard mapping for emergency response after large earthquakes. More information on subduction-type earthquake case studies is also desirable for future research and better modelling of seismic parameters, as well as better distinction of source and runout areas in the confection of earthquake-induced landslide inventories, to allow a clearer distinction between susceptibility and hazard analyses.

Acknowledgements Marisol Lara aided with figure preparation. Many thanks to Bill Murphy, Dave Petley, Matthew Brain, Greg De Pascale, Alejandra Serey, César Pastén, Gonzalo Montalva, Sofía Rebolledo, Marisol Lara, former students and colleagues of JTC1-TR4 group for fruitful discussions on earthquake-induced landslides along the many years that provide the background for this review.

References

Abuzied S, Ibrahim S, Kaiser M, Saleem T (2016) Geospatial susceptibility mapping of earthquake-induced landslides in Nuweiba area, Gulf of Aqaba, Egypt. J Mountain Sci 13(7):1286–1303

Aki K (1988) Local effects in ground motion. In: Earthquake engineering and soil dynamics II: recent advances in ground motion evaluation, special publication, 20. American Association of Civil Engineers, pp 103–155

Aleotti P, Chowdhury R (1999) Landslide hazard assessment: summary review and new perspectives. Bull Eng Geol Environ 58:21–44

Allen TI, Wald DJ, Hotovec AJ, Lin K-W, Earle P, Marano KD (2008) An atlas of ShakeMaps for selected global earthquakes. Geol Survey (US)

Allstadt KE, Jibson RW, Thompson EM, Massey EI, Wald DJ, Godt JW, Rengers FK (2018) improving near-real-time Coseismic landslide models: lessons learned from the 2016 Kaikōura, New Zealand, Earthquake. Bull Seismol Soc Am 108(3B):1649–1664

Alvioli M, Marchesini I, Reichenbach P, Rossi M, Ardizzone F, Fiorucci F, Guzzetti F (2016) Automatic delineation of geomorphological slope units with r.slopeunits v1.0 and their optimization for landslide susceptibility modeling. Geosci Model Dev 9:3975–3991. https://doi.org/10.5194/gmd-9-3975-2016

Ambraseys NN, Menu JM (1988) Earthquake-induced ground displacements. Earthquake Eng Struct Dynam 16:985–1006

Arias A (1970) A measure of earthquake intensity. In: Hansen RJ (ed) Seismic design for nuclear power plants. MIT Press, Cambridge, pp 438–483

Ashford S, Sitar N, Lysmer J, Deng N (1997) Topographic effects on the seismic response of steep slopes. Bull Seismol Soc Am 87(3):701–709
Bai S, Wang J, Zhang Z, Cheng C (2012) Combined landslides susceptibility mapping after Wenchuan earthquake at the Zhouqu segment in the Bailongjiang basin, China. CATENA 99:18–25
Bai S, Cheng C, Wang J, Thiebes B, Zhang Z (2013) Regional scale rainfall- and earthquake-triggered landslide susceptibility assessment in Wudu County, China. J Mountain Sci 10(5):743–753
Bojadjieva J, Sheshov V, Bonnard C (2018) Hazard and risk assessment of earthquake-induced landslides—case study. Landslides 2018(15):161–171
Budimir MEA, Atkinson PM, Lewis HG (2015) Seismically induced landslide hazard and exposure modelling in Southern California based on the 1994 Northridge, California earthquake event. Landslides 2015(12):895–910
Caccavale M, Matano F, Sacchi M (2017) An integrated approach to earthquake-induced landslide hazard zoning basedonprobabilistic seismicscenario forPhlegrean Islands (Ischia, Procida and Vivara), Italy. Geomorphology 295:235–259
Can T, Nefeslioglu HA, Gokceoglu C, Sonmez H, Duman TY (2005) Susceptibility assessments of shallow earthflows triggerered by heavy rainfall at three subcatchments by logistic regression analyses. Geomorphology 72:250–271
Cao J, Zhang Z, Wang C, Liu J, Zhang L (2019) Susceptibility assessment of landslides triggered by earthquakes in the Western Sichuan Plateau. CATENA 175:63–76
CEN (Comité Européen of Normalisation), 2004. Eurocode 8: design of structures for earthquake resistance. Part 1: general rules, seismic action and rules for buildings. In: European Committee for Standardization (Comité Européen de Normalisation), European Standard EN 1998–1:2004(E). COMITÉ EUROPÉEN DE NORMALISATION, Brussels (231 pp)
Chousianitis K, Del Gaudio V, Kalogeras I, Ganas A (2014) Predictive model of Arias intensity and Newmark displacement for regional scale evaluation of earthquake-induced landslide hazard in Greece. Soil Dyn Earthq Eng 65:11–29
Das HO, Sonmez H, Gokceoglu C, Nefeslioglu HA (2013) Influence of seismic acceleration on landslide susceptibility maps: a case study from NE Turkey (the Kelkit Valley). Landslides 10:433–454
Fan X, Hsein JC, Wasowski J, Huang R, Xu Q, Scaringi G, van Westen CJ, Havenith HB (2018) What we have learned from the 2008 Wenchuan Earthquake and its aftermath: a decade of research and challenges. Eng Geol 241:25–32
Gallen SF, Clark MK, Godt JW (2015) Coseismic landslides reveal near-surface rock strength in a highrelief, tectonically active setting. Geology 43:11–14
Gallen SF, Clark MK, Godt JW, Roback K, Niemi NA (2017) Application and evaluation of a rapid response earthquake-triggered landslide model to the 25 April 2015 Mw 7.8 Gorkha earthquake, Nepal. Tectonophysics 714–715:173–187
García M, Pastén C, Sepúlveda SA, Montalva G (2018) Dynamic numerical investigation of a stepped-planar rockslide in the Central Andes, Chile. Eng Geol 237:64–75
García-Rodríguez MJ, Malpica JA, Benito B, Díaz M (2008) Susceptibility assessment of earthquake-triggered landslides in El Salvador using logistic regression. Geomorphology 95:172–191
Gischig VS, Eberhardt E, Moore JR, Hungr O (2015) On the seismic response of deep-seated rock slope instabilities—insights from numerical modeling. Eng Geol 193:1–18
Glade T, Anderson M, Crozier MJ (2005) Landslide hazard and risk. Wiley, Chichester, 793 p
Godt JW, Şener B, Verdin KL, Wald DJ, Earle PS, Harp EL, Jibson RW (2008) Rapid assessment of earthquake-induced landsliding. In: Proceedings of the 1st world landslides forum, United Nations University, Tokyo, Japan, pp 219–222
Grant A, Wartman J, Abou-Jaoude G (2016) Multimodal method for coseismic landslide hazard assessment. Eng Geol 212:146–160

Harp EL, Jibson RW (2002) Anomalous concentrations of seismically triggered rockfalls in Pacoima Canyon: are they caused by highly susceptible slopes or local amplification of seismic shaking? Bull Seismol Soc Am 92(8):3180–3189

Hasegawa S, Dahal RK, Nishimura T, Nonomura A, Yamanaka M (2009) DEM-based analysis of earthquake-induced shallow landslide susceptibility. Geotech Geol Eng 27:419–430

Havenith HB, Jongmans D, Faccioli E, Abdrakhmatov K, Bard PY (2002) Site effect analysis around the seismically induced Ananevo rockslide, Kyrgyzstan. Bull Seismol Soc Am 92:3190–3209

Havenith HB, Vanini M, Jongmans D, Faccioli E (2003) Initiation of earthquake induced slope failure: influence of topographical and other site specific amplification effects. J Seismolog 7:397–412

Hencher SR (1981) Friction parameters for the design of rock slopes to withstand earthquake loading. Dams and earthquake. Thomas Telford Ltd., London, pp 79–87

Hervas J, Bobrowsky P (2009) Mapping: inventories, susceptibility, hazard and risk. In: Sassa K, Canuti P (eds) Landslides: disaster risk reduction. Springer, Berlin, pp 321–349

Hess DN, Leshchinsky BA, Bunn M, Mason HB, Olsen MJ (2017) A simplified three-dimensional shallow landslide susceptibility framework considering topography and seismicity. Landslides 14:1677–1697

Highland LM, Bobrowsky P (2008) The landslide handbook—a guide to understanding landslides. Geol Survey Circular 1325:129. Reston, Virginia, US

Hosmer D, Lemeshow S (2000) Applied logistic regression. Wiley-Interscience Publication, NewYork

Huang Y, Zhao L (2018) Review on landslide susceptibility mapping using support vector machines. CATENA 165:520–529

James N, Sitharam T (2017) Landslide hazard assessment and vulnerability studies for the state of Sikkim, India based on multicriteria analytical hierarchical approach. In: Paper read at 16th world conference on earthquake, Santiago, Chile

Jessee MA (2017) An investigation of seismic hazard: Using geophysical data to predict the location and impact of earthquake-induced landslides around the globe and to estimate earthquake potential in the Philippines. PhD dissertation, Indiana University

Jibson RW (1993) Predicting earthquake-induced landslide displacements using Newmark's sliding block analysis. Transportation Research Record 1411. Transportation Research Board, National Research Council, Washington DC, p.9–17.

Jibson RW (2007) Regression models for estimating coseismic landslide displacement. Eng Geol 91:209–218

Jibson RW, Harp EL, Michael JA (2000) A method for producing digital probabilistic seismic landslide hazard maps. Eng Geol 58:271–289

Kamp U, Growley BJ, Khattak GA, Owen LA (2008) GIS-based landslide susceptibility mapping for the 2005 Kashmir earthquake region. Geomorphology 101:631–642

Katz O, Crouvi O (2007) The geotechnical effects on long human habitation (2000<years): earthquake induced landslide hazard in the city of Zefat, northern Israel. Eng Geol 95:57–78

Keefer DK, Wilson RC (1989) Predicting earthquake-induced landslides, with emphasis on arid and semi-arid environments. In: Sadler PM, Morton DM (eds) Landslides in a semiarid environment, 2. Inland Geological Society of Southern California Publications, Riverside, CA, pp118–149

Khalaj S, BahooToroody F, Mahdi Abaei M, BahooToroody A, De Carlo F, Abbassi R (2020) A methodology for uncertainty analysis of landslides triggered by an earthquake. Comput Geotech 117:103262

Kramer SL (1996) Geotechnical earthquake engineering. Prentice-Hall, Upper Saddle River

Kritikos T, Robinson TR, Davies TRH (2015) Regional coseismic landslide hazard assessment without historical landslide inventories: a new approach. J Geophys Res Earth Surf 120:711–729

Kwag S, Hahm D (2018) Development of an earthquake-induced landslide risk assessment approach for nuclear power plants. Nucl Eng Technol 50:1372–1386

Lara M, Sepúlveda SA (2010) Landslide susceptibility and hazard assessment in San Ramón Ravine, Santiago de Chile, from an engineering geological approach. Environ Earth Sci 60:1227–1243

Lara M, Sepúlveda SA, Celis C, Rebolledo S, Ceballos P (2018) Landslide susceptibility maps of Santiago city Andean foothills, Chile. Andean Geol 45(3):433–442. Lee CT (2014) Statistical seismic landslide hazard analysis: an example from Taiwan. Eng Geol 182:201–212
Lee CT, Huang CC, Lee JF, Pan KL, Lin ML, Dong JY (2008) Statistical approach to earthquake-induced landslide susceptibility. Eng Geol 100:43–58
Lee CT, Hsieh BS, Sung CS, Lin PS (2012) Regional Arias intensity attenuation relationship for Taiwan considering VS30. Bull Seismol Soc Am 102:129–142
Lin PS, Lee CT (2003) Arias intensity in landslide susceptibility analysis. In: Proceeding of 2003 annual meeting of Chinese geophysical society (Taiwan), pp 91–96 (in Chinese with English abstract)
Ma S, Xu C, Shao X (2020) Spatial prediction strategy for landslides triggered by large earthquakes oriented to emergency response, mid-term resettlement and later reconstruction. Int J Disaster Risk Reduction 43:101362
Massey C, Della PF, Holden C, Kaiser A, Richards L, Wartman J, McSaveney MJ, Archibald G, Yetton M, Janku L (2017) Rock slope response to strong earthquake shaking. Landslides 14:249–268
Massey C, Townsend D, Rathje E, Allstadt KE, Lukovic B, Kaneko Y, Bradley B, Wartman J, Jibson RW, Petley DN, Horspool N, Hamling I, Carey J, Cox S, Davidson J, Dellow S, Godt JW, Holden C, Jones K, Kaiser A, Little M, Lyndsell B, McColl S, Morgenstern R, Rengers FK, Rhoades D, Rosser B, Strong D, Singeisen C, Villeneuve M (2018) Landslides triggered by the 14 November 2016 Mw 7.8 Kaikōura earthquake, New Zealand. Bull Seismol Soc Am 108(3B):1630–1648
Miles SB, Keefer DK (2009) Evaluation of CAMEL—comprehensive areal model of earthquake-induced landslides. Eng Geol 104:1–15
Nam K, Wang F (2019) The performance of using an autoencoder for prediction and susceptibility assessment of landslides: a case study on landslides triggered by the 2018 Hokkaido Eastern Iburi earthquake in Japan. Geoenviron Disasters 6:19
Newmark NM (1965) Effects of earthquakes on dams and embankments. Geotechnique 15:139–160. Nguyen BQV, Lee SR, Kim YT (2020) Spatial probability assessment of landslide considering increases in porewater pressure during rainfall and earthquakes: case studies at Atsuma and Mt. Umyeon. CATENA 187:104317
Nian TK, Guo XS, Zheng DF, Xiu ZX, Jiang ZB (2019) Susceptibility assessment of regional submarine landslides triggered by seismic actions. Appl Ocean Res 93:101964
Nonomura A, Hasegawa S (2017) Susceptibility of slopes to earthquake-induced landslides: a new index derived from helicopter-borne electromagnetic resistivity and digital elevation data sets. Landslides 14:2155–2163
Nowicki MA, Wald DJ, Hamburger MW, Hearne M, Thompson EM (2014) Development of a globally applicable model for near-real time prediction of seismically induced landslides. Eng Geol 173:54–65
Nowicki Jessee MA, Hamburger MW, Allstadt K, Wald DJ, Robeson SM, Tanyas H et al (2018) A global empirical model for near-real-time assessment of seismically induced landslides. J Geophys Res Earth Surf 123:1835–1859
Pareek N, Sharma ML, Arora MK (2010) Impact of seismic factors on landslide susceptibility zonation: a case study in part of Indian Himalayas. Landslides 7:191–201
Pareek N, Sharma ML, Arora MK, Pal S (2013) Inclusion of earthquake strong ground motion in a geographic information system-based landslide susceptibility zonation in Garhwal Himalayas. Nat Hazards 65:739–765
Peng WF, Wang CL, Chen ST, Lee ST (2009) Incorporating the effects of topographic amplification and slideing areas in the modeling of earthquake-induced landslide hazards, using the cumulative displacement method. Comput Geosci 35:946–966
Phillips SJ, Anderson RP, Schapire RE (2006) Maximum entropy modelling of species geographic distributions. Ecol Model 190:231–259
Poppeliers C, Pavlis GL (2002) The seismic response of a steep slope: high resolution observations with a dense, three-component seismic array. Bull Seismol Soc Am 92(8):3102–3115

Pradhan B, Lee S (2010) Landslide susceptibility assessment and factor effect analysis: back propagation artificial neural networks and their comparison with frequency ratio and bivariate logistic regression modelling. Environ Model Softw 25(6):747–759. https://doi.org/10.1016/j.envsoft.2009.10.016

Refice A, Capolongo D (2002) Probabilistic modeling of uncertainties in earthquake-induced landslide hazard assessment. Comput Geosci 28:735–749

Regmi AD, Dhital MJ, Zhang JQ, Su LJ, Chen XQ (2016) Landslide susceptibility assessment of the region affected by the 25 April 2015 Gorkha earthquake of Nepal. J Mt Sci 13(11):1941–1957

Rizzitano S, Cascone E, Giovanni B (2014) Coupling of topographic and stratigraphic effects on seismic response of slopes through 2D linear and equivalent linear analyses. Soil Dyn Earthq Eng 67(2014):66–84

Robinson TR, Rosser NJ, Densmore AL, Williams JG, Kincey ME, Benjamin J, Bell HJA (2017) Rapid post-earthquake modelling of coseismic landslide intensity and distribution for emergency response decision support. Nat Hazard 17:1521–1540

Rodríguez-Peces MJ, Pérez García JL, García-Mayordomo J, Azañòn JM, Insua Arévalo JM, Delgado-García J (2011) Applicability of Newmark method at regional, sub-regional and site scales: seismically induced Bullas and La Paca rock-slide cases (Murcia, SE Spain). Nat Hazards 59:1109–1124

Romeo R (2000) Seismically induced landslide displacements: a predictive model. Eng Geol 58:337–351

Rossi M, Reichenbach P (2016) LAND-SE: a software for statistically based landslide susceptibility zonation, version 1.0. Geosci Model Dev 9:3533–3543

Saaty TL (1980) The analytical hierarchy process. McGraw-Hill, New York

Sepúlveda SA (2004) The effect of topographic amplification on seismic rock slope instability. PhD thesis, University of Leeds

Sepúlveda SA, Murphy W, Jibson RW, Petley DN (2005a) Seismically-induced rock slope failures resulting from topographic amplification of strong ground motions: the case of Pacoima Canyon, California. Eng Geol 80:336–348

Sepúlveda SA, Murphy W, Petley DN (2005b) Topographic controls on coseismic rock slides during the 1999 Chi-Chi Earthquake, Taiwan. Q J Eng Geol Hydrogeol 38:189–196

Serey A, Piñero-Feliciangeli L, Sepúlveda SA, Poblete F, Petley DN, Murphy W (2019) Landslides induced by the 2010 Chile megathrust earthquake: a comprehensive inventory and correlations with geological and seismic factors. Landslides 16(6):1153–1165

Serey A, Sepúlveda SA, Murphy W, Petley DN, De Pascale G (2020) Developing conceptual models for the recognition of coseismic landslides hazard for shallow crustal and megathrust earthquakes in different mountain environments—an example from the Chilean Andes. Q J Eng Geol Hydrogeol (in Press). https://doi.org/10.1144/qjegh2020-023

Sharifi-Mood M, Olsen MJ, Gillins DT, Mahalingam R (2017) Performance-based, seismically-induced landslide hazard mapping of Western Oregon. Soil Dyn Earthq Eng 103:38–54

Shinoda M, Miyata Y (2017) Regional landslide susceptibility following the Mid NIIGATA prefecture earthquake in 2004 with NEWMARK'S sliding block analysis. Landslides 14:1887–1899

Shinoda M, Miyata Y, Kurokawa U, Kondo K (2019) Regional landslide susceptibility following the 2016 Kumamoto earthquake using back-calculated geomaterial strength parameters. Landslides 16:1497–1516

Shrestha S, Kang TS (2019) Assessment of seismically-induced landslide susceptibility after the 2015 Gorkha earthquake, Nepal. Bull Eng Geol Environ 78:1829–1842

Song Y, Gong J, Gao S, Wang D, Cui T, Li Y, Wei B (2012) Susceptibility assessment of earthquake-induced landslides using Bayesian network: a case study in Beichuan, China. Comput Geosci 42:189–199

Song J, Gao Y, Feng T (2018) Probabilistic assessment of earthquake-induced landslide hazard including the effects of ground motion directionality. Soil Dyn Earthq Eng 105:83–102

Stafford PJ, Berril JB, Pettinga JR (2009) New predictive equations for Arias intensity from crustal earthquakes in New Zealand. J Seismol 2009(13):31–52

Tang C, Zhu J, Liang J (2009) Emergency assessment of seismic landslide susceptibility: a case study of the 2008 Wenchuan earthquake affected area. Earthquake Eng Eng Vibr 8(2):207–217
Tanoli JI, Ningsheng C, Regmi AD, Jun L (2017) Spatial distribution analysis and susceptibility mapping of landslides triggered before and after Mw7.8 Gorkha earthquake along Upper Bhote Koshi, Nepal. Arab J Geosci 10:277
Tanyas H, Rossi M, Alvioli M, van Westen CJ, Marchesini I (2019) A global slope unit-based method for the real-time prediction of earthquake-induced landslides. Geomorphology 327:126–146
Tanyaş H, vanWesten CJ, Allstadt KE, Anna Nowicki Jessee M, Gorum T, Jibson RW, Godt JW, Sato HP, Schmidt RG, Marc O, Hovius N (2017) Presentation and analysis of a worldwide database of earthquake-induced landslide inventories. J Geophys Res Earth Surf 122:1991–2015
Uchida T, Kataoka S, Iwao T, Matsuo O, Terada H, Nakano Y, Sugiura N, Osanai N (2004) A study on methodology for assessing the potential of slope failures during earthquakes. Technical note of National Institute for Land and Infrastructure Management, p 91 (in Japanese with English summary)
Uchida T, Osanai N, Onoda S, Takayama T, Tomura K (2006) A simple method for producing probabilistic seismic shallow landslide hazard. In: Disaster mitigation of debris flows, slope failures and landslides: proceedings of the INTERPRAEVENT international symposium, 25–29 Sept 2006, Niigata, Japan, pp 529–534
Umar Z, Pradhan B, Ahmad A, Jebur MN, Tehrany MS (2014) Earthquake induced landslide susceptibility mapping using an integrated ensemble frequency ratio and logistic regression models in West Sumatera Province, Indonesia. CATENA 118:124–135
Valagussa A, Frattini P, Crosta GB, Valbuzzi E, Gambini S (2017) Regional landslide susceptibility analysis following the 2015 Nepal earthquake. In: Mikoš M et al (eds) Advancing culture of living with landslides. Springer, pp 1035–1042
van Westen CJ (1997) Statistical landslide hazard analysis. In: Application guide, ILWIS 2.1 for Windows, ITC, Enschede, The Netherlands, pp 73–84
Van Westen CJ, Van Asch TWJ, Soeters R (2006) Landslide hazard and risk zonation—why is it still so difficult? Bull Eng Geol Environ 65(2):167–184
Wang Y, Song C, Lin Q, Li J (2016) Occurrence probability assessment of earthquake-triggered landslides with Newmark displacement values and logistic regression: the Wenchuan earthquake, China. Geomorphology 258:108–119
Wartman J, Dunham L, Tiwari B, Pradel D (2013) Landslides in Eastern Honshu induced by the 2011 off the Pacific Coast of Tohoku earthquake. Bull Seismol Soc Am 103(2B):1503–1521
Wasowski J, Del Gaudio V, Casarano D, Lollino P, Muscillo S (2013) Local scale seismic landslide susceptibility assessment based on historic earthquake records combined with accelerometer monitoring and ambient noise data. In: Ugai K et al (eds) Earthquake-induced landslides. Springer, Berlin, Heidelberg, pp 11–20. https://doi.org/10.1007/978-3-642-32238-9_2
Wilson RC (1993) Relation of Arias intensity to magnitude and distance in California. Open-File report 93–556, US Geological Survey, Reston, Virginia, 42 pp
Wilson RC, Keefer D (1985) Predicting areal limits of earthquake-induced landsliding. In: Ziony JI (ed) Earthquake hazards in the Los Angeles region—an earth-science perspective. U.S. Geological Survey Professional Paper 1360, pp 317–345
Worden CB, Wald DJ (2016) ShakeMap manual online: technical manual, user's guide, and software guide, U.S. Geol Surv. https://doi.org/10.5066/F7D21VPQ
Xu C, Dai F, Xu X, Lee YH (2012a) GIS-based support vector machine modeling of earthquake-triggered landslide susceptibility in the Jianjiang river watershed, China. Geomorphology 145–146:70–80
Xu C, Xu X, Dai F, Saraf AK (2012b) Comparison of different models for susceptibility mapping of earthquake triggered landslide related with the 2008 Wenchuan earthquake in China. Comput Geosci 46:317–329
Xu C, Shen L, Wang G (2016) Soft computing in assessment of earthquake-triggered landslide susceptibility. Environ Earth Sci 75:767

Yalcin A (2008) GIS-based landslide susceptibility mapping using analytical hierarchy process and bivariate statistics in Ardesen (Turkey): comparisons of results and confirmations. Catena 72(1):1–12. https://doi.org/10.1016/j.catena.2007.01.003. Yao X, Tham LG, Dai FC (2008) Landslide susceptibility mapping based on support vector machine: a case study on natural slopes of Hong Kong, China. Geomorphology 101:572–582

Yang Z, Lan H, Gao X, Li L, Meng Y, Wu Y (2015) Urgent landslide susceptibility assessment in the 2013 Lushan earthquake-impacted area, Sichuan Province, China. Nat Hazards 75:2467–2487

Yin Y, Wang F, Sun P (2009) Landslide hazards triggered by the 2008 Wenchuan earthquake, Sichuan, China. Landslides 6:139–151

Yoshimatsu H, Abe S (2006) Are view of landslide hazards in Japan and assessment of their susceptibility using an analytical hierarchic process (AHP) method. Landslides 3:149–158

Youssef AM, Pourghasemi HR, Pourtaghi ZS, Al-Katheeri MM (2016) Landslide susceptibility mapping using random forest, boosted regression tree, classification and regression tree, and general linear models and comparison of their performance at Wadi Tayyah Basin, Asir Region, Saudi Arabia. Landslides 13:839–856

Zhang J, Cui P, Ge Y, Xiang L (2012) Susceptibility and risk assessment of earthquake-induced landslides based on landslide response units in the Subao River basin, China. Environ Earth Sci 65:1037–1047

Zhang S, Li C, Zhang L, Peng M, Zhan L, Xu Q (2020) Quantification of human vulnerability to earthquake-induced landslides using Bayesian network. Eng Geol 265:105436

Numerical Evaluation of the Effect of Drainage Wells on the Stability of the Ichiba No. 1 Landslide During an Mj6.7 Earthquake

Fei Cai, Akihiko Wakai, Naoki Watanabe, Gonghui Wang, and Gen Furuya

Abstract This research evaluated numerically the effect of drainage wells on the response of landslides during earthquakes, taking the Ichiba No. 1 landslide, a coseismic landslide, as an example. The unsaturated–saturated finite element seepage analyses were carried out to evaluate the groundwater level lowered by the drainage wells. The dynamic total stress finite element analyses were used to calculate the response of the slope during the earthquake. The results indicated that the groundwater level can be effectively lowered, especially around the drainage wells. It is important to take the influence of the groundwater level on the response of the slope during the earthquake into account in a dynamic total stress finite element analysis. Whether a collapse with a very large movement occurred could be judged mainly based on the histogram of displacement at some typical points on the slope surface. The deformed finite element mesh and residual shear strain after the earthquake could be used to evaluate where the sliding surfaces developed. For a slope with a large gradient, it may collapse even if it was improved by drainage wells. Thus, such slopes should be cut back to a small slope gradient to prevent them from collapsing in earthquakes.

F. Cai (✉) · A. Wakai
Gunma University, Kiryu, Japan
e-mail: feicai@gunma-u.ac.jp

A. Wakai
e-mail: wakai@gunma-u.ac.jp

N. Watanabe
Niigata University, Niigata, Japan
e-mail: jibanken@cc.niigata-u.ac.jp

G. Wang
Kyoto University, Kyoto, Japan
e-mail: wang.gonghui.3r@kyoto-u.ac.jp

G. Furuya
Toyama Prefectural University, Imizu, Japan
e-mail: gfuruya@pu-toyama.ac.jp

I. Towhata et al. (eds.), *Coseismic Landslides*, Springer Natural Hazards,
https://doi.org/10.1007/978-981-19-6597-5_15

1 Introduction

A large earthquake with a Japan Meteorological Agency (JMA) magnitude (Mj) of 6.7 (Mw 6.2) occurred at 22:08 (Japan Standard Time, JST) on 22 November, 2014 in the northern part of Nagano Prefecture. It was officially named as "The 2014 Nagano Prefecture Kamishiro Fault Earthquake" by Nagano Prefecture. The epicenter of the earthquake was located at 36°41.5′ N, 137°53.4′ E, and the hypocenter was at a depth of 5 km. It ruptured about a 20-km-long northern part of the 31-km-long Kamishiro fault located at the northern end section of the Itoigawa-Shizuoka Tectonic Line active fault system, with a length of about 160 km, as shown in Fig. 1. The focal mechanism suggests the activity of a reverse fault with a strike-slip component and a compression axis that is oriented NNE-SSW (JMA 2014).

The Nagano Prefecture Kamishiro Fault Earthquake resulted in severe damage. Approximately 50 persons were injured and at least 250 residential houses were destroyed or severely damaged in the area hit strongly by the earthquake. Interpretation of aerial photographs and field investigations showed the earthquake also triggered about 100 landslides (Furuya et al. 2015, Katsura et al. 2016). The Ichiba

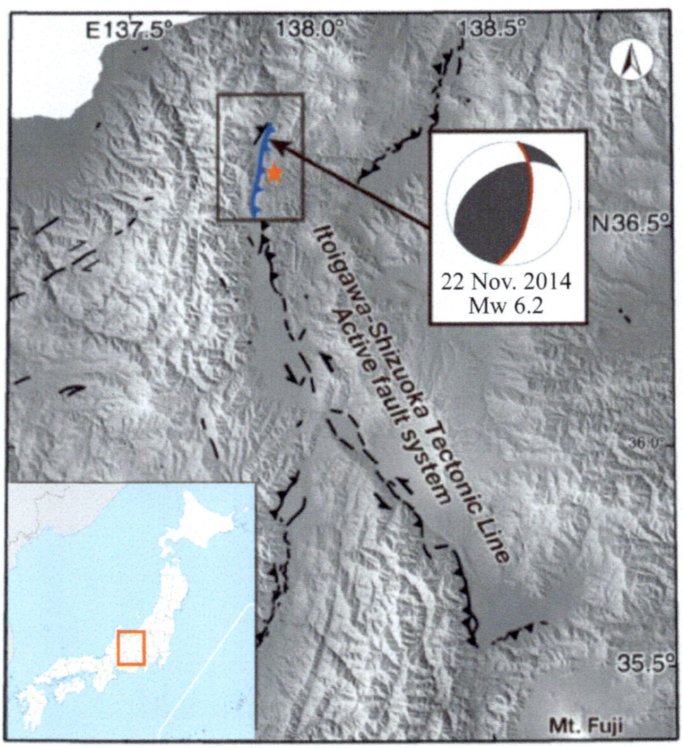

Fig. 1 2014 Nagano Prefecture Kamishiro Fault Earthquake and Itoigawa-Shizuoka Tectonic Line active fault system. The CMT solution is from JMA (2014) (modified from Katsube et al. 2017)

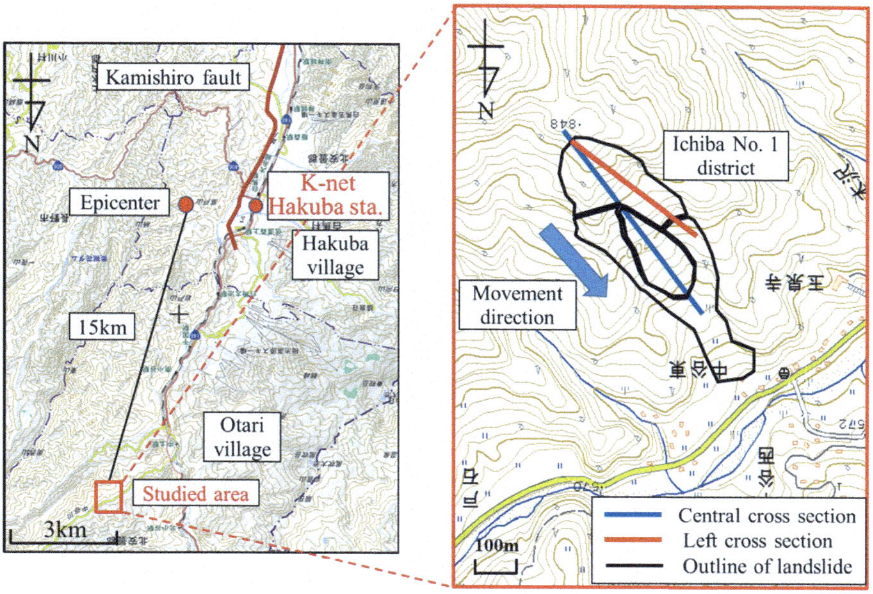

Fig. 2 Ichiba No. 1 landslide area (modified on GSI map)

No. 1 landslide, about 15 km from the epicenter in the NNE direction, is located at Otari Village, Nagano Prefecture, and was one of the largest landslides induced by the Nagano Prefecture Kamishiro Fault Earthquake, as shown in Fig. 2.

The Ichiba No. 1 landslide had a crown elevation of about 800 m, a length of 760 m, and a width of 200 m. The surficial soil layer of the landslide was a residual clayey soil resulting from an old landslide. The surficial clayey soil layer was fluidized and flowed downhill as a debris flow during the earthquake. The landslide had some blocks and scars, and upheaved areas induced by rotational slides were observed on the upper part of the slope. On the other hand, the moved blocks and scars were not observed in the lower part of the slope, but some soils of the debris flow from the upper part of the slope accumulated on the surface of the lower part of the slope. As shown in Fig. 3, the debris flow was separated by the stable lower part of the slope, which was stabilized with drainage well before the earthquake and split into the right and left parts which still flowed downhill for a long distance. The debris flow damaged some protection works, such as drainpipes, but it was stopped at last by a check dam, and the road and river were protected from the debris flow.

In the Ichiba No. 1 area, landslides have occurred frequently since ancient times, and they may damage residential houses and public facilities in the future. Because the structure of the strata of a large area around the Ichiba No. 1 landslide is similar to that of the Ichiba No. 1 landslide site, the area could experience rather extensive damage during a future earthquake. Therefore, six drainage wells were planned to protect the area from damage induced by landslides during earthquakes, and four were constructed at the end of 2017.

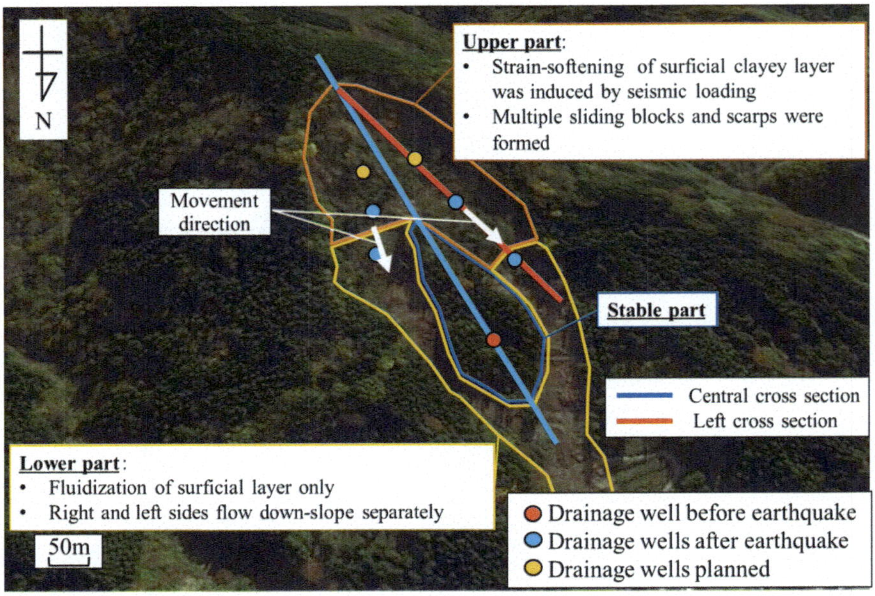

Fig. 3 Ichiba No. 1 landslide area (modified on Google map)

This research takes the Ichiba No. 1 landslide as an example to evaluate the effect of drainage wells on the stability of landslides during an earthquake. The unsaturated–saturated finite element analysis was used to evaluate the lowering of the groundwater level as the effect of the drainage wells. The dynamic finite element analysis, in which the strain-softening constitutive model was used to consider the relationship between the shear stress and shear strain of the surficial clayey soil layer, was used to evaluate the stabilizing effect of the drainage wells by comparing the response of the slope during the earthquake before and after the construction of the drainage wells.

2 Field Investigation

Figure 4 shows a geological cross-section along the centerline (blue line in Fig. 3), of the upper part above the stable part of the slope, based on information from three boreholes. It shows that the surficial layer was brown clayey soil with gravel, which had turned into soil, and was soft, and had high water content. Beneath the surficial layer, gravelly clayey soil with the color of original rock, an angular gravel-like fine sandstone layer, a fine sandstone layer, and alternations of sandstone and mudstone layers were found in the depth direction. A shallow sliding surface, 10 m deep below the ground surface of the landslide triggered by the Nagano Prefecture Kamishiro Fault Earthquake, developed along with the interface between the clayey layer and

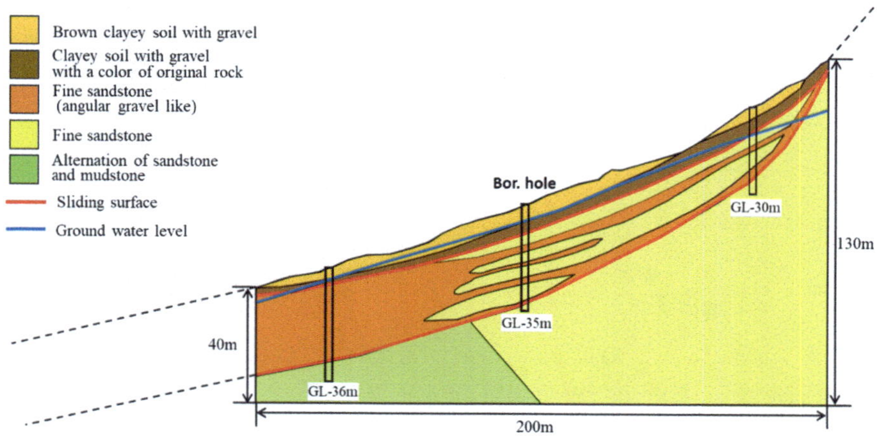

Fig. 4 Geological cross-section of the Ichiba No. 1 landslide constructed based on boreholes (modified from Nippon Koei 2015)

the sandstone layer. Moreover, a deep-seated sliding surface, 25–30 m deep below the ground surface, of an old landslide existed in the sandstone layer; however, it did not move or moved with a negligibly small-displacement during the Nagano Prefecture Kamishiro Fault Earthquake. The groundwater level was 3–10 m below the ground surface.

Besides the boreholes, block sampling, portable dynamic cone penetration tests, surface wave exploration, and electrical prospecting were also conducted at the locations shown in Fig. 5.

Figure 6 shows the resistivity distribution measured by electrical prospecting at three sections, i.e., the upper and lower transverse sections and one longitudinal section, as shown in Fig. 5. Figure 6 indicated that the depth at which the resistivity changed drastically seemed consistent with the depth of the sliding surface of the landslide that occurred during the earthquake.

Figure 7 shows the shear wave velocity measured by surface wave exploration at two transverse sections, indicated by the red and black lines. The measured shear wave velocity was generally consistent with those measured by PS logging and elastic wave exploration conducted by other groups. Figure 7 indicated that the surficial soil layer had a low shear wave velocity: smaller than 200 m/s.

Figure 8 shows the converted N-value distribution in the depth direction measured by portable dynamic cone penetration tests at five locations. The results showed that the surficial clayey soil layer had a low converted N-value of 2–8 counts. The converted N-value of the soil layer beneath the surficial clayey soil layer was significantly increased. The shear wave velocity of 125–215 m/s, calculated by an empirical equation $V_s = 100N^{1/3}$ (Japan Road Association 2012), was consistent with that measured by the surface wave exploration.

Figure 9 shows the measured flow rate of the drainage wells, and Fig. 10 shows the internal structure of a drainage well. The groundwater was lowered mainly by

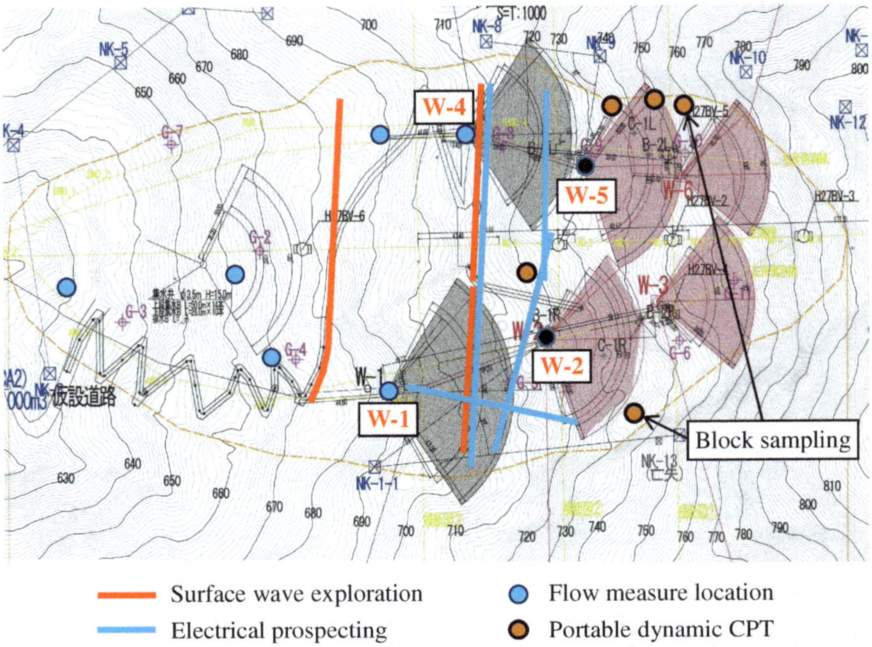

Fig. 5 Field investigation of the Ichiba No. 1 landslide

the drain pipes installed almost horizontally at two or three depth levels from the wall of the well into the ground. As shown in Fig. 10, the shallow drain pipes had water flow, but the middle and deep drain pipes had almost no water flow or the water flow was very small. Even for the shallow drain pipes, the flow rates of drain pipes were significantly different pipe by pipe. For the drainage wells numbered by W-1, the drain pipes No. 1–5 had no water flow, and the drain pipes No. 6–12 had a significantly different water flow rate, as shown in Table 1. This implies that the groundwater flow varied locally.

Block samples were excavated near the sliding surface. Figure 11 shows the results of direct shear tests on the samples of block sampling under three normal stresses. The results show that the peak cohesion and internal friction angle were 38.6 kN/m^2 and 38.0°. These values will be used to calculate the peak shear strength of the strain-softening constitutive model used in dynamic finite element analysis of the response of the slope during the earthquake.

3 Finite Element Seepage Analysis

Two-dimensional unsaturated–saturated finite element seepage analysis was carried out to evaluate the groundwater level lowered by the drainage wells.

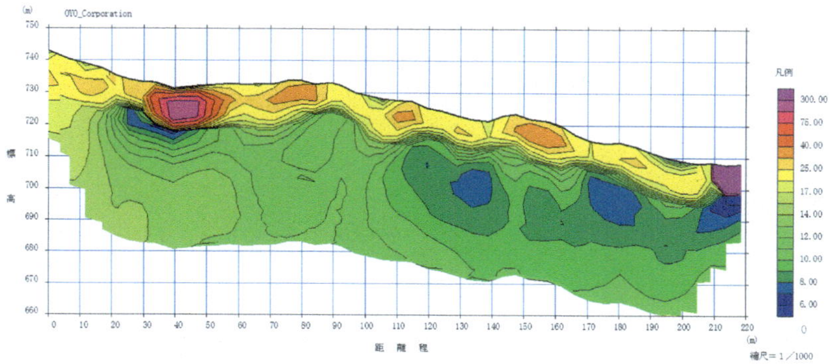

(a) Resistivity distribution calculated by connecting two upper transverse sections

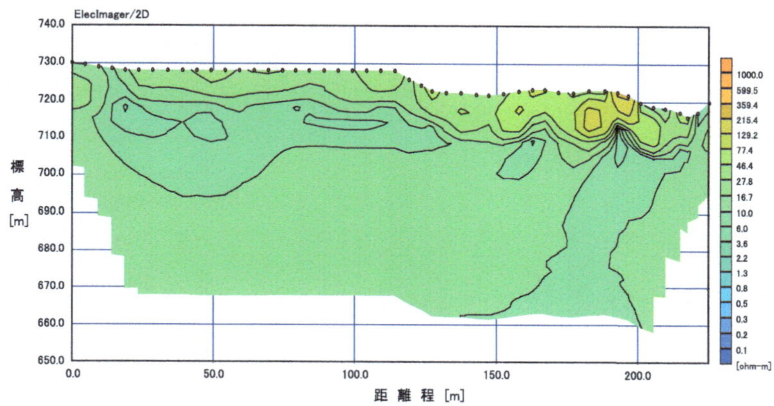

(b) Resistivity distribution of lower transverse section

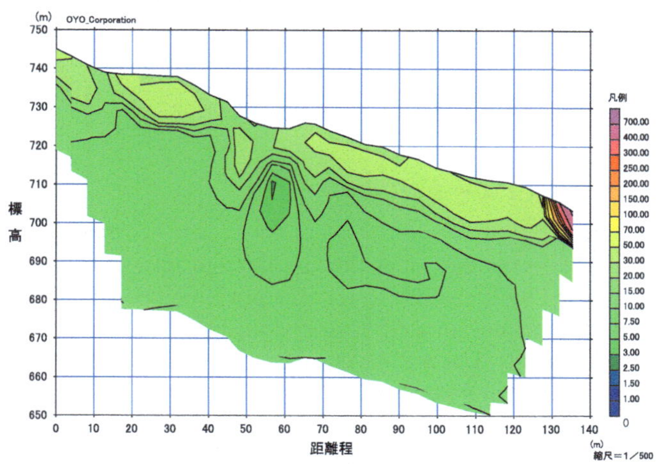

(c) Resistivity distribution of longitudinal section

Fig. 6 Resistivity distribution of electrical prospecting (locations of sections shown in Fig. 5)

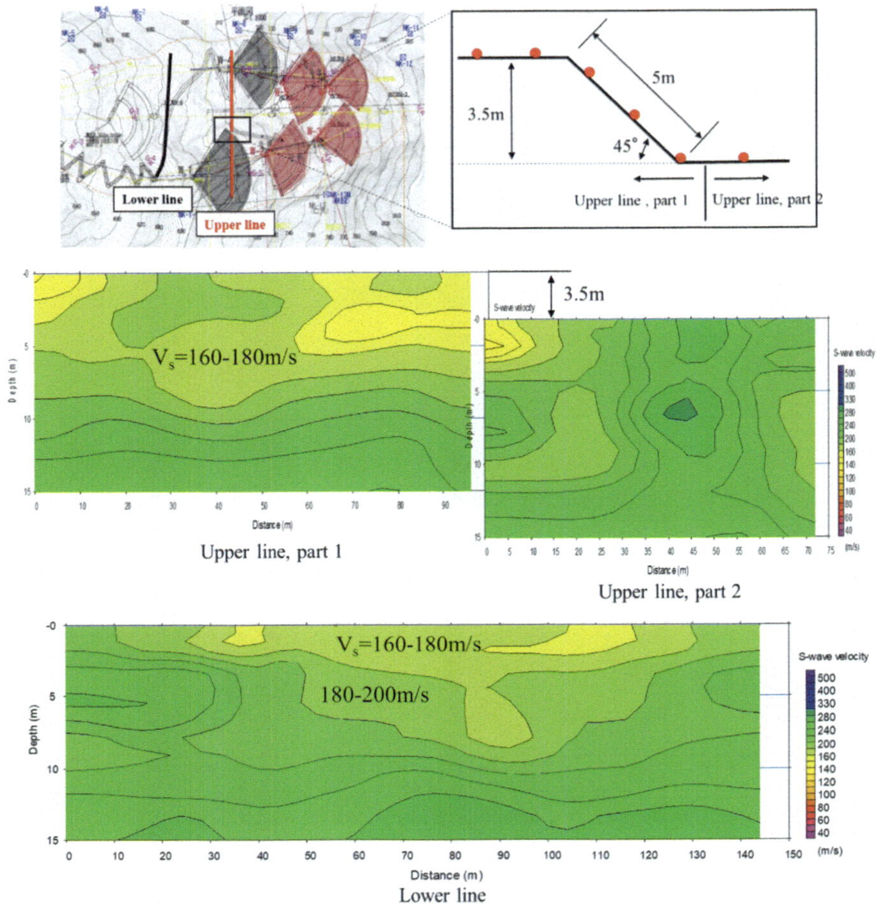

Fig. 7 Shear wave velocity measured by surface wave exploration

3.1 Fundamental Flow Equation

The equation that governs the water flowing through unsaturated–saturated soils was firstly introduced by Richards (1931). The governing equation is based on the mass conservation in an unsaturated–saturated medium together with Darcy's law for water motion. Namely, Darcy's law has been proved to be valid for water flow through both unsaturated and saturated soils. However, the permeability coefficient for saturated soils is assumed to be constant, while for unsaturated soils, it depends on the pore volume occupied by water. In geotechnical engineering, the phase relation is presented by the saturation degree. However, to implement Richards' equation, the permeability coefficient is defined in terms of the volumetric water content. The volumetric water content, θ, is defined as the product of the porosity, n_0, and the degree of saturation, S, i.e., $\theta = n_0 S$.

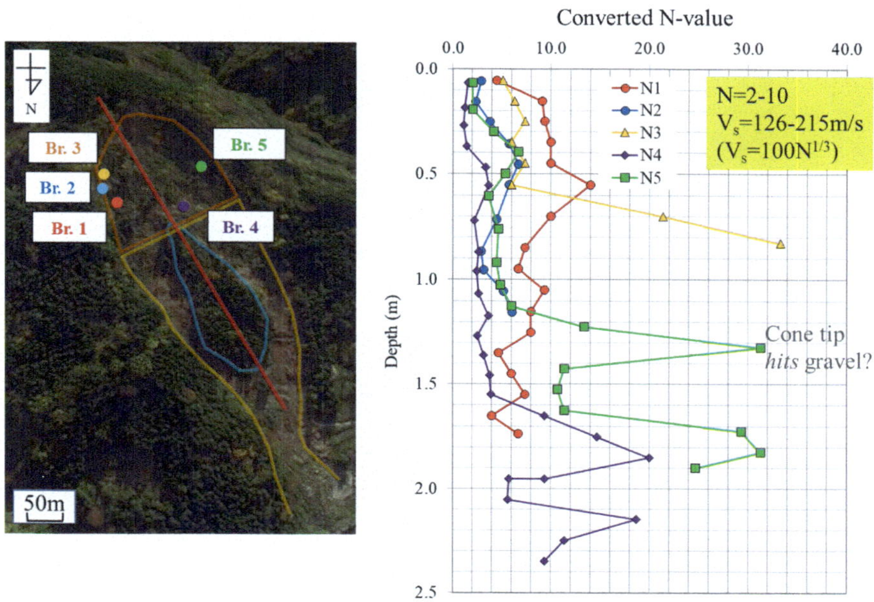

Fig. 8 Converted *N*-value versus depth measured by portable dynamic cone penetration tests

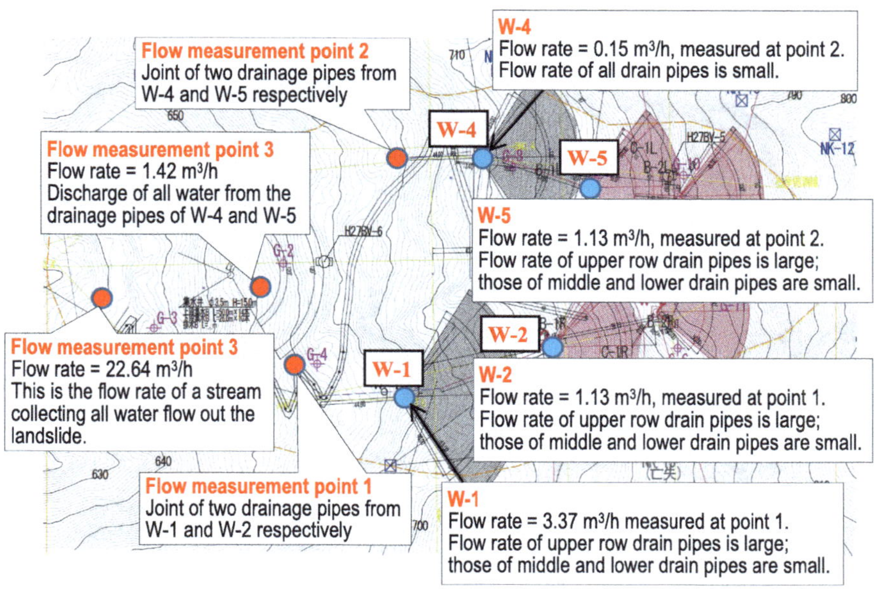

Fig. 9 Flow rates of drainage wells

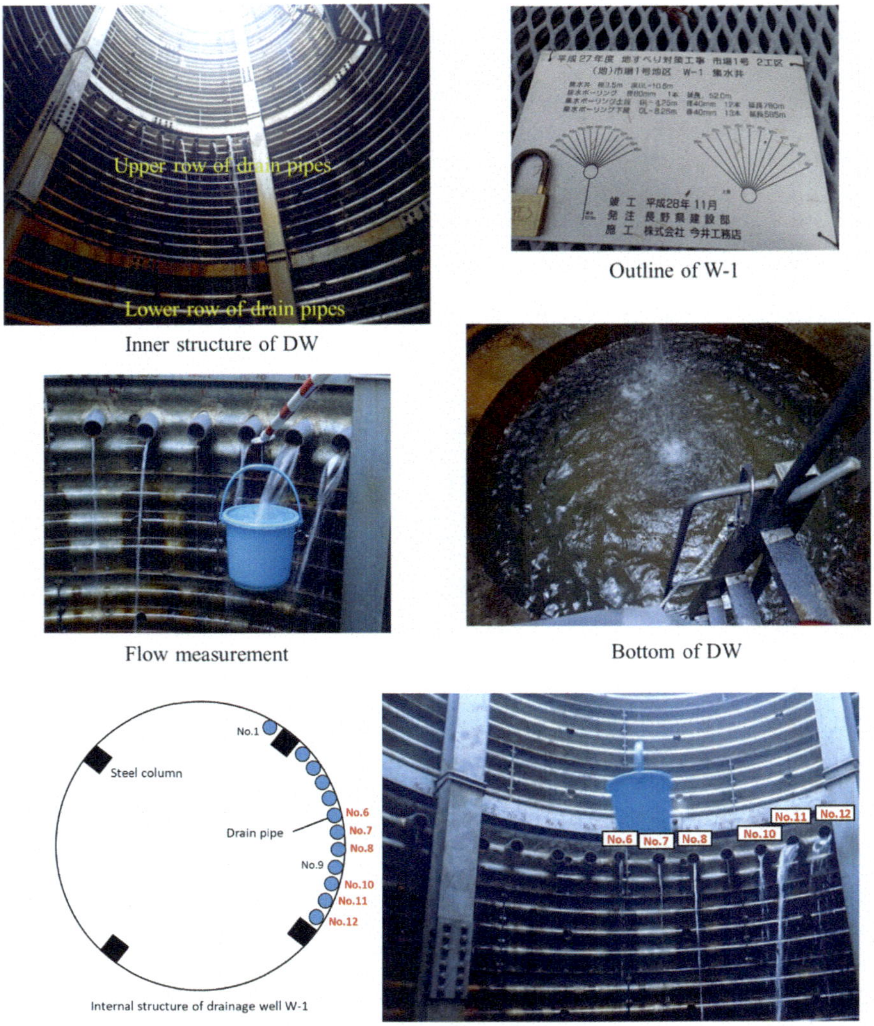

Fig. 10 Internal structure of drainage well (DW) and water flow of drain pipes

Table 1 Water flow rate from the shallow drain pipes of drainage well numbered by W-1

Drain pipe	No. 6	No. 7	No. 8	No. 10	No. 11	No. 12
Flow rate (m^3/h)	0.065	0.112	0.450	0.390	0.556	0.428

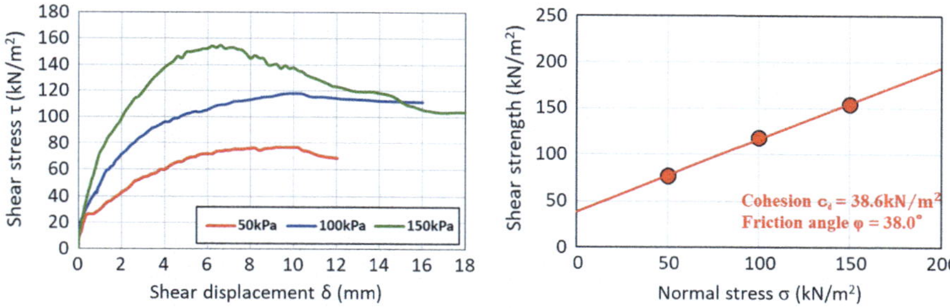

Fig. 11 Direct shear tests on samples from block sampling

Based on Darcy's law and the mass conservation, the governing equation for water flowing through unsaturated–saturated soils under isothermal conditions is written as

$$\frac{\partial}{\partial x}\left[K(\theta)\frac{\partial}{\partial x}(\phi + z)\right] + \frac{\partial}{\partial z}\left[K(\theta)\frac{\partial}{\partial z}(\phi + z)\right] = [c(\phi) + \alpha S_s]\frac{\partial \phi}{\partial t} \tag{1}$$

where $K(\theta)$ is the permeability coefficient, φ is the pressure head, z is the elevation head, S_s is the specific storage coefficient, $\alpha = \begin{cases} 1, \ (S = 1) \\ 0, \ (S < 1) \end{cases}$, t is time, and $c(\varphi)$ is the specific water capacity.

3.2 *Hydraulic Properties*

The permeability coefficient and the specific water capacity in Eq. (1) must be determined. For unsaturated soils, these two parameters depend on the volumetric water content (θ), which in turn is related to the pressure head (φ). This relationship between the water content (w) in soil and the suction can be expressed in a plot of volumetric water content (θ) against suction, which is known as the soil–water characteristic curve. Various equations of the soil–water characteristic curve have been proposed by some investigators. Among them, a set of closed-form equations formulated by van Genuchten (1980), which is based on the capillary model proposed by Mualem (1976), had been widely used. In these equations, the relative degree of saturation, S_e, as defined in Eq. (2), the specific water capacity, $c(\varphi)$, and the permeability coefficient for unsaturated soils, $K(\theta)$, are given by

$$S_e = \frac{\theta - \theta_r}{\theta_s - \theta_r} = \left(1 + |\alpha\phi|^n\right)^{-m} \tag{2}$$

$$c(\phi) = \alpha(n - 1)(\theta_s - \theta_r)S_e^{1/m}\left(1 - S_e^{1/m}\right)^m \tag{3}$$

$$K(\theta) = K_s K_r = K_s S_e^{1/2}\left[1 - \left(1 - S_e^{1/m}\right)^m\right]^2 \quad (4)$$

where $m = 1 - \frac{1}{n}$, and $n > 1$. θ_s is the saturated volumetric water content, i.e., the volumetric water content in the saturated state, and θ_r denotes the residual volumetric water content. As introduced by van Genuchten et al. (1991), the residual volumetric water content represents the volumetric water content at which the slope of the soil–water characteristic curve and permeability coefficient in unsaturated states go to zero when the suction becomes large. K_s is the saturated permeability coefficient, i.e., the permeability coefficient in the saturated state, and K_r (=$K(\theta)/K_s$) is the relative permeability coefficient. α, n, and m are empirical parameters of the hydraulic properties. Thus, the hydraulic functions are determined by a set of 5 parameters, θ_r, θ_s, α, n, and K_s. The van Genuchten model has been proved to be able to provide a better match to the experimental data (Stankovich and Lockinton 1995), although some alternative models, in which the parameters could be obtained more easily, are available (Brooks and Corey 1966). Leong and Rahardjo (1997a, b) made a comprehensive evaluation of the models of the hydraulic properties of soils.

3.3 Effect of Drainage Wells on Groundwater Level

Table 2 gives the parameters of the van Genuchten model, the saturated hydraulic coefficient, and the specific storage coefficient for each stratum used in the unsaturated–saturated finite element seepage analysis to evaluate the effect of drainage wells lowering the groundwater level within the slope. These parameters were determined empirically using the database of typical soils and rocks in Japan (Nishigaki and Takeshita 1993) and the specific storage coefficients compiled by Domenico and Schwartz (1990).

The finite element implementation of the unsaturated–saturated seepage analysis can be found elsewhere (e.g. Cai and Ugai 2004). Concerning the boundary conditions, the upper (right) and lower (left) vertical boundaries (Fig. 12) had known water heads, i.e., the groundwater level was determined based on the groundwater level in the boreholes, and the water level of the river, respectively. The drain pipes can be modeled using three methods: (1) as a thin soil layer with a saturated hydraulic conductivity 100 times larger than that of the geotechnical materials surrounding the drain pipes as given in Table 2. In this case, the water pressure of the center points of drainage wells in the thin soil layer was assumed to be zero; (2) as seepage points, i.e., if the calculated water pressure of the points was larger than zero, then it was assumed to be zero, and if the calculated water pressure was negative, then it was not changed (Cai et al. 1998); and (3) as sinks along the length of the drain pipes below the groundwater level, and the flow rates of sinks were calculated by averaging the measured flow rates of drain pipes of a drainage well. In this research, method (1)

Table 2 Hydraulic properties of materials

Stratum	Saturated hydraulic coefficient K_s ($\times 10^{-7}$ m/s)	Parameters of VG model				Specific storage S_s (1/m)
		θ_s (−)	θ_r (−)	α	n	
Brown clayey soil with gravel	4.00	0.445	0.078	4.76	1.248	1.95×10^{-5}
Clayey soil with gravel with a color of original rock						
Fine sandstone (angular gravel like)	1.05	0.445	0.221	1.47	6.656	7.55×10^{-7}
Fine sandstone						
Alternation of sandstone and mudstone						
Thin layer modeling drain pipes	1.05 (before drainage well constructed) 105 (after drainage well constructed)	0.445	0.221	1.47	6.656	7.55×10^{-7}

was used to model the drain pipes in the unsaturated–saturated finite element seepage analysis.

Figure 12 shows the finite element mesh of the central cross-section for unsaturated–saturated finite element seepage analysis. The ground surface was determined by surveying results before the earthquake. The strata were constructed based on boreholes for a part of the upper part of the slope, and for the other parts of the slope, the strata were constructed by parallel extension of the strata constructed based on boreholes. Figure 13 shows the calculated groundwater level of the central cross-section before and after the construction of the drainage wells in the lower stable part of the slope. The groundwater level was lowered in a larger region approximately centered at the drainage well, which was constructed at the lower part of the slope before the earthquake; however, the groundwater level of the upper part of the slope was almost not influenced by the drainage well. The maximum groundwater level lowering of 16.05 m occurred near the tip of the drain pipes, modeled by a thin soil layer with a higher saturated hydraulic coefficient. The drainage well was constructed before the earthquake, which significantly lowered the groundwater level of the lower part of the slope. This was considered to be a possible reason why the lower part of the slope was stable, but the upper part slope failed or moved significantly during the Nagano Prefecture Kamishiro Fault Earthquake.

Figure 14 shows the finite element mesh of the left cross-section for unsaturated–saturated finite element seepage analysis. The ground surface for the mesh was determined by surveying results before the earthquake, and the strata were constructed

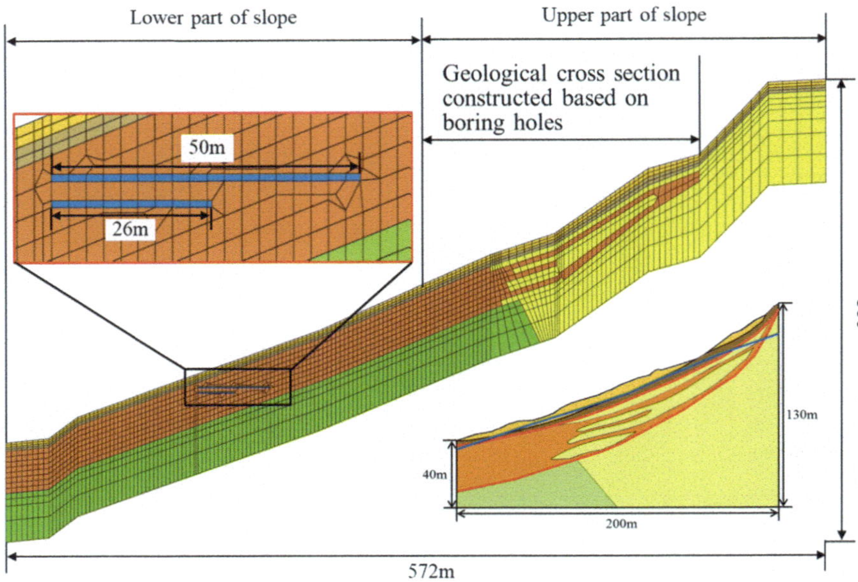

Fig. 12 Finite element mesh for central cross-section

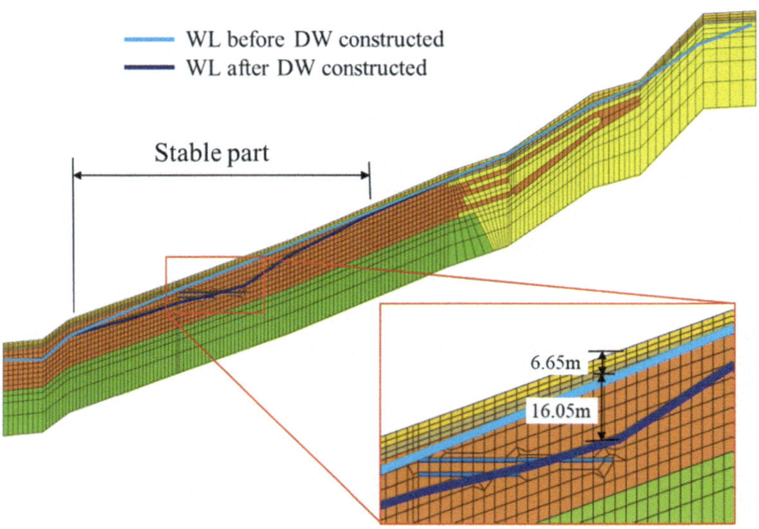

Fig. 13 Results of seepage analysis for central cross-section

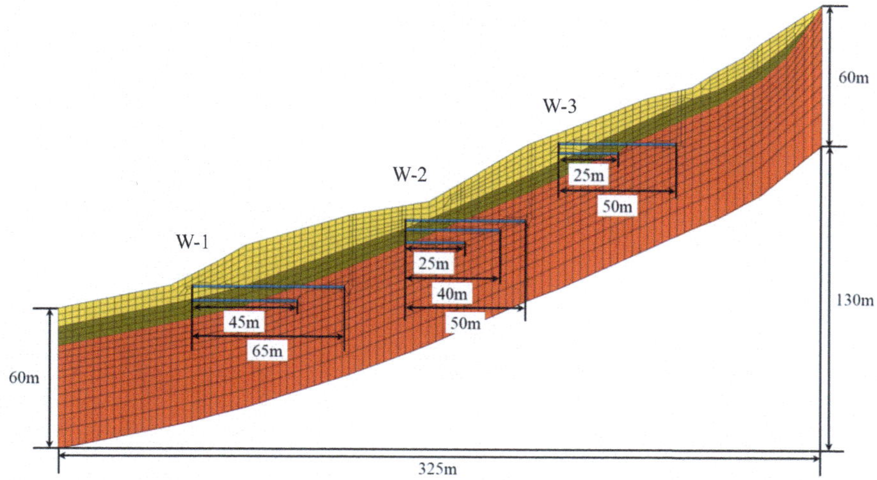

Fig. 14 Finite element mesh for left cross-section (indicated by the red line in Fig. 3)

based on boreholes. Figure 15 shows the calculated groundwater level of the left cross-section before and after the construction of three drainage wells. Figure 15 indicated that the groundwater level was lowered in a large part of the left cross-section, especially around three drainage wells. The maximum lowering of the groundwater level occurred near the tip of the drain pipes, especially for drainage wells W-1 and W-3. The maximum lowering of the groundwater level for W-2 was smaller than those of W-1 and W-3 because W-2 was located between W-1 and W-3.

Figures 13 and 15 show that the length of the shallow drain pipes inserted into the saturated area below the groundwater level was significantly longer than those of the middle and deep drain pipes. The longer the drain pipes were inserted into the saturated area, the more groundwater was drained out by the drain pipes. In addition, as given in Table 2, the saturated hydraulic coefficient of the surficial soil layer in which the shallow drain pipes were installed was about 4 times larger than those of other soil or rock layers in which the middle and deep drain pipes were installed. Thus, the calculated results were qualitatively consistent with the measured flow rate of the drain pipes.

4 Dynamic Finite Element Analysis

Dynamic finite element total stress analyses were performed to simulate the response of the Ichiba No. 1 landslide during the Nagano Prefecture Kamishiro Fault Earthquake. The initial stress of the slope needed by the dynamic finite element analysis was determined using gravity and groundwater levels evaluated by an unsaturated–saturated finite element seepage analysis. A strain-softening UW constitutive model

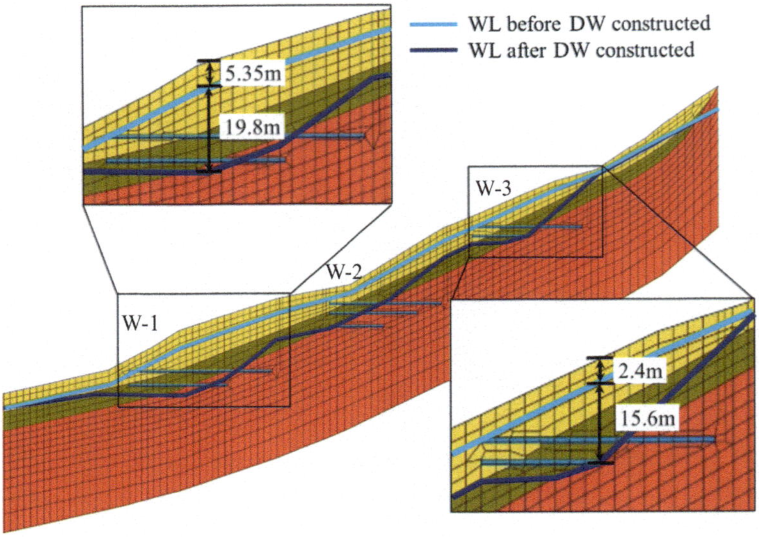

Fig. 15 Results of seepage analysis for left cross-section

(Wakai et al. 2010) and UW constitutive model (Wakai and Ugai 2004) were used to describe the stress–strain relationship and hysteresis loop of gravelly clayey soil layer with a color same as original rock in which the sliding surface of the landslide developed during the earthquake, and the other soil or rock layers, respectively. The skeleton curve of the stress–strain relationship was described by Eq. (5) in the UW constitutive model (Wakai and Ugai 2004) and the strain-softening UW constitutive model (Wakai et al. 2010).

$$\tau = \frac{G_0\gamma}{1 + G_0\gamma/\tau_f} \tag{5}$$

where τ presents the shear stress, γ the shear strain, G_0 the shear modulus at small strain amplitude, and τ_f the shear strength, which is given by Eq. (6) for the UW constitutive model (Wakai and Ugai 2004) and Eq. (7) for the strain-softening UW constitutive model (Wakai et al. 2010) to consider shear strength reduction from peak shear strength τ_{f0} to residual shear strength τ_{fr} with the increase of the shear strain.

$$\tau_f = c\cos\varphi + \frac{\sigma_1 + \sigma_3}{2}\sin\varphi \tag{6}$$

$$\tau_f = \tau_{f0} + \frac{\tau_{fr} - \tau_{f0}}{A + \gamma^p}\gamma^p \tag{7}$$

where c is cohesion, φ is the internal friction angle, σ_1 and σ_3 are major and minor principal stress, A is a model parameter to be determined based on tests, and γ^p plastic

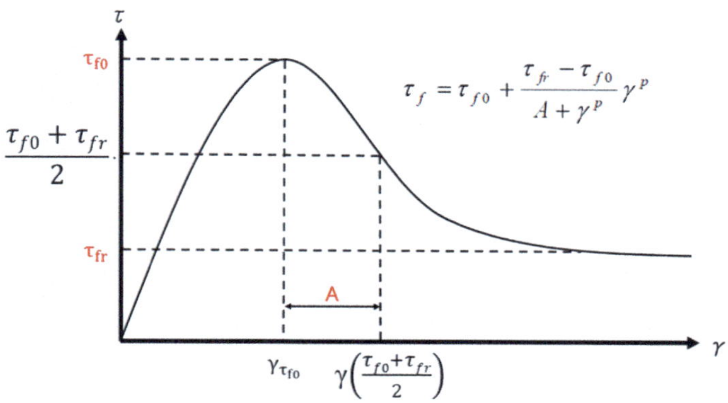

Fig. 16 Shear strain versus shear stress of strain-softening UW model

maximum shear strain. The peak shear strength τ_{f0} and residual shear strength τ_{fr} were calculated using Eq. (6) and peak and residual shear strength parameters c and φ. Figure 16 shows the shear strain and shear stress relationship described by Eq. (7).

The hysteresis loop is given by

$$\tilde{\tau} = \frac{a\tilde{\gamma}^n + G_0\tilde{\gamma}}{1 + b\tilde{\gamma}} \tag{8}$$

where b and n are two constitutive model parameters and a is dependent on other variables (Wakai and Ugai, 2004). The details of the UW constitutive model and strain-softening UW constitutive model, and the finite element implementation of the dynamic finite element analysis can be found in Wakai and Ugai (2004) and Wakai et al. (2010).

Table 3 gives the parameters of the UW constitutive model and strain-softening UW constitutive model used in this research. The deformation modulus E was calculated by $E = 2G_0(1 + \nu)$, here ν is Poisson's ratio and $G_0 = \rho V_s^2$. The shear wave velocity V_s was determined comprehensively based on those measured by the surface wave exploration and portable dynamic cone penetration tests. The empirical values of Poisson's ratio were used. The parameters of peak shear strength of the soil layer in which the sliding surface was developed were determined by direct shear tests on the samples obtained by block sampling. The influence of water pressure on the shear strength was considered using an apparent cohesion, which was equal to the cohesion measured subtracted by water pressure multiplied by tan φ. The residual cohesion was empirically assumed to be zero, and the residual friction angle was determined using Fig. 17 and plastic index $I_p = 17.9$ measured in the laboratory.

The dynamic finite element analyses were conducted to simulate the response of the slope during the earthquake. Table 5 lists the analyzed cases for three geological cross-sections, the lower stable part of the central cross-section, the upper part of the central cross-section, and the left cross-section. For each cross-section, three groundwater conditions were considered, i.e., cases with no groundwater considered,

Table 3 Material properties for dynamic finite element analysis

Stratum	Unit weight γ (kN/m^3)	Deformation modulus E (MN/m^2)	Poisson's ratio ν	Peak shear strength		Softening parameter		Hysteresis damping	
				Cohesion c (kN/m^2)	Friction angle φ (°)	τ_{fr}/τ_{f0}	A	$b\gamma_{G0}$	n
Brown clayey soil with gravel	16.7	90.8	0.35	See Table 4	38	0.36	0.12	0.75	4.2
Clayey soil with gravel with a color of original rock	16.7	146	0.32		38	0.36	0.12	0.78	3.8
Fine sandstone (angular gravel like) Fine sandstone Alternation of sandstone and mudstone	19.7	594	0.25		41	–	–	6.5	1.43

Table 4 Apparent cohesion c (kN/m^2) of materials above or below groundwater level (GWL)

Stratum	Above GWL	Below GWL
Brown clayey soil with gravel	38.6	20.4
Clayey soil with gravel with a color of original rock	38.6	20.0
Fine sandstone (angular gravel like) Fine sandstone Alternation of sandstone and mudstone	200	164

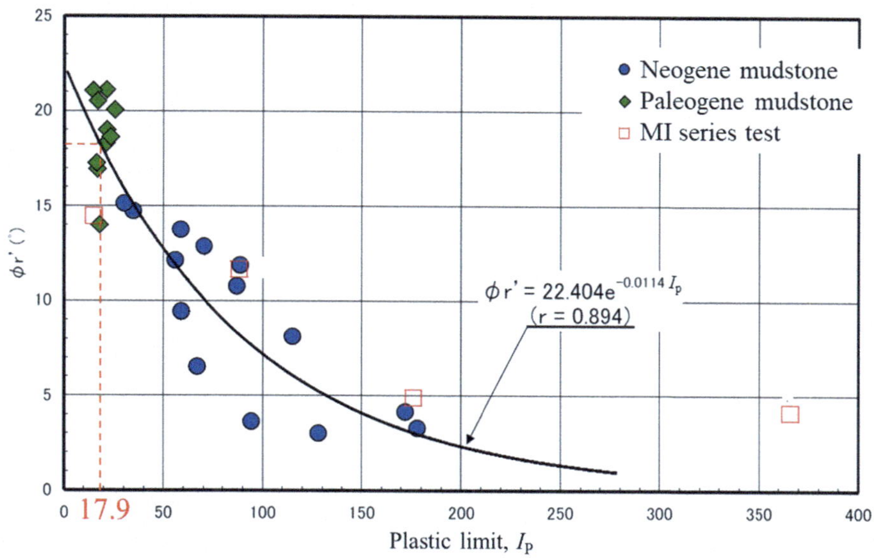

Fig. 17 Relation between plastic limit I_p and residual friction angle φ_r for the clay of the sliding plane originated from Tertiary mudstones (modified from Yamazaki et al. 2000)

and before and after construction of the drainage wells. Figure 18 shows the finite element mesh of the left cross-section used in the dynamic finite element analysis. To eliminate the effect of the left and right vertical boundaries, one cushion area about 300 m long with the same soil or rock layers was connected to the left and right sides, respectively. The left and right cushion areas were not shown in Fig. 18.

Input seismic motions in the direction of the cross-sections, as shown in Fig. 19, were calculated using the observed seismic motions in NS and EW directions at K-net (Hakuba), of which the location was indicated in Fig. 2, and the azimuth of the two cross-sections, respectively. They were slightly different from one another because of the differing azimuths of the two cross-sections. The duration included 35 s of seismic motion and 15 s of no shaking that was added to confirm the displacement of the landslide after the earthquake.

Figure 20 shows the original finite element mesh and deformed finite element mesh after the earthquake in the cases before and after the construction of the drainage

Table 5 Cases of dynamic finite element analysis

Item		Lower part of slope Central cross-section		Upper part of slope Central cross-section		Left cross-section
Softening parameter (τ_{fr}/τ_{f0})		0.13	0.36	0.13	0.36	0.36
GW not considered	Field	Stable	Stable	Two collapses	Two collapses	Two collapses
	Numerical	Stable	Stable	Two collapses	One collapse	One collapse
	Reproducibility	Good	Good	Good	Half & half	Half & half
Before drainage wells constructed	Field	Stable	Stable	Two collapses	Two collapses	Two collapses
	Numerical	Stable	Stable	Two collapses	Two collapses	Two collapses
	Reproducibility	Good	Good	Good	Good	Good
After drainage well constructed	Field	Stable	Stable	NA	NA	NA
	Numerical	Stable	Stable	Two collapses	Two collapses	One collapse
	Reproducibility	Good	Good	NA	NA	NA

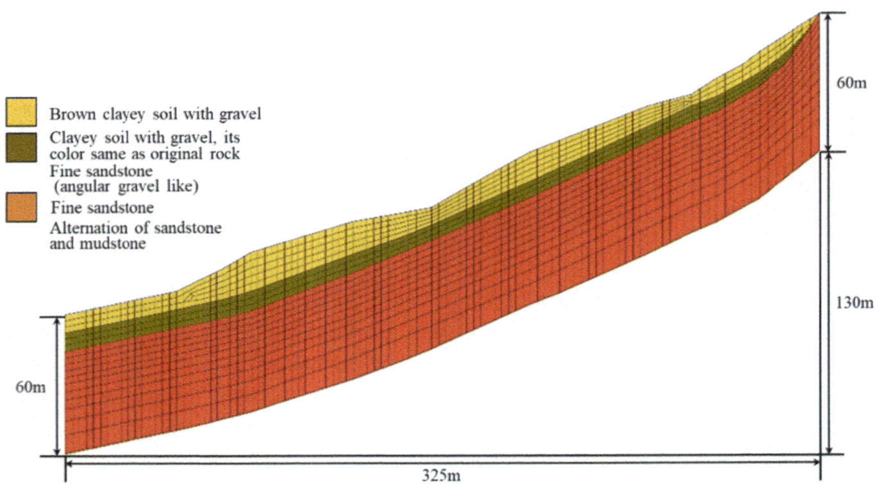

Fig. 18 Finite element mesh for dynamic analysis of left cross-section

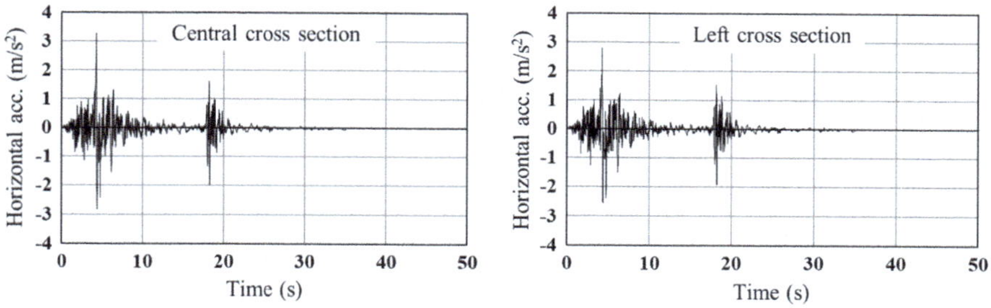

Fig. 19 Input seismic motions for dynamic finite element analysis

well in the lower part of the slope of the central cross-section. The difference of the original and deformed mesh was the residual deformation induced by the earthquake. Figure 20 indicated that the residual displacement of the lower part of the central cross-section was small both before and after the construction of the drainage well, and the difference of the residual displacement between before and after construction of the drainage well deformed mesh was very small anywhere in the lower part of the central cross-section. Figure 21 shows the histogram of the horizontal displacement of point A on the surface of the lower part of the slope, i.e., the red solid circle in Fig. 20. The residual horizontal displacement of point A was only 0.25 m both before and after the construction of the drainage well. The calculated results showed that the residual vertical displacement of point A was only 0.025 m after the construction of the drainage well and 0.03 before construction of the well. These results indicated that the small slope gradient of the lower part of the slope should be the main reason why the lower part of the slope of the central cross-section had not moved during the earthquake. The effect of the drainage well on the lower stable part of the central cross-section was small and can be neglected.

Figure 22 shows the residual shear strain distribution of the upper part of the slope for the central cross-section in the cases where the groundwater was not considered, and where the groundwater was considered but no drainage well was constructed. If the groundwater was not considered in the dynamic finite element analysis, only one collapse was reproduced in the area with higher elevation and a larger slope gradient.

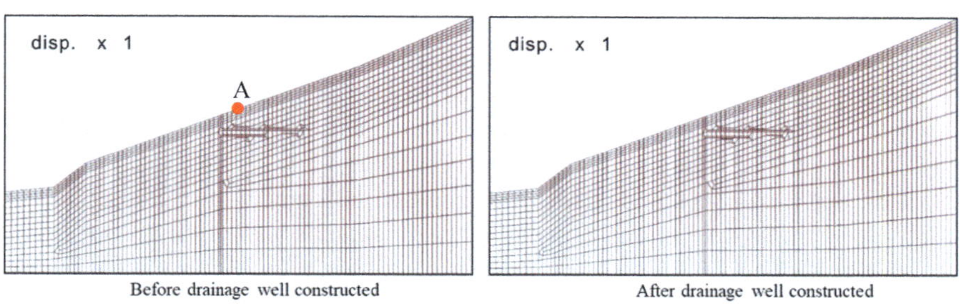

Fig. 20 Deformed finite element mesh after the earthquake for the lower part of the slope of the central cross-section

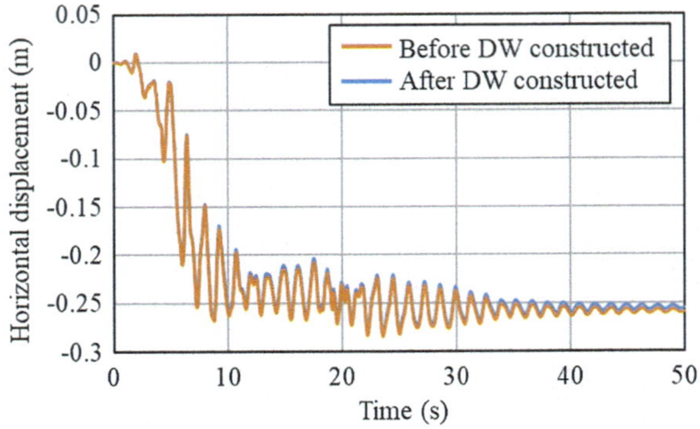

Fig. 21 Histogram of horizontal displacement of point A (red solid circle in Fig. 20) on the surface of the lower part of the central cross-section

However, if the groundwater was considered, two collapses were reproduced: one was the same as that in the case where the groundwater was not considered, and another occurred in the area with a lower elevation and smaller slope gradient. This was because the shear strength in the case where the groundwater was not considered was larger than that in the case where the groundwater was considered, for soil and rock layers below the groundwater level.

Figure 23 shows the histogram of the horizontal displacement of points A and B on the surface, as shown in Fig. 23 (left). The residual horizontal displacement of point B was almost zero if the groundwater was not considered. This indicated that collapse did not occur at the area where point B was located. However, after 35 s with no input seismic motion, i.e., the acceleration equal to zero (see Fig. 19), the horizontal displacements did not stop at point A in the cases whether or not the groundwater was considered and at point B in the case where the groundwater was considered. This indicated that collapse was reproduced by the dynamic finite element analysis at the area where point A was located, whether or not the groundwater was considered, and collapse was reproduced at the area where point B was located only in the case the groundwater was considered. The results in the case where the groundwater was considered were generally consistent with collapses that were observed at the Ichiba No. 1 landslide after the earthquake.

Figure 24 shows the histogram of the horizontal displacement of points C and D on the surface of the left cross-section before and after the construction of the drainage wells if the slope were to be hit by the Nagano Prefecture Kamishiro Fault Earthquake. The horizontal displacement of point D stopped after the earthquake in the case where the drainage wells were constructed. However, the horizontal displacement of point C, whether or not the drainage wells were constructed, and of point C in the case in which the drainage wells were not constructed, did not stop after the earthquake. This indicated that, although a residual horizontal displacement of 0.53 m occurred, collapse with a very large movement did not occur at the lower part of the left cross-section where point D was located if the drainage wells were constructed before the earthquake. However, even if the drainage wells were constructed before the earthquake, collapse still occurred in the upper part of the left cross-section where point C was located because of its larger slope gradient.

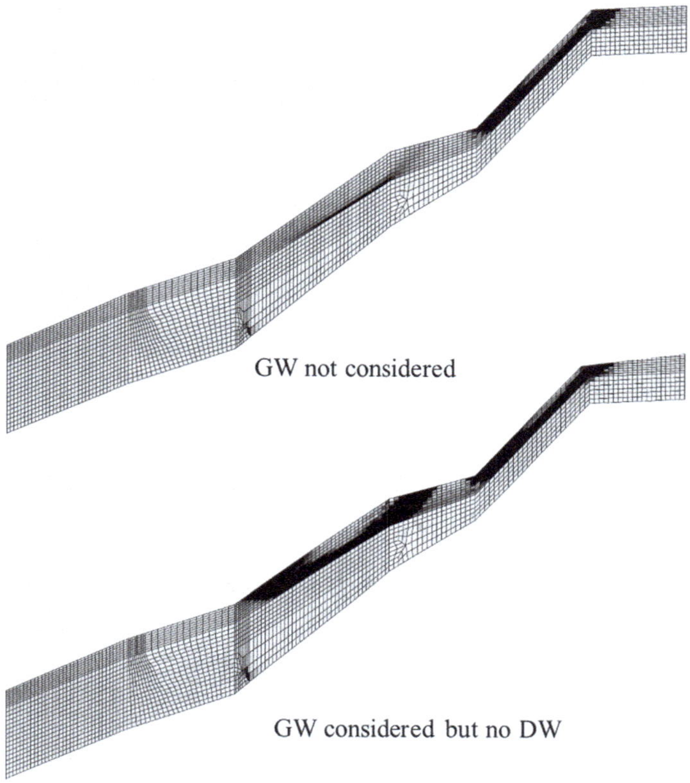

Fig. 22 Residual shear strain after the earthquake for the upper part of the slope for the central cross-section, GW—groundwater, DW—drainage well

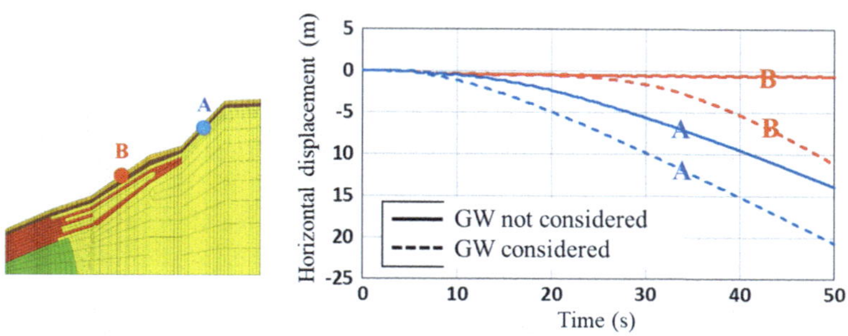

Fig. 23 Histogram of horizontal displacement of points A and B on the slope surface of the upper part of the central cross-section (Fig. 3)

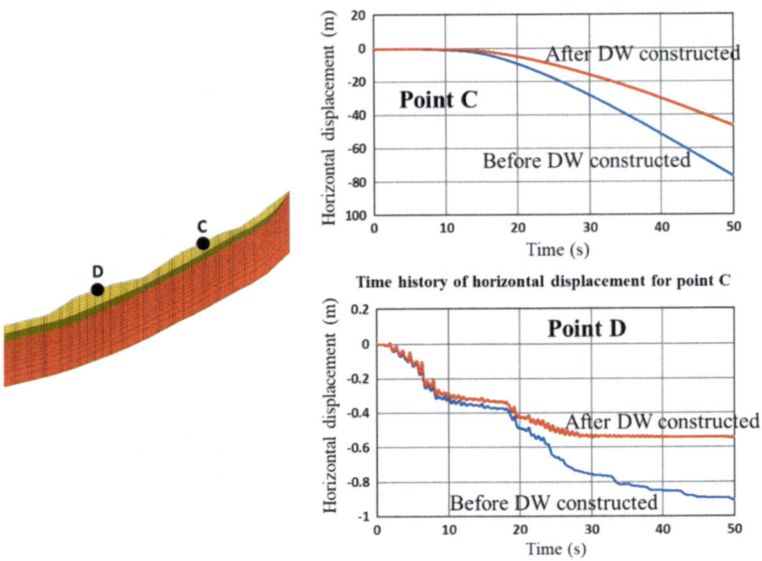

Fig. 24 Histogram of horizontal displacement of points C and D on the slope surface of the left cross-section (Fig. 3)

Figure 25 shows the original finite element mesh and the deformed finite element mesh for the left cross-section before and after the construction of the drainage wells if the slope were to be hit by the Nagano Prefecture Kamishiro Fault Earthquake. It was obvious that even after the construction of the drainage wells, the upper part of the left cross-section still collapsed during the earthquake. However, the residual displacement, even of the upper part, was decreased to some extent after the construction of the drainage wells. Because the residual displacement of the lower part of the left cross-section was smaller than 0.9 m, the difference of the original and deformed finite element mesh for the lower part almost could not be observed, especially after the displacement was multiplied by 0.2 to show clearly the deformed finite element mesh of the upper part of the left cross-section.

Figure 26 shows the residual shear strain distribution for the left cross-section before and after the construction of the drainage wells if the slope were to be hit by the Nagano Prefecture Kamishiro Fault Earthquake. Two clear sliding bands with a large shear strain were observed in the case in which the drainage wells had not been constructed when the slope was hit by the earthquake. However, after the drainage wells were constructed, only one clear sliding band was observed at the upper part of the left cross-section, and the residual shear strain of the lower part was significantly smaller than that in the case before the construction of the drainage wells.

The numerical results for the central and left cross-section indicated that whether a collapse with a very large movement occurred should be judged mainly based on the histogram of displacement at some typical points on the slope surface. The

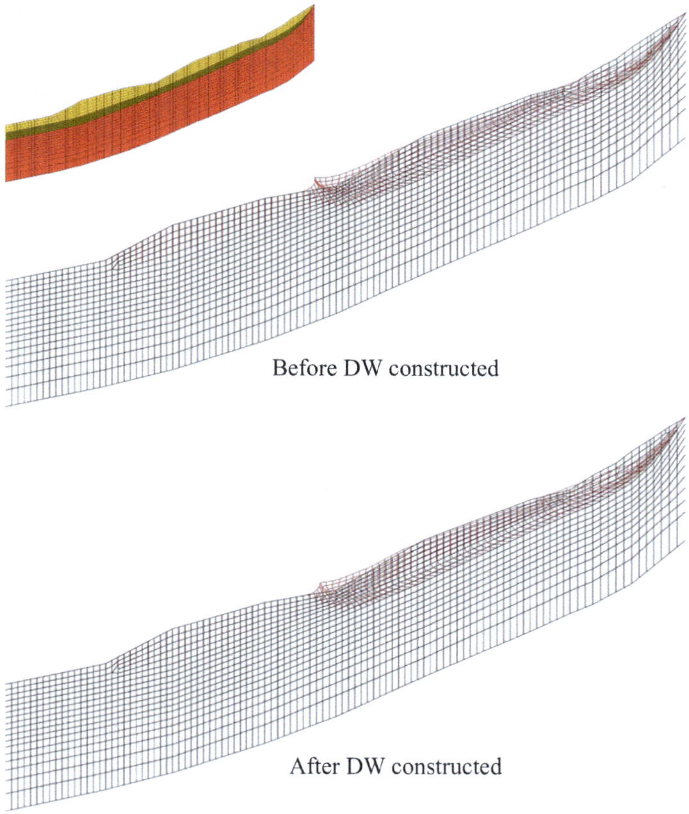

Fig. 25 Deformed finite element mesh (displacement multiplied by 0.2) of left cross-section after the earthquake

deformed finite element and residual shear strain after the earthquake could be used to evaluate where the sliding surfaces developed.

5 Concluding Remarks

This research was conducted using numerical analysis to evaluate the effect of drainage wells on the response of landslides during earthquakes, taking the Ichiba No. 1 landslide that occurred during the Nagano prefecture Kamishiro fault earthquake as an example. A field investigation was conducted to determine the material properties needed for the numerical analysis. The unsaturated–saturated finite element seepage analyses were carried out to evaluate the groundwater level lowered by the drainage wells. The dynamic total stress finite element analyses, in which strain-softening

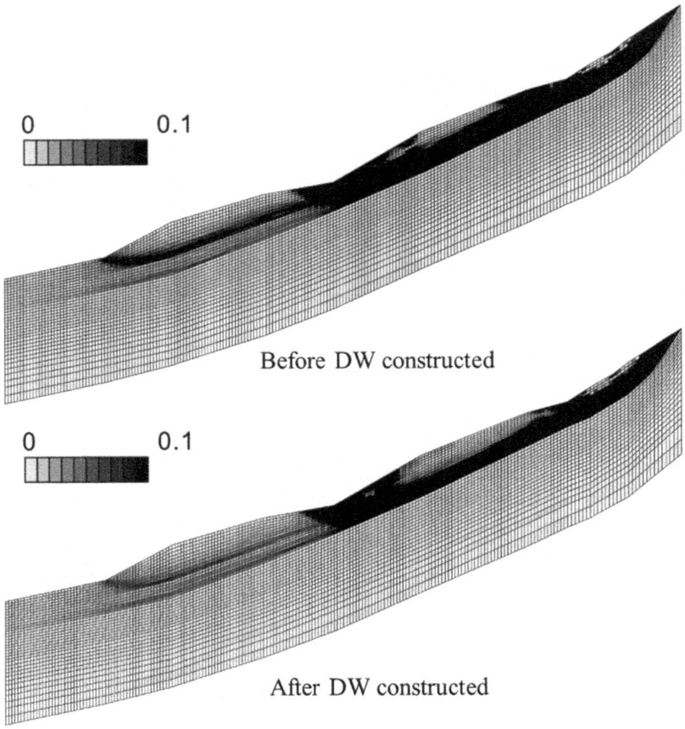

Fig. 26 Residual shear strain distribution of left cross-section after the earthquake

was considered, were used to calculate the response of the slope during the earthquake. From the field investigation and numerical analysis, we made the following conclusions.

(1) The depth at which the resistivity changed drastically, measured by electrical prospecting, seemed consistent with the depth of the sliding surface of the landslide that occurred during the earthquake.
(2) The soil layer that moved largely during the earthquake had a lower shear wave velocity than those of the soil and rock layers below the sliding surface. The shear wave velocity measured by the surface wave exploration was consistent with that calculated by the converted N-values of the portable dynamic cone penetration test using an empirical equation.
(3) The flow rates of drain pipes of drainage wells were significantly different pipe by pipe, and only part of the shallow drain pipes had water flow for the drainage wells constructed to stabilize the Ichiba No. 1 landslide.
(4) Numerical results of finite element seepage analysis indicated that the groundwater level could be effectively lowered, especially around the drainage wells. The maximum lowering of the groundwater level could be locally large—up to 19.8 m for the Ichiba No. 1 landslide.

(5) It is important to consider the influence of the groundwater level on the response of the slope during the earthquake using a dynamic total stress finite element analysis.
(6) Whether a collapse with a very large movement occurred could be judged mainly based on the histogram of displacement at some typical points on the slope surface. The deformed finite element mesh and residual shear strain after the earthquake could be used to evaluate where the sliding surfaces developed.
(7) For a slope with a large gradient, it may collapse even if it was stabilized by drainage wells. Thus, such slopes should be cut back to a small slope gradient to prevent them from collapsing in earthquakes.

Acknowledgements This research was partially supported by the grant-in-aid for collaborative research projects of the Research Institute for Natural Hazards and Disaster, Niigata University. We gratefully thank all those who have helped in conducting this research. Special acknowledgments are also given to the graduate students Ryoichiro Satake, Kazuki Tajima, Yusuke Yamamoto, Kyosuke Hori, and Masato Turuta at Gunma University who helped in conducting the field investigations and finite element simulations.

References

Brooks RH, Corey AT (1966) Properties of porous media affecting fluid flow. J Irrig Drainage Eng Div ASCE 92(2):61–88
Cai F, Ugai K (2004) Numerical analysis of rainfall effects on slope stability. Int J Geomech ASCE 4(2):69–78
Cai F, Ugai K, Wakai A, Li Q (1998) Effects of horizontal drains on slope stability under rainfall by three-dimensional finite element analysis. Comput Geotech 23:255–275
Domenico PA, Schwartz FW (1990) Physical and chemical hydrogeology. John Wiley & Sons
Furuya G, Hata T, Watanabe N, Goto S, Kawasaki I (2015) Characteristics of earthquake-induced landslides in a heavy snowfall region—landslides triggered by the northern Nagano prefecture earthquake, March 12, 2011, Japan. J Jpn Landslide Soc 52(1):40–44
Japan Road Association (2012) Specifications for Highway Bridges, Part IV Substructures
JMA (2014) Available at https://www.jma.go.jp/jma/kishou/books/saigaiji/saigaiji_201406.pdf
Katsube A, Kondo H, Kurosawa H (2017) Surface rupturing earthquakes repeated in the 300 years along the ISTL active fault system, central Japan. Geophys Res Lett 44:6057–6064
Katsura S, Kimura T, Maruyama K, Ishida K (2016) Characteristics of landslides induced by the Kamishiro fault earthquake in Nagano prefecture on November 22, 2014. J Jpn Landslide Soc 53(3):11–20
Leong EC, Rahardjo H (1997a) Review of soil-water characteristic curve equations. J Geotech Geoenvir Eng ASCE 123(12):1106–1117
Leong EC, Rahardjo H (1997b) Permeability functions for unsaturated soils. J Geotech Geoenvir Eng ASCE 123(12):1118–1126
Mualem Y (1976) A new model for predicting the hydraulic conductivity of unsaturated porous media. Water Resour Res 12:513–522
Nippon Koei (2015) Document for joint field investigation of the Ichiba No. 1 landslide area, Otari Village, Nagano Prefecture

Nishigaki M, Takeshita Y (1993) Research on the method of determining unsaturated soil hydraulic properties by in-situ or laboratory tests. Technical Report of Department of Civil Engineering, Okayama University (in Japanese)

Richards LA (1931) Capillary conduction of liquids through porous mediums. Physics 1:318–333

Stankovich JM, Lockinton DA (1995) Brooks-Corey and van Genuchten soil-water-retention models. J Irrig Drainage Eng ASCE 121(1):1–7

van Genuchten MT (1980) A closed-form equation for predicting the hydraulic conductivity of unsaturated soils. Soil Sci Soc Am J 44(5):892–898

van Genuchten MT, Leij F, Yates S (1991) The RETC code for quantifying the hydraulic functions of unsaturated soils. EPA Research Document EPA/600/2–91/065

Wakai A, Ugai K (2004) A simple constitutive model for the seismic analysis of slopes and its applications. Soils Found 44:83–97

Wakai A, Ugai K, Onoue A, Kuroda S, Higuchi K (2010) Numerical modeling of an earthquake-induced landslide considering the strain-softening characteristics at the bedding plane. Soils Found 50(4):533–545

Yamazaki T, Mayumi T, Yoshita E (2000) Ring shear characteristics of high purity clay minerals-Correlation with natural slip surface clay. J Jpn Landslide Soc 37(2):30–39

Displacement Predictive Relationships for Earthquake-Triggered Landslides and Structural Vulnerability Models for Buildings Exposed to Co-seismic Ground Movements

Kyriazis Pitilakis and Stavroula Fotopoulou

Abstract Seismically-induced permanent slope displacement is an essential threat for buildings that are situated close to potentially unstable slopes. The present paper addresses a coherent approach that predicts the slope displacement and assesses the seismic vulnerability of buildings undergoing this type of hazard. The first part of the paper introduces two-dimensional fully nonlinear dynamic parametric analyses on a wide range of slope configurations, soil properties and input seismic motions. The obtained comprehensive database of slope displacement was interpreted by an advanced statistic method to derive an analytical expressions that was able to predict the co-seismic slope displacement by means of the best-correlated scalar and vector of intensity measures. The second part of this paper concerns an analytical assessment of the vulnerability of reinforced concrete buildings that are subject to the earthquake-induced displacement in a neighboring slope with a range of height, gradient and soil properties. The conducted analysis is based on generic probabilistic fragility functions that were derived for typical reinforced concrete buildings. Finally, the developed method was applied to a compound disaster in which the ground shaking and the landslide displacement affect a building. The present paper particularly addresses the vulnerability of a typical low-rise, high-code reinforced concrete buildings that are located very close to a cliff in which the compound disaster mechanism is induced by the topographically amplified ground shaking and the seismically triggered landslide displacement. In more details, two disaster scenarios were considered, i.e., (i) the two hazards do not interact with each other, or (ii) the landslide hazard is conditional on the earthquake that is the primary trigger of the landslide.

Keywords Seismically induced slope displacements · Displacement predictive models · Coupled fragility curves · RC buildings

K. Pitilakis · S. Fotopoulou (✉)
Research Unit of Soil Dynamics and Geotechnical Earthquake Engineering (SDGEE), Aristotle University of Thessaloniki (AUTH), Thessaloniki, Greece
e-mail: sfotopou@civil.auth.gr

I. Towhata et al. (eds.), *Coseismic Landslides*, Springer Natural Hazards,
https://doi.org/10.1007/978-981-19-6597-5_16

1 Predictive Models for Co-Seismic Slope Displacements

The performance-based design of structures is a common trend in the world and is favored by geotechnical earthquake engineering as well. Because the residual displacement is the key factor to govern the seismic performance as is typically shown by slopes undergoing earthquake effects, the practice commonly relies on a displacement-based procedure which assesses the co-seismic slope displacement on the basis of some metric(s) of the seismic ground motion and the slope properties. Typical examples of this procedure are found in the landmark studies by Newmark (1965) and Makdisi and Seed (1978), while several empirical models that followed employed a rigid block model of slopes in which both decoupled and coupled versions were available and provided various indices of the seismic slope performance in terms of displacements (e.g., Bray 2007).

While the well-known Newmark-type analyses rely on a rigid block modelling resting on a frictional slope, many other methods make use of stress–strain models of soil in order to evaluate the seismically induced slope displacements. Among them, Fotopoulou and Pitilakis (2015) used the FLAC2D numerical code (Itasca 2011) and proposed a new computational procedure that is based on two-dimensional fully nonlinear dynamic parametric analyses on a variety of field conditions such as shape of slopes, soil properties and input earthquake motions. Because of the intrinsic uncertainties in the subsurface conditions, twelve typical 2D slope models were analyzed in association with varying geometrical characteristics, material properties of the surface soils and strength and stiffness of the potential slip planes; refer to Table 1.

The proposed method was intended only to evaluate the seismically induced slope displacements, whereas the possible effects of excess pore water pressure due to cyclic shear loading was out of scope. The seismic performance of the analyzed slopes was expressed by the permanent horizontal displacements that sustained in the sliding mass after the earthquake shaking ceased. A family of slopes having different flexibility and resistance in the potential sliding surface were analyzed. As shown in Fig. 1, the horizontal displacements thus obtained range from very small values (<0.01 m) to large values (>1 m). In total, 285 permanent horizontal displacements were calculated from this set of analysis.

The study was initiated with the validation of the reliability of the numerical analyses by comparing the numerically obtained displacements and those obtained by Newmark-type analysis or its empirical simplification. Thereinafter, the computed displacements were correlated with a variety of intensity measures (IMs) by running linear regression analyses. Scalar IMs and two-dimensional vector IMs were optimally selected to take care of proficiency (Padgett et al. 2008) and sufficiency (Luco and Cornell 2007) criteria. The correlation coefficient between the IMs is also considered for the selection of an appropriate IM vectors. Moreover, the yield coefficient k_y, that stands for the overall resistance of the slope, was taken into account in the regression equations as the residuals of the studied IMs were linearly dependent on

Table 1 Analyzed slope soil models

Yielding seismic coefficient k_y	Slope height h (m)	Slope inclination i (degrees)	Soil properties in the surface layer (see Table 3 for details)	Depth of the sliding mass H (m)	Elastic fundamental period T_s (s)
0.05	40	45	Vs = 400 m/s, c = 5 kPa, φ = 44°	5	0.050
	40	45	Vs = 400 m/s, c = 50 kPa, φ = 25°	16	0.160
0.1	20	45	Vs = 400 m/s, c = 5 kPa, φ = 44°	4	0.040
	20	45	Vs = 400 m/s, c = 30 kPa, φ = 25°	10	0.100
0.15	20	30	Vs = 250 m/s, c = 0.5 kPa, φ = 37°	2	0.032
	20	30	Vs = 250 m/s, c = 15 kPa, φ = 25°	8	0.130
	40	30	Vs = 400 m/s, c = 30 kPa, φ = 25°	15	0.150
0.2	20	15	Vs = 150 m/s, c = 15 kPa, φ = 18°	14	0.373
	40	15	Vs = 150 m/s, c = 30 kPa, φ = 18°	26	0.690
0.25	40	15	Vs = 150 m/s, c = 0.5 kPa, φ = 27°	5	0.130

(continued)

Table 1 (continued)

Yielding seismic coefficient k_y	Slope height h (m)	Slope inclination i (degrees)	Soil properties in the surface layer (see Table 3 for details)	Depth of the sliding mass H (m)	Elastic fundamental period T_s (s)
0.3	40	30	Vs = 400 m/s, c = 5 kPa, φ = 44°	4	0.040
	20	15	Vs = 250 m/s, c = 15 kPa, φ = 24°	14	0.224

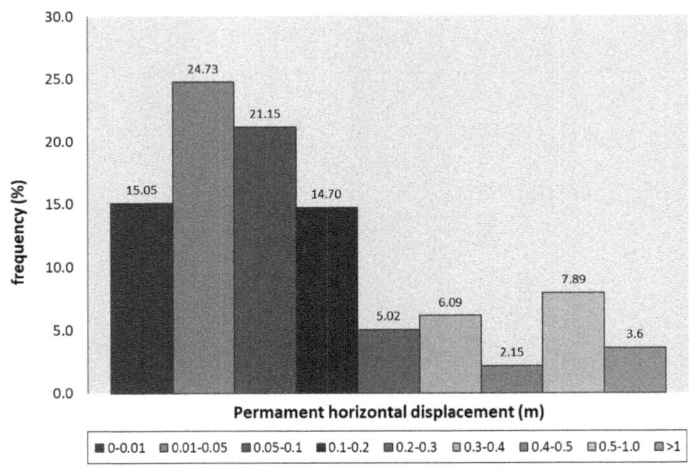

Fig. 1 Histogram of the computed numerical horizontal displacements (the total number of data, $N = 285$) (modified from Fotopoulou and Pitilakis 2015)

k_y. To take care of the sufficiency criterion with respect to the earthquake magnitude M, a magnitude term was added to the suggested scalar analytical expressions, based on the regression analyses of the residuals $\varepsilon|\text{IM}$ from the calculation of the seismic slope displacements on the basis of magnitude M. Based on the optimally selected scalar IMs and vector of two IMs, the final recommendations were made for the scalar (Eqs. 1–3) and vector (Eqs. 4–5) models for displacement prediction (Fotopoulou and Pitilakis 2015):

$$\text{Ln}(D) = -9.891 + 1.873 \cdot \ln(\text{PGV}) - 5.964 \cdot k_y + 0.285 \cdot M \pm \varepsilon \cdot 0.65 \quad (1)$$

$$\mathrm{Ln}(D) = -2.965 + 2.127 \cdot \ln(\mathrm{PGA}) - 6.583 \cdot k_y + 0.535 \cdot M \pm \varepsilon \cdot 0.72 \quad (2)$$

$$\mathrm{Ln}(D) = -10.246 - 2.165 \cdot \ln\left(k_y/\mathrm{PGA}\right) + 7.844 \cdot k_y + 0.654 \cdot M \pm \varepsilon \cdot 0.75 \quad (3)$$

$$\mathrm{Ln}(D) = -8.076 + 1.873 \cdot \ln(\mathrm{PGV}) + 0.200 \cdot \ln(I_a) - 5.964 \cdot k_y \pm \varepsilon \cdot 0.61 \quad (4)$$

$$\mathrm{Ln}(D) = -8.360 + 1.873 \cdot \ln(\mathrm{PGV}) - 0.347 \cdot \ln\left(k_y/\mathrm{PGA}\right) - 5.964 \cdot k_y \pm \varepsilon \cdot 0.64 \quad (5)$$

in which D (m) is the seismic slope displacement, PGA(g) is the peak ground acceleration, PGV (cm/s) is the peak ground velocity, I_a (m/s) is the Arias Intensity (Arias 1970) and ε is the standard normal variant with zero mean and unit standard deviation. The intensity measures (PGA, PGV, I_a) given in the above equations can be estimated through site-specific seismic hazard analyses that consider local site effects. As an alternative, simplified relationships were proposed by Fotopoulou and Pitilakis (2015) in which the IMs at the depth of the potential sliding surface are assessed through a correlation with the corresponding IMs at the rock outcrop.

2 Vulnerability Assessment of Reinforced Concrete Buildings Subject to Co-Seismic Slope Displacement

2.1 Vulnerability Assessment Method

The present study concerns the seismically triggered landslides and the special attention is focused on the co-seismic slope movements. Note that the flow-type massive landslides or lateral spreading after liquefaction are out of scope. The physical or the structural landslide vulnerability of a building is defined as the extent of loss expressed by the percentage of structural damage in a designated structural element or a set of elements that are prone a landslide hazard (e.g., Corominas et al. 2014). The basic framework of the method was developed by the EU research project SAFELAND (www.safeland.no) and have been recently published in Fotopoulou and Pitilakis (2013a; b) and Mavrouli et al. (2014).

The methodology is described in what follows for the derivation of fragility curves of low-rise reinforced concrete buildings that are located next to the ridge of a potentially unstable slope. While the earthquake engineering concerns structures subjected to ground shaking, the present study has to deal with buildings subjected to forced differential displacement that is produced by an earthquake-triggered landslide and results in structural distress and damage. Therefore, the present study eliminates the issue of shaking-induced structural damage, inclusive of the effect of topography, from the evaluation of the building's vulnerability.

The fragility curves were derived in principle by numerical procedures that consist of two consecutive steps and statistical analyses. First, the total and differential permanent displacements at the foundation level were calculated on a studied building located next to the slope crest by using a plane-strain finite difference and nonlinear dynamic analyses. The input seismic motions for the analyses were different progressively scaled acceleration time histories recorded on rock outcrops. Then, detailed analyses were conducted on the structural response of buildings to the ground deformation by running nonlinear static analyses in which the computed ground deformation were put in the building foundation. To be more specific, the differential ground displacement as obtained in the first step at the foundation level was applied quasistatically to one of the building's supports. The reinforced concrete structures were idealized by nonlinear constitutive models that takes reasonably into account the governing effects of cracking and irreversible deformations. The local damage index (or EPD) in terms of maximum strains in steel (ε_s) and concrete (ε_c) is used to evaluate the building's response to the seismically induced slope displacements; refer to Crowley et al. (2004), Bird et al. (2005), Negulescu and Foerster (2010), and Fotopoulou and Pitilakis (2013a, b) among others. Four limit states of damage (LS_1, LS_2, LS_3, LS_4) were defined in order to describe the exceedance of slight, moderate, extensive and complete structural damage of the building. Table 2 described each damage state and the mean limit strain values that address reinforced concrete frame buildings designed with "low" and "high" seismic code provisions.

The intensity or extent of landslide is expressed in terms of PGA at the outcrop or the permanent ground displacement (PGD) within the slope area. More specifically,

Table 2 Structural damage state description and definition of damage limit states for reinforced concrete frame buildings designed with "low" and "high" seismic design level (after Crowley et al. 2004)

Structural damage band	Description
None to slight	Linear elastic response, flexural or shear type hairline cracks (<1.0 mm) in some members and no yielding in any critical section Limit State 1 (LS_1): steel bar yielding
Moderate	Member flexural strengths achieved, limited ductility developed, crack widths reach 1.0 mm and initiation of concrete spalling Limit State 2 (LS_2): $\varepsilon_s = 0.0125$, $\varepsilon_c = 0.0045$
Extensive	Significant repair required to building, wide flexural or shear cracks and buckling of longitudinal reinforcement may occur Limit State 3 (LS_3): "High" code frames: $\varepsilon_s = 0.04$, $\varepsilon_c = 0.010$ "Low" code frames: $\varepsilon_s = 0.025$, $\varepsilon_c = 0.006$
Complete	Repair of building not feasible either physically or economically, demolition required, could be due to shear failure of vertical elements or excessive displacement Limit State 4 (LS_4): "High" code frames: $\varepsilon_s = 0.06$ "Low" code frames: $\varepsilon_s = 0.045$

Table 3 Parameters in the proposed fragility functions in terms of permanent ground displacement (PGD) based on the regression analysis method and the maximum likelihood method

	Parametric models	Median PGD (m)				Dispersion β
		LS1	LS2	LS3	LS4	
Regression analysis method	sand_i30	0.17	0.46	1.09	1.53	0.57
	clay_i30	0.24	0.81	1.97	2.69	0.56
	sand_i15	0.25	0.70	1.50	1.96	0.57
	clay_i15	0.33	1.07	2.62	3.61	0.55
	sandy_i45_h20	0.07	0.22	0.57	0.80	0.55
	sandy_i45_h40	0.05	0.16	0.45	0.67	0.50
	clay_i45	0.80	2.82	–	–	0.73
Maximum likelihood method	sand_i30	0.15	0.39	0.87	1.64	0.43
	clay_i30	0.22	0.67	1.70	3.43	0.43
	sand_i15	0.24	0.60	1.29	2.19	0.40
	clay_i15	0.28	0.94	2.32	–	0.43
	sandy_i45_h20	0.09	0.18	0.47	0.91	0.43
	sandy_i45_h40	0.04	0.16	0.45	0.67	0.50
	clay_i45	0.77	1.72	–	–	0.47

the maximum computed PGD value, wherever it is observed at the slope area, is considered as landslide intensity. The latter is generally better correlated with structural deformation and damage (Fotopoulou 2012). Therefore, the predictive relationships for displacement as presented in Sect. 3 are used for the direct estimation of PGD (or D) considered throughout this study as a landslide intensity parameter. The results of the nonlinear static analysis (pairs of PGD and the maximum strain) of the structure are statistically correlated with the corresponding predefined damage limit states so that probabilistic fragility curves may be derived and expressed as lognormal distribution functions:

$$F(\mathrm{LS} \geq \mathrm{LS}_i|\mathrm{PGD}) = \Phi\left[\frac{1}{\beta}\mathrm{In}\left(\frac{\mathrm{PGD}}{\overline{\mathrm{PGD}_i}}\right)\right] \quad (6)$$

where LS_i designates the damage limit state i (where i = 1, 2, 3, 4), $\Phi[\cdot]$ stands for the standard normal cumulative distribution function, $\overline{\mathrm{PGD}_i}$ means the median value of PGD at which the building reaches the limit state i and β indicates the standard deviation of the natural logarithm of PGD. A single value of β was assigned in common to all the limit states to ensure that the fragility curves do not intersect with each other. Then, the fragility parameters (median and log-standard deviation) were estimated by two different procedures, given the simulated damage data. The first one was based on a regression analysis (e.g., Cornell et al. 2002; Nielson and DesRoches 2007; Argyroudis and Pitilakis 2012; Fotopoulou and Pitilakis 2017), whereas the second one relies on a purely statistical approach; i.e., the Maximum Likelihood Method (e.g., Shinozuka et al. 2000; Fotopoulou and Pitilakis 2013b).

2.2 *Derivation of Fragility Curves Based on Numerical Parametric Analysis*

Parametric investigation was carried out extensively on a reference building that is situated near a slope shoulder (Fig. 2). In total, 24 models were analyzed with different geometry of the slope (slope height h and inclination i), different properties of soil in the slope (soft to stiff clayey and loose to dense sand) corresponding to soil categories B, C and D of Eurocode 8—Part 1 (EN 1998–1 2005), and the distance between the building and the slope shoulder ($L = 3, 5$ m). The soil properties were determined to be consistent with the considered slope angles so that the slope is stable under static (gravity) action. This stability requirement means that a certain extent of cohesion was needed for steep (45°) sandy slopes, i.e., $\varphi = 44°$ and $c =$ 10 kPa. Steep slopes made of soft clay were not analyzed because the slope would be already unstable under static conditions. Moreover, the analysis was limited to those cases that are substantially unstable during earthquakes. For instance, those slopes with small gradient ($i = 15°$) resting on stiff soil were not investigated because the induced permanent deformation and the likelihood of building damage would be negligible.

Note that the present study addresses dry soil conditions, and the effect of groundwater such as altering the seismic response, the slope and the extent of structural damage is out of scope. Fotopoulou and Pitilakis (2013b) described details of the parametric soil models and the input ground motion. The reference studied building is a simplified single bay-single story reinforced concrete bare frame structure resting on a flexible foundation. It was found that this idealization of building adequately reproduced the seismic response of real low-rise reinforced concrete frame buildings designed with "high" seismic code provisions (e.g., Bird et al. 2005; Negulescu and Foerster 2010).

Seven sets of fragility curves in terms of PGD within the slope area were finally proposed for the investigated models based on the most influential slope parameters; i.e., the slope inclination in conjunction with the soil material and the slope height for the 45° inclined sand slopes. Table 3 presents the median and the dispersions of the suggested fragility curves that were obtained on the basis of the method of regression analysis and the maximum likelihood method, respectively. On the other hand, Fig. 3 shows the corresponding graphs of fragility curves when using the latter method. It is noted for the 45° inclined clay slope, no fragility curves parameters for the extensive (LS3) and complete (LS4) damage are provided, as the corresponding damage states are not reached. The different fragility parameters for the two different methods are related with the different assumptions adopted in each method, and clearly demonstrate the influence of epistemic uncertainty on the

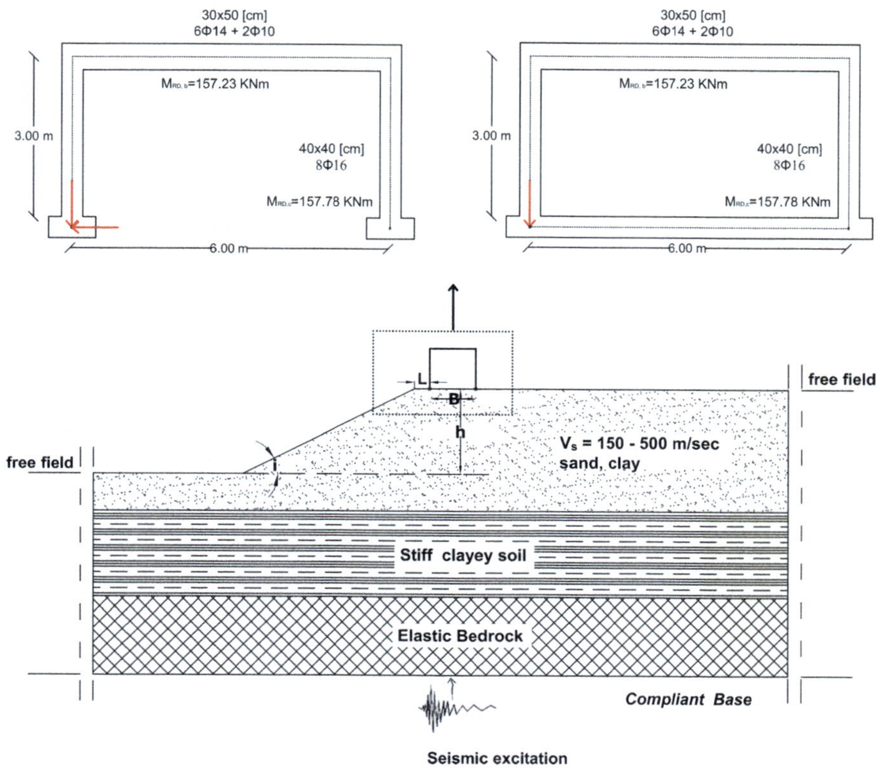

Fig. 2 Model of a single bay-single story reinforced concrete frame building resting on flexible and stiff foundation systems and displacement loading patterns for the nonlinear quasi-static analysis (top), sketch of model for dynamic analyses (bottom) (modified from Fotopoulou and Pitilakis 2013b)

fragility analysis. In particular, the estimated medians in terms of PGD are generally in good agreement between the two methods, while the estimated β values are slightly greater for the regression analysis method, increasing the dispersion of the curves. The derived fragility curves are used to evaluate the vulnerability of low-rise, "high"-code reinforced concrete frame buildings resting on flexible foundations. As stated above, these foundations are prone to seismically induced slope displacements for a variety of slope configurations and soil conditions. Similar curves associated with higher vulnerability are derived for "low-code" reinforced concrete frame structures.

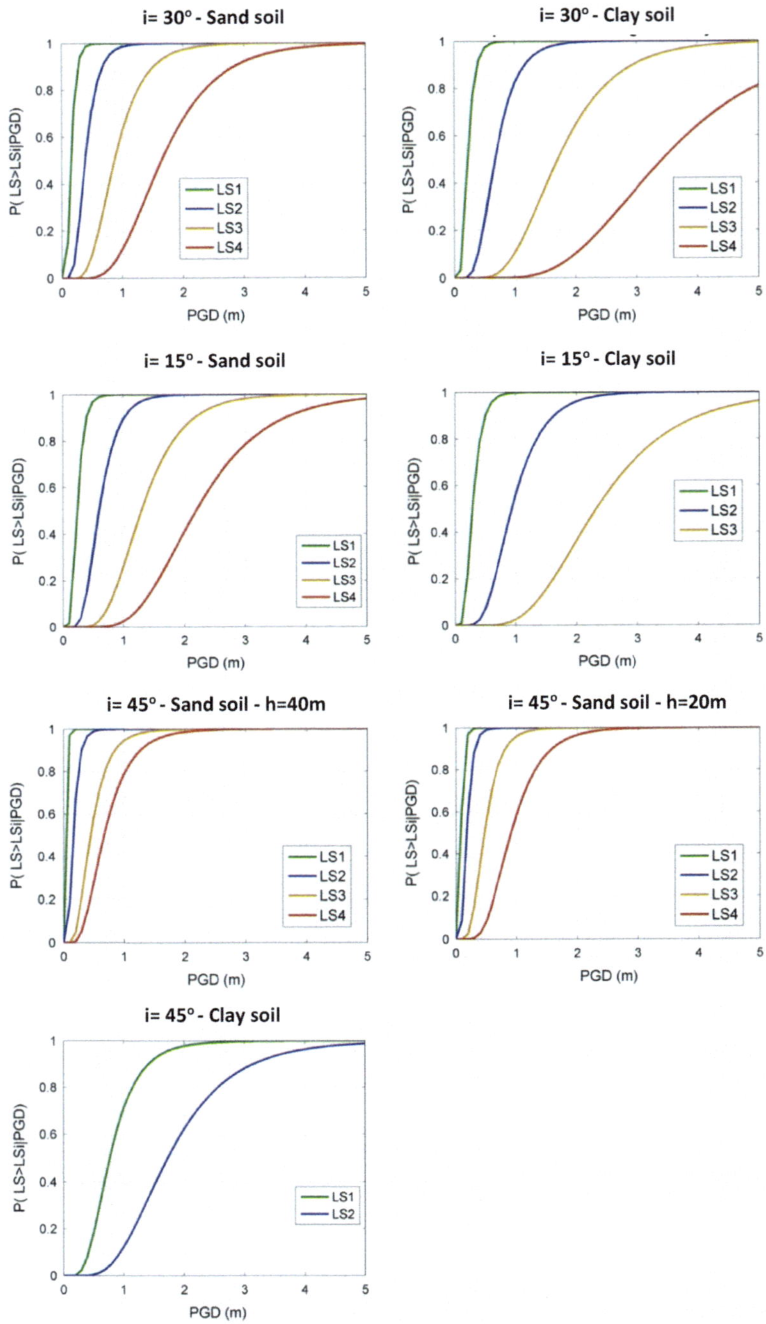

Fig. 3 Example of obtained fragility curves as a function of permanent ground displacement (PGD), addressing high-code, low-rise reinforced concrete frame buildings subjected to permanent slope displacements based on the maximum likelihood method (modified from Fotopoulou and Pitilakis 2013b)

3 Assessment of Combined Vulnerability of Reinforcd Concrete Buildings to Ground Shaking and Co-Seismic Slope Displacement

The discussion in the previous section addressed the building's vulnerability to co-seismic slope displacement due landslide without considering the possible effect of existing damage produced by ground shaking. In this section, the complex issue of the combined damage mechanisms, induced by landslide displacement and shaking-induced existing damage, is discussed in order to propose a combined vulnerability assessment method. In particular, the vulnerability is assessed for a typical low-rise, high-code reinforced concrete building that is situated near the slope's shoulder and subjected to the above-mentioned combined effect of the ground shaking (modified for the topography) and permanent landslide deformation. Details regarding the adopted slope and structural models, as well as to the analyses performed, can be found in Fotopoulou and Pitilakis (2017). In brief, the following two different scenarios were considered:

(I) Those two hazards do not interact, i.e., the damage due to ground shaking is independent of and does not affect the damage due to the co-seismic landsliding (National Institute of Building Sciences 2004) and
(II) Those two hazards do interact, i.e., the extent of landslide hazard is affected by the earthquake that is the initial trigger of the landslide.

Fragility curves were derived by combining the probabilities of damage states due to the landslide (P_L) and ground shaking (P_{GS}). It is noted that the damage state probabilities due to ground shaking (P_{GS}) may be computed using two-dimensional dynamic nonlinear computations and adequate statistical analysis, as described in Fotopoulou and Pitilakis (2017). Otherwise, one may use appropriate generic seismic fragility functions to derive the damage state probabilities due to ground shaking (P_{GS}). Thus, once the independent damage state probabilities are obtained for landslide (P_L) and ground shaking (P_{GS}), respectively, the combined probability (P_{COMB}) of exceeding a given limit state i is given by:

$$P_{COMB}[LS \geq LS_i] = P_{GS}[LS \geq LS_i] + P_{landslide} \cdot P_L[LS \geq LS_i] - P_{GS}[LS \geq LS_i] \cdot P_{landslide} \cdot P_L[LS \geq LS_i] \tag{7}$$

where $P_{landslide}$ is the probability of landslide occurrence during the earthquake. This probability of landslide occurrence ($P_{landslide}$) in Scenario I was set equal to 1.0, because this scenario considers that the two hazards are independent of each other. Although this consideration is rather "simplified", it is reasonably acceptable because it takes into account the uncertainties and complexities involved in the interaction between the two modes of loading and failure (ground shaking and permanent slope displacement).

Scenario II is more realistic because it is reasonable that strong shaking occurs prior to the development of slope deformation towards the ultimate landslide. Nevertheless, we encounter substantial uncertainty in the estimation of the landslide probability. It was decided, therefore, that the probability of landslide occurrence ($P_{\text{landslide}}$) in Scenario II is set equal to 0.75, based on the estimated maximum horizontal displacements as suggested by the entire soil dynamic analyses with the hypothesis that the threshold (critical) horizontal displacement is 7 cm at the initiation of the landslide movement. This threshold displacement of 7 cm pertains principally to thin, shallow landslides; refer to, e.g., Jibson (2011), which is the case expected in this study for the dense sand soil materials.

The above-mentioned equations were used for the given structure located at different distances from the slope's shoulder ($L = 3$ m, 5 m and 10 m) and the combined probabilities were calculated for the two Scenarios (I and II) and for each damage limit state due to ground shaking and seismically induced landsliding (Fig. 4).

The analyses on the above-mentioned conditions exhibited that the building standing at 3 m and 5 m from the crest would suffer higher combined extensive and complete damage due to ground shaking and landsliding compared to the vulnerability of the building located at 10 m from the crest, given that PGA is held identical. Scenario II, which assumes that the landslide hazard is dependent on the preceding earthquake, is less conservative than Scenario I. Generally, lower vulnerability values are obtained for the post-yield limit states for the building standing next to the crest (i.e., 3 m and 5 m) and more subject to the landslide hazard. Thus, unless a reliable estimation of the landslide occurrence is available, the use of the method proposed herein to derive coupled fragility curves without considering the interaction of the two hazards (Scenario I) is recommended for general engineering applications, because their estimation of seismic fragility is rather conservative for the structures standing near the shoulder of potentially hazardous slopes.

4 Conclusion

An analytical approach was presented in this article to assess the seismic performance of slopes in terms of permanent ground displacements due to landslide and to evaluate the seismic vulnerability of typical buildings subjected to the co-seismic ground shaking and displacement. The first part of this article described a set of new analytical relationships used for prediction of seismically induced slope displacements by means of two-dimensional fully nonlinear dynamic parametric analyses with different slope configurations, soil conditions and input motions. Scalar and vector optimally selected analytical models were developed to assess the co-seismic slope displacement that were identified based on proficiency and sufficiency criteria of the scalar and vector of two intensity measures. The developed displacement models can be used for assessing the performance of soil slopes in both deterministic and probabilistic manners at regional or local scales.

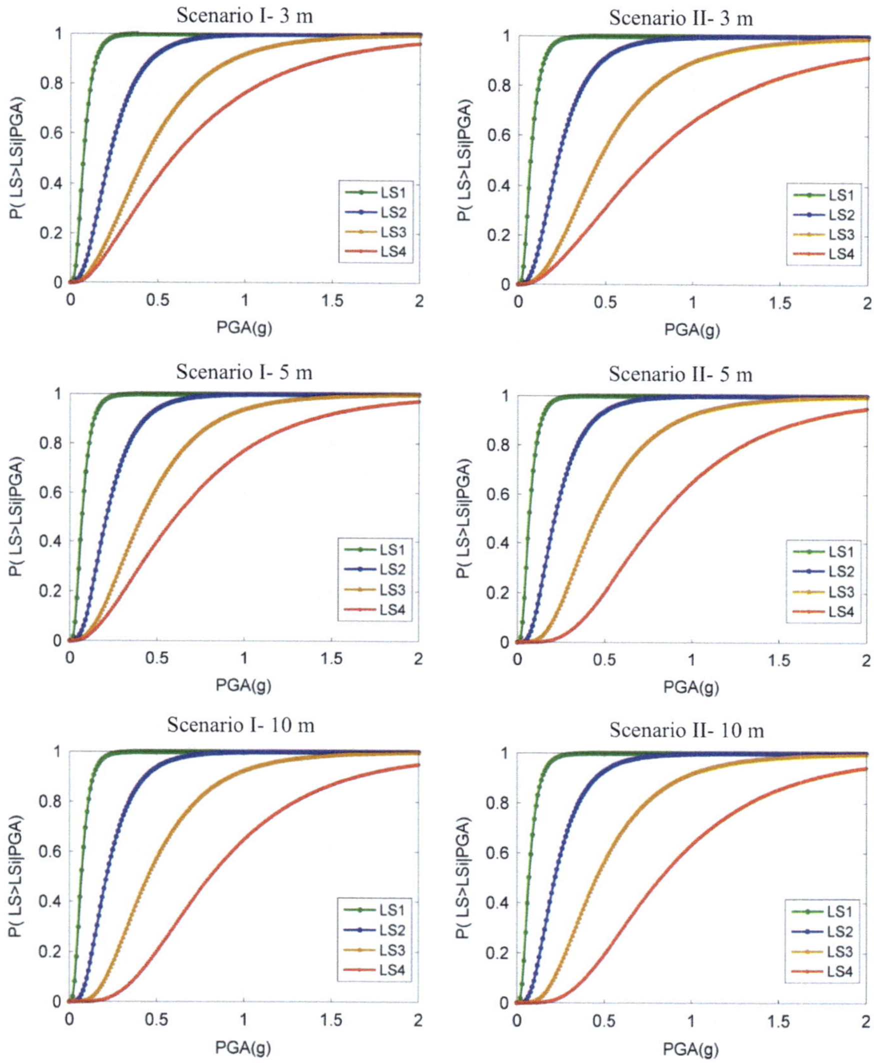

Fig. 4 Fragility curves in terms of PGA for the building located at 3 m, 5 m and 10 m from the slope shoulder and subjected to the combined effect of ground shaking and seismically induced landsliding for Scenarios I and II (modified from Fotopoulou and Pitilakis 2017)

The second part presented an analytical method for vulnerability assessment on reinforced concrete buildings subjected to seismically induced slope displacements. The method is based on the derivation of generic probabilistic fragility functions for typical reinforced concrete buildings that takes into account the slope inclination, the height and the soil material in the precarious and potentially disaster-prone slopes. Seven sets of fragility curves in terms of permanent ground displacement

(PGD) within the slope area were finally recommended for the different investigated models based on the most influential features that are namely the slope inclination in conjunction with the slope soil material as well as the slope height for the 45° inclined sandy slopes. Different statistical approaches are considered to derive the fragility parameters, i.e., the regression analysis method and the maximum likelihood method, indicating the influence of epistemic uncertainty on the fragility analysis. The proposed fragility curves can be applied to evaluate the physical vulnerability of reinforced concrete buildings subjected to earthquake-induced slow-moving slides.

Finally, the complex issue of combined damage due to ground shaking and landslide hazard was investigated by proposing a coupled vulnerability assessment method which combines the two kinds of hazards. In particular, the vulnerability was assessed of a typical low-rise, high-code reinforced concrete building standing next to the slope's shoulder subjected to the combined effect of the modified (due to the topography) ground shaking and the permanent co-seismic ground deformation (landsliding). In this assessment, two different scenarios were considered; (i) the two hazards do not interact, and (ii) the landslide hazard is conditional and governed by the preceding earthquake that is the initial trigger of the landslide. The proposed method develops coupled fragility curves and can be used within a probabilistic risk assessment framework to derive more accurate fragility curves and consequently assess the vulnerability of a typical reinforced concrete frame buildings standing in the vicinity of a slope's crest exposed to the combined effects of ground shaking and the landslide hazard. However, it should be noted that the derived coupled fragility curves are pertinent to the specific case studies in this article, and thus much more effort is needed to extend them to more general situations so that generic fragility curves may be proposed for different slope conditions and structural typologies.

Acknowledgements The authors express their profound appreciation and acknowledgment to Professor Ikuo Towhata for his contribution in getting the paper this final version.

References

Argyroudis S, Pitilakis K (2012) Seismic fragility curves of shallow tunnels in alluvial deposits. Soil Dyn Earthq Eng 35:1–12

Arias A (1970) A measure of earthquake intensity. In: Hansen RJ (ed) Seismic design for nuclear power plants. MIT Press, pp 438–483

Bird JF, Crowley H, Pinho R, Bommer JJ (2005) Assessment of building response to liquefaction induced differential ground deformation. Bul NZ Soc Earthq Eng 38:215–234

Bray JD (2007) Simplified seismic slope displacement procedures. In: Pitilakis KD (ed) Earthquake geotechnical engineering. Springer

Cornell CA, Jalayer F, Hamburger RO, Foutch DA (2002) Probabilistic basis for 2000 SAC federal emergency management agency steel moment frame guidelines. J Struct Eng 128:526–533

Corominas J, Van Westen C, Frattini P, Cascini L, Malet J-P, Fotopoulou S, Catani F, Van Den Eeckhaut M, Mavrouli O, Agliardi F, Pitilakis K, Winter MG, Pastor M, Ferlisi S, Tofani V, Hervás J, Smith JT (2014) Recommendations for the quantitative assessment of landslide risk. Bull Eng Geol Environ 73:209–263

Crowley H, Pinho R, Bommer JJ (2004) A probabilistic displacement-based vulnerability assessment procedure for earthquake loss estimation. Bull Earthq Eng 2:173–219
EN 1998–1 (2005) Eurocode 8: design of structures for earthquake resistance—part 1: general rules, seismic actions and rules for buildings. CEN: Brussels, Belgium
Fotopoulou S (2012) Seismic vulnerability of reinforced concrete buildings in sliding slopes. PhD thesis, Department of Civil Engineering, Aristotle University of Thessaloniki, Greece
Fotopoulou S, Pitilakis K (2013a) Vulnerability assessment of reinforced concrete buildings subjected to seismically triggered slow-moving earth slides. Landslides 10(5):563–582
Fotopoulou S, Pitilakis K (2013b) Fragility curves for reinforced concrete buildings to seismically triggered slow-moving slides. Soil Dyn Earthq Eng 48:143–161
Fotopoulou S, Pitilakis K (2015) Predictive relationships for seismically induced slope displacements using numerical analysis results. Bull Earthq Eng 13(11):3207–3238
Fotopoulou S, Pitilakis K (2017) Vulnerability assessment of reinforced concrete buildings at precarious slopes subjected to combined ground shaking and earthquake induced landslide. Soil Dyn Earthq Eng 93:84–98
Itasca Consulting Group Inc (2011) FLAC (Fast Lagrangian Analysis of Continua), version 7.0. Itasca Consulting Group, Inc., Minneapolis
Jibson RW (2011) Methods for assessing the stability of slopes during earthquakes—a retrospective. Eng Geol 122(1–2):43–50
Luco N, Cornell AC (2007) Structure-Specific Scalar Intensity Measures for Near-Source and Ordinary Earthquake Ground Motions. Earthq Spectra 23(2):357–392
Makdisi F, Seed H (1978) Simplified procedure for estimating dam and embankment earthquake induced deformations. J Geotech Eng 104(7):849–867
Mavrouli O, Fotopoulou S, Pitilakis K, Zuccaro G, Corominas J, Santo A, Cacace F, De Gregorio D, Di Crescenzo G, Foerster E, Ulrich T (2014) Vulnerability assessment for reinforced concrete buildings exposed to landslides. Bull Eng Geol Environ 73:265–289
National Institute of Building Sciences (2004) Direct physical damage—general building stock. HAZUS-MH Technical manual, Chapter 5. Federal Emergency Management Agency, Washington, DC
Negulescu C, Foerster E (2010) Parametric studies and quantitative assessment of the vulnerability of a RC frame building exposed to differential settlements. Nat Hazard 10:1781–1792
Newmark NM (1965) Effects of earthquakes on dams and embankments. Geotechnique 15(2):139–159
Nielson BG, DesRoches R (2007) Analytical seismic fragility curves for typical bridges in the central and southeastern United States. Earthq Spectra 23(3):615–633
Padgett JE, Nielson BG, DesRoches R (2008) Selection of optimal intensity measures in probabilistic seismic demand models of highway bridge portfolios. Earthq Eng Struct Dyn 37:711–725
Shinozuka M, Feng MQ, Lee J, Naganuma T (2000) Statistical analysis of fragility curves. J Eng Mech ASCE 126:1224–1231

A Prototype Earthquake-Induced Landslide Forecast Tool for New Zealand

Chris Massey, Bilijana Lukovic, and Sally Dellow

Abstract This paper outlines a coseismic landslide forecast tool that produces advisory information for emergency managers after a significant earthquake, in near-real time. The tool provides rapid information about the severity, location and impacts of landslides, using near-real-time ground shaking data recorded by the GeoNet strong motion instrument network. The efficacy of the tool is demonstrated using the M_W 7.8 14 November 2016 Kaikoura earthquake, and the landslides it generated, as an example of how the tool would work and the outputs it generates.

Keywords Kaikoura Earthquake · Earthquake-induced landslides · Coseismic landslides · Landslide susceptibility · Forecast modelling · Landslide hazard modelling

1 Introduction

New Zealand maintains a 24-h response capability for advice and investigation of significant landslide occurrences or threats throughout the country (McSaveney et al. 2010). The purpose of the all-hours capability is to ensure that appropriate advice is available to maximize public safety, and to collect reliable, consistent and often ephemeral information for landslide research. This capability is maintained within the GeoNet Project of GNS Science (www.gns.cri.nz) a wholly New Zealand government owned research organization which maintains (among other things) the science functions of the former New Zealand Geological Survey and Seismological Observatory. GNS Science is the lead New Zealand government agency responsible for assessing, monitoring and alerting natural hazard risk (tsunami, earthquakes, volcanic eruptions

C. Massey (✉) · B. Lukovic · S. Dellow
GNS Science, Avalon, New Zealand
e-mail: c.massey@gns.cri.nz

B. Lukovic
e-mail: b.lukovic@gns.cri.nz

S. Dellow
e-mail: s.dellow@gns.cri.nz

I. Towhata et al. (eds.), *Coseismic Landslides*, Springer Natural Hazards,
https://doi.org/10.1007/978-981-19-6597-5_17

and landslides), and for providing advice on mitigating the risk these hazards may pose to people and infrastructure (MCDEM 2017).

This paper describes the details of a prototype earthquake-induced landslide (EIL) forecast tool that will produce outputs (advisory information) for the GeoNet landslide duty officers, which will also have benefits for emergency managers. A "tool" in the context of this paper is defined as software used to carry out a function based on an input. The function of this tool is to provide rapid advisory information about the severity and likely location of landslides following a major earthquake, where the earthquake data is used as the input data for the tool. The prototype EIL forecast tool is the first of several to be developed as part of a larger landslide forecast project being carried out by the GNS Science landslide and social science teams and others. As part of the larger project, it was envisaged that several forecast tools would be developed for two types of landslide trigger (earthquake and rain). The end-users of the information generated by such tools are identified as being technical specialists, specifically the GeoNet landslide duty officers, emergency managers, the public and commercial entities such as infrastructure providers. The aims of the overall project are to: (1) rapidly identify whether an earthquake or a rain event can generate landslides and the severity of landsliding; and (2) rapidly generate advisory information such as a spatial representation (map) of where landslides could occur in a significant earthquake or rainfall event.

1.1 The Need for Such a Tool

New Zealand is prone to earthquakes. Landslides have occurred during many historical seismic events. Between 1840 and 1997, at least 22 earthquakes (about one every seven years on average) have resulted in substantial or widespread and damaging landsliding, documented in Hancox et al. (2002). Since 1997 there have been 11 earthquakes (Table 1) that have caused widespread landsliding (about one every 2 years on average), the most recent being the M_W7.8 14 November 2016, Kaikoura earthquake, which triggered more than 28,000 landslides (Fig. 1) over an area of 10,000 km^2 (Massey et al. 2018, 2020). Many of these more recent earthquakes (Table 1) had their focal depths below 10 km. In New Zealand, the substantial damage caused by earthquake-induced landsliding (EIL) has been second only to building damage caused by strong shaking. In only three major earthquakes (1929 Murchison, 1968 Inangahua and 2016 Kaikoura earthquakes), twenty-five people have been killed. Sixteen deaths resulted from landslides triggered by the 1929 Murchison earthquake alone. Five people were killed by earthquake-induced landslides triggered by earthquakes during the 2010/11 Canterbury earthquake sequence.

In the first few hours and days of the GeoNet landslide team's response to a large magnitude earthquake (typically $\geq$ M6 and PGA $\geq$ 0.2 g), much of the reconnaissance is focused on identifying where landslides might have occurred, and how severe their impacts might be on people and infrastructure. To do this, we rely on information relating to, e.g.: (1) the earthquake epicentre; (2) strong motion instrument data;

Table 1 Historical earthquakes causing substantial landsliding in New Zealand since 1997. This table is an extension of the table in Hancox et al. (2002), which summarizes the period 1840–1997

Name	Date	Magnitude	Depth	Effects*	Key reference
1. Fiordland	22/08/2003	M_W7.2	24 km	MM9 (but 17%g recorded). Remote area only minor damage to buildings. Surficial landsliding over 3000 km^2. More than 400 landslides with volumes up to 700,000 m^3 occurred on steep slopes within 25 km of the underlying earthquake fault plane. Subduction earthquake	Hancox et al. (2003)
2. Fiordland George Sound	16/10/2007	M_W6.7	5 km	MM7. The shaking caused landslides on the steep slopes of Milford Sound. There were no reports of any structural damage	GeoNet (2007a, b)
3. Gisborne	20/12/2007	M_W6.7	33 km	MM8 (28%g). Subduction earthquake	GeoNet (2007a)
4. Dusky Sound	15/07/2009	M_W7.8	12 km	MM7 (15%g). Subduction earthquake. Triggered at least 241 mapped landslides over a main area of 2000 km^2	Fry et al. (2010)
5. Darfield	4/09/2010	M_W7.2	11 km	MM9 (94%g). Only a few landslides were triggered and the epicentral region was relatively flat	Stahl et al. (2014)
6. Canterbury Earthquake Sequence**	22/02/2011 13/06/2011 23/12/2011	M_W6.2 M_W6.2 M_W6.0	~5 km	MM9 to MM10 (220%g, 200%g and 70%g). Triggered many rockfalls and landslides over a 40–60 km^2 area	Massey et al. (2014)

(continued)

Table 1 (continued)

Name	Date	Magnitude	Depth	Effects*	Key reference
7. Cook Strait	21/07/2013	M_W6.5	16 km	MM7 (20%g). Triggered at least 50 mainly small and shallow landslides over an area of about 100 km^2	Van Dissen et al. (2013)
8. Lake Grassmere	16/08/2013	M_W6.6	7 km	MM9 (79%g). Triggered at least 200 landslides over an area of 180 km^2	Van Dissen et al. (2013)
9. Eketahuna	20/01/2014	M_W6.2	34 km	MM7 (26%g). Triggered at least 177 mapped landslides and rockfalls over a main area of 1000 km^2	Rosser et al. (2014)
10. Wilberforce	6/01/2015	M_W6.0	5 km	MM7 (22%g). Triggered at least 200 landslides within an area of 370 km^2	Rosser and Carey (2017)
11. Kaikoura	14/11/2016	M_W7.8	15 km	MM10 (130%g). Triggered more than 28,000 landslides over a main area of 3600 km^2 and a wider area of 10,000 km^2	Massey et al. (2018)

* PGAs are given as %g and are the recorded maximum horizontal component, note these may be lower than the MM intensities, which include "felt" reports

**Only earthquakes $\geq M_W$6.0 are listed. Smaller magnitude earthquakes in the sequence also triggered localized landsliding

and (3) our knowledge of the physiography and geology around the epicentre and area strongly shaken, as landslides tend to occur on steep slopes and in susceptible materials. We typically adopt an instrument peak ground acceleration (PGA) of 0.2 g (20%g or ~2 m/s/s, which is equivalent to MM7 adopting the Moratalla et al. (2020) PGA to MMI relationship), as a threshold above which many landslides tend to be triggered. We then use this information combined with any local on ground reports to guide aerial and ground-based reconnaissance teams. An example of what the GeoNet landslide team does when responding to a major earthquake, in this case, the M_W7.8 14 November 2016 Kaikoura earthquake, is documented in Dellow et al. (2017). Using this example, in the first few hours and days after the earthquake our aerial reconnaissance, via helicopter, was focused around the epicentral region,

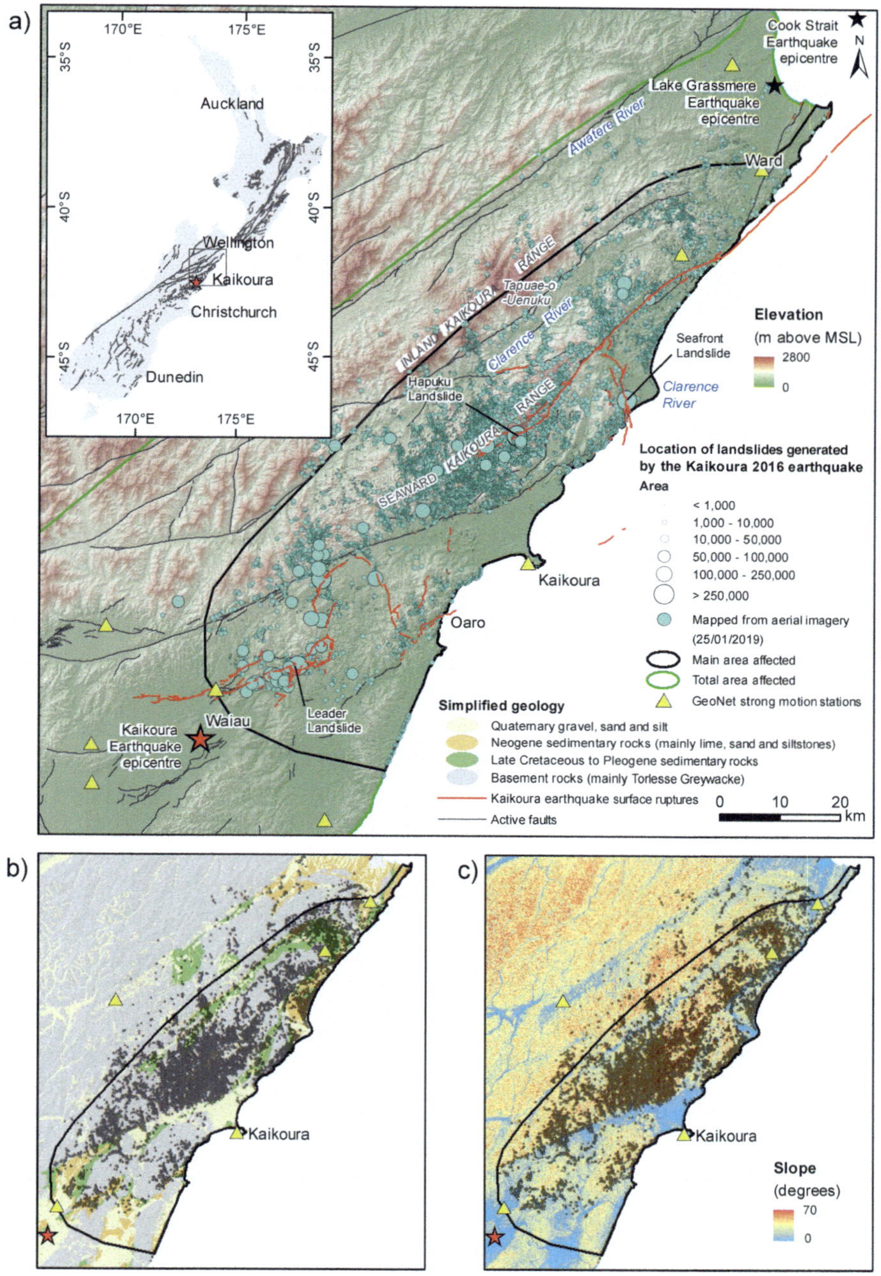

Fig. 1 (Inset) The area of New Zealand affected by coseismic landslides triggered by the M_W 7.8 14 November 2016 Kaikoura earthquake. **a** The mapped >28,000 coseismic landslide source areas and their size (area) triggered by the earthquake—Version 2.0 of the landslide inventory (Massey et al., 2020)—superimposed on the 8 by 8 m digital elevation model (DEM) for New Zealand, classified by elevation in metres above sea level. **b** The landslide source area distribution overlain on the main geological units. **c** Landslide source area distribution shown on the 8 m ground resolution DEM for New Zealand

and in areas shown to have been strongly shaken by the GeoNet strong motion network, which was relatively sparse in this region. Several days were spent looking for landslides in the more populated areas within the epicentral region. However, subsequent reconnaissance showed that most landslides occurred on steep slopes much farther north, in locations far from any strong motion sensor.

1.2 Existing Landslide Forecast Tools and Products and Their Limitations

The U.S. Geological Survey (USGS) has developed near-real-time global earthquake-triggered-landslide products to augment the USGS Prompt Assessment of Global Earthquakes for Response (PAGER) system (Allstadt et al. 2018; Nowicki Jessee et al. 2018). The 14 November 2016 Kaikoura earthquake (Fig. 2) provided a test case for evaluating the performance and near-real-time response applicability of three published global seismically induced landslide models, the results of which are presented in Allstadt et al. (2018).

The statistical models used to forecast landslide probability (shown in Fig. 2) adopt shaking estimates for the earthquake from the USGS ShakeMap. The landslide models are relatively simple, coarse in resolution, and are based solely on input proxies that are globally available. The ShakeMap shaking estimates changed in the hours, days, weeks and months after the earthquake as additional information was included in the models. As a result, the landslide forecasts from the three models also changed with each ShakeMap iteration (Fig. 2). Allstadt et al. (2018) found that once a point source (earthquake epicentre) was replaced with an approximate rupture extent in ShakeMap, about 9 h after the earthquake, the landslide models were all successful at roughly identifying the area of highest landslide occurrence. However, all models dramatically overpredicted the landslide hazard level—in this case the number of landslides likely to have been generated.

2 Methodology

2.1 Designing the Landslide Forecast Tool

Meetings and workshops were held by GNS Science to develop a "story map" for the EIL forecast tool. The initial end-users were identified as being the GeoNet landslide duty officers. Based on discussions with these end-users, their immediate "primary" needs were identified as being: (1) written (advisory) information should be made available as soon as possible (minutes and hours) after an event, including what Territorial Local Authorities (TLAs, which in New Zealand are councils) may have been affected by landslides and the magnitude of the landslides generated; and (2)

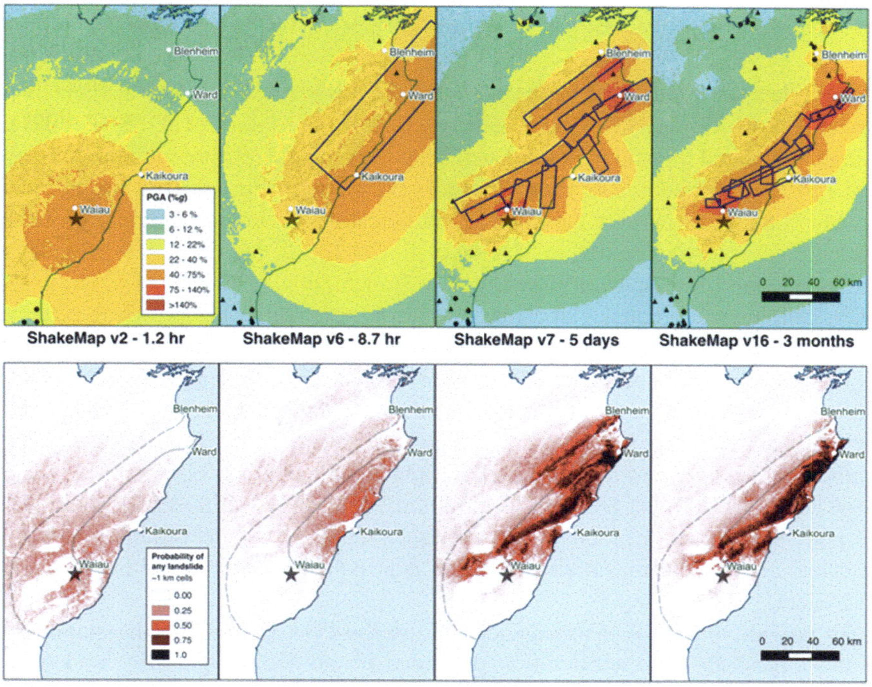

Fig. 2 The upper maps show the evolution of the ShakeMap ground shaking models produced for the Kaikoura Earthquake, at various times after the earthquake. The lower maps show the landslide probability calculated using the Nowicki et al. (2014) landslide model adopting the given ShakeMap model. All maps are taken and modified from Allstadt et al. (2018)

geospatial forecasts (maps) showing the spatial distribution of landslides and the locations of major infrastructure and population centres.

The secondary needs of the GeoNet landslide duty officers were identified as being: (1) the EIL forecast tool will need to be able to host multiple landslide forecast models (the statistical relationship that is used to forecast landslide probability), which would be run, following an event, by the tool to forecast landslide occurrence given a series of input variables; (2) the landslide models should adopt several types of ground shaking data as inputs, e.g. instrument data, ShakeMap models and other modelled data sets as they become available; and (3) the landslide susceptibility models should incorporate a regional-scale runout model to identify likely areas that could be inundated by debris given a landslide occurring.

2.2 The Landslide Model Used for the EIL Forecast Tool

To provide a "minimum viable product" (MVP), it was decided to use the landslide model developed by GNS Science, which is detailed in Massey et al. (2016, 2021), using the updated logistic regression and deep learning models presented in Massey et al. (2022). The models adopted for the tool were trained using three EIL inventories: (1) M_W 7.8 2016 Kaikoura earthquake and landslides adopting Version 2.0 of the landslide inventory presented in Massey et al. (2020); (2) M_W 7.1 1969 Inangahua earthquake landslide inventory presented in Hancox et al. (2014); and (3) M_W 7.3 1929 Murchison earthquake landslide inventory presented in Hancox et al. (2016). The two models estimate the probability (susceptibility) of a landslide occurring at a given location. The location has been defined using a grid. The entire land mass of New Zealand was subdivided into a series of 32 m by 32 m grid cells, and the predictor variables needed for the model are calculated for each grid cell. The predictor variables used in the model are: (1) slope angle; (2) elevation; (3) local slope height (relief); (4) geology; (5) proximity to an active surface fault trace; and (6) peak ground acceleration (PGA). Variables (1–5) are static variables and can be calculated pre-event. Variable (6) PGA is a dynamic variable and is generated post-event.

To provide the MVP, it was decided to use the PGA (defined as the amplitude of the largest absolute acceleration) recorded on an accelerogram by a GeoNet strong motion station during an earthquake. This information is taken from the GeoNet "strong response" data downloaded from GeoNet after a triggering event. The GeoNet strong motion data products for an earthquake event are available in two channels: (1) beta version of the strong motion tool: https://beta-strongmotion.geonet.org.nz/#/event/2021p169083; and (2) strong motion data products that are accessible via ftp: ftp://ftp.geonet.org.nz/strong/processed/2021/03_Mar/2021-03-04_132738/. These data sets contain the continuous peak strong motion data recorded by GeoNet in the past hour, which are calculated in near-real time, and thus can be affected by shaking from non-earthquake sources.

2.3 The Prototype EIL Forecast Tool

Figure 3 is a "story board" which shows the outline design of the current tool based on delivering a MVP. The current EIL forecast tool: (1) is triggered manually, where the landslide duty officer would trigger the EIL tool if the recorded PGAs at two or more sensors were $\geq$0.2 g. They would do this by monitoring the GeoNet strong motion tool, which displays the strong motion instrument data in near-real time over a one hour rolling window. Although this application does not currently allow alerting, the PGA thresholds can be "crudely" linked to earthquake magnitude and depth; (2) uses the currently available GeoNet "Strong Response" instrument PGA data as an input to the landslide model as this is the only earthquake strong motion data currently

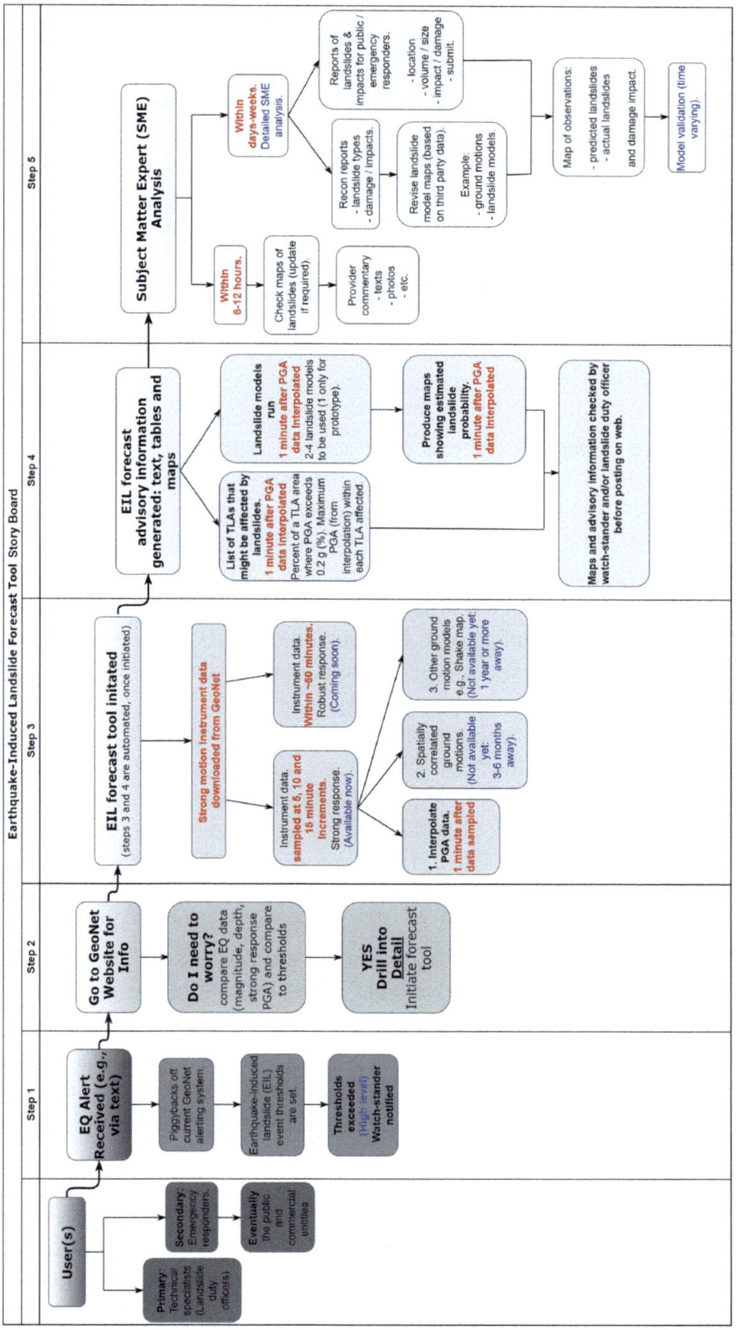

Fig. 3 Earthquake-induced landslide forecast tool "story board"

available for New Zealand in near-real time—the EIL forecast tool creates a grid of PGA by interpolating between the PGAs recorded at the strong motion instrument station using the natural neighbour method; (3) uses the pre-processed data for the entire New Zealand landmass required for the landslide model and calculates the landslide probability within the 0.2 g (20%g) PGA simplified bounding polygon; (4) uses the landslide forecast models, detailed in Massey et al. (2021, 2022) that produce landslide probability metrics; and 5) produces only two outputs, a table of advisory information relating to TLA boundaries and landside probability maps.

Once triggered, the tool runs automatically using ESRI ArcGIS (version 10.5.1) and Python scripting language (version 2.7.8).

3 Example Results

To explore the utility of the EIL forecast tool, we have applied it to the instrument ground motions recorded by the GeoNet strong motion network as recorded five minutes after the M_W7.8 14 November 2016 Kaikoura earthquake. At the time of the earthquake, the GeoNet "Strong Response" data was not available and so GeoNet created a Strong Response data set to show how the instrument data would have appeared in the data set in near-real time after the earthquake. In future, the PGA data in the Strong Response data set will be sampled at 5, 10 and 15-min intervals after an earthquake as there can be delays between the instrument recording the data and it appearing in the Strong Response data set.

The steps in the EIL forecast tool process are as follows: Step (1) the Python script is triggered manually; Step (2) the Strong Response instrument data set is sampled, interpolated and a 0.2 g (20%g) PGA simplified bounding polygon created; Step (3) landslide probabilities are calculated within the 0.2 g bounding polygon using the "static" data sets and the dynamic PGA grid interpolated from the Strong Response data set (Fig. 4); Step (4) results in the form of "advisory" information (Table 2) and maps (Fig. 5) are then produced.

The Python scripts require input data that has been processed pre-event to create the "static" data sets needed for the post-event processing. Some of this data is in raster format, which is used for calculating landslide probability, and some is in vector format, which is used to create the output advisory information and maps. These data are stored in a "pre-event" ESRI File geodatabase. Once the tool is run post-event, a "processing" ESRI File geodatabase is created by the scripts to store intermediate and final outputs created by the EIL tool as it runs.

A results folder, with a time stamp related to the time the GeoNet Strong Response data is sampled by the tool, is created which contains the final outputs, maps and advisory information (Table 2 and Fig. 5). This folder contains a "results" ESRI File geodatabase, ArcGIS Map projects to view the results, landslide probability maps in PDF format and advisory information in table (CSV) format. Once the tool is triggered, it takes between 5 and 7 min to run and produce the advisory information and maps. The speed depends upon the computer specifications on which the tool

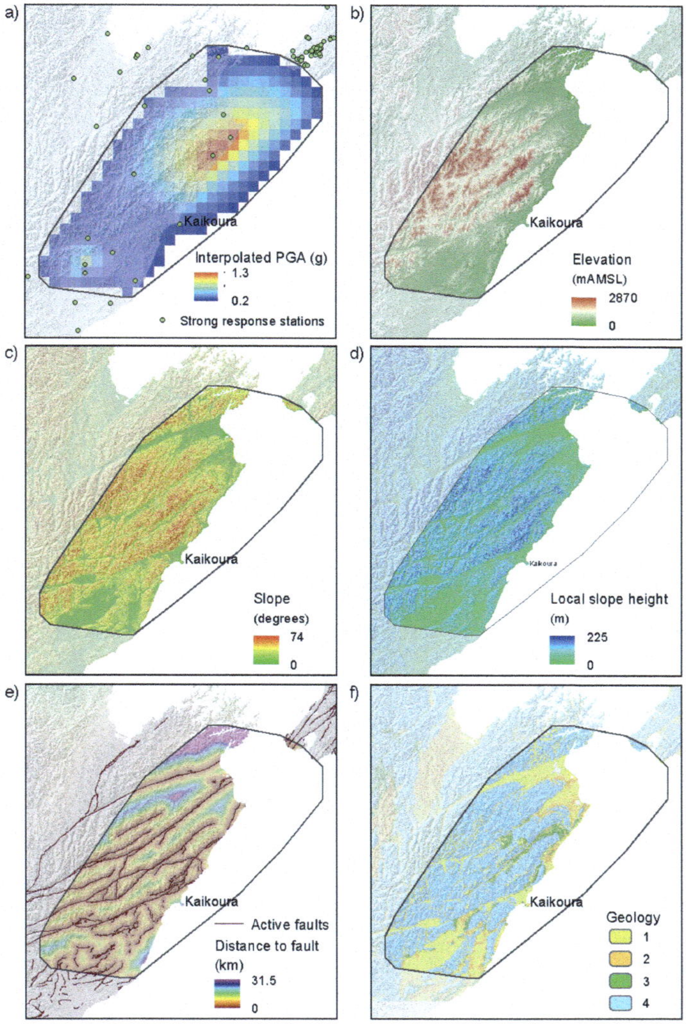

Fig. 4 The input data needed for the EIL forecast tool based on the landslide forecast model described by Massey et al. (2021, 2022)—Kaikoura Earthquake example: **a** PGA model interpolated from the Strong Response data set and clipped to the simplified 0.2 g (20%g) bounding polygon (shown as a black line in all maps **a**–**f**, which is used by the tool to calculate the landslide probabilities, note that the GeoNet strong motion instrument stations used are shown as circles; **b** Slope elevation; **c** Slope angle; **d** Local slope height (relief); **e** Active faults from the New Zealand active faults database, note immediately after a future earthquake we will not know which faults have ruptured and so for forecast purposes we include all active faults within the 0.2 g PGA bounding polygon; and **f** geological materials based on the "Dominant Unit" attribute from the New Zealand Geological QMap (scale 1:250,000)

Table 2 Example advisory information generated by the EIL forecast tool for the Kaikoura earthquake-related Strong Response data

TLA that might be affected by landslides	Area where PGA exceeds 0.2 g (km^2)	Percent of TLA where PGA exceeds 0.2 g	Maximum PGA within TLA (g)	TLA that might be affected by landslides
Hurunui District	4205.08	48.6	0.68	Hurunui District
Kaikoura District	2046.78	100	1.27	Kaikoura District
Lower Hutt City	1.2	0.3	0.22	Lower Hutt City
Marlborough District	6362.82	60.8	1.27	Marlborough District
Wellington City	22.72	7.8	0.26	Wellington City

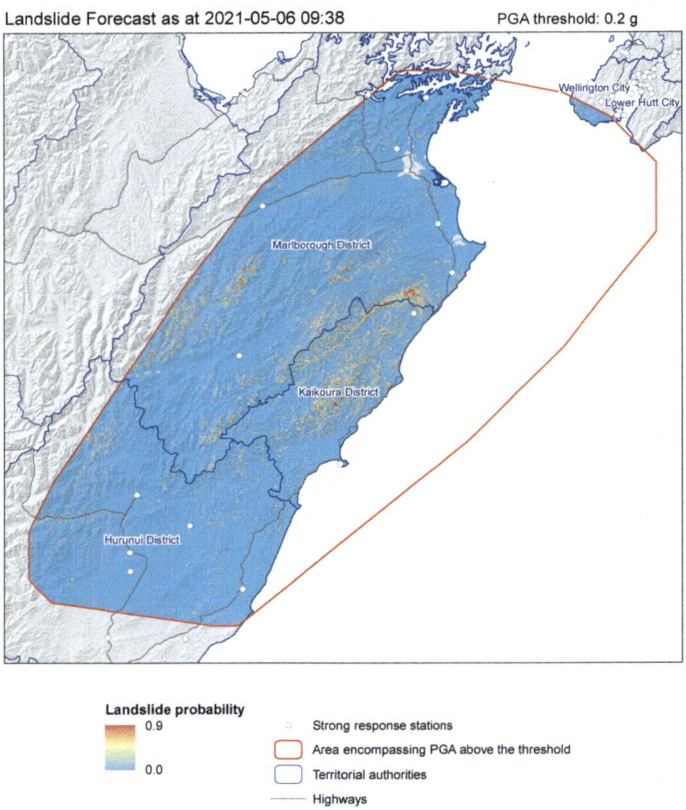

Fig. 5 Example of landslide probability map generated by the EIL forecast tool using the strong motion instrument data from the Mw 7.8 14 Nov 2016 Kaikoura earthquake, adopting the landslide susceptibility logistic regression model of Massey et al. (2021, 2022). The time stamp relates to the date and time that the Strong Response data set was sampled

is being run. The times given here are for the tool to run on a "typical" desktop computer, where the static data sets are installed on the hard drive of the computer.

4 Discussion

The example landslide probability map generated by the EIL forecast tool (Fig. 5) for the Kaikoura earthquake-related ground motions demonstrates the efficacy of such a tool in providing rapid advisory information (5–7 min after being triggered) about the severity and likely location and impacts of landslides following a major earthquake. It is not the intent of the tool to forecast the precise location of landslides, instead, it will provide information that would help "first responders" to target the worst affected areas, therefore, getting aid to those who may need it, as quickly as possible. The information generated will also allow the relative 'intensity' of the event, with regards to the severity of landslides generated, to be analyzed and assessed.

Over the next few years we plan to: (1) establish automatic triggering of the tool, based on ground shaking thresholds being exceeded; and (2) retrain the landslide susceptibility models by incorporating other high-quality landslide inventories from some of the more recent earthquakes that occurred in different geological terrains around New Zealand (listed in Table 1 and Hancox et al. 2002). Eventually, the tool will also adopt multiple ground shaking products produced by different ground motion models as well as estimates of the runout of landslide debris.

5 Conclusions

This report outlines a prototype EIL forecast tool for New Zealand that will produce maps and data, in near-real time, approximately 5–7 min after being triggered. The function of this tool is to provide rapid advisory information about the 'intensity' and location of landslides following a major earthquake in New Zealand.

These maps and data produced by the tool have already been used by emergency managers; regional and district councils; infrastructure managers; major landowners and the public, to help organize their response to the M_W 7.4 Kermadec earthquakes that occurred on 4 March 2021. The tool is not intended to forecast the precise location of landslides but rather to provide responders with information that would help them focus post-earthquake reconnaissance to aid those people affected as quickly as possible.

The current tool uses two deep learning statistical models to forecast the spatial distribution of landslide probability over an area. The models estimate the landslide probability (0–1, where 1 = landslide occurrence) at the regional scale, defined as being 1:25,000 to 1:250,000. The two statistical methods are: (1) logistic regression and (2) a deep learning neural network.

Overall, the statistical models developed here provide a robust forecast of landslide probability at different levels of earthquake PGA. However, the results from the models have a low pseudo R^2 (Massey et al., 2021, 2022), indicating that they cannot give accurate forecasts for individual grid cells. They may nevertheless give good overall estimates of the scale 'intensity' of landslide occurrence to be expected in individual earthquakes.

Although the current tool and models are not the definitive version, the forecast probabilities produced by the models are statistically robust to be used to forecast landslide intensity and locations in the minutes after a major earthquake.

Acknowledgements The authors would like to acknowledge: G. Clitheroe and N. Balfour from GeoNet for providing the Strong Response data set; The Enhanced Geohazards Monitoring project (GeoNet) for funding this work; and K. Allstadt (USGS) for useful discussions and guidance during the development of the EIL forecast tool.

References

Allstadt KE, Jibson RW, Thompson EM, Massey CI, Wald DJ, Godt JW, Rengers FK (2018) Improving near-real-time coseismic landslide models : lessons learned from the 2016 Kaikoura, New Zealand, earthquake. Bull Seismolog Soc Am 108(3B):1649–1664. https://doi.org/10.1785/0120170297

Fry B, Bannister S, Beavan RJ, Bland L, Bradley BA, Cox SC, Cousins WJ, Gale NH, Hancox GT, Holden C, Jongens R, Power WL, Prasetya G, Reyners ME, Ristau J, Robinson R, Samsonov S, Wilson KJ, GeoNet team (2010) The Mw 7.6 dusky sound earthquake of 2009: preliminary report. Bull NZ Soc Earthq Eng 43(1):24–40

GeoNet (2007a) https://www.geonet.org.nz/earthquake/story/2808298

GeoNet (2007a) https://www.geonet.org.nz/earthquake/story/2839343

Hancox GT, Perrin ND, Dellow GD (2002) Recent studies of historical earthquake-induced landsliding, ground damage, and MM intensity in New Zealand. Bull NZ Soc Earthq Eng 35(2):59–95

Hancox GT, Cox SC, Turnbull IM, Crozier MJ (2003) Reconnaissance studies of landslides and other ground damage caused by the Mw7.2 Fiordland earthquake of 22 August 2003. Lower Hutt: Institute of Geological & Nuclear Sciences Limited. Institute of Geological & Nuclear Sciences science report 2003/30. 32 p

Massey CI, McSaveney MJ, Taig T, Richards L, Litchfield NJ, Rhoades DA, McVerry GH, Lukovic B, Heron DW, Ries W, Van Dissen RJ (2014) Determining rockfall risk in Christchurch using rockfalls triggered by the 2010–2011 Canterbury earthquake sequence, New Zealand. Earthq Spectra 30(1):155–181. https://doi.org/10.1193/021413EQS026M [February 2014]

Massey CI, Townsend DB, Rathje E, Allstadt KE, Lukovic B, Kaneko Y, Bradley B, Wartman J, Jibson RW, Petley DM, Horspool NA, Hamling IJ, Carey JM, Cox SC, Davidson J, Dellow GD, Godt GW, Holden C, Jones KE, Kaiser AE, Little M, Lyndsell BM, McColl S, Morgenstern RM, Rengers FK, Rhoades DA, Rosser BJ, Strong DT, Singeisen C, Villeneuve M (2018) Landslides triggered by the 14 November 2016 Mw 7.8 Kaikoura earthquake, New Zealand. Bull Seismolog Soc Am 108(3B):1630–1648. https://doi.org/10.1785/0120170305

Massey CI, Townsend DB, Lukovic B, Morgenstern R, Jones KE, Rosser BJ, de Vilder SJ (2020) Landslides triggered by the 14 November 2016 Mw 7.8 Kaikoura earthquake: an update. Landslides 17(10):2401–2408. https://doi.org/10.1007/s10346-020-01439-x

Massey CI, Lukovic B, Huso R, Buxton R, Potter SH (2021) Earthquake-induced landslide forecast tool for New Zealand: version 2.0. GNS Science, Lower Hutt (NZ), 77 p (GNS Science report; 2018/08). https://doi.org/10.21420/G2TP9V

Massey CI, Wolter A, Huso R, Lukovic B, Brideau M-A (2022) Coseismic landslide susceptibility and triggering analyses. In: I. Towhata et al. (eds.), Coseismic Landslides, pp 633– 679. 10.1007/978-981-19-6597-5_18

McSaveney MJ, Massey CI, Dellow GD (2010) Landslide response and monitoring : the New Zealand GeoNet experience (paper 053). In: Williams AL, Pinches GM, Chin CY, McMorran TJ, Massey CI (eds) Geologically active : delegate papers 11th Congress of the International Association for Engineering Geology and the Environment, Auckland, Aotearoa, 5–10 September 2010. CRC Press Boca Raton, Fla, p 443–450

Ministry of Civil Defence and Emergency Management (MCDEM) (2017) Ministerial review: better responses to natural disasters and other emergencies, 17 November 2017. https://dpmc.govt.nz/sites/default/files/2018-01/ministerial-review-better-responses-natural-disaster-other-emergencies.pdf

Moratalla J, Goded T, Gerstenberger M, Canessa S (2020) New ground motion to intensity conversion equations (GMCIEs) for New Zealand. In: 17th World Conference on Earthquake Engineering; Sendai, Japan [date to be confirmed]

Nowicki MA, Wald DJ, Hamburger MW, Hearne M, Thompson EM (2014) Development of a globally applicable model for near-real time prediction of seismically induced landslides. Eng Geol 173:54–65

Nowicki Jessee MA, Hanburger MW, Allstadt K, Wald DJ, Robeson SM, Tanyas H, Nearne M, Thompson E (2018) A global empirical model for near-real-time assessment of seismically induced landslides. JRG Earth Surf 123(8):1835–1859. https://doi.org/10.1029/2017JF004494

Rosser BJ, Carey JM (2017) Comparison of landslide inventories from the 1994 Mw 6.8 Arthurs Pass and 2015 Mw 6.0 Wilberforce earthquakes, Canterbury, New Zealand. Landslides 14(3):1171–1180. https://doi.org/10.1007/s10346-017-0797-8

Rosser BJ, Townsend DB, McSaveney MJ, Ries W (2014) Landslides and ground damage associated with the M6.2 Eketahuna earthquake, 20 January 2014. GNS Science. GNS Science report 2014/51, Lower Hutt, NZ, 34 p

Stahl T, Bilderback EL, Quigley MC, Nobes DC, Massey CI (2014) Coseismic landsliding during the Mw 7.1 Darfield (Canterbury) earthquake : implications for paleoseismic studies of landslides. Geomorphology 214:114–127. https://doi.org/10.1016/j.geomorph.2014.03.020

Van Dissen RJ, McSaveney MJ, Townsend DB, Hancox GT, Little TA, Ries W, Perrin ND, Archibald GC, Dellow GD, Massey CI, Misra S (2013) Landslides and liquefaction generated by the Cook Strait and Lake Grassmere earthquakes: a reconnaissance report. Bull N Z Soc Earthq Eng 46(4):196–200

Coseismic Landslide Susceptibility and Triggering Analyses

Chris Massey, Andrea Wolter, Rand Huso, Biljana Lukovic, and Marc-Andre Brideau

Abstract This chapter reviews the methods used to carry out coseismic landslide susceptibility and slope stability analyses, from the regional- to the site-scale and based on user needs, data requirements and constraints. Earthquake-induced landslide (EIL) triggering metrics (ground shaking) are discussed and linked to landslide severity and other environmental impacts likely to occur at different levels of ground shaking. Six case studies are used to explore the various methods of determining coseismic landslide susceptibility and slope stability analyses, the metrics they produce, the scale of applicability and data requirements. The key developments and challenges of each method are discussed and summarised. One such challenge is that the efficacy of each method is limited by the availability of the input data and/or the ability to derive it, irrespective of the project scale or need. Possibly the largest challenge for experts is the ability to transfer the results from such models into actionable information (e.g. hazard and risk analyses and engineering design) to support science-based decision-making.

C. Massey (✉) · A. Wolter · R. Huso · B. Lukovic · M.-A. Brideau
GNS Science, Avalon, New Zealand
e-mail: c.massey@gns.cri.nz

A. Wolter
e-mail: a.wolter@gns.cri.nz

R. Huso
e-mail: r.huso@gns.cri.nz

B. Lukovic
e-mail: b.lukovic@gns.cri.nz

M.-A. Brideau
e-mail: MBrideau@bgcengineering.ca

M.-A. Brideau
BGC, Squamish, Canada

I. Towhata et al. (eds.), *Coseismic Landslides*, Springer Natural Hazards,
https://doi.org/10.1007/978-981-19-6597-5_18

1 Introduction

Assessment of the coseismic behaviour of slopes and their susceptibility to failure due to ground shaking requires consideration of complex interactions between seismic waves and slopes (e.g. Ashford et al. 1997; Sepulveda et al. 2005; Rai et al. 2016; Massey et al. 2017). In this chapter, we review some of the methods used to analyse landslide susceptbility from the regional- to the site-specific scale. Initially, earthquake-induced landslide (EIL) triggering metrics are discussed (Sect. 2) and linked to landslide severity and other environmental impacts likely to occur at different levels of ground shaking (Sect. 3). Next, various regional-scale landslide susceptibility techniques are discussed, and two case studies using two statistical methods—logistic regression and deep learning—are applied to three New Zealand-based earthquake-induced landslide inventories (Sect. 4). The results from these case studies are compared to a case study that uses a deterministic, quasi-physics-based method. Following on from these regional-scale analyses, several site-specific numerical analysis techniques are then discussed, from pseudostatic to permanent displacement and stress-deformation approaches, which are then applied to three case studies (Sect. 4). The final part of the chapter summarises the challenges of using different site-specific approaches , and provides guidance on how to choose the most appropriate type of analysis based on the expert's needs and data availability (Sect. 5).

In this chapter, we refer to national scale as being >1:250,000, regional scale as being 1:25,000 to 1: 250,000, local scale as being 1:5000 to 1:25,000 and site-specific scale as <1:5000 (Corominas et al. 2014).

Many national- to regional-scale studies use mapped landslide distributions to explore the relationships between the occurrence and/or displacement of earthquake-triggered landslides and the variables that may control them (e.g. Jibson 2011; Grant et al. 2016; Parker et al. 2015). These variables can broadly be grouped into ***landslide forcing (triggering) variables***, representing the intensity of the event-specific seismic ground motions and their proxies and ***landslide susceptibility (pre-conditioning) variables*** that capture the strength of the hillslope materials at national- to regional-scale, which are typically used as proxies for the static shear stresses at the slope scale. At the site-specific scale, such proxies are usually replaced with earthquake accelerograms, geotechnical material parameters and failure models. Therefore, the choice of scale and technique to be used is ultimately a function of data availability and desired result, e.g. regional-scale landslide probability versus site-specific displacement amounts, for a given level of shaking.

The response of a slope to an earthquake is complex and is thought to be controlled by: (1) the nature of the earthquake source; (2) wave propagation path effects; and (3) local site conditions and their effects on amplifying or de-amplifying ground shaking (Kramer 1996; Sepulveda et al. 2005; Kaiser et al. 2013). At the site- or slope-specific scale, most numerical dynamic slope analyses are concerned with local site conditions and their effects on slope amplification and stability. Amplified ground motions in slopes can result from near surface impedance contrasts associated with material velocity contrasts. These can be caused by: local surficial deposits (fill, colluvium, alluvium, etc.) overlying rock (e.g. Bourdeau and Havenith 2008; Del Gaudio and Wasowski 2011); weathered materials overlying less weathered materials (e.g. Siddiqqi and Atkinson 2002); highly fractured zones within more intact materials and discrete large-scale fracture zones (e.g. Moore et al. 2011; Gischig et al. 2015); or a combination of these factors. Impedance contrasts cause wave amplification, however, the amount of amplification depends on the thickness of the contrasting 'softer' material. If the thickness of the softer material, e.g. the near surface weathered materials overlying the 'stiffer' materials, is typically less than 2 m, the effect of the impedance contrast on amplification is minimal. Focusing of seismic waves by surface morphology—mainly slope inclination, height, and shape (e.g. convex, concave, or planar)—may result in topographic amplification (e.g. Geli et al. 1988; Benites and Haines 1994; Meunier et al. 2008; Hough et al. 2010), at larger 'ridge-scales' and at smaller 'site-scales' (Kaiser et al. 2014). For slopes, the characteristic site period can be influenced by local slope materials, material contrasts and topography and provides an indication of the frequency at which the most significant amplification can be expected (e.g. Kramer 1996). Earthquake ground motions of similar amplitude and duration are thought to affect the slope in different ways, depending on the frequency content of the earthquake (Arnold 2016), which is strongly influenced by earthquake magnitude and source-to-site distance (Kaiser et al. 2014). The resulting acceleration contrasts and the rock/soil mass strength of the slope-forming materials thus control the stability of the slope, which is often expressed using metrics that describe its probability of failure, factor of safety, yield acceleration or permanent displacement.

2 Triggering Metrics

2.1 National- to Regional-Scale

To represent the landslide forcing variables, many national- to regional-scale studies adopt: (1) peak ground velocity models (PGV); (2) peak ground acceleration models (PGA); (3) Arias Intensity models (I_a); (4) the proximity of a landslide to a coseismic fault ruptures; (5) permanent coseismic "fault" displacement derived from, e.g. InSAR and GPS measurements (e.g. Hamling et al. 2017); and/or (6) local slope relief (LSR) and material contrasts to take into account local amplification effects (e.g. Massey et al. 2018). Variables (1–3) are used as direct proxies for ground shaking, and variables (4–6) as indirect proxies for ground shaking. The permanent coseismic displacement variable may also serve as a proxy for other susceptibility factors such as rock mass damage. This is because permanent off-fault ground displacement can lead to rock/soil mass deformation which decreases the shear strength of the rock/soil mass (Bloom et al. 2022). The proximity to a fault is inherently included in the estimation of PGV, PGA and I_a; however, some studies include it separately to examine the influence of local ground deformation (off-fault displacement) and other near-field effects that might not be fully considered in the ground-motion models. It should also be noted that the proximity to fault variable may not account for faults that do not rupture to the surface, but which also contribute to the shaking intensity. Local slope relief (LSR), which can be defined using different metrics, is generally used to characterise the local shape of a slope. It is used to represent a proxy for slopes that could amplify ground shaking due to their 'larger-scale relief', where larger values of LSR represent steeper and higher slopes, which can amplify ground shaking more than smaller and less steep slopes (e.g. Ashford et al. 1997; Massey et al. 2016; Rai et al. 2016; EN1998-5 2004). Some studies also use slope aspect to investigate directivity effects caused by the earthquake-rupture sequence on landslide occurrence (e.g. Fan et al. 2019).

2.2 Local- and Site-Specific Scale

For coseismic landslide studies, the response of a slope to an earthquake is typically assessed using metrics that describe both the amplification effects—the site response—and amounts of permanent displacement. At local- and site-specific scales, there is no need to use proxies to represent the earthquake shaking and amplification effects, as ground motion accelerograms—time-acceleration histories—can be directly input in the software used to estimate such effects. Amplification effects are usually described as a ratio between the measured or modelled PGA, PGV, I_a or spectral acceleration in the free field, compared to those measured or modelled at, e.g. the slope crest, at various locations in the slope, or along the failure surface of an assumed sliding mass (e.g. Makdisi and Seed 1978; Ashford and Sitar 2002; Bray and Travasarou 2007; Athanasopoulos-Zekkos and Seed 2013; Massey et al. 2017; Brideau et al. 2021a).

Several studies have investigated the relative contributions that the earthquake source and path, and local site effects (e.g. topography, material contrasts, and geological structure) have on the amplification of shaking and amount of permanent slope displacement at the slope scale (e.g. Del Gaudio and Wasowski 2011; Jibson 2011; Moore et al. 2011; and Strenk and Wartman 2011). There is, however, no consensus between the authors on what factors are more important, probably because different factors will be important in different situations, and much of the focus of these studies is on the relative importance of topographic versus material factors. Some studies (e.g. Kaiser et al. 2013, 2014; Arnold 2016; and Massey et al. 2015, 2017) based on the observations, monitoring and modelling of rock slope response to the 2010/11 Canterbury earthquake sequence, found that steeper and taller slopes amplified PGAs but also that the frequency content of the earthquake shaking influenced the dominant failure mode (e.g. rockfall versus debris avalanching versus incipient displacement and cracking). Their results showed that accelerations were amplified mainly due to impedance contrasts between geological materials. The resulting acceleration contrasts and rock mass strength controlled slope stability, the amount of permanent slope displacement and the volume of debris falling from the given slope. Similar relationships have been found by others (e.g.: Ashford and Sitar 2002; Sepulveda et al. 2005; Rizzitano et al. 2014; Gischig et al. 2015) for rock slopes in different materials.

3 Triggering Thresholds and Landslide Severity

In general, peak ground acceleration (PGA) contours show a correlation with landslide severity, here defined as landslide density (e.g. Meunier et al. 2008) and have been used to help identify the limit of the area most likely to be affected by earthquake-induced landslides (EIL). Wilson and Keefer (1985) were the first to propose a minimum PGA threshold, of 0.05 g. Typical thresholds reported range from 0.02 to 0.11 g (Keefer and Tannaci 1981; Keefer 1984; Del Gaudio and Wasowski 2011; Jibson and Harp 2016; Tanyaş et al. 2017). Whilst these PGA thresholds relate to the limit of EIL generated by an individual earthquake, they may only represent a few isolated failures, i.e. representing areas where the landslide density is low. Landslide density has been shown to increase nonlinearly with increasing PGA (e.g. Meunier et al. 2008), and with decreasing distance to a surface rupturing fault (e.g. Massey et al. 2018, 2020a, b). Analysis of landslides triggered during earthquake events of varying magnitude and levels of ground shaking in New Zealand have revealed that widespread soil/rock falls and avalanches are typically only triggered for earthquakes with associated ground shaking greater than Modified Mercalli (MM) intensity of MM > 5 and peak ground accelerations (PGA) of $\geq$0.2 g (Dowrick et al. 2008; Hancox et al. 2002, 2014, 2016). However, the intensity of the ground shaking varies with earthquake magnitude and depth; therefore, earthquake magnitude alone is not a good measure of ground shaking intensity. For large slides, longer duration shaking at low frequencies may be more important than PGA for landslide triggering (e.g. Jibson and Tanyas 2020). Landslide and other environmental criteria often observed at different levels of ground shaking are given in Table 1.

4 National- to Regional-Scale Susceptibility Techniques

Landslide susceptibility is a qualitative or quantitative assessment of the spatial distribution of landslides that may occur in an area. Susceptibility may also include a description of the velocity and intensity of existing or potential landslides (Corominas et al. 2014). Landslide susceptibility maps do not represent hazard, as there is no indication of the frequency (annual probability) of landslides occurring or, for rapid landslides, where the debris from the landslide source may travel. However, they are a fundamental input to hazard analysis as they can be used to determine the most probable locations of where landslides may initiate (de Vilder and Massey

Table 1 Landslide and Environmental Criteria often observed at different levels of ground shaking—peak ground acceleration (PGA) and the New Zealand Modified Mercalli intensity scale (MM)—modified from Hancox et al. (2002); and de Vilder and Massey (2020), adopting the PGA to MMI relationship of Moratalla et al. (2021)

PGA and MODIFIED MERCALLI (MM) INTENSITY SCALE Landslide and Environmental Criteria based on New Zealand observations
PGA: < 0.1 g (MM > 5–6) • Trees and bushes shake or are heard to rustle. Loose material dislodged on some slopes, e.g. existing slides, talus and scree slope • A few very small (<1000 m^3) soil and regolith slides and avalanches and occasional soil/rock falls from steep banks and cuts • A few minor cases of liquefaction (sand boils) in highly susceptible alluvial and estuarine deposits
PGA: 0.1–0.2 g (MM 6–7) • Water made turbid by stirred-up mud • A few very small (<1000 m^3) disrupted soil slides, avalanches and falls of sand and gravel banks, and rock falls and avalanches from steep slopes and cuttings common • Fine cracking on some slopes and ridge crests • A few small to moderate (1000–100,000 m^3) slides and avalanches in soil and regolith, and falls of soil/rock falls from steep slopes (>30°) and on coastal cliffs, gorges, road cuts/excavations, etc • Small discontinuous areas of minor shallow sliding/cracking and mobilisation of scree slopes in places. Minor small failures in road cuts and in more susceptible materials • A few instances of liquefaction (small water and sand ejections) in alluvium
PGA: ≥ 0.2–0.5 g (MM 7–8) • Cracks appear on steep slopes and in wet ground • A few pre-existing landslides could re-activate, causing new cracks to form • Significant landsliding likely in susceptible areas • Small to moderate (1000–100,000 m^3) slides widespread; many soil/rock falls on steep slopes (terrace edges, gorges, cliffs, cuts, etc.) • Significant areas of shallow regolith landsliding and reactivation of scree slopes • A few large (100,000–1,000,000 m^3) landslides from coastal cliffs and possibly large to very large (>1,000,000 m^3) rock slides and avalanches from steep mountain slopes • Larger landslides in narrow valleys may form small temporary landslide-dammed lakes • Roads damaged and blocked by small to moderate failures of cuts and slumping of road-edge fills • Evidence of soil liquefaction common, with small sand boils and water ejections in alluvium and localised lateral spreading (fissuring, sand and water ejections) and settlements along banks of rivers, lakes, canals, etc

(continued)

Table 1 (continued)

PGA and MODIFIED MERCALLI (MM) INTENSITY SCALE Landslide and Environmental Criteria based on New Zealand observations
PGA: ≥ 0.5–1.1 g (MM 8–9) • Landsliding widespread and damaging in susceptible terrain, particularly on slopes steeper than 20°. Cracking on flat and sloping ground • Extensive areas of shallow regolith failures and many rock falls and disrupted rock and soil slides on moderate and steep slopes (2000B0–35° or greater), cliffs, escarpments, gorges and man-made cuts • Many small to large (1000–1,000,000 m^3) failures of regolith and bedrock, and some very large landslides (>1,000,000 m^3) on steep, susceptible slopes • Very large failures on coastal cliffs and low-angle bedding planes in, e.g. Neogene sedimentary rocks. Large rock/debris avalanches on steep mountain slopes in well-jointed greywacke and granitic rocks. Landslide-dammed lakes formed by large landslides in narrow valleys • Some pre-existing landslides will re-activate, causing new cracks to form. The debris from some may break down to form avalanches • Damage to road and rail infrastructure widespread, with moderate to large failures of road cuts and slumping of road-edge fills. Small to large cut slope failures and rock falls in open mines and quarries • Liquefaction effects widespread with numerous sand boils and water ejections on alluvial plains, and extensive, potentially damaging, lateral spreading (fissuring and sand ejections) along banks of rivers, lakes, canals, etc. Spreading and settlements of river stop banks likely
PGA: > 1.1 g (MM > 9) • Landsliding is very widespread in susceptible terrain (3) • Similar effects to MM 8–9, but more intensive and severe, with very large rock masses displaced on steep mountain slopes and coastal cliffs. Landslide-dammed lakes formed. Many moderate to large failures of road and rail cuts and slumping of road-edge fills and embankments may cause great damage and closure of roads and railway lines • Many pre-existing landslides will re-activate, causing new cracks to form. The debris from some may break down to form avalanches • Liquefaction effects (as for MM 8–9) widespread and severe. Lateral spreading and slumping may cause rents over large areas, causing extensive damage, particularly along river banks, and affecting bridges, wharfs, port facilities and road and rail embankments on swampy, alluvial or estuarine areas
NOTES: 1. 'Some' or 'a few' indicates that threshold for an effect or response has just been reached at that intensity. At less than MM 5 shaking, landslide effects are insignificant in New Zealand and are therefore not included in this scale 2. Intensity is principally a measure of building damage. Environmental damage (response criteria) also occurs, mainly on susceptible slopes and in certain materials; hence, the effects described above may not occur in all places but can be used to reflect the average or predominant level of damage (or MM intensity) in a given area 3. Environmental response criteria have not been suggested for > MM 9 because the environmental effects at high levels of shaking would be similar to those at MM 8–9, but possibly more widespread and severe

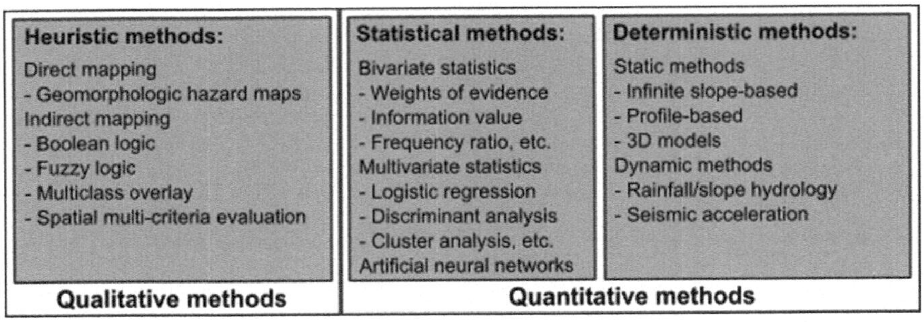

Fig. 1 Different methods for generating landslide susceptibility maps, divided into heuristic (knowledge-driven), statistical (data-driven) and deterministic (physically based modelling) methods (modified from Corominas et al. 2014)

2020). Susceptibility maps are typically determined using information from landslide inventories along with any predisposing 'susceptibility' factors, which can be developed for a specific triggering event. The data is typically analysed using the following three methods (Fig. 1): (1) ***Heuristic techniques***, where expert knowledge and relative ranking scales are used to assign a rank of susceptibility from low to high for a particular mapping unit; (2) ***Statistical techniques***, where combinations of predisposing factors are evaluated statistically, with susceptibility expressed in terms of probability; or (3) ***Deterministic techniques***, where physics-based models are applied either regionally or to individual slopes to determine their susceptibility to failure, adopting metrics such as the yield acceleration (Kramer 1996) or amount of permanent ground (slope) displacement for coseismic landslide susceptibility, which can also be expressed in terms of probability. Comprehensive reviews of landslide susceptibility mapping include Corominas et al. (2014), Guzzetti et al. (2006), Reichenbach et al. (2018) and van Westen et al. (2008). The focus in this chapter is on quantitative methods, and so heuristic methods are not discussed further.

To be useful, such methods should ideally include the ability to link the level of earthquake-induced ground shaking to the estimated landslide probability. These methods could therefore be used on static earthquake shaking data such as PGA/PGV estimates for: (1) different annual exceedance probabilities derived from probabilistic seismic hazard modelling; or (2) in near-real time using instrument or modelled earthquake data available soon after an event (e.g. Nowicki Jessee et al. 2018). As well as forecasting landslide occurrence, such methods have also been successfully used to explore the relationships between earthquake shaking and landslide occurrence, where the landslides can be linked to the triggering event (e.g. Parker et al. 2015; Massey et al. 2018). The majority of these EIL models are either global or regional scale.

4.1 Statistical Methods

The following sections comprise two case studies that show the susceptibility results, expressed as landslide probability (0–1), using two regional-scale 'statistical' methods—logistic regression (LR) and deep learning (DL). The results from these are compared and their main highlights discussed. Both methods adopt variations of the same input EIL data sets: (1) M_W 7.8 2016 Kaikoura earthquake and landslides adopting Version 2.0 of the landslide inventory presented in Massey et al. (2020b); (2) M_W 7.1 1969 Inangahua earthquake landslide inventory presented in Hancox et al. (2014); and (3) M_W 7.3 1929 Murchison earthquake landslide inventory presented in Hancox et al (2016), (Fig. 2).

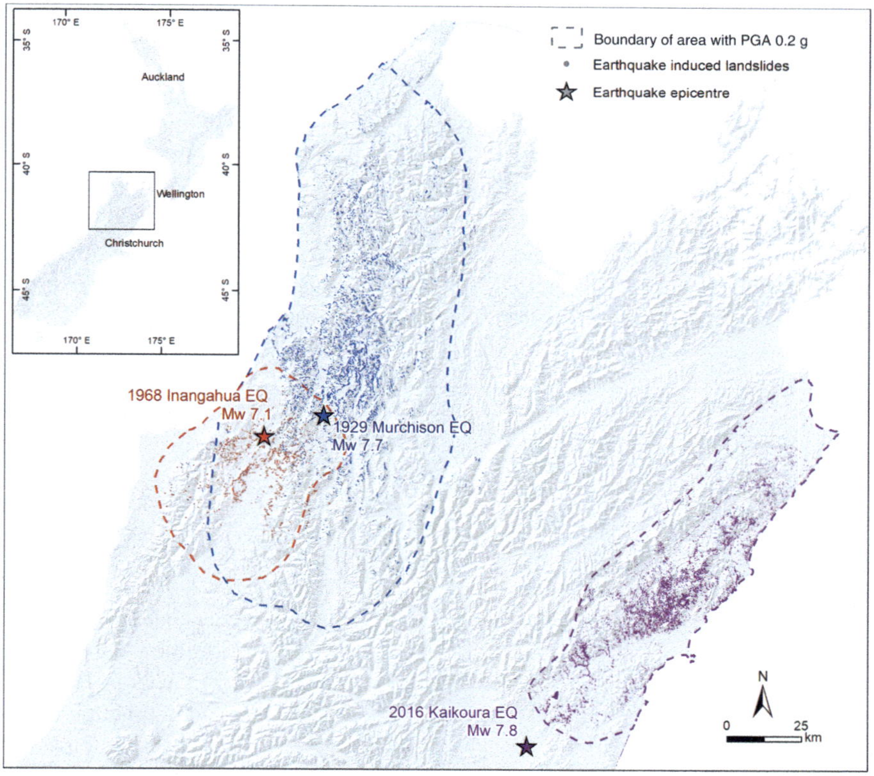

Fig. 2 Earthquake-induced landslide inventories used to train the LR and DL models. The dashed lines show the boundary of the area associated with each earthquake where the peak ground acceleration (PGA) ≥ 0.2 g

4.2 Statistical Method Inputs

To undertake the modelling, a sample grid at 32 m ground resolution was defined, based upon an 8 m ground resolution digital elevation model, resampled from the 2012 version of the Land Information New Zealand (LINZ) digital elevation model for New Zealand. The 32 m grid is much less than the typical hillslope lengths in the regions used for modelling, which can vary from 100 to ≫1000 m. For this analysis, only landslide source areas, not debris trails, from each inventory were used. To reduce sample bias, it was assumed that a grid cell is defined as being a landslide source (LS = 1), if its centroid falls within a landslide source area, even if the grid cell is not fully occupied by the landslide source.

The predictor variables used in the training were chosen based on variables previously found to influence landslide occurrence (e.g. Von Ruette et al. 2011; Parker et al. 2015; Massey et al. 2018, 2020b). The predictor variables used in the model, and the methods to calculate them, are shown in Fig. 3 and described in Table 2. Three proxies of landslide forcing variables were adopted: (1) the PGA, derived from the ShakeMap NZ model (Goded et al. 2021; Horspool et al. 2015); (2) the proximity of a landslide to a coseismic fault rupture; and (3) local slope relief (LSR). The landslide susceptibility variables of (4) geology; (5) elevation; (6) slope angle; and (7) slope curvature were used. Table 2 details how these variables were calculated. These data sets were generated for each earthquake, within an area 'envelope' where the PGAs from the ShakeMap models were ≥0.2 g (Table 1), an example of the input data for the Kaikoura 2016 EIL V2.0 inventory is shown in Fig. 3.

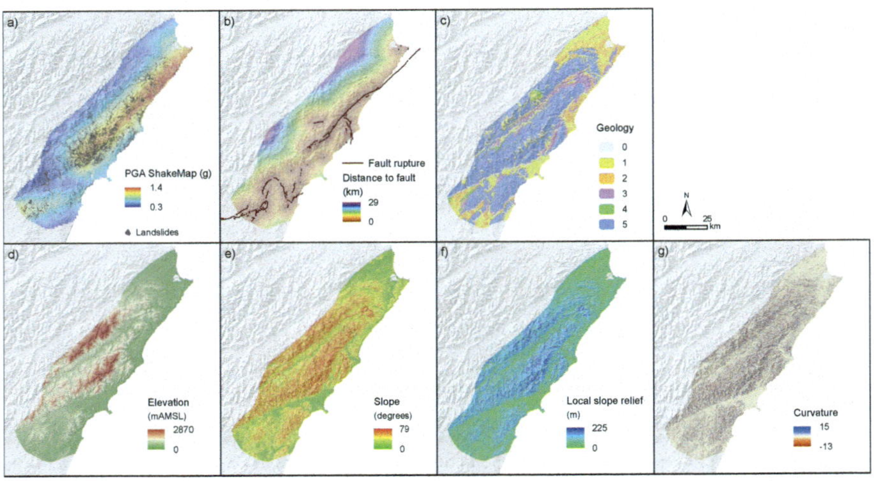

Fig. 3 Example variables used to represent the M_W 7.8 2016 Kaikoura earthquake, and the landslides it generated (from Massey et al. 2020a, b)

Table 2 Predictor variables used in the LR and DL models, their ID codes, descriptions and units

Variable type	Variable ID	Description	Units
Landslide Susceptibility	GeolCode 1	Quaternary sands, silts and gravels. These materials typically form terrace deposits on the top of the steep coastal cliffs as well as inland slopes adjacent to the main rivers of the area. Many of these terraces have been incised by rivers	N/A
	GeolCode 2	Neogene limestones, sandstones and siltstones, which are typically weak. They occur along sections of the coast north of Kaikoura, and along the coast and inland from Greymouth to Karamea on the West Coast	N/A
	GeolCode 3	Upper Cretaceous to Paleogene rocks including limestones, sandstones, siltstones and minor volcanic rocks. These are typically weak (like the Neogene limestones and sandstones), and easily erodible and they can contain thin clay seams, which are volcanic in origin. They are typically exposed in narrow strips overlying the greywacke basement rocks in the Kaikoura region, but are more prevalent within the area affected by the Murchison and Inangahua earthquakes	N/A
	GeolCode 4	All intrusive and extrusive volcanic rocks, ranging in age from Jurassic to Quaternary. Including andesites, basalts, rhyolites, ignimbrites, dacites, granites and granodiorites. These represent the smallest (in land area) geology group within the area affected by the three earthquakes	N/A
	GeolCode 5	Lower Cretaceous Torlesse (Pahau terrane) 'basement' rocks are predominantly sandstones and argillite, also known as greywacke. The greywacke rocks are typically moderately to well bedded and tend to be closely jointed. They form many of the coastal slopes as well as the steeper inland Kaikoura mountain ranges and are the dominant rock type affected by the Kaikoura earthquake	N/A

(continued)

Table 2 (continued)

Variable type	Variable ID	Description	Units
	$\mathrm{Slope}_{\mathrm{MEAN}}$	Local hillslope gradient taken from the 8 m resolution digital elevation model generated by GNS Science, adopting the mean value of all of the 8 m by 8 m cells that fall within each cell of the sample 32 m by 32 m grid. This variable is a proxy for the static shear stresses in the slope	Degree (°)
	$\mathrm{Elev}_{\mathrm{MEAN}}$	Local hillslope elevation taken from the 8 m resolution digital elevation model generated by GNS Science, adopting the mean value of all of the 8 m by 8 m cells that fall within each cell of the sample 32 m by 32 m grid. This variable represents the observation that topography can limit the size of the landslides. For example, slopes that are higher in elevation tend to have larger surface areas and can therefore generate larger landslides than slopes at lower elevations, which tend to have smaller surface areas	Metre above mean seal level (MAMSL)
	Curv	Curvature generated using ArcGIS, taken from the curvature of the surface on a cell-by-cell basis, as fitted through that cell and its eight surrounding neighbours using the 8 m resolution digital elevation model generated by GNS Science. A negative value indicates the surface is upwardly convex at that cell. A positive profile indicates the surface is upwardly concave at that cell. A value of 0 indicates the surface is flat. This variable is a proxy for slope 'sharpness' that represents topographic amplification effects, as amplification of shaking has been recorded at sharp breaks in slope (e.g. Massey et al. 2017; Janku 2017), as well as topographic depressions associated with higher groundwater levels	One hundredth (1/100) of a *z*-unit

(continued)

Table 2 (continued)

Variable type	Variable ID	Description	Units
Landslide (earthquake) forcing	FaultDist	The Euclidean distance from the centroid of each of the 32 m by 32 m sample grid cell to the nearest fault that ruptured using the mapped surface expression, taken from the GNS Science Active Faults database (Langridge et al. 2016), which includes those faults that ruptured during each earthquake. The proximity to a fault is inherently included in the estimation of PGA; however, it was included separately to represent the influence of off-fault ground deformation and other near-field effects that might not be fully taken into account in the ground-motion models	Meters
	PGA_{SM}	Grid of the mean peak ground acceleration (PGA) derived from ShakeMap NZ (Goded et al. 2021, and Horspool et al. 2015), developed by the U.S. Geological Survey calibrated for New Zealand by Horspool et al. (2015). Grid resolution is 1000 m by 1000 m. The PGA values were attributed to the sample grid cell, by taking the PGA value at its centroid	g
	LSR	Local slope relief calculated using focal statistics in ArcGIS. It represents the local height (and angle) of the sample grid cell. It is calculated as the difference in elevation between the lowest in elevation 8 m by 8 m grid cell, within an 80 m (ten (10) 8 m cells) radius from the centroid of the given sample grid cell, and the mean elevation of that grid cell ($Elev_{MEAN}$). This variable represents a proxy for slopes that could amplify ground shaking due to their 'larger-scale steepness' (larger than just a sample grid cell size), where larger values of LSR represent the steeper and higher slopes of the region, which can amplify ground shaking more than lower in height and less steep slopes (Ashford and Sitar 2002; Massey et al. 2017)	Meters (m)

4.3 Case Study 1: Logistic Regression

Logistic regression is routinely used to forecast landslide occurrence (e.g. Von Ruette et al. 2011; Budimir et al. 2015), by calculating landslide probability from a series of input "predictor" variables. The method is used to estimate the coefficients for predicting the probability (P_{LS}) that $Y = 1$, given the values of one or more predictor variables. The condition $Y = 1$ corresponds to the occurrence of a landslide and $Y = 0$ corresponds to no landslide, here within a sample grid cell on the ground. The regression coefficients are estimated using a maximum likelihood criterion. In our case study, models were fitted to each separate GeolCode (Table 2), where landslide probability ($P_{LS:\ Geology\ x}$) for each geology group (GeolCode) is given by the following equation:

$$P_{\mathrm{LS:Geology}\ x} = \frac{1}{1 + \exp\left(-\left(\begin{array}{l} C_{\mathrm{Intercept}} + C_{\mathrm{PGA_{SM}}} \cdot \mathrm{PGA_{SM}} + C_{\mathrm{FaultDist}} \cdot \mathrm{FaultDist} + \\ C_{\mathrm{Elev_{MEAN}}} \cdot \mathrm{Elev_{MEAN}} + C_{\mathrm{Slope_{MEAN}}} \cdot \mathrm{Slope_{MEAN}} + \\ C_{\mathrm{LSR}} \cdot \mathrm{LSR} + \mathrm{Curve} + C_{\mathrm{Curve}} \cdot \mathrm{Curve} \end{array}\right)\right)} \quad (1)$$

The results and coefficients from the LR model are given in Table 3. The models were fitted using all data from all three EIL data sets combined into the five separate GeolCodes, following the method described by Massey et al. (2018). This was done to see how well the LR model performed, trained on all data, against the DL model described below, which was trained by randomly selecting only a small percentage of the data from each GeolCode. It was decided to group the data into GeolCodes because field observations made of landslides triggered by the Kaikoura earthquake (Massey et al. 2018, 2020b) suggested that the landslides within each GeolCode varied in mechanism and occurred in varying physiographic settings. It was also found that an LR model that included geology as a categorical predictor overestimated the cumulative forecast of the number of landslide grid cells ($Y = 1$) compared to the recorded number of landslide grid cells.

4.4 Case Study 2: Deep Learning

A deep learning (DL) model was trained using the Python coding language in the Pandas data analysis toolbox and the Keras interface provided by the TensorFlow artificial neural network library. The model was built and trained adopting the following steps:

- ***Step 1:*** The data from all three EIL data sets was merged into one data set for each GeolCode, to create five data sets. This was done to reduce the number of

Table 3 Summary coefficients from the logistic regression model fitted to all data from three EIL data sets per GeolCode

Variable ID	GeolCode 1		GeolCode 2		GeolCode 3		GeolCode 4		GeolCode 5	
	Est	Std. dev	Est	Std. dev	Est	Std. dev	Est	Std. dev	Est	Std. dev
Intercept	– 6.30	0.04	– 8.16	0.03	– 7.09	0.04	– 7.00	0.04	– 9.11	0.02
$Elev_{MEAN}$	– 0.0011	0.00002	0.0019	0.00002	– 0.00024	0.00003	– 0.00040	0.00003	– 0.00038	0.00001
$Slope_{MEAN}$	0.059	0.001	0.056	0.001	0.035	0.001	0.011	0.001	0.043	0.001
LSR	0.048	0.001	0.021	0.000	0.025	0.001	0.030	0.001	0.035	0.000
Curv	– 0.154	0.012	– 0.127	0.005	– 0.122	0.007	– 0.136	0.007	– 0.104	0.003
FaultDist	– 0.00015	0.000002	– 0.00009	0.000001	– 0.00005	0.000001	– 0.00004	0.000001	– 0.00003	0.0000002
PGA_{SM}	0.15	0.03	1.25	0.02	1.37	0.03	1.75	0.03	1.63	0.01
Pseudo R^2	0.38		0.30		0.16		0.12		0.19	

discontinuities in the hyperplane[1] produced by the training process. There is no smooth transition in the data set between one GeolCode and another, and their order is not related.

- ***Step 2:*** The data limits (range) were calculated for each data set, which are the maximum and minimum values for each variable.
- ***Step 3:*** The input data for each variable was rescaled (normalised) relative to its range, to create a new 'normalised' data set. This step saves the shifting and scaling of values by GeolCode so any new data may be properly scaled and fed into the model to produce a forecast. Neural network models train quickest when the input values are between zero and one, or between negative one and positive one.
- ***Step 4:*** An optimal data structure was created to reduce the time required for model training. This was done in two stages using hypercube sampling and smoothing to create a representative subset of the original data set within each GeolCode:
    - The hypercube sampling used the Pandas 'groupby' function to create a grouping of rows of closely related data. All rows with approximately the same values for all the inputs (by GeolCode) are put in a hypercube and examined for inclusion in the data used for the Bayesian hyperparameter step, and potentially for training the model on a controlled subset of the data. The number of hypercubes is determined by the relative distribution of rows of data in n-space (6 dimensions in this case: every row contains one each of Elevation, Slope, LSR, Curvature, FaultDist and PGA, together with the GeolCode and the Landslide value zero or one) and the total number of rows of data. For the case of a few million rows, the model varied the scaling of the size of the hypercube to allow for somewhere between 5000 and 8000 (configurable) individual hypercubes with data. This resulted in some hypercubes with only a few rows and others with as many as a hundred thousand rows. From each of these hypercubes, the number of randomly selected rows was chosen based on the presence or absence of Landslide events; if there were no events, a subset of the lines up to one hundred was selected for the training, and if there were events, then the number potentially selected was set at a maximum of five hundred. This meant that if there were a hundred thousand rows in a hypercube that did not contain an event, one hundred rows would be randomly selected to represent that cube, and if there were events, then five hundred rows would be randomly selected. Doing this allowed the selection of a representative sample of the input data that would cover the entire range of possibilities. This resulted in 1,684,316 of 28,814,079 total rows selected for the Bayesian step and possibly for the final model training, or 5.85% of the available data (see Table 4 for a breakdown of

[1] In mathematics, a hyperplane H is a linear subspace of a vector space V such that the basis of H has cardinality one less than the cardinality of the basis for V. In other words, if V is an n-dimensional vector space than H is an $(n-1)$-dimensional subspace. Examples of hyperplanes in 2 dimensions are any straight line through the origin. In 3 dimensions, any plane containing the origin. Source: https://deepai.org/machine-learning-glossary-and-terms/hyperplane.

Table 4 Number of grid cells and hypercubes used for model training

	GeolCode 1	GeolCode 2	GeolCode 3	GeolCode 4	GeolCode 5
Number (N) of hypercubes (HC)	5653	5616	6204	5361	5564
Total N grid cells in all HC	252,824	319,956	316,919	308,148	486,469
% of total data set used for training (%)	3.7	13.0	13.7	10.2	3.4

the selected rows by GeolCode). Doing this greatly decreased the execution time for determining the best model structure (the Bayesian Step 5).

- To level and simplify the data for the Bayesian step and to make certain the landslide value was representative for this region of the n-space (the rows randomly sampled from the number available might not otherwise represent the overall likelihood of an event in this hypercube), the selected rows from each hypercube sampling phase could have the Landslide value (initially either a zero or one for all rows in this cube) modified to be the mean value of all the rows in the hypercube. If twenty percent of the rows in a hypercube were events ($Y = 1$) and eighty percent were not ($Y = 0$), a value of 0.2 could be assigned to the Landslide value for each of the selected rows instead of a zero or one. This smoothing of the data made the model much quicker to train and was determined to be beneficial for the Bayesian step and also for training the final models.

- ***Step 5:*** Hyperparameter optimisation. Hyperparameters are parameters whose values are used to control the learning process—they define the model structure as they reflect the number of neurons and layers. The aim of hyperparameter optimization is to find the hyperparameters of the machine learning algorithm that return the best performance as measured on a validation subset of the data. Each test-model structure is tested using three different passes through the data, where a different portion (1/3rd) of the training data is held back for validation. The optimisation implemented the Bayesian model-based optimisation routine in SciKit-Learn, splitting the data into sections (2/3rds for training and 1/3rd for testing) and adopting a sigmoid activation function, to find the best model structure, i.e. the one that returns the best performance. The Bayesian optimisation initially used 32 steps to find the 'best' model structure, followed by a further eight refinements of the initial model results. The performance 'skill' of each model run was 'scored' using the root mean squared error (RMSE). This was calculated by validating the prediction from the training data provided by the k-fold 'split' method, which split the data set into training and testing data, with the test data not used for training the model. Each of the three passes during this process selected a random portion of the data set for training and some for the testing. The RMSE values were summed for the final value used by the Bayesian optimisation routine.

- ***Step 6:*** Determining the optimal epochs for the training. This step split the data into training and validation sets that were both passed to the TensorFlow 'fit' routine to determine how many epochs were required for the model (the best model from the previous Step 5) to 'learn' from the data without loss of performance when applied to independent data. When the model validation started to deviate from the learning (over-training), the optimal number of epochs was determined. This was done three times, and the lowest value was stored for the final training, as it best represented the number of epochs required to train the model on the entire data set.
- ***Step 7:*** The training process. The best model from Step 5 for each GeolCode and the optimal number of epochs value for each GeolCode from Step 6 was applied to train the models using all available data or the subset of data from the hypercube step, and also using the smoothed subset of data. The trained model was then saved. After some experimentation, it was clear that the smoothed subset of data was sufficient for training the models, and model training was significantly faster than using all available data.
- ***Step 8:*** Applying the final model. This step took the saved model from the previous step (step 7), read in the original data sets, scaled them by the numbers derived from Step 2, and created a new data set with an additional column called "Forecast", which represents the forecast landslide probability ($P_{LS} = 0$ to 1) calculated using the preferred model for the given GeolCode.

4.5 Model Performance

The statistical performance of the LR and DL models was compared using two approaches: (1) the Receiver Operator Characteristic (ROC) curve and area under the curve (AUC); and (2) the cumulative forecast probability versus the cumulative landslide record, referred to as the 'CumF-R'.

ROC curves (e.g. Deleo 1993; Fawcett 2006; Lombardo and Mai 2018) are graphical plots based on the analysis of 'non-failing' and 'failing' predictions. The ROC curve is created by plotting the true positive rate (TPR) against the false positive rate (FPR) at various threshold settings. The true positive rate is also known as sensitivity in machine learning. The false positive rate is known as probability of false alarm and can be calculated as (1-specificity), where specificity is true negative rate (TNR). The ROC curves illustrate the diagnostic ability of a binary classifier system as its discrimination threshold is varied. The shape of the curve is used to evaluate the goodness of fit for a binary classifier, in this case whether a grid cell is a landslide or not ($Y = 1$ or $Y = 0$), where the class prediction for each instance is based on the continuous random variable, in this case the landslide probability. The TPR defines how many correct positive results (in this case landslides or cells where $Y = 1$) occur among all positive samples (total number of landslide cells). FPR, on the other hand, defines how many incorrect positive results (non-landslides or cells with $Y = 0$) occur among all negative samples (total number of cells with $Y = 0$). The number of landslide (true positive) and non-landslide (false positive) cells was counted for all the cells with the landslide probability greater than a threshold, where the threshold is a varying parameter.

An ROC curve demonstrates the following: (i) the trade-off between true positive and false positive rates; (ii) that the closer the curve follows the left y-axis border and then the top x-axis border of the ROC space (AUC close to 1), the more informative the prediction may be; and (iii) that the closer the curve comes to the 45° diagonal of the ROC space (AUC $= 0.5$), the less informative the prediction may be. Therefore, a model result with an AUC > 0.9 indicates that the ability of the model to forecast landslide probability is 'excellent', and an AUC < 0.70 indicates that the ability of the model to forecast landslide probability is 'poor' (Lombardo and Mai 2018).

The True Positive Rate (sensitivity), False Positive Rate (1-specificity) and AUC were calculated for each modelled scenario, in the following way:

- ***Step 1:*** Forecasting. Using the LR and DL models fitted to each scenario, the 'forecast' probability was calculated for each grid cell in the data set, based on the variables for the given cell. The data was then ordered from high to low forecast probability;
- ***Step 2:*** True Positive Rate (Sensitivity). This is the cumulative number (forecast probability sorted from high to low, Step 1 of correctly predicted landslides (cells where $Y = 1$), divided by the total number of cells classified as being a landslide ($Y = 1$) in the given data set;
- ***Step 3:*** False Positive Rate (1-Specificity). This is the cumulative number (forecast probability sorted from high to low, Step 1 of non-landslides predicted as landslides (cells where $Y = 0$), divided by the total number of cells classified as not being a landslide ($Y = 0$);
- ***Step 4:*** Area under the curve (AUC). The area was calculated from a histogram created by plotting the TPR on y-axis against the FPR on x-axis for each forecast probability sorted from high to low (Step 1)
- ***Step 5:*** ROC was also plotted, and AUC calculated using the Python SciKit-Learn library.

The CumF-R relationship is a way to show whether the forecast model is under- or over-predicting the number of grid cells that are forecast as being a landslide. It therefore shows the 'consistency' of the modelled forecast probabilities against the actual grid cells recorded as being landslides (Parker et al. 2015; Massey et al. 2018). These are calculated as the cumulative sum of forecast landslide probability values for each sample grid cell sorted from low to high probability. The corresponding recorded Y-values, i.e. $Y = 1$ (landslide) or $Y = 0$ (no landslide) are also summed on the sorted data set, providing the cumulative sum of grid cells classified as being a landslide ($Y = 1$). A well performing model should plot as a "straight" line and close to the line of equality, in cumulative forecast versus cumulative record space. A poorly performing model will not be a straight line and will either over- or under-forecast the cumulative number of landslide cells. The landslide probability results from the LR and DL models, per GeolCode for the entire data set relating to the three EIL inventories, are shown in Fig. 4 and in Table 5.

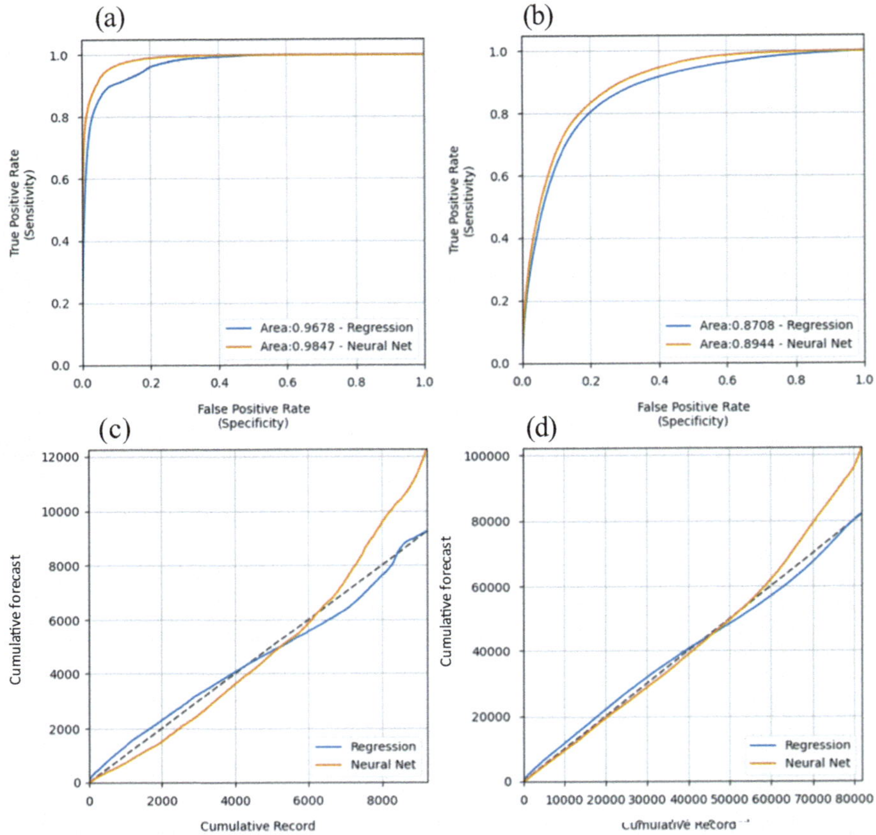

Fig. 4 LR model ('Regression') and DL model ('Neural Net') results. **a** and **b** show ROC curves and AUC (Area under the curve 'Area') and **c** and **d** show the Cumulative Forecast versus Record (Cum-FR) relationship. **a** GeolCode 1; and **b** GeolCode 5. CumF-R plots for: **c** GeolCode 1 and **d** GeolCode 5. The results are based on models fitted to the three EIL data sets

Based on the ROC:AUC results, both the LR and DL models can be classified as 'good' to 'excellent' in their ability to forecast landslide probability for the three EIL inventories, using the input variables listed in Table 2. The results indicate that the DL model outperforms the LR model, when comparing the ROC:AUC values. However, the DL model over- and under-predicts the cumulative number (cumF-R) of landslide grid cells (where $Y = 1$) by up to 22%. This is thought to be due to the hypercube smoothing used in the DL model to reduce training time. These results suggest that the simple LR model, trained on all the data, is good to excellent at forecasting landslide probability for the three EIL inventories, when compared to the more complex DL model, which takes considerably more time to setup and train than the LR model. The results from these models can also be used to investigate the relative importance of each variable on forecasting landslide probability, as well as investigating variability between the EIL data sets. In addition, the relative contribution each variable has

Table 5 Results from the LR and DL models fitted to the three EIL inventories

Scenario		Original data set		Logistic regression			Deep learning			
		Total N cells Y = 0	Total N cells Y = 1 (landslide)	Cumulative forecast total N cells Y = 1	Difference (forecast-actual)	ROC:AUC	ROC:AUC	Cumulative forecast total N cells Y = 1	Difference (forecast-actual)	% difference (%)
GeolCode	1	6,889,309	9239	9239	0.0	0.97	0.98	11,115	1876	20
GeolCode	2	2,431,145	39,263	39,263	0.0	0.91	0.92	44,672	5409	14
GeolCode	3	2,286,075	19,425	19,425	0.0	0.84	0.92	23,608	4183	22
GeolCode	4	2,999,531	14,053	14,053	0.0	0.82	0.89	16,640	2587	18
GeolCode	5	14,044,122	81,917	81,917	0.0	0.87	0.89	98,193	16,276	20

on forecasting landslide probability within each GeolCode may help to elucidate the failure mechanisms of landslides within each GeolCode, as field observations suggest that they are different (Massey et al. 2020a, b). Such analyses are the focus of other research.

These case studies highlight:

- the need for high-quality landslide inventories used for training,
- that relatively simple LR models can perform well (statistically), when comprehensive data sets are available for training,
- the more complex and time-consuming DL models can perform well based on limited training data and as a result could be trained on incomplete inventories and
- the need to use multiple validation approaches to investigate the results (e.g. the ROC:AUC and cum F-R methods), as both methods provide useful insights.

4.6 Deterministic Methods

A good summary of the various regional-scale quasi-physics-based deterministic methods is presented by Jibson (2011). At the regional scale, such methods are only quasi-physics based, as they rarely consider the full earthquake accelerogram, the dynamic response of the slope when excited by it, or different failure modes of the slope (e.g. rockfall, rock/soil avalanche, planar/circular slides, etc.). They are typically used to estimate the yield acceleration (e.g. Newmark 1965; Chien and Tsai 2017) and/or coseismic permanent displacement of rigid body slope-failure modes (e.g. Jibson 2007, 2011; Saygili and Rathje 2008), which can then be used to estimate the probability of the slope (grid cell) being a landslide. Such methods may also require some estimate of the threshold of coseismic permanent displacement of the slope, which will trigger the onset of catastrophic failure, e.g. where the unstable mass may transition from sliding to avalanching. These estimates depend on the material properties, failure mechanisms, depth and thickness of the failure surface(s) as well as the local topography. Grant et al. (2016) report 'threshold' coseismic displacements of 0.05 m for shallow slides and rock falls and 0.15 m for rotational slides/slumps. However, field observations made by the authors suggest that such thresholds vary considerably and thus represent a large uncertainty in such models.

4.7 Deterministic Method Inputs

The challenge with the LR and DL and other data-driven statistical methods to estimate landslide susceptibility is that they require information about past landslides to be able to forecast landslide occurrence with any accuracy. Therefore, such models tend only to be as good as the data sets they are trained on. Deterministic physics-based landslide susceptibility methods bridge such issues as they do not need to be trained on past EIL data sets. This is because they capture the underlying physical processes that lead to slope failure and thus landslide occurrence. However, such methods, at the regional-scale, are also data dependent as they require information on, e.g. the materials present (mapping); the presence of persistent discontinuities (geological structure) along which failure may preferentially occur; their physical (geotechnical) properties such as shear strength; and the depth of water table in the slope, to forecast landslide occurrence with any accuracy. Such information is rarely available at such scales, and so region-wide generalisations of such factors, usually derived from site-specific studies, are made.

4.8 Case Study 3: Multi-Modal Deterministic Method

An example of a regional-scale quasi-physics-based multi-modal (M-M) landslide susceptibility model is the one developed by Grant et al. (2016) and Pollock et al. (2019). Pollock (2020) applied the method to estimate the probability of a slope failing as either a rockfall, shallow rock or soil slide or rotational slide/slump in response to the 2016 Kaikoura earthquake. Pollock (2020) compared the results to a subset of the Version 2.0, Kaikoura EIL inventory (Massey et al. 2020b) in order to validate their model. Pollock (2020) reports ROC:AUC values of 0.88, 0.81 and 0.93 for shallow slides, rotational slides/slumps and rock falls, respectively. These values can be compared to those from the LR and DL models discussed in the previous section (Table 5). However, they only relate to a relatively small region of the area affected by the Kaikoura earthquake. Although not quantified by Pollock (2020), the landslide susceptibility (probability) maps produced suggest the M-M model overestimates the cumulative forecast (cum F-R) of grid cells classified as being a landslide ($Y = 1$). The sensitivity analysis (Pollock 2020) suggests this might be due to a combination of the adopted threshold displacement values and the local topography (slope angles) used to define the susceptibility ranges.

This case study highlights:

- the need for good resolution geological (material and structure) maps across the study area,
- the need for geotechnical properties of the different materials present in the study area, at a spatial resolution that reflects the mapping,
- the need to make assumptions with regards to, e.g. the displacement thresholds and susceptibility values to use, which signify whether a slope has failed or not,
- that such models can be used to gain an appreciation of the types and volumes of landslide that could occur, which can be linked to the geological material and structural controls, and topographic settings and
- that the model performs well when comparing the patterns of the results to the actual mapped landslide distribution from the event, although it does tend to overestimate the severity (number) of landslides produced.

4.9 Developments and Challenges of Regional-Scale Methods

The LR, DL and M-M models all provide useful insights into the relationships between earthquake shaking and landslide susceptibility factors. The statistical methods cannot be directly used to determine the type or volume of landslides produced during an earthquake, as their main output is probability based. These probability outputs, if linked to shaking intensity, can however give an estimate of the severity of landslides as per the cum F-R plots (Fig. 4c and d), from which the total area of landslides could be determined. This estimate of the total area could then be distributed using well-established landslide frequency-area relationships (e.g. Larsen et al. 2010; Massey et al. 2020a) to derive the likely number of landslides of a given area—and ultimately volume—produced.

The efficacy of these regional-scale statistical methods is that once the training data is collated, multiple models can be run on the same data sets, thus providing an estimate of the epistemic uncertainty between models, which is useful when forecasting a chaotic process such as where a landslide might occur. However, they still need to be trained using EIL inventories. Therefore, the best statistically based models will be those trained on EIL data sets: where the landslides can be directly related to the ground motions that caused them; that contain many earthquakes of different types, e.g. subduction and shallow crustal; that have high-quality ground motion (derived from multiple sources and not just, e.g. ShakeMap) and susceptibility data sets available; that have been field mapped/verified for quality purposes; and that have undergone rigorous statistical validation.

The case studies presented here adopt three well-documented landslide inventories. Consequently, such analyses have in the past not been possible where such data is not available (e.g. Kritikos et al. 2015). As a result, rapid post-earthquake modelling of coseismic landslide intensity and distribution for emergency response can be limited, as current methods to identify coseismic landslides immediately after an earthquake using, e.g., optical imagery are too slow to effectively inform emergency response activities (Robinson et al. 2017). Recent developments in DL, LR and fuzzy logic statistical models (e.g. Nowicki Jessee et al. 2018) are however being incorporated into tools allowing near-real-time 'impact' forecasts of landslides to be made (e.g. USGS 2021; Massey et al. 2021, and Chap. 17 in this book) based on rapid post-event (minutes to hours) shaking data. It appears that regional-scale quasi-physics-based methods are lagging behind, as such models are rarely used for impact forecasts, possibly because maps and geotechnical information at the relevant scales are not available. Such models are, however, being used for landscape evolution research (e.g. Fan et al. 2019) and hazard and risk analysis projects (e.g. Pollock et al. 2019) where information on landslide mechanism and volume is needed.

An exciting recent development in this area is the implementation of regional-scale landslide forecast models, like the ones described previously in this section and in Chap. 17 of this book. Such models are being used to produce impact forecasts, e.g. the area of ground or numbers of people affected, by EIL using immediate post-event near-real-time ground shaking data/models. Such models and tools will therefore in future facilitate more targeted emergency response, and hazard- and risk-zone delineation.

5 Site-Scale Numerical Analysis Techniques

At the local- to site-specific scale, coseismic landslide studies typically analyse how a slope may perform in response to seismic shaking, using numerical modelling to determine: (1) amplification effects—the site response to the earthquake shaking amplitude, duration and frequency content; and (2) amounts of permanent displacement in response to the shaking. Site-specific assessments require more detailed data on soil and rock mass characteristics, which the regional-scale landslide susceptibility models discussed previously tend not to use. In this section, the typical modelling objectives are discussed, along with the model inputs required and the typical methods used to achieve the results. The process is illustrated in Fig. 5. Several case studies are used throughout this section to highlight the application of the methods, as well as current limitations and questions.

5.1 Desired Results

Dynamic slope analyses tend to comprise three main methods: ***Pseudostatic analysis***, ***Permanent-displacement analysis*** and ***Stress-deformation analysis*** (Jibson 2011). Despite developments in understanding earthquake mechanisms and the physical processes affecting slope behaviour, analysing the coseismic performance of a slope remains a challenging task as the interaction between seismic waves and the slope includes many variables, such as: (1) the loading history of the site; (2) variability in geotechnical and hydrogeological properties of the material within the slope pre-shaking and during shaking (e.g. strain softening/hardening); (3) earthquake characteristics (epicentre, fault mechanism, magnitude, duration, rupture distance, etc.); and (4) site effects (amplification, directivity, liquefaction, etc.). The desired results depend on the project scope, and range from a preliminary assessment of slope stability given a design ground motion—usually represented as a single value such as PGA—to sophisticated analysis of stress and deformation, possibly under multiple-earthquake loading cycles and/or pore water pressure conditions. Such methods and approaches can also be used to gain insights into site effects and triggering of slope failures for use in risk analysis, as well as research on how stress propagates through a slope during an earthquake and what the post-earthquake 'damage state' of the slope may be as a result of coseismic permanent slope displacement.

In practice, there are several standards and codes that provide practitioners with guidance on how to carry out dynamic slope analyses as well as setting slope performance thresholds regarding, e.g. factors of safety and displacement amounts, for different types of slopes based on the people and assets potentially at risk from

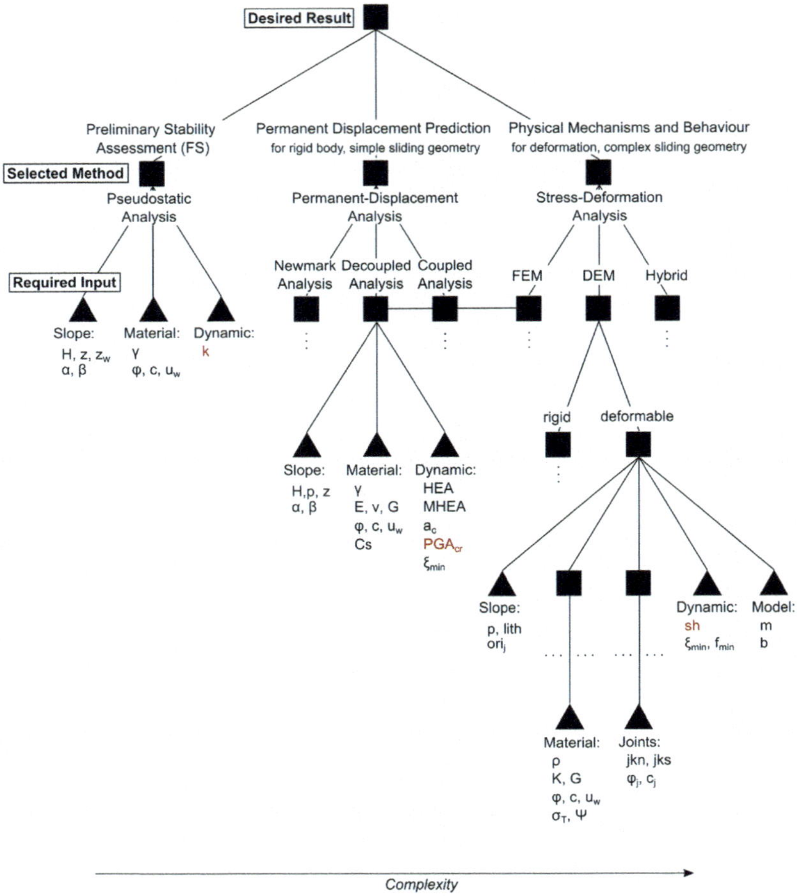

Variables:

Slope:
- H slope/landslide height
- z depth to failure surface
- z_w depth of water table
- α slope angle
- β failure surface angle
- p slope profile
- lith lithological contacts
- ori_j orientation, persistence, spacing of joints

Material:
- γ unit weight
- ρ density
- E Young's modulus
- v Poisson's ratio
- K bulk modulus
- G shear modulus
- φ friction angle
- c cohesion
- u_w pore water pressure
- σ_T tensile strength
- Ψ dilation angle
- Cs shear wave velocity

Dynamic:
- k seismic coefficient
- HEA horizontal equivalent acceleration history
- MHEA maximum HEA
- a_c critical acceleration
- PGA_{cr} peak ground acceleration at crest
- sh stress history
- ξ_{min} damping ratio
- f_{min} centre frequency

Joints:
- jkn joint normal stiffness
- jks joint shear stiffness
- φ_j joint friction angle
- c_j joint cohesion

Model:
- m mesh size
- b model boundaries

Fig. 5 Decision tree of representative approaches to site-specific dynamic slope analysis. The complexity of analyses increases from left to right, and red values indicate higher uncertainty in parameter selection. Ellipses (…) indicate methods or parameter choices not listed. Horizontal line connecting decoupled and FEM methods indicates decoupled analysis requires FEM inputs and simulations failure or displacement of the slope. Monteith (2020) summarises some of these in a New Zealand context, but also includes some of the relevant international codes and standards, which relate to the construction of cut and fill slopes. It is not the intent of this chapter to go into these standards and codes.

5.2 *Input*

Figure 5 indicates the inputs required for dynamic slope analysis, depending on the desired outcomes and method selected. It is shown as a decision tree to emphasise the complexity of even the simplest method, in this case pseudostatic analysis. Values highlighted in red show which inputs tend to have a high level of uncertainty associated with them. Inputs are categorised into slope and sliding surface geometry, static and dynamic material and discontinuity and model parameters. Slope geometry inputs range from slope height, slope angle, assumed average failure surface angle and depth to failure surface to the actual slope profile—with or without an assumed failure surface geometry.

Material (rock and soil) properties are usually assumed to be homogeneous within the material units analysed. Typically, Mohr–Coulomb strength parameters are used for all materials in the slope for dynamic analysis, as indicated in Fig. 5. However, particularly in more complex analyses, other constitutive models may be employed. Scaling properties from laboratory to field in any analysis are challenging. Common, although increasingly criticized (see Yang et al. 2021 and Elmo and Stead 2021), approaches for slope rock masses include the Geological Strength Index (GSI; e.g. Marinos et al. 2007) and deriving Mohr–Coulomb parameter estimates from the Hoek–Brown failure criterion (Hoek 1990; Hoek and Brown 2019). Discrete fracture networks (DFNs) and synthetic rock masses are newer approaches developed mainly for mining applications to estimate field-scale properties and are discussed in, e.g. Exadaktylos and Stavrapoulou (2008), Elmo et al. (2014), Guajardo et al. (2020) and Elmo et al. (2021). Although average material properties can be derived from existing data, these may not reflect actual site conditions and do not consider spatial or temporal variability in characteristics. It is thus imperative to visit the slope investigated to validate properties used in any site-specific analysis. It is also important to understand the failure mechanism(s) of the materials forming the slope. For example, are they expected to behave as brittle or plastic (hardening, softening or perfectly plastic) materials, and are there any structural geological controls that would promote particular types of failure (e.g. translational slides versus rotational slumps; Hungr et al. 2014)?

Ground motion records tend to be the most uncertain input in dynamic slope analysis. They are usually selected based on the type of seismic events expected to occur in a region. Characteristics that should be considered when selecting records for analyses include peak ground acceleration and velocity (PGA and PGV), earthquake magnitude, fault mechanism and rupture distance. If used in stress-deformation analysis, accelerograms—recorded or synthetic time-acceleration histories—tend to be used, which are then typically deconvoluted, filtered and baseline corrected. In some cases, they may need to be converted to stress histories. Other parameters specific to each analysis method are listed in Fig. 5. These include seismic coefficients, damping parameters and model parameters, and they are discussed within each method-specific section below.

5.3 Methods: Pseudostatic Analysis

Pseudostatic analysis is historically the most popular method, largely because of its simplicity. It treats dynamic perturbation of a slope system as a static, permanent body force in a limit equilibrium analysis and provides an index of dynamic stability. As such, it simplifies dynamic motion greatly and is commonly overly conservative. Conversely, it may be unconservative if dynamic pore pressure or significant strength loss is likely, or if an earthquake generates multiple high amplitude acceleration peaks. The method also does not allow for study of post-failure behaviour and is highly dependent on the choice of seismic coefficient to represent the applied body force. The seismic coefficient has traditionally been selected inconsistently based on regulatory guidelines, expert judgement, and pre-existing studies. Bray and Travasarou (2009, 2011) advise that site-specific conditions must be considered in the selection of the coefficient. They have provided guidelines on selecting the seismic coefficient based on variables such as allowable displacement, initial fundamental period of the potential sliding mass, spectral acceleration and dominant frequency of the ground motion, and earthquake moment magnitude.

5.4 Case Study 4: Pseudostatic

Chen et al. (2004) back analyse the Tsao-Ling rockslide in Taiwan, triggered by the 1999 M 7.6 Chi-Chi earthquake. After conducting field investigations and using existing laboratory testing results to characterise the sliding mass and rock formations involved in the event and to determine the failure surface geometry. They selected a nearby seismic record for pseudostatic analysis in a preliminary assessment of dynamic behaviour. The method assumes a sliding surface, which was determined from field observations, as well as one set of parameters to represent the slope properties.

Total unit weight and peak and residual cohesion and friction angles defined the materials involved in the analysis. Pore water pressure was neglected as the slope was assumed to be dry before the earthquake. As materials observed in the field had spatially variable strength, multiple analyses were completed, with parameters reduced stepwise from peak to residual values. The ground motion record was resolved to determine the horizontal acceleration component in the direction of sliding for the analysis and then multiplied by a reduction factor to input a representative value for the record. Six reduction factors were tested assuming an input PGA of 0.66 g, in effect using pseudostatic coefficients (often referred to as k-values) of 0.11 to 0.66. Although the vertical component of acceleration is typically ignored, it was considered in this case. Based on back analysis, Chen et al. (2004) determined that a material strength reduction of 30% of peak values and a ground motion reduction factor of 2/3 best represented the failure.

After back analysing the Tsao-Ling rockslide for calibration, the material remaining on the slope was analysed for stability. Given the design horizontal ground motion (PGA) of 0.33 g for the area (based on the seismic zoning code), the slope was analysed as being stable with peak material properties and no pore pressure. When strength was reduced and static pore water pressure increased, the modelled slope was unstable. The case was subsequently analysed using discontinuum stress-deformation methods, which investigated the failure mechanisms of the rockslide and the importance of pore water pressures in the unstable slope (Tang et al. 2009).

This case study highlights:

- the need to use well-constrained case studies,
- the completion of sensitivity analysis of key parameters to bracket a dynamic stability problem,
- the importance of ground investigations, even for simple dynamic analysis,
- the straightforward interpretation of results (stable or unstable), and
- the significance of considering vertical ground motion in certain cases.

Due to its simplicity, the pseudostatic approach can be used to evaluate the sensitivity of a slope to different forcing and susceptibility factors, however, back analysis is needed to calibrate the site-specific values used in the models.

In New Zealand, there is currently no national standard on how to select the most appropriate pseudostatic coefficient for slope stability analyses. Most New Zealand practitioners use either the New Zealand Transport Agencies Bridge manual (NZTA, 2013), Eurocode 8 (EN1998-5, 2004) or Bray and Travasarou (2007, 2009, 2011), to select such values.

5.5 *Methods: Permanent-Deformation Analysis*

The permanent-deformation method was spearheaded by Newmark (1965). Three types of permanent-deformation analysis have since evolved. The rigid-block approach is the original Newmark method and assumes a rigid block on a sliding plane. Hsieh and Lee (2011) present an example of the method, exploring the relationship between Newmark displacement, Arias intensity and critical acceleration. The decoupled approach separates the dynamic response calculation from the calculation of permanent displacement and requires ground acceleration records to be used as input . In a permanent-deformation analysis, the peak surface acceleration at the crest is the parameter with the greatest source of uncertainty. Makdisi and Seed (1978) developed the most cited decoupled approach, designed for embankment dams and assuming elastic behaviour below yield and perfectly plastic behaviour above yield. Due to these assumptions, decoupled analysis tends to generate overly conservative estimates of permanent displacement. Strenk and Wartman (2011) investigate parameter variability in the Newmark and decoupled approaches and suggest incorporating variability and uncertainty into slope deformation analysis using probabilistic methods. Bray and Rathje (1998) updated the approach for deeper failures. In an example, Massey et al. (2017) analysed the response of several rock slopes in the Port Hills of Christchurch, to the 2010/11 Canterbury earthquake sequence and determined that impedance contrasts related to lithological changes strongly influenced amplification within the slopes. Two of these Port Hills slopes were further analysed by Arnold (2016) using stress-deformation analysis (see below) adopting a particle flow code (PFC) to investigate the observed rock slope failure mechanisms. Finally, coupled analyses, which model the dynamic response and permanent displacement concurrently, are the most sophisticated of the permanent-displacement approaches (Jibson 2011; Bray and Macedo 2019). The coupled models address some of the limitations of the Newmark and decoupled analyses, such as conservative displacements and high uncertainty in crest accelerations. Bray and Travasarou (2007) present a formulation and examples of this method.

5.6 *Case Study 5: Decoupled Permanent-Displacement*

To determine how historic seismic events may have contributed to landslide displacement rates at the slow-moving 22 million m^3 Utiku landslide in New Zealand, decoupled permanent-displacement analysis was completed. Decoupled analysis was considered appropriate, as displacement order of magnitude (not necessarily absolute value) was required. The planar rock slide had been monitored since 1964, with instrumentation comprising: six boreholes intersecting the basal clay sliding zone; five Casagrande piezometer standpipes; two inclinometers; theodolite surveys since 1965; four continuous GPS stations since 2008; a rain gauge; and a strong motion sensor since 2008 (Massey et al. 2016). Geomorphological and engineering geological fieldwork, shear wave velocity measurements of the landslide mass and laboratory tests of borehole and field samples (ring-shear, direct shear, dynamic and index testing) were also completed (Carey et al. 2019). As such, the geological, hydrogeological and geotechnical characteristics of the landslide are well understood.

Seven earthquake accelerograms (time-acceleration histories) were selected based on a four-step selection process (NZS1170, Standards New Zealand 2004; Massey et al. 2016). Both scaled and unscaled vertical and horizontal accelerograms were applied in the analysis. In this study, PGA scaling factors were used to raise the PGA of the accelerograms to the value of the 10,000-year return period PGA estimated from the National Seismic Hazard Model (NSHM) for New Zealand. Numerical model boundaries were far enough from the area of interest to avoid unrepresentative wave reflection. Static limit equilibrium and finite element simulations in

SLIDE and RS2 (Rocscience 2010a, b) first validated the residual shear strength properties of the sliding surface derived from laboratory tests and considering pore water pressures. Then, QUAKE/W (Geo-Slope 2014), finite element method (FEM) software, was used to determine the dynamic response of the translational rockslide to each selected earthquake accelerogram. Finally, permanent displacements were calculated in Slope/W (Geo-Slope 2012) using Newmark's (1965) double integration method based on the average time-acceleration histories along the basal sliding surface—which was determined from borehole and inclinometer measurements.

Results indicated that the displacement rates determined from the decoupled analyses—although relatively large for some of the larger (in PGA) earthquake accelerograms—were too low compared to actual rates, given the estimated return period (annual exceedance probability) of the earthquake PGAs needed to initiate them, as determined from the NSHM (Fig. 6). The authors concluded that changes in pore water pressures within the landslide mass, along with engineering works, were more likely to account for the pre-historic and historic landslide displacement.

This case study highlights:

- the need to use well-constrained case studies,
- the selection of ground motions using clear guidelines and steps,

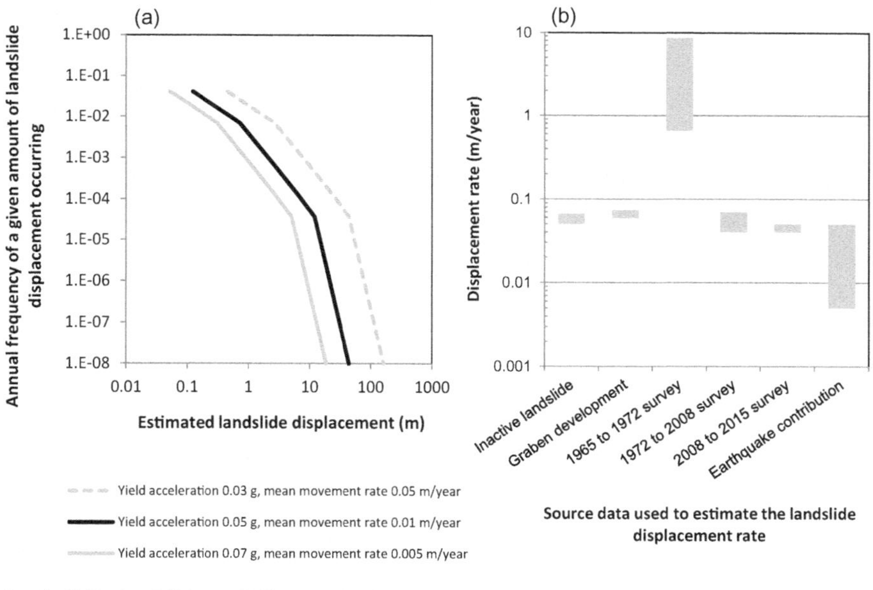

Fig. 6 Utiku landslide modelling results from Massey et al. (2016). **a** The log annual frequency of a given amount of landslide displacement occurring and the mean rate of landslide displacement, for the given range of yield accelerations. **b** The mean landslide displacement rates estimated from monitoring, geomorphological indices and modelling. Note: the earthquake contribution is derived from the decoupled analyses; and the period of increased acceleration was due to engineering works loading the upper part of the landslide (McSaveney and Massey 2017)

- the application of a relatively straightforward analysis technique to determine the significance of seismic activity to slope deformation and
- the validation of the results using ground investigations and understanding of physical processes.

Note that decoupled analysis assumes the landslide mass is a rigid body, which in this case study is a representative assumption given most of the displacement occurs along the basal sliding surface. However, such methods cannot account for progressive failure post-earthquake or at accelerations lower than the yield acceleration. As mentioned, the method does not allow for deformation within the landslide body, and thus may provide a conservative estimate of permanent displacement.

5.7 Methods: Stress-Deformation Analysis

Stress-deformation analysis relies on the construction of sophisticated finite or discrete element (FEM or DEM) or hybrid models. It provides the most accurate representation of individual slopes, but requires site-specific, high-quality data and specialised knowledge. As suggested in Fig. 5, there are many software options available to develop dynamic FEM and DEM models. To interpret the results of the simulations, the user needs to understand the static and dynamic behaviour of the modelled materials (see below). Stress-deformation analysis does not inherently provide simple metrics such as factor of safety or critical displacement for a given slope; rather it calculates the stress distributions in a simulation and suggests how a material may deform within the slope in response to a given loading condition. Three examples of its application are Bozzano et al. (2008), Bazgard et al. (2011) and Rizzitano et al. (2014). Bazgard et al. (2011) derive the boundary conditions for dynamic models in PFC3D and FLAC3D (Itasca 2011), and Bozzano et al. (2008) use FLAC (Itasca 2008) to simulate the reactivation of the Salcito landslide in Italy after the 2002 Molise earthquake. They determine that ground motion amplification is likely to have caused the reactivation of the landslide. Rizzitano et al. (2014) demonstrate the complex amplification effects of topography and stratigraphy using the FEM code QUAKE/W (Geo-Slope 2014).

5.8 Case Study 6: Dynamic DEM

The site of the 1855 Gold's Landslide in Wellington, New Zealand, showcases the application of stress-deformation analysis using the DEM code UDEC v. 7.0 (Itasca 2020). The slope is directly above the main highway connecting Wellington to Lower Hutt, and thus would cause major disruption if it reactivated. As part of a broader project investigating slopes in the Wellington region, the Gold's slope was studied to understand how it would behave in a large earthquake. Ground investigations included engineering geological characterisation of outcrops and two deep boreholes (108 m and 65 m depth, respectively) with geophysical full-waveform sonic logging, downhole imaging, and rock strength, weathering and fracture characterisation.

To model the slope, the following steps were followed:

- ***Step 1:*** A representative two-dimensional cross section was imported. Three greywacke zones with weathering grades of highly weathered at surface to slightly weathered at depth were defined within the slope profile. The basal and lateral boundaries were extended by at least double the slope height and quiet boundaries applied to avoid boundary effects and artificial wave reflection.
- ***Step 2:*** For this analysis, deformable blocks were chosen. Mesh size was set to allow adequate wave propagation through the model. A general guideline is that the element size should be 1/8th to 1/10th of the highest frequency wavelength of the input ground motion (Itasca 2020).
- ***Step 3:*** Most simulations applied Mohr–Coulomb plastic and area contact constitutive models for the greywacke and discontinuities, respectively. The material properties were determined by upscaling laboratory values of jointed greywacke samples from similar greywacke rock masses using the GSI and Hoek–Brown to Mohr–Coulomb approaches.
- ***Step 4:*** The boundaries were changed from velocity in the initial equilibrium step with only gravity to free field quiet boundaries during the dynamic step with the selected ground motion applied.
- ***Step 5:*** An initial state was determined by cycling through to static equilibrium.
- ***Step 6:*** The dynamic analysis was completed by applying the input ground motion as a stress boundary at the base of the model. Rayleigh damping was applied.

Models were run with one of three rock mass inputs: (i) homogeneous, unjointed rock mass (GSI 100 or GSI 50–60); (ii) rock mass with two explicit orthogonal, closely spaced joint sets—derived from field measurements; and (iii) rock mass with closely spaced Voronoi joints—spacing derived from field measurements. The rock mass properties were derived using the approach described by Brideau et al. (2021b), to account for the low persistence discontinuities, which could not be modelled discretely. This approach comprised the following steps:

- ***Step 1:*** The uniaxial compressive strength for a given weathering grade of greywacke was derived from tested cores containing many closely spaced and low persistence discontinuities.
- ***Step 2:*** The GSI was then based on the number, spacing and surface condition of the discontinuities with persistence and spacing greater than the core scale, but typical at the outcrop/meter scale (e.g. 1–5 m). As per the guidance given by Hoek and Brown (2019), those discontinuities at the slope scale, e.g. persistence and spacing greater than a few metres should be modelled explicitly, adopting the appropriate normal- and shear-stiffness and shear strength values for the discontinuity surface.
- ***Step 3:*** The larger persistence discontinuities were modelled discretely or as a Voronoi network.

Ground motion accelerograms from six New Zealand and international crustal and subduction earthquakes (0.25–1.74 g PGA) were selected and run to test the response of the slope, using a similar selection method to the one described in the Utiku landslide case study. The accelerograms, both unscaled and scaled, where the PGA of the given accelerogram was scaled to the PGA of the chosen return period, were derived from the NSHM. The ground motions were converted to stress time histories after deconvolution, baseline correction and filtering. Sensitivity analyses on discontinuity spacing, strain softening behaviour of material and discontinuities and damping parameters were completed to understand the influence of these inputs on slope behaviour.

Results indicate that the Gold's slope will re-activate (displacement > 0.7 m) in the 5 highest magnitude events (Fig. 7), representing earthquake PGAs with return periods in the > 700-year range. The highest modelled displacements occurred in the jointed models, and particularly when discontinuity properties were reduced within the model—accounting for strain softening of the joints in response to displacement. Discontinuities are thus critical to consider in this case. Note that modelled displacements have not been calibrated against actual displacements.

This case study highlights:

- the need to use well-constrained case studies,
- the incorporation of data from field mapping and jointed laboratory samples,
- sensitivity analysis of multiple parameters/scenarios, leading to a better understanding of model and slope behaviour and
- the significance of including discontinuities in dynamic analyses.

The interpretation of results is not as straightforward as pseudostatic or permanent-displacement methods, but the simulations provide much more information on physical mechanisms and evolution of slope behaviour. No assumptions about sliding surface geometry or site effects are needed, as the model indicates strain distributions and amplification metrics as a result. The method is useful as a tool to research slope behaviour, or for the design of critical assets.

5.8.1 Developments and Challenges of Site-Scale Methods

Site-specific analysis of the dynamic behaviour of natural and artificial slopes is becoming more accurate and complete. Although the pseudostatic method has been applied for decades to obtain preliminary assessment of how a given slope may behave during an earthquake, most experts now see it as an overly simplistic tool. The factor of safety metric can be meaningless and misleading in practice. Most popular software packages are slowly replacing pseudostatic analysis with permanent-displacement and/or stress-deformation analysis options. For example, Rocscience recently released a "Dynamic Data Analysis Dialog" that allows processing of seismic signals, the option to model excess dynamic pore water pressures, and two new boundary condition types to simulate seismic loading more accurately (Rocscience 2020). SLIDE, traditionally a popular limit equilibrium tool,

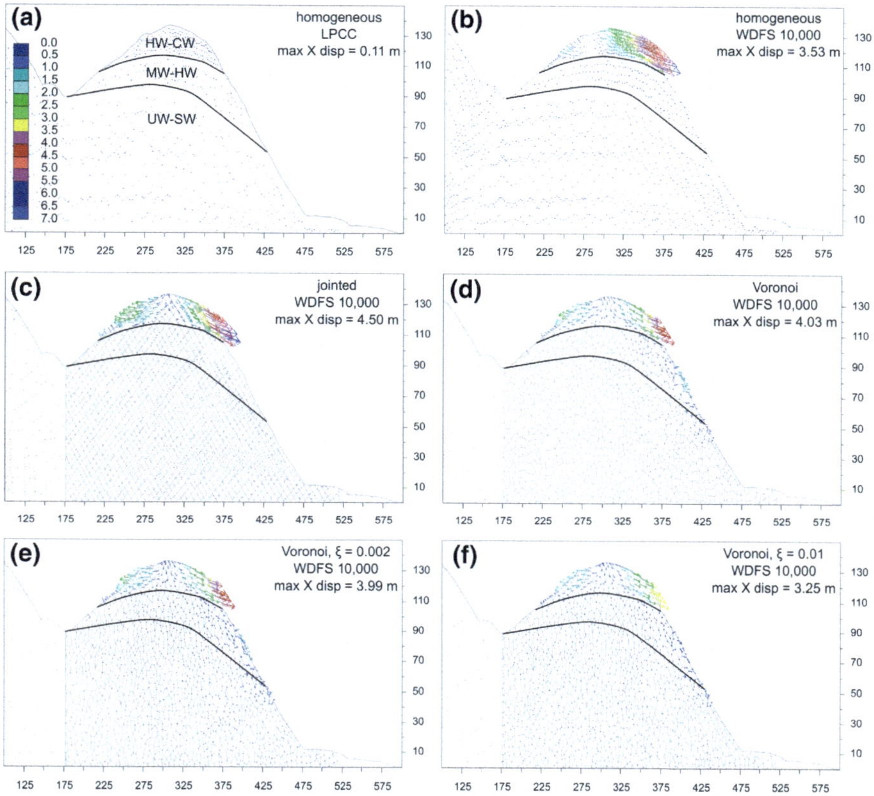

Fig. 7 Modelled horizontal displacement magnitude of the Gold's slope as simulated with: **a** the LPCC earthquake accelerogram and assuming a homogeneous rock mass: **b** the same homogenous rock slope using the WDFS accelerogram, scaled to the 10,000-year return period PGA; **c** as **b** but assuming a discretely jointed rock mass, **d** as per **b** but assuming a Voronoi jointed rock mass; and **e** and **f** as per **d** but with Rayleigh damping sensitivity analysis. Note: model dimensions are in metres and the vectors show the maximum displacement in metres; and black curves represent weathering grade boundaries (UW = unweathered, SW = slightly weathered, MW = moderately weathered, HW = highly weathered, CW = completely weathered)

includes new options for staged pseudostatic analysis as well as Newmark analysis (Rocscience 2020).

Bray (2007) and Jibson (2011) state that applying coupled permanent-displacement tools is more accurate than pseudostatic or decoupled analysis while remaining easier to interpret than stress-deformation analysis. The advance of coupled approaches over the last decade is a significant change in dynamic analysis. However, these coupled tools still require detailed, high resolution material and ground motion parameters that may not be easily acquired. Most analyses have been validated for engineered soil features such as fill and embankment dams. Further studies should be conducted on rock slopes.

Stress-deformation analyses are currently reserved mainly for sites where detailed investigations are warranted, such as where slope failure hazard may pose a risk to life and/or critical infrastructure. These analyses allow stresses and strains in the simulation to develop, focusing on how seismic waves propagate in and affect a given slope. The interpretation of results requires more in-depth understanding of slope processes and how they interact with ground shaking. Displacement can be used as a metric to indicate potential catastrophic failure or used to help define such thresholds for regional-scale studies such as the regional-scale M-M model discussed previously. Such stress-deformation models should, however, be calibrated against actual displacements. Recent developments include discontinuity analysis within dynamic models (e.g. the Gold's case study above) and dynamic analysis in hybrid (DEM/FEM) codes such as Elfen (Rockfield 2021), allowing for brittle fracture. Nonetheless, user expertise and data input remain crucial limiting factors in all analyses. For example, such methods usually assume simplified Mohr–Coulomb material index values for potentially heterogeneous failure surfaces and masses.

Model calibration and validation have improved in the past decade. As the examples above have shown, validation can be in the form of back analysis of existing seismically triggered landslides for stability, comparison between monitoring and modelled displacements or comparison between the modelled site effects (frequency responses, amplification ratios, etc.) and actual recorded effects (e.g. Massey et al. 2017). The development of improved ground motion recording has facilitated validation by providing higher resolution data.

The key developments over the past decade include:

- the replacement of pseudostatic analysis with more accurate methods in common commercial software packages,
- the development and refinement of coupled permanent-displacement analysis,
- hybrid DEM/FEM dynamic analysis,
- improved ground motion recording and processing and
- improved calibration and validation of results.

6 Discussion

The case studies used in this paper still highlight similar issues and challenges to those identified by Wasowski et al. (2011) based on their review of the coseismic landslide literature available at the time (ten years ago). These key challenges are discussed below.

Key challenges: general

- The limitation of input data, as this tends to dictate the choice of method;
- the project needs and budget/scope limitations (including data collection if unavailable prior to project start); and

- the ability to transfer the results from models to the decision makers—from science/engineering to practice to policies—thus allowing more robust decisions to be made (e.g. Quigley et al. 2020); this is arguably the largest challenge.

Key challenges: Regional-scale methods

- collating high-quality global EIL inventories and making them opensource;
- standardising landslide mapping procedures to ensure, e.g. landslide source areas can be separated from their debris trails;
- creating catalogues of ground motion models for landslide-initiating earthquakes;
- sharing of source code and model coefficients for, e.g. deep learning and other machine learning models trained on EIL inventories;
- not forgetting that simple models such as logistic regression, when trained on high-quality EIL inventories and data sets can perform as well as more complex deep learning models;
- landslide scientists working with machine learning scientists and statisticians to take advantage of the truly multi-disciplinary nature of this work; and
- results evolving from the probability of a landslide occurring in each grid cell to the probability of a landslide of given type and volume occurring on a slope.

Key challenges: site-specific scale methods

- not much laboratory testing data exists on the dynamic properties of materials, especially rock; future analyses could incorporate sophisticated dynamic laboratory testing results;
- even sophisticated analyses are based on simplistic deterministic Mohr–Coulomb constitutive models;
- the critical ground motion parameters controlling site effects remain poorly understood and are complex;
- most dynamic analyses are completed in engineered soils; more investigations in slopes and rock masses should be completed;
- dynamic pore water pressure is still not commonly analysed (cf. Biondi et al. 2000; Dey et al. 2016);
- most slope analyses are two-dimensional, but the third and fourth dimensions of ground motions and slopes are significant (cf. Al-Homoud and Tahtamoni 2000);
- models are currently mainly deterministic; probabilistic analysis is capable of improved characterisation of the variability in input parameters and aiding in understanding controlling parameters;
- how rock mass 'damage' and 'fatigue' develop in slopes subjected to multiple seismic loading events as well as other processes such as transient changes in pore water pressures over time, e.g., exploring path dependencies;
- the need for strain-related strength reduction relationships, which would help model rock mass 'damage'; and
- many modelling codes still cannot simulate brittle fracture and crack propagation.

These key challenges reiterate the findings of Wasowski et al. (2011): *'It is apparent that balanced and integrated approaches between the scientific communities focused on local-scale and regional-scale investigations ought to be further encouraged. This should help to foster the practical applicability of the research on seismic landslide hazards'.*

7 Summary and Conclusions

There are many different methods available to investigate the response of slopes to earthquake shaking, at scales from national to site-specific. Their applicability, however, is limited by the availability of input data, or the ability to derive it, irrespective of the project scale or need. Software development is also occurring at a pace that is faster than our ability to capture high-quality EIL data sets—at both regional and local scales—and use them to investigate the fundamental physics-based mechanisms that underpin landslide initiation mechanisms.

Table 6 summarises the needs, data requirements and main constraints of the various coseismic landslide triggering analysis methods discussed in this paper, which become increasingly more data dependent, from the regional- to the local-scale. The information in Table 6 can be used to help choose the most applicable method for a given project's needs and available data.

Possibly the largest challenge for experts is the ability to transfer the results from such models into hazard and risk analyses and engineering design, and ultimately to the decision makers, thus facilitating more science-based decisions. An exciting recent development in this area is the implementation of regional-scale landslide forecast models, like the ones described in this book. Such models are being used to produce impact forecasts, e.g. the area of ground or numbers of people affected by EIL, using immediate post-event near-real-time ground shaking data and models. Such models and tools will facilitate more targeted emergency response and hazard- and risk-zone delineation in future.

Table 6 Summary of needs, data requirements and main constraints of the coseismic landslide triggering methods of analyses discussed

Need (increasing complexity)	Method	Data requirements		Main constraints
		Essential	Desirable	
Regional to local > 1:5,000 to 1:250,000				
Landslide probability	Statistical analysis, e.g. LR, DL, fuzzy logic, etc	• Existing EIL inventories, geology maps at appropriate scales • Digital elevation models at appropriate scales • Dynamic: PGA model of earthquake shaking	• Dynamic: PGA and or PGV from multiple ground motion models, e.g. ShakeMap, Synthetic, etc	• Not able to directly derive landslide types/mechanism, or frequency-magnitude (area/volume) of landslides • Require detailed landslide inventories on which to train the models • Can only be applied, with confidence, to areas in similar geological and physiographical settings to those on which the models were trained • Can only be applied, with confidence, to earthquakes with similar mechanisms to those on which the models were trained

(continued)

Table 6 (continued)

Need (increasing complexity)	Method	Data requirements		Main constraints
		Essential	Desirable	
Probability, landslide type/failure mode and volume, where no existing EIL inventories exist	Deterministic analysis, e.g. quasi-physics based such as the M-M method	Same as statistical models, and: • Material: strength parameters • Failure surface: depth estimates • Landslide: types and failure modes	Same as statistical models, and: • Groundwater: phreatic surface or pore water pressure	• Requires detailed geotechnical material properties, at a resolution that is consistent with the mapping • Requires knowledge of failure types, depths and pore water pressure/phreatic surface depths • The need to make assumptions with respect to displacement thresholds and susceptibility factors
Local to site-specific scale < 1:5,000				
Factor of safety	Pseudostatic analysis	• Slope geometry: height, angle • Failure surface: depth, angle • Material: properties such as unit weight, friction angle and cohesion • Dynamic: seismic coefficient (horizontal)	• Slope geometry: profile section • Groundwater: phreatic surface or pore water pressure Material property variability • Dynamic: vertical seismic coefficient • Calibration via back analysis	• Typically, over-conservative • Average values for slope, failure surface, material, dynamic parameters assumed (no heterogeneity) • Assumed rigid body with simple geometry • Highly dependent on choice of seismic coefficient • No displacement estimates or post-failure analysis

(continued)

Table 6 (continued)

Need (increasing complexity)	Method	Data requirements		Main constraints
		Essential	Desirable	
Permanent displacement, given simple sliding geometry ± rigid body	Permanent-displacement analysis	Same as pseudostatic models, and: • Material properties: bulk and shear moduli • Dynamic: horizontal accelerogram (time-acceleration history) and damping parameters	Same as pseudostatic models, and: • Material properties: strain-related shear modulus and damping ratio relationships • Dynamic: vertical ground motion inputs	• Average values for slope, failure surface, material, dynamic parameters assumed (no heterogeneity) • Calibration of modelled displacement values required • No post-failure analysis • Newmark and decoupled approaches: – Assumed rigid body – Displacements can be too conservative
Physical mechanisms and behaviour, given complex sliding geometry and deformation	Stress-deformation analysis	Same as permanent-displacement analysis models, and: • Material properties: contacts, tensile strength, dilation angle • Joint properties: joint orientation, persistence, spacing, stiffness, joint friction angle and cohesion	Same as permanent-displacement analysis models, and: • Material: other constitutive criteria, if possible • Groundwater: water table, pore water pressure • Material: variability in properties • Dynamic: multiple loading	• Requirements for data and computing high • Experience and judgement required to interpret results (outputs not simple metrics) • Most dynamic studies still use average parameter values and simple strength criteria in two dimensions (i.e. Mohr–Coulomb; see first constraint)

Acknowledgements The authors would like to acknowledge the JTC-1 for facilitating this paper. We would also like to thank the New Zealand governments Strategic Science Investment Fund and their Endeavour Fund (via the Earthquake-induced landscape dynamic programme) and the GeoNet project, which funded this research. We would like to thank D Rhoades, C Mueller, K Leith and P Glassey (GNS Science) for reviewing previous drafts of this work.

References

Al-Homoud A, Tahtamoni W (2000) Reliability analysis of three-dimensional dynamic slope stability and earthquake-induced permanent displacement. Soil Dyn Earthq Eng 19(2):91–114

Arnold L (2016) Seismically-induced rock-slope failure: numerical investigation using the bonded particle model. PhD Thesis, University of Washington, USA. 600 p

Ashford SA, Sitar N (2002) Simplified method of evaluating seismic stability of steep slopes. J Geotech Geoenviron Eng 119. https://doi.org/10.1061/(ASCE)1090-0241(2002)128:2(119)

Ashford SA, Sitar N, Lysmer J, Deng N (1997) Topographic effects on the seismic response of steep slopes. Bull Seism Soc Am 87:701–709

Athanasopoulos-Zekkos A, Seed RB (2013) Simplified methodology for consideration of two-dimensional dynamic response of levees in liquefaction-triggering evaluation. J Geotech Geoenviron Eng 139(11):1911–1922

Bazgard A, Davies T, Cundall P (2011) Dynamic analysis of mountain response using PFC3D and FLAC3D. In: Sainsbury D, Hart R, Detournay C, Nelson M (eds) Continuum and distinct element modelling in geomechanics. Itasca, Minneapolis, MN, USA, pp 1–7

Benites RA, Haines AJ (1994) Quantification of seismic wavefield amplification by topographic features. Institute of Geological & Nuclear Sciences client report, 343901

Biondi G, Cascone E, Maugeri M, Motta E (2000) Pore pressure effect on seismic response of slopes. 12WCEE 2000, pp 1–8

Bloom CK, Howell A, Stahl T, Massey C, Singeisen C (2021) The influence of off-fault deformation zones on the near-fault distribution of coseismic landslides. Submitted to Geology (February 2021)

Bloom CK, Howell A. Stahl T. Massey CI, Singeisen C (2022) The influence of off-fault deformation zones on the nearfault distribution of coseismic landslides. Geology 50(3):272–277.https://doi.org/10.1130/G49429.1

Bourdeau C, Havenith H-B (2008) Site effects modelling applied to the slope affected by the Suusamyr earthquake (Kyrgyzstan, 1992). Eng Geol 97:126–145

Bozzano F, Lenti L, Martino S, Paciello A, Scarascia Mugnozza G (2008) Self-excitation process due to local seismic amplification responsible for the reactivation of the Salcito landslide (Italy) on 31 October 2002. J Geophys Res 113:B10312

Bray J (2007) Chapter 14: Simplified seismic slope displacement procedures. In: Pitilakes K (ed) Earthquake geotechnical engineering: 4th international conference on earthquake geotechnical engineering-invited lectures. Springer, Dordrecht, The Netherlands, pp 327–353

Bray JD, Rathje EM (1998) Earthquake-induced displacements of solid-waste landfills. J Geotech Geoenviron Eng 124(3):242–253

Bray JD, Travasarou T (2007) Simplified procedure for estimating earthquake-induced deviatoric slope displacements. J Geotech Geoenviron Eng 133(4):381–392

Bray JD, Travasarou T (2009) Pseudostatic coefficient for use in simplified seismic slope stability evaluation. J Geotech Geoenviron Eng 135(9):1336–1340

Bray J, Travasarou T (2011) Pseudostatic slope stability procedure. In: 5th international conference on earthquake geotechnical engineering, pp 10–13

Bray JD, Macedo J (2019) Procedure for estimating shear-induced seismic slope displacement for shallow crustal earthquakes. J Geotech Geoenviron Eng 145(12):04019106

Brideau M-A, Massey CI, Carey JM, Kellet RL, Abbott ER, Monteith F, Kupec J (2021a) Geomechanical characterisation and dynamic numerical modelling of two anthropogenic fill slopes. Eng Geol 281(105980). https://doi.org/10.1016/j.enggeo.2020.105980

Brideau M-A, Massey CI, Carey JM, Lyndsell B (2021b) Geomechanical characterisation of discontinuous greywacke from the Wellington region based on laboratory testing. NZ J Geol Geophys. https://doi.org/10.1080/00288306.2020.1853181

Budimir MEA, Atkinson PM, Lewis HG (2015) A systematic review of landslide probability mapping using logistic regression. Landslides 12:419–436. https://doi.org/10.1007/s10346-014-0550-5

Carey JM, Massey CI, Lyndsell B, Petley DN (2019) Displacement mechanisms of slow-moving landslides in response to changes in porewater pressure and dynamic stress. Earth Surf Dyn 7(3):707–722

Chen TC, Lin ML, Hung JJ (2004) Pseudostatic analysis of Tsao-Ling rockslide caused by Chi-Chi earthquake. Eng Geol 71(1–2):31–47

Chien Y-C, Tsai C-C (2017) Immediate estimation of yield acceleration for shallow anddeep failures in slope-stability analyses. Int J Geomech 17:04017009

Corominas J, van Westen C, Frattini P, Cascini L, Malet J, Fotopoulou S, Catani F, van den Eeckaut, Mavrouli O, Agliardi F, Pitilakis K, Winter MG, Pastor M, Ferlisi S, Tofani V, Hervas J, Smith JT (2014) Recommendations for the quantitative analysis of landslide risk. Bull Eng Geol Environ 73(2),209–263

Deleo J (1993) Receiver operating characteristic laboratory (ROCLAB): Software for developing decision strategies that account for uncertainty. In: Proceedings of 2nd international symposium on uncertainty modelling and analysis. Computer Society Press, College Park

Del Gaudio V, Wasowski J (2011) Advances and problems in understanding the seismic response of potentially unstable slopes. Eng Geol 122:73–83. https://doi.org/10.1016/j.enggeo.2010.9.007

de Vilder SJ, Massey CI (2020) Guidelines for natural hazard risk analysis on public conservation lands and waters—Part 4: a commentary on analyzing landslide risk to point and linear sites. Lower Hutt (NZ): GNS Science. 64 p. Consultancy Report 2020/53

Dey R, Hawlader B, Phillips R, Soga K (2016) Modeling of large-deformation behaviour of marine sensitive clays and its application to submarine slope stability analysis. Can Geotech J 53(7):1138–1155

Dowrick DJ, Hancox GT, Perrin ND, Dellow GD (2008) The Modified Mercalli intensity scale: revisions arising from New Zealand experience. Bull NZ Soc Earthq Eng 41(3):193–205. https://doi.org/10.5459/bnzsee.41.3.193-205

Elmo D, Stead D (2021) The role of behavioural factors and cognitive biases in rock engineering. In: Rock Mechanics and Rock Engineering, pp 1–20

Elmo D, Rogers S, Stead D, Eberhardt E (2014) Discrete Fracture Network approach to characterise rock mass fragmentation and implications for geomechanical upscaling. Min Technol 123(3):149–161

Elmo D, Yang B, Stead D, Rogers S (2021) A discrete fracture network approach to rock mass classification. In: International conference of the international association for computer methods and advances in geomechanics. Springer, Cham, pp 854–861

EN1998–5. (2004) Eurocode 8: Design of structures for earthquake resistance Part5: Foundations, retaining structures and geotechnical aspects. https://www.phd.eng.br/wp-content/uploads/2014/11/en.1998.5.2004.pdf

Exadaktylos G, Stavropoulou M (2008) A specific upscaling theory of rock mass parameters exhibiting spatial variability: analytical relations and computational scheme. Int J Rock Mech Min Sci 45(7):1102–1125

Fan X, Scaringi H, Korup O, West JA, Westen CJ, Tanyas H, Hovius N, Hales TC, Jibson RW, Allstadt KE, Zhang L, Evans SG, Xu C, Li G, Pei X, Xu Q, Huang R (2019) Earthquake-induced chains of geologic hazards: patterns, mechanisms, and impacts. Rev Geophys 57(2):421–503.https://doi.org/10.1029/2018RG000626

Fawcett T (2006) An introduction to ROC analysis. Pattern Recogn Lett 27:861–874

Geli L, Bard P-Y, Jullien B (1988) The effect of topography on earthquake ground motion: A review and new results. Bull Seismol Soc Am 78(1):42–63

Geomechanica (2021) Irazu. Geomechanica, Toronto, ON, Canada

Geo-Slope (2012) SLOPE/W. GEOSLOPE, Calgary, AB, Canada

Geo-Slope (2014) QUAKE/W. GEOSLOPE, Calgary, AB, Canada

Gischig VS, Eberhardt E, Moore JR, Hungr O (2015) On the seismic response of deep-seated rock slope instabilities—Insights from numerical modelling. Eng Geol 193(2015):1–18. https://doi.org/10.1016/j.enggeo.2015.04.003

Goded T, Horspool N, Massey C, Kaiser A, Van Dissen R, Litchfield N, Lukovic B, Heron D, Holden C, Hamling I (2021) Shaking intensity maps for 15 moderate-to-large New Zealand earthquakes. Lower Hutt (NZ): GNS Science (GNS Science report; 2020/35). https://doi.org/10.21420/G2CH12

Grant A, Wartman J, Abou-Jaoude G (2016) Multimodal method for coseismic landslide hazard assessment. Eng Geol 212:146–160

Guajardo C, Elmo D, Rogers S (2020) Numerical investigation on rock strength upscaling using synthetic rock approach. 54th U.S. Rock Mechanics/Geomechanics Symposium

Guzzetti F, Reichenbach P, Ardizzone F, Cardinali M, Galli M (2006) Estimating the quality of landslide susceptibility models. Geomorphology 81(1):166–184. https://doi.org/10.1016/j.geomorph.2006.04.007

Hamling I, Hreinsdóttir S, Clark K, Elliott J, Liang C, Fielding E, Litchfield N, Villamore P, Wallace L, Wright TJ et al (2017) Complex multifault rupture during the 2016 MW 7.8 Kaikoura earthquake, New Zealand. Science 356:1–16. https://doi.org/10.1126/science.aam7194

Hancox GT, Perrin ND, Dellow GD (2002) Recent studies of historical earthquake-induced landsliding, ground damage, and MM intensity in New Zealand. Bull NZ Soc Earthq Eng 35(2):59–95. https://doi.org/10.5459/bnzsee.35.2.59-95

Hancox GT, Ries WF, Lukovic B, Parker RN (2014) Landslides and ground damage caused by the MW 7.1 Inangahua earthquake of 24 May 1968 in northwest South Island, New Zealand. GNS Science Report 2014/06

Hancox GT, Ries WF, Parker RN, Rosser B (2016) Landslides caused by the M_W 7.8 Murchison earthquake of 17 June 1929 in northwest South Island, New Zealand. GNS Science Report 2015/42

Hoek E (1990) Estimating Mohr-Coulomb friction and cohesion values from the Hoek-Brown failure criterion. Int J Rock Mech Mining Sci Geomech Abs 12(3):227–229

Hoek E, Brown E (2019) The Hoek-Brown failure criterion and GSI–2018 edition. J Rock Mech Geotech Eng 11(3):445–463

Horspool N, Chadwick M, Ristau J, Salichon J, Gerstenberger MC (2015) ShakeMapNZ: Informing post-event decision making. In: Proceedings conference of the New Zealand society for earthquake engineering, Rotorua (April 2015)

Hough SE, Altidor JR, Anglade D, Given D, Janvier MG, Maharrey JZ, Meremonte M, Mildor BS, Prepetit C, Yong A (2010) Localized damage caused by topographic amplification during the 2010 M 7.0 Haiti earthquake. Nat Geosci 3(11):778–782

Hsieh S-Y, Lee C-T (2011) Empirical estimation of the Newmark displacement from the Arias intensity and critical acceleration. Eng Geol 122(1–2):34–42

Hungr O, Leroueil S, Picarelli L (2014) The Varnes classification of landslide types, an update. Landslides 11(2):167–194

Itasca (2008) FLAC. Itasca Consulting Group, Minneapolis, MN, USA

Itasca (2011) FLAC3D, PFC3D. Itasca Consulting Group, Minneapolis, MN, USA

Itasca (2020) UDEC v. 7.0. Itasca Consulting Group, Minneapolis, MN, USA

Janku L (2017) Quantification of seismic site effects on slopes in Wellington. University of Canterbury PhD thesis. https://doi.org/10.26021/5754

Jibson R (2007) Regression models for estimating coseismic landslide displacement. Eng Geol 91:209–218. https://doi.org/10.1016/j.enggeo.2007.01.013

Jibson R (2011) Methods for assessing the stability of slopes during earthquakes—A retrospective. Eng Geol 122(1–2):43–50

Jibson R, Harp EL (2016) Extraordinary distance limits of landslides triggered by the 2011 Mineral, Virginia, earthquake. Bull Seismol Soc Am. https://doi.org/10.1785/0120120055

Jibson R, Tanyas H (2020) The influence of frequency and duration of seismic ground motion on the size of triggered landslides—A regional view. Eng Geol 273:105671. https://doi.org/10.1016/j.enggeo.2020.105671

Kaiser AE, Oth A, Benites RA (2013) Separating source, path and site influences on ground motion during the Canterbury earthquake sequence, using spectral inversion. Same risks, new realities: New Zealand Society for Earthquake Engineering Technical Conference and AGM, April 26–28 2013, Wellington. Paper no. 18:8p

Kaiser AE, Holden C, Massey CI (2014) Site amplification, polarity and topographic effects in the Port Hills during the Canterbury earthquake sequence, GNS Science Consultancy Report 2014/121

Keefer DK (1984) Landslides caused by earthquakes. Geol Soc Am Bull 95(4):406–421

Keefer DK, Tannaci NE (1981) Bibliography on landslides, soil liquefaction, and related ground failures in selected historic earthquakes. Open file report 81–572, USGS (May 1981)

Kramer SL (1996) Geotechnical earthquake engineering. New Jersey, Prentice-Hall Inc., Upper Saddle River

Kritikos T, Robinson TT, Davies T (2015) Regional coseismic landslide hazard assessment without historical landslide inventories: A new approach. JGR Earth Surface. 120(4):711–729. https://doi.org/10.1002/2014JF003224

Langridge RM, Ries WF, Litchfield NJ, Villamor P, Van Dissen RJ, Barrell DJA, Rattenbury MS, Heron DW, Haubrock S, Townsend DB et al (2016) The New Zealand active faults database. NZ J Geol Geophys 59:86–96. https://doi.org/10.1080/00288306.2015.1112818

Larsen IJ, Montgomery DR, Korup O (2010) Landslide erosion controlled by hillslope material. Nat Geosci 3(4):247–251. https://doi.org/10.1038/ngeo776

Lombardo L, Mai PM (2018) Presenting logistic regression-based landslide susceptibility results. Eng Geol 244:14–24. https://doi.org/10.1016/j.enggeo.2018.07.019.

Makdisi FI, Seed HB (1978) Simplified procedure for estimating dam and embankment earthquake-induced deformations. J Geotech Eng Div 104(7):8449–8867

Marinos P, Marinos V, Hoek E (2007) The geological strength index (GSI): a characterization tool for assessing engineering properties of rock masses. In: Mark, Pakalnis, Tuchman (eds) Proceedings international workshop on rock mass classification for underground mining, 9498, pp 87–94

Massey CI, Richards L, Della-Pasqua FN, McSaveney MJ, Holden C, Kaiser AE, Archibald GC, Wartman J, Yetton M (2015) Performance of rock slopes during the 2010/11 Canterbury earthquakes (New Zealand), paper no. 902. In: Hassani F (ed) (general chair) Proceedings: the 13th international congress of rock mechanics : ISRM congress 2015 : innovations in applied and theoretical rock mechanics ... May 10–13, 2015. Canadian Institute of Mining Metallurgy and Petroleum, Montreal, Canada [Montreal, Quebec]

Massey CI, Abbott E, McSaveney M, Petley DN, Richards L (2016) Earthquake-induced displacement is insignificant in the reactivated Utiku landslide, New Zealand. In: Aversa S, Cascini L, Picarelli L, Scavia C (eds) Landslides and engineered slopes. experience, theory and practice. Associazione Geotecnica Italiana, Rome, Italy, pp. 31–52

Massey C, Della Pasqua F, Holden C, Kaiser A, Richards L, Wartman J, McSaveney MJ, Archibald G, Yetton M, Janku L (2017) Rock slope response to strong earthquake shaking. Landslides 14:249–268

Massey CI, Townsend DB, Rathje E, Allstadt KE, Lukovic B, Kaneko Y, Bradley B, Wartman J, Jibson RW, Petley DM, Horspool NA, Hamling IJ, Carey JM, Cox SC, Davidson J, Dellow GD, Godt GW, Holden C, Jones KE, Kaiser AE, Little M, Lyndsell BM, McColl S, Morgenstern RM, Rengers FK, Rhoades DA, Rosser BJ, Strong DT, Singeisen C, Villeneuve M (2018) Landslides triggered by the 14 November 2016 Mw 7.8 Kaikoura earthquake, New Zealand. Bull Seismol Soc Am 108(3B):1630–1648. https://doi.org/10.1785/0120170305

Massey CI, Townsend DB, Jones KE, Lukovic B, Rhoades DA, Morgenstern R, Rosser BJ, Ries W, Howarth JD, Hamling IJ, Petley DM, Clark M, Wartman J, Litchfield NJ, Olsen M

(2020a) Volume characteristics of landslides triggered by the Mw7.8 2016 Kaikoura Earthquake, New Zealand, derived from digital surface difference modelling. J Geophys Res Earth Surf 125(7):e2019JF005163. https://doi.org/10.1029/2019JF005163
Massey CI, Townsend DB, Lukovic B, Morgenstern R, Jones KE, Rosser BJ, de Vilder SJ (2020b) Landslides triggered by the 14 November 2016 Mw 7.8 Kaikoura earthquake : an update. Landslides 17(10):2401–2408. https://doi.org/10.1007/s10346-020-01439-x
Massey CI, Lukovic B, Huso R, Buxton R (2021) Earthquake-induced landslide forecast tool for New Zealand: Prototype and option assessment. GNS Science, Lower Hutt (NZ). GNS Science report 2018/08. https://doi.org/10.21420/DOI: https://doi.org/10.21420/G2TP9V
McSaveney M, Massey C (2017) Inadvertent engineered activation of Utiku Landslide, New Zealand. In: 4th World Landslide Forum. Ljubljana, Slovenia, pp 563–568
Meunier P, Hovius N, Haines JA (2008) Topographic site effects and the location of earthquake induced landslides. Earth Planet Sci Lett 275:221–232
Monteith F (2020) SLIDE (Wellington): risk remediation approaches for anthropogenically modified slopes in the Wellington region. GNS Science, Lower Hutt, NZ. GNS Science report 2019/51 [209] p. https://doi.org/10.21420/07JP-Q216
Moore JR, Gischig VS, Burjanek J, Loew S, Faeh D (2011) Site effects in unstable rock slopes: dynamic behavior of the Randa instability (Switzerland). Bull Seismol Soc Am 101(6):3110–3116
Moratalla J, Goded T, Gerstenberger M, Canessa S (2021) New ground motion to intensity conversion equations (GMCIEs) for New Zealand. Seismol Res Lett 92(1):448–459
Newmark N (1965) Effects of earthquakes on dams and embankments. Geotechnique 15(2):139–160
New Zealand Transport Agency (NZTA) (2013) Bridge manual (SM061). Version 3, Amendment 3, October 2018. ISBN/ISSN: 978-0-478-37161-1 (online) | 978-0-478-37162-8 (print)
Nowicki Jessee MA, Hanburger MW, Allstadt K, Wald, DJ, Robeson SM, Tanyas H, Nearne M, Thompson E (2018) A Global empirical model for near-real-time assessment of seismically induced landslides. JRG Earth Surf (8):1835–1859. https://doi.org/10.1029/2017JF004494
Parker RN, Hancox GT, Petley DN, Massey CI, Densmore AL, Rosser NJ (2015) Spatial distributions of earthquake-induced landslides and hillslope preconditioning in the northwest South Island, New Zealand. Earth Surf Dyn 3:501–525. https://doi.org/10.5194/esurf-3-501-2015
Pollock W (2020) A framework for regional scale quantitative landslide risk analysis. University of Washington, Seattle. PhD thesis
Pollock W, Grant A, Wartman J, Abou-Jaoude G (2019) Multimodal method for landslide risk analysis. MethodsX 6:827–836. https://doi.org/10.1016/j.mex.2019.04.012
Quigley MC, Saunders WSA, Massey CI, Van Dissen RJ, Villamor P, Jack H, Litchfield NJ (2020) The utility of earth science information in post-earthquake land-use decision-making: the 2010-2011 Canterbury earthquake sequence in Aotearoa New Zealand. Nat Hazards Earth Syst Sci, pp 3361–3385.https://doi.org/10.5194/nhess-20-3361-2020
Rai M, Rodriguez-Marek A, Yong A (2016) An empirical model to predict topographic effects in strong ground motion using California small- to medium-magnitude earthquake database Manisha. Earthq Spectra 32:1033–1054
Reichenbach P, Rossi M, Malamud BD, Mihir M, Guzzetti F (2018) A review of statistically-based landslide susceptibility models. Earth Sci Rev 180:60–91. https://doi.org/10.1016/j.earscirev.2018.03.001
Rizzitano S, Cascone E, Biondi G (2014) Coupling of topographic and stratigraphic effects on seismic response of slopes through 2D linear and equivalent linear analyses. Soil Dyn Earthq Eng 67:66–84
Robinson TR, Rosser NJ, Densmore AL, Williams JG, Kincey ME, Benjamin J, Bell HJA (2017) Rapid post-earthquake modelling of coseismic landslide intensity and distribution for emergency response decision support. Nat Hazards Earth Syst Sci 17(1521–1540):2017. https://doi.org/10.5194/nhess-17-1521-2017
Rockfield (2021) Elfen. Rockfield, Swansea Waterfront, UK
Rocscience (2010a) RS2. Rocscience Inc, Toronto, ON, Canada
Rocscience (2010b) SLIDE. Rocscience Inc, Toronto, ON, Canada

Rocscience (2020) RocNews March 2020. URL: https://www.rocscience.com/rocnews/rocnews-march-2020-our-stance-on-covid-19-new-features-in-our-fem-products-and-a-whole-lot-more. Last accessed: 07 Oct, 2020

Saygili G, Rathje E (2008) Empirical predictive models for earthquake-induced sliding displacements of slopes. Geotech. Geoenviron. Eng. 134(6):790–803. https://doi.org/10.1061/(ASCE)1090-0241(2008)134:6(790)

Sepulveda SA, Murphy W, Petley DN (2005) Topographic controls on coseismic rock slides during the 1999 Chi-Chi earthquake, Taiwan. Q J Eng Geol Hydrogeol 38:189–196. https://doi.org/10.1144/1470-9236/04-062

Siddiqqi J, Atkinson GM (2002) Ground-motion amplification at rock sites across Canada as determined from the horizontal-to-vertical component ratio. Bull Seismol Soc Am 92:877–884

Standards New Zealand (2004) Structural design actions part 5 earthquake actions. New Zealand Standard

Strenk P, Wartman J (2011) Uncertainty in seismic slope deformation model predictions. Eng Geol 122(1–2):61–72

Tang CL, Hu JC, Lin ML, Angelier J, Lu CY, Chan YC, Chu HT (2009) The Tsaoling landslide triggered by the Chi-Chi earthquake, Taiwan: insights from a discrete element simulation. Eng Geol 106(1–2):1–19

Tanyaş H, van Westen CJ, Allstadt KE, Anna Nowicki Jessee M, Görüm T, Jibson RW, Godt JW, Sato HP, Schmitt RG, Marc O et al (2017) Presentation and analysis of a worldwide database of earthquake-induced landslide inventories. J Geophys Res Earth Surf 122(10):1991–2015. https://doi.org/10.1002/2017jf004236

USGS (2021) https://earthquake.usgs.gov/data/ground-failure/background.php. Last accessed 21 Feb, 2021

van Westen CJ, Castellanos E, Kuriakose SL (2008) Spatial data for landslide susceptibility, hazard, and vulnerability assessment: an overview. Eng Geol 102(3):112–131. https://doi.org/10.1016/j.enggeo.2008.03.010

Von Ruette J, Papritz A, Lehmann P, Rickli C, Or D (2011) Spatial statistical modelling of shallow landslides-validating predictions for different landslide inventories and rainfall events. Geomorphology 133:11–22

Wasowski J, Keefer DK, Lee C-T (2011) Toward the next generation of research on earthquake-induced landslides: Current issues and future challenges. Eng Geol 122(1–2):1–8. ISSN 0013-7952. https://doi.org/10.1016/j.enggeo.2011.06.001

Wilson RC, Keefer DK (1985) Predicting areal limits of earthquake induced landsliding. In: Ziony JI (ed) Evaluating earthquake hazards in the Los Angeles region: an earth-science perspective, U.S. Geological Survey Professional Paper 1360, pp 316–345

Yang B, Mitelman A, Elmo D, Stead D (2021) Why the future of rock mass classification systems requires revisiting their empirical past. Q J Eng Geol Hydrogeol

The manufacturer's authorised representative in the EU is Springer Nature Customer Service Centre GmbH, Europaplatz 3, 69115 Heidelberg, Germany. If you have any concerns regarding our products, please contact ProductSafety@springernature.com

Printed and bound by CPI Group (UK) Ltd, Croydon, CR0 4YY
30/09/2026
02225317-0018